A Course in BE-algebras

Sambasiva Rao Mukkamala

A Course in BE-algebras

Sambasiva Rao Mukkamala
Department of Mathematics
MVGR College of Engineering
Vizianagaram, Andhra Pradesh
India

ISBN 978-981-13-4957-7 ISBN 978-981-10-6838-6 (eBook)
https://doi.org/10.1007/978-981-10-6838-6

Softcover re-print of the Hardcover 1st edition 2018

Printed on acid-free paper

This Springer imprint is published by the registered company Springer Nature Singapore Pte Ltd. part of Springer Nature
The registered company address is: 152 Beach Road, #21-01/04 Gateway East, Singapore 189721, Singapore

Dedicated to my beloved father
M. Brahmanandam

Foreword

Sambasiva Rao Mukkamala is a known author of several books on lattice theory and engineering mathematics. The book A Course in BE-algebras starts with a carefully presented foundation of the theory of BE-algebras. After treating the basic concepts of BE-algebras, the concepts filters, prime filters, pseudo-compliments, very true operators, state operators, state filters, state congruences, states, self-mappings, and endomorphisms are thoroughly explained. The logic and reasoning skills of the author are apparent on every page. This book provides an excellent scope for upcoming research in applied abstract algebra, particularly in BE-algebras. It ends with a fuzzification of various implicative filters.

Each chapter of this book includes a reasonable number of examples which justify the arguments and strengthen the newly introduced concepts. A sufficient number of problems are kept in the exercise at the end of each chapter for the additional practice of the readers. The logic presented at each stage enables the earnest student as well as a researcher to assimilate, in depth, the material in the book.

It is expected to prove an excellent book for graduate and postgraduate students and researchers in BE-algebras around the world. I hope that this book is destined to play a major role in exciting, motivating, and educating the next generation of algebraists from all over the world.

Kunming, P.R. China

Professor Kar-Ping Shum
Honorary Director
Institute of Mathematics, Yunnan University

Preface

A glance at the table of contents will reveal that this book treats topics in BE-algebras at the advanced research level. The aim has been to provide the development of the subject which is honest, rigorous, and at the same time not pedantic. The book provides a transition from elementary topics to advanced topics in BE-algebras, and it introduces the reader to some of the abstract thinking that provides modern abstract algebra.

Since logic is part and parcel of science and technical education, it appears in all branches of study. The concepts are being continuously updated and revised in the research structure of many universities around the world. Keeping this in view, 15 chapters are organised in this book with detailed presentations. In continuation of the literature available on this topic, the present results are explained in a lucid manner. Many examples are included in order to justify the arguments. Hope that this book will help the students and researchers to build a strong foundation in applied abstract algebra and related subject areas.

The introduction explains the origin of the present work in BE-algebras. In Chap. 2, all the necessary preliminaries are collected and presented for ready reference of the reader. Chapters 3 and 4 are devoted to the study of basic concepts like closure operators, fuzzy weak subalgebras, soft BE-algebras, filters and weak filters of BE-algebras. In Chap. 5, properties of quasi-filters and pseudo filters are studied including their characterizations. In Chap. 6, the main emphasis is given to the properties of very true operators of BE-algebras. Chapter 7 deals with pseudo-complements and their properties in terms of Boolean, closed and dense elements. In Chap. 8, the notion of stabilizers of BE-algebras is introduced. Some topological properties are studied with the help of left and right stabilizers. Chapter 9 deals with Bosbach states on bounded BE-algebras, which is the most promising fields of investigation. In Chap. 10, properties of state operators of BE-algebras are dealt with to study the properties of state filters and state congruences. Chapters 11 and 12 are devoted to the study of self-mappings and endomorphisms of BE-algebras. Chapters 13–15 are devoted to the fuzzification of filters, weak filters, implicative filters, transitive filters, and semi-transitive filters of BE-algebras. In Chap. 13, we consider the fuzzification of weak filters. In Chaps. 14 and 15, various properties of implicative and transitive filters are investigated. In the final chapter, notions of transitive and semi-transitive filters are introduced in BE-algebras and then triangular normed fuzzification is applied to semi-transitive filters.

The book was motivated by the desire I have had to present a course in BE-algebras and to further the development of the research course in abstract algebra. The centralized themes of all the chapters of this book summarize what I have tried to offer the student in this book. In each chapter, the number of examples presented is made to depend upon the difficulty and importance of the particular concept presented.

I would like to express my gratitude to Prof. G. Chakradhara Rao of Andhra University, India, for his guidance and friendly encouragement throughout this work. I owe special thanks to Prof. K. V. Lakshmipathi Raju, Principal of MVGR College of Engineering, Vizianagaram, for his motivation and encouragement. I wish to thank Springer India for agreeing to publish. In addition, I am grateful to Prof. K. P. Shum of Yunnan University, China, for writing foreword to this book. I would also like to thank the anonymous reviewers of this book for their valuable suggestions. Finally, no words can exactly express the unparalleled support and encouragement given to me by my family members and friends.

The instinct of acknowledging the inspiration due to ingenious works of pioneers gives me immense pleasure, and fills my heart to the brim, for the motivation that dawned on me finds its reflection in this mammoth task. I remain grateful to all the pioneers and my colleague researchers around the world who have contributed to the theory of BE-algebras and the related areas.

I believe that the reader can find not only a number of interesting results and methods but also an inspiration for his own investigations in this area. The publication of this book does not indicate that this field is closed. On the contrary, I know that it is a quickly developing area and hope that this book will serve as an orientation for working algebraists in this special subject.

Vizianagaram, India
August 2017

Sambasiva Rao Mukkamala

Contents

About the Author

Sambasiva Rao Mukkamala, a gold medalist, is an Associate Professor at Maharaj Vijayaram Gajapathi Raj (MVGR) College of Engineering, Andhra Pradesh, India. Earlier, he had been associated with the Department of Information Technology, Al Musanna College of Technology, Sultanate of Oman. He received his PhD degree in Mathematics from Andhra University, India, in 2009. He is an active member of the World Academic-Industry Research Collaboration Organization (WAIRCO) and the Andhra Pradesh Akademi of Sciences (APAS). With over 20 years of teaching experience, he has published two books and over 55 research papers. His areas of research are lattice theory, C-algebras, BE-algebras, and implication algebras. He has delivered invited talks on his research of interest in several countries.

Chapter 1
Introduction

Historical Back Ground

Residuation is one of the most important concepts of the theory of ordered algebraic structures which naturally arises in many other fields of mathematics. The study of abstract residuated structures has originated from the investigation of ideal lattices of commutative rings with 1. In general, a partially ordered monoid is residuated if for all a, b in its universe there exist $a \to b = \max\{c : ca \leq b\}$ and $a \rightsquigarrow b = \max\{c : ac \leq b\}$ in other words, if for every a the translations $x \to xa$ and $x \rightsquigarrow ax$ are residuated mappings. If the multiplicative identity is the greatest element in the underlying order, then the monoid is integral. Residuation structures include lattice order groups and their negative cones as well as algebraic models of various propositional logics. In the logical context,the monoid operation $\cdot$ can be interpreted as conjunction and the residuals $\to$ and $\rightsquigarrow$ as two implications (they coincide if and only if the conjuncture is commutative).

The notion of residuation in integral residuated partially ordered commutative monoids can be characterized by Is̀eki's BCK/BCI-algebras. As a generalization of set-theoretic difference and propositional calculus, K. Is̀eki initiated the study of BCK-algebras in 1966. After the investigations of K. Is̀eki, many researchers continued the study and produced significant number of results on the theory of BCK-algebras. More emphasis has been given to the study of ideal theory of BCK-algebras after its characterization. The class of CI-algebras is known to include properly the class of BCK-algebras. Hilbert algebras are introduced for investigations in intuitionistic and other non-classical logics, and BE-algebra is a generalization of dual BCK-algebras.

Past Work on BE-algebras

In 1983, the theory of BCH-algebras was introduced and then developed by Q.P. Hu and X. Li [113]. It is observed that the class of BCI-algebras properly contained in the class of BCH-algebras. As a generalization of BCK-algebras, J. Neggers and

S. R. Mukkamala, *A Course in BE-algebras*,
https://doi.org/10.1007/978-981-10-6838-6_1

H.S. Kim [190] introduced the notion of a d-algebra. Later, these authors introduced the notion of a B-algebra as a non-empty set X together with a binary operation $*$ and a constant 0 satisfying the following axioms:

$$\text{(I) } x * x = 0;\ \text{(II) } x * 0 = 0;\ \text{(III) } (x * y) * z = x * (z * (0 * y)) \text{ for any } x, y, z \in X,$$

In some sense, the notion of B-algebras is equivalent to the groups. In addition to these results in the literature, a wide class of abstract algebras called BH-algebras was introduced by Y.B. Jun, E.H. Roh, and H.S. Kim [147] which is another generalization of BCH/BCI/BCK-algebras, *i.e.*, (I); (II) and (IV) $x * y = 0$ and $y * x = 0$ imply $x = y$ for any $x, y \in X$. In 2006, H.S. Kim and Y.H. Kim [153] introduced a wide class of BE-algebras. It was observed that these classes of BE-algebras were considered as a generalization of the class of BCK-algebras of K. Iśeki and S. Tanaka [124].

The theory of filters of BE-algebras was established by B.L. Meng in [173]. In 2009, S.S. Ahn and K.S. So [6] investigated some of the properties of filters of BE-algebras. The concept of ideals of BE-algebras was introduced by S.S. Ahn and K.S. So [5, 6] and then derived various characterizations of such ideals. introduced the notion of ideals of BE-algebras and proved several characterizations of such ideals. W.A. Dudek [79], Y.B. Jun [134], and M. Kondo [160] made significant contributions to the theory of ideals of BCI-algebras. The concepts of a fuzzy set and a fuzzy relation on a set were initially defined by L.A. Zadeh [250]. Fuzzy relations on a group have been studied by Bhattacharya and Mukherjee [13]. In 1996, Y.B. Jun and S.M. Hong [135] discussed the fuzzy deductive systems of Hilbert algebras. In [212], the author introduced the concept of fuzzy filters of BE-algebras and discussed some related properties. Recently, the concept of fuzzy implicative filters [213], weak filter are introduced and studied the respective properties.

The Present Work

The main aim of this book is to study the structure of various filters and their fuzzification in BE-algebras. The notions of quasi-filters, associative filters, v-filters, dual annihilator filters, ideals, endomorphic ideals, state filters, and injective filters are introduced in BE-algebras. In the chapter preliminaries, a survey of fundamental notions, and known results on BE-algebras are found. Later, some subclasses like positively ordered BE-algebras, negatively ordered BE-algebras, and ordered BE-algebras are introduced and their structures are observed.

This book comprises 14 chapters with preliminaries in Chap. 2. In Chap. 3, the structure of weak subalgebras of BE-algebras is studied along with basic concepts like closure operators, soft BE-algebras, direct products, and algebra of functions. Chapter 4 is devoted to filters of BE-algebras and their properties with respect to homomorphisms, congruences, irreducibility, final segments, dual atoms, and prime filters. Structure of the set of all weak filters of BE-algebras is also studied in this chapter. Chapter 5 is devoted to quasi-filters and pseudo filters of BE-algebras. An interconnection between the class of quasi-filters and the class of multipliers is studied with the help of congruences. In Chap. 6, the focus is given to very true operators of BE-algebras and the filters generated by these operators. Main emphasis is also given to the properties of the subclasses of υ-filters which are prime υ-filters and injective filters.

In Chap. 7, pseudo-complements are introduced in BE-algebras. A great deal of attention has been paid to study the properties of Boolean elements, closed elements, and dense elements. The notion of ideals is introduced with the help of pseudo-complements, and the characterizations of ideals are derived. Chapter 8 is devoted to stabilizers of BE-algebras. Some topological properties are studied with the help of left and right stabilizers. Chapters 9 and 10 deal with two typical areas of investigations in BE-algebras. Bosbach states on bounded BE-algebras are one of the most promising fields, which are investigated in Chap. 9. In Chap. 9, most of the basic facts about states and pseudo-states are presented along with some more specialized results. Pseudo-states and pseudo state-morphisms are introduced in terms of pseudo-complements of BE-algebras. Chapter 10 grew out of an investigation of state operators on BE-algebras. Main focus is given to study the relationship between state filters and state congruences. Subdirectly irreducible state BE-algebras are characterized in this chapter.

Chapter 11 is devoted to a discussion of right and left self-mappings of BE-algebras. A few characterization theorems of self-distributive BE-algebras, commutative BE-algebras, implicative BE-algebras, and 3-potent BE-algebras are established with the help of right and left self-maps. In Chap. 12, the concept of endomorphisms and the fundamental homomorphic theorems are described to explain endomorphic ideals and endomorphic congruences.

Fuzzy mathematics is an emerging field where the ordinary concepts are getting transferred into fuzzy case. In fuzzification of filters, the elements of BE-algebra are mapped to the real numbers of the interval [0, 1], and hence it is observed that fuzzy filters are different from ordinary filters. In 1965, L.A. Zadeh [250] initiated the study of fuzzy sets. Since then many researchers around the world applied fuzzy concepts to many algebraic structures like rings, semirings, semigroups, modules, matrices, vector spaces, etc. In 1991, Xi [244] applied the concept of fuzzy sets to BCK-algebras which are introduced by Imai and Iśeki [118]. In [109], S.M. Hong and Y.B. Jun applied the idea of fuzzy subalgebras to BCK-algebras. In this work, some of these fuzzy concepts are applied to weak filters, implicative filters, transitive filters of BE-algebras.

Chapters 13–15 are devoted to the fuzzification of various filters and implicative filters of BE-algebras. Chapter 13 starts with a careful development of fuzzy weak filters and their homomorphic images, and normal fuzzy weak filters, which are subsequently used in our study of implicative filters, transitive filters, and semitransitive filters. In Chap. 14, properties of implicative filters are investigated and weak implicative filters are introduced in BE-algebras. Fuzzy concept is applied to the classes of implicative and weak implicative filters. The final chapter gives the reader a leisurely introduction to transitive filters, semitransitive filters of BE-algebras. Fuzzification and triangular normed fuzzification are also considered for semitransitive filters.

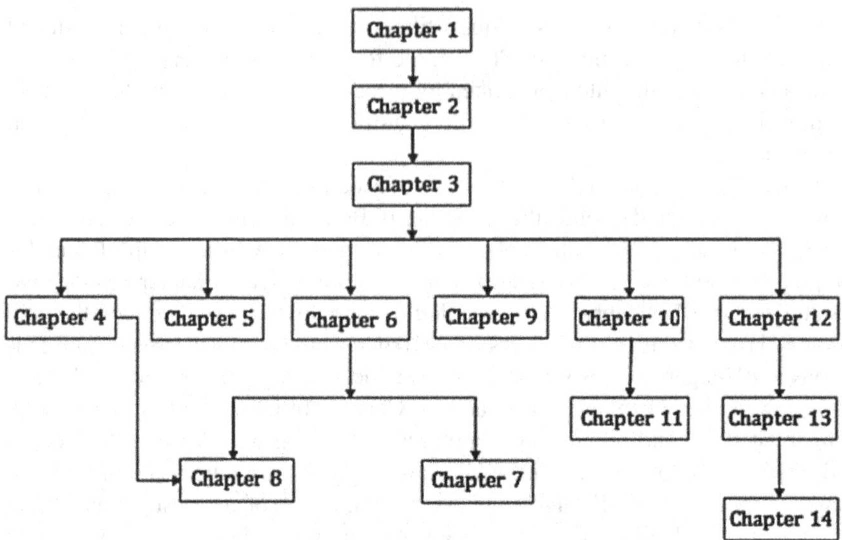

Fig. 1.1 Organization of chapters

The abandonment of the traditional structure of a BE-algebra book means that concepts and notations are more evenly introduced throughout. A very detailed index and the table of notations should help the reader in finding where a concept or notation is first introduced.

Finally, the reader will note that the symbol □ is placed at the end of a proof. Chapters are numbered in Arabic numerals 0 through 15, the sections in a chapter are given by Arabic numerals, 1, 2, etc. Thus Theorem 8.5.10 refers to Theorem 10 of Sect. 5 in Chap. 9. References to the Bibliography are given in the form [24], which refers to a paper with serial number 24 in the References section. "[52]" refers to a paper that had not appeared in print at the time the manuscript of this book was submitted for publication.

Organization of Chapters

Apart from the front matter, preliminaries are the first chapter of this book. In the following figure, the arrangement of the chapters based on their weightage and interdependency is shown (Fig. 1.1).

Chapter 2
Preliminaries

Preliminary pages include everything up to the main body of the text or introduction. In this chapter, a few important definitions and results are collected from various sources for the use in the forthcoming chapters. Some basic and important properties of BE-algebras, BCK-algebras, some special classes of BE-algebras like self-distributive, commutative, transitive, implicative and ordered BE-algebras are observed. Some preliminary properties of ordered relations, partially ordered sets and congruences are quoted for further reference to the reader.

2.1 BE-algebras

In this section, we present certain definitions and results which are taken mostly from the papers [118, 124, 152, 153, 239] for the ready reference of the reader. Hilbert algebras are important tools for certain investigations in algebraic logic since they can be considered as fragments of any propositional logic containing a logical connective implication and the constant 1 which is considered as the logical value true. The concept of Hilbert algebras was introduced by L. Henkin and T. Skolem for investigations in intuitionistic and other non-classical logics.

Y. Imai and K. Iśeki introduced two classes of abstract algebras: BCK-algebras and BCI-algebras [118]. It is known that the class of BCK-algebras is a proper subclass of the class of BCI-algebras. An algebra $(X, *, 0)$ of type $(2, 0)$ is said to be a BCI-algebra if it satisfies the following conditions for all $x, y, z \in X$:

(BCI1) $((x * y) * (x * z)) * (z * y) = 0$,
(BCI2) $(x * (x * y)) * y = 0$,
(BCI3) $x * x = 0$,
(BCI4) $x * y = 0$ and $y * x = 0$ imply $x = y$.

If a BCI-algebra X satisfies the condition,

(BCI5) $0 * x = 0$ for all $x \in X$,

S. R. Mukkamala, *A Course in BE-algebras*,
https://doi.org/10.1007/978-981-10-6838-6_2

then X is called a BCK-algebra. Hence, a BCK-algebra is a BCI-algebra. A BCK-algebra is an important class of logical algebras introduced by Iśeki [121] which was extensively investigated by several researchers. In set theory, there are three basic operations, viz union, intersection, and subtraction. If we consider them all simultaneously, then by generalizing their characteristics, the Boolean algebra comes into being; if we consider merely the first two of them, then by abstracting, the lattice theory is established; if we consider only the last one of them, owing to some characteristics of it being very similar to those of the subtraction in real numbers set R as well as the implication operator in propositional calculus, by abstracting, the BCK-algebra arises.

Example 2.1.1 Let $X = \{0, a, b, c, d\}$. Define a binary operation $*$ on X as follows:

$*$	0	a	b	c	d
0	0	0	0	0	0
a	a	0	a	0	0
b	b	b	0	0	b
c	c	b	a	0	b
d	d	a	d	a	0

It can be easily seen that $(X, *, 0)$ is a BCK-algebra.

In a BCI/BCK-algebra, we can define a binary relation $\leq$ on X by $x \leq y$ if and only if $x * y = 0$. Then $(X, \leq)$ is a partial ordering set. In a BCI-algebra X, the following hold:

(BCI6) $(x * y) * z = (x * z) * y$,

(BCI7) $x * (x * (x * y)) = x * y$,

(BCI8) $((x * z) * (y * z)) * (x * y) = 0$,

(BCI9) $x * 0 = x$,

(BCI10) $0 * (x * y) = (0 * x) * (0 * y)$

(BCI11) $x \leq y$ implies $x * z \leq y * z$ and $z * y \leq z * x$ for all $x, y, z \in X$.

H.S. Kim and Y.H. Kim [153] introduced the concept of BE-algebras. These classes of BE-algebras were introduced, as a generalization of the class of BCK-algebras of K. Iśeki and S. Tanaka [124], as an algebra $(X, *, 1)$ of type $(2, 0)$ which satisfies the following for all $x, y, z \in X$:

(BE1) $x * x = 1$,

(BE2) $x * 1 = 1$,

(BE3) $1 * x = x$,

(BE4) $x * (y * z) = y * (x * z)$.

Example 2.1.2 Let $X = \{1, a, b, c\}$. Define a binary operation $*$ on X as follows:

$*$	1	a	b	c
1	1	a	b	c
a	1	1	a	c
b	1	1	1	c
c	1	a	b	1

It can be easily seen that $(X, *, 1)$ is a BE-algebra.

After the introduction of BCK-algebras by Iśeki in 1966, a great deal of literature has been produced on the theory of BCK-algebras. Some of the preliminaries are presented here.

Theorem 2.1.3 ([152]) *Let* $(X, *, 1)$ *be a BE-algebra. Then we have the following:*
(BE5) $x * (y * x) = 1$
(BE6) $x * ((x * y) * y)) = 1$.

Proof In the presence of (BE1) and (BE4), it is obvious. □

Definition 2.1.4 ([239]) An algebra $(X, *, 1)$ of type $(2, 0)$ is called a *dual BCK-algebra* if it satisfies the following properties for all $x, y, x \in X$:
(BE1) $x * x = 1$,
(BE2) $x * 1 = 1$,
(dBCK1) $x * y = y * x = 1 \Rightarrow x = y$,
(dBCK2) $(x * y) * ((y * z) * (x * z)) = 1$,
(dBCK3) $x * ((x * y) * y) = 1$.

Lemma 2.1.5 ([239]) *Let* $(X, *, 1)$ *be a dual BCK algebra and* $x, y, z \in X$. *Then*
(1) $x * (y * z) = y * (x * z)$,
(2) $1 * x = x$.

Proposition 2.1.6 ([239]) *Any dual BCK algebra is a BE-algebra.*

We introduce a relation $\leq$ on a BE-algebra X by $x \leq y$ if and only if $x * y = 1$ for all $x, y \in X$. A non-empty subset S of a BE-algebra X is called a subalgebra of X if $x * y \in S$ whenever $x, y \in S$. A BE-algebra X is called *self-distributive* if $x * (y * z) = (x * y) * (x * z)$ for all $x, y, z \in X$. A BE-algebra X is called *simple* if $x * (x * y) = x * y$ for all $x, y \in X$. It is clear that every self-distributive BE-algebra is simple.

Definition 2.1.7 ([239]) A BE-algebra X is called a *commutative BE-algebra* if $(x * y) * y = (y * x) * x$ for all $x, y \in X$.

Example 2.1.8 Let $\mathbb{N}_0 = \mathbb{N} \cup \{0\}$. Define the binary operation $*$ on $\mathbb{N}_0$ as follows:

$$x * y = \begin{cases} 0 & \text{if } y \leq x \\ y - x & \text{if } x < y \end{cases}$$

Then, $(\mathbb{N}_0, *, 0)$ is a commutative BE-algebra.

If X is a commutative BE-algebra and $x * y = y * x = 1$, then $x = y$ for all $x, y \in X$. We note that $\leq$ is reflexive by (BE1). If X is transitive, then relation $\leq$ is a transitive order on X. If X is commutative, then it is transitive and hence the relation $\leq$ is transitive. Hence if X is a commutative BE-algebra, then relation $\leq$ is a partial ordering on X. The ordered pair $(X, \leq)$ is called a partially ordered BE-algebra.

In the following, the notion of *ordered BE-algebras* is introduced. The interconnections between ordered BE-algebras and the existing classes of BE-algebras are studied particularly.

Definition 2.1.9 A BE-algebra X is said to be *positively ordered* if $x \leq y$ implies $z * x \leq z * y$ for all $x, y, z \in X$. It is called *negatively ordered* if $x \leq y$ implies $y * z \leq x * z$ for all $x, y, z \in X$. X is called *ordered* BE*-algebra*, if it is both positively ordered and negatively ordered.

Theorem 2.1.10 ([201]) *Every self-distributive BE-algebra is ordered.*

Proposition 2.1.11 *Let* $(X, *, 1)$ *be a negatively ordered BE-algebra. Then for any* $x, y, z \in X$,

$$(x * y) * (x * z) \leq x * (y * z)$$

Proof Let $x, y, z \in X$. Then $y \leq x * y \Rightarrow (x * y) * (x * z) \leq y * (x * z) \Rightarrow (x * y) * (x * z) \leq x * (y * z)$. Therefore $(x * y) * (x * z) \leq x * (y * z)$ for all $x, y, z \in X$. □

Proposition 2.1.12 ([239]) *Let* $(X, *, 1)$ *be a commutative BE-algebra. Then for any* $x, y \in X$,

$$((x * y) * y) * y = x * y$$

It was already observed by many authors that self-distributive BE-algebras and commutative BE-algebras are independent classes. However, the following theorem establishes an equivalent condition for a commutative BE-algebra to become a self-distributive BE-algebra.

Theorem 2.1.13 ([239]) *A commutative BE-algebra is self-distributive if and only if it satisfies the condition:* $x * (x * y) = x * y$ *for all* $x, y \in X$.

Hilbert algebras are important tools for certain investigations in algebraic logic since they can be considered as fragments of any propositional logic containing a logical connective implication and the constant 1 which is the logical value true. The concept of Hilbert algebra was introduced by Henkin and Skolem for investigations in intuitionistic and other non-classical logics.

Definition 2.1.14 ([71]) A *Hilbert algebra* is a triple $(H, *, 1)$, where H is a non-empty set, $*$ is a binary operation on H and 1 is a fixed element of H (i.e., a nullary operation) such that the following three axioms are satisfied, for all $x, y, z \in H$:

(H1) $x * (y * x) = 1$,
(H2) $(x * (y * z)) * ((x * y) * (x * z)) = 1$,
(H3) if $x * y = 1$ and $y * x = 1$, then $x = y$.

We introduce a relation $\leq$ on H by $x \leq y$ if and only if $x * y = 1$. It is easy to see that the relation $\leq$ is a partial order on H which is called the natural ordering on H. We say that H is commutative if $(x * y) * y = (y * x) * x$, for all $x, y \in H$.

Proposition 2.1.15 ([69]) *Let* $(H, *, 1)$ *be a Hilbert algebra. Then the following conditions hold for all* $x, y, z \in H$:

(H4) $x * x = 1$,
(H5) $1 * x = x$,

(H6) $x * 1 = 1$,
(H7) $x * (y * z) = y * (x * z)$,
(H8) $x * (y * z) = (x * y) * (x * z)$,
(H9) *if* $x \leq y$, *then* $z * x \leq z * y$ *and* $y * z \leq x * z$.

It is proved in [200] that every Hilbert algebra is a self-distributive BE-algebra but the converse is not true. It can be seen in the following example.

Example 2.1.16 Let $X = \{1, a, b\}$. Define a binary operation $*$ on X as follows:

$*$	1	a	b
1	1	a	b
a	1	1	1
b	1	1	1

It can be easily seen that $(X, *, 1)$ is a self-distributive BE-algebra. But $(X, *, 1)$ is not a Hilbert algebra, because $a * b = b * a = 1$, while $a \neq b$.

Definition 2.1.17 ([1]) An *implication algebra* is a set X with a binary operation $*$ which satisfies, for all $x, y, z \in X$, the following axioms:
(I1) $(x * y) * x = x$,
(I2) $(x * y) * y = (y * x) * x$,
(I3) $x * (y * z) = y * (x * z)$.

Proposition 2.1.18 ([1]) *In any implication algebra* $(X, *)$, *the following hold for all* $x, y \in X$:
(1) $x * (x * y) = x * y$,
(2) $x * x = y * y$,
(3) *There exists a unique element* $1 \in X$ *such that,*
  (a) $x * x = 1$, $1 * x = x$ *and* $x * 1 = 1$,
  (b) *if* $x * y = 1$ *and* $y * x = 1$, *then* $x = y$.

From the above proposition, it is clear that every implication algebra is a BE-algebra. In [200], equivalency among Hilbert algebras, self-distributive BE-algebras and commutative BE-algebras is derived. In 2002, Halaš [106] showed that commutative Hilbert algebras are implication algebras and Diego [69] proved that implication algebras are Hilbert algebras. Hence in conclusion, it is observed that

commutative self-distributive BE-algebra $\Leftrightarrow$ commutative Hilbert algebra $\Leftrightarrow$ implication algebra

Definition 2.1.19 ([191]) An *implicative semilattice* is an algebra $(X, \leq, \wedge, \rightarrow, 1)$ of type $(2, 2, 2, 1)$ in which X is a non-empty set, $\leq$ is a partial order on X, $\wedge$ is the g.l.b. with respect to $\leq$ and $\rightarrow$ is a binary operation in X such that for any $x, y, z \in X$,

$$z \leq x \rightarrow y \text{ if and only if } z \wedge x \leq y$$

Definition 2.1.20 ([205]) A BE-algebra $(X, *, 1)$ is said to be an *implicative BE-algebra* if it satisfies the implicative condition $x = (x * y) * x$ for all $x, y \in X$.

Example 2.1.21 Let $X = \{1, a, b, c\}$. Define a binary operation $*$ on X as follows:

$*$	1	a	b	c
1	1	a	b	c
a	1	1	b	c
b	1	a	1	c
c	1	a	b	1

Then $(X, *, 1)$ is an implicative BE-algebra.

It is also proved in [205] that every commutative implicative BE-algebra is an implication algebra and a commutative BE-algebra X is self-distributive if and only if it is an implicative BE-algebra.

Proposition 2.1.22 ([1]) *Every self-distributive and commutative BE-algebra is implicative.*

Proposition 2.1.23 *Every positively ordered and implicative BE-algebra is commutative.*

Proof Let X be a positively ordered and implicative BE-algebra. Let $x, y \in X$. Since X is implicative, we get that $(y * x) * y = y$. Clearly $x * ((x * y) * y) = (x * y) * (x * y) = 1$. Hence $x \leq ((x * y) * y)$. Since X is positively ordered BE-algebra, we get

$$\begin{aligned}(y * x) * x &\leq (y * x) * ((x * y) * y) \\ &= (x * y) * ((y * x) * y) \\ &= (x * y) * y.\end{aligned}$$

Hence $(y*x)*x \leq (x*y)*y$. Interchanging x and y, we get $(x*y)*y \leq (y*x)*x$. Thus $(x * y) * y = (y * x) * x$. Therefore X is commutative. □

As every self-distributive BE-algebra is positively ordered, the following is clear.

Corollary 2.1.24 *Let* $(X, *, 1)$ *be a self-distributive BE-algebra. Then* X *is commutative if and only if* X *is implicative.*

Definition 2.1.25 ([5]) A BE-algebra $(X, *, 1)$ is said to be a *transitive BE-algebra* if it satisfies the condition: $y * z \leq (x * y) * (x * z)$ for all $x, y, z \in X$.

Example 2.1.26 Let $X = \{1, a, b, c\}$. Define a binary operation $*$ on X as follows:

$*$	1	a	b	c
1	1	a	b	c
a	1	1	a	a
b	1	1	1	a
c	1	1	a	1

It can be easily seen that $(X, *, 1)$ is a transitive BE-algebra.

Proposition 2.1.27 ([173]) *Every transitive BE-algebra is an ordered BE-algebra.*

Proposition 2.1.28 ([173]) *Every commutative BE-algebra is transitive.*

Proposition 2.1.29 ([173]) *Every self-distributive BE-algebra is transitive.*

For any $x, y \in X$, A. Walendziak [239] defined the operation $\vee$ as $x \vee y = (y * x) * x$. Using the axioms of BE-algebra, we can observe that $x \vee y$ is an upper bound for x and y in X. Moreover, we have the following:

Theorem 2.1.30 ([239]) *Let $(X, *, 1)$ be a commutative BE-algebra. Then X is an upper semilattice with respect to the operation $\vee$.*

Proposition 2.1.31 ([239]) *Let $(X, *, 1)$ be a commutative BE-algebra and $x, y, z \in X$. Then the following conditions hold.*
(1) $x * (y \vee z) = (z * y) * (x * y)$,
(2) $x \leq y$ *implies* $x \vee y = y$,
(3) $z \leq x$ *and* $x * z \leq y * z$ *imply* $y \leq x$.

Theorem 2.1.32 ([203]) *Let X be a commutative BE-algebra. Then for all $x, y \in X$, the greatest lower bound* $\inf\{x, y\} = x \wedge y$ *for brevity, is* $x \wedge y = [(x * u) \vee (y * u)] * u$ *whenever* $u \leq x, y$.

Corollary 2.1.33 *Let X be a commutative BE-algebra. If* $\inf\{x, y\} = x \wedge y$ *exists for all $x, y \in X$, then $(X, \vee, \wedge)$ is a lattice.*

Definition 2.1.34 A BE-algebra X is said to satisfy the condition ***L*** if for all $x, y \in X$, there exists $u \in X$ such that $u \leq x$ and $u \leq y$.

Proposition 2.1.35 ([203]) *Let $(X, *, 1)$ be a commutative BE-algebra with condition **L**. For any $x, y, z \in X$, the following conditions hold.*
(1) $(x \vee y) * z = (x * z) \wedge (y * z)$,
(2) $x * (y \wedge z) = (x * y) \wedge (x * z)$,
(3) $x * (x \wedge y) = x * y$,
(4) $(x * y) \vee (y * x) = 1$,
(5) $(x \wedge y) * z = (x * z) \vee (y * z)$.

Theorem 2.1.36 ([203]) *Let $(X, *, 1)$ be a commutative BE-algebra. Then X is a distributive lattice with respect to the operations $\vee$ and $\wedge$ defined in the above theorems.*

For any commutative BE-algebra with condition ***L***, it is observed that $(X, \wedge)$ is a semilattice which is called BE-semilattice. Clearly every implicative semilattice is a BE-semilattice. We have some facts about BE-semilattices which are observed in the following results.

Theorem 2.1.37 ([203]) *Let* X *be a transitive BE-algebra. Then the following are equivalent.*

(1) X *is commutative;*

(2) $y \leq x$ *implies* $x = (x * y) * y$ *for* $x, y \in X$*;*

(3) $(x \vee y) * (y \vee x) = 1$ *for all* $x, y \in X$.

Definition 2.1.38 ([153]) A non-empty subset F of BE-algebra X is called a filter of X if for all $x, y \in X$

(F1) $1 \in F$,

(F2) $x \in F, x * y \in F$ imply that $y \in F$.

Theorem 2.1.39 ([153]) *Let* F *be a filter of a BE-algebra* X*. Then* $x \in F, x \leq y$ *imply that* $y \in F$.

In the following, we observe the filters of BE-algebras generated by non-empty subsets. The proof of the following result is obtained for self-distributive BE-algebra in [8]. We present this result for ordered BE-algebras (particularly positively ordered BE-algebras).

Theorem 2.1.40 *Let* A *be a non-empty subset of an ordered BE-algebra. Then*

$$B = \{x \in X \mid a_n * (\cdots * (a_1 * x) \cdots) = 1 \text{ for some } a_1, \ldots, a_n \in A\}$$

is the smallest filter of X *containing* A.

Proof Clearly $1 \in B$. Let $x, x * y \in B$. Then there exists $a_1, a_2, \ldots, a_n \in A$ and $b_1, b_2, \ldots, b_m \in A$ such that $a_n * (\cdots * (a_1 * x) \cdots) = 1$ and $b_m * (\cdots * (b_1 * (x * y)) \cdots) = 1$. Hence

$$\begin{aligned} 1 &= b_m * (\cdots * (b_1 * (x * y)) \cdots) \\ &= b_m * (\cdots * (x * (b_1 * y)) \cdots) \\ &\ldots \\ &\ldots \\ &= x * (b_m * (\cdots * (b_1 * y)) \cdots) \end{aligned}$$

Hence $x \leq b_m * (\cdots * (b_1 * y) \cdots)$. Since X is an ordered BE-algebra, by the sequential application of the BE-ordering $\leq$, we get that $1 = a_n * (\cdots * (a_1 * x) \cdots) \leq a_n * (\cdots * (a_1 * (b_m * (\cdots * (b_1 * y) \cdots))) \cdots)$. Hence, we get $a_n * (\cdots * (a_1 * (b_m * (\cdots * (b_1 * y) \cdots))) \cdots) = 1$ where $a_1, a_2, \ldots, a_n, b_1, b_2, \ldots, b_m \in A$. Thus $y \in B$. Therefore B is a filter of X. For any $x \in A$, we get $x * (\cdots * (x * x) \cdots) = 1$. Hence, we get that $x \in B$. Therefore $A \subseteq B$.

We now prove that B is the smallest filter containing A. Let F be a filter of X containing A. Let $x \in B$. Then there exists $a_1, a_2, \ldots, a_n \in A$ such that $a_n * (\cdots * (a_1 * x) \cdots) = 1 \in F$. Since $A \subseteq F$, we get that $a_1, a_2, \ldots, a_n \in F$. Since $a_n \in F$, $a_n * (\cdots * (a_1 * x) \cdots) \in F$ and F is a filter, it implies $a_{n-1} * (\cdots * (a_1 * x) \cdots) \in$

F. Since $a_{n-1} \in F$, $a_{n-1} * (\cdots * (a_1 * x) \cdots) \in F$ and F is a filter, it implies $a_{n-2} * (\cdots * (a_1 * x) \cdots) \in F$. Continuing in this way, we finally get that $x \in F$. Hence $B \subseteq F$. Therefore B is the smallest filter containing A. □

Notation 2.1.41 *For any non-empty subset A of an ordered BE-algebra, the smallest filter containing A is called the filter generated by A, which is denoted by $\langle A \rangle$. Obviously we have $\langle A \rangle \subseteq \langle B \rangle$ whenever $A \subseteq B$ for any two subsets A and B of X. Clearly $\langle A \rangle \cup \{1\} = \langle B \rangle$. If A contains the element 1, then $\langle A \rangle = B$. Moreover, if B is a filter of X, then*

$$\langle B \cup A \rangle = \{x \in X \mid a_n * (\cdots * (a_1 * x) \cdots) \in B \text{ for some } n \in \mathbb{N} \text{ and } a_1, \ldots, a_n \in A\}$$

If $A = \{a\}$, then denote $\langle \{a\} \rangle$ by $\langle a \rangle$ and call this principal filter generated by a. Hence $\langle a \rangle = \{x \in X \mid a^n * x = 1 \text{ for some } n \in \mathbb{N}\}$ where $a^n * x = a * (\cdots * (a * x) \cdots)$ and a occurs n times.

Proposition 2.1.42 *If X is self-distributive, then $\langle a \rangle = \{x \in X \mid a * x = 1\}$.*

Proposition 2.1.43 *Let F and G be two filters of an ordered BE-algebra X. Then*

$$\langle F \cup G \rangle = \{x \in X \mid a * (b * x) = 1 \text{ for some } f \in F \text{ and } g \in G\}$$

Proof Let $A = \{x \in X \mid a * (b * x) = 1$ for some $f \in F$ and $g \in G\}$. We now prove that $A = \langle F \cup G \rangle$. Let $x \in A$. Then $f * (g * x) = 1$ for some $f \in F$ and $g \in G$. Hence by Theorem 2.1.40, it gives that $x \in \langle F \cup G \rangle$. Thus it implies $A \subseteq \langle F \cup G \rangle$.

Conversely, let $x \in \langle F \cup G \rangle$. Then there exists $a_1, a_2, \ldots, a_i, \ldots, a_n \in F \cup G$ such that $a_n * (\cdots * (a_1 * x) \cdots) = 1$. By rearranging the $a_i's$ in the above, we get

$$a_n * (\cdots * (a_{i+1} * (a_i \cdots * (a_1 * x) \cdots)) \cdots) = 1 \in G$$

such that $a_1, \ldots, a_i \in F$ and $a_{i+1}, \ldots, a_n \in G$. Since $a_n \in G$ and G is a filter, we get that

$$a_{n-1} * (\cdots * (a_{i+1} * (a_i \cdots * (a_1 * x) \cdots)) \cdots) \in G.$$

By continuing this, we get $a_i * (\cdots * (a_1 * x) \cdots) \in G$. Put $g = a_i * (\cdots * (a_1 * x) \cdots)$. Then

$$\begin{aligned} a_i * (\cdots * (a_1 * (g * x)) \cdots) &= g * (a_i * (\cdots * (a_1 * x) \cdots)) \\ &= (a_i * (\cdots * (a_1 * x) \cdots)) * (a_i * (\cdots * (a_1 * x) \cdots)) \\ &= 1 \in F \end{aligned}$$

Since $a_i \in F$ and F is a filter, it yields that $a_{i-1} * (\cdots * (a_1 * (g * x))) \in F$. By continuing this argument, we get $g * x \in F$. Put $f = g * x$. Then $f * (g * x) = (g * x) * (g * x) = 1$. Since $f \in F$ and $g \in G$, it implies that $x \in A$. Hence $\langle F \cup G \rangle \subseteq A$. Therefore $A = \langle F \cup G \rangle$. □

In what follows, $\mathcal{F}(X)$ denotes the class of all filters of a BE-algebra X. Then, for any two filters F and G of a BE-algebra, it can be easily seen that $F \cap G$ is the infimum of both F and G. Now, in the following theorem, we obtain that $\mathcal{F}(X)$ forms a complete lattice.

Theorem 2.1.44 *For any ordered BE-algebra X, $\mathcal{F}(X)$ forms a complete distributive lattice.*

Proof For any two filters F and G of an ordered BE-algebra, define

$$F \vee G = \langle F \cup G \rangle = \{a \in X \mid x * (y * a) = 1 \text{ for some } x \in F \text{ and } y \in G\}$$

Then clearly $(\mathcal{F}(X), \cap, \vee)$ is a complete lattice with respect to set inclusion. Let $F, G, H \in \mathcal{F}(X)$. Then clearly $F \cap (G \vee H) \subseteq (F \cap G) \vee (F \cap H)$. Conversely, let $x \in F \cap (G \vee H)$. Then $x \in F$ and $x \in G \vee H$. Then there exists $g \in G$ and $h \in H$ such that $g * (h * x) = 1$. Now letting

$$b_1 = h * x \text{ and } b_2 = b_1 * x$$

it is clear that $b_1 \in F$ and $b_2 \in F$. Now $g * b_1 = g * (h * x) = 1 \in G$. Since $g \in G$ and G is a filter, we get that $b_1 = h * x \in G$. Hence $b_1 \in F \cap G$. Again,

$$\begin{aligned} h * b_2 &= h * (b_1 * x) \\ &= h * ((h * x) * x) \\ &= (h * x) * (h * x) \\ &= 1 \in H \end{aligned}$$

Since $h \in H$ and H is a filter, we get that $b_2 \in H$. Hence $b_2 \in F \cap H$. Now

$$\begin{aligned} b_1 * (b_2 * x) &= (h * x) * ((b_1 * x) * x) \\ &= (h * x) * (((h * x) * x) * x) \\ &= ((h * x) * x) * ((h * x) * x) \\ &= 1 \end{aligned}$$

Since $b_1 \in F \cap G$ and $b_2 \in F \cap H$, it yields that $x \in (F \cap G) \vee (F \cap H)$. Hence $F \cap (G \vee H) \subseteq (F \cap G) \vee (F \cap H)$. Therefore $(\mathcal{F}(X), \cap, \vee)$ is a complete distributive lattice. □

Corollary 2.1.45 *Let X be an ordered BE-algebra. Then the class $\mathcal{F}(X)$ of all filters of X is a complete lattice with respect to the inclusion ordering $\subseteq$ in which for any set $\{F_\alpha\}_{\alpha \in \Delta}$ of filters of X, $inf\{F_\alpha\}_{\alpha \in \Delta} = \bigcap_{\alpha \in \Delta} F_\alpha$ and $sup\{F_\alpha\}_{\alpha \in \Delta} = \langle \bigcup_{\alpha \in \Delta} F_\alpha \rangle$.*

Definition 2.1.46 A filter F of a BE-algebra X is called *proper* if $F \neq X$.

Definition 2.1.47 A proper filter M of a BE-algebra X is said to be a *maximal filter* if $\langle M \cup \{x\}\rangle = X$ for any $x \in X - M$, where $\langle M \cup \{x\}\rangle$ is the filter generated by $M \cup \{x\}$.

In the following theorem, we derive a necessary and sufficient condition for every proper filter of an ordered BE-algebra to become a maximal filter.

Theorem 2.1.48 *A proper filter M of an ordered BE-algebra X is maximal if and only if $M \subseteq F \subseteq X$ implies $M = F$ or $F = X$ for any filter F of X.*

Proof Assume that M is a maximal filter of X. Suppose there is a filter F of X such that $M \subseteq F \subseteq X$. Let $M \neq F$. We now show that $F = X$. Suppose $F \neq X$. Choose $x \in X$ such that $x \notin F$. Then clearly $x \notin M$. Since M is a maximal filter, we get that $\langle M \cup \{x\}\rangle = X$. Let $a \in X$ be arbitrary. Then $a \in \langle M \cup \{x\}\rangle$. Then there exists $n \in \mathbb{N}$ such that

$$x^n * a \in M \subseteq F$$

Since $x \in F$ and F is a filter, we get that $a \in F$. Therefore $F = X$.

Conversely, assume that the condition holds. Let $x \in X - M$. Suppose $\langle M \cup \{x\}\rangle \neq X$. Choose $a \notin \langle M \cup \{x\}\rangle$ and $a \in X$. Hence $M \subseteq \langle M \cup \{x\}\rangle \subset X$. Then by the assumed condition, we get $M = \langle M \cup \{x\}\rangle$. Hence $x \in M$, which is a contradiction. Thus M is a maximal filter of X. □

Definition 2.1.49 ([197]) Let H be a non-empty set and $\circ : H \times H \to \mathcal{P}^*(X)$ be a hyperoperation. Then $(H, \circ, 1)$ is called a hyper BE-algebra if it satisfies the following conditions:

(HBE1) $x < 1$ and $x < x$,

(HBE2) $x \circ (y \circ z) = y \circ (x \circ z)$,

(HBE3) $x \in 1 \circ x$,

(HBE4) $1 < x$ implies $x = 1$, for all $x, y, z \in H$, where the relation $<$ is defined by $x < y$ if and only if $1 \in x \circ y$. For any two non-empty subsets A and B of H, $A < B$ means there exists $a \in A$ and $b \in B$ such that $a < b$ and $A \circ B = \bigcup\limits_{a \in A, b \in B} a \circ b$. In addition, we write $A \ll B$ if for any $a \in A$, there exists $b \in B$ such that $a < b$.

Theorem 2.1.50 ([197]) *Let $(H, \circ, 1)$ be a hyper BE-algebra. Then*

(1) $A \circ (B \circ C) = B \circ (A \circ C)$,

(2) $A < A$,

(3) $1 < A$ *implies* $1 \in A$,

(4) $x < y \circ x$,

(5) $x < y \circ z$ *implies* $y < x \circ z$,

(6) $x < (x \circ y) \circ y$,

(7) $y \in 1 \circ x$ *implies* $y < x$ *for all* $x, y, z \in H$ *and* $A, B, C \subseteq H$.

Definition 2.1.51 ([197]) A non-empty subset S of a hyper BE-algebra H is said to be a hyper subalgebra of H, if $x \circ y \subseteq S$ for all $x, y \in S$.

2.2 Order Relations

In this section, we study various types of ordered relations in BE-algebras. Some basic properties of partially ordered sets and lattices are presented. We present some of the necessary and preliminary concepts for the ready reference of the reader.

To make the definitions formal, let us start with two sets A, B and form the set $A \times B$ of all ordered pairs (a, b) with $a \in A$ and $b \in B$. If $A = B$, we write $A^2 = A \times A$. Then a binary relation R on A can simply be defined as a subset of A^2. The elements $a, b \in A$ are in relation with respect to R if and only if $(a, b) \in R$. For $(a, b) \in R$, we will also write aRb, or $a \equiv b(R)$. Binary relations will be denoted by small Greek letters or by special symbols. The basic properties of the binary relation R on A are as follows:

For all $x, y, z \in A$, we have

Reflexive: $(x, x) \in R$.
Symmetric: $(x, y) \in R$ implies $(y, x) \in R$.
Antisymmetric: $(x, y) \in R$ and $(y, x) \in R$ imply $x = y$.
Transitive: $(x, y) \in R$ and $(y, z) \in R$ imply $(x, z) \in R$.

A *quasi-order* on a non-empty set A is a binary relation that is reflexive and transitive. If R is a quasi-order on A, then the pair (A, R) is called a quasi-ordered set. A binary relation R on a non-empty set A is called an *equivalence relation* if R is reflexive, symmetric, and transitive.

A binary relation on a non-empty set is called a *partial order* if it is satisfying reflexive, antisymmetric, and transitive. If $\leq$ is a partial order on a non-empty set A, then the pair $(A, \leq)$ is called a *partially ordered set* or a *poset* for brevity. As usual, we write $a \leq b$ or $b \leq a$ instead of $(a, b) \in \leq$. It is common to write $a < b$ if $a \leq b$ and $a \neq b$. Partially ordered sets are very useful in higher-level mathematics. It plays an important role in modern algebra as well as abstract algebra. Posets are also studied under topology: an important branch of higher mathematics.

A poset $(A, \leq)$ is called a *chain* or a *totally ordered set* (*linearly ordered set*) if for all $a, b \in A$, $a \leq b$ or $b \leq a$. On the other hand, the poset $(A, \leq)$ is an *antichain* provided any two distinct elements $a, b \in A$ are *incomparable*, that is, if neither $a \leq b$ nor $b \leq a$, in symbols $a||b$.

An element a of a poset P is called *maximal* if there exists no element $x \in P$ such that $a < x$, and a is called the *greatest element* of the poset P if $x \leq a$ for all $x \in P$. Dually, we define the notion of a *minimal element* and of the *least element* of P. Obviously, any poset possesses at most one greatest element and at most one least element. It can be easily seen that every greatest element (least element) is maximal (minimal) but not the converse. The greatest element is denoted by 1 and the least element by 0.

Definition 2.2.1 ([63]) Let $(A, \leq)$ and $(B, \leq)$ be two posets. Then a mapping $f : A \to B$ is called *order preserving* if for all $x, y \in A$, $x \leq y$ implies $f(x) \leq f(y)$.

An order preserving mapping on a poset is also known as **monotone**. A mapping $f : A \to B$ is called *order reversing* if $x \leq y$ implies $f(y) \leq f(x)$ for all $x, y \in A$.

An order reversing map is also known as *antitone*. Furthermore, a bijection f of A onto B satisfying the condition $x \leq y$ if and only if $f(x) \leq f(y)$ is called an *isomorphism*, and a bijection f satisfying $x \leq y$ if and only if $f(y) \leq f(x)$ is called a *dual isomorphism* between $(A, \leq)$ and $(B, \leq)$.

Definition 2.2.2 ([63]) A poset $(A, \leq)$ is called a *join-semilattice* if the supremum $a \vee b$ exists for all $a, b \in A$. If for every $a, b \in A$ the infimum $a \wedge b$ exists, then $(A, \leq)$ is called a *meet-semilattice*. A poset $(A, \leq)$ being both join-semilattice and meet-semilattice is called a lattice.

A poset $(A, \leq)$ possesses the supremum $a \vee b$ and the infimum $a \wedge b$ for all $a, b \in A$ if and only if it satisfies the following conditions:

Idempotency: $x \vee x = x;\ x \wedge x = x.$

Commutativity: $x \vee y = y \vee x;\ x \wedge y = y \wedge x.$

Associativity: $x \vee (y \vee z) = (x \vee y) \vee z;\ x \wedge (y \wedge z) = (x \wedge y) \wedge z.$

Absorption: $x \vee (x \wedge y) = x;\ x \wedge (x \vee y) = x.$

We denote, in the above case, the lattice by $(A, \vee, \wedge)$. Moreover, the partial ordering $\leq$ will obey that $x \leq y$ if and only if $x \vee y = y$ if and only if $x \wedge y = x$. Let $(A, \leq)$ be a lattice and $\emptyset \neq B \subseteq A$. Then B is said to be a *sublattice* of A if $a \vee b \in B$ and $a \wedge b \in B$ whenever $a, b \in B$.

We define a lattice $(A, \vee, \wedge)$ to be *meet-distributive* if for all $x, y, z \in A$,

$$x \wedge y = x \wedge z \text{ implies } x \wedge y = x \wedge (y \vee z)$$

A lattice $(A, \vee, \wedge)$ is called *distributive* if it satisfies either of the following equivalent *distributive laws*:

$$x \vee (y \wedge z) = (x \vee y) \wedge (x \vee z),$$
$$x \wedge (y \vee z) = (x \wedge y) \vee (x \wedge z).$$

In addition, the distributivity is equivalent to the *cancelation law*:

$$\text{if } x \vee z = y \vee z \text{ and } x \wedge z = y \wedge z, \text{ then } x = y.$$

Hence a distributive lattice is a lattice in which the operations of join and meet distribute over each other. The prototypical examples of such structures are collections of sets for which the lattice operations can be given by set union and intersection.

Definition 2.2.3 Let $(X, *, 1)$ be a BE-algebra. By a *BE-ordering* or simply *ordering*, we mean a binary relation $\leq$ on the BE-algebra X defined for all $x, y \in X$ as follows:

$$x \leq y \text{ if and only if } x * y = 1.$$

As a consequence of (BE1), the above relation $\leq$ is reflexive. It is clear that $\leq$ is partial ordering whenever X is dual BCK-algebra due to (dBCK1) and (dBCK2).

Proposition 2.2.4 *If X is a transitive BE-algebra, then the relation $\leq$ is transitive.*

Proof Let $x, y, z \in X$. Suppose $x \leq y$ and $y \leq z$. Then $x * y = 1$ and $y * z = 1$. Since X is transitive, we get $1 = y * z \leq (x * y) * (x * z) = 1 * (x * z) = x * z$. Therefore $x \leq z$. □

Proposition 2.2.5 *If X is a commutative BE-algebra, then the relation $\leq$ is antisymmetric.*

Proof Let $x, y \in X$ be arbitrary. Suppose $x \leq y$ and $y \leq x$. Then, we get that $x * y = 1$ and $y * x = 1$. Since X is commutative, it yields that $x = 1 * x = (y * x) * x = (x * y) * y = 1 * y = y$. Therefore $\leq$ is antisymmetric on X. □

Since every commutative BE-algebra is transitive, the following is an easy consequence.

Corollary 2.2.6 *If X is a commutative BE-algebra, then $\leq$ is a partially ordering on X.*

2.3 Congruences on BE-algebras

In this section, some properties of congruences are studied. A congruence of a BE-algebra is characterized in terms of its unit congruence class. An interconnection between a congruence and a homomorphism of a BE-algebra is observed. The quotient structure of a BE-algebra is studied.

Definition 2.3.1 Let $(X, *, 1)$ be a BE-algebra. Then a binary relation θ on X is called a *congruence* on X if it satisfies the following properties for all $x, y, z \in X$:

(CG1) $(x, x) \in \theta$,

(CG2) $(x, y) \in \theta$ implies $(y, x) \in \theta$,

(CG3) $(x, y) \in \theta$ and $(y, z) \in \theta$ imply $(x, z) \in \theta$,

(CG4) θ is compatible with the operation $*$. That is, if $(a, b) \in \theta$ and $(c, d) \in \theta$, then $(a * c, b * d) \in \theta$.

If the binary relation θ satisfies (CG1), (CG2), and (CG3), then it is know as an equivalence relation on the BE-algebra X. It can be easily seen that an equivalence relation θ on X is a congruence on X if and only if $(a, b) \in \theta$ implies $(c * a, c * b) \in \theta$ as well as $(a * c, b * c) \in \theta$. For any congruence θ on a BE-algebra X and $x \in X$, the corresponding congruence class $[x]_\theta$ is defined as $\{y \in X \mid (x, y) \in \theta\}$. The congruence class $[1]_\theta$ is called the *kernel* of the congruence θ. Obviously, the kernel of any congruence on a BE-algebra X is a subalgebra of X. Every congruence relation has a corresponding quotient structure, whose elements are the equivalence classes (or congruence classes) for the relation. In the following result, we observe that the quotient structure of a BE-algebra is again a BE-algebra, which is a general phenomenon in the algebraic structures.

Definition 2.3.2 Let $(X, *, 1)$ be a BE-algebra and θ is a congruence on X. Then the class of all congruence classes of X is denoted by $X_{/\theta}$ and is given as follows:

$$X_{/\theta} = \{[x]_\theta \mid x \in X\}$$

Theorem 2.3.3 *Let θ be a congruence on a BE-algebra $(X, *, 1)$. Then the quotient algebra $(X_{/\theta}, \bullet, [1]_\theta)$ is a BE-algebra, where the operation $\bullet$ on $X_{/\theta}$ is defined for all $a, b \in X$ as follows:*

$$[a]_\theta \bullet [b]_\theta = [a * b]_\theta$$

Proof Let $x \in X$. Then $[x]_\theta \in X_{/\theta}$. Now $[x]_\theta \bullet [x]_\theta = [x * x]_\theta = [1]_\theta$. Also $[1]_\theta \bullet [x]_\theta = [1 * x]_\theta = [x]_\theta$ and $[x]_\theta \bullet [1]_\theta = [x * 1]_\theta = [1]_\theta$. Let $x, y, z \in X$. Then $[x]_\theta \bullet \{[y]_\theta \bullet [z]_\theta\} = [x]_\theta \bullet [y * z]_\theta = [x * (y * z)]_\theta = [y * (x * z)]_\theta = [y]_\theta \bullet [x * z]_\theta = [y]_\theta \bullet \{[x]_\theta \bullet [z]_\theta\}$. Therefore $(X_{/\theta}, \bullet, [1]_\theta)$ is a BE-algebra. □

In the following theorem, we characterize congruences on a BE-algebra.

Theorem 2.3.4 *Let $(X, *, 1)$ be a commutative BE-algebra. Then for any congruence θ on X,*

$$\theta = \{(x, y) \in X \times X \mid x * y, y * x \in [1]_\theta\}$$

Proof Let$(x, y) \in \theta$. Then $(x * y, 1) = (x * y, y * y) \in \theta$ and $(y * x, 1) = (y * x, x * x) \in \theta$. Hence $x * y \in [1]_\theta$ and $y * x \in [1]_\theta$. Conversely, let $x, y \in X$ be such that $x * y, y * x \in [1]_\theta$. Now

$$\begin{aligned} x * y \in [1]_\theta &\Rightarrow (x * y, 1) \in \theta \\ &\Rightarrow ((x * y) * y, 1 * y) \in \theta \\ &\Rightarrow ((x * y) * y, y) \in \theta \end{aligned}$$

Since $y * x \in [1]_\theta$, similarly we get $((y * x)*, x) \in \theta$. Since X is commutative, we get $((y * x) * x, (x * y) * y) \in \theta$. Hence from the above, we get $(x, y) \in \theta$. □

Theorem 2.3.5 *Let θ_1 and θ_2 be two congruences on a commutative BE-algebra X. Then $\theta_1 \subseteq \theta_2$ if and only if $[1]_{\theta_1} \subseteq [1]_{\theta_2}$.*

Proof Let θ_1 and θ_2 be two congruences on X. Assume that $\theta_1 \subseteq \theta_2$. Let $x \in [1]_{\theta_1}$. Then we get $(x, 1) \in \theta_1 \subseteq \theta_2$. Hence $x \in [1]_{\theta_2}$. Therefore $[1]_{\theta_1} \subseteq [1]_{\theta_2}$. Conversely, assume that $[1]_{\theta_1} \subseteq [1]_{\theta_2}$. Let $(x, y) \in \theta_1$. Then we get that $(x*y, 1) = (x*y, y*y) \in \theta_1$. Hence $x * y \in [1]_{\theta_1} \subseteq [1]_{\theta_2}$. Thus $(x * y, 1) \in \theta_2$. Hence $((x * y) * y, y) = ((x * y) * y, 1 * y) \in \theta_2$. Similarly, we can obtain that $((y * x) * x, x) \in \theta_2$. Since X is commutative, it implies that $(x, y) \in \theta_2$. Therefore $\theta_1 \subseteq \theta_2$. □

Theorem 2.3.6 *Let $(X, *, 1)$ be a BE-algebra and $\{\theta_\alpha\}_{\alpha \in \Delta}$ (where Δ is an indexed family) be a family of congruences on X. Put $\theta = \bigcap\limits_{\alpha \in \Delta} \theta_\alpha$. Then the following conditions hold.*

(1) θ is a congruence on X,

(2) $[1]_\theta = \bigcap\limits_{\alpha \in \Delta} [1]_{\theta_\alpha}$.

Proof (1) It is an easy consequence from the definition of a congruence on BE-algebras.
(2) Since $\theta = \bigcap_{\alpha \in \Delta} \theta_\alpha$, we get that $\theta \subseteq \theta_\alpha$ for each $\alpha \in \Delta$. By the above theorem, we get $[1]_\theta \subseteq [1]_{\theta_\alpha}$ for each $\alpha \in \Delta$. Hence $[1]_\theta \subseteq \bigcap_{\alpha \in \Delta} [1]_{\theta_\alpha}$. Conversely, let $x \in \bigcap_{\alpha \in \Delta} [1]_{\theta_\alpha}$. Then we get $x \in [1]_{\theta_\alpha}$ for each $\alpha \in \Delta$. Thus it yields $(x, 1) \in \theta_\alpha$ for each $\alpha \in \Delta$. Hence $(x, 1) \in \bigcap_{\alpha \in \Delta} \theta_\alpha = \theta$. Thus $x \in [1]_\theta$. Therefore $\bigcap_{\alpha \in \Delta} [1]_{\theta_\alpha} \subseteq [1]_\theta$. □

In the following result, we observe the relation between homomorphisms and congruences of BE-algebras. Let us define the *kernel* of a homomorphism $f : X \to Y$ as the set $Ker(f) = \{x \in X \mid f(x) = 1\}$. The proof of the following theorem is obvious and hence it is omitted.

Theorem 2.3.7 *Let $(X, *, 1)$ and $(Y, \diamond, 1')$ be two BE-algebras and $f : X \to Y$ a homomorphism. For any $x, y \in X$, define a binary relation θ on X by*

$$(x, y) \in \theta \text{ if and only if } f(x) = f(y)$$

Then θ is a congruence on X such that $[1]_\theta = Ker(f)$.

For any two BE-algebras $(X, *, 1)$, $(Y, \diamond, 1')$ and a homomorphism $f : X \to Y$, it is clear that the image $f(X)$ is a subalgebra of Y and hence a BE-algebra. Now in the following theorem, one can observe the generalization of the first isomorphism theorem to the case of BE-algebras.

Theorem 2.3.8 *Let $(X, *, 1)$ and $(Y, \diamond, 1')$ be two BE-algebras and $f : X \to Y$ a homomorphism. Then the quotient algebra $X_{/\theta}$ is isomorphic to the image of f, i.e., $X_{/\theta} \cong f(X)$.*

Proof Define $\psi : X_{/\theta} \to f(X)$ by $\psi([x]_\theta) = f(x)$. Let $[x]_\theta = [y]_\theta$ for $x, y \in X$. Then $(x, y) \in \theta$ and hence $f(x) = f(y)$. Thus it gives $\psi([x]_\theta) = \psi([y]_\theta)$. Therefore ψ is well-defined. To show that ψ is one-to-one, let us consider $\psi([x]_\theta) = \psi([y]_\theta)$ where $x, y \in X$. Then it yields $f(x) = f(y)$. Hence $(x, y) \in \theta$, which implies $[x]_\theta = [y]_\theta$. Therefore ψ is one-to-one. Let $y \in f(X)$. Then there exists $x \in X$ such that $y = f(x)$. For this $x \in X$, we can seen that $\psi([x]_\theta) = f(x) = y$. Hence ψ is onto. Let $[x]_\theta, [y]_\theta \in X_{/\theta}$ where $x, y \in X$. Now $\psi([x]_\theta * [y]_\theta) = \psi([x * y]_\theta) = f(x * y) = f(x) \diamond f(y) = \psi([x]_\theta) \diamond \psi([y]_\theta)$. Thus ψ is a homomorphism. Therefore $X_{/\theta}$ is isomorphic to $f(X)$. □

In the following theorem, we characterize the class of all congruences of a BE-algebra.

Lemma 2.3.9 *Let θ be a congruence on a BE-algebra $(X, *, 1)$. Then for any $a, b, t \in X$ and $m, n \in \mathbb{N}$ (where $\mathbb{N}$ the set of all natural numbers), we have the following:*
(1) *$a \in [1]_\theta$ implies $(a^n * t, t) \in \theta$,*
(2) *$a, b \in [1]_\theta$ implies $(a^m * (b^n * t), t) \in \theta$,*
(3) *$a, b \in [1]_\theta$ implies $(a^m * t, b^n * t) \in \theta$.*

Proof (1) Let $a \in [1]_\theta$. Then we get $(a, 1) \in \theta$. For any $t \in X$, we get $(a * t, 1) = (a * t, 1 * t) \in \theta$. By the repeated application of the operation $*$ with a, we get that $(a^n * t, t) \in \theta$.
(2) and (3) follow from (1). □

Theorem 2.3.10 *Let θ be a congruence on a BE-algebra $(X, *, 1)$. For any $x \in X$, we have that*

$$[x]_\theta = \{t \in X \mid a * (b * x) = a^2 * t \text{ for some } a, b \in [1]_\theta\}$$

where $a^2 * t = a * (a * t)$.

Proof Take $S = \{t \in X \mid a * (b * x) = a^2 * t$ for some $a, b \in [1]_\theta\}$. Let $t \in [x]_\theta$. Then $(x, t) \in \theta$ and hence $(x * t, 1) \in \theta$ and $(t * x, 1) \in \theta$. Put $a = x * t$ and $b = t * x$. Clearly $a, b \in [1]_\theta$. Then

$$b * x = (t * x) * x = (x * t) * t = a * t.$$

Hence $a * (b * x) = a * (a * t) = a^2 * t$. Thus $t \in S$. Therefore $[x]_\theta \subseteq S$. Conversely, let $t \in S$. Then $a * (b * x) = a^2 * t$ for some $a, b \in [1]_\theta$. Since $a, b \in [1]_\theta$, we get $(a, 1) \in \theta$ and $(b, 1) \in \theta$. Now

$$\begin{aligned}
(b, 1) \in \theta &\Rightarrow (b * x, x) \in \theta \\
&\Rightarrow (a * (b * x), a * x) \in \theta \\
&\Rightarrow (a * (b * x), x) \in \theta \qquad \text{since } (a, 1) \in \theta, \text{ we get } (a * x, x) \in \theta \\
&\Rightarrow (a^2 * t, x) \in \theta \\
&\Rightarrow (t, x) \in \theta \qquad \text{By the above lemma}
\end{aligned}$$

Hence $t \in [x]_\theta$. Thus $S \subseteq [x]_\theta$. Therefore the theorem is proved. □

In the following, we now observe the notion of tolerance relations of BE-algebras. These tolerance relations are extensively studied by B. Zelinka [254]. The tolerance relation of lattices is studied by I. Chajda and B. Zelinka [39]. According to I. Chajda and B. Zelinka [39], the definition of a tolerance relation is as follows:

Definition 2.3.11 A *tolerance relation* is a reflexive and symmetric binary relation.

Example 2.3.12 Let X be a BE-algebra and $\emptyset \neq S \subseteq X$. For any $x, y \in X$, define a binary relation R_S on X by $(x, y) \in \theta_S$ if and only if $x*y \in S \cup \{1\}$ and $y*x \in S \cup \{1\}$. It is easy to check that θ_S is a tolerance relation on X. Observe that θ_S need not be transitive relation on X.

A binary operation R on a BE-algebra $(X, *, 1)$ is called compatible with X if and only if $(a, b) \in R$ and $(c, d) \in R$ imply that $(a * c, b * d) \in R$ for all $a, b, c, d \in X$. If R is a binary relation on X, then by R^* we denote the relation $\{(y, x) \in R \mid x \in X, y \in X, (x, y) \in R\}$. If R is a symmetric relation on X

compatible with X, then clearly R^* is a symmetric relation compatible with X. For any relation R on a BE-algebra X, it is trivial that $R \cap R^*$ is symmetric on X. Moreover, if R_1 and R_2 are two relations on X compatible with X, then it is easy to observe that $R_1 \cap R_2$ is a relation compatible with X. Based on these facts, we can have the following:

Theorem 2.3.13 *Let $(X, *, 1)$ be a BE-algebra. If R is a tolerance relation on X compatible with X, then $R \cap R^*$ is a tolerance compatible with X.*

Proof The reflexivity and the symmetry of $R \cap R^*$ are evident. Let $(a, b), (c, d) \in R \cap R^*$. Since R is compatible with X, we get that $(a * c, b * d) \in R$. Since $(b, a), (d, c) \in R$, we get that $(b * d, a * c) \in R$. Hence $(a * c, b * d) \in R^*$. Therefore $R \cap R^*$ is a tolerance compatible with X. □

Since $R \cap R^*$ is symmetric for any relation R on X, the following is a direct consequence.

Corollary 2.3.14 *Let $(X, *, 1)$ be a BE-algebra. If R is a reflexive and transitive relation (i.e. a quasi-ordering) on X compatible with X, then $R \cap R^*$ is a congruence on X.*

Let R_1 and R_2 be two binary relations on a BE-algebra X, then define their product R_1R_2 as $(x, y) \in R_1R_2$ if and only if there exists $z \in X$ such that $(x, z) \in R_1$ and $(z, y) \in R_2$ for all $x, y \in X$. Now, it is proved that the product of compatible relations is again compatible.

Theorem 2.3.15 *Let $(X, *, 1)$ be a BE-algebra. If R_1 and R_2 are two binary relations on X compatible with X, then their product R_1R_2 is also a relation compatible with X.*

Proof Straightforward. □

Theorem 2.3.16 *Let $(X, *, 1)$ be a BE-algebra. If $\{R_i\}_{i=1}^{\infty}$ be a sequence of compatible relations on X such that $R_i \subseteq R_{i+1}$ for every positive integer i, then $\bigcup\limits_{i=1}^{\infty} R_i$ is a compatible relation on X.*

Proof Straightforward. □

Define the nth power of a binary relation R so that $R^n = R$ for $n = 1$ and $R^n = RR^{n-1}$ for $n \geq 2$. For any relation R on a BE-algebra X, the transitive closure of R is defined as follows:

$$transitive(R) = \bigcup_{i=1}^{\infty} R^i$$

Theorem 2.3.17 *Let $(X, *, 1)$ be a BE-algebra. If R is a reflexive relation on X, then $R^i \subseteq R^{i+1}$ for any positive integer i.*

Proof We prove this by using induction on i. Let $(x, y) \in R$. Since R is reflexive $(y, y) \in R$. Hence $(x, y) \in R^2$. Thus the result is true for $i = 1$. Assume that $R^i \subseteq R^{i+1}$. Let $(x, y) \in R^{i+1}$. Then there exists $z \in X$ such that $(x, z) \in R^i$ and $(z, y) \in R$. By the assumption, we get $(x, z) \in R^{i+1}$. Hence $(x, y) \in R^{i+2}$. Thus the result true for $i + 1$. Therefore $R^i \subseteq R^{i+1}$ for all i. □

Theorem 2.3.18 *Let $(X, *, 1)$ be a BE-algebra. If R is a reflexive relation on X compatible with X, then the transitive closure of R is compatible with X.*

Proof By Theorem 2.3.15, it yields that the relation R^i is compatible with X for every positive integer i. Since R is reflexive, by the above theorem, we have $R^i \subseteq R^{i+1}$ for all positive integer i. Thus according to Theorem 2.3.16, the relation $transitive(R) = \bigcup_{i=1}^{\infty} R^i$ is compatible with X. □

Definition 2.3.19 For any filter F of a BE-algebra, define a binary relation θ_F on X by

$$(x, y) \in \theta_F \text{ if and only if } x * y \in F \text{ and } y * x \in F \quad \text{for all } x, y \in X.$$

Theorem 2.3.20 *Let F be a filter of a transitive BE-algebra X. Then θ_F is a congruence on X with the kernel $[1]_{\theta_F} = F$. Moreover θ_F is a unique congruence whose kernel is F.*

Proof Since $x * x = 1$ for all $x \in X$, we get $(x, x) \in \theta_F$ and hence θ_F is reflexive. Clearly θ_F is symmetric. Let $x, y, z \in X$ be such that $(x, y) \in \theta_F$ and $(y, z) \in \theta_F$. Then $x * y \in F, y * x \in F, y * z \in F$ and $z * y \in F$. Since X is transitive, we get $y * z \leq (x * y) * (x * z)$. Hence

$$(y * z) * (x * y) * (x * z) = 1 \in F$$

Since $y * z \in F$ and $x * y \in F$, we get $x * z \in F$. Similarly, we get $z * x \in F$. Thus $(x, z) \in \theta_F$. Therefore θ_F is an equivalence relation on X. Clearly $[1]_{\theta_F} = F$. We now show that θ_F is a congruence on X. Let $(x, y) \in \theta_F$, and $(u, v) \in \theta_F$. Then $x * y \in F, y * x \in F, u * v \in F$ and $v * u \in F$. Since X is transitive, $x * y \leq (u * x) * (u * y)$. Since $x * y \in F$, we get $(u * x) * (u * y) \in F$. Since $y * x \in F$ and $y * x \leq (u * y) * (u * x)$, we get $(u * y) * (u * x) \in F$. Hence

$$(u * x, u * y) \in \theta_F \tag{2.1}$$

Again, since X is transitive, we get $v * y \leq (u * v) * (u * y)$. Hence $u * v \leq (v * y) * (u * y)$. Since $u * v \in F$, we get $(v * y) * (u * y) \in F$. Similarly $(u * y) * (v * y) \in F$. Thus

$$(u * y, v * y) \in \theta_F \tag{2.2}$$

From (2.1) and (2.2), we conclude that $(u * x, v * y) \in \theta_F$. Therefore θ_F is a congruence on X whose kernel is F. Let Φ be a congruence on X such that $F = [1]_\Phi$.

For any $a, b \in X$, we get

$$\begin{aligned}(a, b) \in \theta_F &\Leftrightarrow a * b \in F \text{ and } b * a \in F\\ &\Leftrightarrow a * b \in [1]_\Phi \text{ and } b * a \in [1]_\Phi\\ &\Leftrightarrow [a * b]_\Phi = [1]_\Phi \text{ and } [b * a]_\Phi = [1]_\Phi\\ &\Leftrightarrow [a]_\Phi * [b]_\Phi = [1]_\Phi \text{ and } [b]_\Phi * [a]_\Phi = [1]_\Phi\\ &\Leftrightarrow [a]_\Phi \subseteq [b]_\Phi \text{ and } [b]_\Phi \subseteq [a]_\Phi\\ &\Leftrightarrow [a]_\Phi = [b]_\Phi\\ &\Leftrightarrow (a, b) \in \Phi\end{aligned}$$

Hence $\theta_F = \Phi$. Therefore θ_F is a unique congruence whose kernel is F. □

Proposition 2.3.21 *Let F be a filter of a transitive BE-algebra X. Denote F_x for $[x]_{\theta_F} = \{y \in X \mid (x, y) \in \theta_F\}$. Then $F_1 = [1]_{\theta_F}$ is the smallest filter containing F.*

Proof Clearly $1 \in [1]_{\theta_F}$. Let $x \in [1]_{\theta_F}$ and $x * y \in [1]_{\theta_F}$. Then $(1, x) \in \theta_F$ and $(1, x * y) \in \theta_F$. Hence $(1, x) * (1, y) = (1, x * y) \in \theta_F$. Since $(1, x) \in \theta_F$, we get $(1, y) \in \theta_F$. Hence $y \in [1]_{\theta_F}$. Therefore $[1]_{\theta_F}$ is a filter of X. Let $x \in F$. Clearly $1 * x = x \in F$ and also $x * 1 = 1 \in F$. Hence $(1, x) \in \theta_F$. Thus $x \in [1]_{\theta_F}$. Therefore $F \subseteq [1]_{\theta_F}$. Let G be a filter containing F. Let $x \in F_1 = [1]_{\theta_F}$. Therefore $(x, 1) \in \theta_F$. Hence $x = 1 * x \in F \subseteq G$. Therefore $[1]_{\theta_F} \subseteq G$. □

Definition 2.3.22 Let $f : X \to Y$ be a BE-homomorphism. Define the kernel $Ker(f)$ of the homomorphism f as $Ker(f) = \{x \in X \mid f(x) = 1\}$ which is nothing but $f^{-1}(1)$.

Lemma 2.3.23 *Let X and Y be two BE-algebras and $f : X \to Y$ a BE-homomorphism. Then*

(1) *$Ker(f)$ is a filter of X,*

(2) *f is one-to-one if an only if $Ker(f) = \{1\}$, provided X is partially ordered.*

Proof (1) Since $f(1) = 1$, we get that $1 \in Ker(f)$. Let $x \in Ker(f)$ and $x * y \in Ker(f)$. Then we get that $f(x) = 1$ and $f(x * y) = 1$. Then $f(y) = 1 * f(y) = f(x) * f(y) = f(x * y) = 1$. Hence $y \in Ker(f)$. Therefore $Ker(f)$ is a filter of X.

(2) Assume that f is one-to-one. Clearly $\{1\} \subseteq Ker(f)$. Let $x \in Ker(f)$. Then $f(x) = 1$. Hence $f(x) = 1 = f(1)$. Since f is one-to-one, we get $x = 1$. Therefore $Ker(f) \subseteq \{1\}$. Thus $Ker(f) = \{1\}$. Conversely, assume that $Ker(f) = \{1\}$. Let $x, y \in X$ be such that $f(x) = f(y)$. Then $f(x * y) = f(x) * f(y) = f(x) * f(x) = 1$. Hence $x * y \in Ker(f) = \{1\}$. Thus $x * y = 1$. Similarly, we get $y * x = 1$. Since X is partially ordered, we get $x = y$. Therefore f is one-to-one. □

Theorem 2.3.24 *Let F be a filter of a transitive BE-algebra X. Define $X/F = \{F_x \mid x \in X\}$. Then $(X/F, *, F_1)$ is a BE-algebra with greatest element F_1.*

Proof For any $F_x, F_y \in X/F$ where $x, y \in X$, define $F_x * F_y = F_{x*y}$. Let $F_x \in X/F$ where $x \in X$. Then $F_x * F_x = F_{x*x} = F_1$ and $F_1 * F_x = F_{1*x} = F_x$. Also $F_x * F_1 = F_{x*1} = F_1$. Let $F_x, F_y, F_z \in X/F$ where $x, y, z \in X$. Then

$$\begin{aligned} F_x * (F_y * F_z) &= F_x * F_{y*z} \\ &= F_{x*(y*z)} \\ &= F_{y*(x*z)} \\ &= F_y * (F_{x*z}) \\ &= F_y * (F_x * F_z) \end{aligned}$$

Therefore it concludes that $(X/F, *, F_1)$ is a BE-algebra. □

Theorem 2.3.25 *Let X and Y be two BE-algebras where X is transitive and Y is commutative. Suppose* $f : X \to Y$ *is a BE-epimorphism such that* $F = Ker(f) = F_1$. *Then* $\alpha : X/F \to Y$ *defined by* $\alpha(F_x) = f(x)$ *is a BE-isomorphism.*

Proof Clearly α is well-defined. Let $F_x, F_y \in X/F$ be such that $F_x = F_y$. Then

$$\begin{aligned} F_x = F_y &\Rightarrow (x, y) \in \theta_F \\ &\Rightarrow x * y \in F \text{ and } y * x \in F \\ &\Rightarrow x * y \in F_1 \text{ and } y * x \in F_1 \\ &\Rightarrow f(x) * f(y) = f(x * y) = 1 \text{ and } f(y) * f(x) = f(y * x) = 1 \\ &\Rightarrow f(x) = f(y) \qquad \text{since } Y \text{ is commutative} \\ &\Rightarrow \alpha(F_x) = \alpha(F_y) \end{aligned}$$

Therefore α is one-to-one. Let $y \in Y$. Since f is onto, there exists $x \in X$ such that $f(x) = y$. For this $x \in X$, we get that $F_x \in X/F$ and $\alpha(F_x) = f(x) = y$. Therefore α is onto.

Let $F_x, F_y \in X/F$ where $x, y \in X$. Then $\alpha(F_x * F_y) = \alpha(F_{x*y}) = f(x * y) = f(x) * f(y) = \alpha(F_x) * \alpha(F_y)$. Hence α is a homomorphism. Therefore α is an isomorphism. □

Theorem 2.3.26 *Let F and G be two filters of a transitive BE-algebra X such that* $F \subseteq G$. *Denote* $G/F = \{F_x \in X/F \mid x \in G\}$. *Then we have the following conditions:*

(1) $x \in G$ *if and only if* $F_x \in G/F$ *for all* $x \in X$,

(2) G/F *is a filter in* X/F,

(3) *Let F be a closed filter of X. If S and T are the sets of all filters of X and* X/F *respectively, then the mapping* $g : S \to T$ *defined by* $g(G) = G/F$ *is a bijective map.*

Proof (1) The necessary part is clear. Conversely, assume that $F_x \in G/F$ for all $x \in X$. Then $F_x = F_y$ for some $y \in G$. Then $(x, y) \in \theta_F$. Hence $x * y \in F$ and $y * x \in F \subseteq G$. Since $y \in G$, $y * x \in G$ and G is a filter, we get that $x \in G$.

(2) Since $1 \in G$, we get $[1]_{\theta_F} = F_1 \in G/F$. Let $F_x, F_x * F_y \in G/F$. Then $F_x \in G/F$ and $F_{x*y} \in G/F$. Hence $x \in G$ and $x * y \in G$. Since G is a filter, we get $y \in G$. It implies that $F_y \in G/F$. Therefore G/F is a filter in X/F.
(3) Let G and H be two filters of X such that $g(G) = g(H)$. Then $G/F = H/F$. Let $x \in G$. Then it yields that $F_x \in G/F = H/F$. Hence $x \in H$. Thus it concludes that $G \subseteq H$. Similarly, we can obtain $H \subseteq G$. Therefore $G = H$. Therefore g is one-to-one.

Let G/F be a filter in X/F. We claim that G is a filter in X. Since G/F is a filter in X/F, we get $[1]_{\theta_F} = F_1 \in G/F$. Hence $1 \in G$. Let $x, x * y \in G$. Then $F_x \in G/F$ and $F_x * F_y = F_{x*y} \in G/F$. Since G/F is a filter in X/F, we get $F_y \in G/F$. Hence $y \in G$. Thus G is a filter in X. Clearly $f(G) = G/F$. Thus g is onto. Therefore $g : S \to T$ is a bijection. If $f(G) = \cup\{F_x \mid F_x \in G/F\}$. Clearly $(g \circ f)(G/F) = (f \circ g)(G/F) = F_1$. Therefore f is the inverse of the mapping g. □

2.4 Topological Concepts

In this section, several fundamental topological concepts are discussed. In particular, this section offers an exposition that is seldom covered in many topology texts. We also review some basic notions from topology, the main goal being to set up the language.

Definition 2.4.1 ([185]) Let X be a non-empty set. A collection $\Im$ of subsets of X is said to be a *topology* on X if it satisfies the following conditions:

(1) $\emptyset \in \Im$ and $X \in \Im$,

(2) the arbitrary union of sets of $\Im$ is a set of $\Im$,

(3) the finite intersection of sets of $\Im$ is a set of $\Im$.

In this case, the system $(X, \Im)$ is called a *topological space*. The sets in $\Im$ are called the *open sets* of the topological space, and the elements of X are called points of the topological space. The collection of subsets of a non-empty set X is a topology on X, which is called the *discrete topology* on X, and any topological space whose topology is the discrete topology is called a *discrete space*. Let $\Im_1$ and $\Im_2$ be two topologies on a non-empty set X such that $\Im_1 \subseteq \Im_2$. Then we say that $\Im_1$ is weaker (finer) than or finer than $\Im_2$ (or $\Im_2$ is stronger than $\Im_1$). The topology $\{\emptyset, X\}$ is the finest topology on X, and the discrete topology is the strongest topology on X. Obviously, the class of all topologies on X is a partially ordered set with respect to the relation "is finer than."

Definition 2.4.2 ([185]) If $(X, \Im)$ is a topological space and $x \in X$ is an element in X, a subset N of X is called a *neighborhood* of x if there exists some subset D such that $x \in D \subset N$.

A subset S of X is called *open* if every point in S has a neighborhood lying in the set. The *complement* of a subset A of X is denoted $X - A$ and defined to be the set of $z \in X$ such that $z \notin A$. A subset of a topological space is called *closed* if its complement is open. The *interior* of a set $A \subseteq X$ is denoted $IntA$ and defined to be the set of $x \in A$ for which there is an open set $U \subseteq X$ such that $x \in U$ and $U \subseteq A$. If $U \subseteq X$ is any open set such that $U \subseteq A$, then every element of U is contained in A, so that $U \subseteq IntA$. Thus $IntA$ may be described equivalently as the union of all of the open subsets of X that are contained in A. In particular, $IntA$ is automatically an open subset of X too. Obviously, the complement of the interior of $A \subseteq X$ in X is equal to the closure of the complement of A in X, which is to say that $X - IntA = X - A$.

Definition 2.4.3 A *base* (or *basis*) $\mathcal{B}$ for a topological space X with topology $\Im$ is a collection of open sets in $\Im$ such that every open set in $\Im$ can be written as a union of elements of $\mathcal{B}$.

Two important properties of bases are as follows: (1) The base elements cover X. (2) Let B_1, B_2 be base elements and let I be their intersection. Then for each x in I, there is a base element B_3 containing x and contained in I. If a collection $\mathcal{B}$ of subsets of X fails to satisfy either of these, then it is not a base for any topology on X. (It is a subbase, however, as is any collection of subsets of X.) Conversely, if $\mathcal{B}$ satisfies both of the conditions (1) and (2), then there is a unique topology on X for which $\mathcal{B}$ is a base; it is called the topology generated by $\mathcal{B}$. This topology is the intersection of all topologies on X containing $\mathcal{B}$. Bases are useful because many properties of topologies can be reduced to statements about a base generating that topology, and because many topologies are most easily defined in terms of a base which generates them.

Definition 2.4.4 Let X and Y be two topological spaces and f a mapping of X into Y. Then f is said to be a *continuous mapping* if $f^{-1}(G)$ is open in X whenever G is open in Y. It is called an *open mapping* if $f(G)$ is open in Y whenever G is open in X.

A mapping is continuous if it pulls open sets back to open sets, and open if it carries open sets over to open sets. Any image $f(X)$ of a topological space X under a continuous mapping f is called a *continuous image* of X. A *homeomorphism* is a one-to-one continuous mapping of one topological space onto another which is also an open mapping. Two topological spaces X and Y are said to be *homeomorphic* if there exists a homeomorphism of X onto Y.

Definition 2.4.5 Let X be a topological space. A class $\{G_i\}$ of open subsets of X is said to be an *open cover* of X if each point in X belongs to at least one G_i, that is, if $\cup_i G_i = X$.

A subclass of an open cover which is itself an open cover is called a *subcover*. A *compact space* is a topological space in which every open cover as a finite subcover. A *compact subspace* of a topological space is a subspace which is compact as a topological space on its own. Any closed subspace of a compact space is compact, and any continuous image of s compact space is compact.

Definition 2.4.6 A topological space $(X, \Im)$ is called a T_0-*space* if for any two points $x, y \in X$, there is an open set U such that $x \in U$ and $y \notin U$. A topological space $(X, \Im)$ is called a T_1-*space* if for any two distinct points $x, y \in X$, there exist open sets U and V such that $x \in U - V$ and $y \in V - U$. A *Hausdorff space* or a T_2-space is a topological space if for any two distinct points $x, y \in X$, there exist two open sets U and V such that $x \in U$, $y \in V$ and $U \cap V = \emptyset$.

Obviously every Hausdorff space is a T_1-space and every subspace of a Hausdorff space is a Hausdorff space. A topological space is a T_1-space if and only if each point is a closed set. A topological space $(X, \Im)$ is called a *regular space* if for every point $x \in X$ and each closed set V not containing x, there exist disjoint open sets U_1 and U_2 such that $x \in U_1$ and $V \subset U_2$.

Definition 2.4.7 A topological space $(X, \Im)$ is called a *connected space* if it cannot be represented as the union of two disjoint non-empty open sets. Otherwise, it is called *disconnected*.

If $X = A \cup B$, where A and B are disjoint and open, then A and B are also closed sets, and conversely. Thus X is connected if and only if it can be represented as the union of two disjoint non-empty closed sets. A *connected subspace* of X is a subspace Y which is connected as a topological space in its own right. Any continuous image of a connected space is connected, and the product of any non-empty class of connected spaces is connected.

Definition 2.4.8 A topological space X is called *totally disconnected* if for every pair of points x and y in X such that $x \neq y$, there exists a disconnection $X = A \cup B$ with $x \in A$ and $y \in B$.

The above space is evidently a Hausdorff space, and if it has more than one point, it is disconnected. Oddly enough, a one point space is both connected and totally disconnected. The discrete spaces are simplest totally disconnected spaces. A Hausdorff space is totally disconnected if it has an open base whose sets are also closed.

Definition 2.4.9 A topological space X is called a *locally connected space* if x is any point in X and G any neighborhood of x, then G contains a connected neighborhood of x.

This is evidently equivalent to the condition that each point of the space has an open base whose sets are all connected subspaces. Clearly every Banach space is locally connected. Every connected space is locally connected but not the converse. The union of two disjoint open intervals on the real line is a simple example of a space which is locally connected but not connected.

Definition 2.4.10 Let X be an ordered set with more than one element and $a, b \in X$. Then the collection of sets of the form (a, b), $[a_0, b)$ where a_0 is the smallest element of X (if such exists) and the sets of the form $(a, b_0]$ where b_0 is the largest element of X if such exists is a basis, and it generates a topology on X known as the order topology.

Example 2.4.11 Order topology on $\mathbb{R}$ is just the standard Euclidean topology. The basis for the order topology consists solely of open intervals, as there is no smallest or largest element. But this is the same as the basis for the Euclidean topology, which consists of open balls. As the two bases are the same, the topology that they generate is the same.

Definition 2.4.12 Suppose that X and Y are topological spaces. The product topology on $X \times Y$ is generated by the basis $\mathcal{B} = \{U \times V \mid U \subset X \text{ is open; } V \subset Y \text{ is open}\}$.

Note that this definition asserts without proof that $\mathcal{B}$ is a basis. You can show that $(U_1 \times V_1) \cap (U_2 \times V_2) = (U_1 \cap U_2) \times (V_1 \cap V_2)$, i.e., the intersection of the products is the product of the intersections. This along with the fact that $X \times Y \in \mathcal{B}$ implies that $\mathcal{B}$ is a basis.

Exercise

1. Prove that a Hilbert algebra $(X, *, 1)$ is implicative if and only if it is self-distributive and commutative BE-algebra.
2. Consider the interval $[0, 1]$ of real numbers. Define an operation $*$ on $[0, 1]$ by $x * y = \min\{1 - x + y, 1\}$ for all $x, y \in [0, 1]$. Then show that $([0, 1], *, 1)$ is a bounded BE-algebra (if there exists the smallest element 0 of X (i.e., $0 * x = 1$ for all $x \in X$).
3. Assume that $(X, *, 1)$ be a commutative BE-algebra. Let θ_1 and θ_2 be two congruences on X with $\theta_1 \cap \theta_2 = (1, 1)$. Show that the map $\psi : X \to X_{/\theta_1} \times X_{/\theta_2}$ defined by $\psi(a) = ([a]_{\theta_1}, [a]_{\theta_2})$ is an embedding.
4. Give an example of a binary relation on a BE-algebra X whose reflexive closure and the symmetric closure are not compatible with X.
5. For any binary relation R on a BE-algebra X compatible with X, prove that the set $R_e = \{x \in X \mid (x, 1) \in R\}$ is a subalgebra of X such that $x \in R_e$ implies $a * x \in R_e$ for all $x, a \in X$.
6. Let X be a BE-algebra equipped with a topology $\mathcal{T}$ such that $(x, y) \mapsto x * y$ is a continuous from $X \times X$ with its Cartesian product topology to X with the topology. Prove that $\{1\}$ is an open set of $(X, \mathfrak{T})$ if and only if X is a Hausdorff space.
7. Prove that a negatively ordered BE-algebra X is commutative if and only if $(X, \sqcup)$ is a semilattice where $\sqcup$ is defined on X by $x \sqcup y = (y * x) * x$ for all $x, y \in X$.
8. If $(X, *, 1)$ is a commutative BE-algebra, then show that $(X, \vee, ^a, 1)$, where $x \vee y = (x * y) * y$ for all $x, y \in X$ and $x^a = x * a$ for each $x \in [a, 1] = \{t \in X \mid a \leq t\}$, is a semilattice with sectionally antitone involutions in which we have $x * y = (x \vee y)^y$.

Chapter 3
Some Concepts of BE-algebras

BE-algebras are important tools for certain investigations in algebraic logic since they can be considered as fragments of any propositional logic containing a logical connective implication and the constant 1 which is considered as the logical value "true." The notion of BE-algebras was introduced and extensively studied by H.S. Kim and Y.H. Kim in [153]. These classes of BE-algebras were introduced as a generalization of the class of BCK-algebras of K. Iśeki and S. Tanaka [124]. I. Chajda and J. Kuhr [38] studied the algebraic structure derived from a BCK-algebra. They viewed commutative BCK-algebras as semilattices whose sections have antitone involutions, and it is known that bounded commutative BCK-algebras are equivalent to MV-algebras. In [202], A. Rezaei and A.B. Saeid introduced the concept of fuzzy subalgebras of BE-algebras and studied its nature. They stated and proved the Foster's results on homomorphic images and inverse images in fuzzy topological BE-algebras.

In this chapter, some properties of various weak subalgebras of BE-algebras are investigated. The concept of n-fold weak subalgebras is introduced in BE-algebras. A necessary and sufficient condition is derived for a weak subalgebra of a BE-algebra to become a subalgebra. For a given congruence on a BE-algebra X, a one-to-one correspondence is obtained between the class of all congruences on a BE-algebra and the class of all congruences on the corresponding quotient algebra $X_{/\theta}$. It is proved that every BE-algebra is a subdirect product of subdirectly irreducible BE-algebras. It is proved that the set of all closure operators of a BE-algebra forms a complete lattice with respect to set inclusion. The concept of fuzzy weak subalgebras is introduced in BE-algebras. Some properties of fuzzy weak subalgebras are extensively studied. The theory of soft sets of Molodtsov [181] is applied to BE-algebras and studied the properties of soft weak subalgebras. Falling shadows applied to subalgebras of BE-algebras and the relation between fuzzy subalgebra sand falling subalgebras are obtained. Properties of direct products of BE-algebras are investigated. Two canonical mappings of the direct products of BE-algebras are introduced, and their properties are studied. A commutative BE-algebra is constructed over the set of all real-valued functions defined on a given non-empty set and studied its properties.

S. R. Mukkamala, *A Course in BE-algebras*,
https://doi.org/10.1007/978-981-10-6838-6_3

3.1 Subalgebras of BE-algebras

In this section, the notion of an n-fold weak subalgebra is introduced in BE-algebras. Properties of weak subalgebras are studied. A necessary and sufficient condition is derived for a weak subalgebra to become a subalgebra. The notion of normal weak subalgebras is introduced, and their properties are investigated with respect to congruences and homomorphisms.

Definition 3.1.1 Let $(X, *, 1)$ be a BE-algebra. A non-empty subset S of X is said to be a *subalgebra* of X if $x * y \in S$ for all $x, y \in S$.

Example 3.1.2 Let $X = \{1, a, b, c, d\}$. Define a binary operation $*$ on X as follows:

$*$	1	a	b	c	d
1	1	a	b	c	d
a	1	1	b	c	b
b	1	a	1	b	a
c	1	a	1	1	a
d	1	1	1	b	1

Clearly $(X, *, 1)$ is a BE-algebra. Consider the subsets $S_1 = \{1, a, b\}$; $S_2 = \{1, a, c\}$; and $S_3 = \{1, c, d\}$. It can be easily seen that S_1 and S_2 are subalgebras of X. However, S_3 is not a subalgebra of X because of $d \in S_3$ and $c \in S_3$ but $d * c = b \notin S_3$.

Theorem 3.1.3 *A non-empty subset S of a BE-algebra X is a subalgebra of X if and only if $x \in S, x * y \in X - S$ imply that $y \notin S$ for all $x, y \in X$.*

Proof Let S be a non-empty subset of X. Assume that S is a subalgebra of X. Let $x \in S, x * y \in X - S$. Suppose $y \in S$. Since S is a subalgebra and $x \in S$, we get $x * y \in S$. Thus we obtained a contradiction to that $x * y \notin S$. Therefore $y \in S$, which proves the necessary condition.

Conversely, assume that $x \in S, x * y \in X - S$ imply $y \notin S$ for all $x, y \in X$. Let $x, y \in S$. Suppose $x * y \notin S$. Then by the assumption, we get $y \notin S$. It yields a contradiction to that $y \in S$. Hence $x * y \in S$. Therefore S is a subalgebra of X. □

Definition 3.1.4 Let $(X, *, 1)$ be a BE-algebra. A subset S of X is said to be an *n-fold weak subalgebra* of X if $x \in S$ and $y \in S$ imply that $x^n * y \in S$ where n is a positive integer with $n \geq 2$. Here $x^n * y = x * (x * (\cdots (x * y) \cdots))$ and the operation $*$ is applied n times.

In a self-distributive BE-algebra, it is obvious that every n-fold weak subalgebra is a subalgebra. In general, a subalgebra of any algebra forms again the same algebra, but this is not true in case of n-fold weak subalgebras. However, in the following theorem, we observe that an n-fold weak subalgebra of a BE-algebra also forms a BE-algebra but with a different operation which is defined in terms of the operation of the BE-algebra.

Theorem 3.1.5 *Every n-fold weak subalgebra of a BE-algebra forms a BE-algebra.*

Proof Let $(X, *, 1)$ be a BE-algebra and S is an n-fold weak subalgebra of X. For any $x, y \in S$, define a binary operation $\circledast$ on S as $x \circledast y = x^n * y$. Clearly $\circledast$ is well-defined. For any $x \in X$, we observe the following consequences:

(BE1) $x \circledast x = x^n * x = x * 1 = 1$.
(BE2) $x \circledast 1 = x^n * 1 = x * 1 = 1$.
(BE3) $1 \circledast x = 1^n * x = 1 * x = x$.
(BE4) For any $x, y, z \in X$, we get $x \circledast (y \circledast z) = x \circledast (y^n * z) = x^n * (y^n * z) = y^n * (x^n * z) = y^n * (x \circledast z) = y \circledast (x \circledast z)$. Therefore $(S, \circledast, 1)$ is a BE-algebra. □

It can be easily observed that an n-fold weak subalgebra contains 1. Indeed, if S is n-fold weak subalgebra of a BE-algebra X and $x \in S$. Then $1 = x^n * x \in S$. For $n = 2$, the basic weak subalgebra is called *weak subalgebra* of X. Hence every weak subalgebra of a BE-algebra contains the greatest element 1.

Proposition 3.1.6 *Every subalgebra of a BE-algebra is an n-fold weak subalgebra.*

Proof Let $(X, *, 1)$ be a BE-algebra and S a subalgebra of X. Let $x, y \in S$. Since S is a subalgebra of X, it immediately infers that $x * y \in S$. Since $x, x * y \in S$ and S is a subalgebra of X, we get $x * (x * y) \in S$. Continuing in this way, we get $x^n * y \in S$ for any positive integer $n \geq 2$. Therefore S is an n-fold weak subalgebra of X. □

Though the converse of the above proposition is not true, we derive a sufficient condition for an n-fold weak subalgebra of a BE-algebra to become a subalgebra.

Theorem 3.1.7 *An n-fold weak subalgebra S of a BE-algebra is a subalgebra if $x * (x * y) \in S$ implies $x * y \in S$ for all $x, y \in X$.*

Proof Let S be an n-fold weak subalgebra of a BE-algebra X. Let $x, y \in S$. Since S is an n-fold weak subalgebra of X, we get $x * (x * (x^{n-2} * y)) = x^n * y \in S$. From the given condition, we get $x^{n-1} * y = x * (x^{n-2} * y) \in S$. Again, by the given condition, we get $x^{n-2} * y = x * (x^{n-3} * y) \in S$. Continuing in this way, finally we get $x * y \in S$. Hence S is a subalgebra of X. □

Theorem 3.1.8 *A non-empty subset S of a BE-algebra X is an n-fold weak subalgebra of X if and only if $x \in S, x^n * y \in X - S$ imply that $y \notin S$ for all $x, y \in X$.*

Proof Let S be a non-empty subset of X. Assume that S is an n-fold weak subalgebra of X. Let $x \in S$ and $x^n * y \in X - S$. Suppose $y \in S$. Since S is an n-fold weak subalgebra and $x \in S$, we get that $x^n * y \in S$. Thus we obtained a contradiction to that $x^n * y \notin S$. Therefore $y \in S$, which proves the necessary condition.

Conversely, assume that for all $x, y \in X$, $x \in S, x^n * y \in X - S$ imply $y \notin S$. Let $x, y \in S$. Suppose $x^n * y \notin S$. Then by the assumption, we get $y \notin S$. It results in a contradiction to that $y \in S$. Hence $x^n * y \in S$. Therefore S is an n-fold weak subalgebra of X. □

In the following propositions, we observe some induced n-fold weak subalgebras of BE-algebras.

Proposition 3.1.9 *If $\{S_\alpha \mid \alpha \in \Delta\}$ is an indexed family of n-fold weak subalgebras of a BE-algebra X, then $\bigcap_{\alpha\in\Delta} S_\alpha$ is also an n-fold weak subalgebra of X.*

Proof Let S_α be an n-fold weak subalgebra for each $\alpha \in \Delta$. Let $x, y \in X$. Suppose $x, y \in \bigcap_{\alpha\in\Delta} S_\alpha$. Then $x, y \in S_\alpha$ for each $\alpha \in \Delta$. Since each S_α is an n-fold weak subalgebra, we get $x^n * y \in S_\alpha$ for each $\alpha \in \Delta$. Hence $x^n * y \in \bigcap_{\alpha\in\Delta} S_\alpha$. Therefore $\bigcap_{\alpha\in\Delta} S_\alpha$ is an n-fold weak subalgebra of X. □

Proposition 3.1.10 *Let X and Y be two BE-algebras. If S and T are n-fold weak subalgebras of X and Y, respectively, then $S \times T$ is an n-fold weak subalgebra of the product algebra $X \times Y$.*

Proof Let S and T be n-fold weak subalgebras of X and Y, respectively. Let $x, z \in X$ and $y, w \in Y$ be such that $(x, y) \in S \times T$ and $(z, w) \in S \times T$. Then $x, z \in S$ and $y, w \in T$. Since S and T are n-fold weak subalgebras, we get $x^n * z \in S$ and $y^n * w \in T$. Hence $(x, y)^n * (z, w) = (x^n * z, y^n * w) \in S \times T$. Therefore $S \times T$ is an n-fold weak subalgebra of the product algebra $X \times Y$. □

Proposition 3.1.11 *Let X, Y be two BE-algebras and $f : X \to Y$ is a homomorphism. If S is an n-fold weak subalgebra of X, then $f(S)$ is an n-fold weak subalgebra of X.*

Proof Let S be an n-fold weak subalgebra of X. Let $x, y \in Y$. Suppose $x, y \in f(S)$. Then $x = f(a)$ and $y = f(b)$ for some $a, b \in S \cap X$. Since S is an n-fold weak subalgebra of X, we get $a^n * b \in S$. Since f is a homomorphism, we get $x^n * y = f(a)^n * f(b) = f(a^n * b) \in f(S)$. Hence $x^n * y \in f(S)$. Therefore $f(S)$ is an n-fold weak subalgebra of Y. □

Proposition 3.1.12 *Let X, Y be two BE-algebras and $f : X \to Y$ is a homomorphisms. If S is an n-fold weak subalgebra of Y, then $f^{-1}(S)$ (if f^{-1} exists) is an n-fold weak subalgebra of X.*

Proof Let S be an n-fold weak subalgebra of Y. Let $x, y \in X$. Suppose $x, y \in f^{-1}(S)$. Then $f(x), f(y) \in S$. Since S is an n-fold weak subalgebra of Y, we get $f(x)^n * f(y) \in S$. Since f is a homomorphism, we get $f(x^n * y) = f(x^n) * f(y) = f(x)^n * f(y) \in S$. Hence $x^n * y \in f^{-1}(S)$. Therefore $f^{-1}(S)$ is an n-fold weak subalgebra of X. □

3.2 Subdirectly Irreducible BE-algebras

In this section, some homomorphic theorems are proved. The concept of subdirectly irreducible BE-algebras is introduced, and it is proved that a BE-algebra is a subdirect product of subdirectly irreducible BE-algebras with the help of congruences of BE-algebras.

Theorem 3.2.1 *The homomorphic image of a BE-algebra is also a BE-algebra.*

Proof Straightforward. □

Definition 3.2.2 Let $(X, *, 1)$ be a BE-algebra and θ an equivalence relation on X. We say that $A\theta B$ if and only if for all $a \in A$, there exists $b \in B$ such that $(a, b) \in \theta$ for any $A, B \subseteq X$.

If θ is a congruence relation on a BE-algebra X, then the congruence class of any $x \in X$ modulo θ is denoted by $[x]_\theta = \{y \in X \mid (x, y) \in \theta\}$. Then it was already observed that the class $X_{/\theta}$ of all congruence classes of X forms a BE-algebra with respect to the operation $[x]_\theta * [y]_\theta = [x * y]_\theta$ for all $x, y \in X$ in which the greatest element is $[1]_\theta$.

Definition 3.2.3 Let f be a homomorphism from a BE-algebra X to a BE-algebra Y. Then

$$Ker(f) = \{(x, y) \in X \times Y \mid f(x) = f(y)\}.$$

Theorem 3.2.4 *For any BE-homomorphism $f : X \to X$, $Ker(f)$ is a congruence on X.*

Proof Straightforward. □

Theorem 3.2.5 *Let X and Y be two BE-algebras. If $h : X \to Y$ is a surjective homomorphism, then there exists an isomorphism $f : X_{/Ker(h)} \to Y$.*

Proof Define $f : X_{/Ker(h)} \to Y$ by $f([x]_{Ker(h)}) = h(x)$ for all $x \in X$, where $[x]_{Ker(h)}$ is the congruence class of x with respect to the congruence $Ker(h)$. Clearly f is well-defined. Let $[x]_{Ker(h)}, [y]_{Ker(h)} \in X_{/Ker(h)}$ be such that $f([x]_{Ker(h)}) = f([y]_{Ker(h)})$. Then $h(x) = h(y)$, which implies that $(x, y) \in Ker(h)$. Hence $[x]_{Ker(h)} = [y]_{Ker(h)}$. Therefore f is one-to-one. Let $y \in Y$. Since h is a surjection, there exists $x \in X$ such that $h(x) = y$. Now, for this $x \in X$, $[x]_{Ker(h)} \in X_{/Ker(h)}$ and $f([x]_{Ker(h)}) = h(x) = y$. Therefore f is onto. Now, for any $[x]_{Ker(h)}, [y]_{Ker(h)} \in X_{/Ker(h)}$, we get the following:

$$\begin{aligned} f([x]_{Ker(h)} * [y]_{Ker(h)}) &= f([x * y]_{Ker(h)}) \\ &= h(x * y) \\ &= h(x) * h(y) \\ &= f([x]_{Ker(h)}) * f([y]_{Ker(h)}) \end{aligned}$$

Hence f is a homomorphism. Therefore $X_{/Ker(h)}$ is isomorphic to Y, *i.e.*, $X_{/Ker(h)} \cong Y$. □

Theorem 3.2.6 *Let θ be a congruence on a BE-algebra X. Then $f : X \to X_{/\theta}$ defined by $f(x) = [x]_\theta$, for all $x \in X$, is a surjective homomorphism. The above epimorphism is called canonical epimorphism.*

Proof Straightforward. □

Definition 3.2.7 If θ and ϕ are two congruences on a BE-algebra X with $\theta \subseteq \phi$, then define a relation $\phi_{/\theta}$ on $X_{/\theta}$ by $([x]_\theta, [y]_\theta) \in \phi_{/\theta}$ if and only if $(x, y) \in \phi$.

Theorem 3.2.8 *If θ and ϕ are two congruences on a BE-algebra X with $\theta \subseteq \phi$, then $\phi_{/\theta}$ is a congruence on $X_{/\theta}$ and $f : (X_{/\theta})_{\phi_{/\theta}} \to X_{/\phi}$ is an isomorphism.*

Proof Let θ and ϕ be two congruences on a BE-algebra X with $\theta \subseteq \phi$. Since ϕ is a congruence on X, we get that $\phi_{/\theta}$ is an equivalence relation on $X_{/\theta}$. Let $[x]_\theta, [y]_\theta, [z]_\theta, [w]_\theta \in X_{/\theta}$ be such that $([x]_\theta, [y]_\theta) \in \phi_{/\theta}$ and $([z]_\theta, [w]_\theta) \in \phi_{/\theta}$. Then we get $(x, y) \in \phi$ and $(z, w) \in \phi$. Since ϕ is a congruence on X, it yields $(x * z, y * w) \in \phi$. Hence $([x * z]_\theta, [y * w]_\theta) \in \phi_{/\theta}$. Therefore, it concludes that $\phi_{/\theta}$ is a congruence on $X_{/\theta}$. Now, define $f : (X_{/\theta})_{\phi_{/\theta}} \to X_{/\phi}$ by $[[x]_\theta]_{\phi_{/\theta}} \mapsto [x]_\phi$. For any $[[x]_\theta]_{\phi_{/\theta}}, [[y]_\theta]_{\phi_{/\theta}} \in (X_{/\theta})_{\phi_{/\theta}}$, we get the following consequence:

$$\begin{aligned}
f([[x]_\theta]_{\phi_{/\theta}} * [[y]_\theta]_{\phi_{/\theta}}) &= f([[x]_\theta * [y]_\theta]_{\phi_{/\theta}}) \\
&= f([[x * y]_\theta]_{\phi_{/\theta}}) \\
&= [x * y]_\phi \\
&= [x]_\phi * [y]_\phi \\
&= f([[x]_\theta]_{\phi_{/\theta}}) * f([[y]_\theta]_{\phi_{/\theta}})
\end{aligned}$$

Therefore f is a homomorphism. Let $[[x]_\theta]_{\phi_{/\theta}}, [[y]_\theta]_{\phi_{/\theta}} \in (X_{/\theta})_{\phi_{/\theta}}$ be such that $f([[x]_\theta]_{\phi_{/\theta}}) = f([[y]_\theta]_{\phi_{/\theta}})$. Then $[x]_\phi = [y]_\phi$, which implies $(x, y) \in \phi$. Hence it yields $([x]_\theta, [y]_\theta) \in \phi_{/\theta}$. Thus we get $[[x]_\theta]\phi_{/\theta} = [[y]_\theta]\phi_{/\theta}$. Therefore f is one-to-one. Let $[x]_\theta \in X_{/\theta}$ for some $x \in X$. Then $[[x]_\theta]_{\phi_{/\theta}} \in (X_{/\theta})_{\phi_{/\theta}}$ such that $f([[x]_\theta]_{\phi_{/\theta}}) = [x]_\phi$. Hence f is onto. Therefore f is an isomorphism. □

Definition 3.2.9 Let $\{X_i \mid i \in I\}$ be a family of BE-algebras. Then the *direct product* of $X_i, i \in I$ is the Cartesian product given as $\prod\limits_{i \in I} X_i = \{(x_i)_{i \in I} \mid x_i \in X_i\}$.

It can be easily seen that the direct product of the family $\{X_i \mid i \in I\}$ of BE-algebras is also a BE-algebra with respect to the point-wise operation given by

$$(x_i)_{i\in I} * (y_i)_{i\in I} = (x_i * y_i)_{i\in I}$$

for all $(x_i)_{i\in I}, (y_i)_{i\in I} \in \prod\limits_{i\in I} X_i$. In the following, we introduce the projections of direct products.

Definition 3.2.10 Let $\prod\limits_{i\in I} X_i$ be the direct product of the family $\{X_i \mid i \in I\}$ of BE-algebras. Then the mapping $P_k : \prod\limits_{i\in I} X_i \to X_k$ defined by $(x_i)_{i\in I} \mapsto x_k$ is known as the kth *projection* of the direct product $\prod\limits_{i\in I} X_i$ of the given family of BE-algebras.

Proposition 3.2.11 *Let $\prod_{i\in I} X_i$ be the direct product of the family $\{X_i \mid i \in I\}$ of BE-algebras. Then the projections P_k; $k \in I$ are surjective homomorphisms.*

Proof Define $P_k : \prod_{i\in I} X_i \to X_k$ by $P_k((x_i)_{i\in I}) = x_k$. Let $(x_i)_{i\in I}, (y_i)_{i\in I} \in \prod_{i\in I} X_i$. Then

$$\begin{aligned} P_k((x_i)_{i\in I} * (y_i)_{i\in I}) &= P_k((x_i * y_i)_{i\in I}) \\ &= x_k * y_k \\ &= P_k((x_i)_{i\in I} * P_k((y_i)_{i\in I} \end{aligned}$$

Therefore P_k is a homomorphism. Let $x_k \in X_k$. Since each X_i is non-empty, there exists $x_i \in X_i$ such that $(x_i)_{i\in I} \in \prod_{i\in I} X_i$ and $P_k((x_i)_{i\in I}) = x_k$. Therefore P_k is a surjective homomorphism. □

Definition 3.2.12 Let $\{X_i \mid i \in I\}$ be a family of BE-algebras. Then a subalgebra S of the direct product $\prod_{i\in I} X_i$ is said to be a *subdirect product* of $\{X_i \mid i \in I\}$ if $P_k(S) = X_k$ for all $k \in I$.

Definition 3.2.13 A BE-algebra X is said to be *subdirectly irreducible* if every family $\{\theta_i \mid i \in I, \theta_i \neq \Delta_X\}$ of congruence relations on X satisfies the following property:

$$\bigcap_{i\in I} \theta_i \neq \Delta_X$$

where $\Delta_X = \{(x_i)_{i\in I} \in \prod_{i\in I} \mid \text{ all } x_i, i \in I \text{ are equal }\}$ is the zero congruence on X.

Lemma 3.2.14 *Let X be a BE-algebra and θ a congruence on X. Then there is a one-to-one correspondence between the congruences of X containing θ and the congruences on $X_{/\theta}$.*

Proof Let ϕ be a congruence of X containing θ. Define a relation $\phi_{/\theta}$ on $X_{/\theta}$ by $([x]_\theta, [y]_\theta) \in \phi_{/\theta}$ if and only if $(x, y) \in \phi$. Then by Theorem 3.2.8, $\phi_{/\theta}$ is a congruence on $X_{/\theta}$. Let ϕ' be a congruence on $X_{/\theta}$. Define a relation ϕ_θ on X by $(x, y) \in \phi_\theta$ if and only if $([x]_{\phi_{/\theta}}, [y]_{\phi_{/\theta}}) \in \phi'$. Since ϕ' is a congruence on $X_{/\theta}$, it is clear that ϕ_θ is an equivalence relation on X. To prove that ϕ_θ is a congruence on X, let $(x, y) \in \phi_\theta$ and $(z, w) \in \phi_\theta$. Then we get $([x]_{\phi_{/\theta}}, [y]_{\phi_{/\theta}}) \in \phi'$ and $([z]_{\phi_{/\theta}}, [w]_{\phi_{/\theta}}) \in \phi'$. Hence we get the following consequence:

$$\begin{aligned} ([x]_{\phi_{/\theta}}, [y]_{\phi_{/\theta}}) \in \phi' \text{ and} ([z]_{\phi_{/\theta}}, [w]_{\phi_{/\theta}}) \in \phi' &\Rightarrow ([x]_{\phi_{/\theta}} * [z]_{\phi_{/\theta}}, [y]_{\phi_{/\theta}} * [w]_{\phi_{/\theta}}) \in \phi' \\ &\Rightarrow ([x * z]_{\phi_{/\theta}}, [y * w]_{\phi_{/\theta}}) \in \phi' \\ &\Rightarrow (x * z, y * w) \in \phi_\theta. \end{aligned}$$

Hence ϕ_θ is a congruence on X. Let $(a, b) \in \theta$. Then we get $[a]_\theta = [b]_\theta$. Hence $([a]_\theta, [b]_\theta) \in \phi'$. Thus, we get $(a, b) \in \phi_\theta$. Hence $\theta \subseteq \phi_\theta$. Therefore ϕ_θ is a congruence on $X_{/\theta}$ containing θ.

Let ϕ be a congruence on X containing θ. Then $\phi_{/\theta}$ is a congruence on $X_{/\theta}$ and hence $(\phi_{/\theta})_\theta$ is a congruence on X containing θ. Note that

$$
\begin{aligned}
(x, y) \in \phi &\Leftrightarrow ([x]_\theta, [y]_\theta) \in \phi_{/\theta} \\
&\Leftrightarrow (x, y) \in (\phi_{/\theta})_\theta.
\end{aligned}
$$

Hence $\phi = (\phi_{/\theta})_\theta$. Again, let ϕ' be a congruence on $X_{/\theta}$ containing. Then ϕ_θ is a congruence on X containing θ and hence $(\phi_\theta)_{/\theta}$ is a congruence on $X_{/\theta}$. Note that

$$
\begin{aligned}
([x]_{\phi_\theta}, [y]_{\phi_\theta}) \in \phi' &\Leftrightarrow (x, y) \in \phi_\theta \\
&\Leftrightarrow ([x]_{\phi_{/\theta}}, [y]_{\phi_{/\theta}}) \in (\phi_\theta)_{/\theta}.
\end{aligned}
$$

Hence $\phi' = (\phi_\theta)_{/\theta}$. Therefore the lemma is proved. □

Lemma 3.2.15 *Let $\{\theta_i \mid i \in I\}$ be a family of congruences on a BE-algebra X such that $\bigcap_{i \in I} \theta_i = \Delta_X$. Then X is isomorphic to a subdirect product of BE-algebras $X_{/\theta_i}, i \in I$.*

Proof Define a mapping $f : X \to \prod_{i \in I} X_{/\theta_i}$ by $f(a) = ([a]_{\theta_i})_{i \in I}$. Clearly f is well-defined. Let $a, b \in X$ such that $f(a) = f(b)$. Then $([a]_{\theta_i})_{i \in I} = ([b]_{\theta_i})_{i \in I}$. Hence $[a]_{\theta_i} = [a]_{\theta_i}$ for all $i \in I$. Thus it yields $(a, b) \in \theta_i$ for all $i \in I$. Hence $(a, b) \in \bigcap_{i \in I} \theta_i = \Delta_X$. Hence $a = b$, which yields that f is one-to-one. It is clearly observed that f is onto. Now for any $a, b \in X$, we get

$$
\begin{aligned}
f(a * b) &= ([a * b]_{\theta_i})_{i \in I} \\
&= ([a]_{\theta_i})_{i \in I} * ([b]_{\theta_i})_{i \in I} \\
&= f(a) * f(b).
\end{aligned}
$$

Hence f is a homomorphism and thus it is an isomorphism. We now show that $f(X)$ is a subalgebra of $X_{/\theta_i}, i \in I$. Clearly $f(X)$ is a subalgebra of $\prod_{i \in I} X_{/\theta_i}$. For any $k \in I$,

$$
\begin{aligned}
P_k(f(X)) &= \{P_k(f(a)) \mid f(a) \in f(L)\} \\
&= \{P_k(([a]_{\theta_i})_{i \in I}) \mid a \in X\} \\
&= \{[a]_{\theta_k} \mid a \in X\} \\
&= X_{\theta_k}.
\end{aligned}
$$

Therefore $f(X)$ is a subdirect product of BE-algebras $X_{/\theta_i}, i \in I$. □

Lemma 3.2.16 *Let X be a BE-algebra and $a, b \in X$ such that $a \neq b$. Then there is a congruence $\theta_{(a,b)}$ on X such that $(a, b) \notin \theta_{(a,b)}$ and $\theta_{(a,b)}$ is maximal with respect to this property.*

Proof Let $a, b \in X$ such that $a \neq b$. Then, define $P = \{\phi \in \mathcal{C}(X) \mid (a, b) \notin \phi\}$, where $\mathcal{C}(X)$ is the set of all congruences on X. Then by Zorn's lemma, P has a maximal element, say $\theta_{(a,b)}$. □

In the following lemma, a necessary and sufficient condition is derived for a BE-algebra to become a subdirectly irreducible BE-algebra.

Lemma 3.2.17 *A BE-algebra X is subdirectly irreducible if and only if $\mathcal{C}(X)$ has one and only one element θ_0 such that $\Delta_X \subset \theta_0 \subseteq \theta$ for all $\theta \in \mathcal{C}(X) - \Delta_X$.*

Proof Assume that X is subdirectly irreducible. Let

$$\theta_0 = \bigcap\{\theta \in \mathcal{C}(X) \mid \theta \neq \Delta_X\}.$$

Since X is subdirectly irreducible, it gives that $\Delta_X \subset \theta_0$. Otherwise, if $\theta_0 = \Delta_X$, then there exists some $\theta_i \in \mathcal{C}(X)$ such that $\theta_i = \Delta_X$. Therefore θ_0 is unique element in $\mathcal{C}(X)$ such that $\Delta_X \subset \theta_0 \subseteq \theta$ for all $\theta \in \mathcal{C}(X) - \Delta_X$.

Conversely, assume that $\mathcal{C}(X)$ has one and only one element θ_0 such that $\Delta_X \subset \theta_0 \subseteq \theta$ for all $\theta \in \mathcal{C}(X) - \Delta_X$. Let $\{\theta_i\}_{i \in I}$ be a family of congruences of X such that $\bigcap\limits_{i \in I} \theta_i = \Delta_X$. Suppose $\theta_i \neq \Delta_X$ for all $i \in I$. Hence $\bigcap\limits_{i \in I} \theta_i = \theta' \neq \Delta_X$, which is a contradiction. Hence $\theta_i = \Delta_X$ for some $i \in I$. Therefore X is subdirectly irreducible. □

Theorem 3.2.18 *Every BE-algebra is a subdirect product of subdirectly irreducible BE-algebras.*

Proof Let X be a BE-algebra and $a, b \in X$ with $a \neq b$. Consider the family of congruences $C = \{\theta_{(a,b)} \mid a \neq b\}$ as constructed in Lemma 3.2.16. Suppose $(x, y) \in \bigcap\limits_{a \neq b} \theta_{(a,b)}$. Then, it implies that $(x, y) \in \theta_{(a,b)}$ for all $a, b \in X$ with $a \neq b$. Suppose $x \neq y$. Then $(x, y) \in \theta_{(a,b)}$, which is a contradiction. Hence $x = y$, which implies that $\bigcap\limits_{a \neq b} \theta_{(a,b)} = \Delta_X$. Then by Lemma 3.2.15, X is isomorphic to a subdirect product of BE-algebras $X_{/\theta_{(a,b)}}, a \neq b$.

Now we prove $X_{/\theta_{(a,b)}}, a \neq b$ is subdirectly irreducible. Let $\langle\theta_{(a,b)}\rangle$ be the set of all congruences on X containing $\theta_{(a,b)}$. Let $\psi_{(a,b)}$ be the smallest congruence such that $(a, b) \in \psi_{(a,b)}$. Clearly $\theta_{(a,b)} \subseteq \psi_{(a,b)} \cup \theta_{(a,b)}$. Suppose $\theta_{(a,b)} = \psi_{(a,b)} \cup \theta_{(a,b)}$. Then $\psi_{(a,b)} \subseteq \theta_{(a,b)}$. Hence $(a, b) \in \theta_{(a,b)}$ for $a \neq b$, which is a contradiction. Thus $\theta_{(a,b)} \subset \psi_{(a,b)} \cup \theta_{(a,b)}$. Therefore $\psi_{(a,b)} \cup \theta_{(a,b)} \in \langle\theta_{(a,b)}\rangle$ and $\psi_{(a,b)} \cup \theta_{(a,b)} \neq \theta_{(a,b)}$. Suppose ϕ is another congruence in $\langle\theta_{(a,b)}\rangle$ such that $\phi \neq \theta_{(a,b)}$. Then $(a, b) \in \phi$, which implies that $\psi_{(a,b)} \subseteq \phi$. Hence $\psi_{(a,b)} \cup \theta_{(a,b)} \subseteq \phi$. Therefore $\theta_0 = \psi_{(a,b)} \cup \theta_{(a,b)}$ is the only element such that $\Delta_X \subset \theta_0 \subseteq \theta$ for all congruences θ in $\langle\theta_{(a,b)}\rangle$ with $\theta \neq \theta_{(a,b)}$. By Lemma 3.2.14, there is a one-to-one correspondence between congruences on $X_{/\theta_{(a,b)}}, a \neq b$ and those of $\langle\theta_{(a,b)}\rangle$. Hence $\mathcal{C}(X_{/\theta_{(a,b)}})$ has one and only element θ' such that $\overline{\theta}_{(a,b)} \subset \theta' \subseteq \theta$ for all $\theta \in \mathcal{C}(X_{/\theta_{(a,b)}}) - \overline{\theta}_{(a,b)}$, where $\overline{\theta}_{(a,b)}$ is the smallest(zero) congruence on $X_{/\theta_{(a,b)}}$. Then by Lemma 3.2.17, we get that $X_{/\theta_{(a,b)}}, a \neq b$ is subdirectly irreducible. Hence X isomorphic to a subdirect product of subdirectly irreducible BE-algebras. □

3.3 Closure Operators

In this section, the notion of closure operators is introduced in BE-algebras. Some properties of closure operators of BE-algebras are studied. Throughout this section, X stands for a BE-algebra which satisfies the antisymmetric property with respect to the ordering $\leq$.

Definition 3.3.1 Let $(X, *, 1)$ be a BE-algebra. A mapping $\mathcal{C} : X \to X$ is called a *closure operator* on X if it satisfies the following conditions for all $x, y \in X$:

(C1) $x \leq \mathcal{C}(x)$,
(C2) If $x \leq y$ then $\mathcal{C}(x) \leq \mathcal{C}(y)$,
(C3) $\mathcal{C}(\mathcal{C}(x)) = \mathcal{C}(x)$.

Example 3.3.2 Let $X = \{1, a, b, c, d\}$. Define a binary operation $*$ on X as follows:

$*$	1	a	b	c	d
1	1	a	b	c	d
a	1	1	b	c	b
b	1	a	1	b	a
c	1	a	1	1	a
d	1	1	1	b	1

Clearly $(X, *, 1)$ is a BE-algebra. Define a self-mapping $\mathcal{C} : X \to X$ as follows:

$$\mathcal{C}(a) = \mathcal{C}(1) = 1; \mathcal{C}(b) = \mathcal{C}(d) = b \, and \, \mathcal{C}(c) = c$$

It can be easily seen that $\mathcal{C}$ is a closure operator on X.

Definition 3.3.3 A closure operator $\mathcal{C}$ on a BE-algebra $(X, *, 1)$ is said to be a BE-*closure operator* on X if $\mathcal{C}(x * y) \leq \mathcal{C}(x) * \mathcal{C}(y)$ for all $x, y \in X$.

Example 3.3.4 For any BE-algebra $(X, *, 1)$, define $i_X : X \to X$ by $i_X(x) = x$ for all $x \in X$. Then i_X is a closure operator on X. Moreover it is a BE-closure operator.

Example 3.3.5 For any BE-algebra $(X, *, 1)$, define $\omega_X : X \to X$ by $\omega_X(x) = 1$ for all $x \in X$. Then ω_X is a BE-closure operator on X. For, let $x, y \in X$. Clearly $x \leq 1 = \omega_X(x)$. Suppose $x \leq y$. Then $x * y = 1$ and hence $1 = \omega_X(1) = \omega_X(x * y) \leq \omega_X(x) * \omega_X(y)$. Hence $\omega_X(x) \leq \omega_X(y)$. Clearly $\omega_X(\omega_X(x)) =$

$\omega_X(1) = 1 = \omega_X(x)$. Hence ω_X is a closure operator on X. For $x, y \in X$, we get $\omega_X(x * y) = 1 = 1 * 1 = \omega_X(x) * \omega_X(y)$. Therefore ω_X is a BE-closure operator on X.

Proposition 3.3.6 *Let $(X, *, 1)$ be a self-distributive BE-algebra. For any $a \in X$, the self-map $f_a : X \to X$ defined by $f_a(x) = a * x$ for all $x \in X$ is a BE-closure operator on X.*

Proof Let $a \in X$. Clearly $x \leq a*x = f_a(x)$ for all $x \in X$. Let $x, y \in X$ be such that $x \leq y$. Since X is positively ordered, we get $a * x \leq a * y$. Hence $f_a(x) \leq f_a(y)$. Since X is self-distributive, we get $f_a(f_a(x)) = a * (a * x) = (a * a) * (a * x) = 1 * (a * x) = a * x = f_a(x)$. Therefore f_a is a closure operator on X. Finally $f_a(x * y) = a * (x * y) = (a * x) * (a * y) = f_a(x) * f_a(y)$. Therefore f_a is a BE-closure operator on X. □

Theorem 3.3.7 *Let $(X, *, 1)$ be a BE-algebra and $\mathcal{C} : X \to X$ a BE-closure operator on X. Define $S = \{x \in X \mid \mathcal{C}(x) = x\}$. Then the following conditions hold:*

(1) *$S = \mathcal{C}(X)$ and is a subalgebra of X such that $1 \in S$,*
(2) *$(S, *_c, 1)$ is a BE-algebra defined as $x *_c y = \mathcal{C}(x * y)$ for all $x, y \in S$.*

Proof (1) Clearly $S \subseteq \mathcal{C}(X)$. Conversely, let $x \in \mathcal{C}(X)$. Then $x = \mathcal{C}(a)$ for some $a \in X$. Since $\mathcal{C}(\mathcal{C}(a)) = \mathcal{C}(a)$, we get $x = \mathcal{C}(a) \in S$. Thus $S = \mathcal{C}(X)$. Let $x, y \in S$. Then $\mathcal{C}(x) = x$ and $\mathcal{C}(y) = y$. Hence $\mathcal{C}(x * y) \leq \mathcal{C}(x) * \mathcal{C}(y) = x * y$. Clearly $x * y \leq \mathcal{C}(x * y)$. Thus $\mathcal{C}(x * y) = x * y$. Therefore $x * y \in S$, which implies S is a subalgebra of X. Since $1 \leq \mathcal{C}(1)$, we get $\mathcal{C}(1) = 1$. Hence $1 \in S$.
(2) Let $x, y, z \in S$. Then $1 *_c x = \mathcal{C}(1 * x) = \mathcal{C}(x) = x$. Also $x *_c 1 = \mathcal{C}(x * 1) = \mathcal{C}(1) = 1$ and $x *_c x = \mathcal{C}(x * x) = \mathcal{C}(1) = 1$. Finally, $x *_c (y *_c z) = x *_c \mathcal{C}(y * z) = \mathcal{C}(x * \mathcal{C}(y * z)) = \mathcal{C}(x * (y * z)) = \mathcal{C}(y * (x * z)) = \mathcal{C}(y * \mathcal{C}(x * z)) = y *_c \mathcal{C}(x * z) = y *_c (x *_c z)$. Therefore $(S, *_c, 1)$ is a BE-algebra. □

Corollary 3.3.8 *For any BE-algebra X, $(\mathcal{C}(X), *_c, 1)$ is a BE-algebra.*

Proposition 3.3.9 *For any BE-algebra X, $S = \{x \in X \mid \mathcal{C}(x) = 1\}$ is a subalgebra of X.*

Theorem 3.3.10 *Let $(X, *, 1)$ and $(S, *_c, 1)$ be BE-algebras $\mathcal{C} : X \to X$ be a closure operator on X. Then $\mathcal{C} : S \to \mathcal{C}(X)$ is a homomorphism.*

Proof Let $x, y \in S$. Then $x = \mathcal{C}(x)$ and $y = \mathcal{C}(y)$. From $(C3)$, we get $\mathcal{C}(x), \mathcal{C}(y) \in S$. Hence $\mathcal{C}(x *_c y) = \mathcal{C}(\mathcal{C}(x) * \mathcal{C}(y)) = \mathcal{C}(x * y) \leq \mathcal{C}(x) * \mathcal{C}(y)$. Again from $(C1)$, we get $\mathcal{C}(x) * \mathcal{C}(y) \leq \mathcal{C}(\mathcal{C}(x) * \mathcal{C}(y)) = \mathcal{C}(x *_c y)$. Hence $\mathcal{C}(x *_c y) = \mathcal{C}(x) * \mathcal{C}(y)$. Thus $\mathcal{C} : S \to \mathcal{C}(X)$ is a homomorphism. □

Definition 3.3.11 Let X be a BE-algebra and $\mathcal{C}$ a closure operator of X. Then an element $x \in X$ is said to be *simple* with respect to $\mathcal{C}$, if $\mathcal{C}(x) = x$.

For any $x \in X$, it obvious that $\mathcal{C}(x)$ is simple w.r.t. $\mathcal{C}$. Since $1 \leq \mathcal{C}(1)$, it is clear that 1 is simple with respect to every closure operator $\mathcal{C}$ on a BE-algebra X.

Proposition 3.3.12 *Let $\mathcal{C}$ be a closure operator on a BE-algebra X. Then for each $x \in X$, $\mathcal{C}(x)$ is the least simple element which includes x.*

Proof By the property $\mathcal{C}$, we get $\mathcal{C}(x)$ is simple element of X. By the condition $(C2)$, we get that $x \leq \mathcal{C}(x)$. Suppose y is a simple element such that $x \leq y$. Then, it yields that $\mathcal{C}(x) \leq \mathcal{C}(y) = y$. Hence $\mathcal{C}(x) \leq y$. Therefore $\mathcal{C}(x)$ is the smallest simple element which includes x. □

Let X be a BE-algebra X and S is a subset of X. Then for every $x \in X$, let us denote $S(x] = \{y \in X \mid x \leq y\}$. Instead of $S(x]$ we use the simplified notation $(x] = \{y \in X \mid x \leq y\}$.

Definition 3.3.13 If X is a BE-algebra, by a closure system or Moore family we mean any non-empty subset $C \subseteq X$ such that for each $x \in X$, the set $C(x]$ has the least element.

Proposition 3.3.14 *The set of all simple elements of a closure operator is a closure system.*

Proof Let $\mathcal{C}$ be a closure operator on a BE-algebra X. Denote by S_X the set of all simple elements of X. We now prove that S_X is a closure system. Let $A \subseteq S_X$ and $x \in X$. We prove that $A(x]$ has the least element. Clearly $\mathcal{C}(x)$ is the least element which includes x. Also $\mathcal{C}(x) \in A(x]$. Thus $A(x]$ has the least element, precisely $\mathcal{C}(x)$. Therefore S_X is a closure system. □

Lemma 3.3.15 *Let $\mathcal{C}$ be a closure operator on a BE-algebra $(X, *1)$. Then an element $x \in X$ is simple with respect to $\mathcal{C}$ if and only if $x = \mathcal{C}(a)$ for some $a \in X$.*

Proof Assume that $x \in X$ is simple w.r.t. $\mathcal{C}$. Then clearly $x = \mathcal{C}(x)$. Conversely, assume that $x = \mathcal{C}(a)$ for some $a \in X$. Then $\mathcal{C}(x) = \mathcal{C}(\mathcal{C}(a)) = \mathcal{C}(a) = x$. Therefore x is simple w.r.t. $\mathcal{C}$. □

Lemma 3.3.16 *Let $\mathcal{C}$ be a closure operator on a BE-algebra X. If $\{x_\alpha\}_{\alpha\in\Delta}$ is a family of simple elements of X w.r.t. $\mathcal{C}$, then $\inf\limits_{\alpha\in\Delta} x_\alpha$ is also a simple element w.r.t. $\mathcal{C}$ in X.*

Proof Suppose $\{x_\alpha\}_{\alpha\in\Delta}$ is a family of simple elements of X with respect to $\mathcal{C}$. Let $x = \inf\limits_{\alpha\in\Delta} x_\alpha$. Then clearly $x \in X$ and $x \leq x_\alpha$ for all $\alpha \in \Delta$. Hence, we get that $\mathcal{C}(x) \leq \mathcal{C}(x_\alpha) = x_\alpha$ for all $\alpha \in \Delta$. Therefore $\mathcal{C}(x) \leq \inf\limits_{\alpha\in\Delta} x_\alpha = x$. On the other hand, it is clear that $x \leq \mathcal{C}(x)$. Hence $\mathcal{C}(x) = x$. Therefore $x = \inf\limits_{\alpha\in\Delta} x_\alpha$ is simple w.r.t. $\mathcal{C}$. □

Proposition 3.3.17 *Let $\mathcal{C}$ be a closure operator on a BE-algebra X. For any $x \in X$, $\mathcal{C}(x) = \inf\{y \in X \mid y$ is simple w.r.t. $\mathcal{C}$ and $x * y = 1\}$.*

Proof Let $A = \inf\{y \in X \mid y$ is simple w.r.t. $\mathcal{C}$ and $x * y = 1\}$ and $y_0 = \inf\limits_{y\in A} y$. It is enough to prove that $\mathcal{C}(x) = y_0$. Let $y \in A$. Then $\mathcal{C}(y) = y$ and $x * y = 1$. Thus $x \leq y$ and hence $\mathcal{C}(x) \leq \mathcal{C}(y) = y$ for all $y \in A$. Therefore $\mathcal{C}(x) \leq \inf\limits_{y\in A} y = y_0$. On the other hand, since $\mathcal{C}(\mathcal{C}(x)) = \mathcal{C}(x)$ and $x * \mathcal{C}(x) = 1$. Hence $\mathcal{C}(x) \in A$, which implies $y_0 \leq \mathcal{C}(x)$. Therefore $\mathcal{C}(x) = y_0$. □

Let $\Omega(X)$ be the set of all closure operators on X and for any $\Phi, \Psi \in \Omega(X)$, define $\Phi \leq \Psi$ if and only if $\Phi(x) \leq \Psi(x)$ for all $x \in X$. If X is transitive, then $\leq$ is a partial order on $\Omega(X)$ and hence $(\Omega(X), \leq)$ is a partially ordered set in which ω_X is the greatest element of $(\Omega(X), \leq)$. In the following result, it is observed that $(\Omega(X), \leq)$ is a complete lattice.

Definition 3.3.18 Let $\{\Phi_\alpha\}_{\alpha\in\Delta} \subseteq \Omega(X)$, define $(\inf\limits_{\alpha\in\Delta} \Phi_\alpha)(x) = \inf\limits_{\alpha\in\Delta} \Phi_\alpha(x)$ for all $x \in X$.

Proposition 3.3.19 *Let $\{\Phi_\alpha\}_{\alpha\in\Delta} \subseteq \Omega(X)$. Then $\inf\limits_{\alpha\in\Delta} \Phi_\alpha \in \Omega(X)$ which is the g.l.b. of $\{\Phi_\alpha\}_{\alpha\in\Delta}$ in the partially ordered set $(\Omega(X), \leq)$. Therefore $(\Omega(X), \leq)$ is a complete lattice.*

Proof Write $\Phi = \inf\limits_{\alpha\in\Delta} \Phi_\alpha$. Let $x \in X$. Since each Φ_α is a closure operator, we get that $x \leq \Phi_\alpha(x)$ for all $\alpha \in \Delta$. Hence $x \leq \inf\limits_{\alpha\in\Delta}(\Phi_\alpha(x)) = (\inf\limits_{\alpha\in J} \Phi_\alpha)(x) = \Phi(x)$. Let $x, y \in X$ be such that $x \leq y$. Then $\Phi_\alpha(x) \leq \Phi_\alpha(y)$ for all $\alpha \in \Delta$. Since $\Phi \leq \Phi_\alpha$ for all $\alpha \in \Delta$, we get that $\Phi(x) \leq \Phi_\alpha(x) \leq \Phi_\alpha(y)$ for all $\alpha \in \Delta$ and hence $\Phi(x) \leq \inf\limits_{\alpha\in\Delta}(\Phi_\alpha(y))$. Therefore $\Phi(x) \leq \Phi(y)$. Let $x \in X$. Clearly $\Phi(x) \leq \Phi(\Phi(x))$. Since each Φ_α is a closure operator on X, we get that $\Phi(x) = (\inf\limits_{\alpha\in\Delta} \Phi_\alpha)(x) = \inf\limits_{\alpha\in\Delta} \Phi_\alpha(x) = \inf\limits_{\alpha\in\Delta} \Phi_\alpha(\Phi_\alpha(x)) = (\inf\limits_{\alpha\in\Delta} \Phi_\alpha)(\Phi_\alpha(x)) = \Phi(\Phi_\alpha(x)) \geq \Phi(\inf\limits_{\alpha\in\Delta} \Phi_\alpha(x)) = \Phi((\inf\limits_{\alpha\in\Delta} \Phi_\alpha)(x)) = \Phi(\Phi(x))$. Therefore $\Phi(\Phi(x)) = \Phi(x)$. Now, for $x, y \in X$, we get $\Phi(x * y) = (\inf\limits_{\alpha\in\Delta} \Phi_\alpha)(x * y) = \inf\limits_{\alpha\in\Delta} \Phi_\alpha(x * y) \leq \inf\limits_{\alpha\in\Delta}\{\Phi_\alpha(x) * \Phi_\alpha(y)\} = \{\inf_{\alpha\in\Delta} \Phi_\alpha(x)\} * \{\inf_{\alpha\in\Delta} \Phi_\alpha(y)\} = \{(\inf_{\alpha\in\Delta} \Phi_\alpha)(x)\} * \{(\inf_{\alpha\in\Delta} \Phi_\alpha)(y)\} = \Phi(x) * \Phi(y)$. Therefore Φ is a closure operator on X. Suppose $\Psi \in \Omega(X)$ and $\Psi \leq \Phi_\alpha$ for all $\alpha \in \Delta$. Then $\Psi(x) \leq \Phi_\alpha(x)$ for all $\alpha \in \Delta$. Hence

$$\Psi(x) \leq \inf_{\alpha\in\Delta} \Phi_\alpha(x) = (\inf_{\alpha\in\Delta} \Phi_\alpha)(x) = \Phi(x).$$

Thus $\Psi(x) \leq \Phi(x)$. Hence $\Psi \leq \Phi$ and hence Φ is the g.l.b. of $\{\Phi_\alpha\}_{\alpha\in\Delta}$. Since ω_X is the greatest element of $(\Omega, \leq)$, it yields that $(\Omega, \leq)$ is the complete lattice. □

Theorem 3.3.20 *Let $\{\Phi_\alpha\}_{\alpha\in\Delta} \subseteq \Omega(X)$ and $\Psi = \sup\limits_{\alpha\in\Delta} \Phi_\alpha$. Then, for any $x \in X$, $\Psi(x) = x$ if and only if $\Phi_\alpha(x) = x$ for all $\alpha \in \Delta$.*

Proof Let $\{\Phi_\alpha\}_{\alpha\in\Delta} \subseteq \Omega(X)$ and $\Psi = \sup\limits_{\alpha\in\Delta} \Phi_\alpha$. Suppose $x \in X$ and $\Psi(x) = x$. Since $\Phi_\alpha \leq \Psi$ for all $\alpha \in \Delta$, we get that $\Phi_\alpha(x) \leq \Psi(x) = x$ for all $\alpha \in \Delta$. Since each Φ_α is closure, it is clear that $x \leq \Phi_\alpha(x)$ for all $x \in X$. Hence $\Phi_\alpha(x) = x$. Conversely, assume that $\Phi_\alpha(x) = x$ for all $\alpha \in \Delta$. Define $\eta : X \to X$ by

$$\eta(y) = \begin{cases} x & \text{if } y * x = 1 \\ 1 & \text{otherwise} \end{cases}$$

We first prove that η is a closure operator on X. Since $x * x = 1$, we get that $\eta(x) = x$ for all $x \in X$. Let $y, z \in X$ be such that $y \leq z$. Suppose $z * x = 1$. Then

$1 = z * x \leq z * y$. Hence $z * y = 1$. Therefore $\eta(z) = x = \eta(y)$. If $z * x \neq 1$, then $\eta(z) = 1$. Hence $\eta(y) \leq \eta(z)$. Let $y \in X$. If $y * x = 1$, then $\eta(y) = x$. Now, $\eta(\eta(y)) = \eta(x) = x = \eta(y)$. Suppose $y * x \neq 1$. Then $\eta(y) = 1$. Therefore $\eta(\eta(y)) = \eta(1) = 1 = \eta(y)$. Therefore η is a closure operator on X.

We now prove that $\Phi_\alpha \leq \eta$ for all $\alpha \in \Delta$. Let $y \in X$. Suppose $y * x = 1$. Then $y \leq x$, which implies $\Phi_\alpha(y) \leq \Phi_\alpha(x) = x = \eta(y)$. Suppose $y * x \neq 1$. Then $\Phi_\alpha(y) \leq 1 = \eta(y)$. Therefore $\Phi_\alpha \leq \eta$. Since α is arbitrary, we get that $\Psi(x) \leq \eta(x) = x$. On the other hand, it is clear that $x \leq \Psi(x)$. Therefore $\Psi(x) = x$ for all $x \in X$. □

In the following, the notion of dual atoms is introduced in $\Omega(X)$:

Definition 3.3.21 A closure operator Φ on a BE-algebra X is said to be a *dual atom* of $\Omega(X)$ if there exists no closure operator Ψ on X such that $\Phi < \Psi < \omega_X$. In other words, if Ψ is a closure operator on X such that $\Phi \leq \Psi \leq \omega_X$, then either $\Phi = \Psi$ or $\Psi = \omega_X$.

Proposition 3.3.22 *Let X be a BE-algebra and $1 \neq a \in X$. Define $\Phi_a : X \to X$ by*

$$\Phi_a(x) = \begin{cases} a & if\, x * a = 1 \\ 1 & otherwise \end{cases}$$

for all $x \in X$. Then Φ_a is a closure operator on X. Moreover Φ_a is a dual atom of $\Omega(X)$.

Proof We first show that Φ_a is a closure operator on X.
(C1) Let $x \in X$. Suppose $x * a = 1$. Then $x \leq a = \Phi_a(x)$. If $x * a \neq 1$, then $x \leq 1 = \Phi_a(x)$.
(C2) Let $x, y \in X$ be such that $x \leq y$. If $y * a \neq 1$, then $\Phi_a(x) \leq 1 = \Phi_a(y)$. Suppose $y * a = 1$. Then $1 = y * a \leq x * a$. Hence $\Phi_a(x) = a = \Phi_a(y)$. Therefore $\Phi_a(x) \leq \Phi_a(y)$ whenever $x \leq y$.
(C3) Let $x \in X$. Suppose $x * a = 1$. Then $\Phi_a(x) = a$. Clearly $\Phi_a(a) = a$. Hence $\Phi_a(\Phi_a(x)) = \Phi_a(a) = a = \Phi_a(x)$. Suppose $x * a \neq 1$. Then $\Phi_a(x) = 1$. Then $\Phi_a(\Phi_a(x)) = \Phi_a(1) = 1 = \Phi_a(x)$.

Let Ψ be a closure operator on X such that $\phi_a \leq \Psi \leq \omega_X$. Then we have the following:
Case (i): Suppose $\Psi(a) = 1$. If $x * a = 1$, then we get $a = \Phi_a(x) \leq \Psi(x)$. Now, $1 = \Psi(a) = \Psi(x * a) \leq \Psi(x)$. If $x * a \neq 1$, then $1 = \Phi_a(x) \leq \Psi(x)$. Hence $\Psi(x) = 1$. Therefore $\Psi = \omega_X$.
Case (ii): Suppose $\Psi(a) < 1$. If $x * a = 1$, then $x \leq a$ and hence $\Psi(x) \leq \psi(a)$. Also, $a = \Phi_a(x) \leq \Psi(x)$ implies $\Psi(a) \leq \Psi(x)$. Thus $\Psi(x) = \Psi(a)$. Now, $\Phi_a(\Psi(a)) \leq \Psi(\Psi(a)) = \Psi(a) < 1$. Hence $\Psi(a) * a = 1$. So that, $\Psi(a) = a$. Hence $\Psi(x) = a$ if $x * a = 1$. If $x * a \neq 1$, then $\Phi_a(x) = 1$ and so $\Psi(x) = 1$. Therefore $\Psi = \Phi_a$. Thus $\Psi = \omega$ or $\Psi = \Phi_a$. Hence Φ_a is a dual atom of $\Omega(X)$. □

Theorem 3.3.23 *If Φ is a dual atom of $\Omega(X)$. Then there exists $1 \neq b \in X$ such that $\Phi = \Phi_b$.*

Proof Suppose that Φ is a dual atom of $\Omega(X)$. Then we get $\Phi \neq \omega_X$. Therefore there exists $1 \neq x \in X$ such that $\Phi(x) \neq 1$. If $b = \Phi(x)$, then $b \neq 1$. Now, we prove that $\Phi_b = \Phi$. Since Φ is dual atom, it is clear that $\Phi_b \leq \Phi$. Let $y \in X$. If $y * b = 1$, then $y \leq b$. Hence $\Phi_b(y) = b$ and $\Phi(y) \leq \Phi(b) = b$. Thus $\Phi(y) \leq b = \Phi_b(y)$. Hence we get $\Phi(y) = \Phi_b(y)$. If $y * b \neq 1$, then $\Phi_b(y) = 1$. Therefore $\Phi(y) \leq \Phi_b(y)$. Hence $\Phi(y) = \Phi_b(y)$. Therefore $\Phi = \Phi_b$. □

3.4 Fuzzy Weak Subalgebras

In this section, we study some of the fuzzy concepts such as fuzzy weak subalgebras, fuzzy relations, fuzzy products, etc. The properties of fuzzy weak subalgebras are extensively studied. The properties of triangular normed weak subalgebras and fuzzy dot weak subalgebras are also studied.

Definition 3.4.1 ([250]) Let X be a set. Then a *fuzzy set* in X is a function $\mu : X \to [0, 1]$.

In 1965, Lotfi A. Zadeh introduced the notion of fuzzy sets through his celebrated paper "Fuzzy sets." The theory of fuzzy sets admits the gradual assessment of the membership of elements in a set. Many researchers around the world have been applied the fuzzy concepts to many algebraic structure and derived fuzzy characterization of subalgebras, filters, and ideals.

Definition 3.4.2 Let X be a BE-algebra. A fuzzy set μ of X is said to be a *fuzzy weak subalgebra* if $\mu(x * (x * y)) \geq \min\{\mu(x), \mu(y)\}$ for all $x, y \in X$.

Example 3.4.3 Let $X = \{1, a, b, c\}$. Define a binary operation $*$ on X as follows:

$*$	1	a	b	c
1	1	a	b	c
a	1	1	a	a
b	1	1	1	a
c	1	1	a	1

Then $(X, *, 1)$ is a BE-algebra. Define a fuzzy set $\mu : X \to [0, 1]$ as follows:

$$\mu(x) = \begin{cases} 1 & \text{if } x = 1 \\ 0 & \text{otherwise} \end{cases}$$

Then it can be easily verified that μ is a fuzzy weak subalgebra of X.

Lemma 3.4.4 *If μ is a fuzzy weak subalgebra of a BE-algebra X, then $\mu(1) \geq \mu(x)$ for all $x \in X$.*

Proof Let μ be a fuzzy weak subalgebra of X. For all $x \in X$, we have $x * (x * x) = x * 1 = 1$. Hence $\mu(1) = \mu(x * (x * x)) \geq \min\{\mu(x), \mu(x)\} = \mu(x)$. □

Proposition 3.4.5 *Let μ be a fuzzy weak subalgebra of a BE-algebra X and $n \in \mathbb{Z}^+$. Then*

(1) $\mu(x *^n x) \geq \mu(x)$ *for any odd number n,*
(2) $\mu(x *^n x) = \mu(x)$ *for any even number n*

where $x *^n x = (\cdots((x * x) * x)\cdots x)$ and $*$ is operated n times.

Proof (1) Let $x \in X$ and assume that n is odd. Then $n = 2k - 1$ for some positive integer k. We prove this result by using induction on n. For $k = 1$, the result is clear by above lemma. Assume that $\mu(x *^{2k-1} x) \geq \mu(x)$. Then by the assumption, we get the following:

$$\begin{aligned}\mu(x *^{2(k+1)-1} x) &= \mu(x *^{2k+1} x)\\ &= \mu(((x * x) * x) *^{2k-1} x)\\ &= \mu(x *^{2k-1} x)\\ &\geq \mu(x)\end{aligned}$$

which proves (1). Similarly, (2) can be proved. □

Theorem 3.4.6 *Let X be a BE-algebra and μ a fuzzy weak subalgebra of X. If there exists a sequence $\{x_n\}$ in X such that $\lim\limits_{n\to\infty} \mu(x_n) = 1$, then $\mu(1) = 1$.*

Proof Assume that μ is a fuzzy weak subalgebra of X. By Lemma 3.4.4, we get $\mu(1) \geq \mu(x)$ for all $x \in X$. Thus $\mu(1) \geq \mu(x_n)$ for $n \in \mathbb{Z}^+$. Hence $1 \geq \mu(1) \geq \lim\limits_{n\to\infty} \mu(x_n) = 1$. Thus $\mu(1) = 1$. □

Definition 3.4.7 Let μ_1 and μ_2 be two fuzzy weak subalgebras of a BE-algebra X. For all $x, y \in X$, the intersection of μ_1 and μ_2 is defined as $\mu_1 \cap \mu_2$ which is given by

$$(\mu_1 \cap \mu_2)(x * y) = \min\{\mu_1(x * y), \mu_2(x * y)\}.$$

Theorem 3.4.8 *Let μ_1 and μ_2 be two fuzzy weak subalgebras of a BE-algebra X. Then $\mu_1 \cap \mu_2$ is a fuzzy weak subalgebra of X.*

Proof Let $x, y \in X$. Since μ_1 and μ_2 are fuzzy weak subalgebras, by the above theorem we get

$$\begin{aligned}(\mu_1 \cap \mu_2)(x * (x * y)) &= \min\{\mu_1(x * (x * y)), \mu_2(x * (x * y))\}\\ &\geq \min\{\min\{\mu_1(x), \mu_1(y)\}, \min\{\mu_2(x), \mu_2(y)\}\}\\ &= \min\{\min\{\mu_1(x), \mu_2(x)\}, \min\{\mu_1(y), \mu_2(y)\}\}\\ &= \min\{(\mu_1 \cap \mu_2)(x), (\mu_1 \cap \mu_2)(y)\}.\end{aligned}$$

Therefore $\mu_1 \cap \mu_2$ is a fuzzy weak subalgebra of X. □

Corollary 3.4.9 *Let $\{\mu_i \mid i \in \Delta\}$ be a family of fuzzy weak subalgebras of X. Then $\bigcap_{i\in\Delta} \mu_i$ is also a fuzzy weak subalgebras of X.*

Definition 3.4.10 Let μ be a fuzzy set in a BE-algebra X. For any $\alpha \in [0, 1]$, the set $\mu_\alpha = \{x \in X \mid \mu(x) \geq \alpha\}$ is said to be a *level subset* of μ.

In the following theorem, a necessary and sufficient condition is established for a fuzzy set of a BE-algebra to become a fuzzy weak subalgebra in X. Note that it is a consequence of the transfer principle of fuzzy sets described in [161].

Theorem 3.4.11 *A fuzzy set μ of a BE-algebra X is a fuzzy weak subalgebra of X if and only if for each $\alpha \in [0, 1]$, μ_α is a weak subalgebra of X when $\mu_\alpha \neq \emptyset$.*

Proof Assume that μ is a fuzzy weak subalgebra of X. Let $x, y \in \mu_\alpha$. Then we get that $\mu(x) \geq \alpha$ and $\mu(y) \geq \alpha$. Since μ is a fuzzy weak subalgebra, we get $\mu(x * (x * y)) \geq \min\{\mu(x), \mu(y)\} \geq \alpha$. Thus $x * (x * y) \in \mu_\alpha$. Therefore μ_α is a weak subalgebra of X.

Conversely, assume that μ_α is a weak subalgebra of X for each $\alpha \in [0, 1]$ with $\mu_\alpha \neq \emptyset$. Let $x, y \in X$ be such that $\mu(x) = \alpha_1$ and $\mu(y) = \alpha_2$. Then we get that $x \in \mu_{\alpha_1}$ and $y \in \mu_{\alpha_2}$. Without loss of generality, assume that $\alpha_1 \leq \alpha_2$. Then clearly $\mu_{\alpha_2} \subseteq \mu_{\alpha_1}$. Hence $y \in \mu_{\alpha_1}$. Since μ_{α_1} is a weak subalgebra of X, we get $x * (x * y) \in \mu_{\alpha_1}$. Thus $\mu(x * (x * y)) \geq \alpha_1 = \min\{\alpha_1, \alpha_2\} = \min\{\mu(x), \mu(y)\}$. Therefore μ is a fuzzy weak subalgebra of X. □

Definition 3.4.12 Let μ be a fuzzy weak subalgebra of a BE-algebra X. Then the weak subalgebras μ_α, $\alpha \in [0, 1]$, are called *level weak subalgebras* of X.

Theorem 3.4.13 *Any weak subalgebra of a BE-algebra X can be realized as a level weak subalgebra of some fuzzy weak subalgebra of X.*

Proof Let S be a weak subalgebra of a BE-algebra X. Define a fuzzy set $\mu : X \to [0, 1]$ as follows:

$$\mu(x) = \begin{cases} \alpha & \text{if } x \in S \\ 0 & \text{if } x \notin S \end{cases}$$

where $0 < \alpha < 1$ is fixed. Let $x, y \in X$. Suppose $x, y \in S$. Since S is a weak subalgebra, we get $x * (x * y) \in S$. Then, it gives $\mu(x) = \mu(x * (x * y)) = \mu(y) = \alpha$. Hence $\mu(x * (x * y)) \geq \min\{\mu(x), \mu(y)\}$. Suppose $x \notin S$ and $y \notin S$. Then we get $\mu(x) = \mu(y) = 0$. Hence $\mu(x * (x * y)) \geq \min\{\mu(x), \mu(y)\}$. If exactly one of x and y is in S, then exactly one of $\mu(x)$ and $\mu(y)$ is equal to 0. Hence $\mu(x * (x * y)) \geq \min\{\mu(x), \mu(y)\}$. Thus $\mu(x * (x * y)) \geq \min\{\mu(x), \mu(y)\}$ for all $x, y \in X$. Therefore μ is a fuzzy weak subalgebra of X. Clearly $\mu_\alpha = S$. Hence the theorem is proved. □

Theorem 3.4.14 *Let μ be a fuzzy weak subalgebra of a BE-algebra X. Then two-level weak subalgebras μ_{α_1} and μ_{α_2}(with $\alpha_1 < \alpha_2$) of μ are equal if and only if there is no $x \in X$ such that $\alpha_1 \leq \mu(x) < \alpha_2$.*

Proof Assume that $\mu_{\alpha_1} = \mu_{\alpha_2}$ for $\alpha_1 < \alpha_2$. Suppose there exists some $x \in X$ such that $\alpha_1 \leq \mu(x) < \alpha_2$. Then μ_{α_2} is a proper subset of μ_{α_1}, which is impossible. Conversely, assume that there is no $x \in X$ such that $\alpha_1 \leq \mu(x) < \alpha_2$. Hence $\mu_{\alpha_2} \subseteq \mu_{\alpha_1}$. If $x \in \mu_{\alpha_1}$, then $\mu(x) \geq \alpha_1$. Hence by the assumed condition, $\mu(x) \geq \alpha_2$. Hence $x \in \mu_{\alpha_2}$. Thus $\mu_{\alpha_1} \subseteq \mu_{\alpha_2}$. Therefore $\mu_{\alpha_1} = \mu_{\alpha_2}$. □

Definition 3.4.15 Let X be a BE-algebra and f an endomorphism on X. For any fuzzy set μ in X, define a mapping $\mu^f : X \to [0, 1]$ such that $\mu^f(x) = \mu(f(x))$ for all $x \in X$.

Clearly the above map μ^f is well-defined and fuzzy set in X.

Theorem 3.4.16 *Let X be a BE-algebra and f an onto endomorphism on X. For a fuzzy set μ in X, μ is a fuzzy weak subalgebra of X if and only if μ^f is a fuzzy weak subalgebra of X.*

Proof Assume that μ is a fuzzy weak subalgebra of X. Let $x, y \in X$. Then $\mu^f(x * (x * y)) = \mu(f(x * (x * y))) = \mu(f(x) * (f(x) * f(y))) \geq \min\{\mu(f(x)), \mu(f(y))\} = \min\{\mu^f(x), \mu^f(y)\}$. Hence μ^f is a fuzzy weak subalgebra of X. Conversely, assume that μ^f is a fuzzy weak subalgebra of X. Let $x \in X$. Since f is onto, there exists $y \in X$ such that $f(y) = x$. Let $x, y \in X$. Then there exist $a, b \in X$ such that $f(a) = x$ and $f(b) = y$. Hence we get

$$\begin{aligned}
\mu(x * (x * y)) &= \mu(f(a) * (f(a) * f(b))) \\
&= \mu(f(a * (a * b))) \\
&= \mu^f(a * (a * b)) \\
&\geq \min\{\mu^f(a), \mu^f(b)\} \\
&= \min\{\mu(f(a)), \mu(f(b))\} \\
&= \min\{\mu(x), \mu(y)\}
\end{aligned}$$

Therefore μ is a fuzzy weak subalgebra of X. □

Definition 3.4.17 ([251]) A *fuzzy relation* on a set S is a fuzzy set $\mu : S \times S \to [0, 1]$.

Definition 3.4.18 ([251]) Let μ and ν be two fuzzy sets in a BE-algebra X. Then the Cartesian product of μ and ν is defined as $(\mu \times \nu)(x, y) = \min\{\mu(x), \nu(y)\}$ for all $x, y \in X$.

Let μ and ν be two fuzzy sets in a BE-algebra X. Then it obvious that $\mu \times \nu$ is a fuzzy relation on X. Also $(\mu \times \nu)_\alpha = \mu_\alpha \times \nu_\alpha$ for all $\alpha \in [0, 1]$. For any two BE-algebras X and Y, define an operation $*$ on $X \times Y$ as $(x, y) * (x', y') = (x * x', y * y')$ for all $x, x' \in X$ and $y, y' \in Y$. Then it can be easily observed that $(X \times Y, *, (1, 1))$ is a BE-algebra.

Theorem 3.4.19 *Let μ and ν be two fuzzy weak subalgebras of a BE-algebra X. Then $\mu \times \nu$ is a fuzzy weak subalgebra of $X \times X$.*

Proof Let $a, b \in X \times X$. Then there exist $(x, x'), (y, y') \in X \times X$ be such that $a = (x, x'), b = (y, y')$. Since μ and ν are fuzzy weak subalgebras of X, we can obtain

$$\begin{aligned}(\mu \times \nu)(a * (a * b)) &= (\mu \times \nu)(x * (x * y), x' * (x' * y')) \\ &= \min\{\mu(x * (x * y)), \nu(x' * (x' * y'))\} \\ &\geq \min\{\min\{\mu(x), \mu(y)\}, \min\{\nu(x'), \nu(y')\}\} \\ &= \min\{\min\{\mu(x), \nu(x')\}, \min\{\mu(y), \nu(y')\}\} \\ &= \min\{(\mu \times \nu)(x, x'), (\mu \times \nu)(y, y')\} \\ &= \min\{(\mu \times \nu)(a), (\mu \times \nu)(b)\}\end{aligned}$$

Therefore $\mu \times \nu$ is a fuzzy weak subalgebra of $X \times X$. □

In 1994, Y. Yu et al. [248] introduced and studied about s-norms in L-algebras. Cho et al. [48] introduced the notion of fuzzy subalgebras with respect to a s-norm in BCK-algebras, and they investigated some related results. In the following, the notion of *t-norm* is introduced.

Definition 3.4.20 Let $I = [0, 1]$. Then by a *t-norm* T, we mean a function $T : I \times I \longrightarrow I$ satisfying the following properties:

(1) $T(x, x) = 1$,
(2) $y \leq z$ implies $T(x, y) \leq T(x, z)$,
(3) $T(x, y) = T(y, x)$,
(4) $T(x, T(y, z)) = T(T(x, y), z)$ for all $x, y, z \in I$.

A simple example of such defined t-norm is a function $T(\alpha, \beta) = \min\{\alpha, \beta\}$. In the general case $T(\alpha, \beta) \leq \min\{\alpha, \beta\}$ and $T(\alpha, 1) = 1$ for all $\alpha, \beta \in [0, 1]$. Moreover, $([0, 1]; T)$ may be considered as a commutative semigroup with 1 as the neutral element. Let $I = [0, 1]$ and $T : I \times I \longrightarrow I$ a function defined as follows:

$$T_m(x) = \min\{x, y\} = \begin{cases} x & \text{if } x \leq y \\ y & y < x \end{cases}$$

Then clearly T_m is a t-norm on I. For any t-norm T on I, it can be easily observed that $T(\alpha, \beta) \leq \min\{\alpha, \beta\}$ for all $\alpha, \beta \in I$. For any t-norm T on I, define $\Delta_T = \{\alpha \in I \mid T(\alpha, \alpha) = \alpha\}$. A t-norm T is continuous if T is a continuous function.

Definition 3.4.21 Let T be a t-norm. A fuzzy set μ in X is said to be *sensible* if $Im(\mu) \subseteq \Delta_T$.

Definition 3.4.22 A fuzzy set μ of X is said to be a *fuzzy weak subalgebra* of X w.r.t. a t-norm T (simply called *t-fuzzy weak subalgebra*) if $\mu(x*(x*y)) \geq T(\mu(x), \mu(y))$ for all $x, y \in X$. If a fuzzy weak subalgebra μ of X w.r.t. T is sensible, we say that μ is a *sensible fuzzy weak subalgebra* of X with respect to T.

Theorem 3.4.23 *Let T be a t-norm. Every sensible fuzzy weak subalgebra of a BE-algebra X with respect to T is a fuzzy weak subalgebra of X.*

Proof Let μ be a sensible fuzzy weak subalgebra of X with respect to T. Then $\mu(x*(x*y)) \geq T(\mu(x), \mu(y))$ for all $x, y \in X$. Since μ is sensible, we have

$$\begin{aligned}\min\{\mu(x), \mu(y)\} &= T(\min\{\mu(x), \mu(y)\}, \min\{\mu(x), \mu(y)\})\\ &= T(\min\{\mu(x), \mu(x)\}, \min\{\mu(y), \mu(y)\})\\ &= T(\mu(x), \mu(y))\end{aligned}$$

Hence $\mu(x*(x*y)) \geq T(\mu(x), \mu(y)) = \min\{\mu(x), \mu(y)\}$. Thus μ is fuzzy weak subalgebra of X. □

Definition 3.4.24 Let μ be a fuzzy set in a BE-algebra X. Then μ is said to be a *fuzzy dot weak subalgebra* of X if for all $x, y \in X$, $\mu(x*(x*y)) \geq \mu(x)\cdot\mu(y)$, where $\cdot$ denotes ordinary multiplication.

Clearly every fuzzy weak subalgebra of a BE-algebra X is a fuzzy dot weak subalgebra, but the converse is not true.

Theorem 3.4.25 *If μ is a fuzzy dot weak subalgebra of a BE-algebra X, then μ^n, $n \in \mathbb{Z}^+$ is a fuzzy dot weak subalgebra of X.*

Proof Clearly μ^n is a fuzzy set in X. Suppose μ is a fuzzy dot weak subalgebra of X. Let $x, y \in X$. Then $\mu(x*(x*y)) \geq \mu(x)\cdot\mu(y)$. Hence $\mu^n(x*(x*y)) = [\mu(x*(x*y))]^n \geq [\mu(x)\cdot\mu(y)]^n = [\mu(x)]^n\cdot[\mu(y)]^n = \mu^n(x)\cdot\mu^n(y)$. Therefore μ^n is a fuzzy dot weak subalgebra of X. □

It is well known, the characteristic function of a set is a special fuzzy set. Suppose A is a non-empty subset of X. By χ_A, we denote the characteristic function of A, that is,

$$\chi_A(x) = \begin{cases} 1 & \text{if } x \in A\\ 0 & \text{if } x \notin A\end{cases}$$

Theorem 3.4.26 *Let X be a BE-algebra and A a non-empty subset of X. Then A is a weak subalgebra of X if and only if χ_A is a fuzzy dot weak subalgebra of X.*

Proof Assume that A is a weak subalgebra of X. Suppose $x, y \in A$. Then $x*(x*y) \in A$. Hence $\chi_A(x*(x*y)) = 1 \geq \chi_A(x)\cdot\chi_A(y)$. Suppose $x \notin A$ and $y \notin A$. Then $\chi_A(x) = \chi_A(y) = 0$. Hence $\chi(x*(x*y)) \geq 0 = \chi_A(x)\cdot\chi_A(y)$. Suppose $x \in A$ and $y \notin A$. Then we get $\chi_A(x) = 1$ and $\chi_A(y) = 0$. Therefore, $\chi_A(x*(x*y)) \geq 0 = 1\cdot 0 = \chi_A(x)\cdot\chi_A(y)$. Therefore χ_A is a fuzzy dot weak subalgebra of X. Conversely, assume that χ_A is a fuzzy dot weak subalgebra of X. Let $x, y \in A$. Then $\chi_A(x) = \chi_A(y) = 1$. Hence $\chi_A(x*(x*y)) \geq \chi_A(x)\cdot\chi_A(y) = 1\cdot 1 = 1$. Thus $x*(x*y) \in A$. Therefore A is a weak subalgebra of X. □

3.5 Soft BE-algebras

In this section, the notion of soft sets is introduced in BE-algebras. The soft set theory of Molodtsov [181] is applied to the theory of BE-algebras. The notion of soft BE-algebras and soft subalgebras is introduced, and their basic properties are investigated. The concept of union and intersection soft BE-algebras is introduced, and then their basic properties are studied.

Soft set theory is introduced by D. Molodtsov [181]. In what follows, let U be an initial universe set and X be a set of parameters. Let $\mathcal{P}(U)$ denote the power set of U and $A, B, C, \cdots \subseteq X$. We say that the pair (U, E) is a *soft universe*.

Definition 3.5.1 Let (U, E) be a soft universe. A *soft set* (f_A, A) of X over U is defined to be the set of ordered pairs

$$(f_A, A) = \{(x, f_A(x)) \mid x \in X, f_A(x) \in \mathcal{P}(U)\}$$

where $f_A : X \to \mathcal{P}(U)$ such that $f_A(x) = \emptyset$ if $x \notin A$. The subscript A in the notation f_A indicates that f_A is the *approximation function* of (f_A, A).

Definition 3.5.2 For two soft sets (f_A, A) and (f_B, B) of X over a common universe U, we say that (f_A, A) is a *soft subset* of (f_B, B), denoted by $(f_A, A)\widetilde{\subseteq}(f_B, B)$, if it satisfies the following properties:

(1) $A \subset B$,
(2) For any $\epsilon \in A$, $(f_A, A)(\epsilon)$ and $(f_B, B)(\epsilon)$ are identical approximations.

Definition 3.5.3 For a soft set (f_A, A) of X and a subset τ of U, the *τ-exclusive set* of (f_A, A), denoted by $e(f_A; \tau)$, is defined to be the set

$$e(f_A; \tau) = \{x \in A \mid f_A(x) \subseteq \tau\}.$$

Lemma 3.5.4 *For any subset A of X (the set of parameters), the following hold:*

(1) $e(f_A; U) = A$,
(2) $f_A(x) = \cap\{\tau \in U \mid x \in e(f_A; \tau)\}$,
(3) $(\forall \tau_1, \tau_2 \in U)(\tau_1 \subseteq \tau_2 \Rightarrow e(f_A; \tau_1) \subseteq e(f_A; \tau_2))$.

Proof (1) Let A be a subset of X. Clearly $e(f_A; U) \subseteq A$. Conversely, let $x \in A$. Then $U \supseteq f_A(x)$ and hence $x \in e(f_A; U)$. Thus $A \subseteq e(f_A; U)$. Therefore $e(f_A; U) = A$.
(2) Let $\tau \in U$. Then, for any $x \in e(f_A; \tau)$, it is clear that $f_A(x) \subseteq \tau$. Hence, it yields that $f_A(x) \subseteq \cap\{\tau \in U \mid x \in e(f_A; \tau)\}$. Conversely, let $t \in \cap\{\tau \in U \mid x \in e(f_A; \tau)\}$. Then $t \in \tau$ where $x \in e(f_A; \tau)$ for all $\tau \in U$. Hence $f_A(x) \subseteq \tau$ and $t \in \tau$ for all $\tau \in U$. Hence $t \in f_A(x)$.
(3) Let $\tau_1, \tau_2 \in U$ be such that $\tau_1 \subseteq \tau_2$. Suppose $x \in e(f_A; \tau_1)$. Then we get $f_A(x) \subseteq \tau_1 \subseteq \tau_2$. Hence $x \in e(f_A; \tau_2)$. Therefore $e(f_A; \tau_1) \subseteq e(f_A; \tau_2)$. □

Let X be a BE-algebra and A a non-empty subset of X. Suppose R refers to an arbitrary binary relation between an element of A and an element of X. In other words, R will denote a subset of $A \times X$. A set-valued function $f_A : A \to \mathcal{P}(X)$ can be defined as

$$f_A(x) = \{y \in X \mid (x, y) \in R\} \text{ for all} x \in A.$$

Then the pair (f_A, A) is called a soft set over X.

Definition 3.5.5 Let (U, X) be a soft universe and (f_A, A) a soft set over X. Then (f_A, A) is said to be a *soft* BE-*algebra* over X if $f_A(x)$ is a subalgebra of X for all $x \in A$.

Example 3.5.6 Let $X = \{1, a, b, c\}$. Define a binary operation $*$ on X as follows:

$*$	1	a	b	c
1	1	a	b	c
a	1	1	b	b
b	1	a	1	a
c	1	1	1	1

Then $(X, *, 1)$ is a BE-algebra. Let (f_A, A) be a soft set over X, where $A = X$ and $f_A : A \to \mathcal{P}(X)$ is a set-valued function defined by

$$f_A(x) = \{y \in X \mid (x, y) \in R \Leftrightarrow x * y \in \{1, c\}\}$$

for all $x \in A$. Then $f_A(1) = \{1, c\}$, $f_A(a) = \{1, a\}$, $f_A(b) = \{1, b\}$, and $f_A(c) = X$ are subalgebras of X. Hence (f_A, A) is a soft BE-algebra over X.

Theorem 3.5.7 *Let (f_A, A) be a soft BE-algebra over X. If B is a subset of A, then (f_{A_B}, B) is a soft BE-algebra over X, where f_{A_B} is the restriction of f_A to B.*

Proof Clearly f_{A_B} is a soft set over X. Fix $x \in X$. Let $a, b \in f_{A_B}(x)$. Then $a, b \in f_A(x)$. Since (f_A, A) is a soft BE-algebra, we get $f_A(x)$ is a subalgebra of X. Hence $a * b \in f_A(x)$. Therefore $a * b \in f_{A_B}(x)$. Thus f_{A_B} is a subalgebra of X. Therefore (f_{A_B}, B) is a soft BE-algebra. □

Definition 3.5.8 Let (U, E) be a soft universe and $A, B, C \subseteq E$. If (f_A, A) and (f_B, B) are two soft sets over a common universe U, then the intersection of (f_A, A) and (f_B, B) is defined to be a soft set (f_C, C) satisfying the following conditions:

(1) $C = A \cap B$,
(2) $(\forall t \in C)(f_C(t) = f_A(t)$ or $f_C(t) = f_B(t)$.
In this case, we write $(f_C, C) = (f_A, A)\widetilde{\cap}(f_B, B)$.

Definition 3.5.9 Let (U, E) be a soft universe and $A, B, C \subseteq E$. If (f_A, A) and (f_B, B) are two soft sets over a common universe U, then the union of (f_A, A) and (f_B, B) is defined to be a soft set (f_C, C) satisfying the following conditions:

(1) $C = A \cup B$,
(2) for all $t \in C$,

$$f_C(t) = \begin{cases} f_A(t) & \text{if } t \in A - B \\ f_B(t) & \text{if } t \in B - A \\ f_A(t) \cup f_B(t) & \text{if } t \in A \cap B \end{cases}$$

In this case, we write $(f_C, C) = (f_A, A)\tilde{\cup}(f_B, B)$.

Theorem 3.5.10 *Let (f_A, A) and (f_B, B) be two soft BE-algebras over a BE-algebra X. If $A \cap B \neq \emptyset$, then the intersection $(f_A, A)\tilde{\cap}(f_B, B)$ is a soft BE-algebra over X.*

Proof By the above definition, we have $(f_A, A)\tilde{\cap}(f_B, B) = (f_C, C)$ where $C = A \cap B$ and $f_C(x) = f_A(x)$ or $f_B(x)$ for all $x \in C$. Now $f_C : C \to P(X)$ is a mapping, and thus (f_C, C) is a soft set over X. Since (f_A, A) and (f_B, B) are soft BE-algebras over X, either $f_C(x) = f_A(x)$ or $f_B(x)$ is a subalgebra for all $x \in C$. Hence $(f_C, C) = (f_A, A)\tilde{\cap}(f_B, B)$ is a soft BE-algebra over X. □

The following corollary is a direct consequence of the above theorem.

Corollary 3.5.11 *Let (f_A, A) and (f_B, B) be two soft BE-algebras over X. Then their intersection $(f_A, A)\tilde{\cap}(f_B, B)$ is a soft BE-algebra over X.*

Theorem 3.5.12 *Let (f_A, A) and (f_B, B) be two soft BE-algebras over X. If A and B are disjoint, then the union $(f_A, A)\tilde{\cup}(f_B, B)$ is a soft BE-algebra over X.*

Proof By the above definition, we have $(f_A, A)\tilde{\cup}(_B, B) = (f_C, C)$, where $C = A \cup B$ and for every $t \in C$,

$$f_C(t) = \begin{cases} f_A(t) & \text{if } t \in A - B \\ f_B(t) & \text{if } t \in B - A \\ f_A(t) \cup f_B(t) & \text{if } t \in A \cap B \end{cases}$$

Since $A \cap B = \emptyset$, we get either $x \in A - B$ or $x \in B - A$ for all $x \in C$. Let $x \in A - B$. Since (f_A, A) is soft BE-algebra over X, we get $f_C(x) = f_A(x)$ is a subalgebra of X. Now, suppose $x \in B - A$. Then, since (f_B, B) is a soft BE-algebra over X, it follows that $f_C(x) = f_A(x)$ is a subalgebra of X. Therefore $(f_C, C) = (f_A, A)\tilde{\cup}(f_B, B)$ is a soft BE-algebra over X. □

Definition 3.5.13 If (f_A, A) and (f_B, B) are two soft sets over a common universe U, then "(f_A, A) AND (f_B, B)" denoted by $(f_A, A)\tilde{\wedge}(f_B, B)$ is defined to be a soft set as follows:

$$(f_A, A)\tilde{\wedge}(f_B, B) = (f_{A\times B}, A \times B)$$

where $f_{A\times B}(x, y) = f_A(x) \cap f_B(y)$ for all $(x, y) \in A \times B$.

Definition 3.5.14 If (f_A, A) and (f_B, B) are two soft sets over a common universe U, then "(f_A, A) OR (f_B, B)" denoted by $(f_A, A)\tilde{\vee}(f_B, B)$ is defined to be a soft set as follows:

$$(f_A, A)\widetilde{\vee}(f_B, B) = (f_{A\times B}, A \times B)$$

where $f_{A\times B}(x, y) = f_A(x) \cup f_B(y)$ for all $(x, y) \in A \times B$.

Theorem 3.5.15 *If (f_A, A) and (f_B, B) are two soft BE-algebras over X, then $(f_A, A)\widetilde{\wedge}(f_B, B)$ is a soft BE-algebra over X.*

Proof By the above definition, it is clear that $(f_A, A)\widetilde{\wedge}(f_B, B) = (f_{A\times B}, A \times B)$ where $f_{A\times B}(x, y) = f_A(x) \cap f_B(y)$ for all $(x, y) \in A \times B$. Since $f_A(x)$ and $f_B(y)$ are subalgebras of X, we get that the intersection $f_A(x) \cap f_B(y)$ is also a subalgebra of X. Hence, it gives that $f_{A\times B}(x, y) = f_A(x) \cap f_B(y)$ is a subalgebra of X for all $(x, y) \in A \times B$. Therefore $(f_A, A)\widetilde{\wedge}(f_B, B) = (f_{A\times B}, A \times B)$ is a soft BE-algebra over X. □

Definition 3.5.16 A soft BE-algebra (f_A, A) over X is said to be *trivial* if $f_A(x) = \{1\}$. It is said to be *whole* if $f_A(x) = X$ for all $x \in A$.

Let $\Psi : X \to Y$ be a mapping of BE-algebras. For a soft set (f_A, A) over X, (Ψ_{f_A}, A) is a soft set over Y where $\Psi_{f_A} : A \to \mathcal{P}(Y)$ is defined by $\Psi_{f_A}(x) = \Psi(f_A(x))$ for all $x \in A$.

Lemma 3.5.17 *Let $\Psi : X \to Y$ be a homomorphism of BE-algebras. If (f_A, A) is a soft BE-algebra over X, then (Ψ_{f_A}, A) is a soft BE-algebra over Y.*

Proof Since (f_A, A) is a soft BE-algebra over X, it gives that $f_A(x)$ is a subalgebra of X for every $x \in A$. Hence, for every $x \in X$, $\Psi_{f_A}(x) = \Psi(f_A(x))$ is a subalgebra of Y and its homomorphic image is also a subalgebra of Y. Therefore (Ψ_{f_A}, A) is a soft BE-algebra over Y. □

Theorem 3.5.18 *Let $\Psi : X \to Y$ be a homomorphism of BE-algebras and let (f_A, A) be a soft set over X. If (f_A, A) is a soft BE-algebra over X such that $f_A(x) = Ker(\Psi)$ for all $x \in A$, then (Ψ_{f_A}, A) is the trivial soft BE-algebra over Y.*

Proof Assume that $f_A(x) = Ker(\Psi)$ for all $x \in A$. Then $\Psi_{f_A}(x) = \Psi(f_A(x)) = \{1\}$ for all $x \in A$. Hence by Lemma 3.5.17 and Definition 3.5.16, (Ψ_{f_A}, A) is the trivial soft BE-algebra over Y. □

Definition 3.5.19 Let S be a subalgebra of X. A subset D of X is said to be a subalgebra of X related to S (briefly, S-subalgebra) if $x * y \in D$ for all $x, y \in S$.

Definition 3.5.20 Let (f_A, A) and (f_B, B) be two soft BE-algebras over X. Then we say that (f_A, A) is a *soft subalgebra* of (f_B, B) (we write $(f_A, A)\widetilde{<}(f_B, B)$) if it satisfies the following:

(1) $A \subset B$,
(2) $f_A(x)$ is a $f_B(x)$-subalgebra of X for all $x \in A$.

Theorem 3.5.21 *Let (f_A, A) and (g_A, A) be two soft BE-algebras over X. Then*

$$(\forall x \in A)(f_A(x) \subset g_A(x) \Rightarrow (f_A, A)\widetilde{<}(g_A, A)).$$

Proof It is obvious. □

Theorem 3.5.22 *Let (f_A, A) be a soft BE-algebra over X and let (f_{B_1}, B_1) and (f_{B_2}, B_2) be soft subalgebras of (f_A, A). Then*

(1) $(f_{B_1}, B_1)\tilde{\cap}(f_{B_2}, B_2)\tilde{<}(f_A, A)$,
(2) *If* $B_1 \cap B_2 = \emptyset$, *then* $(f_{B_1}, B_1)\tilde{\cup}(f_{B_2}, B_2)\tilde{<}(f_A, A)$,
(3) $(f_{B_1}, B_1)\tilde{\wedge}(f_{B_2}, B_2)\tilde{<}(f_{A\times A}, A \times A)$ *where* $(f_{A\times A}, A \times A) = (f_A, A)\tilde{\wedge}(f_A, A)$.

Proof (1) By Definition 3.5.8, we have

$$(f_{B_1}, B_1)\tilde{\cap}(f_{B_2}, B_2) = (f_B, B)$$

where $B = B_1 \cap B_2$ and $f_B(x) = f_{B_1}(x)$ or $f_B(x) = f_{B_2}(x)$ for all $x \in B$. Clearly $B \subset A$. Let $x \in B = B_1 \cap B_2$. If $x \in B_1$, then since $(f_{B_1}, B_1)\tilde{<}(f_A, A)$, we get that $f_B(x) = f_{B_1}(x)$ is an $f_A(x)$-subalgebra. If $x \in B_2$, then since $(f_{B_2}, B_2)\tilde{<}(f_A, A)$, we get that $f_B(x) = f_{B_2}(x)$ is an $f_A(x)$-subalgebra. Therefore $(f_{B_1}, B_1)\tilde{\cap}(f_{B_2}, B_2)\tilde{<}(f_A, A)$.
(2) Suppose $B_1 \cap B_2 = \emptyset$. We can write $(f_{B_1}, B_1)\tilde{\cup}(f_{B_2}, B_2) = (f_B, B)$ where $B = B_1 \cup B_2$ and

$$f_B(x) = \begin{cases} f_{B_1}(x) & \text{if } x \in B_1 - B_2 \\ f_{B_2}(x) & \text{if } x \in B_2 - B_1 \\ f_{B_1}(x) \cup f_{B_2}(x) & \text{if } x \in B_1 \cap B_2 \end{cases}$$

for all $x \in B$. Since $(f_{B_i}, B_i)\tilde{<}(f_A, A)$ for $i = 1, 2$, $B = B_1 \cup B_2 \subset A$ and $f_{B_i}(x)$ is an $f_A(x)$-subalgebra for all $x \in B_i$, $i = 1, 2$. Since $B_1 \cap B_2 = \emptyset$, we get that $f_B(x)$ is an $f_A(x)$-subalgebra for all $x \in B$. Therefore $(f_{B_1}, B_1)\tilde{\cup}(f_{B_2}, B_2) = (f_B, B)\tilde{<}(f_A, A)$.
(3) By Theorem 3.5.15, it is clear that $(f_{A\times A}, A \times A)$ is a soft BE-algebra over X. Hence

$$(f_{B_1}, B_1)\tilde{\wedge} f_{B_2}, B_2) = (h_B, B)$$

where $B = B_1 \times B_2$ and $h_B(x, y) = f_{B_1}(x) \cap f_{B_2}(x)$ for all elements $(x, y) \in B$. Obviously $B \subset A \times A$ and $h_B(x, y) = f_{B_1}(x) \cap f_{B_2}(y)$ is a $f_{A\times A}(x, y) = f_A(x) \cap f_A(y)$-subalgebra. Therefore $(f_{B_1}, B_1)\tilde{\wedge}(f_{B_2}, B_2)\tilde{<}(f_{A\times A}, A \times A)$. □

Theorem 3.5.23 *Let $\Psi : X \to Y$ be a morphism of BE-algebras and let (f_A, A) and (f_B, B) be two soft BE-algebras over X. Then $(f_A, A)\tilde{<}(f_B, B) \Rightarrow (\Psi_{f_A}, A)\tilde{<}(\Psi_{f_B}, B)$.*

Proof Assume that $(f_A, A)\tilde{<}(f_B, B)$. Let $x \in A$ be an arbitrary element. Then we get $A \subset B$ and $f_A(x)$ is a subalgebra of $f_B(x)$. Since Ψ is a homomorphism, we get that $\psi_{f_A}(x) = \Psi(f_A(x))$ is a subalgebra of $\Psi(f_B(x)) = \Psi_{f_B}(x)$. Therefore $(\Psi_{f_A}, A)\tilde{<}(\Psi_{f_B}, B)$. □

In the following, union and intersectional soft weak subalgebras are introduced.

Definition 3.5.24 A soft set $(\widetilde{f}, X)$ of a BE-algebra X over the universe U is said to be a *union soft weak subalgebra* of X if $\widetilde{f}(x * (x * y)) \subseteq \widetilde{f}(x) \cup \widetilde{f}(y)$ for all $x, y \in X$.

Definition 3.5.25 A soft set $(\widetilde{f}, X)$ of a BE-algebra X over the universe U is said to be an *intersection soft subalgebra* of X if $\widetilde{f}(x * (x * y)) \supseteq \widetilde{f}(x) \cap \widetilde{f}(y)$ for all $x, y \in X$.

Theorem 3.5.26 *A soft set $(\widetilde{f}, X)$ is a union soft weak subalgebra of X if and only if $e_X(\widetilde{f}; \tau)$ is a weak subalgebra of X whenever $e_X(\widetilde{f}; \tau) \neq \emptyset$ for all $\tau \in \mathcal{P}(U)$.*

Proof Assume that $(\widetilde{f}, X)$ is a union soft weak subalgebra of X. Let $\tau \in \mathcal{P}(U)$ and $e_X(\widetilde{f}; \tau) \neq \emptyset$. Choose $x, y \in e_X(\widetilde{f}; \tau)$. Then $\tau \supseteq \widetilde{f}(x)$ and $\tau \supseteq \widetilde{f}(y)$. Since $(\widetilde{f}, X)$ is a union soft weak subalgebra, we get $\tau \supseteq \widetilde{f}(x) \cup \widetilde{f}(x) \supseteq \widetilde{f}(x * (x * y))$. Hence $e_X(\widetilde{f}; \tau)$ is a weak subalgebra of X.

Conversely, assume that $e_X(\widetilde{f}; \tau)$ is a weak subalgebra of X for all $\tau \in \mathcal{P}(U)$ with $e_X(\widetilde{f}; \tau) \neq \emptyset$. For any $x, y \in X$, let $\widetilde{f}(x) = \tau_x$ and $\widetilde{f}(y) = \tau_y$. Take $\tau = \max\{\tau_x, \tau_y\}$. Then $\tau \supseteq \widetilde{f}(x)$ and $\tau \supseteq \widetilde{f}(y)$. Thus $\widetilde{f}(x) \cup \widetilde{f}(y)$. Hence $x, y \in e_X(\widetilde{f}; \tau)$. Since $e_X(\widetilde{f}; \tau)$ is a weak subalgebra, we get $x * (x * y) \in e_X(\widetilde{f}; \tau)$. Hence $\widetilde{f}(x) \cup \widetilde{f}(y) \supseteq \tau \supseteq \widetilde{f}(x * (x * y))$. Therefore $(\widetilde{f}, X)$ is a union soft weak subalgebra of X. □

Theorem 3.5.27 *Let $(\widetilde{f}, X)$ be a soft set of X over U. Define a soft set $(\widetilde{f}^*, X)$ of X over U where $\widetilde{f}^* : X \to \mathcal{P}(U)$ is given by*

$$\widetilde{f}^*(x) = \begin{cases} \widetilde{f}(x) & \text{if } x \in e_X(\widetilde{f}; \tau) \\ \delta & \text{otherwise} \end{cases}$$

where τ is any subset of U and δ is a subset of U satisfying $\delta \supseteq \bigcup_{x \notin e_X(\widetilde{f}; \tau)} \widetilde{f}(x)$. Then $(\widetilde{f}^*, X)$ is a union soft weak subalgebra of X whenever $(\widetilde{f}, X)$ is a union soft weak subalgebra of X.

Proof Let $x \in X$ be arbitrary. Clearly $\widetilde{f}(x) \subseteq \widetilde{f}^*(x)$. Assume that $(\widetilde{f}, X)$ is a union soft weak subalgebra of X. Then by the above theorem, we get $e_X(\widetilde{f}; \tau)(\neq \emptyset)$ is a union soft weak subalgebra of X for all $\tau \in \mathcal{P}(U)$. Let $x, y \in X$. Suppose that $x, y \in e_X(\widetilde{f}; \tau)$. Then we get $x * (x * y) \in e_X(\widetilde{f}; \tau)$. Hence $\widetilde{f}^*(x * (x * y)) = \widetilde{f}(x * (x * y)) \subseteq \widetilde{f}(x) \cup \widetilde{f}(y) = \widetilde{f}^*(x) \cup \widetilde{f}^*(y)$. Suppose $x \notin e_X(\widetilde{f}; \tau)$ and $y \notin e_X(\widetilde{f}; \tau)$. Then we get $\widetilde{f}^*(x) = \widetilde{f}^*(y) = \delta$. Thus $\widetilde{f}^*(x * (x * y)) \subseteq \delta = \widetilde{f}^*(x) \cup \widetilde{f}^*(y)$. Therefore $(\widetilde{f}^*, X)$ is a union soft weak subalgebra of X. □

3.6 Falling Shadows Applied to (Weak) Subalgebras

In this section, falling shadows representation theory is applied to subalgebras of BE-algebras. The notion of falling (weak) subalgebras of a BE-algebra is introduced. Relations between fuzzy (weak) subalgebras and falling (weak) subalgebras are obtained. A characterization of a falling subalgebra is established. The basic theory of falling shadows is presented in the following:

Definition 3.6.1 A *probability space* is a triple $(\Omega, \mathcal{A}, P)$ consisting of the sample space Ω an arbitrary non-empty set the σ-algebra $\mathcal{A} \subseteq 2^{\Omega}$ (also called σ-field) a set of subsets of Ω, called events, such that $\mathcal{A}$ contains the sample space, i.e., $\Omega \in \mathcal{A}$ and the probability measure $P : \mathcal{A} \to [0, 1]$ a function on $\mathcal{A}$.

Definition 3.6.2 A probability space $(\Omega, \mathcal{A}, P)$ is said to be a *complete probability space* if for all $B \in \mathcal{A}$ with $P(B) = 0$ and all $C \subset B$ one has $C \in \mathcal{A}$. Often, the study of probability spaces is restricted to complete probability spaces.

Example 3.6.3 A number between 0 and 1 is chosen at random, uniformly. Here $\Omega = [0, 1]$, $\mathcal{A}$ is the σ-algebra of Borel sets on Ω, and P is the Lebesgue measure on $[0, 1]$. In this case the open intervals of the form (a, b), where $0 < a < b < 1$, could be taken as the generator sets. Each such set can be assigned the probability of $P((a, b)) = (ba)$, which generates the Lebesgue measure on $[0, 1]$, and the Borel σ-algebra on Ω.

Definition 3.6.4 Given a subset $E \subseteq \mathbb{R}$, with the length of an (open, closed, semi-open) interval $I = [a, b]$ given by $\ell(I) = ba$, the *Lebesgue outer measure* $\lambda^*(E)$ is defined as

$$\lambda^*(E) = \inf \left\{ \Sigma_{k=1}^{\infty} \ell(I_k) \mid (I_k)_{k \in \mathbb{N}} \text{ is a sequence of open intervals with } E \subseteq \bigcup_{k=1}^{\infty} I_k \right\}.$$

The Lebesgue measure is defined on the Lebesgue σ algebra, which is the collections of all the sets E which satisfy the condition that, for every $A \subseteq \mathbb{R}$,

$$\lambda^*(A) \geq \lambda^*(A \cap E) + \lambda^*(A \cap E^c)$$

For any set in the Lebesgue σ-algebra, its Lebesgue measure is given by its Lebesgue outer measure $\lambda^*(E) = \lambda^*(E)$.

P.Z. Wang and E. Sanchez [241] introduced the theory of falling shadows which directly relates probability concepts with the membership function of fuzzy sets. The mathematical structure of the theory of falling shadows is formulated in [240]. In 2012, Y.B. Jun and C.H. Park applied the theory of falling shadows to subalgebras and ideals of BCK/BCI-algebras.

The theory of falling shadows: Given a universe of discourse U, let $\mathcal{P}$ denote the power set of U. For each $u \in U$, let

$$\dot{u} := \{E \subseteq U \mid u \in E\} \tag{3.1}$$

and for each $E \in \mathcal{P}(U)$, let

$$\dot{E} := \{\dot{u} \mid u \in E\} \tag{3.2}$$

An ordered pair $(\mathcal{P}(U), \mathcal{B})$ is said to be a hyper-measurable structure on U if $\mathcal{B}$ is a σ-field in $\mathcal{P}(U)$ and $\dot{U} \subseteq \mathcal{B}$. Given a probability space $(\Omega, \mathcal{A}, P)$ and a hyper-measurable structure $(\mathcal{P}(U), \mathcal{B})$ on U, a random set on U is defined to be a mapping $\xi : \Omega \to \mathcal{P}(U)$ which is $\mathcal{A} - \mathcal{B}$ measurable, that is,

$$(\forall C \in \mathcal{B})(\xi^{-1}(C) := \{\omega \mid \omega \in \Omega \text{ and } \xi(\omega) \in C\} \in \mathcal{A}). \tag{3.3}$$

Suppose that ξ is a random set on U. Let

$$\tilde{H}(u) := P(\omega \mid u \in \xi(\omega) \text{ for each } u \in U. \tag{3.4}$$

Then $\tilde{H}$ is a kind of fuzzy set in U. We call $\tilde{H}$ a *falling shadow* of the random set ξ, and ξ is called a *cloud* of $\tilde{H}$.

Example 3.6.5 Let $(\Omega, \mathcal{A}, P) = ([0, 1], \mathcal{A}, m)$, where $\mathcal{A}$ is a Borel field on $[0, 1]$ and m the usual Lebesgue measure. Let $\tilde{H}$ be a fuzzy set in U and $\tilde{H}_t := \{u \in U \mid \tilde{H}(u) \geq t\}$ be a *t-cut* of $\tilde{H}$. Then

$$\xi : [0, 1] \to \mathcal{P}(U), t \mapsto \tilde{H}_t$$

is a random set and ξ is a cloud of $\tilde{H}$. We shall call ξ defined above as the *cut-cloud* of $\tilde{H}$.

Definition 3.6.6 Let X be a BE-algebra and $(\Omega, \mathcal{A}, P)$ a probability space. Suppose

$$\xi : \Omega \to \mathcal{P}(X)$$

is a random set. If $\xi(\omega)$ is a subalgebra of X for any $\omega \in \Omega$, then the falling shadow $\tilde{H}$ of the random set ξ, i.e.,

$$\tilde{H}(x) = P(\omega \mid x \in \xi(\omega))$$

is said to be a *falling subalgebra* of X.

Example 3.6.7 Let $(\Omega, \mathcal{A}, P)$ be a probability space and let

$$F(X) := \{f \mid f : \Omega \to X \text{ is a mapping }\}$$

where X is a BE-algebra. Define an operation $\circledast$ on $F(X)$ by

$$(\forall \omega \in \Omega)\ ((f \circledast g)(\omega) = f(\omega) * g(\omega))$$

for all $f, g \in F(X)$. Let $\theta \in F(X)$ be defined by $\theta(\omega) = 0$ for all $\omega \in \Omega$. It can be easily to check that $(F(X), \circledast, \theta)$ is a BE-algebra. For any subalgebra (weak subalgebra) S of X and $f \in F(X)$, let

$$S_f := \{\omega \in \Omega \mid f(\omega) \in S\}$$

and

$$\xi : \Omega \to \mathbb{P}(F(X)),\ \omega \mapsto \{f \in F(X) \mid f(\omega) \in S\}.$$

Then $S_f \in \mathbb{A}$ and $\xi(\omega) = \{f \in F(X) \mid f(\omega) \in S\}$ is a subalgebra (weak subalgebra) of $F(X)$. Since

$$\xi^{-1}(f) = \{\omega \in \Omega \mid f \in \xi(\omega)\} = \{\omega \in \Omega \mid f(\omega) \in S\} = S_f = \mathbb{A},$$

ξ is a random set of $F(X)$. Let $\tilde{H}(f) = P(\omega \mid f(\omega) \in S)$. Then $\tilde{H}$ is a falling subalgebra (falling weak subalgebra) of $F(X)$.

Example 3.6.8 Let $X = \{1, a, b, c\}$. Define an operation $*$ on X which is given in the following Cayley table:

$*$	1	a	b	c
1	1	a	b	c
a	1	1	b	c
b	1	a	1	c
c	1	a	b	1

Then $(X, *, 1)$ is a BE-algebra. Let $(\Omega, \mathbb{A}, P) = ([0, 1], \mathbb{A}, m)$ and let $\xi : [0, 1] \to \mathbb{P}(X)$ be defined by

$$\xi(t) = \begin{cases} \{1, a\} & \text{if } t \in [0, 0.4) \\ \{1, b\} & \text{if } t \in [0.4, 0.6) \\ \{1, c\} & \text{if } t \in [0.6, 1] \end{cases}$$

Then $\xi(t)$ is a subalgebra of X for all $t \in [0, 1]$. Hence $\tilde{H}(x) = P(t \mid x \in \xi(t))$ is a falling fuzzy subalgebra of X and $\tilde{H}$ is represented as follows:

$$\tilde{H}(x) = \begin{cases} 1 & \text{if } x = 1 \\ 0.4 & \text{if } x = a, c \\ 0.2 & \text{if } x = b \end{cases}$$

In this case, we can easily check that $\tilde{H}$ is a fuzzy subalgebra of X.

Definition 3.6.9 Let $(\Omega, \mathcal{A}, P)$ be a probability space and let $\xi : \Omega \to P(X)$ be a random set. For any subset S of the BE-algebra X and $f \in F(X)$, let $S_f := \{\omega \in \Omega \mid f(\omega) \in S\}$ and

$$\xi : \Omega \to \mathbb{P}(F(X))$$

$$\omega \mapsto \{f \in F(\omega) \mid f(\omega) \in S\}.$$

Then $S_f \in A$.

Theorem 3.6.10 *Let $(\Omega, \mathcal{A}, P)$ be a probability space and let $\xi : \Omega \to P(X)$ be a random set. Let S be a non-empty subset of X, and $\omega \in \Omega$. If S is a weak subalgebra (subalgebra) of X, then $\xi(\omega) = \{f \in F(X) \mid f(\omega) \in S\}$ is a weak subalgebra (subalgebra) of $F(X)$.*

Proof Suppose that S is a weak subalgebra of X. Since $1(\omega) = 1 \in S$ for any $\omega \in \Omega$, we have that $1 \in \xi(\omega)$. Let $f, g \in F(X)$ be such that $f \in \xi(\omega)$ and $g \in \xi(\omega)$. For any $\omega \in \Omega$, we have $f(\omega) \in S$ and $g(\omega) \in S$. Since S is a weak subalgebra of X, we get $(f \circledast (f \circledast g))(\omega) = f(\omega) * (f(\omega) * g(\omega)) \in S$. Hence $f \circledast (f \circledast g) \in \xi(\omega)$. Therefore $\xi(\omega)$ is a weak subalgebra of $F(X)$. The result is similarly proved for subalgebras of X. □

Theorem 3.6.11 *Let $(\Omega, \mathcal{A}, P)$ be a probability space and $\tilde{H}$ be a falling shadow of a random set $\xi : \Omega \to \mathbb{P}(X)$. Then $\tilde{H}$ is a falling weak subalgebra (subalgebra) of X if and only if*

$$\Omega(x; \xi) \cap \Omega(y; \xi) \subseteq \Omega(x * (x * y); \xi)\ (resp.\ \Omega(x; \xi) \cap \Omega(y; \xi) \subseteq \Omega(x * y); \xi)$$
$$for\ any\ x, y \in X.$$

Proof Assuming that $\tilde{H}$ is a falling fuzzy weak subalgebra of X. Then $\xi(\omega)$ is a weak subalgebra of X for any $\omega \in \Omega$. Let $x, y \in X$ be such that $\omega \in \Omega(x; \xi) \cap \Omega(y; \xi)$. Then we get $\omega \in \Omega(x; \xi)$ and $\omega \in \Omega(y; \xi)$. Hence $x \in \xi(\omega)$ and $y \in \xi(\omega)$. Since $\xi(\omega)$ is a weak subalgebra of X, we get $x*(x*y) \in \xi(\omega)$. Thus $\omega \in \Omega(x*(x*y); \xi)$. Therefore $\Omega(x; \xi) \cap \Omega(y; \xi) \subseteq \Omega(x * (x * y); \xi)$.

Conversely, assume that the condition holds. Let $x, y \in X$ be such that $x \in \xi(\omega)$ and $y \in \xi(\omega)$. Then $\omega \in \Omega(x; \xi)$ and $\omega \in \Omega(y; \xi)$, which implies that $\omega \in \Omega(x; \xi) \cap \Omega(y; \xi) \subseteq \Omega(x * (x * y); \xi)$. Thus $x * (x * y) \in \xi(\omega)$. Hence $\xi(\omega)$ is a weak subalgebra of X. Therefore $\tilde{H}$ is a falling fuzzy weak subalgebra of X. The result is proved similarly for falling subalgebras. □

Theorem 3.6.12 *Let $(X, *, 1)$ be a BE-algebra and $(\Omega, \mathcal{A}, P)$ a probability space. If $\xi : \Omega \to P(X)$ be a random set, then every fuzzy subalgebra of X is a falling fuzzy subalgebra of X.*

Proof Let $\tilde{H}$ be a fuzzy subalgebra of X. Then the level set $\tilde{H}_t$ is a subalgebra of X for all $t \in [0, 1]$. Let

$$\xi : [0, 1] \to \mathbb{P}(X)$$

be a random set and $\xi(t) = \tilde{H}_t$. Then $\tilde{H}$ is a falling fuzzy subalgebra of X. □

The converse of the above theorem need not be true as seen in the following example:

Example 3.6.13 Let $X = \{1, a, b, c\}$. Define an operation $*$ on X which is given in the following Cayley table:

$*$	1	a	b	c
1	1	a	b	c
a	1	1	c	b
b	1	c	1	a
c	1	b	a	1

Then $(X, *, 1)$ is a BE-algebra. Let $(\Omega, \mathbb{A}, P) = ([0, 1], \mathbb{A}, m)$, and let $\xi : [0, 1] \to \mathbb{P}(X)$ be defined by

$$\xi(t) = \begin{cases} \{1, a\} & \text{if } t \in [0, 0.4) \\ \{1, b\} & \text{if } t \in [0.4, 0.6) \\ \{1, c\} & \text{if } t \in [0.6, 1] \end{cases}$$

Then $\xi(t)$ is a subalgebra of X for all $t \in [0, 1]$. Hence $\tilde{H}(x) = P(t \mid x \in \xi(t))$ is a falling fuzzy subalgebra of X and $\tilde{H}$ is represented as follows:

$$\tilde{H}(x) = \begin{cases} 1 & \text{if } x = 1 \\ 0.4 & \text{if } x = a, c \\ 0.2 & \text{if } x = b \end{cases}$$

But $\tilde{H}$ is not a fuzzy subalgebra of X because of $\tilde{H}(a * c) = \tilde{H}(b) = 0.2 < 0.4 = \min\{\tilde{H}(a), \tilde{H}(c)\}$.

Theorem 3.6.14 *Let X be a BE-algebra and $(\Omega, \mathcal{A}, P)$ a probability space. A falling shadow $\tilde{H}$ of a random set $\xi : \Omega \to \mathbb{P}(X)$ is a falling weak subalgebra of X if and only if for each $\omega \in \Omega$, the following condition holds:*

$$(\forall x \in \xi(\omega))(\forall x * (x * y) \in X \setminus \xi(\omega))\ (y \in X \setminus \xi(\omega)).$$

Proof Assume that $\tilde{H}$ is a falling weak subalgebra of X. Then $\xi(\omega)$ is a weak subalgebra of X for all $\omega \in \Omega$. Let $x, y \in X$ be such that $x \in \xi(\omega)$ and $x * (x * y) \in X \setminus \xi(\omega)$. Suppose that $y \in \xi(\omega)$. Since $\xi(\omega)$ is a weak subalgebra of X and $x \in \xi(\omega)$, we get that $x * (x * y) \in \xi(\omega)$, which is a contradiction to that $x * (x * y) \in X \setminus \xi(\omega)$. Hence $y \notin \xi(\omega)$, which means $y \in X \setminus \xi(\omega)$.

Conversely, assume that the condition holds. Let $x, y \in X$ be such that $x \in \xi(\omega)$ and $y \in \xi(\omega)$. Suppose that $x * (x * y) \notin \xi(\omega)$. Then by the assumed condition, we get that $y \in X \setminus \xi(\omega)$ which is a contradiction to that $y \in \xi(\omega)$. Hence $x * (x * y) \in \xi(\omega)$. Thus $\xi(\omega)$ is a weak subalgebra of X. Therefore $\tilde{H}$ is a falling weak subalgebra of X. $\square$

The following theorem can be verified similarly.

Theorem 3.6.15 *Let X be a BE-algebra and* $(\Omega, \mathcal{A}, P)$ *a probability space. A falling shadow* $\tilde{H}$ *of a random set* $\xi : \Omega \to \mathbb{P}(X)$ *is a falling subalgebra of X if and only if for each* $\omega \in \Omega$*, the following condition holds:*

$$(\forall x \in \xi(\omega))(\forall x * y \in X \setminus \xi(\omega))\ (y \in X \setminus \xi(\omega)).$$

Definition 3.6.16 Let $(\Omega, \mathcal{A}, P)$ be a probability space and $\xi : \Omega \to P(X)$ a random set. A falling subalgebra (resp. weak subalgebra) $\tilde{H}$ is called *order preserving* if for all $x, y \in X$,

$$x \le y \Rightarrow \Omega(x; \xi) \subseteq (y; \xi)$$

Proposition 3.6.17 *Let* $(\Omega, \mathcal{A}, P)$ *be a probability space and* $\xi : \Omega \to P(X)$ *a random set. Let* $\tilde{H}$ *be a falling weak subalgebra of a BE-algebra X. Then*

(1) $\Omega(x; \xi) \cap \Omega(y; \xi) \subseteq \Omega(x * (x * y); \xi)$ *for all* $x, y \in X$,
(2) $\Omega(x; \xi) \subseteq \Omega(1; \xi)$ *for all* $x \in X$,
(3) $\Omega(x; \xi) \cap \Omega(x^n * y; \xi) \subseteq \Omega(x^{n+2} * y; \xi)$ *for all* $x, y \in X$ *and* $n \in \mathbb{N}$.

Proof (1) Suppose $\omega \in \Omega(x; \xi) \cap \Omega(y; \xi)$. Then $x \in \xi(\omega)$ and $y \in \xi(\omega)$. Since $\xi(\omega)$ is a weak subalgebra of X, we get $x * (x * y) \in \xi(\omega)$. Hence $\omega \in \Omega(x * (x * y); \xi)$. Therefore $\Omega(x; \xi) \cap \Omega(y; \xi) \subseteq \Omega(x * (x * y); \xi)$.
(2) Putting $x = y$ in (1), we get $\Omega(x; \xi) = \Omega(x; \xi) \cap \Omega(x; \xi) \subseteq \Omega(x * (x * x); \xi) = \Omega(1; \xi)$.
(3) From (1) is it clear. □

Theorem 3.6.18 *Let* $(\Omega, \mathcal{A}, P)$ *be a probability space and* $\xi : \Omega \to P(X)$ *be a random set. Suppose* $\tilde{H}$ *be a falling weak subalgebra of a BE-algebra X. Then*

$$\tilde{H}(x * (x * y)) \ge T(\tilde{H}(x), \tilde{H}(y))$$

for all $x, y \in X$, *where* $T(s, t) = \min\{s + t - 1, 1\}$ *for any* $s, t \in [0, 1]$.

Proof From the definition of falling weak subalgebra, we get $\xi(\omega)$ is a weak subalgebra of X for any $\omega \in \Omega$. Let $x, y \in X$. Then by the above proposition, we get $\Omega(x; \xi) \cap \Omega(y; \xi) \subseteq \Omega(x * (x * y); \xi)$. Hence

$$\begin{aligned}
\tilde{H}(x * (x * y)) &= P(\omega \mid x * y \in \xi(\omega)) \\
&\ge P(\omega \mid x \in \xi(\omega)) \cap P(\omega \mid y \in \xi(\omega)) \\
&\ge P(\omega \mid x \in \xi(\omega)) + P(\omega \mid y \in \xi(\omega)) - P(\omega \mid x \in \xi(\omega) \text{ or } y \in \xi(\omega)) \\
&\ge \tilde{H}(x) + \tilde{H}(y) - 1
\end{aligned}$$

Hence $\tilde{H}(x * y) \ge \min\{\tilde{H}(x) + \tilde{H}(y) - 1, 1\} = T(\tilde{H}(x), \tilde{H}(y))$. □

Corollary 3.6.19 *Every falling fuzzy weak subalgebra is a T-fuzzy weak subalgebra.*

Definition 3.6.20 Let Ω be a sample space and ξ and η be two functions, each assigning a real number $\xi(\omega)$ and $\eta(\omega)$ to every outcome $\omega \in \Omega$, that is, $\xi : \Omega \to \mathcal{P}(U)$ and $\eta : \Omega \to \mathcal{P}(U)$. Then the pair $X = (\xi, \eta)$ is said to be a *two-dimensional random variable*. The induced sample space (range) of the two-dimensional random variable is

$$\mathfrak{F} = \{(x, y) \mid x \in \mathcal{A}_1;\ y \in \mathcal{A}_2\}$$

Let $(\Omega, \mathcal{A}, P)$ be a probability space, where P is a probability distribution of two-dimensional random variables (ξ, η) on $[0, 1]^2$. There are many types of probability distributions, but only three types are the most classic ones.

(1) If the whole probability P is concentrated and uniformly distributed on the main diagonal of the unit square $[0, 1]^2$, then P is called a diagonal distribution.

(2) If the whole probability P is concentrated and uniformly distributed on the antidiagonal of the unit square $[0, 1]^2$, then P is called an antidiagonal distribution.

(3) If the whole probability P is uniformly distributed on the unit square $[0, 1]^2$, then P is called an independent distribution (Figs. 3.1 and 3.2).

Definition 3.6.21 Let $(\Omega, \mathcal{A}, P)$ be a probability space, and let $\tilde{H}$, $\tilde{H}_1$, and $\tilde{H}_2$ be falling shadows of random sets $\xi, \xi_1, \xi_2 : \Omega \to \mathcal{P}(X)$, respectively. Then the product of $\tilde{H}_1$ and $\tilde{H}_2$ is defined by

$$(\tilde{H}_1 \circledast \tilde{H}_2)(z) = \inf\{P([\tilde{H}_1(x), 1] \times [\tilde{H}_2(y), 1]) \mid z = x * y\}.$$

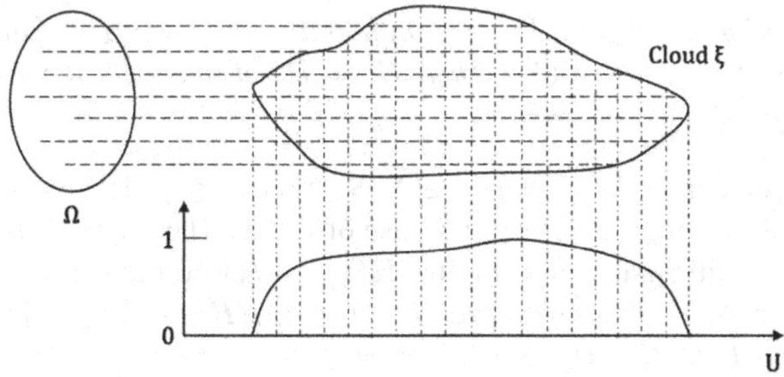

Fig. 3.1 Falling shadow

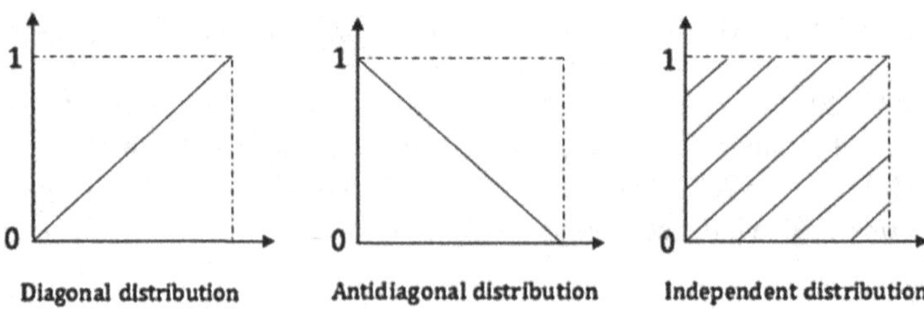

Fig. 3.2 Types of probability distributions

The down product between $\tilde{H}_1$ and $\tilde{H}_2$ is defined by

$$(\tilde{H}_1 \odot \tilde{H}_2)(z) = \inf\{P([\tilde{H}_1(x), 1] \times [\tilde{H}_2(y), 1]) \mid z \leq x * y\}.$$

The down set of $\tilde{H}$ is defined by

$$\tilde{H}^{\downarrow}(z) = \inf\{P([\tilde{H}(x), 1]) \mid z \leq x\}.$$

Remark 3.6.22 From the above definition, it is easy to see that

$$\tilde{H}^{\downarrow}(z) = \inf\{P([\tilde{H}(x), 1]) \mid z \leq x\} = \inf\{\tilde{H}(x) \mid z \leq x\} \leq \tilde{H}(z)$$

for any $z \in X$, and therefore $\tilde{H}^{\downarrow} \subseteq \tilde{H}$.
If the probability distribution P in the above definition is diagonal, antidiagonal, and independent, respectively, then $(\tilde{H}_1 \circledast \tilde{H}_2)(z) = \inf\{T(\tilde{H}_1(x), \tilde{H}_2(y)) \mid z = x * y\}$, and $(\tilde{H}_1 \odot \tilde{H}_2)(z) = \inf\{T(\tilde{H}_1(x), \tilde{H}_2(y)) \mid z \leq x * y\}$.

Proposition 3.6.23 *Let $(\Omega, \mathcal{A}, P)$ be a probability space and $\tilde{H}$ a falling shadow of a random set $\xi : \Omega \to \mathcal{P}(X)$. Then $\tilde{H} = \tilde{H}^{\downarrow}$ if and only if $x \leq y$ implies $\tilde{H}(x) \leq \tilde{H}(y)$ for any $x, y \in X$.*

Proof Assume that $\tilde{H} = \tilde{H}^{\downarrow}$. Let $x, y \in X$. Suppose $x \leq y$. Then we get $\tilde{H}(x) = \tilde{H}^{\downarrow}(x) = \inf\{\tilde{H}(z) \mid x \leq z\} \leq \tilde{H}(y)$ because of $x \leq y$. Therefore $\tilde{H}(x) \leq \tilde{H}(y)$.

Conversely, assume that $x \leq y$ implies $\tilde{H}(x) \leq \tilde{H}(y)$ for any $x, y \in X$. Clearly $\tilde{H}^{\downarrow} \subseteq \tilde{H}$. Let $z \in X$. Then we have $\tilde{H}^{\downarrow}(z) = \inf\{\tilde{H}(x) \mid z \leq x\} \geq \tilde{H}(z)$ by hypothesis. Thus $\tilde{H} \subseteq \tilde{H}^{\downarrow}$. Therefore $\tilde{H} = \tilde{H}^{\downarrow}$. □

Theorem 3.6.24 *Let $(\Omega, \mathcal{A}, P)$ be a probability space and $\tilde{H}$ a falling shadow of a random set $\xi : \Omega \to \mathcal{P}(X)$. If the probability distribution P of two-dimensional random variables is diagonal (antidiagonal or independent, resp.), then $\tilde{H}$ is a T-fuzzy subalgebra of X if and only if $\tilde{H} \circledast \tilde{H} \subseteq \tilde{H}$.*

Proof We first consider the case where the probability distribution P is diagonal. The other cases can be proved similarly. Suppose $\tilde{H}$ is a T-fuzzy subalgebra of X. Let $z \in X$. Then there exist $x, y \in X$ such that $z = x * y$. Hence $\tilde{H}(z) = \tilde{H}(x * y) \geq$

$T(\tilde{H}(x), \tilde{H}(y))$. Thus $\tilde{H}(z) \geq \inf\{T(\tilde{H}(x), \tilde{H}(y)) \mid z = x * y\} = (\tilde{H} \circledast \tilde{H})(z)$. Therefore $(\tilde{H} \circledast \tilde{H}) \subseteq \tilde{H}$.

Conversely, assume that $(\tilde{H} \circledast \tilde{H}) \subseteq \tilde{H}$. Let $x, y \in X$. Then we have

$$\tilde{H}(x * y) \geq (\tilde{H} \circledast \tilde{H})(x * y) \geq T((\tilde{H}(x), \tilde{H}(y)).$$

Therefore $\tilde{H}$ is a T-fuzzy subalgebra of X. □

Proposition 3.6.25 *Let $(\Omega, \mathcal{A}, P)$ be a probability space, where the probability distribution P of two-dimensional random variables is diagonal. Let $\tilde{H}_1$ and $\tilde{H}_2$ be falling shadows of random sets $\xi_1, \xi_2 : \Omega \to P(X)$, respectively. If $\tilde{H}_1$ and $\tilde{H}_2$ are T-fuzzy subalgebras of X, then $\tilde{H}_1 \odot \tilde{H}_2$ is a T-fuzzy subalgebra of X.*

Proof Suppose $\tilde{H}_1$ and $\tilde{H}_2$ are T-fuzzy subalgebras of X. Let $x, y \in X$. Then

$$\begin{aligned}(\tilde{H}_1 \odot \tilde{H}_2)(x * y) &\geq \inf\{\tilde{H}_1(x_1 * y_1) \wedge \tilde{H}_2(x_2 * y_2) \mid x \leq x_1 * x_2, y \leq y_1 * y_2\}\\ &\geq \inf\{\tilde{H}_1(x_1) \wedge \tilde{H}_1(y_1) \wedge \tilde{H}_2(x_2) \wedge \tilde{H}_2(y_2) \mid x \leq x_1 * x_2, y \leq y_1 * y_2\}\\ &= \inf\{\tilde{H}_1(x_1) \wedge \tilde{H}_2(x_2) \mid x \leq x_1 * x_2\} \wedge \inf\{\tilde{H}_1(y_1) \wedge \tilde{H}_2(y_2) \mid y \leq y_1 * y_2\}\\ &= (\tilde{H}_1 \odot \tilde{H}_2)(x) \wedge (\tilde{H}_1 \odot \tilde{H}_2)(y)\end{aligned}$$

Therefore $\tilde{H}_1 \odot \tilde{H}_2$ is a T-fuzzy subalgebra of X. □

3.7 Direct Products of BE-algebras

In this section, the notion of direct products of BE-algebras is introduced. The properties of commutativity, self-distributivity, and transitivity of BE-algebras are studied in terms of direct products of BE-algebras. Two canonical mappings of the direct product of BE-algebras are introduced, and then we obtain some of their properties.

Definition 3.7.1 Let $A = (A, *, 1_A)$ and $B = (B, *, 1_B)$ be two BE-algebras. Define the *direct product* of A and B as the algebraic structure $(A \times B, \circledast, 1)$, where the underlying set $A \times B$ is given by $\{(a, b) \mid a \in A \text{ and } b \in B\}$, $1 = (1_A, 1_B)$ and whose binary operation $\circledast$ is given by

$$(a_1, b_1) \circledast (a_2, b_2) = (a_1 * a_2, b_1 * b_2)$$

where $a_1, a_2 \in A$ and $b_1, b_2 \in B$. Note that the binary operation $\circledast$ is componentwise.

Theorem 3.7.2 *The direct product of two BE-algebras is again a BE-algebra.*

Proof The properties (BE1), (BE2), (BE3), and (BE4) of $A \times B$ follow from those of A and B. Hence the proof is a routine verification. □

Now, we extend the concept of direct products to any finite family of BE-algebras. Let $I_n = \{1, 2, \ldots, n\}$ and $\{A_i = (A_i, *, 1_i) \mid i \in I_n\}$ be a finite family of BE-algebras. Define the direct product of BE-algebras $A_1, A_2, \ldots, A_n$ as the algebraic structure

$$\prod_{i=1}^{n} A_i = \left(\prod_{i=1}^{n} A_i, \circledast, (1_1, 1_2, \ldots, 1_n)\right)$$

where

$$\prod_{i=1}^{n} A_i = A_1 \times A_2 \times \cdots A_n = \{(a_1, a_2, \ldots, a_n) \mid a_i \in A_i \text{ and } i \in I_n\}$$

and whose operation $\circledast$ is defined by $(a_1, a_2, \ldots, a_n) \circledast (b_1, b_2, \ldots, b_n) = (a_1 * b_1, a_2 * b_2, \ldots, a_n * b_n)$. Clearly, $\circledast$ is a binary operation on $\prod_{i=1}^{n} A_i$.

Theorem 3.7.3 *Let $\{A_i \mid i \in I_n\}$ be a finite family of BE-algebras. Then $\prod_{i=1}^{n} A_i$ is a BE-algebra.*

Proof It is routine. □

Proposition 3.7.4 *Let $\{A_i = (A_i, *, 1_i) \mid i \in I_n\}$ be a finite family of BE-algebras. Then each A_i is commutative if and only if $\prod_{i=1}^{n} A_i$ is commutative.*

Proof Assume that each A_i; $i \in I_n$ is a commutative BE-algebra. Let $(a_1, \ldots, a_n) \in \prod_{i=1}^{n} A_i$ and $(b_1, \ldots, b_n) \in \prod_{i=1}^{n} A_i$. Then $a_i, b_i \in A_i$ for each $i \in I_n$. Since each $A_i, i \in I_n$ is commutative, we get $(a_i * b_i) * b_i = (b_i * a_i) * a_i$ for each $i \in I_n$. Then we get

$$\begin{aligned}((a_1, \ldots, a_n) \circledast (b_1, \ldots, b_n)) \circledast (b_1, \ldots, b_n) &= (a_1 * b_1, \ldots, a_n * b_n) \circledast (b_1, \ldots, b_n)\\ &= ((a_1 * b_1) * b_1, \ldots, (a_n * b_n) * b_n)\\ &= ((b_1 * a_1) * a_1, \ldots, (b_n * a_n) * a_n)\\ &= ((b_1, \ldots, b_n) \circledast (a_1, \ldots, a_n)) \circledast (a_1, \ldots, a_n)\end{aligned}$$

Therefore $\prod_{i=1}^{n} A_i$ is commutative.

Conversely, assume that $\prod_{i=1}^{n} A_i$ is a commutative BE-algebra. Suppose $a_i, b_i \in A_i$ for all $i \in I_n$. Then $(a_1, \ldots, a_n), (b_1, \ldots, b_n) \in \prod_{i=1}^{n} A_i$ and

$$((a_1, \ldots, a_n) \circledast (b_1, \ldots, b_n)) \circledast (b_1, \ldots, b_n) = ((b_1, \ldots, b_n) \circledast (a_1, \ldots, a_n)) \circledast (a_1, \ldots, a_n)$$

Hence we have

$$\begin{aligned}((a_1 * b_1) * b_1, \ldots, (a_n * b_n) * b_n) &= (a_1 * b_1, \ldots, a_n * b_n) \circledast (b_1, \ldots, b_n)\\ &= ((a_1, \ldots, a_n) \circledast (b_1, \ldots, b_n)) \circledast (b_1, \ldots, b_n)\\ &= ((b_1, \ldots, b_n) \circledast (a_1, \ldots, a_n)) \circledast (a_1, \ldots, a_n)\\ &= (b_1 * a_1, \ldots, b_n * a_n) \circledast (a_1, \ldots, a_n)\\ &= ((b_1 * a_1) * a_1, \ldots, (b_n * a_n) * a_n)\end{aligned}$$

Thus $(a_i * b_i) * b_i = (b_i * a_i) * a_i$ for each $i \in I_n$. Therefore each A_i is commutative. □

Definition 3.7.5 For any $(a_1, \ldots, a_n), (b_1, \ldots, b_n) \in \prod_{i=1}^{n} A_i$, define an ordering $\leq_{\circledast}$ on $\prod_{i=1}^{n} A_i$ as

$$(a_1, \ldots, a_n) \leq_{\circledast} (b_1, \ldots, b_n) \text{ if and only if } a_i \leq_i b_i \text{ for each } i \in I_n.$$

where $\leq_i$ is the natural BE-ordering on each $A_i, i \in I_n$.

Since each $\leq_i$ is reflexive on A_i respectively, it is clear that $\leq_{\circledast}$ is reflexive on $\prod_{i=1}^{n} A_i$. It can be easily seen that $(a_1, \ldots, a_n) \leq_{\circledast} (b_1, \ldots, b_n)$ if and only if $(a_1, \ldots, a_n) \circledast (b_1, \ldots, b_n) = (1_1, \ldots, 1_n)$.

Theorem 3.7.6 *Let $\{A_i = (A_i, *, 1_i) \mid i \in I_n\}$ be a finite family of BE-algebras. If $\prod_{i=1}^{n} A_i$ is commutative, then $\leq_{\circledast}$ is antisymmetric.*

Proof Assume that $\prod_{i=1}^{n} A_i$ is a commutative BE-algebra. Then by the above theorem, we get that each $A_i; i \in I_n$ is a commutative BE-algebra. Let $(a_1, \ldots, a_n) \in \prod_{i=1}^{n} A_i$ and $(b_1, \ldots, b_n) \in \prod_{i=1}^{n} A_i$ be such that $(a_1, \ldots, a_n) \leq_{\circledast} (b_1, \ldots, b_n)$ and $(b_1, \ldots, b_n) \leq_{\circledast} (a_1, \ldots, a_n)$. Then we get $a_i \leq b_i$ and $b_i \leq a_i$ for each $i \in I_n$. Since each $A_i, i \in I_n$ is commutative, we get $a_i = b_i$ for each $i \in I_n$. Hence $(a_1, \ldots, a_n) = (b_1, \ldots, b_n)$. Therefore $\leq_{\circledast}$ is antisymmetric. □

Proposition 3.7.7 *Let $\{A_i = (A_i, *, 1_i) \mid i \in I_n\}$ be a finite family of BE-algebras. Then each A_i is self-distributive if and only if $\prod_{i=1}^{n} A_i$ is self-distributive.*

Proof Assume that each A_i is a self-distributive. Let $(a_1, \ldots, a_n); (b_1, \ldots, b_n); (c_1, \ldots, c_n) \in \prod_{i=1}^{n} A_i$. Then $a_i, b_i, c_i \in A_i$ for each $i \in I_n$. Since each $A_i, i \in I_n$ is self-distributive, we get $a_i * (b_i * c_i) = (a_i * b_i) * (a_i * c_i)$ for each $i \in I_n$. Put $a = (a_1, \ldots, a_n)$; $b = (b_1, \ldots, b_n)$; and $c = (c_1, \ldots, c_n)$. Then we get

$$
\begin{aligned}
a \circledast (b \circledast c) &= (a_1, \ldots, a_n) \circledast ((b_1, \ldots, b_n) \circledast (c_1, \ldots, c_n)) \\
&= (a_1, \ldots, a_n) \circledast (b_1 * c_1, \ldots, b_n * c_n) \\
&= (a_1 * (b_1 * c_1), \ldots, a_n * (b_n * c_n)) \\
&= ((a_1 * b_1) * (a_1 * c_1), \ldots, (a_n * b_n) * (a_n * c_n)) \\
&= (a_1 * b_1, \ldots, a_n * b_n) \circledast (a_1 * c_1, \ldots, a_n * c_n) \\
&= ((a_1, \ldots, a_n) \circledast (b_1, \ldots, b_n)) \circledast ((a_1, \ldots, a_n) \circledast (c_1, \ldots, c_n)) \\
&= (a \circledast b) \circledast (a \circledast c)
\end{aligned}
$$

Therefore $\prod_{i=1}^{n} A_i$ is self-distributive.

Conversely, assume that $\prod_{i=1}^{n} A_i$ is a self-distributive BE-algebra. Suppose $a_i, b_i, c_i \in A_i$ for all $i \in I_n$. Then $a = (a_1, \ldots, a_n), b = (b_1, \ldots, b_n), c = (c_1, \ldots, c_n) \in \prod_{i=1}^{n} A_i$ and

$$
\begin{aligned}
a \circledast (b \circledast c) &= (a_1, \ldots, a_n) \circledast ((b_1, \ldots, b_n) \circledast (c_1, \ldots, c_n)) \\
&= ((a_1, \ldots, a_n) \circledast (b_1, \ldots, b_n)) \circledast ((a_1, \ldots, a_n) \circledast (c_1, \ldots, c_n)) \\
&= (a \circledast b) \circledast (a \circledast c)
\end{aligned}
$$

Hence we have

$$
\begin{aligned}
(a_1 * (b_1 * c_1), \ldots, a_n * (b_n * c_n)) &= (a_1, \ldots, a_n) \circledast (b_1 * c_1, \ldots, b_n * c_n) \\
&= (a_1, \ldots, a_n) \circledast ((b_1, \ldots, b_n) \circledast (c_1, \ldots, c_n)) \\
&= a \circledast (b \circledast c) \\
&= (a \circledast b) \circledast (a \circledast c) \\
&= ((a_1, \ldots, a_n) \circledast (b_1, \ldots, b_n)) \circledast ((a_1, \ldots, a_n) \circledast (c_1, \ldots, c_n)) \\
&= (a_1 * b_1, \ldots, a_n * b_n) \circledast (a_1 * c_1, \ldots, a_n * c_n) \\
&= ((a_1 * b_1) * (a_1 * c_1), \ldots, (a_n * b_n) * (a_n * c_n))
\end{aligned}
$$

Thus $a_i * (b_i * c_i) = (a_i * b_i) * (a_i * c_i)$ for each $i \in I_n$. Therefore each A_i is self-distributive. □

Proposition 3.7.8 *Let $\{A_i = (A_i, *, 1_i) \mid i \in I_n\}$ be a finite family of BE-algebras. Then each A_i is transitive if and only if $\prod_{i=1}^{n} A_i$ is transitive.*

Proof Assume that each A_i is transitive. Let $(a_1, \ldots, a_n); (b_1, \ldots, b_n); (c_1, \ldots, c_n) \in \prod_{i=1}^{n} A_i$. Then $a_i, b_i, c_i \in A_i$ for each $i \in I_n$. Since each A_i is transitive, we get $a_i * b_i \leq_i (c_i * a_i) * (c_i * b_i)$ for each $i \in I_n$. Put $a = (a_1, \ldots, a_n)$; $b = (b_1, \ldots, b_n)$ and $c = (c_1, \ldots, c_n)$. Then we get

$$\begin{aligned} a \circledast b &= (a_1, \ldots, a_n) \circledast (b_1, \ldots, b_n) \\ &= (a_1 * b_1, \ldots, a_n * b_n) \\ &\leq_{\circledast} ((c_1 * a_1) * (c_1 * b_1), \ldots, (c_n * a_n) * (c_n * b_n)) \\ &= (c_1 * a_1, \ldots, c_n * a_n) \circledast (c_1 * b_1, \ldots, c_n * b_n) \\ &= ((c_1, \ldots, c_n) \circledast (a_1, \ldots, a_n)) \circledast ((c_1, \ldots, c_n) \circledast (b_1, \ldots, b_n)) \\ &= (c \circledast a) \circledast (c \circledast b) \end{aligned}$$

Therefore $\prod_{i=1}^{n} A_i$ is transitive.

Conversely, assume that $\prod_{i=1}^{n} A_i$ is a transitive BE-algebra. Suppose $a_i, b_i, c_i \in A_i$ for all $i \in I_n$. Then $a = (a_1, \ldots, a_n), b = (b_1, \ldots, b_n), c = (c_1, \ldots, c_n) \in \prod_{i=1}^{n} A_i$ and

$$\begin{aligned} a \circledast b &= (a_1, \ldots, a_n) \circledast (b_1, \ldots, b_n) \\ &\leq_{\circledast} ((c_1, \ldots, c_n) \circledast (a_1, \ldots, a_n)) \circledast ((c_1, \ldots, c_n) \circledast (b_1, \ldots, b_n)) \\ &= (c \circledast a) \circledast (c \circledast b) \end{aligned}$$

Hence we have

$$\begin{aligned} (a_1 * b_1, \ldots, a_n * b_n) &= (a_1, \ldots, a_n) \circledast (b_1, \ldots, b_n) \\ &= a \circledast b \\ &\leq_{\circledast} (c \circledast a) \circledast (c \circledast b) \\ &= ((c_1, \ldots, c_n) \circledast (a_1, \ldots, a_n)) \circledast ((c_1, \ldots, c_n) \circledast (b_1, \ldots, b_n)) \\ &= (c_1 * a_1, \ldots, c_n * a_n) \circledast (c_1 * b_1, \ldots, c_n * b_n) \\ &= ((c_1 * a_1) * (c_1 * b_1), \ldots, (c_n * a_n) * (c_n * b_n)) \end{aligned}$$

Thus $a_i * b_i \leq_i (c_i * a_i) * (c_i * b_i)$ for each $i \in I_n$. Therefore each A_i is transitive. □

Theorem 3.7.9 *Let $\{A_i = (A_i, *, 1_i) \mid i \in I_n\}$ be a finite family of BE-algebras. If $\prod_{i=1}^{n} A_i$ is transitive, then $\leq_{\circledast}$ is transitive.*

Proof Assume that $\prod_{i=1}^{n} A_i$ is a transitive BE-algebra. Then by the above theorem, we get that each $A_i, i \in I_n$ is a transitive BE-algebra. Let $(a_1, \ldots, a_n) \in \prod_{i=1}^{n} A_i$; $(b_1, \ldots, b_n) \in \prod_{i=1}^{n} A_i$; and $(c_1, \ldots, c_n) \in \prod_{i=1}^{n} A_i$ be such that $(a_1, \ldots, a_n) \leq_{\circledast} (b_1, \ldots, b_n)$ and $(b_1, \ldots, b_n) \leq_{\circledast} (c_1, \ldots, c_n)$. Then we get $a_i \leq b_i$ and $b_i \leq c_i$ for each $i \in I_n$. Since each $A_i, i \in I_n$ is transitive, we get $a_i \leq_i c_i$ for each $i \in I_n$. Hence $(a_1, \ldots, a_n) \leq_{\circledast} (c_1, \ldots, c_n)$. Therefore $\leq_{\circledast}$ is transitive. □

Since every commutative BE-algebra is transitive, the following corollary is obvious.

Corollary 3.7.10 *Let* $\{A_i = (A_i, *, 1_i) \mid i \in I_n\}$ *be a finite family of BE-algebras. If* $\prod_{i=1}^{n} A_i$ *is commutative, then* $\leq_{\circledast}$ *is a partial ordering on* $\prod_{i=1}^{n} A_i$.

Theorem 3.7.11 *Let* $\{\varphi_i : A_i \to B_i \mid i \in I_n\}$ *be a family of BE-homomorphisms. If* $\varphi : \prod_{i=1}^{n} A_i \to \prod_{i=1}^{n} B_i$ *is the map given by* $(a_1, \ldots, a_n) \mapsto (\varphi_1(a_1), \ldots, \varphi_n(a_n))$, *then* φ *is a BE-homomorphism with* $Ker(\varphi) = \prod_{i=1}^{n} Ker(\varphi_i)$, $\varphi(\prod_{i=1}^{n} A_i) = \prod_{i=1}^{n} \varphi(A_i)$. *Furthermore,* φ *is a monomorphism (respectively, epimorphism) if and only if each* φ_i *is monomorphism (respectively, epimorphism).*

Proof Let $\{\varphi_i : A_i \to B_i \mid i \in I_n\}$ be a family of BE-homomorphisms and $\varphi : \prod_{i=1}^{n} A_i \to \prod_{i=1}^{n} B_i$ a map given by $(a_1, \ldots, a_n) \mapsto (\varphi_1(a_1), \ldots, \varphi_n(a_n))$. Let $(a_1, \ldots, a_n); (b_1, \ldots, b_n) \in \prod_{i=1}^{n} A_i$. Then

$$\begin{aligned}
\varphi((a_1, \ldots, a_n) \circledast (b_1, \ldots, b_n)) &= \varphi(a_1 * b_1, \ldots, a_n * b_n) \\
&= (\varphi_1(a_1 * b_1), \ldots, \varphi_n(a_n * b_n)) \\
&= (\varphi_1(a_1) * \varphi_1(b_1), \ldots, \varphi_n(a_n) * \varphi_n(b_n)) \\
&= (\varphi_1(a_1), \ldots, \varphi_n(a_n)) \circledast (\varphi_1(b_1), \ldots, \varphi_n(b_n)) \\
&= \varphi(a_1, \ldots, a_n) \circledast \varphi(b_1, \ldots, b_n)
\end{aligned}$$

This shows that φ is a BE-homomorphism. Moreover,

$$\begin{aligned}
(a_1, \ldots, a_n) \in Ker(\varphi) &\Leftrightarrow \varphi(a_1, \ldots, a_n) = (1_1, \ldots, 1_n) \\
&\Leftrightarrow (\varphi_1(a_1), \ldots, \varphi_n(a_n)) = (1_1, \ldots, 1_n) \\
&\Leftrightarrow \varphi_i(a_i) = 1_i \text{ for each } i \in I_n \\
&\Leftrightarrow a_i \in Ker(\varphi_i) \text{ for each } i \in I_n \\
&\Leftrightarrow (a_1, \ldots, a_n) \in \prod_{i=1}^{n} Ker(\varphi_i)
\end{aligned}$$

Hence $Ker(\varphi) = \prod_{i=1}^{n} Ker(\varphi_i)$. Now

$$\begin{aligned}
(b_1, \ldots, b_n) \in \varphi\left(\prod_{i=1}^{n} A_i\right) \Leftrightarrow &\text{ there exists } (a_1, \ldots, a_n) \in \prod_{i=1}^{n} A_i \text{ such that} \\
&(b_1, \ldots, b_n) = \varphi(a_1, \ldots, a_n)
\end{aligned}$$

$$\begin{aligned}&\Leftrightarrow \text{ there exists } (a_1, \ldots, a_n) \in \prod_{i=1}^{n} A_i \text{ such that}\\ &\qquad (b_1, \ldots, b_n) = (\varphi_1(a_1), \ldots, \varphi_n(a_n))\\ &\Leftrightarrow \text{ there exists } a_i \in A_i \text{ such that } b_i = \varphi_i(a_i) \in \varphi_i(A_i) \text{ for each } i \in I_n\\ &\Leftrightarrow (b_1, \ldots, b_n) \in \prod_{i=1}^{n} \varphi_i(A_i)\end{aligned}$$

Therefore $\varphi\Big(\prod_{i=1}^{n} A_i\Big) = \prod_{i=1}^{n} \varphi_i(A_i)$.

Assume that φ is one-to-one. Suppose $\varphi_i(a_i) = \varphi_i(b_i)$ for each $i \in I_n$. Then

$$\begin{aligned}\varphi(a_1, \ldots, a_n) &= (\varphi_i(a_1), \ldots, \varphi_n(a_n))\\ &= (\varphi_i(b_1), \ldots, \varphi_n(b_n))\\ &= \varphi(b_1, \ldots, b_n)\end{aligned}$$

Since φ is one-to-one, we get $(a_1, \ldots, a_n) = (b_1, \ldots, b_n)$. Hence $a_i = b_i$ for each $i \in I_n$. Therefore, φ_i is one-to-one for each $i \in I_n$. Conversely, assume that φ_i is one-to-one for each $i \in I_n$. Suppose $\varphi(a_1, \ldots, a_n) = \varphi(b_1, \ldots, b_n)$ for $(a_1, \ldots, a_n), (a_1, \ldots, a_n) \in \prod_{i=1}^{n} A_i$. Then we get

$$\begin{aligned}(\varphi_i(a_1), \ldots, \varphi_n(a_n)) &= \varphi(a_1, \ldots, a_n)\\ &= \varphi(b_1, \ldots, b_n)\\ &= (\varphi_i(b_1), \ldots, \varphi_n(b_n))\end{aligned}$$

Hence $\varphi_i(a_i) = \varphi_i(b_i)$ for each $i \in I_n$. Since each φ_i is one-to-one, we get $a_i = b_i$ for each $i \in I_n$. Hence $(a_1, \ldots, a_n) = (b_1, \ldots, b_n)$. Therefore φ is one-to-one.

We now prove that φ is onto if and only if each φ_i is onto. Assume that φ is onto. Suppose $b_i \in B_i$ for each $i \in I_n$. Then $(b_1, \ldots, b_n) \in \prod_{i=1}^{n} B_i$. Since φ is onto, there exists $(a_1, \ldots, a_n) \in \prod_{i=1}^{n} A_i$ such that $(b_1, \ldots, b_n) = \varphi(a_1, \ldots, a_n) = (\varphi_1(a_1), \ldots, \varphi_n(a_n))$. Hence $b_i = \varphi_i(a_i)$ for each $i \in I_n$. Therefore φ_i is onto for each $i \in I_n$. Conversely, assume that φ_i is onto for each $i \in I_n$. Suppose $(b_1, \ldots, b_n) \in \prod_{i=1}^{n} B_i$. Then $b_i \in B_i$ for each $i \in I_n$. Since each φ_i is onto, there exists $a_i \in A_i$ such that $b_i = \varphi_i(a_i)$ for each $i \in I_n$. Hence $(b_1, \ldots, b_n) = (\varphi_1(a_1), \ldots, \varphi_n(a_n)) = \varphi(a_1, \ldots, a_n)$. Thus φ is onto. Therefore the proof is completed. □

Corollary 3.7.12 *Let $\{(A_i, *, 1_i) \mid i \in I_n\}$ and $\{(B_i, *, 1_i) \mid i \in I_n\}$ be any two families of BE-algebras such that $A_i \cong B_i$ for each $i \in I_n$. Then $\prod_{i=1}^{n} A_i \cong \prod_{i=1}^{n} B_i$.*

Theorem 3.7.13 *Let* $\{(A_i, *, 1_i) \mid i \in I_n\}$ *be a family of BE-algebras. Then* $f_k : \prod_{i=1}^{n} A_i \to A_k$ *given by* $(a_1, \ldots, a_k, \ldots, a_n) \mapsto a_k$ *is an epimorphism of BE-algebras for each* $k \in I_n$.

Proof Let I_n be an indexed set. For each $k \in I_n$, define $f_k : \prod_{i=1}^{n} A_i \to A_k$ by $f_k(a_1, \ldots, a_k, \ldots, a_n) = a_k$; for all $(a_1, \ldots, a_k, \ldots, a_n) \in \prod_{i=1}^{n} A_i$. Let $(a_1, \ldots, a_k, \ldots, a_n), (b_1, \ldots, b_k, \ldots, b_n) \in \prod_{i=1}^{n} A_i$. Suppose $(a_1, \ldots, a_k, \ldots, a_n) = (a_1, \ldots, a_k, \ldots, a_n)$. Then $a_i = b_i$ for each $i \in I_n$. Hence

$$\begin{aligned} f_k(a_1, \ldots, a_k, \ldots, a_n) &= a_k \\ &= b_k \\ &= f_k(b_1, \ldots, b_k, \ldots, b_n) \end{aligned}$$

Thus f_k is well-defined for each $k \in I_n$. Let $(a_1, \ldots, a_k, \ldots, a_n), (b_1, \ldots, b_k, \ldots, b_n) \in \prod_{i=1}^{n} A_i$. Then

$$\begin{aligned} f_k((a_1, \ldots, a_k, \ldots, a_n) \circledast (b_1, \ldots, b_k, \ldots, b_n)) &= f_k(a_1 * b_1, \ldots, a_k * b_k, \ldots, a_n * b_n) \\ &= a_k * b_k \\ &= f_k(a_1, \ldots, a_k, \ldots, a_n) * f_k(b_1, \ldots, b_k, \ldots, b_n) \end{aligned}$$

Therefore f_k is a homomorphism of BE-algebras. Let $c_k \in A_k$. Then $(1_1, \ldots, c_k, \ldots, 1_n) \in \prod_{i=1}^{n} A_i$ and $f_k(1_1, \ldots, c_k, \ldots, 1_n) = c_k$. Therefore f_k is onto and hence f_k is an epimorphism. □

The mappings f_k in the above theorem are called the canonical projections of the direct product. In the following theorem, we can observe the relation between direct product and its canonical projections.

Theorem 3.7.14 *Let* $\{(A_i, *, 1_i) \mid i \in I_n\}$ *be a family of BE-algebras. Then there exists a BE-algebra D, together with a family of homomorphisms* $\{f_i : D \to A_i \mid i \in I_n\}$ *with the following property: for any BE-algebra X and a family of homomorphisms* $\{\varphi_i : X \to A_i \mid i \in I_n\}$, *there exists a unique homomorphism* $\varphi : X \to D$ *such that* $f_i \circ \varphi = \varphi_i$ *for all* $i \in I_n$. *Furthermore, D is uniquely determined up to isomorphism.*

Proof Let $\{(A_i, *, 1_i) \mid i \in I_n\}$ be a family of BE-algebras. Then the direct product $\prod_{i=1}^{n} A_i$ is a BE-algebra. Let $D = \prod_{i=1}^{n} A_i$, and let $\{f_i : D \to Ai \mid i \in I_n\}$ be the family of canonical projections. Suppose that X is any BE-algebra and $\{\varphi_i : X \to A_i \mid i \in I_n\}$ a family of homomorphisms. Define $\varphi : X \to D$ by $\varphi(x) = (\varphi_1(x), \ldots, \varphi_i(x), \ldots, \varphi_n(x))$ for all $x \in X$. If $x, y \in X$, then

$$\begin{aligned}\varphi(x * y) &= (\varphi_1(x * y), \ldots, \varphi_i(x * y), \ldots, \varphi_n(x * y)) \\ &= (\varphi_1(x) * \varphi_1(y), \ldots, \varphi_i(x) * \varphi_i(y), \ldots, \varphi_n(x) * \varphi_n(y)) \\ &= (\varphi_1(x), \ldots, \varphi_i(x), \ldots, \varphi_n(x)) \circledast (\varphi_1(y), \ldots, \varphi_i(y), \ldots, \varphi_n(y)) \\ &= \varphi(x) \circledast \varphi(y)\end{aligned}$$

Thus φ is a homomorphism. Moreover, for any $x \in X$, we get

$$(f_i \circ \varphi)(x) = f_i(\varphi(x)) = f_i((\varphi_1(x), \ldots, \varphi_i(x), \ldots, \varphi_n(x))) = \varphi_i(x).$$

Hence $f_i \circ \varphi = \varphi_i$ for all $i \in I_n$. We now show that φ is unique. Suppose $\varphi' : X \to D$ be another homomorphism such that $f_i \circ \varphi' = \varphi_i$ for each $i \in I_n$. If $x \in X$, then $(f_i \circ \varphi)(x) = \varphi_i(x) = (f_i \circ \varphi')(x)$. By the definition of φ, we get $\varphi(x) = (\varphi_1(x), \ldots, \varphi_i(x), \ldots, \varphi_n(x))$ and assume that $\varphi'(x) = (a_1, \ldots, a_i, \ldots, a_n)$. Thus, for each $i \in I_n$, we get

$$\begin{aligned}a_i &= f_i(a_1, \ldots, a_i, \ldots, a_n) \\ &= f_i(\varphi'(x)) \\ &= (f_i \circ \varphi')(x) \\ &= (f_i \circ \varphi)(x) \\ &= f_i(\varphi(x)) \\ &= f_i(\varphi_1(x), \ldots, \varphi_i(x), \ldots, \varphi_n(x)) \\ &= \varphi(x)\end{aligned}$$

Hence $\varphi(x) = (\varphi_1(x), \ldots, \varphi_i(x), \ldots, \varphi_n(x)) = (a_1, \ldots, a_n) = \varphi'(x)$. Therefore, φ is unique.

Suppose that a BE-algebra D' has the same property as D with the family of homomorphisms $\{f_i' : D' \to A_i \mid i \in I_n\}$. If we apply this property for D to the family of homomorphisms $\{f_i' : D' \to A_i \mid i \in I_n\}$ and also apply it for D' to the family of homomorphisms $\{f_i : D \to A_i \mid i \in I_n\}$, then we obtain unique homomorphisms $\alpha : D' \to D$ and $\beta : D \to D'$ such that $f_i \circ \alpha = f_i'$ and $f_i' \circ \beta = f_i$ for all $i \in I_n$. Thus, $\alpha \circ \beta : D \to D$ is a unique homomorphism such that $f_i \circ (\alpha \circ \beta) = f_i$ for all $i \in I_n$. Since $id_D : D \to D$ is a homomorphism such that $f_i \circ id_D = f_i$ for all $i \in I_n$, $\alpha \circ \beta = id_D$ by uniqueness. A similar argument shows that $\beta \circ \alpha = id_{D'}$. Therefore, β is an isomorphism; that is, D is uniquely determined up to isomorphism. □

Theorem 3.7.15 *Let* $\{(A_i, *, 1_i) \mid i \in I_n\}$ *be a family of BE-algebras. Then* $g_k : A_k \to \prod_{i=1}^{n} A_i$ *given by* $a_k \mapsto (1_1, \ldots, a_k, \ldots, 1_n)$ *is a monomorphism of BE-algebras for each* $k \in I_n$.

Proof For each $k \in I_n$, define $g_k : A_k \to \prod_{i=1}^{n} A_i$ by $g_k(a_k) = (1_1, \ldots, 1_{k-1}, ak, 1_{k+1}, \ldots, 1_n)$ for all $a_k \in A_k$. Let $a_k, b_k \in A_k$. If $a_k = b_k$, then

$$\begin{aligned} g_k(a_k) &= (1_1, \ldots, 1_{k-1}, a_k, 1_{k+1}, \ldots, 1_n) \\ &= (1_1, \ldots, 1_{k-1}, b_k, 1_{k+1}, \ldots, 1_n) \\ &= g_k(b_k) \end{aligned}$$

Hence g_k is well-defined. Let $a_k, b_k \in A_k$. Then we get

$$\begin{aligned} g_k(a_k * b_k) &= (1_1, \ldots, 1_{k-1}, a_k * b_k, 1_{k+1}, \ldots, 1_n) \\ &= (1_1 * 1_1, \ldots, 1_{k-1} * 1_{k-1}, a_k * b_k, 1_{k+1} * 1_{k+1}, \ldots, 1_n * 1_n) \\ &= (1_1, \ldots, 1_{k-1}, a_k, 1_{k+1}, \ldots, 1_n) \circledast (1_1, \ldots, 1_{k-1}, b_k, 1_{k+1}, \ldots, 1_n) \\ &= g_k(a_k) \circledast g_k(b_k) \end{aligned}$$

Therefore, g_k is a homomorphism. Suppose $g_k(a_k) = g_k(b_k)$. Then

$$\begin{aligned} (1_1, \ldots, 1_{k-1}, a_k, 1_{k+1}, \ldots, 1_n) &= g_k(a_k) \\ &= g_k(b_k) \\ &= (1_1, \ldots, 1_{k-1}, b_k, 1_{k+1}, \ldots, 1_n) \end{aligned}$$

Hence, $a_k = b_k$. Thus, g_k is one-to-one and so g_k is a monomorphism. □

3.8 BE-algebra of Functions

In this section, a BE-algebra structure is constructed on a set of real-valued functions. Some properties of these BE-algebras are observed with the help of the properties of functions. It is proved that the set of all real-valued function can be made into a commutative BE-algebra.

Definition 3.8.1 Let $\mathbb{R}_0 = \mathbb{R} \cup \{0\}$ be the non-empty subset of positive real numbers which contains zero. For any non-empty set X, define the set $\mathbb{R}_0^X$ as follows:

$$\mathbb{R}_0^X = \{f \mid f : X \to \mathbb{R}_0 \text{ is a function }\}.$$

Theorem 3.8.2 *For the set $\mathbb{R}_0^X$ defined above, define an operation $*$ on $\mathbb{R}_0^X$ as follows:*

$$(f * g)(x) = 1 + g(x) - \max\{f(x), g(x)\} \textit{for all } x \in X.$$

Then $*$ is a closure operation or binary operation on $\mathbb{R}_0^X$.

Proof It is obvious. □

For any function $f \in \mathbb{R}_0^X$, denote the function $f(x) = 0$ for all $x \in X$ by $\overline{0}$ and the function $f(x) = 1$ for all $x \in X$ by $\overline{1}$. We consider the natural ordering $\leq_{\mathbb{R}}$ on

$\mathbb{R}_0^X$ as follows:

$$f \leq_{\mathbb{R}} g \text{ if and only if } f(x) \leq_{\mathbb{R}} g(x)$$

for all $x \in X$ and for all $f, g \in \mathbb{R}_0^X$. Then a few properties of $\mathbb{R}_0^X$ are observed in the following:

Lemma 3.8.3 *Let X be a non-empty set and $\leq_{\mathbb{R}}$ is an ordering on $\mathbb{R}_0^X$. Then*

(1) *$\leq_{\mathbb{R}}$ is a partial ordering on $\mathbb{R}_0^X$,*
(2) *for $f, g \in \mathbb{R}_0^X$, $f \leq_{\mathbb{R}} g$ implies $f * g = \overline{1}$,*
(3) *for $f, g \in \mathbb{R}_0^X$, $f \leq_{\mathbb{R}} g$ implies $(g * f) * f = g$.*

Proof (1) It is clear.
(2) Let $f, g \in \mathbb{R}_0^X$. Suppose $f \leq_{\mathbb{R}} g$. Then we get that $f(x) \leq_{\mathbb{R}} g(x)$ for all $x \in X$. Hence $(f * g)(x) = 1 + g(x) - \max\{f(x), g(x)\} = 1 + g(x) - g(x) = 1 = \overline{1}(x)$. Therefore $f * g = \overline{1}$.
(3) Let $f, g \in \mathbb{R}_0^X$. Suppose $f \leq_{\mathbb{R}} g$. For any $x \in X$, we get $(g * f)(x) = 1 + f(x) - \max\{f(x), g(x)\} = 1 + f(x) - g(x)$. Hence $g * f = 1 + f - g$. Thus, we get the following:

$$\begin{aligned}(g * f) * f &= \overline{1} + f - \max\{f, g * f\} \\ &= \overline{1} + f - \max\{f, \overline{1} + f - g\} \\ &= \overline{1} + f - (\overline{1} + f - g) \\ &= g.\end{aligned}$$

Therefore $(g * f) * f = g$. □

Proposition 3.8.4 *Let X be a non-empty set. For any $f, g, h \in \mathbb{R}_0^X$, we have:*

(1) *$f * f = \overline{1}$,*
(2) *$\overline{1} * f = f$,*
(3) *$f * \overline{1} = \overline{1}$,*
(4) *$f * g = g * f$ implies $f = g$,*
(5) *$(f * g) * ((g * h) * (f * h)) = \overline{1}$.*

Proof (1) For any $x \in X$, we have $(f * f)(x) = 1 + f(x) - \max\{f(x), f(x)\} = 1 + f(x) - f(x) = 1$.
(2) For any $x \in X$, we have $(\overline{1} * f)(x) = 1 + f(x) - \max\{\overline{1}(x), f(x)\} = 1 + f(x) - 1 = f(x)$.
(3) For any $x \in X$, we have $(f * \overline{1})(x) = 1 + \overline{1}(x) - \max\{\overline{1}(x), f(x)\} = 1 + 1 - 1 = 1 = \overline{1}(x)$.
(4) Suppose $f * g = g * f$. Then, for any $x \in X$, we get $1 + g(x) - \max\{f(x), g(x)\} = (f * g)(x) = (g * f)(x) = 1 + f(x) - \max\{f(x), g(x)\}$. Hence $f(x) = g(x)$. Therefore $f = g$.
(5). Let $f, g, h \in \mathbb{R}_0^X$. To prove (5), we set

$$K = (\overline{1} + g - \max\{f, g\}) * ((\overline{1} + h - \max\{g, h\}) * (\overline{1} + h - \max\{f, h\})).$$

We now deal with all the 6 possible cases: (i) $f \leq_{\mathbb{R}} g \leq_{\mathbb{R}} h$; (ii) $f \leq_{\mathbb{R}} h \leq_{\mathbb{R}} g$; (iii) $g \leq_{\mathbb{R}} f \leq_{\mathbb{R}} h$; (iv) $g \leq_{\mathbb{R}} h \leq_{\mathbb{R}} f$; (v) $h \leq_{\mathbb{R}} f \leq_{\mathbb{R}} g$; (vi) $h \leq_{\mathbb{R}} g \leq_{\mathbb{R}} f$.
Case (i): Suppose $f \leq_{\mathbb{R}} g \leq_{\mathbb{R}} h$. Then clearly $K = (\overline{1} + g - g) * ((\overline{1} + h - h) * (\overline{1} + h - h) = \overline{1} * (\overline{1} * \overline{1}) = \overline{1}$.
Case (ii): Suppose $f \leq_{\mathbb{R}} h \leq_{\mathbb{R}} g$. Then $K = (\overline{1}+g-g)*((\overline{1}+h-h)*(\overline{1}+h-h)) = \overline{1}$.
Case (iii): Suppose $g \leq_{\mathbb{R}} f \leq_{\mathbb{R}} h$. Then $K = (\overline{1} + g - f) * ((\overline{1} + h - h) * (\overline{1} + h - h)) = \overline{1}$.
Case (iv): Suppose $g \leq_{\mathbb{R}} h \leq_{\mathbb{R}} f$. Then $K = (\overline{1} + g - f) * ((\overline{1} + h - h) * (\overline{1} + h - f)) = (\overline{1} + g - f) * (\overline{1} + h - f)$. Since $f(x), g(x)$, and $h(x) \in \mathbb{R}_0$, we get $g - f \leq_{\mathbb{R}} h - f$ and hence $(\overline{1} + g - f) \leq_{\mathbb{R}} (\overline{1} + h - f)$. Hence $K = (\overline{1} + g - f) * (\overline{1} + h - f) = \overline{1}$.
Case (v): Suppose $h \leq_{\mathbb{R}} f \leq_{\mathbb{R}} g$. Since $f(x), g(x) \in \mathbb{R}_0$, we get $h - g \leq_{\mathbb{R}} h - f$. Then $K = (\overline{1} + g - g) * ((\overline{1}+h - g) * (\overline{1} + h - f)) = (\overline{1}+h - g) * (\overline{1} + h - f) = \overline{1}$.
Case (vi): Suppose $h \leq_{\mathbb{R}} g \leq_{\mathbb{R}} f$. Since $g \leq_{\mathbb{R}} f$, we get $h - f \leq_{\mathbb{R}} h - g$. Hence we get the following:

$$\begin{aligned}
K &= (\overline{1} + g - f) * ((\overline{1} + h - g) * (\overline{1} + h - f)) \\
&= (\overline{1} + g - f) * (\overline{1} + (\overline{1} + h - f) - \max\{\overline{1} + h - f, \overline{1} + h - g\}) \\
&= (\overline{1} + g - f) * (2 + h - f - (\overline{1} + h - g) \\
&= (\overline{1} + g - f) * (\overline{1} + g - f) \\
&= \overline{1}.
\end{aligned}$$

Therefore $(f * g) * ((g * h) * (f * h)) = \overline{1}$. □

Proposition 3.8.5 *Let X be a non-empty set. For any $f, g, h \in \mathbb{R}_0^X$, we have:*

(1) $f * ((f * g) * g) = \overline{1}$,
(2) $f * g = \overline{1}, g * f = \overline{1} \Rightarrow f = g$,
(3) $(g * h) * ((f * g) * (f * h)) = \overline{1}$.

Proof (1) Let $f, g \in \mathbb{R}_0^X$. Suppose $f \leq_{\mathbb{R}} g$. Then $f * g = \overline{1}$. Hence $f * ((f * g) * g) = f * (\overline{1} * g) = f * g = \overline{1}$. Suppose $g \leq_{\mathbb{R}} f$. Then by (5), we get $(f * g) * g = f$. Hence $f * ((f * g) * g) = f * f = \overline{1}$.
(2) Let $f, g \in \mathbb{R}_0^X$ be such that $f * g = \overline{1}$ and $g * f = \overline{1}$. Then

$$\begin{aligned}
1 + f(x) - \max\{f(x), g(x)\} &= (g * f)(x) \\
&= \overline{1} \\
&= (f * g)(x) \\
&= 1 + g(x) - \max\{f(x), g(x)\}
\end{aligned}$$

Therefore $f(x) = g(x)$ for all $x \in X$. Therefore $f = g$.
(3) Let $f, g, h \in \mathbb{R}_0^X$. To prove (3), we set

$$K = (\bar{1} + h - \max\{g, h\}) * ((\bar{1} + g - \max\{f, g\}) * (\bar{1} + h - \max\{f, h\})).$$

We now deal with all the 6 possible cases: $(i)\ f \leq_{\mathbb{R}} g \leq_{\mathbb{R}} h$; $(ii)\ f \leq_{\mathbb{R}} h \leq_{\mathbb{R}} g$; $(iii)\ g \leq_{\mathbb{R}} f \leq_{\mathbb{R}} h$; $(iv)\ g \leq_{\mathbb{R}} h \leq_{\mathbb{R}} f$; $(v)\ h \leq_{\mathbb{R}} f \leq_{\mathbb{R}} g$; $(vi)\ h \leq_{\mathbb{R}} g \leq_{\mathbb{R}} f$.
Case (i): Suppose $f \leq_{\mathbb{R}} g \leq_{\mathbb{R}} h$. Then clearly $K = (\bar{1} + h - h) * ((\bar{1} + g - g) * (\bar{1} + h - h) = \bar{1} * (\bar{1} * \bar{1}) = \bar{1}$.
Case (ii): Suppose $f \leq_{\mathbb{R}} h \leq_{\mathbb{R}} g$. Then $K = (\bar{1}+h-g)*((\bar{1}+g-g)*(\bar{1}+h-h)) = \bar{1}$.
Case (iii): Suppose $g \leq_{\mathbb{R}} f \leq_{\mathbb{R}} h$. Then $K = (\bar{1} + h - h) * ((\bar{1} + g - f) * (\bar{1} + h - h)) = \bar{1}$.
Case (iv): Suppose $g \leq_{\mathbb{R}} h \leq_{\mathbb{R}} f$. Then $K = (\bar{1}+h-h)*((\bar{1}+g-f)*(\bar{1}+h-f)) = (\bar{1}+g-f)*(\bar{1}+h-f)$. Since $f(x), g(x)$ and $h(x) \in \mathbb{R}_0$, we get $g - f \leq_{\mathbb{R}} h - f$ and hence $(\bar{1}+g-f) \leq_{\mathbb{R}} (\bar{1}+h-f)$. Hence $K = (\bar{1}+g-f)*(\bar{1}+h-f) = \bar{1}$.
Case (v): Suppose $h \leq_{\mathbb{R}} f \leq_{\mathbb{R}} g$. Since $f \leq_{\mathbb{R}} g$, we get $h - g \leq_{\mathbb{R}} h - f$. Hence $K = (\bar{1} + h - g) * ((\bar{1} + g - g) * (\bar{1} + h - f)) = (\bar{1} + h - g) * (\bar{1} + h - f) = \bar{1}$.
Case (vi): Suppose $h \leq_{\mathbb{R}} g \leq_{\mathbb{R}} f$. Since $g \leq_{\mathbb{R}} f$, we get $h - f \leq_{\mathbb{R}} g - f$. Hence we get the following:

$$\begin{aligned} K &= (\bar{1} + h - g) * ((\bar{1} + g - f) * (\bar{1} + h - f)) \\ &= (\bar{1} + h - g) * (\bar{1} + (\bar{1} + h - f) - \max\{\bar{1} + g - f, \bar{1} + h - f\}) \\ &= (\bar{1} + h - g) * (2 + h - f - (\bar{1} + g - f) \\ &= (\bar{1} + h - g) * (\bar{1} + h - g) \\ &= \bar{1}. \end{aligned}$$

Therefore $(g * h) * ((f * g) * (f * h)) = \bar{1}$. It concludes the proof. □

The following result is an easy consequence of the above two results.

Corollary 3.8.6 *Let X be a non-empty set. For any $f, g, h \in \mathbb{R}_0^X$, we have:*

(1) *$f \leq_{\mathbb{R}} g$ implies $g * h \leq_{\mathbb{R}} f * h$,*
(2) *$f \leq_{\mathbb{R}} g$ implies $h * f \leq_{\mathbb{R}} h * g$,*
(3) *$f \leq_{\mathbb{R}} g$ and $g \leq_{\mathbb{R}} h$ imply $f \leq_{\mathbb{R}} h$.*

Proof (1) Let $f, g \in \mathbb{R}_0^X$ be such that $f \leq_{\mathbb{R}} g$. Then by Proposition 3.8.4(5), we get $\bar{1} = f * g \leq_{\mathbb{R}} ((g * h) * (f * h))$. Hence $(g * h) * (f * h) = \bar{1}$. Therefore $(g * h) \leq_{\mathbb{R}} (f * h)$.
(2) Let $f, g \in \mathbb{R}_0^X$ be such that $f \leq_{\mathbb{R}} g$. Then by Proposition 3.8.4(5), we get $\bar{1} = f * g \leq_{\mathbb{R}} ((g * h) * (f * h))$. Hence $(g * h) * (f * h) = \bar{1}$. Therefore $(g * h) \leq_{\mathbb{R}} (f * h)$.
(3) It is direct from Proposition 3.8.5(3). □

Theorem 3.8.7 *For any non-empty set X, the algebra $(\mathbb{R}_0^X, *, \bar{1})$ is a BE-algebra.*

Proof The properties (BE1) (BE2) and (BE3) are clear from (1), (2), (3) of Proposition 3.8.4. We now prove (BE4). Let $f, g, h \in \mathbb{R}_0^X$. From Proposition 3.8.5(1), it is clear that $g \leq ((g * h) * h)$. Hence $((g * h) * h) * (f * h) \leq g * (f * h)$ by Corollary

3.8.6(1). Again by Proposition 3.8.4 (5), we get $f * (g * h) \leq ((g * h) * h) * (f * h)$. Then by the transitivity of $\leq$, it implies $f * (g * h) \leq g * (f * h)$. For ($\geq$), we simply interchange the role of f and g in ($\leq$) to get $g * (f * h) \leq f * (g * h)$. Hence by Proposition 3.8.5(2), we get $f * (g * h) = g * (f * h)$. Therefore $(\mathbb{R}_0^X, *, \overline{1})$ is a BE-algebra. □

Corollary 3.8.8 $(\mathbb{R}_0^X, *, \overline{1})$ *is a transitive BE-algebra.*

Proof It is clear by Proposition 3.8.5(3). □

Theorem 3.8.9 *Let $\leq$ be the usual BE-order and $\leq_{\mathbb{R}}$ be the ordinary order on real numbers, then $f \leq g$ if and only if $f \leq_{\mathbb{R}} g$ for all $f, g \in \mathbb{R}_0^X$.*

Proof Let $f, g \in \mathbb{R}_0^X$. Then $f \leq g$ if and only if $f * g = \overline{1}$ if and only if $\overline{1} + g - \max\{f, g\} = \overline{1}$ if and only if $g = \max\{f, g\}$ if and only if $f \leq_{\mathbb{R}} g$. □

Theorem 3.8.10 *Let $(\mathbb{R}_0^X, *, \overline{1})$ be a BE-algebra and $f, g \in \mathbb{R}_0^X$. Then*

$$f \vee g = (g * f) * f = \max\{f, g\}.$$

Proof Let $f, g \in \mathbb{R}_0^X$. To prove the result, we set $K = 1 + f - \max\{f, g * f\} = 1 + f - \max\{f, 1 + f - \max\{f, g\}\}$. We consider the two cases: (1). $f = \max\{f, g\}$ and (2). $g = \max\{f, g\}$. In the case (1), we have $K = \overline{1} + f - \max\{f, \overline{1} + f - f\} = \overline{1} + f - \overline{1} = f$. If case (2) holds, then we have $K = \overline{1} + f - \max\{f, \overline{1} + f - g\} = \overline{1} + f - (\overline{1} + f - g) = g$ and so $K = \max\{f, g\}$. □

Theorem 3.8.11 $(\mathbb{R}_0^X, *, \overline{1})$ *is a commutative BE-algebra.*

Proof Let $f, g \in \mathbb{R}_0^X$. Then by the above theorem, we get $(f * g) * g = \max\{f, g\} = (g * f) * f$. Therefore $(\mathbb{R}_0^X, *, \overline{1})$ is commutative. □

Corollary 3.8.12 *The BE-algebra $\mathbb{R}_0^X$ is an upper semilattice with respect to BE-ordering $\leq$.*

Exercise

1. For any congruence θ on a BE-algebra $(X, *, 1)$, prove that there exists a one-to-one correspondence between the set of all subalgebras of X and the set of all subalgebras of the quotient algebra $X_{/\theta}$.
2. For any partially ordered BE-algebra X, prove that the self-map $\mathcal{C} : \wp(X) \to \wp(X)$ defined by $\mathcal{C}(P) = \{y \in X \mid y \leq x$ for some $x \in P\}$ is a closure operator on $\wp(X)$.
3. Give an example of a self-map on a BE-algebra X which is closure but not BE-closure.
4. Let Φ_1, Φ_2 be two closure operators on a BE-algebra X and $x \in X$. Then prove that $\Phi_1 \leq \Phi_2$ if and only if $\Phi_2(x) = x$ implies $\Phi_1(x) = x$ for all $x \in X$.
5. Let X be a BE-algebra and C a closure system of X. Then prove that the assignment $x \mapsto \mathcal{C}(x) =$ least element of $C(x]$ is a closure on X.

6. Let T be a t-norm and μ a fuzzy set in a BE-algebra X. If the level subset μ_α of μ is a subalgebra of X for $\alpha \in [0, 1]$, then prove that μ is a fuzzy subalgebra of X w.r.t. T.
7. Suppose T is a t-norm satisfying $T(\alpha, \alpha) = \alpha$ for all $\alpha \in [0, 1]$. If a fuzzy set μ in a BE-algebra X is a fuzzy subalgebra of X with respect to T, then prove that the non-empty level subset μ_α is a subalgebra of X for every $\alpha \in [0, 1]$.
8. For any two fuzzy dot subalgebras μ and ν of a BE-algebra X, define their intersection as $(\mu \cap \nu)(x) = \mu(x) \cdot \nu(x)$ for all $x \in X$. Prove that $\mu \cap \nu$ is also a fuzzy dot subalgebra of X.
9. Let ν be a fuzzy set in a BE-algebra X. Then the strongest fuzzy relation μ_ν is a fuzzy relation on X defined by $\mu_\nu(x, y) = \min\{\nu(x), \nu(y)\}$ for all $x, y \in X$. If ν is a fuzzy subalgebra in X, then prove that μ_ν is a fuzzy subalgebra of $X \times X$.
10. Prove that a soft set $(\widetilde{f}, X)$ is a union soft subalgebra of X if and only if $e_X(\widetilde{f}; \tau)$ is a subalgebra of X whenever $e_X(\widetilde{f}; \tau) \neq \emptyset$ for all $\tau \in \mathcal{P}(U)$.
11. Prove that a soft set $(\widetilde{f}, X)$ is an intersection soft subalgebra of X if and only if $e_X(\widetilde{f}; \tau)$ is a subalgebra of X whenever $e_X(\widetilde{f}; \tau) \neq \emptyset$ for all $\tau \in \mathcal{P}(U)$.
12. Let $\{A_i = (A_i, *, 1_i) \mid i \in I_n\}$ be a finite family of BE-algebras. Then prove that each A_i is implicative if and only if $\prod_{i=1}^{n} A_i$ is implicative.

Chapter 4
Filters of BE-algebras

In a BE-algebra, filters are important substructures which play an important role in the characterization of many special classes of BE-algebras. Also, as it is well known that filters are exactly the kernels of congruences, many authors tried to define various filters of BE-algebras in order to construct quotient BE-algebras and investigate some of their properties. In [155], Kim et al. investigated several properties of upper sets in BE-algebras. Using the notion of upper sets, many authors derived equivalent conditions describing filters of BE-algebras. In [5], S.S. Ahn and K.S. So introduced the notion of ideals in BE-algebras and then proved several characterizations of such ideals. They introduced more extended upper sets of BE-algebras and obtained some relations with filters of BE-algebras. Concepts of ideals and filters are identical in BCK-algebras, and one can find many related properties of filters of BCC-algebras in [83, 133], BCI-algebras in [79], lattice implication algebras in [160, 257], BCK-algebras in [81], BL-algebras in [162], and BZ-algebras in [87]. I. Chajda et al. [41] considered properties of deductive systems in Hilbert algebras which are upper semilattices as posets. They showed that every maximal deductive system is prime and gave a condition for a deductive system to be prime. In [173], B.L. Meng deeply studied the filter theory of BE-algebras. He gave a procedure by which he generated a filter by a subset of a transitive BE-algebra. He presented the construction of a quotient algebra of a transitive BE-algebra via a filter.

The goal of this chapter is to consider some types of filters of BE-algebras and show the relationship among them. The notion of weak filters is introduced, and the properties of weak filters are studied. For a given quasi-ordering, a method of construction is proposed to convert the respective quasi-ordered set into BE-algebras. Some isomorphism theorems involving filters are proposed and proved in transitive BE-algebras. The concept of θ-filters of BE-algebras is introduced and the primeness of θ-filters is investigated. The theory of soft sets is applied to the filters of BE-algebras. Properties of irreducible filters of BE-algebras are investigated. An equivalent condition is derived for a filter to become irreducible.

S. R. Mukkamala, *A Course in BE-algebras*,
https://doi.org/10.1007/978-981-10-6838-6_4

Some studies are made on segments to study the relation between final segments and filters. The notions of associative filters and absorbent filters are introduced in BE-algebras. Characterization theorems of associative filters and absorbent filters of BE-algebras are derived. The notions of dual atomistic BE-algebras and dual atomistic filters are introduced. Dual atomistic BE-algebras are characterized in terms of dual atoms of BE-algebras. Properties of prime filters of BE-algebras are studied. The notion of relative dual annihilators is introduced and studied the relation between prime filters and dual annihilator filter.

4.1 Definition and Properties

In this section, properties of filters are studied with respect to Cartesian products and homomorphisms. Some characterization theorems are derived for the elements of filters. A sufficient condition is derived for a subalgebra of a BE-algebra to become a filter.

Definition 4.1.1 A non-empty subset F of a BE-algebra X is called a *filter* if it satisfies the following conditions for all $x, y \in X$:

(1) $1 \in F$,
(2) $x \in F$ and $x * y \in F$ imply that $y \in F$.

It is obvious that $\{1\}$ is a filter of a BE-algebra X. Suppose that $\mathcal{F}(X)$ denote the set of all filters of a BE-algebra X. Then it can be easily observed that $\bigcap_{F \in \mathcal{F}(X)} F$ is also a filter of X.

Example 4.1.2 Let $X = \{1, a, b, c, d\}$. Define a binary operation $*$ on X as follows:

$*$	1	a	b	c	d
1	1	a	b	c	d
a	1	1	b	c	b
b	1	a	1	b	a
c	1	a	1	1	a
d	1	1	1	b	1

Obviously $(X, *, 1)$ is a BE-algebra. It is easy to check that $F = \{b, c, 1\}$ is a filter of X.

Denote by $\uparrow a$ the set $\uparrow a = \{x \in X \mid a * x = 1\}$. Note that, for $a \in X$, the set $\uparrow a$ may not be a filter in general. However, we prove that it is a filter in a self-distributive BE-algebra.

Proposition 4.1.3 *Let X be self-distributive. For any $a \in X$, the set $\uparrow a$ is a filter of X.*

Proof Clearly $1 \in \uparrow a$. Let $x, x * y \in \uparrow a$. Then $a * x = 1$ and $a * (x * y) = 1$. Now $a * y = 1 * (a * y) = (a * x) * (a * y) = a * (x * y) = 1$. Hence $y \in \uparrow a$. Therefore $\uparrow a$ is a filter in X. □

In the following, we observe some properties of filters of BE-algebras.

Theorem 4.1.4 ([153]) *Every filter of a BE-algebra satisfies the following conditions:*

(1) $x \in F$ *and* $x \leq y$ *imply that* $y \in F$ *for all* $x, y \in X$,
(2) $a \in F$ *implies* $(a * x) * x \in F$ *for all* $x \in X$,
(3) $x, y \in F$, $x * (y * z) = 1$ *imply that* $z \in F$ *for all* $x, y, z \in X$.

Proof (1) Assume that F is a filter of the BE-algebra X. Let $x, y \in X$ be such that $x \in F$ and $x \leq y$. Then $x * y = 1 \in F$. Since F is a filter and $x \in F$, we get that $y \in F$.
(2) Assume that F is a filter of X. Let $a \in F$. Then $a*((a*x)*x) = (a*x)*(a*x) = 1 \in F$. Since $a \in F$ and F is a filter, we get that $(a * x) * x \in F$.
(3) Let $x, y \in F$. Suppose $x * (y * z) = 1 \in F$. Since $x, y \in F$, we get $z \in F$. □

In the following theorems, some necessary and sufficient conditions are derived for a non-empty subset of a BE-algebra to become a filter.

Theorem 4.1.5 *A non-empty set* F *a BE-algebra* X *is a filter of* X *if and only if it satisfies the following condition for all* $x, y \in X$:

(1) $x \in F$ *and* $x \leq y$ *imply that* $y \in F$,
(2) $a, b \in F$ *implies* $(a * (b * x)) * x \in F$.

Proof Assume that F is a filter of X. Let $x \in F$ and $x \leq y$. Then $x*y = 1 \in F$. Since $x \in F$ and F is a filter, we get $y \in F$. Let $a, b \in F$. Since $a*((a*(b*x))*(b*x)) = (a*(b*x))*(a*(b*x)) = 1$, we get that $a \leq ((a*(b*x))*(b*x))$. Since $a \in F$, it yields that $(a * (b * x)) * (b * x) \in F$. Hence $b * ((a * (b * x)) * x) \in F$. Since $b \in F$, it implies that $(a * (b * x)) * x \in F$.

Conversely, assume that F satisfies the given conditions. For any $x \in F$, we have $x \leq 1$. Hence by condition (1), we get $1 \in F$. Let $x, y \in X$ be such that $x, x * y \in F$. Now

$$\begin{aligned}(x * ((x * y) * y)) * y &= ((x * y) * (x * y)) * y \\ &= 1 * y \\ &= y\end{aligned}$$

By the condition (2), we get $y \in F$. Therefore F is a filter of X. □

Corollary 4.1.6 *A non-empty set* F *a BE-algebra* X *is a filter of* X *if and only if it satisfies the following condition for all* $x, y \in X$:

(1) $1 \in F$,
(2) $a, b \in F$ *implies* $(a * (b * x)) * x \in F$.

The following theorem is a direct consequence of above results.

Theorem 4.1.7 *A non-empty subset F of a BE-algebra X is a filter of X if and only if it satisfies the following conditions for all $x, y \in X$:*

(1) $x \in F$ *implies* $y * x \in F$,
(2) $a, b \in F$ *implies* $(a * (b * x)) * x \in F$.

In a BE-algebra X, it was already observed that a filter is a subalgebra of X but the converse is not true in general. However, in the following theorem, we derive an equivalent condition for a subalgebra of a BE-algebra to become a filter.

Theorem 4.1.8 *Let X be a BE-algebra. A subalgebra S of X is a filter if and only if it satisfies the following condition for all $x, y \in X$:*

$$x \in S \text{ and } x * (x * y) \in S \text{ imply that } y \in S.$$

Proof Let S be a subalgebra of X. Assume that S is a filter of X. Clearly $1 \in S$. Let $x, y \in X$ be such that $x \in S$ and $x*(x*y) \in S$. Since S is a filter, we get that $y \in F$.

Conversely, assume that S satisfies the given condition. Since S is a subalgebra of F, we get $1 \in S$. Let $x, y \in X$ be such that $x \in S$ and $x * y \in S$. Since S is a subalgebra, we get $x * (x * y) \in S$. Then by the given condition, we get $y \in F$. Therefore S is a filter of X. □

Let $X_1, X_2, \ldots, X_n$ be some BE-algebras. Then their Cartesian product $X_1 \times X_2 \times \ldots.. \times X_n$ is also a BE-algebra with respect to the following operation:

$$(a_1, a_2, \ldots, a_n) * (b_1, b_2, \ldots, b_n) = (a_1 * b_1, a_2 * b_2, \ldots, a_n * b_n)$$

where $a_i, b_i \in X_i$ for $1 \leq i \leq n$ and $n \in N$. The greatest element of this Cartesian product is nothing but $(1, 1, \ldots, 1)$. In the following, we observe the property of Cartesian product of filters.

Theorem 4.1.9 *If F_1 and F_2 are two filters of BE-algebras X_1 and X_2, respectively, then $F_1 \times F_2$ is a filter of the product algebra $X_1 \times X_2$. Conversely, every filter F of $X_1 \times X_2$ can be expressed as $F = F_1 \times F_2$ where F_i is a filter of X_i for $i = 1, 2$.*

Proof Let F_1 and F_2 be the filters of X_1 and X_2, respectively. Since $1 \in F_1$ and $1 \in F_2$, we get $(1, 1) \in F_1 \times F_2$. Let $(x_1, x_2) \in F_1 \times F_2$ and $(x_1 * y_1, x_2 * y_2) = (x_1, x_2) * (y_1, y_2) \in F_1 \times F_2$ where $x_1, y_1 \in X_1$ and $x_2, y_2 \in X_2$. Hence $x_1 * y_1 \in F_1$ and $x_2 * y_2 \in F_2$. Since $x_1 \in F_1, x_1 * y_1 \in F_1$ and F_1 is a filter in X_1, we get that $y_1 \in F_1$. Similarly, we get $y_2 \in F_2$. Hence $(y_1, y_2) \in F_1 \times F_2$. Therefore $F_1 \times F_2$ is a filter of $X_1 \times X_2$.

Conversely, let F be any filter of $X_1 \times X_2$. Consider the projections $\Pi_i : X_1 \times X_2 \longrightarrow X_i$ for $i = 1, 2$. Let F_1 and F_2 be the projections of F on X_1 and X_2, respectively.

That is $\Pi_i(F) = F_i$ for $i = 1, 2$. We now prove that F_1 and F_2 are filters of X_1 and X_2, respectively. Since $(1, 1) \in F$, we get $1 = \Pi_1(1, 1) \in F_1$. Let $x_1 \in F_1$ and $x_1 * y_1 \in F_1$. Then $(x_1, 1) \in F$ and $(x_1, 1) * (y_1, 1) = (x_1 * y_1, 1 * 1) \in F$. Since F is a filter, we get $(y_1, 1) \in F$. Thus $y_1 = \Pi_1(y_1, 1) \in \Pi_1(F) = F_1$. Therefore F_1 is a filter of X_1. Similarly, we can get F_2 is a filter of X_2.

We now prove that $F = F_1 \times F_2$. For any $a \in F$, we can express that $a = (\Pi_1(a), \Pi_2(a))$ and hence $F \subseteq F_1 \times F_2$. Let $(x, y) \in F_1 \times F_2$. Then $x \in F_1 = \Pi_1(F)$ and $y \in F_2 = \Pi_2(F)$. Hence $(x, 1) \in F$ and $(1, y) \in F$. Now, we get the following consequence:

$$\begin{aligned}(x, 1) * (x, y) &= (x * x, 1 * y)\\ &= (1, y) \in F\end{aligned}$$

Since $(x, 1) \in F$ and F is a filter, we get $(x, y) \in F$. Thus $F_1 \times F_2 \subseteq F$. Therefore $F = F_1 \times F_2$. □

Corollary 4.1.10 *Let $X = X_1 \times X_2 \times \cdots \times X_n$ be the Cartesian product of the BE-algebra $X_1, X_2, \ldots, X_n$ and F is a subset of X. Then F is a filter of X if and only if F is of the form $F_1 \times F_2 \times \cdots \times F_n$ where F_i is a filter of X_i for $1 \leq i \leq n$ and $n \in N$.*

Let X_1 and X_2 be two BE-algebras. A mapping $f : X_1 \to X_2$ is called a *BE-homomorphism* if it satisfies the following condition:

$$f(x * y) = f(x) * f(y)$$

for all $x, y \in X_1$. If f is a homomorphism, then $f(1) = f(1 * 1) = f(1) * f(1) = 1$. A one-to-one homomorphism is called a injective, and an onto homomorphism is called surjective.

Theorem 4.1.11 *Let $(X_1, *, 1)$ and $(X_2, *, 1)$ be two BE-algebras and $f : X_1 \to X_2$ a BE-homomorphism. Then $f^{-1}(F)$ (if exists) is a filter in X_1 whenever F is a filter of X_2.*

Proof Since $f(1) = 1 \in F$, we get that $1 \in f^{-1}(F)$. Let $x \in f^{-1}(F)$ and $x * y \in f^{-1}(F)$. Hence $f(x) \in F$ and $f(x) * f(y) = f(x * y) \in F$. Since F is a filter of X_2, we get that $f(y) \in F$. It concludes that $y \in f^{-1}(F)$. Therefore $f^{-1}(F)$ is a filter of X_1. □

Definition 4.1.12 An element of a partially ordered set X is called a *node* if it is comparable *w.r.t.* the *BE*-ordering $\leq$ with every element of X.

It is clear that 1 is a node in every BE-algebras.

Proposition 4.1.13 *An element a of a BE-algebra X is node if and only if for every $x \in X$ either $a * x = 1$ or $x * a = 1$.*

Let X be a linearly ordered BE-algebras. It is clear that every element a of X is a node of X.

Example 4.1.14 Let $X = \{1, a, b, c\}$. Define a binary operation $*$ on X as follows:

$*$	a	b	c	1
a	1	b	1	1
b	a	1	1	1
c	a	b	1	1
1	a	b	c	1

Obviously $(X, *, 1)$ is a BE-algebra. It is easy to check that the element c is a node in X. Also the element b is not a node, because $a * b = b \neq 1$ and $b * a = a \neq 1$.

If $\mathcal{F}(X)$ is the set of all filters of a BE-algebra, then clearly $\mathcal{F}(X)$ is a partially ordered set with respect to the ordering set-inclusion $\subseteq$.

Theorem 4.1.15 *Let X be a self-distributive BE-algebra and $F \in \mathcal{F}(X)$. Then the following are equivalent:*

(1) *F is a node of $\mathcal{F}(X)$,*
(2) *for any $x \in F$ and $y \notin F$ imply $y \nleq x$,*
(3) *for any $G \in \mathcal{F}(X)$, either $F \subseteq G$ or $G \subseteq F$.*

Proof (1) $\Rightarrow$ (2): Assume that F is a node of $\mathcal{F}(X)$. Let $x \in F$ and $y \notin F$. Then $\uparrow y \nsubseteq F$. Since $\uparrow y$ is a filter, we get $F \subset\uparrow y$. Hence $\uparrow x \subset F \subset\uparrow y$. Therefore $y \lneqq x$.
(2) $\Rightarrow$ (3): Assume the condition (2). Let $G \in \mathcal{F}(X)$ be such that $F \nsubseteq G$ and $G \nsubseteq F$. Then there exists $x \in F - G$ and $y \in G - F$. Hence $y \nleq x$, which is a contradiction.
(3) $\Rightarrow$ (1): It is clear from the definition of node. □

In the following definition, the notion of n-fold weak filters is introduced in BE-algebras.

Definition 4.1.16 Let $(X, *, 1)$ be a BE-algebra. A subset F of X is said to be an *n-fold weak filter* of X if it satisfies the following conditions:

(WF1) $1 \in F$,
(WF2) $x \in F$ and $x * y \in F$ imply that $x^n * y \in F$.

where n is a positive integer with $n \geq 2$. Here $x^n * y = x * (x * (\cdots (x * y) \cdots))$ and the operation $*$ is applied n times. For $n = 2$, the base 2-fold weak filter is simply called a weak filter.

It is obvious that the sets $\{1\}$ and X of a BE-algebra X are n-fold weak filters of X. Moreover, the intersection of an indexed family of weak filters of a BE-algebra is again a weak filter.

Proposition 4.1.17 *Every subalgebra of a BE-algebra is an n-fold weak filter.*

Proof Let S be a subalgebra of a BE-algebra X. Let $x, x * y \in S$. Since S is a subalgebra, we get $x^2 * y = x * (x * y) \in S$. Since $x \in S$ and $x^2 * y \in S$, we get $x^3 * y \in S$. Continuing in this way, we get $x^n * y \in S$ for any $n \in \mathbb{Z}^+$. Therefore S is an n-fold weak filter of X. □

Since every filter of a BE-algebra is a subalgebra, the following corollary is direct.

Corollary 4.1.18 *Every filter of a BE-algebra is an n-fold weak filter.*

Example 4.1.19 Let $X = \{1, a, b, c, d\}$. Define a binary operation $*$ on X as follows:

$*$	1	a	b	c	d
1	1	a	b	c	d
a	1	1	a	c	c
b	1	1	1	c	c
c	1	a	b	1	a
d	1	1	a	1	1

It can be routinely verified that $(X, *, 1)$ is a BE-algebra. Consider the set $F = \{1, a, c\}$. It can be easily seen that F is a weak filter but not a filter, because $a \in F$ and $a * d = c \in F$ but $d \notin F$.

Though every n-fold weak filter of a BE-algebra is not a filter, in the following theorem, a necessary and sufficient condition is derived for every n-fold weak filter to become a filter.

Theorem 4.1.20 *An n-fold weak filter of a BE-algebra is a filter if and only if $x \in F$ and $x^n * y \in F$ imply that $y \in F$ for all $x, y \in F$.*

Proof Let F be an n-fold weak filter of a BE-algebra X. Assume that F is a filter of X. Let $x, y \in X$ be such that $x \in F$ and $x^n * y \in F$. Since F is a filter and $x \in F$, we get $x^{n-1} * y \in F$. Again, since $x \in F$, we get $x^{n-2} * y \in F$. Continuing in this way, finally we get $y \in F$.

Conversely assume that F satisfies the condition. Let $x, y \in X$. Suppose $x \in F$ and $x * y \in F$. Since F is an n-fold weak filter of X, we get that $x^n * y \in F$. Since $x \in F$ and $x^n * y \in F$, by the assumed condition, we get $y \in F$. Therefore F is a filter of X. □

Theorem 4.1.21 *A non-empty subset F of a BE-algebra X is an n-fold weak filter of X if and only if $x \in F, x^n * y \in X - F$ imply that $x * y \notin F$ for all $x, y \in X$.*

Proof Let F be a non-empty subset of X. Assume that F is an n-fold weak filter of X. Let $x \in F$ and $x^n * y \in X - F$. Suppose $x * y \in F$. Since F is an n-fold weak filter and $x \in F$, we get that $x^n * y \in F$. Thus we obtained a contradiction to that $x^n * y \notin F$. Therefore $x * y \notin F$, which proves the necessary condition.

Conversely, assume that for all $x, y \in X$, $x \in F, x^n * y \in X - F$ imply $x * y \notin F$. Let $x, x * y \in F$. Suppose $x^n * y \notin F$. Then by the assumption, we get $x * y \notin F$. It results in a contradiction to that $x * y \in F$. Hence $x^n * y \in F$. Therefore F is an n-fold weak filter of X. □

In the following propositions, we observe some induced n-fold weak subalgebras of BE-algebras.

Proposition 4.1.22 *Let X and Y be two BE-algebras. If F and G are n-fold weak filters of X and Y, respectively, then $F \times G$ is an n-fold weak filter of the product algebra $X \times Y$.*

Proof Let F and G be n-fold weak filters of X and Y, respectively. Let $x, z \in X$ and $y, w \in Y$ be such that $(x, y) \in F \times G$ and $(x * z, y * w) = (x, y) * (z, w) \in F \times G$. Then $x*z \in F$ and $y*w \in G$. Since F and G are n-fold weak filters, we get $x^n*z \in F$ and $y^n * w \in G$. Hence $(x, y)^n * (z, w) = (x^n * z, y^n * w) \in F \times G$. Therefore $F \times G$ is an n-fold weak filter of $X \times Y$. □

Proposition 4.1.23 *Let X, Y be two BE-algebras and $f : X \to Y$ is a homomorphisms. If F is an n-fold weak filter of Y, then $f^{-1}(F)$ (if f^{-1} exists) is an n-fold weak filter of X.*

Proof Suppose F is an n-fold weak filter of Y. Let $x, y \in X$. Suppose $x, x * y \in f^{-1}(F)$. Then $f(x) \in F$ and $f(x) * f(y) = f(x * y) \in F$. Since F is an n-fold weak filter of Y, we get $f(x)^n * f(y) \in F$. Since f is a homomorphism, we get $f(x^n * y) = f(x^n) * f(y) = f(x)^n * f(y) \in F$. Hence $x^n * y \in f^{-1}(F)$. Therefore $f^{-1}(F)$ is an n-fold weak filter of X. □

4.2 Quasi-ordering and Filters

In this section, a special method is presented to make a BE-algebra from a quasi-ordered set. The connections between filters and congruences are studied. A characterization is derived for the filters of quotient algebras corresponding to congruences.

Definition 4.2.1 Let X be a BE-algebra. A relation R on X is called a *quasi-ordering* if R is both reflexive and transitive on X. Then the ordered pair (X, R) is called a *quasi-ordered set*.

Proposition 4.2.2 *Let X be a transitive BE-algebra. For any $x, y \in X$, define a relation R_X on X by $(x, y) \in R_X$ if and only if $x * y = 1$. Then R_X is a quasi-ordering on X. Moreover*

(1) $(x, 1) \in R_X$,
(2) *If $x \in X$ is such that $(1, x) \in R_X$, then $x = 1$.*

Proof Let $x \in X$. Since $x * x = 1$, we get $(x, x) \in R_X$. Hence R_X is reflexive. Let $x, y, z \in X$ be such that $(x, y) \in R_X$ and $(y, z) \in R_X$. Then $x * y = 1$ and $y * z = 1$. Since X is transitive, we get $1 = y * z \leq (x * y) * (x * z) = 1 * (x * z) = x * z$. Hence $(x, z) \in R_X$. Thus R_X is transitive. Therefore R_X is a quasi-ordering on X. We now prove the remaining two identities. Since $x * 1 = 1$, we get that $(x, 1) \in R_X$. Again, let $x \in X$ be such that $(1, x) \in R_X$. Then $x = 1 * x = 1$. □

Lemma 4.2.3 *Let R_X be the induced quasi-ordering of a commutative BE-algebra X and $x, y \in X$. If $(x, y) \in R_X$, then $(z * x, z * y) \in R_X$ and $(y * z, x * z) \in R_X$ for all $z \in X$.*

Proof Let $x, y \in X$ be such that $(x, y) \in R_X$. Then we get $x * y = 1$. Since X is transitive, we get $1 = x * y \leq (z * x) * (z * y)$. Hence $(z * x) * (z * y) = 1$. Therefore $(z * x, z * y) \in R_X$ for all $z \in X$. Since X is commutative, we get $(y * z) * z = (z * y) * y$. Hence

$$\begin{aligned}(y * z) * (x * z) &= x * ((y * z) * z)\\ &= x * ((z * y) * y)\\ &= (z * y) * (x * y)\\ &= (z * y) * 1\\ &= 1.\end{aligned}$$

Therefore $(y * z, x * z) \in R_X$ for all $z \in X$. □

For every quasi-ordering R of X, denote by $\mathfrak{C}_R$ the relation on X given by

$$(x, y) \in \mathfrak{C}_R \text{ if and only if } (x, y) \in R \text{ and } (y, x) \in R$$

Obviously, $\mathfrak{C}_R$ is an equivalence relation on X, which is called an equivalence relation induced by R. Denote by $[a]_{\mathfrak{C}_R}$ the equivalence class containing a and by $X/\mathfrak{C}_R$ the set of all equivalence classes of X with respect to $\mathfrak{C}_R$, that is, $[a]_{\mathfrak{C}_R} = \{x \in X \mid (x, a) \in \mathfrak{C}_R\}$ and $X/\mathfrak{C}_R = \{[a]_{\mathfrak{C}_R} \mid a \in X\}$. For any $ab \in X$ define a relation $\leq_R$ on $X/\mathfrak{C}_R$ by

$$[a]_{\mathfrak{C}_R} \leq_R [b]_{\mathfrak{C}_R} \text{ if and only if } (a, b) \in R.$$

Then $\leq_R$ is a partial order on $X/\mathfrak{C}_R$, and so $(X/\mathfrak{C}_R, \leq_R)$ is a poset, which is called a poset assigned to the quasi-ordered set (X, R). A relation R on X is said to be compatible if $(u * x, v * y) \in R$ whenever $(x, y) \in R$ and $(u, v) \in R$ for all $x, y, u, v \in X$. A compatible equivalence relation on X is called a congruence on X. The set $[1]_R = \{x \in X \mid (1, x) \in R\}$ is called the *kernel* of R.

Theorem 4.2.4 *Let R_X be the induced quasi-ordering of a BE-algebra X and $\Theta = \mathfrak{C}_{R_X}$ be the equivalence relation induced by R_X. Then Θ is a congruence relation on X with kernel $[1]_\Theta = \{1\}$.*

Proof Clearly Θ is reflexive and symmetric on X. Let $(x, y) \in \Theta$ and $(y, z) \in \Theta$. Since $(x, y) \in \Theta$, we get $(x, y) \in R_X$ and $(y, x) \in R_X$. Since $(y, z) \in \Theta$, we get $(y, z) \in R_X$ and $(z, y) \in R_X$. Since $(x, y) \in R_X$ and $(y, z) \in R_X$, we get $(x, z) \in R_X$. Similarly, we can obtain $(z, x) \in R_X$. Hence $(x, z) \in \Theta$. Thus Θ is transitive, and so it is an equivalence relation on X.

Let $x, y, u, v \in X$ be such that $(x, y) \in \Theta$ and $(u, v) \in \Theta$. Then $(x, y) \in R_X$, $(y, x) \in R_X$, $(u, v) \in R_X$, and $(v, u) \in R_X$. Using the above lemma, we obtain $(x * u, x * v) \in R_X$ and $(x * v, y * v) \in R_X$. By the transitivity of R_X, we get $(x * u, y * v) \in R_X$. Similarly, we have $(y * v, x * u) \in R_X$. Hence $(x * u, y * v) \in \Theta$; that is, Θ is a congruence relation on X. Now if $x \in [1]_\Theta$, then $(1, x) \in \Theta$ and so $(x, 1) \in R_X$. It follows from Proposition 4.2.2 that $x = 1$. Therefore $[1]_\Theta = \{1\}$. □

Let X be a BE-algebra and $\emptyset \neq K \subseteq X$. Denote by θ_K the relation on X given by

$$(x, y) \in \theta_K \text{ if and only if } x * y \in K \text{ and } y * x \in K \qquad \text{for all } x, y \in X.$$

Lemma 4.2.5 *Let K be a non-empty subset K of a BE-algebra X. If the relation θ_K is reflexive relation on the BE-algebra X, then $[1]_{\theta_K} = K$.*

Proof Let θ_K be reflexive for every non-empty subset K of X. For any $x \in X$, we get $(x, x) \in \theta_K$. Hence $1 = x * x \in K$. Let $a \in K$. Then $a * 1 = 1 \in K$ and $1 * a = a \in K$. Hence $(1, a) \in \theta_K$, that is, $a \in [1]_{\theta_K}$. Conversely, let $a \in [1]_{\theta_K}$. Then we get $(1, a) \in \theta_K$. Therefore $a = 1 * a \in K$. □

In the following, we can observe the method of converting a quasi-ordered set into a BE-algebra.

Theorem 4.2.6 *Let (X, R) be a quasi-ordered set. Suppose $1 \notin X$ and let $X_0 = X \cup \{1\}$. Define a binary operation $*$ on X_0 as follows:*

$$x * y = \begin{cases} 1 & \text{if } (x, y) \in R \\ y & \text{otherwise} \end{cases}$$

Then $(X_0, *, 1)$ is a BE-algebra.

Proof Let $x \in X$. Since R is reflexive, we get $(x, x) \in R$. Hence $x * x = 1$. Suppose $(1, x) \in R$ for some $x \in X$. Then $1 * x = 1$, which is a contradiction. Hence $(1, x) \notin R$ for every $x \in X$. Thus, we have $1 * x = x$ for all $x \in X$. Observe that $x * 1 = 1$ for all $x \in X$.
Case I: Suppose $(y, z) \notin R$ and $(x, z) \notin R$. Then $x * (y * z) = x * z = z = y * z = y * (x * z)$.
Case II: Suppose $(y, z) \in R$ and $(x, z) \notin R$. Then $x * (y * z) = x * 1 = 1 = y * z = y * (x * z)$.
Case III: Suppose $(y, z) \notin R$ and $(x, z) \in R$. Then $x * (y * z) = x * z = 1 = y * 1 = y * (x * z)$.

Case IV: Suppose $(y, z) \in R$ and $(x, z) \in R$. Then $x * (y * z) = x * 1 = 1 = y * 1 = y * (x * z)$. Hence $x * (y * z) = y * (x * z)$ for all $x, y, z \in X$. Therefore $(X_0, *, 1)$ is a BE-algebra. □

Corollary 4.2.7 *$(X_0, *, 1)$ is self-distributive.*

Proof Let $x, y, z \in X_0$. We again consider the following cases.

Case I: Suppose $(x, y) \notin R$ and $(y, z) \notin R$. If $(x, z) \in R$, then $x * (y * z) = x * z = 1 = y * 1 = (x * y) * (x * z)$. If $(x, z) \notin R$, then we get $x * (y * z) = x * z = z = y * z = (x * y) * (x * z)$.
Case II: Suppose $(x, y) \in R$ and $(y, z) \notin R$. If $(x, z) \in R$. Then $x*(y*z) = x*z = 1 = 1*1 = (x*y)*(x*z)$. If $(x, z) \notin R$, then $x*(y*z) = x*z = z = 1*z = (x*y)*(x*z)$.
Case III: Suppose that $(x, y) \notin R$ and $(y, z) \in R$. If $(x, z) \in R$, then $x * (y * z) = x * 1 = 1 = y * 1 = (x * y) * (x * z)$. If $(x, z) \notin R$, then $x * (y * z) = x * 1 = 1 = y * z = (x * y) * (x * z)$.
Case IV: Suppose $(x, y) \in R$ and $(y, z) \in R$. Then by transitivity of R, we get $(x, z) \in R$. Thus, it yields that $x * (y * z) = x * 1 = 1 = 1 * 1 = (x * y) * (x * z)$. Hence $x * (y * z) = (x * y) * (x * z)$ for all $x, y, z \in X$. By summing he above cases, it concludes that $(X_0, *, 1)$ is self-distributive. □

4.3 Homomorphisms and Filters

In this section, some properties of filters are studied in terms of homomorphisms of BE-algebras. Some isomorphism theorems involving filters are proved in BE-algebras. Throughout this section, we consider the congruence θ which is defined in Definition 2.3.19 and all BE-algebras are transitive and partially ordered sets.

Theorem 4.3.1 *Let X and Y be two self-distributive BE-algebras and $f : X \to Y$ a homomorphism. Then for any filter F of X, $F/(F \cap Ker(f)) \cong f(A)$.*

Proof Clearly $F \cap Ker(f)$ is a filter of X. Let $G = F \cap Ker(f)$. Define a mapping $\phi : F/G \to Y$ by $\phi(G_x) = f(x)$ for all $x \in F$. Then for any $G_x, G_y \in F/G$, we get the following consequence:

$$\begin{aligned} G_x = G_y &\Leftrightarrow (x, y) \in \theta_G \\ &\Leftrightarrow x * y \in G \text{ and } y * x \in G \\ &\Leftrightarrow f(x * y) = 1 \text{ and } f(y * x) = 1 \\ &\Leftrightarrow f(x) * f(y) = 1 \text{ and } f(y) * f(x) = 1 \\ &\Leftrightarrow f(x) = f(y) \\ &\Leftrightarrow \phi(G_x) = \phi(G_y) \end{aligned}$$

Hence ϕ is well-defined and one-to-one. Moreover, for all $G_x, G_y \in F/G$, we get

$$\phi(G_x * G_y) = \phi(G_{x*y}) = f(x * y) = f(x) * f(y) = \phi(G_x) * \phi(G_y).$$

Hence ϕ is a homomorphism. Thus we obtain $Img(\phi) = \{\phi(G_x) \mid x \in F\} = \{f(x) \mid x \in F\} = f(A)$. Therefore $F/(F \cap Ker(f)) \cong f(A)$. □

By taking $F = X$, the following is a direct consequence of the above theorem.

Corollary 4.3.2 *Let X and Y be two self-distributive BE-algebras and $f : X \to Y$ an epimorphism. Then $X/Ker(f)$ is isomorphic to Y.*

Theorem 4.3.3 *Let F and G be two filters of a self-distributive BE-algebra X such that $G \subseteq F$. Then X/F is isomorphic to $(X/G)/(F/G)$.*

Proof Let $G \subseteq F$. Then clearly $\theta_G \subseteq \theta_F$. Define a mapping $\psi : X/G \to X/F$ by $\psi(G_x) = F_x$ for all $x \in X$. Let $G_x, G_y \in X/G$ be such that $G_x = G_y$. Then $(x, y) \in \theta_G \subseteq \theta_F$. Hence $F_x = F_y$. Thus $\psi(G_x) = \psi(G_y)$, which gives ϕ is well-defined. Now, let $G_x, G_y \in X/G$ and consider $\psi(G_x * G_y) = \psi(G_{x*y}) = F_{x*y} = F_x * F_y = \psi(G_x) * \psi(G_y)$. Therefore ψ is a homomorphism. We now show that $Ker(\psi) = F/G$. Since F_1 is the unit element in X/F, we get

$$\begin{aligned} Ker\ \psi &= \{G_x \in X/G \mid \psi(G_x) = F_1\} \\ &= \{G_x \in X/G \mid F_x = F_1\} \\ &= \{G_x \in X/G \mid x \in F\} \\ &= F/G. \end{aligned}$$

Then by Theorem 4.3.1, $(X/G)/Ker(\psi)$ is isomorphic to X/F. Hence $(X/G)/(F/G) \cong X/F$. □

Definition 4.3.4 Let X and Y be two BE-algebras. Define $X^\Delta = \{(a, 1) \mid a \in X\}$ and $Y^\Delta = \{(1, b) \mid b \in Y\}$.

Proposition 4.3.5 *Let X and Y be two BE-algebras. Then X^Δ and Y^Δ are filters of $X \times Y$.*

Proof Clearly $(1, 1) \in X^\Delta$. Let $a \in X^\Delta$ and $r \in X \times Y$. Then $a = (x, 1)$ and $r = (y, z)$ where $x, y \in X$ and $z \in Y$. Hence, it infers that $r * a = (y, z) * (x, 1) = (y * x, z * 1) = (y * x, 1) \in X^\Delta$. Let $a, b \in X^\Delta$ and $r \in X \times Y$. Then $a = (x, 1), b = (y, 1)$ and $r = (p, q)$. Hence $(a * (b * r)) * r = ((x, 1) * ((y, 1) * (p, q))) * (p, q) = ((x * (y * p)) * p, (1 * (1 * q)) * q) = ((x * (y * p)) * p, 1) \in X^\Delta$. From Theorem 4.1.7, we get X^Δ is a filter in $X \times Y$. Similarly, Y^Δ is a filter of $X \times Y$. □

Theorem 4.3.6 *Let X and Y be two self-distributive BE-algebras. Then X^Δ is isomorphic to X and Y^Δ is isomorphic to Y.*

Proof Define $f : X \to X^\Delta$ by $f(x) = (x, 1)$ for all $x \in X$. Clearly f is well-defined. Let $x, y \in X$ be such that $f(x) = f(y)$. Hence $(x, 1) = (y, 1)$, which implies that $x = y$. Therefore f is one-to-one. Let $a \in X^\Delta$. Then $a = (x, 1)$ for some

$x \in X$. For this x, we have $f(x) = (x, 1)$. Hence f is onto. Let $x, y \in X$. Then $f(x * y) = (x * y, 1) = (x * y, 1 * 1) = (x, 1) * (y, 1) = f(x) * f(y)$. Therefore f is an isomorphism. Similarly, it will be proved that Y is isomorphic to Y^{Δ}. □

Definition 4.3.7 Let F be a filter of a self-distributive BE-algebra X and θ a congruence on X. Then F is said to satisfy the θ-condition if $F_x * F_y = F_1$ implies $F_x = F_y$.

Theorem 4.3.8 *Let F and G be two filters of a self-distributive BE-algebra X which satisfy the θ-condition. Then the following statements hold:*

(1) $(F \vee G)/G \cong F/(F \cap G)$,

(2) $(F\vee G)/(F\cap G) \cong [(F\vee G)/F]\times[(F\vee G)/G] \cong [G/(F\cap G)]\times[F/(F\cap G)]$.

In particular, if $F \vee G = X$ *(i.e., F and G are co-maximal), then* $X/(F \cap G) \cong (X/F) \times (X/G)$.

Proof (1) Define $g : F \to (F \vee G)/G$ by $g(x) = G_x$ for all $x \in F$. Let $x, y \in F$ be such that $x = y$. Then $g(x) = G_x = G_y = g(y)$. Hence g is well-defined. Let $G_x \in (F \vee G)/G$. Then $x \in F \vee G$. Hence $a * (b * x) = 1$ for some $a \in F$ and $b \in G$. Thus $b * (a * x) = 1 \in G$. Since $b \in G$, we get $a * x \in G$ and thus $G_a * G_x = G_{a*x} = G_1$. Since G satisfies θ-condition, we get $G_a = G_x$. Thus for each $G_x \in (F \vee G)/G$, there exists $a \in F$ such that $g(a) = G_a = G_x$. Therefore g is onto. For any $x, y \in F$, $g(x * y) = G_{x*y} = G_x * G_y = g(x) * g(y)$. Thus g is epimorphism. Now

$$\begin{aligned}
Ker(g) &= \{x \in F \mid g(x) = G_1, G_1 \text{ is the unit element in } (F \vee G)/G\} \\
&= \{x \in F \mid G_x = G_1\} \\
&= \{x \in F \mid (x, 1) \in \theta_G\} \\
&= \{x \in F \mid x \in G\} \\
&= F \cap G
\end{aligned}$$

Therefore by Theorem 4.3.1, we get $(F \vee G)/G \cong F/(F \cap G)$.

(2) Define $h : F \vee G \to [(F \vee G)/F] \times [(F \vee G)/G]$ by $h(x) = (F_x, G_x)$ for all $x \in F \vee G$. Clearly h is well-defined. Let $(F_x, G_1) \in [(F \vee G)/F] \times [(F \vee G)/G]$. Then $x \in F \vee G$. Hence $a * (b * x) = 1$ for some $a \in F$ and $b \in G$. Since $b \in G$, we get $G_b = G_1$. Since $a \in F$, we get $b * x \in F$. Since F satisfies θ-condition, we get $F_b = F_x$. Thus $h(b) = (F_b, G_b) = (F_x, G_1)$. Thus each element of type (F_x, G_1) has a pre-image in $F \vee G$ w.r.t. h. Similarly, elements of the type (F_1, G_y) has a pre-image in $F \vee G$ w.r.t. h. Therefore h is onto. For $x, y \in F \vee G$, we get $h(x * y) = (F_{x*y}, G_{x*y}) = (F_x * F_y, G_x * G_y) = (F_x, G_x) * (F_y, G_y) = h(x) * h(y)$. Hence h is a homomorphism. Now

$$\begin{aligned} Ker(h) &= \{x \in F \vee G \mid h(x) = (F_1, G_1)\} \\ &= \{x \in F \vee G \mid (F_x, G_x) = (F_1, G_1)\} \\ &= \{x \in F \vee G \mid F_x = F_1 \text{ and } G_x = G_1\} \\ &= \{x \in F \vee G \mid x \in F \text{ and } x \in G\} \\ &= \{x \in F \vee G \mid x \in F \cap G\} \\ &= F \cap G. \end{aligned}$$

Therefore by Theorem 4.3.1, we get $(F \vee G)/F \cap G$ is isomorphic to $[(F \vee G)/F] \times [(F \vee G)/G]$. From (1), we have $(F \vee G)/F \cong G/(F \cap G)$ and $(F \vee G)/G \cong F/(F \cap G)$. Thus

$$(F \vee G)/(F \cap G) \cong [(F \vee G)/F] \times [(F \vee G)/G] \cong [G/(F \cap G)] \times [F/(F \cap G)].$$

In particular if $X = F \vee G$, then $X/(F \cap G) \cong (X/F) \times (X/G)$. □

The following theorem is a routine verification.

Theorem 4.3.9 *Let $F_i, i = 1, 2, 3, \ldots, n$ be the filters of the BE-algebra $X_i, i = 1, 2, 3, \ldots, n$, respectively. Then $F = F_1 \times F_2 \times F_3 \times \cdots \times F_n$ is a filter of $X = X_1 \times X_2 \times X_3 \times \cdots \times X_n$.*

Theorem 4.3.10 *Let F and G be two filters of a self-distributive BE-algebra X and 0 is the least element of X with respect to the BE-ordering $\leq$. Then the mapping $f : X \to (X/F) \times (X/G)$ defined by $f(x) = (F_x, G_x)$ for all $x \in X$ is a homomorphism. Moreover, the following hold:*

(1) *f is injective if and only if $F \cap G = \{1\}$,*
(2) *If f is surjective, then $F \vee G = X$.*

Proof Clearly f is well-defined. Let $x, y \in X$. Then $f(x * y) = (F_{x*y}, G_{x*y}) = (F_x * F_y, G_x * G_y) = (F_x, G_x) * (F_y, G_y) = f(x) * f(y)$. Therefore f is a homomorphism. Moreover,

(1) We have already observed that f is injective if and only if $Ker(f) = \{1\}$. Now

$$\begin{aligned} x \in Ker(f) &\Leftrightarrow f(x) = \overline{1}, \text{ the unit element in } (X/F) \times (X/G) \\ &\Leftrightarrow (F_x, G_x) = (F_1, G_1) \\ &\Leftrightarrow F_x = F_1 \text{ and } G_x = G_1 \\ &\Leftrightarrow x \in F \text{ and } x \in G \\ &\Leftrightarrow x \in F \cap G \end{aligned}$$

Thus, it gives that $Ker(f) = F \cap G$. Therefore f is injective if and only if $F \cap G = \{1\}$.
(2) Assume that f is surjective. If X has the least element 0, then $(F_0, G_1) \in (X/F) \times (X/G)$. Since f is surjective, there exists $x \in X$ such that $f(x) = (F_0, G_1)$. Hence

$$\begin{aligned} f(x) = (F_0, G_1) &\Leftrightarrow (F_x, G_x) = (F_0, G_1) \\ &\Leftrightarrow F_x = F_0 \text{ and } G_x = G_1 \\ &\Leftrightarrow x * 0 \in F \text{ and } x \in G \end{aligned}$$

Since $(x * 0) * (x * 0) = 1$; $x * 0 \in F$ and $x \in G$ imply that $0 \in F \vee G$. Hence $F \vee G = X$. □

The following is an extension of the above theorem.

Corollary 4.3.11 *Let $F^i, i = 1, 2, 3, \dots, n$ be the filters of a self-distributive BE-algebra X and 0 is the least element of X with respect to the BE-ordering $\leq$. Then the mapping $f : X \to (X/F^1) \times (X/F^2) \times (X/F^3) \times \cdots \times (X/F^n)$ defined by $f(x) = (F_x^1, F_x^2, F_x^3, \dots, F_x^n)$ for all $x \in X$ is a homomorphism. Moreover, the following hold:*

(1) *f is injective if and only if $\bigcap_{i=1}^{n} F^i = \{1\}$,*
(2) *If f is surjective, then $F^i \vee F^j = X$ for $i \neq j$.*

4.4 Congruences and Filters

In this section, the notion of θ-filters is introduced and these θ-filters are characterized. The concept of θ-prime filters is introduced in BE-algebras. The famous Stone's theorem is generalized to the case of θ-filters and θ-prime filters of BE-algebras. Let us recall that the smallest congruence on X is given by $0_X = \{(a, a) \mid a \in X\}$.

Definition 4.4.1 Let θ be a congruence on a BE-algebra X. For any filter F of X, define the extension of F as F^θ as $F^\theta = \{ x \in X \mid (x, a) \in \theta \text{ for some } a \in F\}$.

Lemma 4.4.2 *For any congruence θ and any filter F of X, the set F^θ is a filter of X.*

Proof Clearly $1 \in F^\theta$. Let $a \in F^\theta$ and $x \in X$. Then $(a, t) \in \theta$ for some $t \in F$. Since F is a filter, we get $x * t \in F$. Since $(a, t) \in \theta$, we get $(x * a, x * t) \in \theta$. Since $x * t \in F$, it yields that $x * a \in F^\theta$. Let $a, b \in F^\theta$. Then $(a, x) \in \theta$ and $(b, y) \in \theta$ for some $x, y \in F$. Now, for any $t \in X$, we get

$$\begin{aligned} (b, y) \in \theta &\Rightarrow (b * t, y * t) \in \theta \\ &\Rightarrow (a * (b * t), x * (y * t)) \in \theta \qquad \text{since } (a, x) \in \theta \\ &\Rightarrow ((a * (b * t)) * t, (x * (y * t)) * t) \in \theta \end{aligned}$$

Since $x, y \in F$ and F is a filter, by Theorem 4.1.7, we get that $(x * (y * t)) * t \in F$. Hence $(a * (b * t)) * t \in F^\theta$. Therefore, by Theorem 4.1.7, we get F^θ is a filter of X. □

Lemma 4.4.3 *Let θ be a congruence on a commutative BE-algebra X. Then for any two filters F, G of X, we have the following:*

(1) $F \subseteq F^\theta$,
(2) $F \subseteq G$ *implies* $F^\theta \subseteq G^\theta$,
(3) $(F \cap G)^\theta = F^\theta \cap G^\theta$,
(4) $(F^\theta)^\theta = F^\theta$.

Proof (1) For any $a \in F$, we get that $(a, a) \in \theta$. Thus $a \in F^\theta$. Therefore $F \subseteq F^\theta$.
(2) Suppose that $F \subseteq G$. Let $x \in F^\theta$. Then we get that $(x, a) \in \theta$ for some $a \in F \subseteq G$. Therefore it can be concluded that $x \in G^\theta$. Hence $F^\theta \subseteq G^\theta$.
(3) Clearly $(F \cap G)^\theta \subseteq F^\theta \cap G^\theta$. Conversely, let $x \in F^\theta \cap G^\theta$. Then we get $(x, a) \in \theta$ and $(x, b) \in \theta$ for some $a \in F$ and $b \in G$. Since $a \in F$, we get $(b * a) * a \in F$. Similarly, we get $(a * b) * b \in G$. Since X is commutative, we get $(b * a) * a = (a * b) * b \in F \cap G$. Now, we get the following:

$$\begin{aligned}(x, a) \in \theta &\Rightarrow (1, a * b) = (x * x, a * b) \in \theta && \text{since } (x, b) \in \theta\\ &\Rightarrow (1 * x, (a * b) * b) \in \theta && \text{since } (x, b) \in \theta\\ &\Rightarrow (x, (a * b) * b) \in \theta\end{aligned}$$

Since $(a * b) * b \in F \cap G$, we get that $x \in (F \cap G)^\theta$. Therefore $(F \cap G)^\theta = F^\theta \cap G^\theta$.
(4) Clearly $F^\theta \subseteq (F^\theta)^\theta$. Conversely, let $x \in (F^\theta)^\theta$. Then $(x, a) \in \theta$ for some $a \in F^\theta$. Hence $(a, b) \in \theta$ for some $b \in F$. Thus $(x, b) \in \theta$ and $b \in F$. Hence $x \in F^\theta$. Thus $(F^\theta)^\theta \subseteq F^\theta$. □

Definition 4.4.4 Let θ be a congruence on a BE-algebra X. A filter F of X is said to be a *θ-filter* of X if $a \in F$ implies $[a]_\theta \subseteq F$ for all $a \in X$.

For any congruence θ on a BE-algebra X, it can be easily observed that the unit filter $\{1\}$ is a θ-filter if and only if $[1]_\theta = \{1\}$. In the following, we now characterize the θ-filters of BE-algebras.

Theorem 4.4.5 *Let θ be a congruence on a BE-algebra X. For any filter F of X, the following conditions are equivalent:*

(1) *F is a θ-filter,*
(2) *For any $x, y \in X$, $(x, y) \in \theta$ and $x \in F$ imply that $y \in F$*
(3) $F = \bigcup_{x \in F} [x]_\theta$.

Proof (1) ⇒ (2): Assume that F is a θ-filter of X. Let $x, y \in X$ be such that $(x, y) \in \theta$. Suppose $x \in F$. Therefore we get that $y \in [x]_\theta \subseteq F$.
(2) ⇒ (3): Assume the condition (2). Let $x \in F$. Since $x \in [x]_\theta$, we get $F \subseteq \bigcup_{x \in F} [x]_\theta$. Conversely, let $a \in \bigcup_{x \in F} [x]_\theta$. Then $(a, x) \in \theta$ for some $x \in F$. By the condition (2), we get that $a \in F$ and hence $\bigcup_{x \in F} [x]_\theta \subseteq F$. Therefore $F = \bigcup_{x \in F} [x]_\theta$.
(3) ⇒ (1): Assume that the condition (3) holds. Let $a \in F$. Then $(x, a) \in \theta$ for some $x \in F$. Let $t \in [a]_\theta$. Thus $t \in [a]_\theta \subseteq \bigcup_{x \in F} [x]_\theta$. Therefore F is a θ-filter. □

In the following theorem, a set of equivalent conditions is established for every filter of a BE-algebra to become a θ-filter.

Theorem 4.4.6 *Let θ be a congruence on a partially ordered BE-algebras X. Then the following are equivalent:*

(1) *θ is the smallest congruence,*
(2) *Every filter is a θ-filter,*
(3) *Every principal filter is a θ-filter.*

Proof (1) $\Rightarrow$ (2): Assume that θ is the smallest congruence on X. Let F be a filter of X and $x \in F$. If $t \in [x]_\theta$, then $(t, x) \in \theta$. Hence $t = x \in F$. Thus $[x]_\theta \subseteq F$. Thus F is a θ-filter of X.
(2) $\Rightarrow$ (3): It is obvious.
(3) $\Rightarrow$ (1): Assume that every principal filter is a θ-filter. Let $x, y \in X$ and $(x, y) \in \theta$. Then $[x]_\theta = [y]_\theta$. Since $\langle y\rangle$ is a θ-filter, we get $x \in [x]_\theta = [y]_\theta \subseteq \langle y\rangle$. Thus $x*y = 1$. Similarly, we get $y * x = 1$. Since X is partially ordered, we get $x = y$. Hence θ is the smallest congruence on X. □

In the following definition, the concept of θ-prime filters is now introduced in a commutative BE-algebra (a BE-algebra X in which $a \vee b = (b*a)*a = (a*b)*b = b \vee a$ for all $a, b \in X$).

Definition 4.4.7 Let θ be a congruence on a commutative BE-algebra X. A proper θ-filter P of X is said to be a θ-prime if for all $a, b \in X$, $a \vee b \in [1]_\theta$ implies either $a \in P$ or $b \in P$.

Proposition 4.4.8 *Every prime θ-filter of commutative BE-algebras X is a θ-prime filter.*

Proof Let P be a prime θ-filter of X. Let $a, b \in X$ be such that $a \vee b \in [1]_\theta$. Since P is θ-filter, we get $a \vee b \in [1]_\theta \subseteq P$. Thus we get $a \in P$ or $b \in P$. Therefore P is a θ-prime filter of X. □

Corollary 4.4.9 *If θ is the smallest congruence on a commutative BE-algebra X, then every prime filter is a θ-prime filter.*

Proof Let P be a prime filter of a commutative BE-algebra X. Then by the above theorem, it yields that P is a θ-filter. Hence by the main proposition, we get P is a θ-prime filter of X. □

Theorem 4.4.10 *Let θ be a congruence on a commutative and self-distributive BE-algebra X and P a θ-filter of X. Then P is a θ-prime if and only if for any filters F, G of X, $F \cap G \subseteq [1]_\theta$ implies that $F \subseteq P$ or $F \subseteq P$.*

Proof Assume that P is a θ-prime filter of X. Let F and G be two filters of X such that $F \cap G \subseteq [1]_\theta$. Let $a \in F$ and $b \in G$. Since F and G are filters of X, we get that $a \vee b = (b * a) * a \in F$ and $b \vee a = (a * b) * b \in G$. Since X is commutative, we get $a \vee b \in F \cap G$. Then $a \vee b \in F \cap G \subseteq [1]_\theta$. Since P is θ-prime, we get either $a \in P$ or $b \in P$. Thus we get either $F \subseteq P$ or $G \subseteq P$.

Conversely, assume the condition. Let $a, b \in X$ be such that $a \vee b \in [1]_\theta$. Since P is a θ-filter and $1 \in P$, we get $a \vee b \in [1]_\theta \subseteq P$. Hence we get $\langle a\rangle \cap \langle b\rangle = \langle a \vee b\rangle \subseteq P$. Thus by condition (2), we get that either $a \in P$ or $b \in P$. Therefore P is a θ-prime filter of X. □

Proposition 4.4.11 *Let θ be a congruence on a commutative BE-algebra X. For any filter F of X, F^θ is the smallest θ-filter of X such that $F \subseteq F^\theta$.*

Proof From Lemmas 4.4.2 and 4.4.3(1), we get that F^θ is a θ-filter of X containing the filter F. Let K be a θ-filter of X such that $F \subseteq K$. Let $x \in F^\theta$. Then we get $(x, a) \in \theta$ for some $a \in F \subseteq K$. Hence $x \in [x]_\theta = [a]_\theta \subseteq K$. Therefore $F^\theta \subseteq K$. □

For any self-distributive BE-algebra X, it is already observed that the set $\mathcal{F}(X)$ of all filter of X forms a complete distributive lattice. For any BE-algebra X, let us denote that $\mathcal{F}_\theta(X)$ is the set of all θ-filters of X. Keeping in view of the operation ^{e} depicted for θ-filters, it can be observed that $\mathcal{F}_\theta(X)$ can be made into a distributive lattice with respect to the operations: $F \wedge G = F \cap G$ and $F \sqcup G = (F \vee G)^\theta$ for any $F, G \in \mathcal{F}_\theta(X)$.

Theorem 4.4.12 *Let θ be a congruence on a commutative and self-distributive BE-algebra X. For any proper θ-filter F of X, $F = \bigcap\{ P \mid P$ is a θ-prime filter, $F \subseteq P \}$.*

Proof Let $F_0 = \bigcap\{ P \mid P$ is a θ-prime filter, $F \subseteq P \}$. Clearly $F \subseteq F_0$. Let $a \notin F$. Consider $\Im = \{ G \mid G$ is a θ-filter, $F \subseteq G$ and $a \notin G \}$. Clearly $F \in \Im$. Let $\{G_\alpha | \alpha \in \Delta\}$ be chain of θ-filters in $\Im$. Clearly $\bigcup\limits_{\alpha\in\Delta} G_\alpha$ is a θ-filter such that $F \subseteq \bigcup\limits_{\alpha\in\Delta} G_\alpha$ and $a \notin \bigcup\limits_{\alpha\in\Delta} G_\alpha$. Let M be a maximal element of $\Im$. Suppose that $x \notin M$ and $y \notin M$. Then $M \subset M \vee \langle x\rangle \subseteq \{M \vee \langle x\rangle\}^\theta$. Similarly $M \subset \{M \vee \langle y\rangle\}^\theta$. By the maximality of M, we get $a \in \{M \vee \langle y\rangle\}^\theta$ and $a \in \{M \vee \langle x\rangle\}^\theta$. Hence $a \in \{M \vee \langle x\rangle\}^\theta \bigcap \{M \vee \langle y\rangle\}^\theta = \{M \vee \langle x \vee y\rangle\}^\theta$. If $x \vee y \in [1]_\theta$, then $x \vee y \in [1]_\theta \subseteq M$. Hence $a \in M$, which is a contradiction. Hence M is θ-prime. Therefore for any $a \notin F$, there exists a θ-prime filter M such that $F \subseteq M$ and $a \notin M$. Thus $a \notin F_0$. Therefore $F_0 \subseteq F$. □

Corollary 4.4.13 $[1]_\theta = \bigcap\{ P \mid P$ *is a θ-prime filter* $\}$

Corollary 4.4.14 *If θ is the smallest congruence on X, then we have the following:*

$$\{1\} = \bigcap\{ P \mid P \text{ is a } \theta\text{-prime filter} \}$$

Corollary 4.4.15 *Let θ be a congruence on a self-distributive BE-algebra X. If $a \notin [1]_\theta$, then there exists a θ-prime filter P of X such that $a \notin P$.*

In the following theorem, we generalize the celebrated Stone's theorem for ideals to the case of θ-filters of a commutative and self-distributive BE-algebra.

Theorem 4.4.16 *Let θ be a congruence relation on a commutative and self-distributive BE-algebra X. Suppose F is a θ-filter and A a $\vee$-closed subset($a, b \in A$ implies $a \vee b \in A$) of X such that $F \cap A = \emptyset$. Then there exists a θ-prime filter P such that $F \subseteq P$ and $P \cap A = \emptyset$.*

Proof Let F be a θ-filter and A a $\vee$-closed subset of X such that $F \cap A = \emptyset$. Consider $\Gamma = \{ G \mid G$ is a θ-filter, $F \subseteq G$ and $G \cap A = \emptyset \}$. Clearly $F \in \Gamma$. Let $\{G_i | i \in \Delta\}$ be a chain of θ-filters in Γ. Clearly $\cup_{i\in\Delta} G_i$ is a θ-filter such that $F \subseteq \cup_{i\in\Delta} G_i$ and $(\cup_{i\in\Delta} G_i) \cap A = \emptyset$. Let M be a maximal element of Γ. Suppose that $x \notin M$ and $y \notin M$. Then $M \subset M \vee \langle x \rangle \subseteq \{M \vee \langle x \rangle\}^\theta$ and $M \subset M \vee \langle x \rangle \subseteq \{M \vee \langle x \rangle\}^\theta$. By the maximality of M, we get $\{M \vee \langle x \rangle\}^\theta \cap A \neq \emptyset$ and $\{M \vee \langle y \rangle\}^\theta \cap A \neq \emptyset$. Choose $a \in \{M \vee \langle x \rangle\}^\theta \cap A$ and $b \in \{M \vee \langle y \rangle\}^\theta \cap A$. Hence

$$\begin{aligned} a \vee b &\in \{M \vee \langle x \rangle\}^\theta \cap \{M \vee \langle y \rangle\}^\theta \\ &= \{M \vee \langle x \vee y \rangle\}^\theta \end{aligned}$$

If $x \vee y \in [1]_\theta$, then $x \vee y \in [1]_\theta \subseteq M$. Since M is a θ-filter, we get that $a \vee b \in M^\theta = M$. Hence $a \vee b \in M \cap A$, which is a contradiction. Therefore M is a θ-prime filter of X. □

4.5 Irreducible Filters of BE-algebras

In this section, some properties of irreducible filters of BE-algebras are studied extensively. Some necessary conditions of irreducible filters and sufficient conditions for a filter to become an irreducible filter are derived. An equivalent condition is derived for every proper filter of a BE-algebra to become an irreducible filter.

Definition 4.5.1 A proper filter F of a BE-algebra X is said to be *irreducible* if for any two proper filters A and B of X, $F = A \cap B$ implies $F = A$ or $F = B$.

Example 4.5.2 Let $X = \{1, a, b, c\}$. Define a binary operation $*$ on X as follows:

$*$	1	a	b	c
1	1	a	b	c
a	1	1	b	b
b	1	a	1	a
c	1	1	1	1

Clearly $(X, *, 1)$ is a BE-algebra. All the filters of the BE-algebra are $\{1\}$, $\{1, a\}$, $\{1, b\}$ and X. It can be easily seen that the filters $\{1, a\}$ and $\{1, b\}$ are irreducible filters but the filter $\{1\}$ is not irreducible, because $\{1\} = \{1, a\} \cap \{1, b\}$ and also neither $\{1\} = \{1, a\}$ nor $\{1\} = \{1, b\}$.

From the above example, it can also be observed that the intersection of irreducible filters of a BE-algebra need not necessarily be an irreducible filter. For, consider the irreducible filters $F_1 = \{1, a\}$ and $F_2 = \{1, b\}$ of the BE-algebra X. Now, their intersection $F_1 \cap F_2 = \{1\}$ is not an irreducible filter of F.

Theorem 4.5.3 *Let F be a proper filter of a BE-algebra X. Then F is an irreducible filter of X if it satisfies the following condition for all $x, y \in X$:*

$$\text{For } x, y \notin F, \text{ there exists } z \notin F \text{ such that } x \leq z \text{ and } y \leq z.$$

Proof Assume that F satisfies the given condition. We prove the theorem by contradiction. Suppose F is not an irreducible filter of X. Then there exist two proper filters A and B of X such that $F = A \cap B$, $F \neq A$ and $F \neq B$. Then clearly $F \subset A$ and $F \subset B$. Let $x \in A - F$ and $y \in B - F$. Then by the assumed condition, there exists $z \in X - F$ such that $x \leq z$ and $y \leq z$. Since $x \in A, x \leq z$ and A is a filter of X, we get $z \in A$. Since $y \in B$ and $y \leq z$, we get $z \in B$. Hence $z \in A \cap B = F$, which is a contradiction. Therefore F is an irreducible filter of X. □

Theorem 4.5.4 *A proper filter F of a BE-algebra X is irreducible if and only if for every $x, y \notin F$, there exists $z \notin F$ such that $x * z \in F$ and $y * z \in F$.*

Proof The necessary condition follows from the above theorem. To prove the sufficient condition, assume that F satisfies the condition. Suppose F is not an irreducible filter of X. Then there exist two proper filters A and B of X such that $F = A \cap B$, $F \neq A$ and $F \neq B$. Clearly $F \subset A$ and $F \subset B$. Choose $x \in A - F$ and $y \in B - F$. Then by the assumed condition, there exists $z \in X - F$ such that $x * z \in F$ and $y * z \in F$. Since $F \subset A$, we get $x, x * z \in A$. Since A is a filter, we get $z \in A$. Since $F \subset B$, we get $y, y * z \in B$. Since B is a filter, we get $z \in B$. Hence $z \in A \cap B = F$, which is a contradiction. Therefore F is an irreducible filter of X. □

Theorem 4.5.5 *Let F be a proper filter of a BE-algebra X and $x \in X$. Suppose $x \notin F$. Then there exists an irreducible filter M of X such that $F \subseteq M$ and $x \notin M$.*

Proof Consider the following collection of filters of X:

$$\mathcal{IR} = \{G \mid G \text{ is a filter of } X, F \subseteq G, x \notin G\}.$$

Clearly $F \in \mathcal{IR}$ and hence $\mathcal{IR} \neq \emptyset$. Let $\{G_\alpha\}_{\alpha \in \Delta}$ be a collection of elements of $\mathcal{IR}$ which is forming a chain. Clearly $F \subseteq G_\alpha$ and $x \notin G_\alpha$ for each $\alpha \in \Delta$. Hence $F \subseteq \bigcup_{\alpha \in \Delta} G_\alpha$ and $x \notin \bigcup_{\alpha \in \Delta} G_\alpha$. Thus $\bigcup_{\alpha \in \Delta} G_\alpha$ is an upper bound for $\{G_\alpha\}_{\alpha \in \Delta}$. Therefore, by Zorn's lemma, there exists a maximal element M in $\mathcal{IR}$. Then $F \subseteq M$ and $x \notin M$. We now prove that M is an irreducible filter of X. Let A and B be two proper filters of X such that $M = A \cap B$. Suppose $M \neq A$ and $M \neq B$. Then clearly $M \subset A$ and $M \subset B$. Also $F \subseteq M \subset A$ and $F \subseteq M \subset B$. By the maximality of M, we get $x \in A$ and $x \in B$. Hence $x \in A \cap B = M$, which is a contradiction. Therefore M is an irreducible filter of X such that $F \subseteq M$ and $x \notin M$. □

Corollary 4.5.6 *Let X be a BE-algebra and $1 \neq x \in X$. Then there exists an irreducible filter M such that $x \notin M$.*

Proof Let $1 \neq x \in X$ and $F = \{1\}$. Clearly F is a proper filter and $x \notin F$. Then by the main theorem, there exists an irreducible filter M such that $x \notin M$. □

Theorem 4.5.7 *Let F be a proper filter of a BE-algebra X. Then*

$$F = \cap\{M \mid M \text{ is an irreducible filter of } X \text{ such that } F \subseteq M\}.$$

Proof Clearly $F \subseteq \cap\{M \mid M$ is an irreducible filter of X such that $F \subseteq M\}$. Conversely, let $x \notin F$. Then by the above theorem, there exists an irreducible filter M_x such that $F \subseteq M_x$ and $x \notin M_x$. Therefore

$$x \notin \cap\{M \mid M \text{ is an irreducible filter of } X \text{ such that } F \subseteq M\}.$$

Therefore $\cap\{M \mid M$ is an irreducible filter of X such that $F \subseteq M\} \subseteq F$. □

The following corollary is a direct consequence of the above results.

Corollary 4.5.8 *If $\{1\}$ is an irreducible filter of a BE-algebra X, then the intersection of all irreducible filters of a BE-algebra is equal to $\{1\}$.*

Theorem 4.5.9 *Let X and Y be two BE-algebras where X is commutative and $f : X \to Y$ a homomorphism which maps filters to filters. If F is an irreducible filter of Y and $f^{-1}(F) \neq X$, then $f^{-1}(F)$ is an irreducible filter of X.*

Proof Let F be an irreducible filter of Y. Then clearly $f^{-1}(F)$ is a filter of X. Let A and B be two filters of X such that $f^{-1}(F) = A \cap B$. Then $F = f(A \cap B)$. We first prove that $f(A \cap B) = f(A) \cap f(B)$. Clearly $f(A \cap B) \subseteq f(A) \cap f(B)$. Conversely, let $x \in f(A) \cap f(B)$. Then $x = f(a)$ and $x = f(b)$ for some $a \in A$ and $b \in B$. Since A and B are filters of X, we get $(b * a) * a \in A$ and $(a * b) * b \in B$. Since X is commutative, we get $(a * b) * b = (b * a) * a \in A \cap B$. Hence

$$\begin{aligned} x &= (x * x) * x \\ &= (f(a) * f(b)) * f(b) \\ &= f((a * b) * b) \in f(A \cap B). \end{aligned}$$

Hence $f(A) \cap f(B) \subseteq f(A \cap B)$. Thus $F = f(A \cap B) = f(A) \cap f(B)$. Since $f(A)$ and $f(B)$ are filters of Y and F is irreducible in Y, we get $F = f(A)$ or $F = f(B)$. Hence $f^{-1}(F) = A$ or $f^{-1}(F) = B$. Therefore $f^{-1}(F)$ is an irreducible filter of X. □

In the following theorem, a necessary and sufficient condition is derived for the class $\mathcal{F}(X)$ of all filters of a BE-algebra X to become a totally ordered set.

Theorem 4.5.10 *Let X be a BE-algebra. Then $\mathcal{F}(X)$ is a totally ordered set or a chain if and only if every proper filter of X is an irreducible filter.*

Proof Assume that $\mathcal{F}(X)$ is a totally ordered set. Let F be a proper filter of X. Let A and B be two proper filters of X such that $F = A \cap B$. Since $\mathcal{F}(X)$ is totally ordered and $A, B \in \mathcal{F}(X)$, we get that either $A \subseteq B$ or $B \subseteq A$. Hence $F = A \cap B = A$ or $F = A \cap B = B$. Therefore F is an irreducible filter of X.

Conversely assume that every proper filter of X is an irreducible filter. Let F and G be two proper filters of X. Clearly $F \cap G$ is a proper filter of X and hence by the assumed condition $F \cap G$ is irreducible. Since $F \cap G = F \cap G$, we get

$$F = F \cap G \text{ or } G = F \cap G$$

Hence $F \subseteq G$ or $G \subseteq F$. Therefore $\mathcal{F}(X)$ is a totally ordered set. □

4.6 Segments of BE-algebras

In this section, the notion of final segments is introduced in BE-algebras. Some studies are made on segments to obtain the relation between final segments and filters of BE-algebras. Some properties of final segments of self-distributive BE-algebras are studied.

Definition 4.6.1 Let $(X, *, 1)$ be a BE-algebra and $a, b \in X$. Then the set

$$[a, b] = \{x \in X \mid a \leq x \leq b\} = \{x \in X \mid a * x = x * b = 1\}$$

is called a *segment* of X. In the case of $b = 1$, the segment

$$[a, 1] = \{x \in X \mid a \leq x\} = \{x \in X \mid a * x = 1\}$$

is called the *final segment* of X. A BE-algebra which is linearly ordered by the *BE*-relation ($a \leq b$ if and only if $a * b = 1$) is called a *BE*-chain.

Example 4.6.2 Let $X = \{1, a, b, c, d\}$ be a BE-algebra with the following Caley table:

$*$	1	a	b	c	d
1	1	a	b	c	d
a	1	1	b	a	d
b	1	1	1	a	d
c	1	1	b	1	1
d	1	1	a	a	1

Then X is a BE-algebra. Its final segments have the form $[a, 1] = \{1, a\}$, $[b, 1] = \{1, a, b\}$, $[c, 1] = \{1, a, c, d\}$, and $[d, 1] = \{1, a, d\}$. All the above final segments are *BE*-chains.

Proposition 4.6.3 *Every final segment of a transitive BE-algebra is a BE-subalgebra.*

Proof Let $a \in X$. Clearly $1 \in [a, 1]$. Let $x, y \in [a, 1]$. Then, we get $a \leq x$ and $a \leq y$. Hence $a \leq x * a \leq x * y$. Thus $x * y \in [a, 1]$. Therefore $[a, 1]$ is a subalgebra of X for each $a \in X$. □

Proposition 4.6.4 *The set-theoretic union of any two final segments of a given transitive BE-algebra is a subalgebra.*

Proof Let X be a transitive BE-algebra and $a, b \in X$. Clearly $1 \in [a, 1] \cup [b, 1]$. Let $x, y \in [a, 1] \cup [b, 1]$. Suppose $x, y \in [a, 1]$. Then $x * y \in [a, 1]$ and so $x * y \in [a, 1] \cup [b, 1]$. If $x, y \in [b, 1]$, then also $x * y \in [b, 1] \subseteq [a, 1] \cup [b, 1]$. Suppose $x \in [a, 1]$ and $y \in [b, 1]$. Then $a \leq x \leq y * x$, which implies $y * x \in [a, 1] \subseteq [a, 1] \cup [b, 1]$. Hence $[a, 1] \cup [b, 1]$ is a subalgebra of X. □

In the following example, we observe a fact that the final segments are not filters of BE-algebras in general.

Example 4.6.5 Let $X = \{1, a, b, c\}$. Define a binary operation $*$ on X as follows:

$*$	1	a	b	c
1	1	a	b	c
a	1	1	a	a
b	1	1	1	a
c	1	1	a	1

Then $(X, *, 1)$ is a BE-algebra. Consider the final segments $[b, 1] = \{1, a, b\}$. Clearly $[b, 1]$ is not a filter of X, because $a \in [b, 1]$ and $a * c = b \in [b, 1]$ but $c \notin [b, 1]$.

Theorem 4.6.6 *Let F be a filter of a BE-algebra. Then* $F = \bigcup_{a \in F} [a, 1]$.

Proof Let $a \in F$. Choose $x \in X$ such that $x \in [a, 1]$. Then $a \leq x$. Since F is a filter, we get $x \in F$. Hence, it implies $[a, 1] \subseteq F$. Therefore $\bigcup_{a \in F} [a, 1] \subseteq F$. Conversely, let $x \in F$. Since $x * x = 1$, we get $x \leq x$. Hence, it yields that $x \in [x, 1]$. Hence $F \subseteq [x, 1] \subseteq \bigcup_{a \in F} [a, 1]$. □

In general, the converse of the above theorem is not true. For consider the following example.

Example 4.6.7 Let $X = \{1, a, b, c, d\}$. Define a binary operation $*$ on X as follows:

$*$	1	a	b	c	d
1	1	a	b	c	d
a	1	1	b	c	d
b	1	a	1	c	c
c	1	1	b	1	b
d	1	1	1	1	1

Then $(X, *, 1)$ is a BE-algebra. Consider the set $A = \{1, a, b, c\}$. It is observed that $[a, 1] = \{1, a\}$; $[b, 1] = \{1, b\}$; and $[c, 1] = \{1, a, c\}$. Clearly $A = [a, 1] \cup [b, 1] \cup [c, 1] \cup \{1\}$. It can also be observed that A is not a filter of X, because $c \in A$ and $c * d = b \in A$ but $d \neq A$.

Moreover, the BE-algebra mentioned in the above example is observed to be a self-distributive. Therefore, it is concluded that a non-empty set satisfying the necessary condition of the above theorem need not be a filter even in a self-distributive BE-algebra.

Theorem 4.6.8 *Every filter F of a transitive BE-algebra X satisfies the following conditions:*

(1) $F = \bigcup\limits_{a \in F} [a, 1]$,
(2) *for all $a, b \in X$, $[a, 1] = [b, 1]$ and $a \in F$ imply that $b \in F$,*
(3) *$a \in F$ implies $[a, 1] \subseteq F$ for all $a \in X$.*

Proof (1) It is proved in Theorem 4.7.4.
(2) Let $a, b \in X$ be such that $[a, 1] = [b, 1]$. Suppose that $a \in F$. Then by (1), we get $[b, 1] = [a, 1] \subseteq \bigcup\limits_{a \in F} [a, 1] = F$. Since $b \in [b, 1]$, it yields that $b \in F$.
(3) It is clear from (1). □

Proposition 4.6.9 *The Cartesian product of two final segments is also a final segment.*

Proof Let X_1 and X_2 be two BE-algebras. For any $a \in X_1$ and $b \in X_2$, we intend to prove that $[(a, b), (1, 1)] = [a, 1] \times [b, 1]$. For $a \in X_1$ and $b \in X_2$, we get

$$
\begin{aligned}
[(a, b), (1, 1)] &= \{(x, y) \in X_1 \times X_2 \mid (a, b) \leq (x, y)\} \\
&= \{(x, y) \in X_1 \times X_2 \mid a \leq x, b \leq y\} \\
&= \{(x, y) \in X_1 \times X_2 \mid x \in [a, 1] \text{ and } y \in [b, 1]\} \\
&= [a, 1] \times [b, 1]
\end{aligned}
$$

Therefore $[(a, b), (1, 1)]$ is a final segment in the product algebra $X_1 \times X_2$. □

It is known that the principal filter $\langle a \rangle$ generated by an element $a \in X$ is defined as the set $\{x \in X \mid a^n * x = 1 \text{ for some } n \in \mathbb{N}\}$. We now establish a relation between the final segments and principal filters of BE-algebras.

Lemma 4.6.10 *Let X be a BE-algebra and $a \in X$. Then $[a, 1] \subseteq \langle a \rangle$.*

Proof Let $a \in X$. Suppose $x \in [a, 1]$. Then $a \leq x$. Thus $a * x = 1$. Hence $a * (a * x) = 1$. Continuing in this way, we get $a^n * x = 1$ for any $n \in \mathbb{N}$. Hence $x \in \langle a \rangle$. Thus $[a, 1] \subseteq \langle a \rangle$. □

In general, the converse of the above lemma is not true. However, in the following theorem, a set of equivalent conditions is derived for the existence of the reverse inclusion.

Theorem 4.6.11 *Let X be a transitive BE-algebra and $a \in X$. Then the following conditions are equivalent.*

(1) $a \leq a * x$ *implies that* $a \leq x$ *for all* $x \in X$;
(2) $[a, 1] = \langle a \rangle$;
(3) $[a, 1]$ *is a filter of* X.

Proof (1) $\Rightarrow$ (2): Assume that the implication (1) holds. Let $a \in X$. Then clearly $[a, 1] \subseteq \langle a \rangle$. Conversely, let $x \in \langle a \rangle$. Then $a^n * x = 1$ for some $n \in \mathbb{N}$. Hence we get

$$
\begin{aligned}
a^n * x = 1 &\Rightarrow a \leq a^{n-1} * x \\
&\Rightarrow a \leq a * (a^{n-2} * x) \\
&\Rightarrow a \leq a^{n-2} * x \qquad \text{by condition (1)} \\
&\Rightarrow a \leq a * (a^{n-3} * x) \\
&\Rightarrow a \leq a^{n-3} * x \qquad \text{by condition (1)} \\
&\quad \ldots \\
&\Rightarrow a \leq a * x \\
&\Rightarrow a \leq x \qquad \text{by condition (1)}
\end{aligned}
$$

Hence $x \in [a, 1]$. Thus $\langle a \rangle \subseteq [a, 1]$. Therefore $\langle a \rangle = [a, 1]$.
(2) $\Rightarrow$ (3): Let $a \in X$. Since $\langle a \rangle$ is a filter, it is clear.
(3) $\Rightarrow$ (1): Assume that the final segment $[a, 1]$ is a filter of X. Suppose $a \leq a * x$ for all $x \in X$. Then $a * x \in [a, 1]$. Since $a \in [a, 1]$ and $[a, 1]$ is a filter, it yields that $x \in [a, 1]$. Hence $a \leq x$. □

Corollary 4.6.12 *Let $(X, *, 1)$ be a transitive BE-algebra. If a non-trivial final segment $[a, 1]$ is a filter of X, then $a * x \neq a$ for all $1 \neq x \in X$.*

Proof Let $[a, 1]$, where $a \neq 1$, be a filter of X. Suppose $a*x = a$ for some $1 \neq x \in X$. Then $a * x = a \in [a, 1]$. Hence $a \leq a * x$. Then by the above theorem, it yields that $a \leq x$. Hence $a = a * x = 1$, which is a contradiction. Therefore $a * x \neq a$ for all $1 \neq x \in X$. □

Proposition 4.6.13 *Let $(X, *, 1)$ be a commutative BE-algebra. Then for any $x, y \in X$,*

$$((x * y) * y) * y \leq x * y$$

Proof For any $x, y \in X$, consider $x * ((x * y) * y) = (x * y) * (x * y) = 1$. Hence $x \leq (x * y) * y$. Since X is negatively ordered, we get $((x * y) * y) * y \leq x * y$. □

Theorem 4.6.14 *A final segment $[a, 1]$ of a negatively ordered BE-algebra X is a filter if and only if for all $z \in X$*

$$a \leq a * z \text{ implies } a \leq z.$$

Proof Assume that the above implication holds. Let $x, y \in X$ be such that $x \in [a, 1]$ and $x * y \in [a, 1]$. Then $a \leq x$ and $a \leq x * y$. Hence

$$\begin{aligned} a \leq x &\Rightarrow x * y \leq a * y \\ &\Rightarrow a \leq x * y \leq a * y \qquad \text{since } X \text{ is negatively ordered} \\ &\Rightarrow a \leq y \qquad \text{by the given condition} \end{aligned}$$

which implies that $y \in [a, 1]$. Therefore the final segment $[a, 1]$ is a filter of X.

Conversely, assume that the final segment $[a, 1]$ is a filter of X. Suppose $a \leq a*z$ for all $z \in X$. Then $a * z \in [a, 1]$. Since $a \in [a, 1]$ and $[a, 1]$ is a filter, it yields that $z \in [a, 1]$. Hence $a \leq z$. □

Theorem 4.6.15 *In a self-distributive BE-algebra, every final segment is a filter.*

Proof Let X be a self-distributive BE-algebra and $a \in X$. Let $x, y \in X$ be such that $x \in [a, 1]$ and $x * y \in [a, 1]$. Then $a \leq x$ and $a \leq x * y$. Hence

$$\begin{aligned} a \leq x &\Rightarrow x * y \leq a * y \\ &\Rightarrow a \leq x * y \leq a * y \end{aligned}$$

Hence $a * (a * y) = 1$. Since X is self-distributive, we get that $a * y = 1 * (a * y) = (a * a) * (a * y) = a * (a * y) = 1$. Hence $a \leq y$, which implies that $y \in [a, 1]$. Therefore $[a, 1]$ is a filter of X. □

Definition 4.6.16 Let $(X, *, 1)$ be a BE-algebra. For any $a \in X$, define a relation θ_a on X by

$$(x, y) \in \theta_a \text{ if and only } a \leq x * y \text{ and } a \leq y * x$$

for all $x, y \in X$.

Theorem 4.6.17 *In a self-distributive BE-algebra, every θ_a is an equivalence relation on X.*

Proof Let X be a self-distributive BE-algebra. Then by Theorem 4.6.15, $[a, 1]$ is a filter for any $a \in X$. Clearly θ_a is reflexive and symmetric. Let $(x, y) \in \theta_a$ and $(y, z) \in \theta_a$. Then $a \leq x * y$, $a \leq y * x$ and $a \leq y * z$, $a \leq z * y$. Now, we have the following consequence.

$$\begin{aligned} a &\leq x * y \\ &\leq (y * z) * (x * z) \\ &\leq a * (x * z) \qquad \text{since } a \leq y * z \end{aligned}$$

Since $[a, 1]$ is filter, by Theorem 4.6.11, it yields $a \leq x * z$. Also

$$\begin{aligned} a &\leq z * y \\ &\leq (y * x) * (z * x) \\ &\leq a * (z * x) \qquad \text{since } a \leq y * x \end{aligned}$$

Since $[a, 1]$ is filter, by Theorem 4.6.11, we get $a \leq z * x$. Hence $(x, z) \in \theta_a$ and so θ_a is transitive. Therefore θ_a is an equivalence relation on X. □

Theorem 4.6.18 *If X is a transitive BE-algebra, then every θ_a is a congruence on X.*

Proof Since X is transitive, we get that X is negatively ordered. Let $a \in X$. By the above theorem, θ_a is an equivalence relation on X. Let $(x, y) \in \theta_a$ and $(u, v) \in \theta_a$. Then we get $a \leq x * y, a \leq y * x$ and $a \leq u * v, a \leq v * u$. Hence $a \leq x*y \leq u*(x*y) = (u*x)*(u*y)$. Similarly, we can obtain $a \leq (u*y)*(u*x)$. Hence $(u * x, u * y) \in \theta_a$.

Since X is negatively ordered, we get $1 = (u * v) * (u * v) \leq (u * v) * ((v * y) * (u * y)) = (v * y) * ((u * v) * (u * y))$. Hence $(v * y) * ((u * v) * (u * y)) = 1$. Now, we get

$$\begin{aligned} a &\leq 1 \\ &= (v * y) * ((u * v) * (u * y)) \\ &\leq (v * y) * (a * (u * y)) \qquad \text{since } a \leq u * v \\ &= a * ((v * y) * (u * y)) \end{aligned}$$

Then by Theorem 4.6.11, it yields $a \leq (v * y) * (u * y)$. Similarly, we get $a \leq (u * y) * (v * y)$. Hence $(u * y, v * y) \in \theta_a$. Since $(u * x, u * y) \in \theta_a$, by the transitivity, it implies that $(u * x, v * y) \in \theta_a$. Therefore θ_a is a congruence on X. □

Definition 4.6.19 If θ_a is a congruence on a BE-algebra X for each $a \in X$, then the kernel of θ_a is defined by $Ker(\theta_a) = \{x \in X \mid (x, 1) \in \theta_a\}$.

Proposition 4.6.20 *Let X be a BE-algebra and $a \in X$. If θ_a is a congruence on X, then $Ker(\theta_a)$ is a filter of X such that $Ker(\theta_a) = [a, 1]$.*

Proof Clearly $1 \in Ker(\theta_a)$. Let $x \in Ker(\theta_a)$ and $x * y \in Ker(\theta_a)$. Then $(1, x) \in \theta_a$ and $(1, x * y) \in \theta_a$. Since $(1, x) \in \theta_a$, we get $(y, x * y) = (1 * y, x * y) \in \theta_a$. Since $(1, x * y) \in \theta_a$, it yields $(1, y) \in \theta_a$. Hence $y \in Ker(\theta_a$. Therefore $Ker(\theta_a)$ is a filter of X. Now, for any $x \in X$, it is clear that $(1, x) \in \theta_a$ if and only if $x \in [a, 1]$. Hence $Ker(\theta_a) = [a, 1]$. □

The following corollary is a direct consequence of the above proposition.

Corollary 4.6.21 *Let X be a BE-algebra and $a \in X$. If θ_a is a congruence on X, then the final segment $[a, 1]$ is a filter of X.*

The summation of all the above results yields the following:

Theorem 4.6.22 *Let X be a transitive BE-algebra and $a \in X$. Then θ_a is a congruence on X such that $Ker(\theta_a) = [a, 1]$.*

4.7 Associative Filters of BE-algebras

In this section, the notion of associative filters is introduced in BE-algebras. A set of equivalent conditions is derived for every filter to become an associative filter. The notion of absorbent filters is introduced. A set of equivalent condition is derived for every filter to become an absorbent filter

Definition 4.7.1 A non-empty subset F of a BE-algebra X is said to be an *associative filter* of X if, for all $x, y, z \in X$, it satisfies the following conditions:

(A1) $1 \in F$,
(A2) $x * (y * z) \in F$ and $x * y \in F$ imply $z \in F$.

Example 4.7.2 Let $X = \{1, a, b, c, d\}$. Define a binary operation $*$ on X as follows:

$*$	1	a	b	c	d
1	1	a	b	c	d
a	1	1	a	c	c
b	1	1	1	c	c
c	1	a	b	1	a
d	1	1	a	1	1

It is easy to check that $(X, *, 1)$ is a BE-algebra. Consider the set $F = \{1, a, b\}$. Clearly F is a filter of X. It can be routinely verified that F is an associative filter of X.

Proposition 4.7.3 *Every associative filter of a BE-algebra is a filter.*

Proof Let F be an associative filter of a BE-algebra X. Let $x, y \in X$ be such that $x \in F$ and $x*y \in F$. Then, it yields that $1*x = x \in F$ and also $1*(x*y) = x*y \in F$. Since F is an associative filter of X, it implies that $y \in F$. Therefore F is a filter of X. □

The converse of the above proposition is not true. In the Example 4.7.2, the set $\{1\}$ is a filter but not an associative filter. For $a, b, c \in X$, it is observed that $d * (d * c) = d * 1 = 1 \in \{1\}$ and $d * d = 1 \in \{1\}$, but $c \notin \{1\}$. Hence $F = \{1\}$ is not an associative filter of X. However, in the following, a set of equivalent conditions is derived for a filter of a BE-algebra to become an associative filter.

Theorem 4.7.4 *Let F be a filter of a commutative BE-algebra X and $x, y, z \in X$. Then the following conditions are equivalent.*

(1) *F is associative,*
(2) *$x * (y * z) \in F$ implies $(x * y) * z \in F$,*
(3) *$x * (x * y) \in F$ implies $y \in F$.*

Proof (1) $\Rightarrow$ (2): Assume that F is associative. Let $x * (y * z) \in F$. Since X is a commutative BE-algebra, we get the following consequence:

$$\begin{aligned} x * ((y * z) * ((x * y) * z)) &= (y * z) * (x * ((x * y) * z)) \\ &= (y * z) * ((x * y) * (x * z)) \\ &= (x * y) * ((y * z) * (x * z)) \\ &= (x * y) * (x * ((y * z) * z)) \\ &= (x * y) * (x * ((z * y) * y)) \\ &= (x * y) * ((z * y) * (x * y)) \\ &= (z * y) * ((x * y) * (x * y)) \\ &= (z * y) * 1 \\ &= 1 \in F \end{aligned}$$

Since F is an associative filter and $x * (y * z) \in F$, it yields that $(x * y) * z \in F$.
(2) $\Rightarrow$ (3): Put $x = y$ in the condition (2), we get the condition (3).
(3) $\Rightarrow$ (1): Assume the condition (3). Let $x, y, z \in X$ be such that $x * (y * z) \in F$ and $x * y \in F$. Since X is transitive, we get $y * z \leq (x * y) * (x * z)$. Since X is ordered, we get $x * (y * z) \leq x * ((x * y) * (x * z))$. Therefore we get the following consequence.

$$\begin{aligned} (x * (y * z)) * (x * (x * ((x * y) * z))) &= (x * (y * z)) * (x * ((x * y) * (x * z))) \\ &\geq (x * (y * z)) * (x * (y * z)) \\ &= 1 \end{aligned}$$

which implies $x * (y * z) \leq x * (x * ((x * y) * z))$. Since F is a filter, we get $x * (x * ((x * y) * z)) \in F$. Thus by the condition (3), it yields $(x * y) * z \in F$. Since $x * y \in F$ and F is a filter, we get $z \in F$. Therefore F is an associative filter of X. □

Proposition 4.7.5 *Let $(X, *, 1)$ be a commutative BE-algebra with condition* **L**. *Then every associative filter of X is a lattice filter.*

Proof Let F be an associative filter of a commutative BE-algebra X with condition **L**. Then clearly F is a filter of X. Let $x, y \in F$. Then, it yields $y \leq x * y = 1 \wedge (x * y) = (x * x) \wedge (x * y) = x * (x \wedge y)$. Since F is a filter, we get $x * (x \wedge y) \in F$. Again by the property of filters, it yields that $x \wedge y \in F$. Therefore F is a lattice filter of X. □

The converse of the above proposition is not true. However, in the following theorem, a set of equivalent conditions is derived for every lattice filter of a BE-algebra to become an associative filter. In the absence of transitivity and the presence of condition **L**, we observe this fact.

Theorem 4.7.6 *Let F be a filter of a commutative BE-algebra X with condition* **L** *and $x, y, z \in X$. Then the following conditions are equivalent.*

(1) *F is associative,*
(2) *$x * (y * z) \in F$ implies $(x * y) * z \in F$,*
(3) *$x * y \in F$ implies $y \in F$.*

Proof (1) $\Rightarrow$ (2): Since X is commutative, this part is the same as in the above theorem.
(2) $\Rightarrow$ (3): Assume that F satisfies the condition (2). Let $x * y \in F$. Then we get that $x * (1 * y) = 1 * (x * y) = x * y \in F$. Then by the assumed condition (2), it yields that $y = 1 * y = (x * 1) * y \in F$. Therefore the condition (3) holds.
(3) $\Rightarrow$ (1): Assume the condition (3). Suppose that $x * (y * z) \in F,\ x * y \in F$. Since F is a filter and by the condition (3), we have the following consequence:

$$\begin{aligned}(x * (y * z)) \wedge (x * y) \in F &\Rightarrow x * (y \wedge (y * z)) \in F\\ &\Rightarrow x * (y \wedge z) \in F\\ &\Rightarrow (x * y) \wedge (x * z) \in F\\ &\Rightarrow x * z \in F\\ &\Rightarrow z \in F\end{aligned}$$

Therefore F is an associative filter of X. □

Theorem 4.7.7 *If F_1 and F_2 are associative filters of X_1 and X_2, respectively, then $F_1 \times F_2$ is an associative filter of the product algebra $X_1 \times X_2$. Conversely, every associative filter F of $X_1 \times X_2$ can be produced two associative filters F_1 and F_2 of X_1 and X_2, respectively.*

Proof Let F_1 and F_2 be associative filters of X_1 and X_2, respectively. Since $1 \in F_1$ and $1 \in F_2$, we get $(1, 1) \in F_1 \times F_2$. Let $(x_1, y_1) * ((x_2, y_2) * (x_3, y_3)) \in F_1 \times F_2$ and $(x_1, y_1) * (x_2, y_2) \in F_1 \times F_2$. Then $(x_1 * (x_2 * x_3), y_1 * (y_2 * y_3)) \in F_1 \times F_2$ and $(x_1 * x_2, y_2 * y_2) \in F_1 \times F_2$. Hence $x_1 * (x_2 * x_3) \in F_1; x_1 * x_2 \in F_1$ and $y_1 * (y_2 * y_3) \in F_2; y_1 * y_2 \in F_2$. Since F_1 and F_2 are associative filters of X_1 and X_2, respectively, we can get $x_3 \in F_1$ and $y_3 \in F_2$. Hence $(x_3, y_3) \in F_1 \times F_2$. Therefore $F_1 \times F_2$ is an associative filter of $X_1 \times X_2$.

Conversely, let F be any associative filter of $X_1 \times X_2$. Consider $\Pi_i : X_1 \times X_2 \longrightarrow X_i$ for $i = 1, 2$. Let F_1 and F_2 be the projections of F on X_1 and X_2, respectively. That is $\Pi_i(F) = F_i$ for $i = 1, 2$. We now prove that F_1 and F_2 are associative filters of X_1 and X_2, respectively. Since $(1, 1) \in F$, we get $1 = \Pi_1(1, 1) \in F_1$. Let $x * (y * z) \in F_1$ and $x * y \in F_1$. Then $(x, 1) * ((y, 1) * (z, 1)) = (x * (y * z), 1 * (1 * 1)) = (x * (y * z), 1) \in F$ and$(x, 1) * (y, 1) = (x * y, 1 * 1) = (x * y, 1) \in F$. Since F is an associative filter of $X_1 \times X_2$, we get $(z, 1) \in F$. Thus it yields $z = \Pi_1(z, 1) \in \Pi_1(F) = F_1$. Therefore F_1 is an associative filter of X_1. Similarly, we get F_2 is an associative filter of X_2. □

Theorem 4.7.8 (Extension property for associative filters) *Let F, G be two filters of a commutative BE-algebra X such that $F \subseteq G$. If F is an associative filter, then so is G.*

Proof Since X is transitive, it is ordered. Assume that F is an associative filter of X such that $F \subseteq G$. Let $x, y, z \in X$ and $x * (y * z) \in G$. Since X is transitive, we get $y * z \leq (x * y) * (x * z)$. Since X is ordered, it implies $x * (y * z) \leq (x * y) * (x * z)$. Hence, we get the following:

$$\begin{aligned} x * (x * ((x * (y * z)) * ((x * y) * z))) &= (x * (y * z)) * (x * (x * ((x * y) * z))) \\ &= (x * (y * z)) * (x * ((x * y) * (x * z))) \\ &\geq (x * (y * z)) * (x * (y * z)) \\ &= 1 \in F \end{aligned}$$

Since F is associative and X is commutative, it follows from Theorem 4.8.4(3) that

$$(x * (y * z)) * ((x * y) * z)) \in F \subseteq G.$$

Since G is a filter and $x * (y * z) \in G$, it yields that $(x * y) * z \in G$. Therefore, by Theorem 4.7.4, it concludes that G is an associative filter of X. This completes the proof. □

In the following, the concept of *absorbent filters* is introduced in BE-algebras.

Definition 4.7.9 A non-empty subset F of a BE-algebra X is said to be an *absorbent filter* if it satisfies the following properties, for all $x, y \in X$:

(AB1) $1 \in F$
(AB2) $(x * y) * x \in F$ imply that $x \in F$.

It is clear that every filter of an implicative BE-algebra is an absorbent filter.

Theorem 4.7.10 *Every associative filter of a BE-algebra X is an absorbent filter.*

Proof Let F be an associative filter of a BE-algebra X. Let $x, y \in X$ be such that $(x * y) * x \in F$. Then $(x * y) * (1 * x) = 1 * ((x * y) * x) = (x * y) * x \in F$. Since F is associative, it yields that $x = 1 * x = ((x * y) * 1) * x \in F$. Therefore F is an absorbent filter of X. □

Example 4.7.11 Let $X = \{a, b, c, 1\}$. Define a binary relation $*$ on X as follows:

$*$	a	b	c	1
a	1	c	1	1
b	c	1	1	1
c	c	c	1	1
1	a	b	c	1

Then $F = \{1, c, b\}$ is an absorbent filter of X. But F is not an associative filter, because $b * (c * a) = b * c = 1 \in F$ and $(b * c) * a = 1 * a = a \notin F$.

In general, every filter of a BE-algebra need not be an absorbent filter.

Example 4.7.12 Let $X = \{a, b, c, d, 1\}$. Define a binary operation $*$ on X as follows:

$*$	a	b	c	d	1
a	1	b	c	b	1
b	a	1	b	a	1
c	a	1	1	a	1
d	1	1	b	1	1
1	a	b	c	d	1

Then $(X, *, 1)$ is a BE-algebra. It is easy to check that $F = \{c, 1\}$ is a filter of X but not an absorbent filter. For $c, d \in X$, it is observed that $(d * c) * d = b * d = c \in F$ and $d \notin F$.

Theorem 4.7.13 *Let F be a filter of a BE-algebra X. Then the following are equivalent.*

(1) *F is absorbent,*
(2) *for all $a, x, y \in X$, $a \in F$, $(x * y) * (a * x) \in F$ implies $x \in F$,*
(3) *for all $x, y, z \in X$, $x * ((y * z) * y) \in F$ and $y \in F$ imply that $x \in F$.*

Proof (1) $\Rightarrow$ (2): Assume that F is an absorbent filter of X. Let $a, x, y \in X$. Suppose $a \in F$ and $(x * y) * (a * x) \in F$. Then $a * ((x * y) * x) = (x * y) * (a * x) \in F$. Since $a \in F$, we get that $(x * y) * x \in F$. Since F is absorbent, it concludes that $x \in F$.
(2) $\Rightarrow$ (3): Assume the condition (2). Let $x, y, z \in X$. Suppose $x * ((y * z) * y) \in F$ and $x \in F$. Then $(y * z) * (x * y) = x * ((y * z) * y) \in F$. Since $x \in F$, by (2), we get that $y \in F$.
(3) $\Rightarrow$ (1): Assume that F satisfies the condition (3). Let $x, y \in X$ and $(x*y)*x \in F$. Then clearly $1 * ((x * y) * x) = (x * y) * x \in F$. Since $1 \in F$, by the condition (3), we get that $x \in F$. Therefore F is an absorbent filter of X. □

4.8 Dual Atomistic Filters of BE-algebras

In this section, the notion of dual atoms is introduced in BE-algebras and their properties are studied with the help of filters. A necessary and sufficient condition is derived for every subalgebra of a BE-algebra to become a filter.

Definition 4.8.1 An element $a \neq 1$ of a BE-algebra X is said to be a *dual atom* of X if $a \leq x$ implies either $a = x$ or $x = 1$ for all $x \in X$.

Obviously, dual atoms of BE-algebras are the elements which are just covered by 1 with respect to the *BE*-ordering $\leq$. In the following example, we observe the non-dual atoms of BE-algebras.

Example 4.8.2 Let $X = \{a, b, c, 1\}$. Define a binary operation $*$ on X as follows:

$*$	1	a	b	c
1	1	a	b	c
a	1	1	a	b
b	1	1	1	b
c	1	1	1	1

Then clearly $(X, *, 1)$ is a BE-algebra. It is easy to check that the element a is a dual atom in X. Also the elements b and c are not dual atoms because $b \leq a(\neq 1)$ and $c \leq b(\neq 1)$.

Proposition 4.8.3 *Let X be a BE-algebra and $1 \neq a \in X$. Then a is a dual atom whenever $\{a, 1\}$ is a filter of X.*

Proof Suppose $\{a, 1\}$ is a filter of X. Let $x \in X$ be such that $a \leq x$. Since $\{a, 1\}$ is a filter and $a \leq x$, we get $x \in \{a, 1\}$. Hence either $x = a$ or $x = 1$. Therefore a is a dual atom in X. □

The converse of the above proposition is not true. For, consider the BE-algebra X in the above example. It is observed that a is a dual atom of X but $\{a, 1\}$ is not a filter of X because $a, a * b = 1 \in \{a, 1\}$ but $b \notin \{a, 1\}$. However, in the following theorem, a necessary and sufficient condition is derived for every subalgebra of a BE-algebra to become a filter.

Theorem 4.8.4 *If every element $a \neq 1$ of a BE-algebra X is a dual atom, then every subalgebra of X is a filter of X.*

Proof Assume that every element $1 \neq a \in X$ is a dual atom. Let S be a subalgebra of X. Clearly $1 \in S$. Suppose that $x \in S$ and $x * y \in S$. Since $y \leq x * y$ and y is a dual atom, we get that $x * y = 1$ or $y = x * y \in S$. If $x * y = 1$, then it yields $x \leq y$. Since x is a dual atom, we get that $y = x \in S$ or $y = 1$. Hence $y \in S$. Therefore S is a filter of X. □

Since every final segment is a subalgebra, the following result is clear.

Corollary 4.8.5 *If every element $a \neq 1$ of a BE-algebra X is a dual atom, then every initial segment is a filter of X.*

In the following, the notion of an *dual atomistic BE-algebras* is introduced.

Definition 4.8.6 A BE-algebra X is called dual atomistic if every non-unit element of X is a dual atom in X, i.e., $X = \mathcal{A}(X)$

A set of equivalent conditions is derived now for a BE-algebra to become dual atomistic.

Theorem 4.8.7 *The following conditions are equivalent in a BE-algebra X:*

(1) *X is dual atomistic,*
(2) *for all $x, y \in X$ with $x \neq y$, $x * y = y$,*
(3) *for all $x, y \in X$ with $x \neq y$, $(x * y) * y = 1$.*

Proof (1) $\Rightarrow$ (2): Assume that X is atomistic. Let $x, y \in X$ be such that $x \neq y$. Since $y \leq x * y$ and y is an atom of X, we get either $y = x * y$ or $x * y = 1$. Suppose $x * y = 1$. Then we get that $x < y$. Since x is an atom, it gives a contradiction. Therefore $x * y = y$.
(2) $\Rightarrow$ (3): It is obvious.
(3) $\Rightarrow$ (1): Assume the condition (3). Let $1 \neq x \in X$. Suppose $x \leq a$. Then $x*a = 1$. If $a = 1$, then we are through. Suppose $a \neq 1$. We claim that $x = a$. Suppose $x \neq a$. Then by condition (3), we get that $(x * a) * a = 1$. Hence $x * a \leq a$. Thus we get $1 = x * a \leq a$. Therefore $a = 1$, which is a contradiction. Hence $x = a$. Thus x is an atom of X. Therefore X is atomistic. □

In what follows, $\mathcal{A}(X)$ denotes the set of all dual atoms of X unless otherwise mentioned. Hence $\mathcal{A}(X) = \{x \in X \mid x \neq 1 \text{ and } x \text{ is a dual atom }\}$. Consider $\mathcal{A}_1(X) = \mathcal{A}(X) \cup \{1\}$.

Proposition 4.8.8 *For any BE-algebra X, $\mathcal{A}(X)$ is a subalgebra of X.*

Proof Let a, b be two distinct dual atoms of X, i.e., $a, b \in \mathcal{A}(X)$. Clearly $b \leq a * b$. Since b is a dual atom, we get either $b = a * b$ or $a * b = 1$. Suppose $a * b = 1$. Then $a < b$. Since a is a dual atom, it gives a contradiction. Hence $a * b = b \in \mathcal{A}(X)$. Therefore $\mathcal{A}(X)$ is a subalgebra of X. □

Theorem 4.8.9 *If X is dual atomistic, then we have the following:*

(1) *X is implicative,*
(2) *X is commutative,*
(3) *X is simple,*
(4) *X is self-distributive.*

Proof (1) Assume that X is dual atomistic. Let $x, y \in X$. If $x = 1$ or $y = 1$, then the result is clear. Suppose $x \neq 1$ and $y \neq 1$. Since X is dual atomistic, we get $x, y \in \mathcal{A}(X)$. It is observed in the above theorem that $x * y = y$. Hence $(x * y) * x = y * x = x$. Therefore X is implicative.
(2) Let $x, y \in X$. If $x = 1$ or $y = 1$, then the result is clear. Suppose $x \neq 1$ and $y \neq 1$. Then $x, y \in \mathcal{A}(X)$. Hence $(x * y) * y = y * y = 1 = x * x = (y * x) * x$. Therefore X is commutative.
(3) Let $x, y \in X$. If $x = 1$ or $y = 1$, then the result is clear. Suppose $x \neq 1$ and $y \neq 1$. Since X is atomistic, we get that $x, y \in \mathcal{A}(X)$. Hence $x * (x * y) = x * y$. Therefore X is simple.
(4) From (2) and (3), it is clear. □

Definition 4.8.10 A filter F of a BE-algebra X is called *dual atomistic* if $\mathcal{A}_1(F) = F$.

It is clear that every filter of a dual atomistic BE-algebra is a dual atomistic filter.

Theorem 4.8.11 *Let X be a BE-algebra and F a filter of X. Then F is a dual atomistic filter of X if and only if for any $a, b \in F$, $a \neq b$ implies $a * b = b$.*

Proof The necessity part is clear. For sufficiency, assume that for any $a, b \in F$, $a \neq b$ implies $a * b = b$. Let $x \in F$ with $x \neq 1$. Suppose $1 \neq x_0 \in X$ be such that $x \leq x_0$. Since F is a filter, we get $x_0 \in F$. Since $x \leq x_0$, we get either $x = x_0$ or $x \neq x_0$. If $x \neq x_0$, then by the assumed condition we get $x_0 = x * x_0 = 1$. Hence $x_0 = 1$, which is a contradiction to that $x_0 \neq 1$. Thus $x = x_0$ and x is a dual atom of X. The proof is completed. □

Theorem 4.8.12 *A filter F a self-distributive BE-algebra X is a maximal dual atomistic filter if and only if for each $x \notin F$, there exists a dual atom a of X such that $x * a = 1$.*

Proof Assume that F is a maximal dual atomistic filter of X. Let $x \notin F$. Suppose $x * a \neq 1$ for all $a \in \mathcal{A}(X)$. Since $a \leq x * a$ and a is a dual atom, it yields $a = x * a$ for all atoms a. Let $b, c \in \langle F \cup \{x\} \rangle$ with $b \neq c$. Then $x * b, x * c \in F$. Since F is atomistic, we get

$$x * (b * c) = (x * b) * (x * c) = x * c.$$

By the hypothesis, we get $x * b = x * (x * b) \neq 1$ and $x * c = x * (x * c) \neq 1$. Hence $b * c = c$. Therefore by Example 4.7.11, it concludes that $\langle F \cup \{x\} \rangle$ is a dual atomistic filter. It is a contradiction to that F is maximal. Therefore for each $x \notin F$, there exists a dual atom a such that $x * a \neq 1$.

Conversely, assume the condition. Let G be a dual atomistic filter such that $F \subset G$. Choose $x \in G - F$. Clearly $x \neq 1$. Since $x \notin F$, there exists a dual atom a such that $x \leq a$. Hence x is not a dual atom of G, which is a contradiction to that G is dual atomistic. □

For any $a \in X$, define the self-map f_a on X by $f_a(x) = a * x$ for all $x \in X$. If X is self-distributive, then clearly f_a is an idempotent homomorphism. If $\mathcal{S}(X)$ denotes the set of all self-maps of the above type, then $\mathcal{S}(X)$ becomes a commutative monoid under composition of maps and f_1 is the identity. Define a relation $\leq_s$ on $\mathcal{S}(X)$ by $f_a \leq_s f_b$ if and only if $(a * x) * (b * x) = 1$ for all $x \in X$. Then clearly $\leq_s$ is an equivalence relation on $\mathcal{S}(X)$.

Lemma 4.8.13 *Let X be a BE-algebra and $f_a \in \mathcal{S}(X)$ where $a \in \mathcal{A}(X)$. If the subalgebra $\mathcal{A}(X)$ of dual atoms is a filter of X, then $Ker(f_a) = \{1, a\}$.*

Proof Let $a \in \mathcal{A}(X)$. Clearly $\{1, a\} \subseteq Ker(f_a)$. Conversely, let $x \in Ker(f_a)$. Then $a * x = f_a(x) = 1$. Hence $a \leq x$. Since a is a dual atom, we get $x = a$ or $x = 1$. Hence $x \in \{a, 1\}$. □

Theorem 4.8.14 *Let X be a BE-algebra and $a \in X$. Then X is dual atomistic if and only if for all $f_a \in \mathcal{S}(X)$, $Ker(f_a) = \{a, 1\}$.*

Proof Assume that X is dual atomistic. Then $\mathcal{A}(X) = X$. Then by Theorem 4.8.4, we get $\mathcal{A}(X)$ is a filter of X. Hence by the above lemma, we get that $Ker(f) = \{a, 1\}$ for any $a \in X$.

Conversely, assume that $Ker(f_a) = \{a, 1\}$ for all $f_a \in \mathcal{S}(X)$. Let $b \in X$ with $b \neq 1$. Suppose there exists $x \in X$ such that $b \leq x$. Then $f_b(x) = b * x = 1$. Hence $x \in Ker(f_b) = \{b, 1\}$, which implies $x = b$ or $x = 1$. Thus b is a dual atom of X. Therefore X is dual atomistic. □

Definition 4.8.15 An element $a \neq 1$ of a BE-algebra X is said to be a *pure dual atom* if $a \leq x$ implies $a = x$ for all $x \in X - \{1\}$.

Clearly every pure dual atom is a dual atom. However, in the following theorem, a set of equivalent conditions is derived for an element of a BE-algebra to become a pure dual atom.

Theorem 4.8.16 *Let X is a BE-algebra and $1 \neq a \in X$. Then the following are equivalent:*

(1) *a is a pure dual atom,*
(2) *$a = (a * x) * x$ for all $x \in X - \{1\}$,*
(3) *$(a * x) * (y * x) = y * a$ for all $x \in X - \{1\}$ and $y \in X$.*

Proof (1) $\Rightarrow$ (2): Assume that a is a pure dual atom. For any $x \in X$, we get $a * ((a * x) * x) = (a * x) * (a * x) = 1$. Hence $a \leq (a * x) * x$. Since a is pure, it gives that $a = (a * x) * x$.
(2) $\Rightarrow$ (3): Assume that $a = (a * x) * x$ for all $1 \neq x \in X$. For any $y \in X - \{1\}$, we get $(a * x) * (y * x) = y * ((a * x) * x) = y * a$.
(3) $\Rightarrow$ (1): Let $x \in X - \{1\}$ be such that $a \leq x$. Then $a * x = 1$. By taking $y = 1$ in (3), we get $x = 1 * (1 * x) = (a * x) * (y * x) = y * a = 1 * a = a$. Therefore a is a pure dual atom. □

Theorem 4.8.17 *The following conditions are equivalent in a BE-algebra X.*

(1) *every $a \neq 1$ is a pure dual atom,*
(2) *$x * y = y$ for all $x, y \in X$ with $x \neq y$,*
(3) *$(x * y) * y = 1$ for all $x, y \in X$ with $x \neq y$.*

Proof (1) $\Rightarrow$ (2): Assume that every $a \neq 1$ is a pure dual atom. Let $x, y \in X$ with $x \neq y$. Since $y \leq x * y$ and y is a pure dual atom, we get $x * y = y$.
(2) $\Rightarrow$ (3): It is obvious.
(3) $\Rightarrow$ (1): Assume that the condition (3) holds. Let $1 \neq a \in X$. Suppose $a \leq x$ for some $x \in X - \{1\}$. If $a \neq x$, then by (3), we get $x = 1 * x = (a * x) * x = 1$. Hence $x = 1$, which is a contradiction. Thus $a = x$. Therefore a is a pure dual atom of X. □

4.9 Prime Filters of BE-algebras

In this section, some properties of prime filters and maximal filters of BE-algebras are studied. A necessary and sufficient condition is derived for a proper filter of a BE-algebra to become a prime filter. Throughout this section, X stands for a BE-algebra unless otherwise mentioned.

Definition 4.9.1 A proper filter P of a BE-algebra X is said to be *prime* if $F \cap G \subseteq P$ implies $F \subseteq P$ or $G \subseteq P$ for any two filters F and G of X.

Theorem 4.9.2 ([220]) *A proper filter P of a BE-algebra X is prime if and only if $\langle x \rangle \cap \langle y \rangle \subseteq P$ implies $x \in P$ or $y \in P$ for all $x, y \in X$.*

Proof Assume that P is a prime filter of X. Let $x, y \in X$ be such that $\langle x \rangle \cap \langle y \rangle \subseteq P$. Since P is prime, it implies that $x \in \langle x \rangle \subseteq P$ or $y \in \langle y \rangle \subseteq P$. Conversely, assume that the condition holds. Let F and G be two filters of X such that $F \cap G \subseteq P$. Let $x \in F$ and $y \in G$ be the arbitrary elements. Then $\langle x \rangle \subseteq F$ and $\langle y \rangle \subseteq G$. Hence $\langle x \rangle \cap \langle y \rangle \subseteq F \cap G \subseteq P$. Then by the assumed condition, we get $x \in P$ or $y \in P$. Thus $F \subseteq P$ or $G \subseteq P$. Therefore P is a prime filter of X. □

Theorem 4.9.3 *Let X be an ordered BE-algebra and F a filter of X. Then for any $a, b \in X$,*

$$\langle a \rangle \cap \langle b \rangle \subseteq F \text{ if and only if } \langle F \cup \{a\} \rangle \cap \langle F \cup \{b\} \rangle = F.$$

Proof Let F be a filter of X. Assume that $\langle F \cup \{a\} \rangle \cap \langle F \cup \{b\} \rangle = F$ for any $a, b \in X$. Since $a \in \langle F \cup \{a\} \rangle$ and $b \in \langle F \cup \{b\} \rangle$, we get that $\langle a \rangle \subseteq \langle F \cup \{a\} \rangle$ and $\langle b \rangle \subseteq \langle F \cup \{b\} \rangle$. Hence $\langle a \rangle \cap \langle b \rangle \subseteq \langle F \cup \{a\} \rangle \cap \langle F \cup \{b\} \rangle = F$. Therefore $\langle a \rangle \cap \langle b \rangle \subseteq F$.

Conversely, assume that $\langle a \rangle \cap \langle b \rangle \subseteq F$. Clearly $F \subseteq \langle F \cup \{a\} \rangle \cap \langle F \cup \{b\} \rangle$. Let $x \in \langle F \cup \{a\} \rangle \cap \langle F \cup \{b\} \rangle$. Since F is a filter, there exists $m, n \in \mathbb{N}$ such that $a^m * x \in F$ and $b^n * x \in F$. Hence, there exists $m_1, m_2 \in F$ such that $a^m * x = m_1$ and $b^n * x = m_2$. Hence

$$a^m * (m_1 * x) = m_1 * (a^m * x) = m_1 * m_1 = 1$$

Hence $m_1 * x \in \langle a \rangle$. Similarly, we get $m_2 * x \in \langle b \rangle$. Since

$$m_1 * x \leq m_2 * (m_1 * x) = m_1 * (m_2 * x) \text{ and } m_2 * x \leq m_1 * (m_2 * x)$$

we get that $m_1 * (m_2 * x) \in \langle a \rangle$ and $m_1 * (m_2 * x) \in \langle b \rangle$. Hence

$$m_1 * (m_2 * x) \in \langle a \rangle \cap \langle b \rangle \subseteq F$$

As $m_1, m_2 \in F$, we get $x \in F$. Hence $\langle F \cup \{a\} \rangle \cap \langle F \cup \{b\} \rangle \subseteq F$. Thus $\langle F \cup \{a\} \rangle \cap \langle F \cup \{b\} \rangle = F$. □

Theorem 4.9.4 *Every maximal filter of an ordered BE-algebra is a prime filter.*

Proof Let M be a maximal filter of a BE-algebra X. Let $\langle x\rangle \cap \langle y\rangle \subseteq M$ for some $x, y \in X$. Suppose $x \notin M$ and $y \notin M$. Then $\langle M \cup \{x\}\rangle = X$ and $\langle M \cup \{y\}\rangle = X$. Hence

$$\langle M \cup \{x\}\rangle \cap \langle M \cup \{y\}\rangle = X$$

Hence, by the Theorem 4.9.3, it yields that $\langle x\rangle \cap \langle y\rangle \nsubseteq M$, which is a contradiction. Hence $x \in M$ or $y \in M$. Therefore M is a prime filter of X. □

Corollary 4.9.5 *Let X be an ordered BE-algebra. If $M_1, M_2, \ldots, M_n$ and M are maximal filters of X such that $\bigcap_{i=1}^{n} M_i \subseteq M$. Then there exists $j \in \{1, 2, \ldots, n\}$ such that $M_j = M$.*

Proof Let $M_1, M_2, \ldots, M_n$ and M be maximal filters of X such that $\bigcap_{i=1}^{n} M_i \subseteq M$. By the above theorem, M is a prime filter of X. Hence there exists $j \in \{1, 2, \ldots, n\}$ such that $M_j \subseteq M$. Since M_j is a maximal filter, it gives that $M_j = M$. □

Theorem 4.9.6 *Let X ba an ordered BE-algebra and $a \in X$. If F is a filter of X such that $a \notin F$, then there exists a prime filter P such that $a \notin P$ and $F \subseteq P$.*

Proof Let F be a filter of X such that $a \notin F$. Consider $\Im = \{G \in \mathcal{F}(X) \mid a \notin G$ and $F \subseteq G\}$. Clearly $F \in \Im$. Then by the Zorn's lemma, $\Im$ has a maximal element, say M. Clearly $a \notin M$. We now prove that M is prime. Let $x, y \in X$ be such that $\langle x\rangle \cap \langle y\rangle \subseteq M$. Then by Theorem 4.9.3, we get $\langle M \cup \{x\}\rangle \cap \langle M \cup \{y\}\rangle = M$. Since $a \notin M$, we can obtain that $a \notin \langle M \cup \{x\}\rangle$ or $a \notin \langle M \cup \{y\}\rangle$. By the maximality of M, we get that $\langle M \cup \{x\}\rangle = M$ or $\langle M \cup \{y\}\rangle = M$. Hence $x \in M$ or $y \in M$. Therefore M is a prime filter of X. □

Corollary 4.9.7 *Let F be a proper filter of a positively ordered BE-algebra X. Then*

$$F = \cap\{P \mid P \text{ is a prime filter of} X \text{ such that } F \subseteq P\}$$

Proof Clearly $F \subseteq \cap\{P \mid P$ is a prime filter of X such that $F \subseteq P\}$. Conversely, let $x \notin F$. Then by the main theorem, there exists a prime filter P_x such that $x \notin P_x$ and $F \subseteq P_x$. Therefore

$$x \notin \cap\{P \mid P \text{ is a prime filter of } X \text{ such that } F \subseteq P\}.$$

Therefore $\cap\{P \mid P$ is a prime filter of X such that $F \subseteq P\} \subseteq F$. □

Corollary 4.9.8 *Let X be an ordered BE-algebra and $1 \neq x \in X$. Then there exists a prime filter P such that $x \notin P$.*

Proof Let $1 \neq x \in X$ and $F = \{1\}$. Clearly F is a filter and $x \notin F$. Then by the main theorem, there exists a prime filter P such that $x \notin P$. □

The following corollary is a direct consequence of the above results.

Corollary 4.9.9 *The intersection of all prime filters of an ordered BE-algebra is equal to* $\{1\}$.

Theorem 4.9.10 *Let X and Y be two ordered BE-algebras and $f : X \to Y$ a homomorphism such that $f(X)$ is a filter in Y. If F is a prime filter of Y and $f^{-1}(F) \neq X$, then $f^{-1}(F)$ is a prime filter of X.*

Proof Let F be a filter of Y. Then by Theorem 4.1.11, $f^{-1}(F)$ is a filter of X. Let $x, y \in X$ be such that $\langle x\rangle \cap \langle y\rangle \subseteq f^{-1}(F)$. Let $u \in \langle f(x)\rangle \cap \langle f(y)\rangle$. Then there exists $m, n \in \mathbb{N}$ such that $f(x)^n * u = 1 \in F$ and $f(y)^m * u = 1 \in F$. Since $f(x) \in f(X)$ and $f(X)$ is a filter, it implies that $u \in f(X)$. Hence $u = f(a)$ for some $a \in X$. Moreover, $f(x^n * a) = f(y^m * a) = 1 \in F$ because f is a homomorphism. Hence

$$x^n * a \in f^{-1}(F) \text{ and } y^m * a \in f^{-1}(F)$$

Hence

$$a \in \langle f^{-1}(F) \cup \{x\}\rangle \cap \langle f^{-1}(F) \cup \{y\}\rangle$$

Since $\langle x\rangle \cap \langle y\rangle \subseteq f^{-1}(F)$, then by Theorem 4.9.3, we get $a \in f^{-1}(F)$. Hence $u = f(a) \in F$. It concludes that $\langle f(x)\rangle \cap \langle f(y)\rangle \subseteq F$. Since F is a prime filter of Y, we get that $\langle f(x)\rangle \subseteq F$ or $\langle f(y)\rangle \subseteq F$. Thus it yields that $f(x) \in F$ or $f(y) \in F$. Therefore $x \in f^{-1}(F)$ or $y \in f^{-1}(F)$, which concludes that $f^{-1}(F)$ is a prime filter of X. □

In the following theorem, a necessary and sufficient condition is derived, in terms of primeness of filters, for the class $\mathcal{F}(X)$ of all filters of a BE-algebra X to become a chain.

Theorem 4.9.11 *Let X be a BE-algebra. Then $\mathcal{F}(X)$ is a totally ordered set or a chain if and only if every proper filter of X is a prime filter.*

Proof Assume that $\mathcal{F}(X)$ is a totally ordered set. Let F be a proper filter of X. Let $a, b \in X$ be such that $\langle a\rangle \cap \langle b\rangle \subseteq F$. Since $\langle a\rangle$ and $\langle b\rangle$ are filters of X, we get that either $\langle a\rangle \subseteq \langle b\rangle$ or $\langle b\rangle \subseteq \langle a\rangle$. Hence $a \in F$ or $b \in F$. Therefore F is a prime filter of X.

Conversely assume that every proper filter of X is a prime filter. Let F and G be two proper filters of X. Since $F \cap G$ is a proper filter of X, we get that

$$F \subseteq F \cap G \text{ or } G \subseteq F \cap G$$

Hence $F \subseteq G$ or $G \subseteq F$. Therefore $\mathcal{F}(X)$ is a totally ordered set. □

The notion of prime ideals of commutative BCK-algebras was introduced by J. Meng, Y.B. Jun, and X.L. Xin [180]. Later, the properties of prime ideals of BCI and BCK-algebras were studied by Y. Huang [115]. In [18], R.A. Borzooei and O. Zahiri studied the structure of prime ideals of commutative BCI-algebra. Since every commutative BE-algebras is a commutative dual BCK-algebra, we can generalize the following crucial results:

Theorem 4.9.12 ([180]) *Let X be a commutative BE-algebra and F a filter of X. Then F is prime if and only if $x * y \in F$ or $y * x \in F$ for all $x, y \in X$.*

Lemma 4.9.13 ([115]) *Let X be a commutative BE-algebra or BE-semilattice. Then for any $a, b \in X$*

$$\langle a \vee b \rangle = \langle a \rangle \cap \langle b \rangle.$$

Theorem 4.9.14 ([115]) *Let X be a commutative BE-algebra and P a proper filter of X. Then the following conditions are equivalent.*

(1) *P is prime,*
(2) *for any $x, y \in X$, $x \vee y \in P$ implies $x \in P$ or $y \in P$,*
(3) *for any $x, y \in X$, $\langle x \rangle \cap \langle y \rangle \subseteq P$ implies $x \in P$ or $y \in P$.*

Theorem 4.9.15 *Let X be a commutative BE-algebra. Then the following are equivalent:*

(1) *every filter is a prime filter,*
(2) *the filter $\{1\}$ is a prime filter,*
(3) *X is a totally ordered set with respect to BE-ordering.*

Proof (1) $\Rightarrow$ (2): It is obvious.
(2) $\Rightarrow$ (3): Assume that $\{1\}$ is a prime filter. Let $x, y \in X$. Since $\{1\}$ is prime, we get that either $x * y \in \{1\}$ or $y * x \in \{1\}$. Hence $x \leq y$ or $y \leq x$. Therefore X is totally ordered.
(3) $\Rightarrow$ (1): Assume that X is a totally ordered set with respect to BE-ordering. Let F be a filter of X. Let $x, y \in X$. Since X is totally ordered, it implies $x \leq y$ or $y \leq x$ and thus $x * y = 1 \in F$ or $y * x = 1 \in F$. Therefore F is a prime filter of X. □

For any prime filter P of a commutative BE-algebra X, it can be easily observed that $X - P$ is a $\vee$-closed subset of X.

Theorem 4.9.16 ([115]) *Let X be a commutative BE-algebra and A a $\vee$-closed subset of X (i.e., $x \vee y \in A$ for all $x, y \in A$). If F is a filter of X such that $F \cap A = \emptyset$, then there exists a prime filter P of X such that $F \subseteq P$ and $A \cap P = \emptyset$.*

4.10 Dual Annihilators of BE-algebras

In this section, the notion of relative dual annihilators is introduced in a commutative BE-algebra. Some elementary properties of the class of relative dual annihilators of BE-algebras are studied. Some sufficient condition is derived for a dual annihilator of a BE-algebra to become a prime filter. Homomorphic images and inverse images of dual annihilators are studied.

Definition 4.10.1 Let A and B be any two non-empty subsets of a commutative BE-algebra X. Define $(A : B) = \{x \in X \mid x \vee a \in B \text{ for all } a \in A\}$. The set $(A : B)$ is said to be a *dual annihilator* of A relative to B or *relative dual annihilator* of A with respect to B.

For a singleton set $A = \{a\}$, we simply consider the relative annihilator $(\{a\} : B)$ of a relative to B as $(a : B)$. For $B = \{1\}$, we denote for brevity $(A : \{1\})$ by A^+ where $A^+ = \{x \in X \mid x \vee a = 1 \text{ for all } a \in A\}$. Here A^+ is called the *dual annihilator* of A. For any $\emptyset \neq A \subseteq X$, the set $(a : A)$ need not be a filter of X, which is observed in the following:

Example 4.10.2 Let $X = \{1, a, b, c\}$. Define a binary operation $*$ on X as follows:

$*$	1	a	b	c
1	1	a	b	c
a	1	1	b	b
b	1	a	1	a
c	1	1	1	1

Then $(X, *, 1)$ is a BE-algebra. Consider $A = \{a, b\}$ and $c \in X$. It is observed that $a \vee c = a \in A$; $b \vee c = b \in A$; $1 \vee c \notin A$; and $c \vee c = c \notin A$. Hence $(c : A) = \{a, b\}$, which is not a filter of X.

Though the relative dual annihilator $(a : A)$ is not a filter, in the following proposition, it is proved that $(a : A)$ is a filter whenever A is a filter of X. For this, we first observe the following lemma:

Lemma 4.10.3 *Let X be a commutative BE-algebra and $x, y, z \in X$. Then*

$$y * z \leq (z * x) * (y * x).$$

Proof Let $x, y, z \in X$. Since X is a commutative BE-algebra, we get that $(y * z) * ((z * x) * (y * x)) = (y * z) * (y * ((z * x) * x)) = (y * z) * (y * ((x * z) * z)) = (y * z) * ((x * z) * (y * z)) = (x * z) * ((y * z) * (y * z)) = (x * z) * 1 = 1$. Therefore $y * z \leq (z * x) * (y * x)$. □

Proposition 4.10.4 *Let F be a filter of a commutative BE-algebra X. Then for any $a \in X$, $(a : F)$ is a filter of X containing F.*

Proof Let $a \in X$. Clearly $1 \in (a : F)$. Let $x \in (a : F)$ and $x * y \in (a : F)$. Then we get $(x * a) * a = x \vee a \in F$ and $(((x * y) * a) * a = (x * y) \vee a \in F$. Since X is commutative, we get

$$\begin{aligned}((x*y)\vee a)*((x\vee a)*(y\vee a)) &= ((((x*y)*a)*a)*(((x*a)*a)*((y*a)*a))\\ &= ((((x*y)*a)*a)*((y*a)*(((x*a)*a)*a))\\ &= ((((x*y)*a)*a)*((y*a)*(x*a))\\ &= (y*a)*((((x*y)*a)*a)*(x*a))\\ &= (y*a)*(x*((((x*y)*a)*a)*a))\\ &= (y*a)*(x*((x*y)*a))\\ &= (y*a)*((x*y)*(x*a))\\ &= (x*y)*((y*a)*(x*a))\\ &= 1 \qquad\qquad \text{since } (x*y)\le(y*a)*(x*a)\end{aligned}$$

From the above observation, it immediately infers that $(x*y)\vee a \le (x\vee a)*(y\vee a)$. Since $(x*y)\vee a \in F$ and F is a filter, we get $(x\vee a)*(y\vee a)\in F$. Since $x\vee a\in F$, it yields that $y\vee a\in F$. Thus we get $y\in(a:F)$. Therefore $(a:F)$ is a filter of X. Let $x\in F$. Since $x\le x\vee a$, we get $x\vee a\in F$. Therefore $x\in(a:F)$, which implies that $F\subseteq(a:F)$. □

Corollary 4.10.5 *For any subset A of a commutative BE-algebra X, A^+ is a filter of X.*

Corollary 4.10.6 *Let F be an associative filter of a commutative BE-algebra X. Then for any $a\in X$, $(a:F)$ is an associative filter of X.*

Proof Let $x,y,z\in X$. Suppose $x*(y*z)\in(a:F)$ and $x*y\in(a:F)$. Then we get $(x*(y*z))\vee a\in F$ and $(x*y)\vee a\in F$. From the observation made in Proposition 4.10.4, we get $(x*(y*z))\vee a\le(x\vee a)*((y\vee a)*(z\vee a))$ and $(x*y)\vee a\le(x\vee a)*(y\vee a)$. Since F is a filter, we get $(x\vee a)*((y\vee a)*(z\vee a))\in F$ and $(x\vee a)*(y\vee a)\in F$. Since F is an associative filter, we get $z\vee a\in F$. Hence $z\in(a:F)$. Therefore $(a:F)$ is an associative filter of X. □

Proposition 4.10.7 *Let X be a commutative BE-algebra. If F is a filter of X, then $F=(A:F)$ for some singleton subset A of $(a:F)$.*

Proof Let F be a filter of X. Clearly $F\subseteq(a:F)$. Let $x\in(a:F)$ and consider $A=\{x\}$. Then we get $x=x\vee x\in F$. Hence $(a:F)\subseteq F$. Therefore $F=(a:F)$. □

For any filter F of a commutative BE-algebra X, the equality $F=(A:F)$ does not hold in general for any non-empty subset A of X as shown in the following example.

Example 4.10.8 Let $X=\{1,a,b,c,d\}$. Define a binary operation $*$ on X as follows:

$*$	1	a	b	c	d
1	1	a	b	c	d
a	1	1	a	c	d
b	1	1	1	c	d
c	1	a	b	1	d
d	1	1	1	c	1

Clearly $(X, *, 1)$ is a commutative BE-algebra. Consider $F = \{1, a, b\}$ and $A = \{c\}$. Then

$$(F : A) = \{x \in X \mid x \vee a \in F \text{ for all } a \in A\} = \{1, a, b, d\} \neq F.$$

Therefore the condition of the above proposition is not true.

Some elementary properties of relative dual annihilators are observed in the following:

Proposition 4.10.9 *Let F, G be two filters of a commutative BE-algebra X and $a, b \in X$. Then we have the following:*

(1) $(X : F) = F$ *and* $(1 : F) = X$,
(2) *if* $a \leq b$, *then* $(a, F) \subseteq (b : F)$,
(3) *if* $F \subseteq G$, *then* $(a : F) \subseteq (a : G)$,
(4) $(a : F \cap G) = (a : F) \cap (a : G)$,
(5) $\langle a \rangle \subseteq ((a : F) : F)$,
(6) $(\langle a \rangle : F) = (a : F)$,
(7) $(a : F) = (((a : F) : F) : F)$,
(8) $(a : F) = X$ *if and only if* $a \in F$,
(9) $(a : F) \cap ((a : F) : F) = F$.

Proof (1) Clearly $F \subseteq (X : F)$. Conversely, let $x \in (X : F)$. Then $x \vee a \in F$ for all $a \in X$. In particular, for $x \in X$, we get $x = x \vee x \in F$. Hence $(X : F) \subseteq F$. Therefore $(X : F) = F$.
(2) Suppose $a \leq b$. Then $a \vee b = b$. Let $x \in (a : F)$. Then $x \vee a \in F$. Since F is a filter, we get $x \vee a \vee b \in F$. Thus $x \vee b = x \vee a \vee b \in F$, which means $x \in (b : F)$. Therefore $(a, F) \subseteq (b : F)$.
(3) Suppose F and G are two filters of X such that $F \subseteq G$. Let $x \in (a : F)$. Then we get that $x \vee a \in F \subseteq G$. Hence $x \in (a : G)$. Therefore $(a : F) \subseteq (a : G)$.
(4) For any $x \in X$, we get $x \in (a : F \cap G) \Leftrightarrow x \vee a \in F \cap G \Leftrightarrow x \vee a \in F$ and $x \vee a \in G \Leftrightarrow x \in (a : F)$ and $x \in (b : G) \Leftrightarrow x \in (a : F) \cap (a : G)$. Therefore $(a : F \cap G) = (a : F) \cap (a : G)$.
(5) Let $x \in (a : F)$. Then $a \vee x \in F$, which is true for all $x \in (a : F)$. Hence $x \in ((a : F) : F)$. Let $x \in \langle a \rangle$. Then $a^n * x = 1 \in ((a : F) : F)$ for some positive integer n. Since $a \in ((a : F) : F)$ and $((a : F) : F)$ is a filter, we get $x \in ((a : F) : F)$. Therefore $\langle a \rangle \subseteq ((a : F) : F)$.
(6) Since $\{a\} \subseteq \langle a \rangle$, we get $(\langle a \rangle : F) \subseteq (\{a\} : F) = (a : F)$. Conversely, let $x \in (a : F)$ and $t \in \langle a \rangle$. Then $a^n * t = 1$ for some positive integer n. Hence $1 = (a * (a * (\cdots * (a * t))) \vee x \leq (a \vee x) * ((a \vee x) * (\cdots * ((a \vee x) * (t \vee x))$. Thus $(a \vee x) * ((a \vee x) * (\cdots * ((a \vee x) * (t \vee x)) = 1 \in F$. Since $a \vee x \in F$, we get $t \vee x \in F$. This is true for all $t \in \langle a \rangle$. Hence $x \in (\langle a \rangle : F)$.
(7) From (5), we get $\langle a \rangle \subseteq ((a : F) : F)$. Then from (6), we get $(((a : F) : F) : F) \subseteq (\langle a \rangle : F) = (a : F)$. Conversely, let $x \in (a : F)$. Then $x \vee y \in F$ for all $y \in ((a : F) : F)$. Hence $x \in (((a : F) : F) : F)$. Therefore $(a : F) \subseteq (((a : F) :$

$F) : F)$.
(8) Suppose $(a : F) = X$. Since $a \in X = (a : F)$, we get $a = a \vee a \in F$. Conversely, let $a \in F$. Since F is a filter of X, we get $a \vee x \in F$ for any $x \in X$. Hence $X \subseteq (a : F)$.
(9) Clearly $F \subseteq (a : F)$ and $F \subseteq ((a : F) : F)$. Hence $F \subseteq (a : F) \cap ((a : F) : F)$. Conversely, let $x \in (a : F) \cap ((a : F) : F)$. Then $x \in (a : F)$ and $x \in ((a : F) : F)$. Hence $x \vee y \in F$ for all $y \in (a : F)$. In particular, $x = x \vee x \in F$. Therefore $(a : F) \cap ((a : F) : F) \subseteq F$. □

For any filter F of a commutative BE-algebra X, denote the set $\{((a : F) : F) \mid a \in X\}$ of all dual annihilators with respect to F by $\mathcal{A}_F(X)$.

Theorem 4.10.10 *For any filter F of a commutative BE-algebra X, the set $\mathcal{A}_F(X)$ forms a complete semilattice with respect to set intersection. Moreover, the mapping $\Psi_F : X \to \mathcal{A}_F(X)$ defined by $\Psi_F(x) = ((x : F) : F)$ for all $x \in X$ is a dual epimorphism up to semilattices.*

Proof Let F be an arbitrary filter of X. Let $a, b \in X$. Since $a, b \leq a \vee b$, we get $(a : F), (b : F) \subseteq (a \vee b : F)$. Hence $((a \vee b : F) : F) \subseteq ((a : F) : F), ((b : F) : F)$. Therefore $((a \vee b : F) : F) \subseteq ((a : F) : F) \cap ((b : F) : F)$. Conversely, let $x \in ((a : F) : F) \cap ((b : F) : F)$, $y \in (a \vee b : F)$, $f \in F$ and $g \in G$. Clearly $f \vee g \in F \cap G$. Now we have

$$\begin{aligned}
y \in (a \vee b : F) &\Rightarrow y \vee a \vee b \in F \\
&\Rightarrow y \vee a \in (b : F) \\
&\Rightarrow x \vee y \vee a \in F \qquad \text{since } x \in ((b : F) : F) \\
&\Rightarrow x \vee y \in (a : F) \\
&\Rightarrow x \vee y \in (a : F) \cap ((a : F) : F) = F \qquad \text{since } x \in ((a : F) : F) \\
&\Rightarrow x \vee y \in F \qquad \text{for all } y \in (a \vee b : F) \\
&\Rightarrow x \in ((a \vee b : F) : F)
\end{aligned}$$

Hence $((a : F) : F) \cap ((b : F) : F) \subseteq ((a \vee b : F) : F)$. Thus $((a \vee b : F) : F) = ((a : F) : F) \cap ((b : F) : F)$. Therefore $\langle \mathcal{A}_F(X), \cap \rangle$ is a semilattice which is a partially ordered set with respect to set inclusion. Hence $\langle \mathcal{A}_F(X), \cap \rangle$ is a complete semilattice. It is remaining to prove that the mapping $\Psi_F : X \to \mathcal{A}_F(X)$ defined by $\Psi_F(x) = ((x : F) : F)$ for all $x \in X$ is a dual epimorphism up to semilattices. Clearly Ψ_F is well-defined. Let $a, b \in X$. Then

$$\begin{aligned}
\Psi_F(a \vee b) &= ((a \vee b : F) : F) \\
&= ((a : F) : F) \cap ((b : F) : F) \\
&= \Psi_F(a) \cap \Psi_F(b).
\end{aligned}$$

Let $((a : F) : F) \in \mathcal{A}_F(X)$. For this $a \in X$, $\Psi_F(a) = ((a : F) : F)$. Hence Ψ_F is surjective. Therefore Ψ_F is a dual epimorphism up to semilattices. □

Proposition 4.10.11 *Suppose that F is a filter of a simple and commutative BE-algebra X. Let G be a filter that is maximal among all relative dual annihilators of non-unit elements of X with respect to the filter F. Then G is a prime filter of X.*

Proof Suppose $G = (x : F)$ for some $1 \neq x \in X$. Clearly G is a filter of X. Let $a, b \in X$ be such that $a \vee b \in G$ and $a \notin G$. Then $x \vee a \neq 1$, otherwise $a \in (x : F) = G$. Since $x \leq x \vee a$, we get $(x : F) \subseteq (x \vee a : F)$. Since G is maximal among all relative dual annihilators of non-unit elements of X, we get that $(x \vee a : F) = (x : F) = G$. Since $a \vee b \in G$, we get $a \vee b \vee x \in F$ and hence $b \in (x \vee a : F) = G$. Therefore G is a prime filter of X. □

Theorem 4.10.12 *Let F be a filter of a simple and commutative BE-algebra X and $x \in X$. If F is a prime filter of X, then $(x : F)$ is a prime filter of X.*

Proof Assume that F is a prime filter of X. Let $a, b \in X$ be such that $a \vee b \in (x : F)$. Then $(a \vee x) \vee (b \vee x) \in F$. Since F is prime, we get either $a \vee x \in F$ or $b \vee x \in F$. Hence $a \in (x : F)$ or $b \in (x : F)$. Therefore $(x : F)$ is a prime filter of X. □

Proposition 4.10.13 *If X is a totally ordered, simple, and commutative BE-algebra, then A^+ is a prime filter for any $\emptyset \neq A \subseteq X$.*

Proof Suppose X is totally ordered. Let $a, b \in X$ and $a \vee b \in A^+$. Since X is totally ordered, we get either $a \leq b$ or $b \leq a$. Hence $b = a \vee b \in A^+$ or $a = a \vee b \in A^+$. Therefore A^+ is prime. □

For any commutative BE-algebras X, it was already observed that X is a semilattice with respect to the operation $\vee$. An element $a \in X$ is said to be *join-irreducible* if $c = a \vee b$ implies $c = a$ or $c = b$ for any $a, b \in X$. Using these facts, a set of equivalent conditions is derived for the dual annihilator A^+ of a commutative BE-algebra to become a prime filter.

Theorem 4.10.14 *The following are equivalent in a simple and commutative BE-algebra X:*

(1) $\{1\}$ *is a prime filter,*
(2) 1 *is join-irreducible,*
(3) *for* $\emptyset \neq A \subseteq X, A^+$ *is prime.*

Proof (1) $\Rightarrow$ (2): Assume that $\{1\}$ is prime. Let $a, b \in X$ be such that $a \vee b = 1$. Then $a \vee b \in \{1\}$. Since $\{1\}$ is prime, we get $a \in \{1\}$ or $b \in \{1\}$. Hence $a = 1$ or $b = 1$. Therefore 1 is join-irreducible.
(2) $\Rightarrow$ (3): Assume that 1 is join-irreducible. Let $\emptyset \neq A \subseteq X$. Let $x, y \in X$ be such that $x \vee y \in A^+$. Then $(x \vee a) \vee (y \vee a) = x \vee y \vee a = 1$ for all $a \in A$. Since 1 is join-irreducible, we get either $x \vee a = 1$ or $y \vee a = 1$. Hence $x \in A^+$ or $y \in A^+$. Therefore A^+ is a prime filter of X.
(3) $\Rightarrow$ (1): Assume that A^+ is prime for each $\emptyset \neq A \subseteq X$. By putting $A = X$, we get that $X^+ = \{1\}$ is a prime filter of X. □

Lemma 4.10.15 *Let* $(X, *, 1)$ *and* $(Y, *, 1')$ *bet two BE-algebras and* $f : X \to Y$ *be a homomorphism. Then* $f(x \vee y) = f(x) \vee f(y)$ *for all* $x, y \in X$ *where* $x \vee y = (y * x) * x$.

Proof Clearly $f(x \vee y) = f((y * x) * x) = f(y * x) * f(x) = (f(y) * f(x)) * f(x) = f(x) \vee f(y)$. □

Lemma 4.10.16 *Let* $(X, *, 1)$ *and* $(Y, *, 1')$ *bet two BE-algebras. If* $f : X \to Y$ *is a homomorphism, then* $f(A^+) \subseteq \{f(A)\}^+$ *for any non-empty subset* A *of* X.

Proof Let $a \in f(A^+)$ and $y \in f(A)$. Then there exists $b \in A^+$ and $x \in A$ such that $a = f(b)$ and $y = f(x)$. Now $a \vee y = f(b) \vee f(x) = f(b \vee x) = f(1) = 1'$. Therefore $a \in \{f(A)\}^+$. □

In general, the equality $\{f(A)\}^+ = f(A^+)$ does not hold for any non-empty subset A of a BE-algebra. For, consider the following example:

Example 4.10.17 Let $(X, *, 1)$ be a totally ordered and commutative BE-algebras. Define a mapping $f : X \to X$ by $f(x) = 1$ for all $x \in X$. Then clearly f is a homomorphism on X. Consider a non-empty subset A. Suppose $x \in A^+$. Then $x \vee a = 1$ for all $1 \neq a \in A$. Since X is totally ordered, we get either $x \leq a$ or $a \leq x$. If $x \leq a$, then $1 = x \vee a = a$. Since $a \neq 1$, we get $x = 1$. Hence $A^+ = \{1\}$. Clearly $f(A) = \{1\}$. Hence $f(A^+) = f(\{1\}) = \{1\}$ and $\{f(A)\}^+ = X$. Therefore $\{f(A)\}^+ \neq f(A^+)$.

The concept of annihilator preserving homomorphisms is now introduced.

Definition 4.10.18 Let $(X, *, 1)$ and $(Y, *, 1')$ be two commutative BE-algebras. Then a homomorphism $f : X \to Y$ is called dual annihilator preserving if $f(A^+) = \{f(A)\}^+$ for any $\emptyset \neq A \subseteq X$.

Example 4.10.19 Let X and Y be two commutative BE-algebras and $A^+ = \{1\}$ for all $\emptyset \neq A \subseteq X$. Then every homomorphism from X into Y preserves dual annihilators.

Theorem 4.10.20 *Let* $(X, *, 1)$ *and* $(Y, *, 1')$ *be two commutative BE-algebras and* $f : X \to Y$ *an isomorphism. Then both* f *and* f^{-1} *preserve dual annihilators.*

Proof Assume that f is an isomorphism. Let A be a non-empty subset of X. We have always $f(A^+) \subseteq \{f(A)\}^+$. Let $x \in \{f(A)\}^+ \subseteq Y$. Since f is onto, there exists $y \in X$ such that $f(y) = x$. Then we get the following consequence:

$$
\begin{aligned}
f(y) \in \{f(A)\}^+ &\Rightarrow f(y) \vee f(a) = 1' && \text{for all } a \in A \\
&\Rightarrow f(y \vee a) = 1' = f(1) \\
&\Rightarrow y \vee a = 1 && \text{since } f \text{ is one-to-one} \\
&\Rightarrow y \vee a = 1 && \text{for all } a \in A \\
&\Rightarrow y \in A^+ \\
&\Rightarrow x = f(y) \in f(A^+)
\end{aligned}
$$

Hence $\{f(A)\}^+ \subseteq f(A^+)$. Thus $\{f(A)\}^+ = f(A^+)$. Therefore f is preserving dual annihilators. Let B be a non-empty subset of Y. It is enough to prove that $f^{-1}(B^+) = \{f^{-1}(B)\}^+$. Let $x \in \{f^{-1}(B)\}^+$. Then $x \vee b = 1$ for all $b \in f^{-1}(B)$. Hence $f(x) \vee f(b) = f(x \vee b) = f(1) = 1'$ for all $b \in f^{-1}(B)$. Since f is onto, we get $f(x) \in B^+$. Hence $x \in f^{-1}(B^+)$. Hence $\{f^{-1}(B)\}^+ \subseteq f^{-1}(B^+)$. Conversely, let $x \in f^{-1}(B^+)$ and $b \in f^{-1}(B)$. Then $f(x) \in B^+$ and $f(b) \in B$. Hence $f(x \vee b) = f(x) \vee f(b) = 1$, which means that $f(x \vee b) = 1' = f(1)$. Since f is one-to-one, we get $x \vee b = 1$ for all $b \in f^{-1}(B)$. Thus $x \in \{f^{-1}(B)\}^+$. Hence $f^{-1}(B^+) = \{f^{-1}(B)\}^+$. Therefore f^{-1} preserves dual annihilators. □

The notion of relative dual annihilator filters is now introduced in a commutative BE-algebra.

Definition 4.10.21 Let F and G be two filters of a commutative BE-algebra X. Then the filter G is said to be a *relative dual annihilator filter* or *dual annihilator filter* with respect to F if $G = ((G : F) : F)$ or equivalently $G = (S : F)$ for some non-empty subset S of X.

For any filter F of a commutative BE-algebra X, clearly each $(a : F), a \in X$ is a relative dual annihilator filter of X. A filter F of a commutative BE-algebra X is called simply dual annihilator filter of X if $F = F^{++}$. In the following theorem, the images and the inverse images of dual annihilator filters of commutative BE-algebras are observed under the light of a dual annihilator preserving homomorphism.

Theorem 4.10.22 *Let $(X, *, 1)$ and $(Y, *, 1')$ be two commutative BE-algebras. If $f : X \to Y$ is an epimorphism which preserves dual annihilators, then $f(F)$ is a dual annihilator filter of Y for every dual annihilator filter F of X.*

Proof Let F be a dual annihilator filter of X. Then $F = F^{++}$. Since f is onto, we get that $f(F)$ is a filter of Y. Since f preserves dual annihilators, we get $\{f(F)\}^{++} = f(F^{++}) = f(F)$. Therefore $f(F)$ is a dual annihilator filter of Y. □

Theorem 4.10.23 *Let $(X, *, 1)$ and $(Y, *, 1')$ be two simple and commutative BE-algebras. If $f : X \to Y$ is a homomorphism such that f^{-1} preserves dual annihilators, then $f^{-1}(G)$ is a prime dual annihilator filter of X, for every prime dual annihilator filter G of Y.*

Proof Let G be a prime dual annihilator filter of Y. Clearly $f^{-1}(G)$ is a filter of X. Since f^{-1} preserves annihilators, we get that $\{f^{-1}(G)\}^{++} = f^{-1}(G^{++}) = f^{-1}(G)$. Therefore $f^{-1}(G)$ is a dual annihilator filter of X. Suppose G is a prime filter of Y. Let $x, y \in X$ such that $x \vee y \in f^{-1}(G)$. Then $f(x) \vee f(y) = f(x \vee y) \in G$. Since G is a prime filter of Y, we get $f(x) \in G$ or $f(y) \in G$. Hence $x \in f^{-1}(G)$ or $y \in f^{-1}(G)$. Therefore $f^{-1}(G)$ is a prime dual annihilator filter of X. □

Exercise

1. For any non-empty subset A of a BE-algebra X, prove that $\langle A \rangle$ is the intersection of all filters containing A. Prove that the mapping $A \to \langle A \rangle$ is a closure operator and hence $\mathcal{F}(X)$ is a complete lattice with respect to set inclusion.

2. Let F_1, F_2 are two filters of a BE-algebra $(X, *, 1)$. If θ_{F_1} and θ_{F_2} are the congruences generated by F_1 and F_2, respectively, then show that $F_1 \subseteq F_2$ if and only if $\theta_{F_1} \subseteq \theta_{F_2}$.
3. Let $(X, *, 1)$ be a simple BE-algebras (a BE-algebra in which $\{1\}$ and X are the only filters). Then prove that $\{(1, 1)\}, L \times L$ are the only congruence relations on X.
4. Consider the product of congruences θ and ϕ on a BE-algebra X as $(a, b) \in \theta \circ \phi$ if and only if there exists $x \in X$ such that $(a, x) \in \theta$ and $(x, b) \in \phi$. If θ_1 and θ_2 are two congruence relations on X, then prove that $\theta_1 \circ \theta_2$ is a congruence on X.
5. Prove that the homomorphic image of an associative filter is again an associative filter.
6. Let X be a BE-algebra. If A is a subalgebra of X such that $A = F \cup G$ where F, G are two filters, then prove that A is a filter of X.
7. Let X be a partially ordered BE-algebras and $\mathcal{C}$ a closure operator on X. Then show that the set $F = \{x \in X \mid \mathcal{C}(x) = 1\}$ is a filter of X. Prove that the map $\mathcal{C}_0 : X/F \to \mathcal{C}(X)$ defined by $\mathcal{C}_0(F_x) = \mathcal{C}(x)$ is an isomorphism.
8. Let X_1 and X_2 be two BE-algebras and $h : X_1 \to X_2$ a homomorphism. Let $\mathcal{C}_1 : X_1 \to \mathcal{C}(X_1), \mathcal{C}_2 : X_2 \to \mathcal{C}(X_2)$ be two closure operator and $F^1 = \{x \in X_1 \mid \mathcal{C}_1(x) = 1\}, F^2 = \{x \in X_2 \mid \mathcal{C}_2(x) = 1\}$. If h is surjective such that $h(\mathcal{C}_1(x)) = \mathcal{C}_2(h(x))$ for all $x \in X_1$ and $h(\mathcal{C}_1(x)) = 1$ implies $\mathcal{C}_1(x) = 1$, then prove that $h_0 : X_1/F^1 \to X_2/F^2$ is isomorphism.
9. Let F be a filter of a self-distributive BE-algebra. If $\overline{K}$ is a filter of the quotient algebra X/F, then show that $K = \cup\{F_x \mid F_x \in \overline{K}\}$ is a filter of X and $F \subseteq K$, where F_x is the congruence class of x w.r.t. the congruence θ defined as $(x, y) \in \theta$ if and only if $x * y \in F$ and $y * x \in F$. Then there exists a bijection between the set of all filters of X/F and the set of all filters of X which are containing F.
10. For any subalgebra K of a self-distributive BE-algebra X and a filter F of X, prove that $K \cap F$ is a filter and the set $F_K = \bigcup\limits_{x \in K} F_x$ is a subalgebra of X (where F_x is the usual congruence class with respect to θ_F). Furthermore, prove that $K/(K \cap F) \cong F_K/F$.
11. Prove that any finite proper *BE*-chain may be extended to at least two non-isomorphic proper *BE*-chains of the same order.
12. Prove that any final segment $[c, 1]$ is isomorphic to a maximal filter of some BE-algebras.
13. In a BE-algebra X, define the set X_0 as $X_0 = \{x \in \mathcal{A}(X) \mid a * x = x \text{ for all } a \in X - \{x\}\}$ and let $X' = X_0 \cup \{1\}$. Then prove that X' is a dual atomistic filter of X. Furthermore, if $u \in X - X'$ is a minimal element of X, then prove that $b * u = u$ for any $b \in X'$.
14. For any BE-algebra $(X, *, 1)$, prove that X is a dual atomistic BE-algebra if and only if every final segment of X is a two element chain.
15. If X_1 is a subalgebra of a commutative BE-algebra X and P_1 is a prime filter of X_1, then prove that there exists a prime filter P of X such that $P_1 = X_1 \cap P$.
16. Let $\phi : X \to Y$ be a *BE*-homomorphism. If F is a prime filter of the BE-algebra X such that $Ker(\phi) \subseteq F$, then prove that $\phi(G)$ is a prime filter of Y.

17. A non-empty subset F of a BE-algebra X is called a finite $\cap$-structure, if $(\langle x\rangle \cap \langle y\rangle) \cap F \neq \emptyset$ for all $x, y \in F$, and X is called a finite $\cap$-structure if $X - \{0\}$ is a finite $\cap$-structure. Then prove that a filter F of X is prime if and only if $I = X - F$ is a finite $\cap$-structure.
18. Suppose that F is a filter of a commutative BE-algebra X. Let G be a filter that is maximal among all dual annihilators of non-unit elements of F. Then prove that G is prime.

Chapter 5
Quasi-filters of BE-algebras

Several types of algebraic structures have been studied so far in the literature of mathematics. In abstract algebras, the main emphasis is given to the study of algebraic structures and their properties. Though there are many different views on algebraic structures, the central idea of an algebraic structure is a system involving one or more sets equipped with one or more n-ary operations satisfying a given set of axioms. Universal algebras is another branch of mathematics that studies algebraic structures in general. In light of universal algebras, many algebraic structures can be divided into varieties and quasi-varieties depending on the axioms used.

One of the specific objectives of studying weak structures and weak substructures in abstract algebras and universal algebras is to obtain the required outcomes to the full extent but with limited postulates. So far many authors studied the notions of weak ideals, semi-ideals, quasi-filters, etc., in many algebraic structures such as rings, lattices, and implication algebras. In [189], A.S. Nasab and A.B. Saeid introduced the concept of semimaximal filters in Hilbert algebras as a closed set of that closure operator and studied it in detail. Some properties of weak hyper filters were studied in hyper BE-algebras by A. Radfar et al. [197]. Smarandache weak BE-algebras were introduced and studied by A.B. Saeid in [208].

In this chapter, the notion of quasi-filters is introduced in BE-algebras. An equivalent condition is derived for a quasi-filter of a BE-algebra to become a filter. The notion of multipliers is introduced, and some properties of quasi-filters are studied with the help of multipliers. An interconnection is obtained between idempotent multipliers and weak congruences. A bijection is obtained between the class of all quasi-filters of a BE-algebra and the set of all quasi-filters of the respective quotient algebra w.r.t. a congruence on it. The notions of simple filters and pseudo filters are introduced in BE-algebras, and the theory of soft sets is applied to the pseudo filters.

S. R. Mukkamala, *A Course in BE-algebras*,
https://doi.org/10.1007/978-981-10-6838-6_5

5.1 Quasi-filters of BE-algebras

In this section, the notion of quasi-filters is introduced in BE-algebras. Properties of quasi-filters are studied with respect to homomorphisms and Cartesian products. An equivalent condition is derived for a quasi-filter of a BE-algebra to become a filter.

Definition 5.1.1 Let $(X, *, 1)$ be a BE-algebra. A non-empty subset F of X is said to be a *quasi-filter* if it satisfies the following condition for all $x, y \in X$.

$$x \in X \text{ and } y \in F \text{ imply that } x * y \in F.$$

It is obvious that $\{1\}$ is a quasi-filter of a BE-algebra X. Suppose that $\mathcal{QF}(X)$ denote the set of all quasi-filters of a BE-algebra X. It is obvious that $\bigcap\limits_{F \in \mathcal{QF}(X)} F$ is also a quasi-filter of X.

Example 5.1.2 Let $X = \{1, a, b, c\}$. Define a binary operation $*$ on X as follows:

$*$	1	a	b	c
1	1	a	b	c
a	1	1	a	a
b	1	1	1	a
c	1	1	a	1

Then it can be easily verified that $(X, *, 1)$ is a BE-algebra. Consider $F = \{1, a\}$. Clearly F is a quasi-filter of X. It is clear that F is not a filter because $a, a * b = a \in F$ but $b \notin F$.

In the following, we provide some more examples of quasi-filters of BE-algebras.

Proposition 5.1.3 *For any $a \in X$, the set $\uparrow a = \{x \in X \mid a * x = 1\}$ is a quasi-filter of X.*

Proof Clearly $1 \in \uparrow a$. Let $x \in X$ and $y \in \uparrow a$. Then $a * y = 1$. Now $a * (x * y) = x * (a * y) = x * 1 = 1$. Hence $x * y \in \uparrow a$. Therefore $\uparrow a$ is a quasi-filter of X. □

Proposition 5.1.4 *Let X be a BE-algebra and $a \in X$ be a fixed element. Then the set $S_a = \{a * x \mid x \in X\}$ is a quasi-filter of X.*

Proof Clearly $1 = a * a \in S_a$. Let $r \in X$ and $b \in S_a$. Then $b = a * x$ for some $x \in X$. Hence $r * b = r * (a * x) = a * (r * x) \in S_a$. Thus $r * b \in S_a$. Therefore S_a is a quasi-filter of X. □

Proposition 5.1.5 *Let F be a quasi-filter of a BE-algebra X. Then for any $a \in X$, the set $a * F = \{a * x \mid x \in F\}$ is a quasi-filter of X.*

Proof Let F be a quasi-filter of X. Let $a \in X$. Since $1 \in F$, we get that $1 = a * 1 \in a * F$. Let $x \in X$ and $c \in a * F$. Then $c = a * b$ for some $b \in F$. Since $b \in F$ and F is a quasi-filter, we get $x * b \in F$. Now $x * c = x * (a * b) = a * (x * b) \in a * F$. Therefore $a * F$ is a quasi-filter of X. □

Theorem 5.1.6 *Let F and G be two quasi-filters of a BE-algebra X. Then $F \cap G$ and $F \cup G$ are also quasi-filters of X.*

Proof Let $x \in X$ and $a \in F \cap G$. Since $x \in X, a \in F$, and F is a quasi-filter of X, it yields that $x * a \in F$. Similarly, we obtain $x * a \in G$. Hence it concludes that $x * a \in F \cap G$. Therefore $F \cap G$ is a quasi-filter of X. Again, let $x \in X$ and $a \in F \cup G$. Then $a \in F$ or $a \in G$. If $a \in F$, then $x * a \in F \subseteq F \cup G$ because F is a quasi-filter of X. Therefore $F \cup G$ is a quasi-filter of X. If $a \in G$, then similarly we can obtain that $F \cup G$ is a quasi-filter of X. □

The following result is a generalization of the above theorem.

Corollary 5.1.7 *If $\{F_\alpha \mid \alpha \in \Delta\}$ is an indexed family of quasi-filters of a BE-algebra X, then $\bigcap_{\alpha\in\Delta} F_\alpha$ and $\bigcup_{\alpha\in\Delta} F_\alpha$ are also quasi-filters of X.*

In general, the union of two filters of a BE-algebra X may not be a filter of X. It can be seen in the following example.

Example 5.1.8 Let $X = \{1, a, b, c\}$. Define a binary operation $*$ on X as follows:

$*$	1	a	b	c
1	1	a	b	c
a	1	1	b	b
b	1	a	1	a
c	1	c	c	1

Then it can be easily verified that $(X, *, 1)$ is a BE-algebra. Consider the sets $F = \{1, a\}$ and $G = \{1, b\}$. Clearly F and G are filters of X, but $F \cup G = \{1, a, b\}$ is not a filter of X because $a \in F \cup G$ and $a * c = b \in F \cup G$ but $c \notin F \cup G$.

Since every filter is a quasi-filter, however, we have the following result.

Corollary 5.1.9 *The union of two filters of a BE-algebra X is a quasi-filter of X.*

Let F be a quasi-filter and G a subalgebra of a BE-algebra X. Then $F \cup G$ is not a quasi-filter of X in general as seen in the following example.

Example 5.1.10 Let $X = \{1, a, b, c, d\}$. Define a binary operation $*$ on X as follows:

$*$	1	a	b	c	d
1	1	a	b	c	d
a	1	1	b	b	d
b	1	a	1	a	d
c	1	1	1	1	d
d	1	a	b	d	1

Then it can be easily seen that $(X, *, 1)$ is a BE-algebra. It is easy to check that $F = \{1, a, b\}$ is a quasi-filter of X and $G = \{1, c\}$ is a subalgebra of X. But $F \cup G = \{1, a, b, c\}$ is not a quasi-filter of X because $c \in F \cup G$ and $d * c = d \notin F \cup G$.

Theorem 5.1.11 *Let $(X, *, 1)$ be a BE-algebra. If F is a quasi-filter and G a subalgebra of X, then $F \cap G$ is a quasi-filter of G.*

Proof Let $x \in G$ and $a \in F \cap G$. Then $a \in F$ and $a \in G$. Since F is a quasi-filter and $a \in F$, we get that $x * a \in F$. Since $x, a \in G$ and G is a subalgebra, it yields that $x * a \in G$. Hence, it implies $x * a \in F \cap G$. Therefore, it concludes that $F \cap G$ is a quasi-filter of G. □

It can be easily observed that every filter of a BE-algebra is a quasi-filter but not the converse. However, in the following theorem, we derive a necessary and sufficient condition for every quasi-filter of a BE-algebra to become a filter.

Theorem 5.1.12 *A quasi-filter F of a BE-algebra X is a filter if and only if $a, b \in F$ implies $(a * (b * x)) * x \in F$ for all $x, a, b \in X$*

Proof Let F be a quasi-filter of X. Assume that F is a filter of X. Let $a, b \in F$. Then for any $x \in X$, we get $a * ((a * (b * x)) * (b * x)) = (a * (b * x)) * (a * (b * x)) = 1$. Hence, it yields that $a \leq ((a * (b * x)) * (b * x))$. Since $a \in F$, we get that $(a * (b * x)) * (b * x) \in F$. Hence $b * ((a * (b * x)) * x) \in F$. Since $b \in F$, it implies that $(a * (b * x)) * x \in F$.

Conversely, assume that F satisfies the given condition. For any $a \in F$, clearly $1 = a * a \in F$. Let $x, y \in X$ be such that $x, x * y \in F$. We first observe the following:

$$\begin{aligned}(y * ((y * x) * y)) * y &= ((y * x) * (y * y)) * y \\ &= ((y * x) * 1) * y \\ &= 1 * y \\ &= y\end{aligned}$$

By the assumed condition, we get $y \in F$. Therefore F is a filter in X. □

Theorem 5.1.13 *If F_1, F_2 are two quasi-filters of X_1 and X_2 respectively, then $F_1 \times F_2$ is a quasi-filter of the product $X_1 \times X_2$. Conversely, every quasi-filter F of $X_1 \times X_2$ can be expressed as $F = F_1 \times F_2$ where F_i is a quasi-filter of X_i for $i = 1, 2$ provided $a * b = a$ for any $a \in F_1, b \in F_2$.*

Proof Let F_1 and F_2 be quasi-filters of X_1 and X_2, respectively. Since $1 \in F_1$ and $1 \in F_2$, we get $(1, 1) \in F_1 \times F_2$. Let $(x, y) \in X_1 \times X_2$ and $(x_1, y_1) \in F_1 \times F_2$. Since F_1 and F_2 are quasi-filters of X_1 and X_2 respectively, we can get $x * x_1 \in F_1$ and $y * y_1 \in F_2$. Hence $(x, y) * (x_1, y_1) = (x * x_1, y * y_1) \in F_1 \times F_2$. Therefore $F_1 \times F_2$ is a quasi-filter of $X_1 \times X_2$.

Conversely, let F be any quasi-filter of $X_1 \times X_2$. Consider the projections $\Pi_i : X_1 \times X_2 \to X_i$ for $i = 1, 2$. Let F_1 and F_2 be the projections of F on X_1 and X_2 respectively. That is $\Pi_i(F) = F_i$ for $i = 1, 2$. We now prove that F_1 and F_2 are quasi-filters of X_1 and X_2 respectively. Since $(1, 1) \in F$, we get $1 = \Pi_1(1, 1) \in F_1$. Let

$x_1 \in X_1$ and $a \in F_1$. Then $(x_1, 1) \in X_1 \times X_2$ and $(a, 1) \in F$. Since F is a quasi-filter, we get $(x_1 * a, 1 * 1) \in F$. Thus $x_1 * a = \Pi_1(x_1 * a, 1) \in \Pi_1(F) = F_1$. Therefore F_1 is a quasi-filter of X_1. Similarly, we can get F_2 is a quasi-filter of X_2.

We now prove that $F = F_1 \times F_2$. Clearly $F \subseteq F_1 \times F_2$. Let $(x, y) \in F_1 \times F_2$. Then $x \in F_1 = \Pi_1(F)$ and $y \in F_2 = \Pi_2(F)$. Hence $(x, 1) \in F$ and $(y, y) \in F$. Since F is a quasi-filter, we get $(x, y) = (x * y, 1 * y) = (x, 1) * (y, y) \in F$. Thus $F_1 \times F_2 \subseteq F$. Therefore $F = F_1 \times F_2$. □

Theorem 5.1.14 *Let $\phi : X_1 \to X_2$ be an epimorphism from a BE-algebra X_1 onto a BE-algebra X_2. If F is a quasi-filter of X_1, then $\phi(F)$ is a quasi-filter of X_2.*

Proof Clearly $1 = \phi(1) \in \phi(F)$. Let $x \in X_2$ and $\phi(a) \in \phi(F)$. Since ϕ is onto, there exists $t \in X_1$ such that $\phi(t) = x$. Since $a \in F$ and F is a quasi-filter of X_1, we get $t * x \in F$. Hence $x * \phi(a) = \phi(t) * \phi(a) = \phi(t * a) \in \phi(F)$. Therefore $\phi(F)$ is a quasi-filter of X_2. □

For any congruence θ on a BE-algebra X, it is already observed that the quotient algebra $X_{/\theta}$ is a BE-algebra with respect to the operation defined by $[x]_\theta * [y]_\theta = [x * y]_\theta$ for all $x, y \in X$, where $X_{/\theta} = \{[x]_\theta \mid x \in X\}$. Then clearly $\phi : X \to X_{/\theta}$ is a natural homomorphism.

Theorem 5.1.15 *Let θ be a congruence on a BE-algebra X and F a non-empty subset of X. Then F is a quasi-filter of X if and only if $\phi(F) = F_{/\theta}$ is a quasi-filter in $X_{/\theta}$.*

Proof Assume that F is a quasi-filter of X. Since $1 \in F$, we get that $[1]_\theta \in F_{/\theta}$. Let $[x]_\theta \in X_{/\theta}$ and $[a]_\theta \in F_{/\theta}$. Then, it yields that $a \in F$. Since F is a quasi-filter of X, it implies that $x * a \in F$. Hence we get $[x]_\theta * [a]_\theta = [x * a]_\theta \in F_{/\theta}$. Therefore $F_{/\theta}$ is a quasi-filter of $X_{/\theta}$.

Conversely, assume that $F_{/\theta}$ is a quasi-filter of $X_{/\theta}$. Since $[1]_\theta \in F_{/\theta}$, we get that $1 \in F$. Let $x \in X$ and $a \in F$. Then $[x]_\theta \in X_{/\theta}$ and $[a]_\theta \in F_{/\theta}$. Since $F_{/\theta}$ is a quasi-filter of $X_{/\theta}$, we get $[x * a]_\theta = [x]_\theta * [a]_\theta \in F_{/\theta}$. Hence $x * a \in F$. Therefore F is a quasi-filter of X. □

The following results can be proved similarly.

Theorem 5.1.16 *Let $\phi : X_1 \to X_2$ be an epimorphism from a BE-algebra X_1 into a BE-algebra X_2. If G is a quasi-filter of X_2, then $\phi^{-1}(G)$ is a quasi-filter of X_1.*

Corollary 5.1.17 *Let θ be a congruence on a BE-algebra X and $\phi : X \to X_{/\theta}$ be a natural homomorphism. If G is a quasi-filter of $X_{/\theta}$, then $\phi^{-1}(G)$ is a quasi-filter of X.*

Theorem 5.1.18 *Let F be a quasi-filter of a BE-algebra X. If S is a subalgebra of X which does not meet F, then there exists a subalgebra of X containing S which is maximal with respect to the property of not meeting F.*

Proof Consider the collection $\Im = \{T \subseteq X \mid T$ is a subalgebra of X and $T \cap F = \emptyset\}$. Clearly $S \in \Im$ and hence $\Im \neq \emptyset$. Let $\{T_\alpha\}_{\alpha\in\Delta} \subseteq \Im$ be a chain. Let $x, y \in \bigcup\limits_{\alpha\in\Delta} T_\alpha$. Then there exist $\alpha, \beta \in \Delta$ such that $x \in T_\alpha$ and $y \in T_\beta$. Since $\{T_\alpha\}_{\alpha\in\Delta}$ is a chain, we get either $x, y \in T_\alpha$ or $x, y \in T_\beta$. Since each T_α is a subalgebra, we get either $x * y \in T_\alpha$ or $x * y \in T_\beta$. Hence $x * y \in \bigcup\limits_{\alpha\in\Delta} T_\alpha$. Therefore $\bigcup\limits_{\alpha\in\Delta} T_\alpha$ is a subalgebra of X. Suppose $\big(\bigcup\limits_{\alpha\in\Delta} T_\alpha\big) \cap F \neq \emptyset$. Choose $x \in \big(\bigcup\limits_{\alpha\in\Delta} T_\alpha\big) \cap F$. Then there exists $\alpha \in \Delta$ such that $x \in T_\alpha$ and $x \in F$. Hence $x \in T_\alpha \cap F$, which is a contradiction to that $T_\alpha \in \Im$. Thus $\big(\bigcup\limits_{\alpha\in\Delta} T_\alpha\big) \cap F = \emptyset$. Therefore $\bigcup\limits_{\alpha\in\Delta} T_\alpha$ is an upper bound for the chain $\{T_\alpha\}_{\alpha\in\Delta}$. Thus by Zorn's lemma, $\Im$ has a maximal element, say M. Clearly M is a subalgebra of X such that $S \subseteq M$ and $M \cap F = \emptyset$. Hence the theorem is proved. □

Since every filter of a BE-algebra is a quasi-filter, the following result is clear.

Corollary 5.1.19 *Let F be a filter of a BE-algebra X. If S is a subalgebra of X which does not meet F, then there exists a subalgebra of X containing S which is maximal with respect to the property of not meeting F.*

Definition 5.1.20 Let $(X, *, 1)$ be a BE-algebra and $a \in X$. Then define the sets F_a and F^a as $F_a = \{x \in X \mid a * x = x\}$ and $F^a = \{x \in X \mid a * x = 1\}$.

Proposition 5.1.21 *For any $a \in X$ be a fixed element of a BE-algebra X, both F_a and F^a are quasi-filters of X. If X is self-distributive, then F^a is a filter of X.*

Proof Since $a * 1 = 1$, it is clear that $1 \in F_a$. Let $y \in X$ and $x \in F_a$. Then we get $a * x = x$. Hence $a * (y * x) = y * (a * x) = y * x$. Hence we get $y * x \in F_a$. Therefore F_a is a quasi-filter of X. Clearly $1 \in F^a$. Let $y \in X$ and $x \in F^a$. Then we get $a * x = 1$. Hence $a * (y * x) = y * (a * x) = y * 1 = 1$. Thus $y * x \in F^a$. Therefore F^a is a quasi-filter of X.

Moreover, Suppose that X is a self-distributive BE-algebra. Clearly $1 \in F^a$. Let $x, x * y \in F^a$. Then we get that $a * x = 1$ and $a * (x * y) = 1$. Hence we get $a * y = 1 * (a * y) = (a * x) * (a * y) = a * (x * y) = 1$. Thus $y \in F^a$. Therefore F^a is a filter of X. □

We now introduce two operations α and β on the class of quasi-filters.

Definition 5.1.22 For any quasi-filter F of a BE-algebra X, define two operations α and β as

$$\alpha(F) = \{x \in X \mid \text{ to each } s \in F, x * s = 1 \text{ implies } s = 1\} \text{ and}$$
$$\beta(F) = \{x \in X \mid x * s = 1 \text{ and } s \neq 1 \text{ for some } s \in F\}.$$

For any quasi-filter F of a BE-algebra X, it can be easily observed from the above definition that $\alpha(F)$ and $\beta(F)$ are complements to each other in X. Now we prove, in the following theorem, that $\alpha(F)$ is itself a quasi-filter of the BE-algebra X.

Theorem 5.1.23 *If F is a quasi-filter of a BE-algebra X, then $\alpha(F)$ is also a quasi-filter of X.*

Proof It is clear that $1 \in \alpha(F)$. Let $x \in \alpha(F)$ and $y \in X$. Assume that $y * x \leq s$ for all $s \in F$. Then we get that $x \leq y * x \leq s$. Hence $x * s = 1$, which implies $s = 1$, since $x \in \alpha(F)$. Thus we get that $y * x \in \alpha(F)$. Therefore, it concludes that $\alpha(F)$ is a quasi-filter of X. □

In the following lemma, we now prove some basic properties of $\alpha(F)$ and $\beta(F)$.

Lemma 5.1.24 *Let F_1, F_2 and F be quasi-filters of a BE-algebra X. Then the following hold.*

(1) $F_1 \subseteq F_2$ *implies* $\beta(F_1) \subseteq \beta(F_2)$,
(2) $F_1 \subseteq F_2$ *implies* $\alpha(F_2) \subseteq \alpha(F_1)$,
(3) $F_a \cap F^a = \{1\}$,
(4) $F \cap \alpha(F) \subseteq F_a \cap F^a$,
(5) $F \cap \alpha(F) = F_a \cap F^a \cap F$.

Proof (1) Assume that $F_1 \subseteq F_2$. Let $x \in \beta(F_1)$. Then there exists some $s \in F_1$ such that $s \neq 1$ and $x * s = 1$. Hence $x \in \beta(F_2)$ because $s \in F_2$. Therefore $\beta(F_1) \subseteq \beta(F_2)$.
(2) Assume that $F_1 \subseteq F_2$. Then by condition (1), we get that $\beta(F_1) \subseteq \beta(F_2)$. Since $\alpha(F)$ and $\beta(F)$ are complements to each other in X, it yields that $\alpha(F_2) \subseteq \alpha(F_1)$.
(3) Let $x \in F_a \cap F^a$. Then $a * x = x$ and $a * x = 1$. Hence $x = 1$. Therefore $F_a \cap F^a = \{1\}$.
(4) Let $x \in F \cap \alpha(F)$. Then by the definition of $\alpha(F)$, it is clear that $x = 1$. Hence, by the condition (3), we get that $x \in F_a \cap F^a$. Therefore $F \cap \alpha(F) \subseteq F_a \cap F^a$.
(5) Let $x \in F_a \cap F^a \cap F$. Then $x \in F$ and $x = 1$. Suppose $x * s = 1$ and $s \in F$. Then it yields that $s = 1$. Thus $x \in \alpha(F)$, which implies that $x \in F \cap \alpha(F)$. Therefore $F_a \cap F^a \cap F \subseteq F \cap \alpha(F)$. On the other hand, suppose $x \in F \cap \alpha(F)$. Then $x \in F$ and $x = 1$, which implies that $x \in F_a \cap F^a \cap F$. Thus $F \cap \alpha(F) \subseteq F_a \cap F^a \cap F$. Therefore $F \cap \alpha(F) = F_a \cap F^a \cap F$. □

Lemma 5.1.25 *Let F be a quasi-filter of a self-distributive BE-algebra X. If $a \in \alpha(F)$, then for every $s \in F$, we have $a * s = s$.*

Proof Suppose $a \in \alpha(F)$. Let $s \in F$. Then $a * ((a * s) * s) = (a * s) * (a * s) = 1$. Since $s \in F$ and F is a quasi-filter, we get that $(a * s) * s \in F$. Since $a \in \alpha(F)$, it concludes that $(a * s) * s = 1$. Clearly $s \leq a * s$. Since X is self-distributive, it gives $a * s \leq s$. Therefore $a * s = s$. □

Finally, we conclude this section with the following theorem which characterizes the elements of the quasi-filter $\alpha(F)$ of a self-distributive BE-algebra X.

Theorem 5.1.26 *Let F be a quasi-filter of a self-distributive BE-algebra X. Then the following are equivalent:*

(1) $a \in \alpha(F)$,
(2) $F \subseteq F_a$,
(3) $\beta(F) \subseteq \beta(F_a)$.

Proof (1) ⇒ (2): Let F be a quasi-filter of X. Assume $a \in \alpha(F)$. Let $s \in F$. Then, by above lemma, we get that $a * s = s$. Thus $s \in F_a$. Therefore $F \subseteq F_a$.
(2) ⇒ (3): Assume that $F \subseteq F_a$. Since F and F_a are quasi-filters, by above Lemma 5.1.24(1), we get that $\beta(F) \subseteq \beta(F_a)$.
(3) ⇒ (1): Assume that $\beta(F) \subseteq \beta(F_a)$. Let $s \in F$ and $a * s = 1$. Suppose $s \neq 1$. Then $a \in \beta(F)$. This implies that $a \in \beta(F_a)$ and hence there exists $t \in F_a$ such that $t \neq 1$ and $a * t = 1$. Thus $t = a * t = 1$, which is a contradiction. Therefore $s = 1$ and so $a \in \alpha(F)$. □

5.2 Multipliers and Quasi-filters

In this section, the concept of multipliers is introduced in BE-algebras. Some properties of quasi-filters are studied with the help of multiplier. The notion of weak congruences is introduced and obtained an interconnection between idempotent multipliers and weak congruences.

Definition 5.2.1 Let $(X, *, 1)$ be a BE-algebra. Then a self-mapping $f : X \to X$ is said to be a *multiplier* of X if it satisfies the following condition, for all $x, y \in X$.

$$f(x * y) = x * f(y)$$

Lemma 5.2.2 *Let f be a multiplier of a BE-algebra $(X, *, 1)$. Then we have*
(1) $f(1) = 1$,
(2) $x \leq f(x)$ *for all* $x \in X$.

Proof (1) Substituting $f(1)$ for x and 1 for y in the definition of multipliers, we obtain

$$f(1) = f(f(1) * 1) = f(1) * f(1) = 1$$

(2) Replace x by y in the definition, we get $1 = f(1) = f(x * x) = x * f(x)$. Hence $x \leq f(x)$. □

Lemma 5.2.3 *Let $(X, *, 1)$ be a BE-algebra. A multiplier $f : X \to X$ is an identity map if it satisfies $x * f(y) = f(x) * y$ for all $x, y \in X$.*

Proof Let $x, y \in X$ be such that $x * f(y) = f(x) * y$. Now $f(x) = f(1 * x) = f(1) * x = 1 * x = x$. Therefore f is an identity mapping. □

Proposition 5.2.4 *Let X_1 and X_2 be two BE-algebras. Define a map $f : X_1 \times X_2 \to X_1 \times X_2$ by $f(x, y) = (x, 1)$ for all $(x, y) \in X_1 \times X_2$. Then f is a multiplier of $X_1 \times X_2$ with respect to point-wise operations.*

Proof Let $(x_1, x_2), (y_1, y_2) \in X_1 \times X_2$. Then we get

$$\begin{aligned} f((x_1, x_2) * (y_1, y_2)) &= f(x_1 * y_1, x_2 * y_2) \\ &= (x_1 * y_1, 1) \\ &= (x_1 * y_1, x_2 * 1) \\ &= (x_1, x_2) * (y_1, 1) \\ &= (x_1, x_2) * f(y_1, y_2) \end{aligned}$$

Therefore f is a multiplier on the direct product $X_1 \times X_2$. □

In general, every multiplier of a BE-algebra need not be identity. However, in the following, we derive a set of conditions which are all together equivalent to that f being an identity multiplier.

Theorem 5.2.5 *Let $(X, *, 1)$ be a BE-algebra. A multiplier f of X is identity if and only if the following conditions are satisfied for all $x, y \in X$:*
(a) $f^2(x) = f(x)$,
(b) $f(x * y) = f(x) * f(y)$,
(c) $f^2(x) * y = f(x) * f(y)$.

Proof The condition for necessary is trivial. For sufficiency, assume the conditions (a), (b), and (c). Then for any $x, y \in X$, we can get $f(x) * y = f^2(x) * y = f(x) * f(y) = f(x * y)$. Also by the definition of multiplier, we have $f(x * y) = x * f(y)$. Hence

$$f(x * y) = x * f(y) = f(x) * y$$

Therefore by above lemma, f is an identity multiplier of X. □

Theorem 5.2.6 *Let f be a multiplier of a BE-algebra X. For any quasi-filter F of X, both $f(F)$ and $f^{-1}(F)$ are quasi-filters of X.*

Proof Clearly $1 = f(1) \in f(F)$. Let $x \in X$ and $a \in f(F)$. Then we get $a = f(s)$ for some $s \in F$. Now $x * a = x * f(s) = f(x * s) \in f(F)$, because $x * s \in F$. Therefore $f(F)$ is a quasi-filter of X. Since F is a quasi-filter, we get $f(1) = 1 \in F$. Hence $1 \in f^{-1}(F)$. Let $x \in X$ and $a \in f^{-1}(F)$. Then, it yields that $f(a) \in F$. Since F is a quasi-filter, we get $f(x * a) = x * f(a) \in F$. Hence $x * a \in f^{-1}(F)$. Therefore $f^{-1}(F)$ is a quasi-filter of X. □

In the following, the kernel of a multiplier is defined in a usual way.

Definition 5.2.7 Let f be a multiplier of a BE-algebra X. Then define the *kernel* of the multiplier f by $Ker(f) = \{x \in X \mid f(x) = 1\}$.

Proposition 5.2.8 *For any multiplier f of a BE-algebra X, $Ker(f)$ is a quasi-filter.*

Proof Since $f(1) = 1$, we get that $1 \in Ker(f)$. Let $a \in Ker(f)$ and $x \in X$. Then we get $f(x * a) = x * f(a) = x * 1 = 1$. Hence $x * a \in Ker(f)$. Therefore $Ker(f)$ is a quasi-filter of X. □

Definition 5.2.9 Let f be a multiplier of a BE-algebra X. An element $a \in X$ is said to be a *fixed element* if $f(a) = a$.

Let us denote the set of all fixed elements of a BE-algebra X by $Fix_f(X)$ and the image of X under the multiplier f by $f(X)$. Then we have the following:

Theorem 5.2.10 *Let f be a multiplier of a BE-algebra X. Then we have the following:*

(a) *$Fix_f(X)$ is a quasi-filter of X,*
(b) *$f(X)$ is a quasi-filter of X.*

Proof (a) Since $f(1) = 1$, we get $1 \in Fix_f(X)$. Let $x \in X$ and $a \in Fix_f(X)$. Then $f(a) = a$. Now $f(x * a) = x * f(a) = x * a$. Hence $x * a \in Fix_f(X)$. Therefore $Fix_f(X)$ is a quasi-filter of X.
(b) It follows from Theorem 5.2.6. □

Definition 5.2.11 Let $(X, *, 1)$ be a BE-algebra. A multiplier f of X is said to *idempotent* if $f^2(x) = f(x)$ for all $x \in X$.

Definition 5.2.12 Let $(X, *, 1)$ be a BE-algebra. An equivalence relation θ on X is called a *weak congruence* if $(x, y) \in \theta$ implies that $(a * x, a * y) \in \theta$ for any $a \in X$.

Clearly every congruence on a BE-algebra X is a weak congruence on X. In the following theorem, we have an example for a weak congruence in terms of multipliers.

Theorem 5.2.13 *Let f be a multiplier of a BE-algebra X. Define a binary relation θ_f on X as*

$$(x, y) \in \theta_f \text{ if and only if } f(x) = f(y) \text{ for all } x, y \in X$$

Then θ_f is a weak congruence on X.

Proof Clearly θ_f is an equivalence relation on the BE-algebra $(X, *, 1)$. Let $(x, y) \in \theta_f$. Then we get that $f(x) = f(y)$. Now, for any $a \in X$, we have the following:

$$\begin{aligned} f(a * x) &= a * f(x) \\ &= a * f(y) \\ &= f(a * y) \end{aligned}$$

Hence $(a * x, a * y) \in \theta_f$. Therefore θ_f is a weak congruence on X. □

Lemma 5.2.14 *Let f be an idempotent multiplier of a BE-algebra X. Then we have*

(1) *$f(x) = x$ for all $x \in f(X)$,*
(2) *If $(x, y) \in \theta_f$ and $x, y \in f(X)$, then $x = y$.*

Proof (1) If $x \in f(X)$, then $x = f(a)$ for some $a \in X$. Now $f(x) = f^2(x) = f(f(x)) = f(a) = x$.
(2) Let $(x, y) \in \theta_f$ and $x, y \in f(X)$. Then by (1), $x = f(x) = f(y) = y$. □

Theorem 5.2.15 *Let X be a BE-algebra and F a quasi-filter of X. Then there exists an idempotent multiplier f of X such that $f(X) = F$ if and only if $F \cap \theta_f(x)$ is a singleton set for all $x \in X$, where $\theta_f(x)$ is the congruence class of x with respect to θ_f.*

Proof Let f be an idempotent multiplier of X such that $f(X) = F$. Then clearly θ_f is a weak congruence on X. Let $x \in X$ be an arbitrary element. Since $f(x) = f^2(x)$, we get $(x, f(x)) \in \theta_f$. Hence $f(x) \in \theta_f(x)$. Also $f(x) \in f(X) = F$. Hence $f(x) \in F \cap \theta_f(x)$. Therefore $F \cap \theta_f(x)$ is non-empty. Suppose a, b be two elements of $F \cap \theta_f(x)$. Then by the above lemma, we get that $a = b$. Therefore $F \cap \theta_f(x)$ is a singleton set for all $x \in X$.

Conversely, assume that $F \cap \theta_f(x)$ is a singleton set, for all $x \in X$. Let x_0 be the single element of $F \cap \theta_f(x)$. Now define a self-mapping as follows:

$$f : X \to X \text{ by } f(x) = x_0 \quad \text{for all } x \in X$$

Clearly f is idempotent. By the definition of the mapping f, we get $f(b) \in F$ and $f(f(b)) = f(b)$. Since F is a quasi-filter, we get that $a * f(b) \in F$ and hence

$$\begin{aligned} f(f(b)) = f(b) &\Rightarrow (f(b), b) \in \theta_f \\ &\Rightarrow (a * f(b), a * b) \in \theta_f \\ &\Rightarrow a * f(b) \in \theta_f(a * b) \\ &\Rightarrow a * f(b) \in F \cap \theta_f(a * b) \qquad \text{since } a * f(b) \in F \end{aligned}$$

Since $f(a * b) \in F \cap \theta_f(a * b)$ and $F \cap \theta_f(a * b)$ is a singleton set, we get $f(a * b) = a * f(b)$. Therefore f is a multiplier of X. □

5.3 Congruences and Quasi-filters

In this section, the notion of θ-quasifilters is introduced. A bijection is obtained between the set of all θ-quasifilters of a BE-algebra and the set of all quasi-filters of the BE-algebra of all congruence classes. In the following, we first introduce two operations.

Definition 5.3.1 Let θ be a congruence on a BE-algebra X. Define operations α and β as

(1) For any quasi-filter F of X, define $\alpha(F) = \{ [x]_\theta \mid (x, y) \in \theta \text{ for some } y \in F \}$.

(2) For any quasi-filter $\widehat{F}$ of $X_{/\theta}$, define $\beta(\widehat{F}) = \{x \in X \mid (x, y) \in \theta \text{ for some } [y]_\theta \in \widehat{F}\}$.

In the following lemma, some basic properties of the above two operations are observed.

Lemma 5.3.2 *For any congruence θ on a BE-algebra X, the following hold:*

(1) *For any quasi-filter F of X, $\alpha(F)$ is a quasi-filter in $X_{/\theta}$,*
(2) *For any quasi-filter $\widehat{F}$ of $L_{/\theta}$, $\beta(\widehat{F})$ is a quasi-filter in X,*
(3) *α and β are isotone,*
(4) *For any quasi-filter F of X, $x \in F$ implies $[x]_\theta \in \alpha(F)$,*
(5) *For any quasi-filter $\widehat{F}$ of $X_{/\theta}$, $[x]_\theta \in \widehat{F}$ implies $x \in \beta(\widehat{F})$.*

Proof (1) Let $[x]_\theta \in X_{/\theta}$ and $[a]_\theta \in \alpha(F)$. Then we get $(a, y) \in \theta$ for some $y \in F$. Hence, it yields $(x * a, x * y) \in \theta$ and $x * y \in F$ because F is a quasi-filter. Thus we get $[x]_\theta * [a]_\theta = [x * a]_\theta \in \alpha(F)$. Therefore $\alpha(F)$ is a quasi-filter of $X_{/\theta}$.
(2) Let $x \in X$ and $a \in \beta(\widehat{F})$. Then $(a, y) \in \theta$ for some $[y]_\theta \in \widehat{F}$. Hence $(x * a, x * y) \in \theta$. Since $\widehat{F}$ is quasi-filter, $[x * y]_\theta = [x]_\theta * [y]_\theta \in \widehat{F}$. Thus $x * a \in \beta(\widehat{F})$. Hence $\beta(\widehat{F})$ is a quasi-filter of X.
(3) Let F_1, F_2 be two quasi-filters in X such that $F_1 \subseteq F_2$. Let $[x]_\theta \in \alpha(F_1)$. Then $(x, y) \in \theta$ for some $y \in F_1 \subseteq F_2$. Consequently, we get $[x]_\theta \in \alpha(F_2)$. Therefore $\alpha(F_1) \subseteq \alpha(F_2)$. Again, let $\widehat{F_1}, \widehat{F_2}$ be two filters of $X_{/\theta}$ such that $\widehat{F_1} \subseteq \widehat{F_2}$. Suppose $x \in \beta(\widehat{F_1})$. Then $(x, y) \in \theta$ for some $[y]_\theta \in \widehat{F_1} \subseteq \widehat{F_2}$. Hence $y \in \beta(\widehat{F_2})$. Therefore $\beta(\widehat{F_1}) \subseteq \beta(\widehat{F_2})$.
(4) For any $x \in F$, we have $(x, x) \in \theta$. Therefore $[x]_\theta \in \alpha(F)$.
(5) For any $[x]_\theta \in \widehat{F}$, we have $(x, x) \in \theta$. Therefore $x \in \beta(\widehat{F})$. □

Lemma 5.3.3 *Let θ be a congruence on a BE-algebra X. For any quasi-filter F of X, $\alpha\beta\alpha(F) = \alpha(F)$.*

Proof Let F be a quasi-filter of X. Suppose $[x]_\theta \in \alpha(F)$. Then we get $(x, y) \in \theta$ for some $y \in F$. Since $y \in F$, by Lemma 5.3.2(4), we get $[y]_\theta \in \alpha(F)$. Since $(x, y) \in \theta$ and $[y]_\theta \in \alpha(F)$, we get $x \in \beta\alpha(F)$. Hence $[x]_\theta \in \alpha\beta\alpha(F)$. Therefore $\alpha(F) \subseteq \alpha\beta\alpha(F)$.

Conversely, let $[x]_\theta \in \alpha\beta\alpha(F)$. Then $(x, y) \in \theta$ for some $y \in \beta\alpha(F)$. Since $y \in \beta\alpha(F)$, there exists $[a]_\theta \in \alpha(F)$ such that $(y, a) \in \theta$. Hence $[x]_\theta = [a]_\theta \in \alpha(F)$. Therefore $\alpha\beta\alpha(F) \subseteq \alpha(F)$. □

It is now intended to show that the composition $\beta\alpha$ is a closure operator on the set $\mathcal{QF}(X)$ of all filters of a BE-algebra X.

Proposition 5.3.4 *For any quasi-filter F of a BE-algebra, the map $F \to \beta\alpha(F)$ is a closure operator on $\mathcal{QF}(X)$. That is, for any two quasi-filters F, G of X, we have the following:*

(a) $F \subseteq \beta\alpha(F)$,
(b) $\beta\alpha\beta\alpha(F) = \beta\alpha(F)$,
(c) $F \subseteq G \Rightarrow \beta\alpha(F) \subseteq \beta\alpha(G)$.

Proof (a) Let $x \in F$. Then by Lemma 5.3.2(4), we get $[x]_\theta \in \alpha(F)$. Since $(x, x) \in \theta$ and $\alpha(F)$ is a quasi-filter in $X_{/\theta}$, we get $x \in \beta\alpha(F)$. Therefore $F \subseteq \beta\alpha(F)$.
(b) Since $\beta\alpha(F)$ is a quasi-filter in X, by above condition (a), we get $\beta\alpha(F) \subseteq \beta\alpha[\beta\alpha(F)]$. Conversely, let $x \in \beta\alpha[\beta\alpha(F)]$. Then we obtain $(x, y) \in \theta$ for some $[y]_\theta \in \alpha\beta\alpha(F)$. Thus by above Lemma 5.3.2, we get $[y]_\theta \in \alpha(F)$. Hence $x \in \beta\alpha(F)$. Therefore $\beta\alpha[\beta\alpha(F)] \subseteq \beta\alpha(F)$.
(c) Suppose F and G are two quasi-filters of X such that $F \subseteq G$. Let $x \in \beta\alpha(F)$. Then we get $[x]_\theta \in \alpha(F)$. Hence $[x]_\theta = [y]_\theta$ for some $y \in F \subseteq G$. Since $y \in G$, we get that $[x]_\theta = [y]_\theta \in \alpha(G)$. Hence we get $x \in \beta\alpha(G)$. Therefore $\beta\alpha(F) \subseteq \beta\alpha(G)$. □

Denoting by $\mathcal{QF}(X_{/\theta})$ the set of all quasi-filters of $X_{/\theta} = \{[x]_\theta \mid x \in X\}$, we can therefore define a mapping $\alpha : \mathcal{QF}(X) \to \mathcal{QF}(X_{/\theta})$ by $F \mapsto \alpha(F)$ also another mapping $\beta : \mathcal{QF}(X_{/\theta}) \to \mathcal{QF}(X)$ by $F \mapsto \beta(F)$. Then we have the following proposition.

Proposition 5.3.5 *Let θ be congruence on a BE-algebra X. Then α is a residuated map with residual map β.*

Proof For every $F \in \mathcal{QF}(X)$, by Proposition 5.3.4(a), we have $F \subseteq \beta\alpha(F)$. Let $F \in \mathcal{F}(L_{/\theta})$. Suppose $[x]_\theta \in F$. Then we get that $x \in \beta(F)$. Since $\beta(F)$ is a quasi-filter in L, we get that $[x]_\theta \in \alpha\beta(F)$. Hence $F \subseteq \alpha\beta(F)$. Conversely, let $[x]_\theta \in \alpha\beta(F)$. Then $[x]_\theta = [y]_\theta$ for some $y \in \beta(F)$. Since $y \in \beta(F)$, we get that $[x]_\theta = [y]_\theta \in F$. Hence $\alpha\beta(F) \subseteq F$. Therefore for every $F \in \mathcal{F}(X_{/\theta})$, we obtain that $\alpha\beta(F) = F$. Since α and β are isotone, it follows that α is residuated and that the residual of α is nothing but β. □

The notion of θ-quasifilters is now introduced in a BE-algebra.

Definition 5.3.6 Let θ be a congruence on a BE-algebra X. A quasi-filter F of X is called a *θ-quasifilter* if $\beta\alpha(F) = F$.

For any congruence θ on a BE-algebra X, it can be easily observed that the quasi-filter $\{1\}$ is a θ-quasifilter if and only if $[1]_\theta = \{1\}$. Moreover, we have the following:

Lemma 5.3.7 *Let θ be a congruence on a bounded BE-algebra X. For any quasi-filter F of X, the following conditions hold:*
(1) *If F is a θ-quasifilter then $[1]_\theta \subseteq F$,*
(2) *If F is a proper θ-quasifilter then $F \cap [0]_\theta = \emptyset$.*

In the following theorem, a set of sufficient conditions is derived for a proper quasi-filter of a BE-algebra to become a θ-quasifilter.

Theorem 5.3.8 *Let θ be a congruence on a BE-algebra X. A proper filter F of X is a θ-quasifilter if it satisfies the following conditions:*
(1) *For $x, y \in X$ with $x \neq y$, either $x \in F$ or $y \in F$,*
(2) *To each $x \in F$, there exists $x' \notin F$ such that $(x, x') \in \theta$.*

Proof Let F be a proper filter of X. Clearly $F \subseteq \beta\alpha(F)$. Conversely, let $x \in \beta\alpha(F)$. Then $(x, y) \in \theta$ for some $[y]_\theta \in \alpha(F)$. Hence $[y]_\theta = [a]_\theta$ for some $a \in F$. Since $a \in F$, there exists $a' \notin F$ such that $(a, a') \in \theta$. Since $(x, y) \in \theta$, $(y, a) \in \theta$ and $(a, a') \in \theta$, by the transitive property, we get $(x, a') \in \theta$. Since $a' \notin F$, by conditions (1) and (2), we get $x \in F$. Therefore $\beta\alpha(F) = F$. □

We now characterize θ-quasifilters in the following:

Theorem 5.3.9 *Let θ be a congruence on a BE-algebra X. For any quasi-filter F of X, the following conditions are equivalent:*

(1) *F is a θ-quasifilter,*
(2) *For any $x, y \in X$, $[x]_\theta = [y]_\theta$ and $x \in F$ imply that $y \in F$,*
(3) $F = \bigcup\limits_{x \in F} [x]_\theta$,
(4) *$x \in F$ implies $[x]_\theta \subseteq F$.*

Proof (1) $\Rightarrow$ (2): Assume that F is a θ-quasifilter of X. Let $x, y \in X$ be such that $[x]_\theta = [y]_\theta$. Then $(x, y) \in \theta$. Suppose $x \in F = \beta\alpha(F)$. Then $(x, a) \in \theta$ for some $[a]_\theta \in \alpha(F)$. Thus $(a, y) \in \theta$ and $[a]_\theta \in \alpha(F)$. Therefore $y \in \beta\alpha(F) = F$.
(2) $\Rightarrow$ (3): Assume the condition (2). Let $x \in F$. Since $x \in [x]_\theta$, we get $F \subseteq \bigcup\limits_{x \in F} [x]_\theta$. Conversely, let $a \in \bigcup\limits_{x \in F} [x]_\theta$. Then $(a, x) \in \theta$ for some $x \in F$. Hence $[a]_\theta = [x]_\theta$. By the condition (2), we get $a \in F$. Therefore $F = \bigcup\limits_{x \in F} [x]_\theta$.
(3) $\Rightarrow$ (4): Assume the condition (3). Let $a \in F$. Then $(x, a) \in \theta$ for some $x \in F$. Let $t \in [a]_\theta$. Then $(t, a) \in \theta$. Hence $(x, t) \in \theta$. Thus it yields that $t \in [x]_\theta \subseteq F$. Therefore $[a]_\theta \subseteq F$.
(4) $\Rightarrow$ (1): Assume the condition (4). Clearly $F \subseteq \beta\alpha(F)$. Conversely, let $x \in \beta\alpha(F)$. Then $(x, y) \in \theta$ for some $[y]_\theta \in \alpha(F)$. Hence $[y]_\theta = [a]_\theta$ for some $a \in F$. Since $a \in F$, by condition (4), we get that $x \in [y]_\theta = [a]_\theta \subseteq F$. Thus $\beta\alpha(F) \subseteq F$. Therefore F is a θ-quasifilter of X. □

Theorem 5.3.10 *Let θ be a congruence on a BE-algebra X. Then there exists a bijection between the set $\mathcal{QF}_\theta(X)$ of all θ-quasifilters of X and the set of all quasi-filters of the BE-algebra $X_{/\theta}$ of all congruence classes.*

Proof Define a mapping $\psi : \mathcal{QF}(X) \to \mathcal{QF}(X_{/\theta})$ by $\psi(F) = \alpha(F)$ for all $F \in \mathcal{F}_\theta(X)$. Let $F, G \in \mathcal{F}_\theta(X)$. Then we get the following consequence:

$$\begin{aligned} \psi(F) = \psi(G) &\Rightarrow \alpha(F) = \alpha(G) \\ &\Rightarrow \beta\alpha(F) = \beta\alpha(G) \\ &\Rightarrow F = G \qquad \text{since } F, G \in \mathcal{F}_\theta(X) \end{aligned}$$

Hence ψ is one-to-one. We now show that ψ is onto. Let $\widehat{F}$ be a quasi-filter of $X_{/\theta}$. Then $\beta(\widehat{F})$ is a quasi-filter in X. We now show that $\beta(\widehat{F})$ is a θ-quasifilter in X. We have always $\beta(\widehat{F}) \subseteq \beta\alpha\beta(\widehat{F})$. Let $x \in \beta\alpha\beta(\widehat{F})$. Then we get $(x, y) \in \theta$ for some $[y]_\theta \in \alpha\beta(\widehat{F}) = \widehat{F}$. Hence $x \in \beta(\widehat{F})$. Therefore $\beta(\widehat{F}) = \beta\alpha\beta(\widehat{F})$. Now for this $\beta(\widehat{F}) \in X$, we get $\psi[\beta(\widehat{F})] = \alpha\beta(\widehat{F}) = \widehat{F}$. Therefore ψ is onto. Therefore ψ is a bijection between $\mathcal{F}_\theta(X)$ and $\mathcal{F}(X_{/\theta})$. □

5.4 Pseudo Filters of BE-algebras

In this section, the concepts of simple filters and pseudo filters are introduced in BE-algebras. Some properties of pseudo filters are studied. A set of equivalent conditions is derived for every pseudo filter of a BE-algebra to become a filter. The theory of soft sets is applied to pseudo filters of BE-algebras and their properties are studied.

Definition 5.4.1 Let $(X, *, 1)$ be a BE-algebra. A subset A of X is called a *simple filter* of X if it satisfies the following conditions for all $x, y \in X$.

(1) $1 \in A$,
(2) $x * y \in A$ implies $x * (x * y) \in A$.

It is clear that $\{1\}$ and X are simple filters of a BE-algebra X. By the set F_1, we mean a non-empty subset F of a BE-algebra X such that $1 \in F$. It is then clear that F_1 is a simple filter of X whenever X is a simple BE-algebra.

Proposition 5.4.2 *Every non-empty subset F of a BE-algebra which is satisfying the following condition is a simple filter:*

$$x \in F \text{ and } x \leq y \text{ imply that } y \in F \text{ for all } x, y \in X.$$

Proof Let X be a BE-algebra and $x, y \in X$. Let F be a non-empty subset of X satisfying the given condition. Let $x, y \in X$ be such that $x * y \in F$. Since $x * y \leq x * (x * y)$, we get that $x * (x * y) \in F$. Therefore F is a simple filter of X. □

Proposition 5.4.3 *Every quasi-filter of a BE-algebra is a simple filter.*

Proof Let X be a BE-algebra and F a quasi-filter of X. Let $x, y \in X$ be such that $x * y \in F$. Since F is a quasi-filter, we get $x * (x * y) \in F$. Therefore F is a simple filter of X. □

Since every filter is a quasi-filter, the following corollary is a direct consequence.

Corollary 5.4.4 *Every filter of a BE-algebra is a simple filter.*

The notion of pseudo filters is now introduced in BE-algebras.

Definition 5.4.5 Let $(X, *, 1)$ be a BE-algebra. A subset A of X is said to be a *pseudo filter* of X if it satisfies the following condition for all $x, y \in X$:

$$x \in A \text{ and } x * (x * y) \in A \text{ imply that } y \in A.$$

Example 5.4.6 Let $X = \{1, a, b, c, d\}$. Define a binary operation $*$ on X as follows:

$*$	1	a	b	c	d
1	1	a	b	c	d
a	1	1	b	c	d
b	1	a	1	c	c
c	1	1	b	1	b
d	1	1	1	1	1

Then it can be easily verified that $(X, *, 1)$ is a BE-algebra. Clearly $F = \{a, c\}$ is a pseudo filter of X but not a filter of X, since $1 \notin F$.

There is no guaranty that a pseudo filter contains the greatest element 1. If it contains 1, then the pseudo filter will become a filter.

Theorem 5.4.7 *Every pseudo filter of a BE-algebra which contains 1 is a filter.*

Proof Let F be a pseudo filter of a BE-algebra X. Suppose $1 \in F$. Let $a, b \in X$. Suppose $a \in F$ and $a \leq b$. Then $a * b = 1$. Hence $a * (a * b) = a * 1 = 1 \in F$. Since F is a pseudo filter, we get $b \in F$. We now show that F is a filter of X. Let $x, y \in X$ be such that $x \in F$ and $x * y \in F$. Since $x * y \leq x * (x * y)$, by the above observation, we get $x * (x * y) \in F$. Since F is a pseudo filter, we get $y \in F$. Therefore F is a filter of X. □

Corollary 5.4.8 *Every filter of BE-algebra is a pseudo filter.*

The converse of the above corollary is not true. For, the set $\{a, c\}$ in Example 5.4.6 is a pseudo filter of X but not the filter of X. However, in the following theorem, we derive a set of equivalent conditions for a pseudo filter of a BE-algebra to become a filter.

Theorem 5.4.9 *Let F be a pseudo filter of BE-algebra X. Then the following are equivalent:*

(1) *F is a filter of X,*
(2) *$x \in F$ and $x \leq y$ imply that $y \in F$ for all $x, y \in X$,*
(3) *F is a simple filter of X.*

Proof (1) $\Rightarrow$ (2): It is clear.
(2) $\Rightarrow$ (3): Assume that the condition (2) holds. Let $x, y \in X$ be such that $x * y \in F$. Since $x * y \leq x * (x * y)$, by the condition (2), we get $x * (x * y) \in F$. Therefore F is a simple filter of X.
(3) $\Rightarrow$ (1): Assume that F is a simple filter of X. Let $x, y \in X$. Suppose $x \in F$ and $x * y \in F$. Since $x * y \in F$, by the assumed condition, we get $x * (x * y) \in F$. Since $x \in F$ and F is a pseudo filter, we get $y \in F$. Therefore F is a filter of X. □

Corollary 5.4.10 *A pseudo filter of a BE-algebra is a filter if and only if it is a simple filter.*

Proposition 5.4.11 *Let $\{F_\alpha \mid \alpha \in \Delta\}$ be an indexed family of pseudo filters of a BE-algebra X. Then $\bigcap_{\alpha\in\Delta} F_\alpha$ is also a pseudo filter of X.*

Proof Let $x, x * (x * y) \in \bigcap_{\alpha\in\Delta} F_\alpha$ for $x, y \in X$. Then we get $x, x * (x * y) \in F_\alpha$ for each $\alpha \in \Delta$. Since each $F_\alpha; \alpha \in \Delta$ is a pseudo filter of X, we get $y \in F_\alpha$ for each $\alpha \in \Delta$. Hence $y \in \bigcap_{\alpha\in\Delta} F_\alpha$. Therefore $\bigcap_{\alpha\in\Delta} F_\alpha$ is a pseudo filter of X. □

Proposition 5.4.12 *Let $(X, *, 1)$ and $(Y, *, 1)$ be two BE-algebras and $f : X \to Y$ a homomorphism. If G is a pseudo filter of Y, then $f^{-1}(G)$ is a pseudo filter of X.*

Proof Let $x \in f^{-1}(G)$ and $x * (x * y) \in f^{-1}(G)$. Then we get $f(x) \in G$ and $f(x) * (f(x) * f(y)) = f(x * (x * y)) \in G$. Since G is a pseudo filter of Y, we get $f(y) \in G$. Hence $y \in f^{-1}(G)$. Therefore $f^{-1}(G)$ is a pseudo filter of X. □

Theorem 5.4.13 *If F_1 and F_2 are pseudo filters of X_1 and X_2 respectively, then $F_1 \times F_2$ is a pseudo filter of the product algebra $X_1 \times X_2$. Conversely, every pseudo filter F of $X_1 \times X_2$ can be expressed as Cartesian product of two pseudo filters of X_1 and X_2.*

Proof Let F_1 and F_2 be pseudo filters of BE-algebras X_1 and X_2, respectively. Let $(x_1, y_1) \in F_1 \times F_2$ and $(x_1, y_1) * ((x_1, y_1) * (x_2, y_2)) \in F_1 \times F_2$. Then we get $(x_1 * (x_1 * x_2), y_1 * (y_1 * y_2)) \in F_1 \times F_2$. Hence $x_1, x_1 * (x_1 * x_2) \in F_1$ and $y_1, y_1 * (y_1 * y_2) \in F_2$. Since F_1 and F_2 are pseudo filters of X_1 and X_2 respectively, we get $x_2 \in F_1$ and $y_2 \in F_2$. Hence $(x_2, y_2) \in F_1 \times F_2$. Therefore $F_1 \times F_2$ is a pseudo filter of $X_1 \times X_2$.

Conversely, let F be any pseudo filter of $X_1 \times X_2$. Consider the projections $\Pi_i : X_1 \times X_2 \longrightarrow X_i$ for $i = 1, 2$. Let F_1 and F_2 be the projections of F on X_1 and X_2, respectively. That is $\Pi_i(F) = F_i$ for $i = 1, 2$. We now prove that F_1 and F_2 are pseudo filters of X_1 and X_2, respectively. Let $x \in F_1$ and $x * (x * y) \in F_1$. Then $(x, 1) \in F$ and $(x, 1) * ((x, 1) * (y, 1)) = (x * (x * y), 1 * (1 * 1)) \in F$. Since F is a pseudo filter of $X_1 \times X_2$, we get $(y, 1) \in F$. Thus $y = \Pi_1(y, 1) \in \Pi_1(F) = F_1$. Therefore F_1 is a pseudo filter of X_1. Similarly, we can get F_2 is a pseudo filter of X_2. □

In what follows, we take a BE-algebra X as a set of parameters unless otherwise specified. In the following, we introduce the concept of soft pseudo filters of BE-algebras.

Definition 5.4.14 A soft set $(\widetilde{f}, X)$ of X over U is called a *soft pseudo filter* of X if it satisfies the following condition for all $x, y \in X$:

$$\widetilde{f}(x) \cap \widetilde{f}(x * (x * y)) \subseteq \widetilde{f}(y).$$

Example 5.4.15 Let $X = \{1, a, b, c\}$. Define a binary operation $*$ on X as follows:

$*$	1	a	b	c
1	1	a	b	c
a	1	1	1	c
b	1	1	1	c
c	1	c	c	1

Then it can be easily verified that $(X, *, 1)$ is a BE-algebra. Let $(\widetilde{f}, X)$ be a soft set of X over $U = \mathbb{Z}$ (the set of integers), where $\widetilde{f} : X \to \mathcal{P}(U)$ defined as follows:

$$\widetilde{f}(x) = \begin{cases} 3\mathbb{N} & \text{if } x \in \{a, b\} \\ 3\mathbb{Z} & \text{if } x = c \end{cases}$$

It is easy to check that $(\widetilde{f}, X)$ is a soft pseudo filter of X.

Definition 5.4.16 A soft set $(\widetilde{f}, X)$ of X over U is called *order preserving* if it satisfies the following condition for all $x, y \in X$:

$$x * y = 1 (x \leq y) \text{ implies } \widetilde{f}(x) \subseteq \widetilde{f}(y).$$

Proposition 5.4.17 *Every order preserving soft pseudo filter $(\widetilde{f}, X)$ of X over U satisfies the following property for all $x, y, z \in X$:*

$$\widetilde{f}(y) \cap \widetilde{f}(x * (y * z)) \subseteq \widetilde{f}(x * (y * (y * z)).$$

Proof Let $x, y, z \in X$. Since $x * z \leq x * (x * z)$ and $(\widetilde{f}, X)$ is a soft pseudo filter, we get

$$\begin{aligned} \widetilde{f}(y) \cap \widetilde{f}(x * (y * (y * z)) &= \widetilde{f}(y) \cap \widetilde{f}(y * (y * (x * z))) \\ &\subseteq \widetilde{f}(x * z) \\ &\subseteq \widetilde{f}(x * (x * z)). \end{aligned}$$

Therefore $\widetilde{f}(y) \cap \widetilde{f}(x * (y * (y * z)) \subseteq \widetilde{f}(x * (x * z))$ for all $x, y, z \in X$. □

Definition 5.4.18 For a soft set $(\widetilde{f}, X)$ of X and a subset τ of U, the *τ-inclusive set* of $(\widetilde{f}, X)$, denoted by $i_X(\widetilde{f}; \tau)$, is defined to be the set

$$i_X(\widetilde{f}; \tau) = \{x \in A \mid \tau \subseteq \widetilde{f}(x)\}.$$

Theorem 5.4.19 *An soft set $(\widetilde{f}, X)$ is a soft pseudo filter of X if and only if $i_X(\widetilde{f}; \tau)$ is a pseudo filter of X whenever $i_X(\widetilde{f}; \tau) \neq \emptyset$ for all $\tau \in \mathcal{P}(U)$.*

Proof Assume that $(\widetilde{f}, X)$ is a soft weak filter of X. Let $\tau \in \mathcal{P}(U)$ and $i_X(\widetilde{f}; \tau) \neq \emptyset$. Let $x \in i_X(\widetilde{f}; \tau)$ and $x * (x * y) \in i_X(\widetilde{f}; \tau)$. Then $\tau \subseteq \widetilde{f}(x)$ and $\tau \subseteq \widetilde{f}(x * (x *$

$y)$). Since $(\widetilde{f}, X)$ is a soft pseudo filter, we get $\tau \subseteq \widetilde{f}(x) \cap \widetilde{f}(x * (x * y)) \subseteq \widetilde{f}(y)$. Thus $y \in i_X(\widetilde{f}; \tau)$. Therefore $i_X(\widetilde{f}; \tau)$ is a pseudo filter of X.

Conversely, assume that $i_X(\widetilde{f}; \tau)$ is a pseudo filter of X for all $\tau \in \mathcal{P}(U)$ with $i_X(\widetilde{f}; \tau) \neq \emptyset$. For any $x, y \in X$, let $\widetilde{f}(x) = \tau_x$ and $\widetilde{f}(x * (x * y)) = \tau_{x*(x*y)}$. Take $\tau = \tau_x \cap \tau_{x*(x*y)}$. Then we get $\tau \subseteq \widetilde{f}(x)$ and $\tau \subseteq \widetilde{f}(x * (x * y))$. Hence $x \in i_X(\widetilde{f}; \tau)$ and $x * (x * y) \in i_X(\widetilde{f}; \tau)$. Since $i_X(\widetilde{f}; \tau)$ is a pseudo filter of X, we get $y \in i_X(\widetilde{f}; \tau)$. Hence $\widetilde{f}(x) \cap \widetilde{f}(x * (x * y)) = \tau_x \cap \tau_{x*(x*y)} = \tau \subseteq \widetilde{f}(y)$. Therefore $(\widetilde{f}, X)$ is a soft pseudo filter of X. □

Remark 5.4.20 The above filter $i_X(\widetilde{f}; \tau)$ is called an inclusive filter of X.

Theorem 5.4.21 *Let $(\widetilde{f}, X)$ be a soft set of X over U. Define a soft set $(\widetilde{f}^*, X)$ of X over U where $\widetilde{f}^* : X \to \mathcal{P}(U)$ is given by*

$$\widetilde{f}^*(x) = \begin{cases} \widetilde{f}(x) & \text{if } x \in i_X(\widetilde{f}; \tau) \\ \delta & \text{otherwise} \end{cases}$$

where τ is any subset of U and δ is a subset of U satisfying $\delta \subseteq \bigcap_{x \notin i_X(\widetilde{f};\tau)} \widetilde{f}(x)$. Then we have

(1) *$\widetilde{f}^*(x) \subseteq \widetilde{f}(x)$ for all $x \in X$,*
(2) *If $(\widetilde{f}, X)$ is a soft pseudo filter of X, then so is $(\widetilde{f}^*, X)$.*

Proof (1) Let $x \in X$. Suppose $x \in i_X(\widetilde{f}; \tau)$. Then clearly $\widetilde{f}^*(x) = \widetilde{f}(x)$. Let $x \notin i_X(\widetilde{f}; \tau)$. Then $\widetilde{f}^*(x) = \delta \subseteq \bigcap_{x \notin i_X(\widetilde{f};\tau)} \widetilde{f}(x) \subseteq \widetilde{f}(x)$. Hence $\widetilde{f}^*(x) \subseteq \widetilde{f}(x)$ for all $x \in X$.

(2) Assume that $(\widetilde{f}, X)$ is a soft pseudo filter of X. Then by Definition 5.4.18, we get that $i_X(\widetilde{f}; \tau)(\neq \emptyset)$ is a pseudo filter of X for all $\tau \in \mathcal{P}(U)$. Let $x, y \in X$. Suppose $x \in i_X(\widetilde{f}; \tau)$ and $x * (x * y) \in i_X(\widetilde{f}; \tau)$. Since $i_X(\widetilde{f}; \tau)$ is a pseudo filter, we get that $y \in i_X(\widetilde{f}; \tau)$. Hence

$$\begin{aligned} \widetilde{f}^*(y) &= \widetilde{f}(y) \\ &\supseteq \widetilde{f}(x * (x * y)) \cap \widetilde{f}(x) \\ &= \widetilde{f}^*(x * (x * y)) \cap \widetilde{f}^*(x) \end{aligned}$$

Suppose $x \notin i_X(\widetilde{f}; \tau)$ or $x * (x * y) \notin i_X(\widetilde{f}; \tau)$. Then we get that $\widetilde{f}^*(x) = \delta$ or $\widetilde{f}^*(x * (x * y)) = \delta$. Thus

$$\widetilde{f}^*(y) \supseteq \delta = \widetilde{f}^*(x * (x * y)) \cap \widetilde{f}^*(x).$$

Therefore $(\widetilde{f}^*, X)$ is a soft pseudo filter of X. □

Theorem 5.4.22 *If $(\widetilde{f}, X)$ is a soft pseudo filter of X, then the set $X_{\widetilde{f}}$ defined by*

$$X_{\widetilde{f}} = \{x \in X \mid \widetilde{f}(x) \supseteq \widetilde{f}(a) \text{ for some } a \in X\}$$

is a pseudo filter of X.

Proof Let $x, y \in X$ be such that $x \in X_{\widetilde{f}}$ and $x * (x * y) \in X_{\widetilde{f}}$. Then $\widetilde{f}(a) \subseteq \widetilde{f}(x)$ and $\widetilde{f}(a) \subseteq \widetilde{f}(x * (x * y))$. Since $(\widetilde{f}, X)$ is a soft pseudo filter of X, we get that

$$\begin{aligned} \widetilde{f}(y) &\supseteq \widetilde{f}(x) \cap \widetilde{f}(x * (x * y)) \\ &= \widetilde{f}(a) \cap \widetilde{f}(a) \\ &= \widetilde{f}(a). \end{aligned}$$

Hence $\widetilde{f}(y) \supseteq \widetilde{f}(a)$. Thus $y \in X_{\widetilde{f}}$. Therefore $X_{\widetilde{f}}$ is a pseudo filter of X. □

Theorem 5.4.23 *Any pseudo filter can be realized as an inclusive filter of some soft pseudo filter of a BE-algebra.*

Proof Let F be a pseudo filter of X. For any non-empty subset τ of U, let $(\widetilde{f}, X)$ be a soft set of X over U where $\widetilde{f} : X \to \mathcal{P}(U)$ defined by

$$\widetilde{f}(x) = \begin{cases} \tau & \text{if } x \in F, \\ U & \text{if } x \notin F \end{cases}$$

For any $x, y \in X$, let $x \in F$ and $x * (x * y) \in F$. Since F is a pseudo filter, we get that $y \in F$. Hence $\widetilde{f}(x) = \widetilde{f}(x * (x * y)) = \widetilde{f}(y) = \tau$. Hence $\widetilde{f}(x) \cap \widetilde{f}(x * (x * y)) = \tau = \widetilde{f}(y)$. If $x \notin F$ or $x * (x * y) \notin F$, then $\widetilde{f}(x) = U$ or $\widetilde{f}(x * (x * y)) = U$. Hence

$$\widetilde{f}(x) \cap \widetilde{f}(x * (x * y)) = \tau \subseteq \widetilde{f}(y).$$

Therefore $(\widetilde{f}, X)$ is a soft pseudo filter of X. Clearly $i_X(\widetilde{f}; \tau) = F$. □

Exercise

1. Derive a sufficient condition for the homomorphic image of a quasi-filter (skew filter) of a BE-algebra that is again a quasi-filter (skew filter).
2. Let X be a BE-algebra. If A is a quasi-filter of X such that $A = F \cup G$ where F, G are two filters, then prove that either $A = F$ or $A = G$.
3. Let X be a partially ordered BE-algebra and $\mathcal{C}$ a closure operator on X. Then show that the set $F = \{x \in X \mid \mathcal{C}(x) = 1\}$ is a quasi-filter of X.
4. In any BE-algebra $(X, *, 1)$, prove that, if every non-unit element of the BE-algebra is a pure dual atom, then $x * y = y$ for all $x, y \in X$ with $x \neq y$ and hence every subalgebra with 1 is a quasi-filter of X.

5. For any BE-algebra X, define $\overline{X} = \{a \in X \mid a \leq x$ implies $x = a$ for all $x \in X\}$. Then prove that a non-empty subset F with 1 is a quasi-filter if and only if $\overline{X} \subseteq F$.
6. For any BE-algebra $(X, *, 1)$, prove that a quasi-filter F of X is a filter if and only if every element $1 \neq a \in X$ is a pure dual atom of X.

Chapter 6
Very True Operators

Inspired by the considerations of Zadeh [252], Hajek in [105] formalized the fuzzy truth-value very true. He enriched the language of the basic fuzzy logic BL by adding a new unary connective vt and introduced the propositional logic BL_{vt}. The completeness BL_{vt} was proved in [168] by using the so-called BL_{vt}-algebra, an algebraic counterpart of BL_{vt}. In 2006, Vychodil [237] proposed an axiomatization of unary connectives like slightly true and more or less true and introduced $BL_{vt,st}$-logic which extends BL_{vt}-logic by adding a new unary connective "slightly true" denoted by "st." Noting that bounded commutative $R\ell$-monoids are algebraic structures which generalize, e.g., both BL-algebras and Heyting algebras (an algebraic counterpart of the intuitionistic propositional logic), Rachunek and Salounova taken bounded commutative $R\ell$-monoids with a vt-operator as an algebraic semantics of a more general logic than Hajeks fuzzy logic and studied algebraic properties of $R\ell_{vt}$-monoids in [196].

In this chapter, the notion of weak very true operators or simply wvt-operators is introduced on BE-algebras. Properties of weak vt-operators are studied in terms of Cartesian products and strongest self-maps. Corresponding to a wvt-operator v, a wvt-operator is introduced on the quotient algebra with respect to the kernel of the wvt-operator v. The concept of v-filters is introduced in BE-algebras. Characterization theorems of prime v-filters and maximal v-filters are derived. The smallest v-filter generated by a non-empty set is constructed, and then the celebrated Stone's theorem is generalized to the case of v-filters.

The concept of v-congruences is introduced in BE-algebras. A necessary and sufficient condition is derived for every congruence of a BE-algebra to become a v-congruence. v-filters are characterized in terms of v-congruences. A one-to-one correspondence is obtained between the set of all v-filters of a BE-algebra and the set of all v-filters of the quotient algebra corresponding to a v-congruence. A bijection is obtained between the set of all v-congruences and the set of all v-filters of a BE_{wvt}-algebra. The notion of prime v-filters is introduced, and some properties of

S. R. Mukkamala, *A Course in BE-algebras*,
https://doi.org/10.1007/978-981-10-6838-6_6

prime v-filters are studied. Stone's theorem of prime filters of distributive lattices is generalized to the case of prime v-filters. Subdirectly irreducible BE_{wvt}-algebras are characterized.

6.1 Properties of Very True Operators

In this section, the notion of weak very true operators is introduced in BE-algebras. Properties of weak very true operators obtained with respect to Cartesian products and strongest operators. Relations among kernel, image set, and fixed sets of very true operators of BE-algebras are investigated.

Definition 6.1.1 For any BE-algebra X, a self-mapping $v : X \to X$ is said to be a *weak very true operator* or *weak vt-operator* (wvt-operator in brief) on X if for any $x, y \in X$:

(1) $v(1) = 1$,
(2) $v(x) \leq x$, i.e., v is subdiagonal,
(3) $v(x * y) \leq v(x) * v(y)$.

Moreover, if X is a bounded and commutative BE-algebra and v a wvt-operator which satisfies $v(x \vee y) \leq v(x) \vee v(y)$ for any $x, y \in X$, then v is called a vt-operator on X. A wvt-operator v is called hedge on X if it satisfies $v(v(x)) = v(x)$ for any $x \in X$. Any commutative BE-algebra admits vt-operators, e.g., the identity and the globalization v, where $v(x) = 0$ for $x \neq 1$ and $v(1) = 1$. Globalization can be seen as an interpretation of a connective absolutely/fully true.

BE_{vt}-algebras (BE_{wvt}-algebras) are BE-algebras with vt-operator v (weak vt-operator). We denote these BE_{vt}-algebra (weak vt-operator) by (X, v).

Lemma 6.1.2 *Let* (X, v) *be a* BE_{wvt}*-algebra. Then the following hold:*

(1) $v^n(0) = 0$ *for any* $n \in \mathbb{N}$,
(2) *for all* $x, y \in X$, $x \leq y$ *implies* $v(x) \leq v(y)$,
(3) *for any* $x \in X$, $n \in \mathbb{N}$, $v^n(x) \leq x$,
(4) *for any* $m, n \in \mathbb{N}$, $m \leq n$ *implies* $v^n(x) \leq v^m(x)$,
(5) *for any* $x, y \in X$, $v^n(x * y) \leq v^n(x) * v^n(x)$.
Moreover, if v *is a* vt*-operator on a commutative BE-algebra* X*, then*
(6) $v^n(x \vee y) = v^n(x) \vee v^n(y)$.

Proof (1) From the property (2) of wvt-operator, we get $v(0) \leq 0$. Hence $v(0) = 0$. Thus $v^n(0) \leq v^{n-1}(0) \leq \cdots \leq v(0) \leq 0$. Therefore $v^n(0) = 0$.
(2) Suppose $x \leq y$. Hence $1 = v(1) = v(x * y) \leq v(x) * v(y)$. Thus $v(x) \leq v(y)$.
(3) Let $n \in \mathbb{N}$. By the property (2), we get $v(x) \leq x$ for any $x \in X$. Hence $v^2(x) \leq v(x) \leq x$. Continuing in this way, we get $v^n(x) \leq x$.
(4) Let $m, n \in \mathbb{N}$ be such that $m \leq n$. Then for any $x \in X$, from (3), we get

$$v^n(x) \leq v^{n-1}(x) \leq \ldots \leq v^m(x).$$

(5) Let $x, y \in X$. Then for any $n \in \mathbb{N}$, we get

$$\begin{aligned} v^n(x * y) &= v^{n-1}(v(x * y)) \\ &\leq v^{n-1}(v(x) * v(y)) \\ &= v^{n-2}(v(v(x) * v(y))) \\ &\leq v^{n-2}(v^2(x) * v^2(y)) \\ &\leq \cdots \\ &\leq v^n(x) * v^n(y) \end{aligned}$$

(6) Suppose X is commutative and v a vt-operator on X. Let $x, y \in X$. Since $x, y \leq x \vee y$, we get $v(x), v(y) \leq v(x \vee y)$. Hence $v(x) \vee v(y) \leq v(x \vee y)$. By the definition, we get $v(x \vee y) \leq v(x) \vee v(y)$. Therefore $v(x \vee y) = v(x) \vee v(y)$. Hence $v^2(x \vee y) = v(v(x \vee y)) = v(v(x) \vee v(y)) = v(v(x)) \vee v(v(y)) = v^2(x) \vee v^2(y)$. Thus $v^2(x \vee y) = v^2(x) \vee v^2(y)$. Similarly, we get $v^3(x \vee y) = v^3(x) \vee v^3(y)$. Continuing in this way, we get $v^n(x \vee y) = v^n(x) \vee v^n(y)$. □

Theorem 6.1.3 *Let (X, v_1) and (X, v_2) be two BE_{wvt}-algebras. Then the mapping $v_1 \times v_2$ defined, for all $a \in X$ and $b \in Y$, on $X \times Y$ as follows:*

$$(v_1 \times v_2)(a, b) = (v_1(a), v_2(b))$$

is a weak vt-operator on $X \times Y$ and hence $(X \times Y, v_1 \times v_2)$ is a BE_{wvt}-algebra.

Proof Clearly $X \times Y$ is a BE-algebra with the greatest element $(1, 1)$. Since v_1 and v_2 are weak vt-operators on X and Y, respectively, we get $(v_1 \times v_2)(1, 1) = (v_1(1), v_2(1)) = (1, 1)$. Let $(a, b) \in X \times Y$. Then $(v_1 \times v_2)(a, b) = (v_1(a), v_2(b)) \leq (a, b)$. Let $(a, b), (c, d) \in X \times Y$. Since v_1 and v_2 are weak vt-operators on X and Y, respectively, we get the following consequence:

$$\begin{aligned} (v_1 \times v_2)((a, b) * (c, d)) &= (v_1 \times v_2)(a * c, b * d) \\ &= (v_1(a * c), v_2(b * d)) \\ &\leq (v_1(a) * v_1(c), v_2(b) * v_2(d)) \\ &= (v_1(a), v_2(b)) * (v_1(c), v_2(d)) \\ &= (v_1 \times v_2)(a, b) * (v_1 \times v_2)(c, d) \end{aligned}$$

Therefore $v_1 \times v_2$ is a weak vt-operator on $X \times Y$. □

The following corollary is an immediate consequence of the above theorem.

Corollary 6.1.4 *Let v_1 and v_2 be two weak vt-operators on X. Then $v_1 \times v_2$ is a weak vt-operator on X^2 and hence $(X^2, v_1 \times v_2)$ is a BE_{wvt}-algebra.*

Definition 6.1.5 Let μ be a self-map on a BE-algebra X. Then the *strongest self-map* v_μ is a self-map on X^2 defined by

$$v_\mu(x, y) = (\mu(x), \mu(y)) \text{ for all } x, y \in X$$

.

Theorem 6.1.6 *Let μ be a self-map on a BE-algebra X and v_μ the strongest self-map on X^2. Then μ is weak vt-operator on X if and only if v_μ is a weak vt-operator on X^2.*

Proof Assume that μ is a wvt-operator on X. Then $v_\mu(1, 1) = (\mu(1), \mu(1)) = (1, 1)$. Let $(a, b) \in X \times X$. Since μ is a wvt-operator on X, we get $v_\mu(a, b) = (\mu(a), \mu(b)) \leq (a, b)$. Again, let $(a, b), (c, d) \in X \times X$. Then

$$\begin{aligned} v_\mu((a, b) * (c, d)) &= v_\mu(a * c, b * d) \\ &= (\mu(a * c), \mu(b * d)) \\ &\leq (\mu(a) * \mu(c), \mu(b) * \mu(d)) \\ &= (\mu(a), \mu(b)) * (\mu(c), \mu(d)) \\ &= v_\mu(a, b) * v_\mu(c, d) \end{aligned}$$

Hence v_μ is a wvt-operator on X^2. Therefore (X^2, v_μ) is a BE_{wvt}-algebra.

Conversely, assume that v_μ is a wvt-operator on X^2. Then $(\mu(1), \mu(1)) = v_\mu(1, 1) = (1, 1)$. Hence $\mu(1) = 1$. For any $a \in X$, $(\mu(a), \mu(a)) = v_\mu(a, a) \leq (a, a)$. Hence $\mu(a) \leq a$. For $a, b \in X$, we have $(a, a), (a, b) \in X^2$. Now

$$\begin{aligned} (1, \mu(a * b)) &= (\mu(1), \mu(a * b)) \\ &= v_\mu(1, a * b) \\ &= v_\mu(a * a, a * b) \\ &= v_\mu((a, a) * (a, b)) \\ &\leq v_\mu(a, a) * v_\mu(a, b) \\ &= (\mu(a), \mu(a)) * (\mu(a), \mu(b)) \\ &= (\mu(a) * \mu(a), \mu(a) * \mu(b)) \\ &= (1, \mu(a) * \mu(b)) \end{aligned}$$

Hence $\mu(a * b) \leq \mu(a) * \mu(b)$. Therefore μ is a wvt-operator on X. □

Corollary 6.1.7 *Let $(X, *, 1)$ be a BE-algebra. Then (X, μ) is a BE_{wvt}-algebra if and only if (X^2, v_μ) is a BE_{wvt}-algebra.*

For any BE_{wvt}-algebra (X, v), define the kernel of the wvt-operator v as $Ker(v) = \{x \in X \mid v(x) = 1\}$. The image of the BE-algebra X is defined as $v(X) = \{v(x) \mid x \in X\}$. Also the identity set of X is defined as $Id_X(v) = \{x \in X \mid v(x) = x\}$.

Proposition 6.1.8 *Let (X, v) be a BE_{wvt}-algebra. Then $Ker(v)$ is a filter of X such that $Ker(v) \cap Id_X(v) = \{1\}$.*

Proof Clearly $1 \in Ker(v)$. Let $x, x*y \in Ker(v)$. Then we get $v(x) = v(x*y) = 1$. Hence $v(y) = 1 * v(y) = v(x) * v(y) \geq v(x*y) = 1$. Hence $v(y) = 1$ and thus we get $y \in Ker(v)$. Therefore $Ker(v)$ is a filter of X. Let $x \in Ker(v) \cap Id_X(v)$. Then we get $v(x) = 1$ and $v(x) = x$. Hence, it implies that $x = v(x) = 1$. Therefore $Ker(v) \cap Id_X(v) = \{1\}$. □

Proposition 6.1.9 *Let (X, v) be a BE_{wvt}-algebra. If v is idempotent, then $Ker(v) \cap VX = \{1\}$.*

Proof Suppose v is idempotent. Let $x \in Ker(v) \cap vX$. Then we get $x \in Ker(v)$ and $x \in v(X)$. Hence $x = v(a)$ for some $a \in X$. Since v is idempotent, we get $v(x) = v(v(x)) = v(x) = 1$. Therefore $Ker(v) \cap v(X) = \{1\}$. □

6.2 v-filters of BE-algebras

In this section, the concept of v-filters with respect to very true operators is studied. A set of equivalent conditions is derived for every filter of a transitive BE-algebra to become a v-filter. Element-wise characterization is obtained for the smallest v-filters generated by a non-empty sets.

Definition 6.2.1 Let v be a weak vt-operator on a BE-algebra X and F a filter of X. Then F is called a *v-filter* of X if $v(x) \in F$ for every $x \in F$.

Obviously, the filters $\{1\}$ and X are v-filters of the BE_{wvt}-algebra (X, v). For any BE_{wvt}-algebra (X, v), it is clear that the filter $Ker(v)$ is a v-filter in X. It can be easily seen that the intersection of any two v-filters of a BE_{wvt}-algebra is again a v-filter. Moreover, we have $\bigcap\limits_{F \in \mathcal{F}_v(X)} F = \{1\}$, where $\mathcal{F}_v(X)$ is the set of all v-filters of a BE_{wvt}-algebra (X, v).

Example 6.2.2 Let $(X, *, 1)$ be a BE-algebra and F a filter of X. For any $a \in X - F$, define a self-map $v : X \to X$ as follows:

$$v_a(x) = \begin{cases} 1 \text{ if } x \in F \\ a \text{ if } x \notin F \end{cases}$$

for all $x \in X$. Then clearly (X, v) is a BE_{wvt}-algebra. For any $x \in X$ with $x \in F$, we get $v(x) = 1 \in F$. Hence $v(F) \subseteq F$. Therefore F is a v-filter of (X, v).

Proposition 6.2.3 *Let $\{\mathcal{F}_\alpha\}_{\alpha \in \Delta}$ be an indexed family of v-filters of a BE_{wvt}-algebra (X, v). Then the set intersection $\bigcap\limits_{\alpha \in \Delta} F_\alpha$ is a v-filter of (X, v).*

Proof Since each $F_\alpha, \alpha \in \Delta$ is a filter of X, it is clear that $\bigcap_{\alpha\in\Delta} F_\alpha$ is a filter of X. Let $x \in \bigcap_{\alpha\in\Delta} F_\alpha$. Then $x \in F_\alpha$ for each $\alpha \in \Delta$. Since each $F_\alpha, \alpha \in \Delta$ is a v-filter of (X, v), we get $v(x) \in F_\alpha$ for each $\alpha \in \Delta$. Hence $v(x) \in \bigcap_{\alpha\in\Delta} F_\alpha$. Therefore $\bigcap_{\alpha\in\Delta} F_\alpha$ is a v-filter of (X, v). □

Since $\{1\}$ is a v-filter of a BE_{wvt}-algebra (X, v), it is obvious that $\bigcap_{F\in\mathcal{F}_v(X)} F = \{1\}$, where $\mathcal{F}_v(X)$ is the set of all v-filters of a BE_{wvt}-algebra (X, v). In the following, we can observe the properties of v-filters under homomorphisms.

Proposition 6.2.4 *Let (X, v) be a BE_{wvt}-algebra where the operator v is idempotent. Then for any v-filter F of (X, v), we have the following conditions:*

(1) *$v(F)$ is a v-filter of X provided $v(F)$ is a filter of X.*
(2) *$v^{-1}(F)$ is a v-filter of X provided v^{-1} exists.*

Proof (1) Let $x \in v(F)$. Then $x = v(a)$ for some $a \in F$. Since F is a v-filter, we get $v(a) \in F$. Hence $v(x) = v^2(a) \in v(F)$. Therefore $v(F)$ is a v-filter of X.
(2) Let F be a v-filter of (X, v) and suppose $x \in v^{-1}(F)$. Then $v(x) \in F$. Since F is a v-filter, we get $v^2(x) = v(x) \in F$. Hence $v(x) \in v^{-1}(F)$. Therefore $v^{-1}(F)$ is a v-filter of (X, v). □

In the following theorem, a set of equivalent conditions is derived for every filter of a self-distributive BE_{wvt}-algebra to become a v-filter.

Theorem 6.2.5 *The following are equivalent in a self-distributive BE_{wvt}-algebra (X, v):*

(1) *every filter is a v-filter,*
(2) *for each $x \in X$, $\langle x\rangle$ is a v-filter,*
(3) *for each $x \in X$, $x * v(x) = 1$.*

Proof (1) ⇒ (2): It is clear.
(2) ⇒ (3): Assume that $\langle x\rangle$ is a v-filter for each $x \in X$. Since $x \in \langle x\rangle$ and $\langle x\rangle$ is a v-filter of X, we get that $v(x) \in \langle x\rangle$. Since X is self-distributive, it yields that $x * v(x) = 1$.
(3) ⇒ (1): Let F be a filter of X. Let $x \in F$. Then we get $x * v(x) = 1$, which implies $x \leq v(x)$. Hence it yields that $v(x) \in F$. Therefore F is a v-filter of X. □

Theorem 6.2.6 *Let A be a non-empty subset of a transitive BE_{wvt}-algebra (X, v). Then*

$$\langle A\rangle_v = \{x \in X \mid v^{r_1}(a_1)*(\cdots*(v^{r_n}(a_n)*x)\cdots) = 1 \text{ for some } a_1, \ldots, a_n \in A \text{ and } r_1, \ldots, r_n \in \mathbb{N}\}$$

is the smallest v-filter containing A.

Proof Clearly $1 \in \langle A\rangle_v$. Let $x, x * y \in \langle A\rangle_v$. Then there exist $a_1, \ldots, a_n, b_1, \ldots, b_m \in A$ and $r_1, \ldots, r_n, s_1, \ldots, s_m \in \mathbb{N}$ such that $v^{r_1}(a_1)*(\cdots*(v^{r_n}(a_n)*x)\cdots) = 1$ and $v^{s_1}(b_1) * (\cdots * (v^{s_m}(b_m) * (x * y))\cdots) = 1$. Hence

$$\begin{aligned}
1 &= v^{s_1}(b_1) * (\cdots * (v^{s_m}(b_m) * (x * y))\cdots) \\
&= v^{s_1}(b_1) * (\cdots * (x * (v^{s_m}(b_m) * y))\cdots) \\
&\ldots\ldots \\
&\ldots\ldots \\
&= x * (v^{s_1}(b_1) * (\cdots * (v^{s_m}(b_m) * y)\cdots)
\end{aligned}$$

Hence $x \leq v^{s_1}(b_1) * (\cdots * (v^{s_m}(b_m) * y)\cdots)$. By the sequential application of BE-ordering $\leq$ with the elements $v^{r_1}(a_1), \ldots, v^{r_n}(a_n)$, we get

$$\begin{aligned}
1 &= v^{r_1}(a_1) * (\cdots * (v^{r_n}(a_n) * x)\cdots) \\
&\leq v^{r_1}(a_1) * (\cdots * (v^{r_n}(a_n) * (v^{s_1}(b_1) * (\cdots * (v^{s_m}(b_m) * y)\cdots)))\cdots)
\end{aligned}$$

Hence $v^{r_1}(a_1) * (\cdots * (v^{r_n}(a_n) * (v^{s_1}(b_1) * (\cdots * (v^{s_m}(b_m) * y)\cdots)))\cdots) = 1$, which gives that $y \in \langle A\rangle_v$. Therefore $\langle A\rangle_v$ is a filter of X. For any $x \in A$, we get $v(x) * (\cdots * (v(x) * x)\cdots) = 1$ because of $v(x) \leq x$. Hence $x \in \langle A\rangle_v$. Therefore $A \subseteq \langle A\rangle_v$. Let $x \in \langle A\rangle_v$. Then there exist $a_1, \ldots, a_n \in A$ and $r_1, \ldots, r_n \in \mathbb{N}$ such that $v^{r_1}(a_1) * (\cdots * (v^{r_n}(a_n) * x)\cdots) = 1$. Hence $v(v^{r_1}(a_1) * (\cdots * (v^{r_n}(a_n) * x)\cdots)) = v(1) = 1$. Since v is a weak vt-operator, by its definition, we get $1 = v(v^{r_1}(a_1) * (\cdots * (v^{r_n}(a_n) * x)\cdots)) \leq v^{r_1+1}(a_1) * (\cdots * (v^{r_n+1}(a_n) * v(x))\cdots)$. Hence $v(x) \in \langle A\rangle_v$. Therefore $\langle A\rangle_v$ is a v-filter in X such that $A \subseteq \langle A\rangle_v$.

We now prove that $\langle A\rangle_v$ is the smallest v-filter containing A. Let F be a v-filter of X containing A. Let $x \in \langle A\rangle_v$. Then there exist $a_1, \ldots, a_n \in A$ and $r_1, \ldots, r_n \in \mathbb{N}$ such that

$$v^{r_1}(a_1) * (\cdots * (v^{r_n}(a_n) * x)\cdots) = 1 \in F$$

Since $A \subseteq F$, we get $a_1, \ldots, a_n \in F$. Since F is a v-filter, we get $v^{r_1}(a_1), \ldots, v^{r_n}(a_n) \in F$. Since $v^{r_1}(a_1) \in F$ and $v^{r_1}(a_1) * (\cdots * (v^{r_n}(a_n) * x)\cdots) \in F$, we get $v^{r_2}(a_2) * (\cdots * (v^{r_n}(a_n) * x)\cdots) \in F$. Continuing in this way finally we get $x \in F$. Hence $\langle A\rangle_v \subseteq F$. Therefore $\langle A\rangle_v$ is the smallest v-filter of X containing A. □

Corollary 6.2.7 *Let (X, v) be a transitive BE_{wvt}-algebra and F a v-filter of X. If $\emptyset \neq B \subseteq X$, then $\langle F \cup B\rangle_v = \{x \in X \mid v^{r_1}(a_1) * (\cdots * (v^{r_n}(a_n) * x)\cdots) \in F$ for some $a_1, \ldots, a_n \in B$ and $r_i \in \mathbb{N}\}$.*

Corollary 6.2.8 *Let (X, v) be a transitive BE_{wvt}-algebra where v is idempotent. If F is a v-filter of X and $\emptyset \neq B \subseteq X$, then $\langle F \cup B\rangle_v = \{x \in X \mid v(a_1)*(\cdots*(v(a_n)*x)\cdots) \in F$ for some $a_1, \ldots, a_n \in B\}$.*

Corollary 6.2.9 *Let (X, v) be a transitive BE_{wvt}-algebra and F a v-filter of X. If $\emptyset \neq B \subseteq X$, then $\langle F \cup \{a\}\rangle_v = \{x \in X \mid v^{r_1}(a) * (\cdots * (v^{r_n}(a) * x)\cdots) \in I$ for some $r_1, r_2, \ldots, r_n \in \mathbb{N}\}$.*

Corollary 6.2.10 *Let (X, v) be a transitive BE_{wvt}-algebra where v is idempotent. If F is a v-filter of X and $\emptyset \neq B \subseteq X$, then $\langle F \cup \{a\}\rangle_v = \{x \in X \mid v^n(a) * x \in F$ for some $n \in \mathbb{N}\}$.*

Since $\{1\}$ and X are v- filters of (X, v), the following is a direct consequence of the above.

Corollary 6.2.11 *Let (X, v) be a transitive BE_{wvt}-algebra and $a \in X$. Then $G = \{x \in X \mid v^n(a) * x = 1$ for some $n \in \mathbb{N}\}$ is a v-filter of X.*

For any transitive BE_{wvt}-algebra (X, v), the above v-filter G in X is the principal v-filter generated by a and is denoted by $\langle a\rangle_v$. Hence $\langle a\rangle_v = \{x \in X \mid v^n(a) * x = 1$ for some $n \in \mathbb{N}\}$. If (X, v) is self-distributive, then we get $\langle a\rangle_v = \{x \in X \mid v(a) * x = 1\}$.

Theorem 6.2.12 *Let F, G be two v-filters of a transitive BE_{wvt}-algebra X. Then*

$$F \vee G = \{x \in X \mid v^m(a) * (v^n(b) * x) = 1 \text{ for some } a \in F, b \in G \text{ and } m, n \in \mathbb{N}\}$$

is the smallest v-filter of X which contains both F and G.

Proof Clearly, $1 \in F \vee G$. Let $x \in F \vee G$ and $x * y \in F \vee G$. Then there exists $a, c \in F$ and $b, d \in G$ such that $v^{m_1}(a) * (v^{n_1}(b) * x) = 1$ and $v^{m_2}(c) * (v^{n_2}(d) * (x * y)) = 1$. Then

$$x * (v^{m_2}(c) * (v^{n_2}(d) * y)) = v^{m_2}(c) * (v^{n_2}(d) * (x * y)) = 1.$$

Hence $x \leq v^{m_2}(c) * (v^{n_2}(d) * y)$. Since X is transitive, we get

$$\begin{aligned} 1 &= v^{m_1}(a) * (v^{n_1}(b) * x) \\ &\leq v^{m_1}(a) * (v^{n_1}(b) * (v^{m_2}(c) * (v^{n_2}(d) * y))) \\ &= v^{m_1}(a) * (v^{m_2}(c) * (v^{n_1}(b) * (v^{n_2}(d) * y))). \end{aligned}$$

Hence $v^{m_1}(a) * (v^{m_2}(c) * (v^{n_1}(b) * (v^{n_2}(d) * y))) = 1 \in F$ with $a, c \in F$ and $b, d \in G$. Since $a, c \in F$ and F is a v-filter, we get $v^{m_1}(a) \in F$ and $v^{m_2}(c) \in F$. Hence $v^{n_1}(b)*(v^{n_2}(d)*y) = v^{n_1}(b)*(v^{n_2}(d)*y) \in F$. Put $f = v^{n_1}(b)*(v^{n_2}(d)*y)$. Then we get

$$\begin{aligned} v^{n_1}(b) * (v^{n_2}(d) * (f * y)) &= f * (v^{n_1}(b) * (v^{n_2}(d) * y)) \\ &= (v^{n_1}(b) * (v^{n_2}(d) * y)) * (v^{n_1}(b) * (v^{n_2}(d) * y)) \\ &= 1. \end{aligned}$$

Hence $v^{n_1}(b) * (v^{n_2}(d) * (f * y)) = 1 \in G$. Since $b, d \in G$ and G is a v-filter, we get $v^{n_1}(b) \in G$ and $v^{n_2}(d) \in G$. Hence $f * y \in G$. Put $g = f * y$. Then we get

$$f * (g * y) = g * (f * y) = (f * y) * (f * y) = 1$$

Hence $f*(g*y)=1$, which gives $1=\upsilon(1)=\upsilon((f*(g*y)))\leq \upsilon(f)*(\upsilon(g)*\upsilon(y))$. Since $\upsilon(y)\leq y$, we get $\upsilon(f)*(\upsilon(g)*y)=1$. Since $f\in F, g\in G$, we get $y\in F\vee G$. Therefore $F\vee G$ is a filter of X. Let $x\in F$. Since F is a υ-filter, we get $\upsilon(x)\in F$. Clearly $\upsilon(x)*(\upsilon(1)*\upsilon(x))=\upsilon(x)*\upsilon(x)=1$. Since $\upsilon(x)\leq x$, we get $\upsilon(x)*(\upsilon(1)*x)=1$. Since $x\in F$ and $1\in G$, we get $x\in F\vee G$. Hence $F\subseteq F\vee G$. Similarly, we get $G\subseteq F\vee G$. Let K be a υ-filter of X such that $F\subseteq K$ and $G\subseteq K$. Let $x\in F\vee G$. Then there exists $a\in F\subseteq K$ and $b\in G\subseteq K$ such that $\upsilon^m(a)*(\upsilon^n(b)*x)=1\in K$ for some $m,n\in\mathbb{N}$. Since $a\in K$ and K is a υ-filter, we get $\upsilon^m(a)\in K$. Hence $\upsilon^m(b)*x\in K$. Since $b\in K$ and K is a υ-filter, we get $\upsilon^n(b)\in K$. Hence $x\in K$. Thus $F\vee G\subseteq K$. Therefore $F\vee G$ is the smallest υ-filter which contains both F and G. □

Corollary 6.2.13 *For any transitive $BE_{w\upsilon t}$-algebra (X,υ), the set $\mathcal{F}_\upsilon(X)$ of all υ-filters of X forms a complete distributive lattice.*

6.3 Congruences and υ-filters

In this section, the concept of υ-congruences is introduced and then some properties of υ-filters are studied with respect to υ-congruences. An isotone bijection is obtained between the set of all υ-filters and the set of all υ-congruences of a transitive $BE_{w\upsilon t}$-algebra.

Definition 6.3.1 Let (X,υ) be a $BE_{w\upsilon t}$-algebra. A congruence θ on X is said to be a *υ-congruence* on (X,υ) if it satisfies the following property for all $x,y\in X$:

$$(x,y)\in\theta \text{ implies } (\upsilon(x),\upsilon(y))\in\theta$$

For any self-distributive BE-algebra X, it is proved in Theorem 2.3.20 that the relation $\theta_F=\{(x,y)\in X\times X \mid x*y\in F \text{ and } y*x\in F\}$ is a congruence on X. In the following, we show that it is a υ-congruence on a $BE_{w\upsilon t}$-algebra (X,υ).

Proposition 6.3.2 *Let (X,υ) be a self-distributive $BE_{w\upsilon t}$-algebra. If F is a υ-filter of (X,υ), then $\theta_F=\{(x,y)\in X\times X \mid x*y\in F$ and $y*x\in F\}$ is a υ-congruence on (X,υ).*

Proof Since X is self-distributive, it is clear that θ_F is a congruence on X. Let $(x,y)\in\theta_F$. Then $x*y\in F$ and $y*x\in F$. Since F is a υ-filter, we get $\upsilon(x)*\upsilon(y)\geq \upsilon(x*y)\in F$ and $\upsilon(y)*\upsilon(x)\geq\upsilon(y*x)\in F$. Hence $(\upsilon(x),\upsilon(y))\in\theta_F$. Therefore θ_F is a υ-congruence on (X,υ). □

Let θ be a υ-congruence on a $BE_{w\upsilon t}$-algebra (X,υ). Then clearly $(X_{/\theta},*,[1]_\theta)$ is a BE-algebra with respect to the operation $*$ defined as $[x]_\theta*[y]_\theta=[x*y]_\theta$ for all $x,y\in X$. Define a quasi-order $\leq_\theta$ on X as $a\leq_\theta b$ if and only if $[a]_\theta\leq[b]_\theta$ for all $a,b\in X$.

Theorem 6.3.3 *Let θ be a v-congruence on a BE_{wvt}-algebra (X, v). Then the map $\overline{v} : X_{/\theta} \to X_{/\theta}$ defined by*

$$\overline{v}([x]_\theta) = [v(x)]_\theta \quad \textit{for all } x \in X$$

is a wvt-operator on $X_{/\theta}$. Moreover, $(X_{/\theta}, \overline{v})$ is a BE_{wvt}-algebra.

Proof We first observe that $\overline{v}$ is well-defined. Suppose $[x]_\theta = [y]_\theta$ where $x, y \in X$. Then $(x, y) \in \theta$. Since θ is a v-congruence on X, we get that $(v(x), v(y)) \in \theta$. Therefore

$$\overline{v}([x]_\theta) = [v(x)]_\theta = [v(y)]_\theta = \overline{v}([y]_\theta).$$

Therefore the map $\overline{v}$ is well-defined. $\overline{v}([1]_\theta) = [v(1)]_\theta = [1]_\theta$. For any $x \in X$, we get $\overline{v}([x]_\theta) = [v(x)]_\theta \leq [x]_\theta$. Let $x, y \in X$. Then we get the following:

$$\begin{aligned}
\overline{v}([x]_\theta * [y]_\theta) &= \overline{v}([x * y]_\theta)\\
&= [v(x * y)]_\theta\\
&\leq [v(x) * v(y)]_\theta\\
&= [v(x)]_\theta * [v(y)]_\theta\\
&= \overline{v}([x]_\theta) * \overline{v}([y]_\theta)
\end{aligned}$$

Therefore $\overline{v}$ is a wvt-operator on $X_{/\theta}$ and thus $(X_{/\theta}, *, [1]_\theta)$ is a BE_{wvt}-algebra. □

For any BE_{wvt}-algebra (X, v), it is clear that $Ker(v)$ is a v-filter of X. Hence the relation $\theta_{Ker(v)}$ defined by $(x, y) \in \theta_{Ker(v)}$ if and only if $x * y \in Ker(v)$, $y * x \in Ker(v)$ is a congruence on X. Now consider $X_{/Ker(v)} = \{[x]_{/Ker(v)} \mid x \in X\}$ the set of all congruence classes where $[x]_{Ker(v)} = \{y \in X \mid (x, y) \in \theta_{Ker(v)}\}$. Note that the largest element of the BE_{wvt}-algebra $X_{/Ker(v)}$ is the unit congruence class $[1]_{Ker(v)}$. Then the following corollary is a direct consequence of the above theorem.

Corollary 6.3.4 *Let (X, v) be a self-distributive BE_{wvt}-algebra. Then the map $\overline{v}$: $X_{/Ker(v)} \to X_{/Ker(v)}$ defined by $\overline{v}([x]_{Ker(v)}) = [v(x)]_{Ker(v)}$ for all $x \in X$ is a wvt-operator on $X_{/Ker(v)}$.*

Using the above theorem, we now obtain a one-to-one correspondence between the set of all v-filter of BE_{wvt}-algebra (X, v) and the set of all v-filters of the BE_{wvt}-algebra $(X_{/\theta}, \overline{v})$. For any subset S of a BE-algebra X, let us denote $\overline{S} = \{[x]_\theta \mid x \in S\}$.

Theorem 6.3.5 *Let θ a v-congruence on a BE_{wvt}-algebra (X, v) and F a non-empty subset of (X, v). Then F is a v-filter of X if and only if $\overline{F}$ is a v-filter of $X_{/\theta}$.*

Proof Assume that F is a v-filter of (X, v). Since $1 \in F$, we get that $[1]_\theta \in \overline{F}$. Let $[x]_\theta$ and $[x * y]_\theta = [x]_\theta * [y]_\theta \in \overline{F}$. Then $x, x * y \in F$. Since F is a filter, we get that

$y \in F$. Hence $[y]_\theta \in \overline{F}$. Therefore $\overline{F}$ is a filter of $X_{/\theta}$. Let $[x]_\theta \in \overline{F}$. Then $x \in F$. Since F is a υ-filter, we get that $\upsilon(x) \in F$. Hence $\overline{\upsilon}([x]_\theta) = [\upsilon(x)]_\theta \in \overline{F}$. Therefore $\overline{F}$ is a υ-filter of $(X_{/\theta}, \overline{\upsilon})$.

Conversely, assume that $\overline{F}$ is a υ-filter of $X_{/\theta}$. Then $[1]_\theta \in \overline{F}$, which implies that $1 \in F$. Let $x, x * y \in F$. Then $[x]_\theta, [x]_\theta * [y]_\theta = [x * y]_\theta \in \overline{F}$. Since $\overline{F}$ is a filter of $X_{/\theta}$, we get $[y]_\theta \in \overline{F}$. Hence $y \in F$. Therefore F is a filter of X. Let $x \in F$. Then we get $[x]_\theta \in \overline{F}$. Since $\overline{F}$ is a υ-filter of X_θ, we get $[\upsilon(x)]_\theta = \overline{\upsilon}([x]_\theta) \in \overline{F}$. Hence $\upsilon(x) \in F$. Therefore F is a υ-filter of X. □

The following theorem is a routine verification.

Theorem 6.3.6 *If F_1 and F_2 are two υ-filters of BE_{wvt}-algebras (X_1, μ) and (X_2, μ), respectively, then $F_1 \times F_2$ is a υ-filter of the product algebra $X_1 \times X_2$. Conversely, every υ-filter F of $X_1 \times X_2$ can be expressed as $F = F_1 \times F_2$ where F_i is a υ-filter of X_i for $i = 1, 2$.*

In the following, we derive a characterization theorem for υ-filters of BE_{wvt}-algebras.

Theorem 6.3.7 *Let (X, υ) be a self-distributive BE_{wvt}-algebra. If F is a filter of X, then the following conditions are equivalent:*

(1) *F is a υ-filter,*
(2) *there exists a υ-congruence on (X, υ) whose kernel is F,*
(3) *$x, y \in F$ implies $\upsilon(x) * \upsilon(y) \in F$.*

Proof (1) ⇒ (2): Assume that F is a υ-filter of (X, υ). Then by Proposition 6.3.2, θ_F is a υ-congruence on (X, υ). It is enough to show that $Ker(\theta_F) = F$. Let $x \in Ker(\theta_F)$. Then $(1, x) \in \theta_F$. Hence $x = 1 * x \in F$. Therefore $Ker(\theta_F) \subseteq F$. Again, let $x \in F$. Then $1 * x = x \in F$ and $x * 1 = 1 \in F$. Hence $(1, x) \in \theta_F$. Therefore $F \subseteq Ker(\theta_F)$.
(2) ⇒ (3): Assume that the condition (2) holds. Let $x, y \in F = Ker(\theta)$. Then $(1, x) \in \theta$ and $(1, y) \in \theta$. Since θ is a υ-congruence, we get that $(1, \upsilon(x)) \in \theta$ and $(1, \upsilon(y)) \in \theta$. Hence $(\upsilon(x) * \upsilon(y), 1) = (\upsilon(x) * \upsilon(y), 1 * 1) \in \theta$. Therefore $\upsilon(x) * \upsilon(y) \in Ker(\theta) = F$.
(3) ⇒ (1): Let $x \in F$. Since $1, x \in F$, by the condition (3), it yields that $\upsilon(x) = 1 * \upsilon(x) = \upsilon(1) * \upsilon(x) \in F$. Therefore F is a υ-filter of (X, υ). □

Corollary 6.3.8 *Let (X, υ) be a self-distributive and commutative BE_{wvt}-algebra. If F is a υ-filter of (X, υ), then θ_F is the smallest υ-congruence whose kernel is F.*

Proof Since F is a υ-filter of (X, υ), by the main theorem θ_F is a υ-congruence on (X, υ) such that $Ker(\theta_F) = F$. It is enough to prove that θ_F is the smallest congruence on X. Suppose R is a congruence on X such that $F = Ker(R)$.

$$
\begin{aligned}
(x, y) \in \theta_F &\Rightarrow x * y \in F = Ker(R) \text{ and } y * x \in F = Ker(R) \\
&\Rightarrow (x * y, 1) \in R \text{ and } (y * x, 1) \in R \\
&\Rightarrow ((x * y) * y, 1 * y) \in R \text{ and } ((y * x) * x, 1 * x) \in R \\
&\Rightarrow ((x * y) * y, y) \in R \text{ and } ((y * x) * x, x) \in R \\
&\Rightarrow (x, y) \in R \qquad\qquad \text{since } X \text{ is commutative}
\end{aligned}
$$

Therefore θ_F is the smallest υ-congruence on (X, υ) whose kernel is F. □

Theorem 6.3.9 *Let (X, υ) be a commutative $BE_{w\upsilon t}$-algebra. Then there exists an isotone bijection between the set $\mathcal{C}_\upsilon(X, \upsilon)$ of all υ-congruences of (X, υ) and the set $\mathcal{F}_\upsilon(X, \upsilon)$ of all υ-filters of (X, υ).*

Proof Define $\Psi : \mathcal{C}_\upsilon(X, \upsilon) \to \mathcal{F}_\upsilon(X, \upsilon)$ by $\Psi(\theta) = Ker(\theta)$. Since $Ker(\theta)$ is a υ-filter of (X, υ), it is clear that the map Ψ is well-defined. Let $\theta_1, \theta_2 \in \mathcal{C}_\upsilon(X, \upsilon)$ be such that $\Psi(\theta_1) = \Psi(\theta_2)$. Hence $Ker(\theta_1) = Ker(\theta_2)$. We now prove that $\theta_1 = \theta_2$. Let $(x, y) \in \theta_1$. Then we get the following:

$$
\begin{aligned}
(x, y) \in \theta_1 &\Rightarrow (x * y, y * y) \in \theta_1 \\
&\Rightarrow (x * y, 1) \in \theta_1 \\
&\Rightarrow x * y \in Ker(\theta_1) = Ker(\theta_2) \\
&\Rightarrow (x * y, 1) \in \theta_2 \\
&\Rightarrow ((x * y) * y, 1 * y) \in \theta_2 \\
&\Rightarrow ((x * y) * y, y) \in \theta_2
\end{aligned}
$$

Similarly, we get $((y * x) * x, x) \in \theta_2$. Hence, it implies $(x, y) \in \theta_2$. Therefore $\theta_1 \subseteq \theta_2$. Similarly, we can obtain that $\theta_2 \subseteq \theta_1$. Therefore Ψ is one-to-one. We now claim that Ψ is onto. Let F be a υ-filter of (X, υ). Then by Proposition 6.3.2, θ_F is a υ-congruence on (X, υ) such that $Ker(\theta_F) = F$. For this υ-congruence θ_F, we get

$$\Psi(\theta_F) = Ker\ \theta_F = F$$

Hence, it yields that Ψ is onto. Therefore, it concludes that Ψ is a bijection. To show that Ψ is an isotone, let us assume that $\theta_1 \subseteq \theta_2$ for $\theta_1, \theta_2 \in \mathcal{C}_\upsilon(X, \upsilon)$. Let $x \in \Psi(\theta_1) = Ker\ \theta_1$. Then we get $(x, 1) \in \theta_1 \subseteq \theta_2$. Hence $x \in Ker\ \theta_2 = \Psi(\theta_2)$. Therefore $\Psi(\theta_1) \subseteq \Psi(\theta_2)$. Thus Ψ is isotone. □

6.4 Prime υ-filters of BE-algebras

In this section, the notion of prime υ-filters is introduced in BE-algebras. A characterization theorem for prime υ-filters is derived. It is proved that the class of all maximal υ-filters of a $BE_{w\upsilon t}$-algebra is properly contained in the class of all prime υ-filters.

Definition 6.4.1 Let (X, υ) be a transitive $BE_{w\upsilon t}$-algebra. A proper υ-filter P of X is said to be *prime* if $F \cap G \subseteq P$ implies $F \subseteq P$ or $G \subseteq P$ for all υ-filters F and G of X.

In the following, we characterize the prime υ-filters of a $BE_{w\upsilon t}$-algebra. For any $a \in X$, let us recall that the principal υ-filter $\langle a\rangle_\upsilon$ of a BE-algebra is the set $\{x \in X \mid \upsilon(a)^n * x = 1\}$.

Proposition 6.4.2 *Let (X, υ) be a transitive $BE_{w\upsilon t}$-algebra. A proper υ-filter P of X is prime if and only if $\langle x\rangle_\upsilon \cap \langle y\rangle_\upsilon \subseteq P$ implies $x \in P$ or $y \in P$ for all $x, y \in X$.*

Proof Assume that P is a prime υ-filter of X. Let $x, y \in X$ be such that $\langle x\rangle_\upsilon \cap \langle y\rangle_\upsilon \subseteq P$. Since P is prime, it implies that $x \in \langle x\rangle_\upsilon \subseteq P$ or $y \in \langle y\rangle_\upsilon \subseteq P$. Conversely, assume the condition. Let F and G be two υ-filters of X such that $F \cap G \subseteq P$. Let $x \in F$ and $y \in G$ be the arbitrary elements. Then $\langle x\rangle_\upsilon \subseteq F$ and $\langle y\rangle_\upsilon \subseteq G$. Hence $\langle x\rangle_\upsilon \cap \langle y\rangle_\upsilon \subseteq F \cap G \subseteq P$. Then by the assumed condition, we get $x \in P$ or $y \in P$. Thus $F \subseteq P$ or $G \subseteq P$. Therefore P is prime. □

Definition 6.4.3 Let (X, υ) be a transitive $BE_{w\upsilon t}$-algebra. A proper υ-filter M of (X, υ) is said to be a *maximal υ-filter* if $\langle M \cup \{x\}\rangle_\upsilon = X$ for any $x \in X - M$.

Theorem 6.4.4 *Let (X, υ) be a transitive $BE_{w\upsilon t}$-algebra. A proper υ-filter M of X is maximal if and only if $M \subseteq F \subseteq X$ implies $M = F$ or $F = X$ for any υ-filter F of X.*

Proof Assume that M is a maximal υ-filter of X. Suppose there is a υ-filter F of X such that $M \subseteq F \subseteq X$. Let $M \neq F$. Then $M \subset F$. Choose $x \in F$ such that $x \notin M$. Since M is a maximal υ-filter, we get $\langle M \cup \{x\}\rangle_\upsilon = X$. Let $a \in X$. Hence $a \in \langle M \cup \{x\}\rangle_\upsilon$. Then there exist $m_1, m_2, \ldots m_t \in \mathbb{N}$ such that

$$\upsilon^{m_1}(x) * (\upsilon^{m_2}(x) * (\cdots (\upsilon^{m_t}(x) * a) \cdots)) \in M \subseteq F$$

Since $x \in F$ and F is a υ-filter, we get $\upsilon^{m_1}(x), \upsilon^{m_1}(x), \ldots, \upsilon^{m_1}(x) \in F$. Since F is a filter, it implies that $a \in F$. Hence $X \subseteq F$. Therefore $F = X$.

Conversely, assume that the condition holds. Let $x \in X - M$. Suppose $\langle M \cup \{x\}\rangle_\upsilon \neq X$. Choose $a \in X$ such that $a \notin \langle M \cup \{x\}\rangle_\upsilon$. Hence $M \subseteq \langle M \cup \{x\}\rangle_\upsilon \subset X$. Then by the assumed condition, we get $M = \langle M \cup \{x\}\rangle_\upsilon$. Hence $x \in M$, which is a contradiction. Therefore M is maximal. □

In the following, we prove that the class of all prime υ-filters properly contains the class of all maximal υ-filters of a transitive $BE_{w\upsilon t}$-algebra. To prove this, we need the following result.

Proposition 6.4.5 *Let F be a υ-filter of a transitive $BE_{w\upsilon t}$-algebra (X, υ). Then*

$$\langle a\rangle_\upsilon \cap \langle b\rangle_\upsilon \subseteq F \text{ if and only if } \langle F \cup \{a\}\rangle_\upsilon \cap \langle F \cup \{b\}\rangle_\upsilon = F \text{ for any } a, b \in X.$$

Proof Assume that $\langle F \cup \{a\}\rangle_v \cap \langle F \cup \{b\}\rangle_v = F$. Since $a \in \langle F \cup \{a\}\rangle_v$, we get that $\langle a\rangle_v \subseteq \langle F \cup \{a\}\rangle_v$. Similarly, we get $\langle b\rangle_v \subseteq \langle F \cup \{b\}\rangle_v$. Hence $\langle a\rangle_v \cap \langle b\rangle_v \subseteq \langle F \cup \{a\}\rangle_v \cap \langle F \cup \{b\}\rangle_v = F$.

Conversely, assume that $\langle a\rangle_v \cap \langle b\rangle_v \subseteq F$ for all $a, b \in X$. Clearly $F \subseteq \langle F \cup \{a\}\rangle_v \cap \langle F \cup \{b\}\rangle_v$. Let $x \in \langle F \cup \{a\}\rangle_v \cap \langle F \cup \{b\}\rangle_v$. Then there exist $m_1, m_2, \ldots, m_s, n_1, n_2, \ldots, n_t \in \mathbb{N}$ such that $v^{n_1}(a) * (\cdots (v^{n_s}(a) * x) \cdots) \in F$ and $v^{m_1}(b) * (\cdots (v^{m_t}(b) * x) \cdots) \in F$. Hence there exists $u, v \in F$ such that $v^{n_1}(a) * (\cdots (v^{n_s}(a) * x) \cdots) = u$ and $v^{m_1}(a) * (\cdots (v^{m_t}(b) * x) \cdots) = v$. Hence

$$\begin{aligned} v^{n_1}(a) * (\cdots (v^{n_s}(a) * (u * x)) \cdots) &= u * (v^{n_1}(a) * (\cdots (v^{n_s}(a) * x) \cdots)) \\ &= u * u \\ &= 1 \in \langle a\rangle_v. \end{aligned}$$

Since $\langle a\rangle_v$ is a v-filter and $a \in \langle a\rangle_v$, we get $u * x \in \langle a\rangle_v$. By the similar argument, we get that $v * x \in \langle b\rangle_v$. Also $u * x \le v * (u * x) = u * (v * x)$ and $v * x \le u * (v * x)$. Since $u * x \in \langle a\rangle_v$, we get that $u * (v * x) \in \langle a\rangle_v$. Since $v * x \in \langle b\rangle_v$, we get that $u * (v * x) \in \langle b\rangle_v$. Hence $u * (v * x) \in \langle a\rangle_v \cap \langle b\rangle_v \subseteq F$. Since $u, v \in F$ and F is a filter, we get that $x \in F$. Therefore $\langle F \cup \{a\}\rangle_v \cap \langle F \cup \{b\}\rangle_v \subseteq F$. □

Theorem 6.4.6 *Every maximal v-filter of a transitive BE_{wvt}-algebra (X, v) is a prime v-filter.*

Proof Let M be a maximal v-filter of (X, v). Suppose $\langle x\rangle_v \cap \langle y\rangle_v \subseteq M$ for some $x, y \in X$. Suppose $x \notin M$ and $y \notin M$. Then $\langle M \cup \{x\}\rangle_v = X$ and $\langle M \cup \{y\}\rangle_v = X$. Hence

$$\langle M \cup \{x\}\rangle_v \cap \langle M \cup \{y\}\rangle_v = X$$

Hence, by the above proposition, it yields that $\langle x\rangle_v \cap \langle y\rangle_v \nsubseteq M$, which is a contradiction. Hence $x \in M$ or $y \in M$. Therefore M is a prime v-filter of X. □

Theorem 6.4.7 *Let (X, v) ba a transitive BE_{wvt}-algebra and $a \in X$. If F is a v-filter of X such that $a \notin F$, then there exists a prime v-filter P such that $a \notin P$ and $F \subseteq P$.*

Proof Let F be a v-filter of X such that $a \notin F$. Consider $\Im = \{G \in \mathcal{F}_v(X) \mid a \notin G \text{ and } F \subseteq G\}$, where $\mathcal{F}_v(X)$ is the class of all v-filters of X. Clearly $F \in \Im$. Then by Zorn's Lemma, $\Im$ has a maximal element, say M. Clearly $a \notin M$. We now prove that M is prime. Let $x, y \in X$ be such that $\langle x\rangle_v \cap \langle y\rangle_v \subseteq M$. Then by Proposition 6.4.5, we get

$$\langle M \cup \{x\}\rangle_v \cap \langle M \cup \{y\}\rangle_v = M$$

Since $a \notin M$, we can obtain that $a \notin \langle M \cup \{x\}\rangle_v$ or $a \notin \langle M \cup \{y\}\rangle_v$. By the maximality of M, we get that $\langle M \cup \{x\}\rangle_v = M$ or $\langle M \cup \{y\}\rangle_v = M$. Hence $x \in M$ or $y \in M$. Therefore M is prime. □

Corollary 6.4.8 *Let (X, v) be a transitive BE_{wvt}-algebra and $1 \neq x \in X$. Then there exists a prime v-filter P of (X, v) such that $x \notin P$.*

Proof Let $1 \neq x \in X$ and $F = \{1\}$. Then F is a v-filter of X such that $x \notin F$. Hence by the main theorem, there exists a prime v-filter P such that $x \notin P$. □

The following corollary is a direct consequence of the above results.

Corollary 6.4.9 *Let (X, v) be a transitive BE_{wvt}-algebra. Then the intersection of all prime v-filters of X is equal to $\{1\}$.*

Corollary 6.4.10 *Let (X, v) be a transitive BE_{wvt}-algebra and F a proper v-filter of X. Then*

$$F = \cap\{P \mid P \text{ is a prime } v\text{-filter of } X \text{ such that } F \subseteq P\}$$

Proof Clearly $F \subseteq \cap\{P \mid P$ is a prime v-filter of X, $F \subseteq P\}$. Conversely, let $x \notin F$. Then by the main theorem, there exists a prime v-filter P_x such that $x \notin P_x$ and $F \subseteq P$. Therefore $x \notin \cap\{P \mid P$ is a prime v-filter of $X, F \subseteq P\}$. Hence the corollary is proved. □

Theorem 6.4.11 *Let (X, v) be a BE_{wvt}-algebra such that $v(X)$ is a filter. If F is a prime v-filter of X and $v^{-1}(F) \neq X$, then $v^{-1}(F)$ is a prime v-filter.*

Proof Since $v(1) = 1 \in F$, we get $1 \in v^{-1}(F)$. Let $x, x * y \in v^{-1}(F)$. Hence $v(x) \in F$ and $v(x * y) \in F$. Since $v(x * y) \leq v(x) * v(y)$, it yields $v(x) * v(y) \in F$. Since $v(x) \in F$, we get $v(y) \in F$ and hence $y \in v^{-1}(F)$. Therefore $v^{-1}(F)$ is a filter of X. Let $x \in v^{-1}(F)$. Then $v(x) \in F$. Since F is a v-filter, we get $v(v(x)) \in F$. Hence $v(x) \in v^{-1}(F)$. Therefore $v^{-1}(F)$ is a v-filter of X. Let $x, y \in X$ be such that $\langle x\rangle_v \cap \langle y\rangle_v \subseteq v^{-1}(F)$. Let $u \in \langle v(x)\rangle_v \cap \langle v(y)\rangle_v$. Then there exists $m_1, \ldots, m_s, n_1, \ldots, n_t \in \mathbb{N}$ such that

$$v^{m_1+1}(x) * (\cdots (v^{m_s+1}(x) * u) \cdots) = v^{m_1}(v(x)) * (\cdots * (v^{m_s}(v(x)) * u) \cdots) = 1 = v(1) \in v(X)$$

and

$$v^{n_1+1}(y) * (\cdots (v^{n_t+1}(y) * u) \cdots) = v^{n_1}(v(y)) * (\cdots * (v^{n_t}(v(y)) * u) \cdots) = 1 = v(1) \in v(X).$$

Since $v(X)$ is a filter and $v^{m_1+1}(x), \ldots, v^{m_s+1}(x) \in v(X)$, we get $u \in v(X)$. Thus $u = v(a)$ for some $a \in X$. Since $v(a) \leq a$, we get

$$\begin{aligned} 1 &= v^{m_1+1}(x) * (v^{m_2+1}(x) * (\cdots * (v^{m_s+1}(x) * u) \cdots)) \\ &= v^{m_1+1}(x) * (v^{m_2+1}(x) * (\cdots * (v^{m_s+1}(x) * v(a)) \cdots)) \\ &\leq v^{m_1+1}(x) * (v^{m_2+1}(x) * (\cdots * (v^{m_s+1}(x) * a) \cdots)) \end{aligned}$$

Hence $v(v^{m_1}(x) * (v^{m_2}(x) * (\cdots * (v^{m_s}(x) * a) \cdots))) = v(1) = 1 \in F$. Thus, we get $v^{m_1}(x) * (v^{m_2}(x) * (\cdots * (v^{m_s}(x) * a) \cdots)) \in v^{-1}(F)$. Therefore $a \in \langle v^{-1}(F) \cup \{x\}\rangle_v$. Similarly, we can get $a \in \langle v^{-1}(F) \cup \{y\}\rangle_v$. Hence

$$a \in \langle v^{-1}(F) \cup \{x\}\rangle_v \cap \langle v^{-1}(F) \cup \{y\}\rangle_v$$

Since $\langle x\rangle_v \cap \langle y\rangle_v \subseteq v^{-1}(F)$, then by Proposition 6.4.5, we get $a \in \langle v^{-1}(F) \cup \{x\}\rangle_v \cap \langle v^{-1}(F) \cup \{y\}\rangle_v = v^{-1}(F)$. Hence $u = v(a) \in F$. Thus $\langle v(x)\rangle_v \cap \langle v(y)\rangle_v \subseteq F$. Since F is a prime v-filter of X, we get $v(x) \in F$ or $v(y) \in F$. Hence $x \in v^{-1}(F)$ or $y \in v^{-1}(F)$. Therefore $v^{-1}(F)$ is a prime v-filter of X. □

Theorem 6.4.12 *Let (X, v) be a BE_{wvt}-algebra. Then $\mathcal{F}_v(X)$ is a totally ordered set (chain) if and only if every proper v-filter of (X, v) is prime.*

Proof Assume that $\mathcal{F}_v(X)$ is a totally ordered set. Let F be a proper v-filter of (X, v). Let $a, b \in X$ be such that $\langle a\rangle_v \cap \langle b\rangle_v \subseteq F$. Since $\langle a\rangle_v$ and $\langle b\rangle_v$ are v-filters of X, we get either $\langle a\rangle_v \subseteq \langle b\rangle_v$ or $\langle b\rangle_v \subseteq \langle a\rangle_v$. Hence $a \in F$ or $b \in F$. Therefore F is prime.

Conversely assume that every proper v-filter of (X, v) is prime. Let F and G be two proper v-filters of (X, v). Since $F \cap G$ is a proper v-filter of (X, v), we get that $F \subseteq F \cap G$ or $G \subseteq F \cap G$. Hence $F \subseteq G$ or $G \subseteq F$. Therefore $\mathcal{F}_v(X)$ is a totally ordered set. □

6.5 v-injective Filters of BE-algebras

In this section, the notion of v-injective filters is introduced in BE-algebras. It is derived that every filter is a v-injective filter if and only if the respective v-operator is injective. It is observed that every v-injective filter is a v-filter but not the converse. However we derive a necessary and sufficient condition for every v-filter to become a v-injective filter.

Definition 6.5.1 Let (X, v) be a transitive BE_{wvt}-algebra. A filter F of X is said to be a *v-injective filter* if it satisfies the following condition for all $x, y \in X$

$$\text{if there exist } m, n \in \mathbb{N} \text{ such that } v^m(x) = v^n(y) \text{ and } x \in F, \text{ then } y \in F.$$

For any transitive BE_{wvt}-algebra (X, v), it is clear that the filter $Ker(v)$ is a v-injective filter of (X, v) and it is the smallest v-filter of (X, v). It can be easily seen that the intersection of any two v-injective filters of a transitive BE_{wvt}-algebra is again a v-injective filter. Moreover, we have that $\bigcap_{F \in \mathcal{F}^v(X)} F = Ker(v)$, where $\mathcal{F}^v(X)$ is the set of all v-injective filters of a transitive BE_{wvt}-algebra (X, v). Though the unit filter $\{1\}$ is a v-filter of a BE_{vt}-algebra (X, v), it is not sure that $\{1\}$ is an injective filter with respect to v. However a set of equivalent conditions is established for the filter $\{1\}$ of a transitive BE_{wvt}-algebra to become a v-injective filter.

Theorem 6.5.2 *Let (X, v) be a transitive BE_{wvt}-algebra. Then the following are equivalent:*

(1) *$\{1\}$ is a v-injective filter,*
(2) *$Ker(v) = \{1\}$,*
(3) *for all $x \in X, n \in \mathbb{N}$, $v^n(x) = 1$ implies $x = 1$.*

Proof (1) $\Rightarrow$ (2): Assume that $\{1\}$ is a υ-injective filter with respect to υ. Clearly $\{1\} \subseteq Ker(\upsilon)$. Conversely, let $x \in Ker(\upsilon)$. Then $\upsilon(x) = 1 = \upsilon(1)$. Since $\{1\}$ is υ-injective, we get $x \in \{1\}$. Hence $Ker(\upsilon) \subseteq \{1\}$. Therefore $Ker(\upsilon) = \{1\}$.
(2) $\Rightarrow$ (3): Assume that $Ker(\upsilon) = \{1\}$. Let $x \in X$ and $n \in \mathbb{N}$. Suppose $\upsilon(\upsilon^{n-1}(x)) = \upsilon^n(x) = 1$. Then $\upsilon^{n-1}(x) \in Ker(\upsilon) = \{1\}$. Hence $\upsilon^{n-2}(x) \in Ker(\upsilon) = \{1\}$. Hence $\upsilon^{n-2}(x) = 1$. Continuing in this way, finally we get $x = 1$.
(3) $\Rightarrow$ (1): Assume the condition (3). Let $x, y \in X$. Suppose there exist $m, n \in \mathbb{N}$ such that $\upsilon^m(x) = \upsilon^n(y)$. Suppose $x \in \{1\}$, which means $x = 1$. Then $\upsilon^n(y) = \upsilon^m(x) = \upsilon^m(1) = 1$. Thus by condition (3), we get $y = 1$. Hence $y \in \{1\}$. Therefore $\{1\}$ is a υ-injective filter of (X, υ). □

Theorem 6.5.3 *Let (X, υ) be a self-distributive and commutative BE_υ-algebra. Then the following conditions are equivalent:*

(1) *υ is injective,*
(2) *every filter is υ-injective,*
(3) *every principle filter set is υ-injective.*

Proof (1) $\Rightarrow$ (2) and (2) $\Rightarrow$ (3) are clear.
(3) $\Rightarrow$ (1): Assume that every principle filter of X is υ-injective. Let $x, y \in X$ be such that $\upsilon^m(x) = \upsilon^n(y)$ for some $m, n \in \mathbb{N}$. Since X is self-distributive, we get $\langle x \rangle$ and $\langle y \rangle$ are filters of X. Hence $\langle x \rangle$ and $\langle y \rangle$ are υ-injective filters. Since $x \in \langle x \rangle$ and $y \in \langle y \rangle$, we get that $y \in \langle x \rangle$ and $x \in \langle y \rangle$. Hence $x * y = 1$ and $y * x = 1$. Since X is commutative, we get that $x = y$. Therefore υ is an injective map. □

Theorem 6.5.4 *Let (X, υ) be a transitive $BE_{w\upsilon t}$-algebra. Then every υ-injective filter is a υ-filter.*

Proof Let F be a υ-injective filter of (X, υ). Let $x \in F$. Then we get $\upsilon(x) \in \upsilon(F)$. Hence $\upsilon(x) = \upsilon(a)$ for some $a \in F$. Thus we get $\upsilon(\upsilon(x)) = \upsilon^2(x) = \upsilon^2(a)$. Since F is υ-injective and $a \in F$, we get that $\upsilon(x) \in F$. Therefore F is a υ-filter of (X, υ). □

The converse of the above theorem is not true. Hence every υ-filter of a transitive $BE_{w\upsilon t}$-algebra need not be a υ-injective filter with respect to υ, which can be seen in the following example.

Example 6.5.5 Let $X = \{1, a, b, c, d\}$. Define a binary operation $*$ on X as follows:

$*$	1	a	b	c	d
1	1	a	b	c	d
a	1	1	b	c	b
b	1	a	1	b	a
c	1	a	1	1	a
d	1	1	1	b	1

Then it can be easily verified that $(X, *, 1)$ is a BE-algebra. It is easy to check that $F = \{b, c, 1\}$ is a filter in X. Define a mapping $\upsilon : X \to X$ as follows:

$$v(a) = v(1) = 1; v(b) = v(d) = b \text{ and } v(c) = c$$

It can be easily seen that F is a v-filter of (X, v). But F is not a v-injective filter of (X, v) because of $v(b) = v(d)$, $b \in F$ and $d \notin F$.

However, in the following theorem, we derive a necessary and sufficient condition for every v-filter of a transitive BE_{wvt}-algebra to become a v-injective filter.

Theorem 6.5.6 *Let (X, v) be a transitive BE_{wvt}-algebra and F a v-filter of (X, v). Then the following conditions are equivalent:*

(1) *F is v-injective,*
(2) *for all $x \in X, n \in \mathbb{N}$, $v^n(x) \in F \Leftrightarrow x \in F$.*

Proof (1) $\Rightarrow$ (2): Let F be a v-filter of (X, v). Assume that F is injective. Let $x \in X$ and $n \in \mathbb{N}$. Clearly $x \in F$ implies $v^n(x) \in F$. Suppose that $v^n(x) \in F$. Clearly $v(v^n(x)) = v^{n+1}(x)$. Since F is v-injective and $v^n(x) \in F$, we get $x \in F$.

Conversely, assume that $v^n(x) \in F$ if and only if $x \in F$ for all $x \in X$ and $n \in \mathbb{N}$. Let $x, y \in X$ be such that $v^m(x) = v^n(y)$ for some $m, n \in \mathbb{N}$. Now

$$\begin{aligned} x \in F &\Rightarrow v(x) \in v(F) \subseteq F \\ &\Rightarrow v^m(y) = v^n(x) \in F \\ &\Rightarrow y \in F \qquad \text{by the assumed condition} \end{aligned}$$

Therefore F is v-injective. □

Theorem 6.5.7 *Let $F_1, F_2, \ldots F_k$ be v-injective filters of a BE_{wvt}-algebras (X_i, v_i) for $i = 1, 2, \ldots k$, respectively. Then their direct product $F_1 \times F_2 \times \cdots \times F_k$ is a v-injective filter of the product algebra $\prod_{i=1}^{k}(X_i, v_i)$.*

Proof Since each (X_i, v_i) for $i = 1, 2, \ldots, k$ is a BE_{wvt}-algebra, we get that $\prod_{i=1}^{k}(X_i, v_i)$ is also a BE_{wvt}-algebra with the weak very true operator $v_1 \times v_2 \times \cdots \times v_k$. Let F_i be a v-injective filter of (X_i, v_i) for $i = 1, 2, \ldots, k$. It is already verified that $F_1 \times F_2 \times \cdots \times F_k$ is a filter of $\prod_{i=1}^{k} X_i$. Let $(a_1, a_2, \ldots a_k), (b_1, b_2, \ldots, b_k) \in \prod_{i=1}^{k} X_i$. Suppose there exists $m, n \in \mathbb{N}$ such that

$$(v_1 \times v_2 \times \cdots \times v_k)^m(a_1, \ldots, a_k) = (v_1 \times v_2 \times \cdots \times v_k)^n(b_1, \ldots, b_k).$$

and $(a_1, a_2, \ldots, a_k) \in F_1 \times F_2 \times \cdots \times F_k$. Then $a_i \in F_i$ for $i = 1, 2, \ldots, k$. Then

$$\begin{aligned} (v_1 \times v_2 \times \cdots \times v_k)^m(a_1, \ldots, a_k) &= (v_1 \times v_2 \times \cdots \times v_k)^{m-1}(v_1(a_1), \ldots, v_k(a_k)) \\ &= (v_1 \times v_2 \times \cdots \times v_k)^{m-2}(v_1^2(a_1), \ldots, v_k^2(a_k)) \\ &\cdots \\ &= (v_1^m(a_1), \ldots, v_k^m(a_k)). \end{aligned}$$

Similarly $(\upsilon_1 \times \upsilon_2 \times \cdots \times \upsilon_k)^n(b_1, \ldots, b_k) = (\upsilon_1^n(b_1), \ldots, \upsilon_k^n(b_k))$. Hence $(\upsilon_1^m(a_1), \ldots, \upsilon_k^m(a_k)) = (\upsilon_1^n(b_1), \ldots, \upsilon_k^n(b_k))$. Thus $\upsilon_i^m(a_i) = \upsilon_i^n(b_i)$ for $i = 1, 2, \ldots, k$. Since $a_i \in F_i$ and F_i is a υ-injective filter of X_i for each $i = 1, 2, \ldots, k$, we get $b_i \in F_i$ for $i = 1, 2, \ldots, k$. Hence $(b_1, \ldots, b_n) \in F_1 \times \cdots \times F_k$. Hence $F_1 \times F_2 \times \cdots F_k$ is a υ-injective filter of the BE_{wvt}-algebra $\prod_{i=1}^{k}(X_i, \upsilon_i)$. □

6.6 Subdirectly Irreducible BE_{wvt}-algebras

In this section, we observe some of the properties of the kernel of a congruence on BE-algebras. The subdirectly irreducible BE_{wvt}-algebras are characterized in this section. Throughout this section, the weak very true operator υ is idempotent.

Definition 6.6.1 Let X and Y be two BE-algebras and $\alpha : X \to Y$ a homomorphism. Then the *kernel* of α, written $Ker(\alpha)$, is defined by $Ker(\alpha) = \{(a, b) \in X \times X \mid \alpha(a) = \alpha(b)\}$.

Theorem 6.6.2 *Let $\alpha : X \to Y$ be a homomorphism. Then $Ker(\alpha)$ is a congruence on X^2.*

Proof Clearly $Ker(\alpha)$ is an equivalence relation on X^2. Let $(a, b), (c, d) \in Ker(\alpha)$. Then clearly $\alpha(a * c) = \alpha(a) * \alpha(c) = \alpha(b) * \alpha(d) = \alpha(b * d)$. Hence $Ker(\alpha)$ is a congruence on X^2. □

Remark 6.6.3 Let X be a BE-algebra and θ a congruence on X. Then the mapping $\nu : X \to X_{/\theta}$ is a natural homomorphism.

Theorem 6.6.4 *(Homomorphism theorem) Suppose $\alpha : X \to Y$ is a homomorphism of X onto Y. Then there exists an isomorphism $\beta : X_{/Ker(\alpha)} \to Y$ defined by $\alpha = \beta o \nu$, where ν is the natural homomorphism from X to $X_{/Ker(\alpha)}$.*

Theorem 6.6.5 *Let $\alpha : X \to Y$ be a homomorphism and $A \subseteq X$. Then $(a, b) \in \theta(A)$ implies $(\alpha(a), \alpha(b)) \in \theta(\alpha(A))$, where $\theta(A)$ is a congruence generated by $X \times X$.*

Proof It is clear. □

Definition 6.6.6 The mapping $\pi_i : X_1 \times X_2 \to X_i; i \in \{1, 2\}$ defined by $\pi_i(a_1, a_2) = a_i$ is said to be the *projection map* on the ith coordinate of $X_1 \times X_2$.

Theorem 6.6.7 *For $i = 1, 2$, the map $\pi : X_1 \times X_2 \to X_i$ is a surjective homomorphism from $X_1 \times X_2$ to X_i. Furthermore, in $Con(X_1 \times X_2)$, $Ker(\pi_1) \cap Ker(\pi_2) = \Delta$.*

Proof Let $a \in X_1$. Then we get $(a, x) \in X_1 \times X_2$ for any $x \in X_2$. Hence $\pi_1(a, x) = a$. Thus π_1 is surjective. Let $(a_1, b_1), (a_2, b_2) \in X_1 \times X_2$. Then

$$\begin{aligned}\pi_1((a_1, b_1) * (a_2, b_2)) &= \pi_1(a_1 * a_2, b_1 * b_2)\\ &= a_1 * a_2\\ &= \pi_1(a_1, b_1) * \pi_2(a_2, b_2)\end{aligned}$$

Therefore π_1 is a homomorphism. Similarly, we get π_2 is a surjective homomorphism. Let $((a, b), (c, d)) \in Ker(\pi_1) \cap Ker(\pi_2)$. Then we get $a = \pi_1(a, b) = \pi(c, d) = c$. Similarly, we get $b = d$. Hence $(a, b) = (c, d)$. Therefore $((a, b), (c, d)) \in \Delta$. Conversely, let $((a, b), (c, d)) \in \Delta$. Then $(a, b) = (c, d)$, which implies that $a = c$ and $b = d$. Now $a = c$ implies $\pi_1(a, b) = \pi_1(c, d)$. Hence $((a, b), (c, d)) \in Ker(\pi_1)$. Since $b = d$, we similarly get that $((a, b), (c, d)) \in Ker(\pi_2)$. Hence $((a, b), (c, d)) \in Ker(\pi_1) \cap Ker(\pi_2)$. Therefore $Ker(\pi_1) \cap Ker(\pi_2) = \Delta$. □

Definition 6.6.8 Let X, Y be two BE-algebras and $\alpha : X \to Y$ a homomorphism. Let $a, b \in X$. Then we say that α separates the points a and b if $\alpha(a) \neq \alpha(b)$.

Example 6.6.9 Let $X = \{1, a, b, c\}$. Define a binary operation $*$ on X as follows:

$*$	1	a	b	c
1	1	a	b	c
a	1	1	a	a
b	1	1	1	b
c	1	1	1	1

Then $(X, *, 1)$ is a BE-algebra. Define $\alpha : X \to X$ as follows:

$$\alpha(1) = \alpha(a) = 1 \text{ and } \alpha(b) = \alpha(c) = b$$

Obviously α is an endomorphism on X. Clearly α separates a and b but not the points b and c.

Example 6.6.10 Let $(X, *, 1)$ be a BE-algebra, and $Id_X : X \to X$ is a mapping defined by $Id_X(x) = x$ for all $x \in X$. Then α separates every two points of X. The unity map $1_X : X \to X$ defined by $1_X(x) = x$ for all $x \in X$ separates no points.

Theorem 6.6.11 *Let X, X_1, X_2 be three BE-algebras and $\alpha_i : X \to X_i$ for $i \in \{1, 2\}$ a mapping. Then the following conditions are equivalent:*

(1) *The maps α_i separate points for some $i = 1, 2$,*
(2) *$\alpha : X \to X_1 \times X_2$ defined by $\alpha(a) = (\alpha_1(a), \alpha_2(a)$ for all $a \in X$ is injective,*
(3) *$Ker(\alpha_1) \cap Ker(\alpha_2) = \Delta$.*

Proof (1) $\Rightarrow$ (2): Assume that α_i separate points for some $i = 1, 2$. Let $a, b \in X$ and $a \neq b$. Then we get either $\alpha_1(a) \neq \alpha_1(b)$ or $\alpha_2(a) \neq \alpha_2(b)$. Hence $\alpha(a) = (\alpha_1(a), \alpha_2(a)) \neq (\alpha_1(b), \alpha_2(b)) = \alpha(b)$. Therefore α is injective mapping.
(2) $\Rightarrow$ (3): Assume that α is an injective mapping. Clearly $\Delta \subseteq Ker(\alpha_1) \cap Ker(\alpha_2)$. Let $a, b \in X$ and $(a, b) \neq \Delta$. Then we get the following conclusion:

$$\begin{aligned}(a,b) \neq \Delta &\Rightarrow a \neq b \\ &\Rightarrow \alpha(a) \neq \alpha(b) \qquad \text{since } \alpha \text{ is injective} \\ &\Rightarrow (\alpha_1(a), \alpha_2(a)) \neq (\alpha_1(b), \alpha_2(b)) \\ &\Rightarrow \alpha_1(a) \neq \alpha_1(b) \text{ or } \alpha_2(a) \neq \alpha_2(b) \\ &\Rightarrow (a,b) \notin Ker(\alpha_1) \text{ or } (a,b) \notin Ker(\alpha_2) \\ &\Rightarrow (a,b) \notin Ker(\alpha_1) \cap Ker(\alpha_2).\end{aligned}$$

Hence $Ker(\alpha_1) \cap Ker(\alpha_2) \subseteq \Delta$. Therefore $Ker(\alpha_1) \cap Ker(\alpha_2) = \Delta$.
(3) $\Rightarrow$ (1): Let α_1, α_2 be the given maps. Assume that $Ker(\alpha_1) \cap Ker(\alpha_2) = \Delta$. Let $a, b \in X$ with $a \neq b$. Then we get that $(a,b) \neq \Delta = Ker(\alpha_1) \cap Ker(\alpha_2)$. Hence $\alpha_i(a) \neq \alpha_i(b)$ for some $i = 1, 2$. Therefore α_i separate points for some $i = 1, 2$. □

The above theorem can be generalized for an indexed family of BE-algebras.

Corollary 6.6.12 *Let $\{X_i\}_{i \in I}$ be an indexed family of BE-algebras and $\alpha_i : X \to X_i$ for $i \in I$ a mapping. Then the following conditions are equivalent:*

(1) *The maps α_i separate points for some $i \in I$,*
(2) *$\alpha : X \to \prod_{i \in I} X_i$ defined by $\alpha(a) = \alpha_i(a)(i)$ for all $a \in X$ is injective,*
(3) $\bigcap_{i \in I} Ker(\alpha_i) = \Delta$.

The following corollary is a direct consequence of the above theorem.

Corollary 6.6.13 *Let $\{X_i\}_{i \in I}$ be an indexed family of BE-algebras and $\alpha_i : X \to X_i$ for $i \in I$ be an indexed family of homomorphisms. Then the following conditions are equivalent:*

(1) *The maps α_i separate points for some $i \in I$,*
(2) *$\alpha : X \to \prod_{i \in I} X_i$ defined by $\alpha(a) = \alpha_i(a)(i)$ for all $a \in X$ is an embedding,*
(3) $\bigcap_{i \in I} Ker(\alpha_i) = \Delta$.

Definition 6.6.14 [S.Burris and Sankappannavar] An algebra A of type $\mathcal{F}$ is a *subdirect product* of an indexed family $\{A_i\}_{i \in I}$ of algebras of type $\mathcal{F}$ if it satisfies the following properties:

(1) A is a subalgebra of $\prod_{i \in I} A_i$,
(2) $\pi_i(A) = A_i$ for any $i \in I$, where $\pi_i : \prod_{i \in I} A_i \to A_i$ is a natural projection map.

It is know that, for any algebra A of type $\mathcal{F}$, a one-to-one homomorphism $\alpha : A \to \prod_{i \in I} A_i$ is called a *subdirect embedding* if $\alpha(A)$ is a subdirect product of the family $\{A_i\}_{i \in I}$. An algebra A of type $\mathcal{F}$ is called *subdirectly irreducible* if, for every subdirect embedding $\alpha : A \to \prod_{i\in I} A_i$, there exists $i \in I$ such that $\pi_i \circ \nu : A \to A_i$ is an isomorphism.

Theorem 6.6.15 *If $\theta_i \in Con A$ for $i \in I$ and $\bigcap_{i\in I} \theta_i = \Delta$, then the natural homomorphism $\nu : A \to \prod_{i\in I} A_{/\theta_i}$ defined by $\nu(a)(i) = [a]_{\theta_i}$ is a subdirect embedding.*

Proof Let ν_i be the natural homomorphism from A to $A_{/\theta_i}$ for $i \in I$. Since $Ker(\nu_i) = \theta_i$, by the above theorem, ν is an embedding. Since each ν_i is surjective, ν is a subdirect embedding. □

It is already observed that for any v-congruence θ on a BE_{wvt}-algebra (X, v), the quotient algebra $(X_{/\theta}, \overline{v})$ is also a BE_{wvt}-algebra with weak very true operator $\overline{v}$ defined by $\overline{v}[x]_\theta = [v(x)]_\theta$ for all $x \in X$. A characterization is now obtained for a subdirectly irreducible BE_{wvt}-algebras.

Theorem 6.6.16 *Let (X, v) be a BE_{wvt}-algebra where X is commutative. Then (X, v) is subdirectly irreducible if and only if $\mathcal{C}_v(X, v) - \Delta$ has a minimal element.*

Proof Suppose $\mathcal{C}_v(X, v) - \Delta$ has no minimal element. Then we get $\bigcap\{\mathcal{C}_v(X, v) - \Delta\} = \Delta$. Let $I = \mathcal{C}_v(X, v) - \Delta$. Then by above theorem, we get that the natural map $\alpha : (X, v) \to \prod_{\theta\in I}(X_{/\theta}, \overline{v})$ is an embedding. Since the natural map $(X, v) \to (X_{/\theta}, \overline{v})$ is not injective for $\theta \in I$, it concludes that (X, v) is not subdirectly irreducible.

Conversely, assume that $\mathcal{C}_v(X, v) - \Delta$ has a minimal element. Therefore, let $\theta = \bigcap\{\mathcal{C}_v(X, v) - \Delta\} \neq \Delta$. Choose $(a, b) \in \theta \neq \Delta$. Hence $a \neq b$. Suppose $\alpha : (X, v) \to \prod_{i\in I}(X_i, v_{\nu_i})$ is a subdirect embedding. Then $\alpha(a)(i) = \alpha(b)(i)$ for some i. Hence $(\pi_i \circ \alpha)(a) \neq (\pi_i \circ \alpha)(b)$. Thus it implies $(a, b) \neq Ker(\pi_i \circ \alpha)$. Therefore $\theta \nsubseteq Ker(\pi_i \circ \alpha)$, which implies that $Ker(\pi_i \circ \alpha) = \Delta$. Thus $\pi_i \circ \alpha : (X, v) \to (X_i, v_{\nu_i})$ is an isomorphism. Therefore (X, v) is subdirectly irreducible. □

Remark 6.6.17 Let F and G be two filters of X such that $F \subseteq G$. Then clearly $\theta_F \subseteq \theta_G$. Let (X, v) be subdirectly irreducible. Then by the above theorem, $\mathcal{C}_v(X, v) - \Delta$ has a least element. Then by Theorem 6.6.16, there exists a v-filter of (X, v) such that $\theta = \theta_F$. Hence F is the least v-filter $(F \neq \{1\})$ of (X, v). Hence (X, v) is subdirectly irreducible if and only if $\mathcal{F}_v(X, v) - \{1\}$ has the least element.

Theorem 6.6.18 *The following conditions hold in a subdirectly irreducible BE_{wvt}-algebra:*

(1) *If $Ker(v) = \{1\}$, then $v(X)$ is a subdirectly irreducible subalgebra of X.*

(2) *If $Ker(v) \neq \{1\}$, then $Ker(v)$ is a subdirectly irreducible subalgebra of X and $Ker(v) \cap \langle a\rangle \neq \{1\}$ for each element $a \neq 1$ of $v(X)$.*

Proof (1) Let (X, v) be a subdirectly irreducible and $Ker(v) = \{1\}$. Then by the above remark, $\mathcal{F}_v(X, v) - \{1\}$ has the least element, say F. Suppose $F \cap v(X) = \{1\}$. Since F is a v-filter, we get $v(F) = v(F) \cap v(X) \subseteq F \cap v(X) = \{1\}$. Hence

$v(x) = 1$ for all $x \in F$. Thus $v(X) \subseteq Ker(v) = \{1\}$, which is a contradiction. Therefore $F \cap v(X) \neq \{1\}$. We now intend to show that $F \cap v(X)$ is the least element of $v(X) - \{1\}$. Suppose G is a filter in $v(X)$. Let $\langle G\rangle_X$ be the filter of X generated by G and $x \in \langle G\rangle_X$. Then there exists $a_1, a_2, \ldots, a_n \in G$ such that $a_n * (\ldots * (a_1 * x)\ldots) = 1$. Then by the property of the operator v, we get

$$1 = v(1) = v(a_n * (\ldots * (a_1 * x)\ldots)) \leq v(a_n) * (\ldots * (v(a_1) * v(x))\ldots)$$

Since $a_1, a_2, \ldots, a_n \in G \subseteq v(X)$ and $v^2 = v$, we get $v(a_1), v(a_2), \ldots, v(a_n) \in v(X)$. Hence $a_n * (\ldots * (a_1 * v(x))\ldots) = 1$. Thus $v(x) \in G \subseteq \langle G\rangle_X$. Therefore $\langle G\rangle_X$ is a v-filter of (X, v). Clearly $G = \langle G\rangle_X \cap \mu(X)$. Hence from the above, we get $F \subseteq \langle G\rangle_X$. Thus $F \cap v(X) \subseteq \langle G\rangle_X \cap v(X) = G$. Hence $F \cap v(X)$ is the least non-unit filter ($F \cap v(X) \neq \{1\}$) of $v(X)$. By Theorem 6.6.16, it concludes that $v(X)$ is a subdirectly irreducible subalgebra of X.
(2) Let $v(X) \neq \{1\}$. Again, let $F \neq \{1\}$ be the least v-filter of the subdirectly irreducible BE_{wvt}-algebra (X, v). Clearly $Ker(v)$ is a v-filter of (X, v). Hence $F \subseteq Ker(v)$. It is enough to show that F is the least filter of $Ker(v)$. Let $G \neq \{1\}$ be a filter of $Ker(v)$. Then $v(G) \subseteq v(Ker(v)) = \{1\} \subseteq G$. Suppose $x, x * y \in G$ where $x, y \in X$. Then

$$v(y) = 1 * v(y) = v(x) * v(y) \geq v(x * y) = 1.$$

Hence $y \in Ker(v)$. Since G is a filter of $Ker(v)$, we get that $y \in G$. Thus G is a v-filter of X, which implies that $F \subseteq G$. Then by the above theorem, it yields $Ker(v)$ which is subdirectly irreducible. Now, let $1 \neq a \in v(X)$ and $\langle a\rangle_X$ the filter generated by a in X. Then $v(a) = a$. Let $t \in \langle a\rangle_X$. Then there exists $n \in \mathbb{N}$ such that $a^n * t = 1$. Hence

$$1 = v(1) = v(a^n * t) \leq v(a)^n * v(t) = a^n * v(t).$$

Hence $v(t) \in \langle a\rangle_X$ and $v(\langle a\rangle_X) \subseteq \langle a\rangle_X$. Thus $\langle a\rangle_X$ is a v-filter of (X, v) and $\langle a\rangle_X \neq \{1\}$. Therefore $F \subseteq \langle a\rangle_X$. Since $F \subseteq Ker(v)$, it implies that $\{1\} \neq F \subseteq Ker(v) \cap \langle a\rangle_X$. □

Theorem 6.6.19 *Let (X, v) be a transitive BE_{wvt}-algebra. If it satisfies the condition (1) or (2) of the above theorem, then (X, v) is subdirectly irreducible.*

Proof Suppose (X, v) satisfies the condition (1) of Theorem 6.6.18. That is $Ker(v) = \{1\}$ and $v(X)$ is subdirectly irreducible subalgebra of X. Since $v(X)$ is subdirectly irreducible, by Remark 6.6.17, we get $\bigcap\{\mathcal{F}(v(X)) - \{1\}\}$ is a non-trivial($\neq \{1\}$) filter of $v(X)$. Since $\mathcal{F}(v(X)) = \{F \cap v(X) \mid F \in \mathcal{F}(X)\}$, we get $\bigcap(\{F \cap v(X) \mid F \in \mathcal{F}(X)\} - \{1\}$ is non-trivial and so $\bigcap(\mathcal{F}(X) - \{1\}) \neq \{1\}$. Hence the intersection of all non-trivial v-filters of (X, v) is a non-trivial v-filter of X. Therefore (X, v) is subdirectly irreducible.

Suppose (X, v) satisfies the condition (2) of Theorem 6.6.18. That is $Ker(v) \neq \{1\}$ and $Ker(v)$ is subdirectly irreducible subalgebra of X. Let F be the least non-trivial filter of $Ker(v)$. Since $v(F) = \{1\}$, we get that F is a v-filter. We intend to show that F is the least non-trivial v-filter of (X, v). Suppose G is a v-filter of (X, v). Then $v(G) \subseteq G$. If $v(G) = \{1\}$, then $G \subseteq Ker(v)$ and hence $F \subseteq G$. Otherwise, there exists $a \in v(G) - \{1\}$. Hence $\{1\} \neq Ker(v) \cap \langle a \rangle_X \subseteq Ker(v) \cap G$. Hence $F \subseteq G \cap Ker(v) \subseteq G$. Thus F is the least non-trivial v-filter of (X, v). Therefore (X, v) is subdirectly irreducible. □

Exercise

1. Let θ be a v-congruence on a BE_{wvt}-algebra X and $\overline{v}$ a self-map on $X_{/\theta}$ by $[x]_\theta \mapsto [v(x)]_\theta$. Then prove that (X, v) is a BE_{wvt}-algebra if and only if $(X_{/\theta}, \overline{v})$ is a BE_{wvt}-algebra.
2. Let F be a filter of a BE_{wvt}-algebra (X, v). If F is a prime v-filter of (X, v), then prove that $\overline{F}$ is a prime v-filter of the quotient algebra $(X_{/\theta}, \overline{v})$ where θ is the v-congruence on X defined by $(x, y) \in \theta$ such that $x * y \in F$ and $y * x \in F$.
3. Let θ be a v-congruence and F a filter of a BE_{wvt}-algebra (X, v) such that $[1]_\theta = \{1\}$. Then $\overline{F}$ is a prime v-filter of $(X_{/\theta}, \overline{v})$ whenever F is a prime v-filter of (X, v).
4. Let (X, v) be a negatively ordered BE_{wvt}-algebra. Then prove that $(v(x) * v(y)) * (v(x) * v(z)) \leq v(x) * (v(y) * v(z))$ for any $x, y, z \in X$.
5. For any subdirectly irreducible BE_{wvt}-algebra (X, v), prove that $Ker(v)$ is a prime v-filter of the BE_{wvt}-algebra.
6. Let S be a join-closed subset of a commutative BE-algebra X. If F is a v-filter of X such that $F \cap S = \emptyset$, then prove that there exists a prime v-filter P such that $F \subseteq P$ and $P \cap S = \emptyset$.

Chapter 7
Pseudo-complements

In Mathematics, particularly in order theory, a pseudo-complement is one generalization of the notion of complement. In a lattice L with bottom element 0, an element $x \in L$ is said to have a pseudo-complement if there exists a greatest element $x^* \in L$, disjoint from x, with the property that $x \wedge x^* = 0$. More formally, $x^* = \max\{y \in L \mid x \wedge y = 0\}$. The lattice L itself is called a pseudo-complemented lattice if every element of L is pseudo-complemented. Every pseudo-complemented lattice is necessarily bounded; i.e., it has a 1 as well. Since the pseudo-complement is unique by definition (if it exists), a pseudo-complemented lattice can be endowed with a unary operation * mapping every element to its pseudo-complement. The theory of pseudo-complements in lattices, and particularly in distributive lattices was developed by M.H. Stone [228], O. Frink [97] and G. Gratzer [103]. Later many authors like R. Balbes [12], O. Frink [97] extended the study of pseudo-complements to characterize Stone lattices. In 2013, Cilo$\breve{g}$lu and Ceven [53] studied the properties of the elements $x * 0$ in a commutative and bounded BE-algebras. Recently in 2014, R. Borzooei et. al [23] studied some structural properties of bounded and involutory BE-algebras and investigate the relationship between them.

In this chapter, the notion of pseudo-complements is generalized in BE-algebras. Some subclasses of BE-algebras like implicative BE-algebras, commutative BE-algebras, and involutory BE-algebras are characterized with the help of these pseudo-complements. Several interesting properties of Boolean center of a BE-algebra are studied in terms of pseudo-complements. The notion of closed elements is introduced and proved that the class of all closed elements forms a Boolean algebra. The notions of dense elements, D-filters, and dense BE-algebras are introduced. Some characterization theorems are derived for dense BE-algebras with the help of pseudo-complements.

S. R. Mukkamala, *A Course in BE-algebras*,
https://doi.org/10.1007/978-981-10-6838-6_7

The concept of ideals is introduced in BE-algebras, and a relation between ideals and lower sets is derived with the help of pseudo-complements. Some properties of ideals are studied with the help of congruences and quotient algebras. The notion of θ-ideals is introduced and characterized. Properties of some filters generated by pseudo-complements of BE-algebras are investigated.

7.1 Definition and Properties

In this section, the notion of pseudo-complements is introduced in BE-algebras. A few subclasses of BE-algebras are characterized with the help of pseudo-complements. A set of equivalent conditions is derived for a BE-algebra to become a Heyting semilattice. Throughout this section X stands for a partially ordered BE-algebra unless otherwise mentioned.

Definition 7.1.1 By a bounded BE-algebra, we mean an algebra $(X, *, 0, 1)$ of type $(2, 0, 0)$ such that its $\{*, 1\}$-reduct $X^* = (X, *, 1)$ is a BE-algebra and 0 is a constant satisfying the identity $0 * x = 1$ for all $x \in X$(equivalently $0 \leq x$ for all $x \in X$, where $\leq$ is the *BE*-ordering).

Let X be a self-distributive BE-algebra and $a \in X$. Since $\langle a \rangle = \{x \in X \mid a \leq x\}$ is closed under $*$, it is obvious that $\mathcal{X}_a = (\langle a \rangle, *, a, 1)$ is a bounded BE-algebra. Moreover, if $0 \notin X$ and

$$a * b = \begin{cases} a * b & \text{if } a, b \in X, \\ 1 & \text{if } a = 0, b \in X \\ 0 & \text{if } a \in X, b = 1 \end{cases}$$

Then $X^0 = (\{0\} \cup X, *, 0, 1)$ is also a bounded BE-algebra. Observe that a bounded BE-algebra X is finitely subdirectly irreducible, if and only if the reduct X^* is finitely subdirectly irreducible. Hence the results for BE-algebras also hold for bounded BE-algebras.

Definition 7.1.2 Let $(X, *, 0, 1)$ be a bounded BE-algebra, where 0 is the smallest element of X with respect to the ordering $\leq$. For any $x \in X$, define a unary operation $\diamond$ on X as follows:

$$x^{\diamond} = x * 0$$

Here $x^{\diamond}$ is said to be the *pseudo-complement* of x. Then the algebraic structure $(X, *, ^{\diamond}, 0, 1)$ is said to be the pseudo-complemented BE-algebra.

Clearly every finite BE-algebra is a pseudo-complemented BE-algebra. In a BE-algebra X, the element $x^{\diamond}$ may not even a complement of $x \in X$. But, when x is a Boolean element, the element $x^{\diamond}$ shall become a complement of x in X. It is also going to be observed that in a self-distributive and commutative BE-algebra every element is complemented where $x^\diamond$ is the complement of x. Keeping in view of these facts, the element $x^\diamond$ is named the pseudo-complement of x in X. In the following lemma, some of the properties of pseudo-complements of BE-algebras are observed. Observe that $x \vee y = (y * x) * x$ for all $x, y \in X$.

Lemma 7.1.3 ([53]) *Let* $(X, *, 0, 1)$ *be a bounded BE-algebra. Then for any* $x, y \in X$*, the following conditions hold.*

(1) $0^\diamond = 1$ *and* $1^\diamond = 0$,
(2) $x \leq x^{\diamond\diamond}$,
(3) $x * y^\diamond = y * x^\diamond$,
(4) $x * y^{\diamond\diamond} = y^\diamond * x^\diamond$,
(5) $0 \vee x = x^{\diamond\diamond}$ *and* $x \vee 0 = x$.

Some properties of pseudo-complements of transitive BE-algebras are observed in the following:

Proposition 7.1.4 *Let* $(X, *, 0, 1)$ *be a bounded and transitive BE-algebra. Then the following properties hold for all* $x, y \in X$.

(1) $x \leq y$ *implies* $y^\diamond \leq x^\diamond$,
(2) $x^{\diamond\diamond\diamond} = x^\diamond$,
(3) $x * y \leq y^\diamond * x^\diamond$,
(4) $x * y^\diamond = x^{\diamond\diamond} * y^\diamond$,
(5) $(x * y^{\diamond\diamond})^{\diamond\diamond} = x * y^{\diamond\diamond}$,
(6) $x * y^{\diamond\diamond} = x^{\diamond\diamond} * y^{\diamond\diamond}$,
(7) $(x * y)^{\diamond\diamond} \leq x^{\diamond\diamond} * y^{\diamond\diamond}$,
(8) $(x^\diamond * y^\diamond)^{\diamond\diamond} = x^\diamond * y^\diamond$.

Proof (1) Let X be a transitive BE-algebra. It was already observed that X is positively ordered. Let $x, y \in X$ be such that $x \leq y$. Then we get that $x * y = 1$. Since $y \leq y^{\diamond\diamond}$ and X is positively ordered, we get $x * y \leq x * y^{\diamond\diamond}$. Hence, we get that $y^\diamond * x^\diamond = (y * 0) * (x * 0) = x * ((y * 0) * 0) = x * y^{\diamond\diamond} \geq x * y = 1$. Hence $y^\diamond * x^\diamond = 1$. Therefore $y^\diamond \leq x^\diamond$.
(2) Let $x \in X$. From (2) of the above lemma, we get $x^\diamond \leq (x^\diamond)^{\diamond\diamond} = x^{\diamond\diamond\diamond}$. Since $x \leq x^{\diamond\diamond}$, by the condition (1), it yields that $x^{\diamond\diamond\diamond} \leq x^\diamond$. Therefore $x^{\diamond\diamond\diamond} = x^\diamond$.
(3) Let $x, y \in X$. Since X is transitive, we get $y * 0 \leq (x * y) * (x * 0)$. Hence

$$\begin{aligned} 1 &= y^\diamond * y^\diamond \\ &= (y * 0) * (y * 0) \\ &\leq (y * 0) * ((x * y) * (x * 0)) \\ &= (x * y) * ((y * 0) * (x * 0)) \\ &= (x * y) * (y^\diamond * x^\diamond). \end{aligned}$$

Hence $(x * y) * (y^\diamond * x^\diamond) = 1$. Therefore $x * y \leq y^\diamond * x^\diamond$.

(4) Let $x, y \in X$ be two arbitrary elements. Since X is transitive, it yields that

$$\begin{aligned} x^{\diamond\diamond} * y^{\diamond} &= ((x * 0) * 0) * (y * 0) \\ &= y * (((x * 0) * 0) * 0) \\ &= y * (x * 0) \\ &= x * (y * 0) \\ &= x * y^{\diamond}. \end{aligned}$$

(5) Let $x, y \in X$. Clearly $x * y^{\diamond\diamond} \leq (x * y^{\diamond\diamond})^{\diamond\diamond}$. On the other hand, we have

$$\begin{aligned} (x * y^{\diamond\diamond})^{\diamond\diamond} * (x * y^{\diamond\diamond}) &= x * ((x * y^{\diamond\diamond})^{\diamond\diamond} * y^{\diamond\diamond}) \\ &= x * (y^{\diamond} * (x * y^{\diamond\diamond})^{\diamond\diamond\diamond}) \\ &= x * (y^{\diamond} * (x * y^{\diamond\diamond})^{\diamond}) \\ &= x * ((x * y^{\diamond\diamond}) * y^{\diamond\diamond}) \\ &= (x * y^{\diamond\diamond}) * (x * y^{\diamond\diamond}) \\ &= 1. \end{aligned}$$

Hence $(x * y^{\diamond\diamond})^{\diamond\diamond} \leq x * y^{\diamond\diamond}$. Therefore $(x * y^{\diamond\diamond})^{\diamond\diamond} = x * y^{\diamond\diamond}$.
(6) By replacing y by $y^{\diamond}$ in (4), the result follows immediately.
(7) Since X is transitive and $y \leq y^{\diamond\diamond}$, we get that $x * y \leq x * y^{\diamond\diamond}$. Hence form the conditions (5) and (6), it yields that $(x * y)^{\diamond\diamond} \leq (x * y^{\diamond\diamond})^{\diamond\diamond} = x * y^{\diamond\diamond} = x^{\diamond\diamond} * y^{\diamond\diamond}$.
(8) From (5), we get $(x^{\diamond} * y^{\diamond})^{\diamond\diamond} = (x^{\diamond} * (y^{\diamond})^{\diamond\diamond})^{\diamond\diamond} = x^{\diamond} * y^{\diamond\diamond\diamond} = x^{\diamond} * y^{\diamond}$. □

Definition 7.1.5 ([23]) A bounded BE-algebra $(X, *, 0, 1)$ is said to be an *involutory* BE-algebra if $x^{\diamond\diamond} = x$ for all $x \in X$.

Obviously, every commutative BE-algebra is involutory.

Example 7.1.6 Let $X = \{1, a, b, c, d, 0\}$. Define a binary operation $*$ on X as follows:

$*$	1	a	b	c	d	0
1	1	a	b	c	d	0
a	1	1	a	c	c	d
b	1	1	1	c	c	c
c	1	a	b	1	a	b
d	1	1	a	1	1	a
0	1	1	1	1	1	1

Then it can be easily verified that $(X, *, 0, 1)$ is a bounded BE-algebra with smallest element 0. Observe that $a^{\diamond\diamond} = d^{\diamond} = a$; $b^{\diamond\diamond} = c^{\diamond} = b$; $c^{\diamond\diamond} = b^{\diamond} = c$; $d^{\diamond\diamond} = a^{\diamond} = d$; and $1^{\diamond\diamond} = 0^{\diamond} = 1$. Therefore X is an involutory BE-algebra.

In the following theorem, a necessary and sufficient condition is derived for every bounded BE-algebra to become an involutory BE-algebra.

Theorem 7.1.7 *A bounded BE-algebra X is involutory if and only if* $x^\diamond * y^\diamond = y * x$ *for* $x, y \in X$.

Proof Assume that X is an involutory BE-algebra. Let $x, y \in X$. Then by Lemma 7.1.3(3), we get that $x * y = x * y^{\diamond\diamond} = y^\diamond * x^\diamond$. Conversely, assume that $x^\diamond * y^\diamond = y * x$ for all $x, y \in X$. Let $x \in X$. Then $x^{\diamond\diamond} = (x*0)*0 = (x*0)*1^\diamond = x^\diamond * 1^\diamond = 1*x = x$. Therefore X is involutory. □

For any BE-algebra X, consider the set $C(X) = \{x \in X \mid x^{\diamond\diamond} = x\}$. Clearly $0, 1 \in C(X)$.

Corollary 7.1.8 *If* $(X, *, 0, 1)$ *is an involutory BE-algebra, then* $C(X) = X$.

For any two bounded BE-algebras X_1 and X_2, define the pseudo-complementation on the product algebra $X_1 \times X_2$ as $(x, y)^\diamond = (x^\diamond, y^\diamond)$ for $x \in X_1$ and $y \in X_2$. Then we have the following result about the involutory property.

Theorem 7.1.9 *If* X_1 *and* X_2 *are two involutory BE-algebras, then so is* $X_1 \times X_2$.

In the following theorem, a set of equivalent conditions is established for every bounded and transitive BE-algebra to become an involutory BE-algebra.

Theorem 7.1.10 *Let* $(X, *, 0, 1)$ *be a bounded and transitive BE-algebra. Then the following conditions are equivalent.*

(1) *X is involutory,*
(2) *for any* $x, y \in X$, $x^\diamond = y^\diamond$ *implies* $x = y$,
(3) *for all* $x \in X$, $(x * 0) * 0 = (0 * x) * x$.

Proof (1) ⇒ (2): Assume that X is involutory. Let $x, y \in X$ be such that $x^\diamond = y^\diamond$. Then $x^{\diamond\diamond} = y^{\diamond\diamond}$. Since X is involutory, we get $x = x^{\diamond\diamond} = y^{\diamond\diamond} = y$.
(2) ⇒ (3): Assume that $x^\diamond = y^\diamond$ implies $x = y$ for all $x, y \in X$. For any $x \in X$, it is clear that $((x * 0) * 0)^\diamond = (x^{\diamond\diamond})^\diamond = x^{\diamond\diamond\diamond} = x^\diamond$. Therefore, by the assumed condition (2), it yields that $(x * 0) * 0 = x = 1 * x = (0 * x) * x$.
(3) ⇒ (1): It is obvious. □

Some more properties of involutory BE-algebras are observed in the following:

Proposition 7.1.11 *The following conditions hold in an involutory BE-algebra.*

(1) $x^\diamond * y = y^\diamond * x$,
(2) $x^\diamond * y^\diamond = y * x$.

Proof (1) Let $x, y \in X$. Then by Lemma 7.1.3(3), we get $x^\diamond * y = x^\diamond * y^{\diamond\diamond} = y^\diamond * x^{\diamond\diamond} = y^\diamond * x$.

(2) Let $x, y \in X$ be two arbitrary elements. Since X is involutory, we get that $x^\diamond * y^\diamond = (x * 0) * (y * 0) = y * ((x * 0) * 0) = y * ((0 * x) * x) = y * (1 * x) = y * x$. □

Let X be a commutative BE-algebra. For any $x, y \in X$, define $x \wedge y = (x^\diamond \vee y^\diamond)^\diamond$, where $x \vee y = (y * x) * x$ for all $x, y \in X$. Then some of the facts are observed in the following:

Theorem 7.1.12 ([53]) *Let $(X, *, 0, 1)$ be a bounded and commutative BE-algebra. For any $x, y \in X$, the following conditions hold:*

(1) $(x \vee y)^\diamond = x^\diamond \wedge y^\diamond$,
(2) $(x \wedge y)^\diamond = x^\diamond \vee y^\diamond$.

Corollary 7.1.13 *Let $(X, *, 0, 1)$ be a bounded and commutative BE-algebra. Then X is a lattice with respect to the operations $\vee$ and $\wedge$ defined as $x \vee y = (y * x) * x$ and $x \wedge y = (x^\diamond \vee y^\diamond)^\diamond$.*

Though every transitive and commutative BE-algebra need not be an implicative BE-algebra, in the following theorem, a set of equivalent conditions is derived for every transitive and commutative BE-algebra to become an implicative BE-algebra with the help of pseudo-complements.

Theorem 7.1.14 *Let $(X, *, 0, 1)$ be a bounded and commutative BE-algebra and $x, y \in X$. Then the following conditions are equivalent.*

(1) *X is implicative,*
(2) $x = x^\diamond * x$,
(3) $x \vee y = x^\diamond * y$.

Proof (1) $\Rightarrow$ (2): Assume that X is implicative. Let $x \in X$. Then we get $x^\diamond * x = (x * 0) * x = x$.
(2) $\Rightarrow$ (3): Assume that $x = x^\diamond * x$ for all $x \in X$. For any $y \in X$, we get $0 \leq y$ and $x*0 \leq x*y$. Since X is also negatively ordered, it implies $x \leq (x*y)*y \leq (x*0)*y = x^\diamond * y$. Since $y \leq x^\diamond * y$, it yields that $x^\diamond * y$ is an upper bound of x and y. Hence $x \vee y \leq x^\diamond * y$. Also, by (2), we have that $x^\diamond * y \leq (y*x)*(x^\diamond * x) = (y*x)*x = x \vee y$. Therefore $x \vee y = x^\diamond * y$.
(3) $\Rightarrow$ (1): Assume that $x \vee y = x^\diamond * y$ for all $x, y \in X$. Clearly $x \leq (x * y) * x$. Since $0 \leq y$, we get $x * 0 \leq x * y$ and hence $(x * y) * x \leq (x * 0) * x = x^\diamond * x = x \vee x = x$. Therefore X is implicative. □

Remark 7.1.15 Let $(X, *, 0, 1)$ be a bounded commutative BE-algebra satisfying $x = (x * y) * x$ for all $x, y \in X$. Then, for all $x, y \in X$, we get that $x \wedge y = y \wedge x = (x * y^\diamond)^\diamond = (y * x^\diamond)^\diamond$, since $x \wedge y = (x^\diamond \vee y^\diamond)^\diamond = (x * y^\diamond)^\diamond$ and in the similar way, it yields that $y \wedge x = (y * x^\diamond)^\diamond$.

In the following, the notion of bounded *BE*-morphisms is introduced.

Definition 7.1.16 Let $(X, *, 0, 1)$ be a bounded BE-algebra. A *BE*-morphism $f : X \to X$ is called *bounded* if $f(0) = 0$.

For any *BE*-morphism f of a BE-algebra X, it is obvious that $f(1) = 1$. In the following, a necessary and sufficient condition is obtained for a *BE*-morphism to become bounded.

Theorem 7.1.17 *Let $(X, *, 0, 1)$ be a bounded BE-algebra. Then a BE-morphism $f : X \to X$ is bounded if and only if $f(x^\diamond) = f(x)^\diamond$ for all $x \in X$.*

Proof Assume that $f(0) = 0$. Let $x \in X$. Then $f(x^\diamond) = f(x * 0) = f(x) * f(0) = f(x) * 0 = f(x)^\diamond$. Conversely, assume that $f(x^\diamond) = f(x)^\diamond$ for all $x \in X$. Then $f(0) = f(1^\diamond) = f(1)^\diamond = 1^\diamond = 0$. □

The following corollary is an immediate consequence of the above theorem.

Corollary 7.1.18 *Let $(X, *, 0, 1)$ be a bounded BE-algebra and $f : X \to X$ a bounded BE-morphism. If X is involutory, then $f(X)$ is involutory.*

The following result is of intrinsic result. For any $a \in X$, define a self-map $L_a : X \to X$ by $L_a(x) = a^\diamond * x$ for all $x \in X$. Also define $R_a : X \to X$ by $R_a(x) = x * a^\diamond$ for all $x \in X$. Note that L_0 and R_0 are homomorphisms.

Theorem 7.1.19 *Let X be a transitive BE-algebra and $a \in X$. For the left map $L_a : X \to X$ and $x, y \in X$, the following conditions are equivalent.*

(1) $L_a(x * y) \leq L_a(x) * L_a(y)$,
(2) L_a *is a homomorphism,*
(3) $L_a^2 = L_a$.

Proof (1) ⇒ (2): Assume that $L_a(x * y) \leq L_a(x) * L_a(y)$ for all $x, y \in X$. Clearly $x \leq a^\diamond * x$. Since X is transitive, we get $L_a(x) * L_a(y) = (a^\diamond * x) * (a^\diamond * y) \leq x * (a^\diamond * y) = a^\diamond * (x * y) = L_a(x * y)$. Hence L_a is a homomorphism.
(2) ⇒ (3): Assume that L_a is a homomorphism. For any $x \in X$, we get $L_a^2(x) = L_a(L_a(x)) = L_a(a^\diamond * x) = L_a(a^\diamond) * L_a(x) = (a^\diamond * a^\diamond) * L_a(x) = 1 * L_a(x) = L_a(x)$. Hence $L_a^2 = L_a$.
(3) ⇒ (1): Assume that $L_a^2 = L_a$. Let $x, y \in X$. Then $x * y \leq (a^\diamond * x) * (a^\diamond * y)$. Hence

$$\begin{aligned} L_a(x * y) &= a^\diamond * (x * y) \\ &\leq a^\diamond * ((a^\diamond * x) * (a^\diamond * y)) \\ &= (a^\diamond * x) * (a^\diamond * (a^\diamond * y)) \\ &= L_a(x) * L_a^2(y) \\ &= L_a(x) * L_a(y) \end{aligned}$$

Therefore $L_a(x * y) \leq L_a(x) * L_a(y)$. □

Theorem 7.1.20 *Let X be a transitive BE-algebra and $a \in X$. If R_a is injective, then X is involutory.*

Proof Assume that R_a is injective. Let $x \in X$. Then $R_a(x) = x * a^\diamond = x^{\diamond\diamond} * a^\diamond = R_a(x^{\diamond\diamond})$. Since R_a is injective, it yields $x = x^{\diamond\diamond}$. Therefore X is involutory. □

Theorem 7.1.21 *Let $(X, *, 0, 1)$ be a transitive BE-algebra. For any $x, y \in X$, define a binary relation θ on X as $(x, y) \in \theta$ if and only if $x^\diamond = y^\diamond$. Then θ is an equivalence relation on X, and for any $a \in X$ the following holds:*

(1) *The element $a^{\diamond\diamond}$ is the greatest one in the class $[a]_\theta$,*
(2) *The class $[a]_\theta$ contains just one element from $C(X)$ which is $a^{\diamond\diamond}$.*

Proof (1) Clearly θ is an equivalence relation on X. Since $a^{\diamond\diamond\diamond} = a^\diamond$, we get $(a, a^{\diamond\diamond}) \in \theta$. Thus $a^{\diamond\diamond} \in [a]_\theta$. Let x be an element of $[a]_\theta$. Then $x^\diamond = a^\diamond$, which implies $x \le x^{\diamond\diamond} = a^{\diamond\diamond}$. Therefore $a^{\diamond\diamond}$ is the greatest element of the class $[a]_\theta$.
(2) Let $b \in C(X)$ be such that $b \in [a]_\theta$. Then we get $a^\diamond = b^\diamond$. Hence $b = b^{\diamond\diamond} = a^{\diamond\diamond}$. Therefore $[a]_\theta$ contains just one element from $C(X)$ which is $a^{\diamond\diamond}$. □

Theorem 7.1.22 *Let $(X, *, 0, 1)$ be a transitive BE-algebra which satisfies the condition $(x * y)^{\diamond\diamond} = x^{\diamond\diamond} * y^{\diamond\diamond}$ for all $x, y \in X$. Then we get the following:*

(1) *θ is a congruence on X,*
(2) *$C(X)$ is a retract of X.*

Proof (1) Let $(x, y) \in \theta$ and $(z, w) \in \theta$. Then we get $x^\diamond = y^\diamond$ and $z^\diamond = w^\diamond$. Hence $(x * z)^{\diamond\diamond} = x^{\diamond\diamond} * z^{\diamond\diamond} = y^{\diamond\diamond} * w^{\diamond\diamond} = (y * w)^{\diamond\diamond}$. Thus $(x * z)^\diamond = (y * w)^\diamond$. Hence $(x * z, y * w) \in \theta$. Therefore θ is a congruence on X.
(2) Clearly $C(X)$ is a subalgebra of X. Define $\phi : X \to X$ by $\phi(x) = x^{\diamond\diamond}$ for all $x \in X$. Then by the given condition, we get $\phi(x) = x$ for all $x \in C(X)$. Clearly $\phi(x) = x^{\diamond\diamond} \in C(X)$ for all $x \in X - C(X)$. Therefore $C(X)$ is a retract of X. □

Definition 7.1.23 Let X be a bounded BE-algebra. Define a binary operation $\oplus$ on X as $a \oplus b = a^\diamond * b$ for all $a, b \in X$.

It can be easily observed that $a \oplus 1 = 1 = 1 \oplus a$ for all $a \in X$.

Lemma 7.1.24 *Let X be a bounded and transitive BE-algebra. For any $a, b \in X$, we have*

(1) *$a \oplus a^\diamond = 1$,*
(2) *$a \oplus (a * b) = a \oplus (a * b)$,*
(3) *$b \le a * (a \oplus b)$,*
(4) *$a \le b$ implies $c \oplus a \le c \oplus b$ for any $c \in X$,*
(5) *$a \le b$ implies $a \oplus c \le b \oplus c$ for any $c \in X$.*

Proof (1) It is clear.
(2) Let $a, b \in X$. Then $a \oplus (a * b) = a^\diamond * (a * b) = a * (a^\diamond * b) = a * (a \oplus b)$.
(3) Let $a, b \in X$. Clearly $b \leq a^\diamond * (a * b) = a \oplus (a * b) = a * (a \oplus b)$.
(4) Let $a, b \in X$ be such that $a \leq b$. Since X is transitive, we get $b^\diamond \leq a^\diamond$. Hence $a^\diamond * c \leq b^\diamond * c$ for any $c \in X$. Therefore $a \oplus c \leq b \oplus c$.
(5) Let $a, b \in X$ be such that $a \leq b$. Since X is transitive, we get $c^\diamond * a \leq c^\diamond * b$. Hence $c \oplus a \leq c \oplus b$. □

Definition 7.1.25 Let X be a transitive BE-algebra and A be a non-empty subset of X. Then A is said to be a $\oplus$-closed subset of X if $a \oplus b \in A$ whenever $a, b \in A$

Theorem 7.1.26 *Let X be a bounded BE-algebra with $x^0 = 0$ for all $x \neq 0$. Then every subset containing* 1 *is a $\oplus$-closed set.*

Proof Let X be a bounded BE-algebra with the given condition. Let S be a subset of X such that $1 \in S$. For any $a, b \in S$, we get $a^\diamond * b = 0 * b = 1 \in S$. Hence S is $\oplus$-closed subset of X. □

Theorem 7.1.27 *In a transitive BE-algebra, every bounded subalgebra S (i.e., $x^\diamond \in S$ whenever $x \in S$) is a $\oplus$-closed subset.*

Proof Let S be a bounded subalgebra of a transitive BE-algebra X. Let $a, b \in S$. Since S is bounded, we get $a^\diamond \in S$. Since S is a subalgebra, we get $a \oplus b = a^\diamond * b \in S$. Hence S is $\oplus$-closed. □

The converse of the above theorem is not true; i.e., every $\oplus$-closed subset need not to be a subalgebra. For consider the following example:

Example 7.1.28 Let $X = \{0, a, b, c, 1\}$. Define a binary operation $*$ on X as follows:

$*$	0	a	b	c	1
0	1	1	1	1	1
a	c	1	1	1	1
b	b	c	1	1	1
c	a	b	c	1	1
1	0	a	b	c	1

(*a*)

$\oplus$	0	a	b	c	1
0	1	a	b	c	1
a	a	b	c	1	1
b	b	c	1	1	1
c	c	1	1	1	1
1	1	1	1	1	1

(*b*)

Clearly X is a transitive and bounded BE-algebra. Consider the set $A = \{0, b, c, 1\}$. It can be routinely verified that A is $\oplus$-closed. Observe that A is not a subalgebra of X, because $c, 0 \in A$ but $c * 0 = a \notin A$.

Definition 7.1.29 Let X and Y be two bounded BE-algebra and $f : X \to Y$ be a mapping. Then f is said to be $\oplus$-*homomorphism* if for all $a, b \in X$,

$$f(a \oplus b) = f(a) \oplus f(b)$$

A *BE*-homomorphism (resp. $\oplus$-homomorphism) of a bounded BE-algebra X is said to be *bounded* if $f(a^\diamond) = f(a)^\diamond$ for all $a \in X$.

Theorem 7.1.30 *Let X and Y be two bounded BE-algebras and $f : X \to Y$ be a $\oplus$-homomorphism. If $f^{-1}(1) = \{x \in X \mid f(x) = 1\} \neq \emptyset$, then $f^{-1}(1)$ is a $\oplus$-closed subset of X.*

Proof Let $a, b \in f^{-1}(0)$. Then $f(a) = 1$ and $f(b) = 1$. Hence $f(a \oplus b) = f(a) \oplus f(b) = 1 \oplus 1 = 1^\diamond * 1 = 1$. Thus $a \oplus b \in f^{-1}(1)$. Therefore $f^{-1}(1)$ is a $\oplus$-closed. □

Theorem 7.1.31 *Let X, Y be two bounded BE-algebras and $f : X \to Y$ be a surjective $\oplus$-homomorphism. If A is a $\oplus$-closed subset of X, then $f(A)$ is a $\oplus$-closed subset of Y.*

Proof Assume that A is a $\oplus$-closed subset of X. Let $a, b \in f(A)$ where $a, b \in Y$. Then there exists $x, y \in A$ such that $a = f(x)$ and $b = f(y)$. Then we ge $a \oplus b = f(x) \oplus f(b) = f(x \oplus y)$. Since $x, y \in A$ and A is $\oplus$-closed in X, we get $x \oplus y \in A$. Hence $a \oplus b = f(x \oplus y) \in f(A)$. Therefore $f(A)$ is a $\oplus$-closed subset in Y. □

Theorem 7.1.32 *Let X, Y be two bounded BE-algebras and $f : X \to Y$ be a $\oplus$-homomorphism. If A is a $\oplus$-closed subset of Y, then $f^{-1}(A)$ is a $\oplus$-closed subset of X.*

Proof Assume that A is a $\oplus$-closed subset of Y. Let $a, b \in f^{-1}(A)$ where $a, b \in X$. Then $f(a) \in A$ and $f(b) \in A$. Since A is $\oplus$-closed in Y, we get $f(a \oplus b) = f(a) \oplus f(b) \in A$. Hence $a \oplus b \in f^{-1}(A)$. Therefore $f^{-1}(A)$ is a $\oplus$-closed subset in X. □

Theorem 7.1.33 *Let X and Y be two bounded BE-algebras and $f : X \to Y$ be a mapping. If f is a bounded BE-homomorphism, then f is a $\oplus$-homomorphism.*

Proof Suppose f is a bounded *BE*-homomorphism. Let $a, b \in X$. Then $f(a \oplus b) = f(a^\diamond * b) = f(a^\diamond) * f(b) = f(a)^\diamond * f(b) = f(a) \oplus f(b)$. Hence f is $\oplus$-homomorphism. □

Theorem 7.1.34 *Let X and Y be two involutory BE-algebras and $f : X \to Y$ be a mapping. If f is a bounded $\oplus$-homomorphism, then f is a bounded BE-homomorphism.*

Proof Suppose X and Y are involutory and $f : X \to Y$ is a bounded $\oplus$-homomorphism. For any $a, b \in X$, we get

$$\begin{aligned} f(a * b) &= f(a^{\diamond\diamond} * b) \\ &= f(a^\diamond \oplus b) \\ &= f(a^\diamond) \oplus f(b) \\ &= f(a)^\diamond \oplus f(b) \\ &= f(a)^{\diamond\diamond} * f(b) \\ &= f(a) * f(b) \end{aligned}$$

Therefore f is abounded *BE*-homomorphism. □

Corollary 7.1.35 *Let X and Y be two involutory BE-algebras and $f : X \to Y$ be a mapping. Then f is a bounded $\oplus$-homomorphism if and only if f is a bounded BE-homomorphism.*

Lemma 7.1.36 *Let X be an involutory and transitive BE-algebra. For any $a, b \in X$, we have*

(1) $a^\diamond \oplus a = 1$,
(2) $a, b \leq a \oplus b$,
(3) $a \oplus b = b \oplus a$,
(4) $a \leq b$ *if and only if* $a \oplus b^\diamond = 1$,
(5) $a * (a \oplus b) = a \oplus (a * b) = 1$.

Proof (1) Let $a \in X$. Then $a^\diamond \oplus a = a^{\diamond\diamond} * a = a * a = 1$.
(2) Let $a, b \in X$. Clearly $b \leq a^\diamond * b = a \oplus b$. Since X is involutory, we get $a \leq b^\diamond * a = a^\diamond * b = a \oplus b$.
(3) Let $a, b \in X$. Since X is involutory, we get $a \oplus b = a^\diamond * b = b^\diamond * a = b \oplus a$.
(4) Let $a, b \in X$. Assume that $a \leq b$. Then $b^\diamond \leq a^\diamond$. Hence $1 = b^\diamond \oplus b \leq a^\diamond \oplus b = b \oplus a^\diamond$. Conversely, assume that $b \oplus a^\diamond = 1$. Then $b^\diamond * a^\diamond = 1$, which means $b^\diamond \leq a^\diamond$. Since X is involutory, we get $a = a^{\diamond\diamond} \leq b^{\diamond\diamond} = b$.
(5) Let $a, b \in X$. Then $a * (a \oplus b) = a * (a^\diamond * b) = a * (b^\diamond * a) = b^\diamond * (a * a) = 1$. Also $a \oplus (a * b) = a^\diamond * (a * b) = a * (a^\diamond * b) = a * (b^\diamond * a) = b^\diamond * (a * a) = b^\diamond * 1 = 1$. □

Theorem 7.1.37 *If X is a transitive and involutory BE-algebra, then the algebraic structure $\langle X, \oplus, 1\rangle$ forms an abelian group.*

Proof Clearly the operation $\oplus$ is well-defined. Let $a \in X$. Clearly $a \oplus 1 = 1 \oplus a = 1$. Hence 1 is the identity element of X with respect to $\oplus$. Let $a, b, c \in X$. Since X is involutory, we get

$$\begin{aligned}
a \oplus (b \oplus c) &= a^\diamond * (b \oplus c) \\
&= a^\diamond * (b^\diamond * c) \\
&= a^\diamond * (c^\diamond * b) \\
&= c^\diamond * (a^\diamond * b) \\
&= (a^\diamond * b)^\diamond * c \\
&= (a \oplus b)^\diamond * c \\
&= (a \oplus b) \oplus c
\end{aligned}$$

Therefore $\oplus$ is associative on X. Let $a \in X$. Clearly $a \oplus a^\diamond = 1$. Since X is involutory, by (1) of the lemma, we get $a^\diamond \oplus a = 1$. Hence $a^\diamond$ is the inverse of a for each $a \in X$. Therefore $\langle X, \oplus, 1\rangle$ is a group. Let $a, b \in X$. Since X is involutory, by (3) of the above lemma, we get $a \oplus b = b \oplus a$. Therefore $\langle X, \oplus, 1\rangle$ is an abelian group. □

7.2 Boolean Elements of a BE-algebra

In this section, the notion of Boolean elements is introduced in BE-algebras. Several interesting properties of a Boolean center of a BE-algebra are studied. A congruence is introduced on a commutative BE-algebra, and then, the properties of the quotient algebra are studied. Throughout this section X stands for a partially ordered BE-algebra unless otherwise mentioned.

Proposition 7.2.1 *Let X be a BE-algebra and $a, b \in X$. Then $a*c = 1$ and $b*c = 1$ imply $c = 1$ for all $c \in X$ if and only if $\langle a\rangle \cap \langle b\rangle = \{1\}$.*

Proof Assume that $a * c = 1$ and $b * c = 1$ imply $c = 1$ for all $c \in X$. Let $c \in \langle a\rangle \cap \langle b\rangle$. Then there exist positive integers n, m such that $a^n * c = 1$ and $b^m * c = 1$. Hence, it yields $a^n * (b^m * c) = a^n * 1 = 1$. Assume that $k = \min\{m \in \mathbb{Z}^+ \mid b^m * c = 1\}$. If $n \geq 1$, then we get $a \leq a^{n-1} * c$ because $a * (a^{n-1} * c) = a^n * c = 1$. Hence, we get the following:

$$a \leq a^{n-1} * c \leq b^{k-1} * (a^{n-1} * c) = a^{n-1} * (b^{k-1} * c).$$

Also $a^{n-1} * (b^k * c) = 1$. Hence $b * (a^{n-1} * (b^{k-1} * c)) = 1$, which results in $b \leq a^{n-1} * (b^{k-1} * c)$. Thus $a \leq a^{n-1} * (b^{k-1} * c)$ and $b \leq a^{n-1} * (b^{k-1} * c)$. Therefore by the assumption $a^{n-1} * (b^{k-1} * c) = 1$. By iterating the process, it yields that $b^{k-1} * c = a^0 * (b^{k-1} * c) = 1$, which is a contradiction to the assumption. Hence $k = 0$ and $c = b^0 * c = 1$. The converse is an immediate consequence. □

The following corollaries are immediate consequences of the above proposition.

Corollary 7.2.2 *Let X be a BE-algebra in which $\{1\}$ is a prime filter. Then, for all $a, b \in X$, the following conditions are equivalent:*

(1) $\langle a\rangle \cap \langle b\rangle = \{1\}$,
(2) $a = 1$ *or* $b = 1$,
(3) *for any* $c \in X$, $(a * c) * ((b * c) * c) = 1$.

Corollary 7.2.3 *Let $(X, *, 0, 1)$ be a bounded and commutative BE-algebra. Then for any $x, y \in X$, $x \vee y = 1$ if and only if $\langle x\rangle \cap \langle y\rangle = \{1\}$.*

In a bounded BE-algebra X, we can observe that $a * (a^{\diamond} * 0) = a * a^{\diamond\diamond} = 1$. Hence, for any positively ordered BE-algebra, it implies $0 \in \langle a\rangle \vee \langle a^{\diamond}\rangle$. Thus it infers that $\langle a\rangle \vee \langle a^{\diamond}\rangle = X$. On the other hand, if $\langle a\rangle \cap \langle a^{\diamond}\rangle = \{1\}$, then $\langle a\rangle$ is complemented in the distributive lattice of principal filters with $\langle a^{\diamond}\rangle$ as its unique complement. Thus the properties of these kinds of elements can be studied for special purpose and so the notion of *Boolean elements* is introduced in the following:

Definition 7.2.4 Let $(X, *, 0, 1)$ be a bounded BE-algebra. An element $x \in X$ is called a *Boolean element* if $\langle x\rangle \cap \langle x^{\diamond}\rangle = \{1\}$.

It can be easily observed that $\langle x\rangle \cup \langle x^\diamond\rangle = X$ for all $x \in X$. Let us denote the set of all Boolean elements of a bounded BE-algebra X by $\mathcal{B}(X)$ (*i.e.*, $\mathcal{B}(X) = \{x \in X \mid \langle x\rangle \cap \langle x^\diamond\rangle = \{1\}\}$) which is called the *Boolean center* of X. Clearly 0 and 1 are Boolean elements of X and so $0, 1 \in \mathcal{B}(X)$.

Lemma 7.2.5 *Let $(X, *, 0, 1)$ be a bounded BE-algebra. An element $a \in X$ is a Boolean element if and only if for all $c \in X$, $a * c = 1$ and $a^\diamond * c = 1$ imply $c = 1$, that is,* $\sup\{a, a^\diamond\} = 1$.

Proof Replacing b of the Proposition 7.2.1 by $a^\diamond$, it follows immediately. □

Lemma 7.2.6 *Let $(X, *, 0, 1)$ be a bounded BE-algebra and $x, y \in X$ such that $x \vee y$ exists in X. Then there exists $x^\diamond \wedge y^\diamond$ such that $x^\diamond \wedge y^\diamond = (x \vee y)^\diamond$.*

Proof Since $x, y \leq x \vee y$, we get $(x \vee y)^\diamond \leq x^\diamond, y^\diamond$. Thus $(x \vee y)^\diamond$ is a lower upper bound for $x^\diamond$ and $y^\diamond$. Again, let $t \in X$ such that $t \leq x^\diamond, y^\diamond$. Then $x \leq x^{\diamond\diamond} \leq t^\diamond$ and $y \leq y^{\diamond\diamond} \leq t^\diamond$, which implies $x \vee y \leq t^\diamond$. Hence $t \leq t^{\diamond\diamond} \leq (x \vee y)^\diamond$. Therefore $(x \vee y)^\diamond = x^\diamond \wedge y^\diamond$. □

Remark 7.2.7 If $x \in \mathcal{B}(X)$, then $x \vee x^\diamond = 1$, and hence by the above lemma we get that $x^\diamond \wedge x^{\diamond\diamond} = (x \vee x^\diamond)^\diamond = 1^\diamond = 0$. Hence, it yields that $x \wedge x^\diamond \leq x^{\diamond\diamond} \wedge x^\diamond = 0$. Thus, it infers that $x \wedge x^\diamond = 0$. Therefore, it concludes that $x^\diamond$ is the complement of x in X.

Boolean elements of a bounded BE-algebras also satisfy several interesting properties which can be proved using above lemmas and some arithmetical calculus.

Proposition 7.2.8 *Let $(X, *, 0, 1)$ be a bounded and transitive BE-algebra. Then for every $a \in \mathcal{B}(X)$ and $x, y \in X$, the following implications hold.*

(1) $a^\diamond \in \mathcal{B}(X)$,
(2) $a * (a * x) = a * x$,
(3) $a * (x * y) = (a * x) * (a * y)$,
(4) $a * a^\diamond = a^\diamond$ *and* $a^\diamond * a = a$,
(5) $a^{\diamond\diamond} = a$,
(6) $(a * x) * a = a$,
(7) $(a * x) * (x * a) = x * a$,
(8) $(a * x) * x \leq (x * a) * a$,
(9) $((a * x) * a^\diamond) * a^\diamond = a * x^{\diamond\diamond}$,
(10) *if* $b \in \mathcal{B}(X)$, *then* $(a * b) * b = (b * a) * a$.

Proof (1) Let $a \in \mathcal{B}(X)$, i.e., $\langle a\rangle \cap \langle a^\diamond\rangle = \{1\}$. Since $a \leq a^{\diamond\diamond}$, we get $\langle a^{\diamond\diamond}\rangle \subseteq \langle a\rangle$. Hence $\langle a^\diamond\rangle \cap \langle a^{\diamond\diamond}\rangle \subseteq \langle a\rangle \cap \langle a^\diamond\rangle = \{1\}$. Therefore $a^\diamond \in \mathcal{B}(X)$.
(2) Let $a \in \mathcal{B}(X)$, i.e., $\langle a\rangle \cap \langle a^\diamond\rangle = \{1\}$. For any $x \in X$, we get $a^\diamond = a * 0 \leq a * x \leq (a*(a*x))*(a*x)$. Thus $(a*(a*x))*(a*x) \in \langle a^\diamond\rangle$. Since $a \leq (a*(a*x))*(a*x)$, we get $(a * (a * x)) * (a * x) \in \langle a\rangle$. Hence $(a * (a * x)) * (a * x) \in \langle a^\diamond\rangle \cap \langle a\rangle = \{1\}$. Therefore $a * (a * x) \leq a * x$. Clearly $a * x \leq a * (a * x)$. Therefore $a * (a * x) = a * x$.

(3) Let $a \in \mathcal{B}(X)$. Since X is transitive, we get $x * y \leq (a * x) * (a * y)$ for all $x, y \in X$. Since $a \in \mathcal{B}(X)$, by (2), we get $a * (x * y) \leq a * ((a * x) * (a * y)) = (a * x) * (a * (a * y)) = (a * x) * (a * y)$. Therefore $a * (x * y) \leq (a * x) * (a * y)$. On the other hand, we can obtain that $(a * x) * (a * y) \leq a * (x * y)$ because X is transitive. Therefore $a * (x * y) = (a * x) * (a * y)$.
(4) Let $a \in \mathcal{B}(X)$. Then $\langle a \rangle \cap \langle a^\diamond \rangle = \{1\}$. By (2), we get $a * a^\diamond = a * (a * 0) = a * 0 = a^\diamond$. We now prove the remaining. Clearly $a \leq (a^\diamond * a) * a$. Hence $(a^\diamond * a) * a \in \langle a \rangle$. Also we have $a^\diamond * ((a^\diamond * a) * a) = (a^\diamond * a) * (a^\diamond * a) = 1$. Thus $a^\diamond \leq (a^\diamond * a) * a$, which implies $(a^\diamond * a) * a \in \langle a^\diamond \rangle$. Hence $(a^\diamond * a) * a \in \langle a \rangle \cap \langle a^\diamond \rangle = \{1\}$. Thus $a^\diamond * a \leq a$. Therefore $a^\diamond * a = a$.
(5) Let $a \in \mathcal{B}(X)$. Clearly $a \leq a^{\diamond\diamond}$. Since $a \leq a^{\diamond\diamond} * a$, we get $a^{\diamond\diamond} * a \in \langle a \rangle$. Since $a^\diamond \in \mathcal{B}(X)$, by (3), we get $a^\diamond * (a^{\diamond\diamond} * a) = a^{\diamond\diamond} * (a^\diamond * a) = (a^\diamond * 0) * (a^\diamond * a) = a^\diamond * (0 * a) = a^\diamond * 1 = 1$. Hence $a^{\diamond\diamond} * a \in \langle a^\diamond \rangle$. Thus $a^{\diamond\diamond} * a \in \langle a \rangle \cap \langle a^\diamond \rangle = \{1\}$. Hence $a^{\diamond\diamond} \leq a$. Therefore $a^{\diamond\diamond} = a$.
(6) Let $a \in \mathcal{B}(X)$. Then by condition (4), we have $a^\diamond * a = a$. Let $x \in X$. Since X is a transitive BE-algebra, we get that $a^\diamond = a * 0 \leq a * x$ and hence it infers $(a * x) * a \leq a^\diamond * a = a$. Hence, it yields that $(a * x) * a \leq a$. Clearly $a \leq (a * x) * a$. Therefore, it concludes that $(a * x) * a = a$.
(7) Let $a \in \mathcal{B}(X)$. Then by condition (6), we get that $(a * x) * (x * a) = x * ((a * x) * a) = x * a$.
(8) Let $a \in \mathcal{B}(X)$. Then by condition (6), we get that $(a * x) * a = a$ for any $x \in X$. Clearly $x * ((x * a) * a = (x * a) * (x * a) = 1$, and thus it infers $x \leq ((x * a) * a$. Since X is transitive, it yields that $(a * x) * x \leq (a * x) * ((x * a) * a) = (x * a) * ((a * x) * a) = (x * a) * a$.
(9) Let $a \in \mathcal{B}(X)$. For any $x \in X$, by the condition (3), it immediately infers that

$$\begin{aligned} ((a * x) * a^\diamond) * a^\diamond &= ((a * x) * (a * 0)) * (a * 0) \\ &= (a * (x * 0)) * (a * 0) \\ &= (a * x^\diamond) * (a * 0) \\ &= a * (x^\diamond * 0) \\ &= a * x^{\diamond\diamond} \end{aligned}$$

(10) Since $a \in \mathcal{B}(X)$, by (8), we get $(a * b) * b \leq (b * a) * a$. If $b \in \mathcal{B}(X)$, then again by (8), it yields $(b * a) * a \leq (a * b) * b$. Therefore $(a * b) * b = (b * a) * a$. □

It is observed from Proposition 7.2.8(5) that every Boolean element of a BE-algebra is a closed element. It is evident from the following example that every closed element need not be Boolean.

Example 7.2.9 Let $X = \{1, a, b, c, d, 0\}$. Define a binary operation $*$ on X as follows:

$*$	1	a	b	c	d	0
1	1	a	b	c	d	0
a	1	1	a	c	c	d
b	1	1	1	c	c	c
c	1	a	b	1	a	b
d	1	1	a	1	1	a
0	1	1	1	1	1	1

Clearly $(X, *, 0, 1)$ is a bounded BE-algebra. Observe that $a^{\diamond\diamond} = d^{\diamond} = a$; $b^{\diamond\diamond} = c^{\diamond} = b$; $c^{\diamond\diamond} = b^{\diamond} = c$; and $d^{\diamond\diamond} = a^{\diamond} = d$. Therefore $\mathcal{C}(X) = X$. But the elements a and d of the BE-algebra X are not Boolean because $\langle a\rangle \cap \langle a^{\diamond}\rangle = \langle a\rangle \cap \langle d\rangle = \{1, a\} \cap \{1, d, a, c\} = \{1, a\} \neq \{1\}$ also $\langle d\rangle \cap \langle d^{\diamond}\rangle \neq \{1\}$.

Theorem 7.2.10 *In an implicative BE-algebra, every closed element is Boolean.*

Proof Let X be an implicative BE-algebra. Let $a \in \mathcal{C}(X)$. Let $x \in X$ and suppose $a * x = 1$ and $a^{\diamond} * x = 1$. Then $x^{\diamond} \leq a^{\diamond\diamond} = a \leq x$. Since X is implicative, we get $x = (x * 0) * x = x^{\diamond} * x = 1$. Hence by Proposition 7.2.1, we get $\langle a\rangle \cap \langle a^{\diamond}\rangle = \{1\}$. Therefore a is a Boolean element. □

In the following theorem, some equivalent conditions are derived for every element of a BE-algebra to become a Boolean element.

Theorem 7.2.11 *Let X be a transitive BE-algebra X which satisfies the condition: $(x * y) * y = x^{\diamond} * y$ for all $x, y \in X$. Then the following conditions are equivalent:*

(1) *Every element is Boolean,*
(2) *for $x, y \in X$, $x * (x * y) = x * y$,*
(3) *for any $a \in X$, $\langle a\rangle = \{x \in X \mid a * x = 1\}$.*

Proof (1) ⇒ (2): Assume that every element of X is a Boolean element. Let $x, y \in X$. Then by Proposition 7.2.8(2), it is clear that $x * (x * y) = x * y$.
(2) ⇒ (3): Assume that the condition (2) holds. Let $a \in X$. Clearly $\{x \in X \mid a * x = 1\} \subseteq \langle a\rangle$. Conversely, let $x \in \langle a\rangle$. Then $a^n * x = 1$ for some positive integer n. By the repeated use of condition (2), it yields that $a * x = 1$. Therefore $\langle a\rangle = \{x \in X \mid a * x = 1\}$.
(3) ⇒ (1): Assume that the condition (3) holds. Let $a \in X$ and $x \in \langle a\rangle \cap \langle a^{\diamond}\rangle$. Then $a * x = 1$ and $a^{\diamond} * x = 1$. Then by the assumption, we get

$$\begin{aligned} x &= 1 * x \\ &= (a * x) * x \\ &= a^{\diamond} * x \qquad \text{by the given condition on } X \\ &= 1 \end{aligned}$$

Hence $x = 1$, which implies that $\langle a\rangle \cap \langle a^{\diamond}\rangle = \{1\}$. Therefore a is Boolean. □

It is observed that every self-distributive is both commutative and simple but not the converse. Though every element is not Boolean, in the following theorem, a set of equivalent is derived for every element of a BE-algebra to become Boolean, which provides an evidence for the existence of the above converse.

Theorem 7.2.12 *The following conditions are equivalent in a commutative BE-algebra X.*

(1) *X is simple,*
(2) *every element is Boolean,*
(3) *X is self-distributive.*

Proof (1) $\Rightarrow$ (2): Assume that X is simple. Let $a \in X$. Since X is simple, we get $a * a^\diamond = a * (a * 0) = a * 0 = a^\diamond$. Thus, it yields $a \vee a^\diamond = (a * a^\diamond) * a^\diamond = 1$. Then by Corollary 7.2.3, we get $\langle a \rangle \cap \langle a^\diamond \rangle = \{1\}$. Therefore a is a Boolean element.
(2) $\Rightarrow$ (3): Assume that every element of X is Boolean. Let $x, y, z \in X$. By Proposition 7.2.8(3), we get $x * (y * z) = (x * y) * (x * z)$. Therefore X is self-distributive.
(3) $\Rightarrow$ (1): It is obvious. □

Theorem 7.2.13 *Let X be a transitive BE-algebra and $a \in \mathcal{B}(X)$. Define a binary relation θ_a on X as follows:*

$$(x, y) \in \theta_a \text{ if and only if } a * x = a * y$$

for all $x, y \in X$. Then θ_a is a congruence on X such that $[1]_{\theta_a} = \langle a \rangle$.

Proof Clearly θ_a is an equivalence relation on X. Let $(x, y) \in \theta_a$ and $(z, w) \in \theta_a$. Then $a * x = a * y$ and $a * z = a * w$. Since $a \in \mathcal{B}(X)$, we get that $a * (x * z) = (a*x)*(a*z) = (a*y)*(a*w) = a*(y*w)$. Thus $(x*z, y*w) \in \theta_a$. Therefore θ_a is a congruence on X. Now $[1]_{\theta_a} = \{x \in X \mid a*x = a*1\} = \{x \in X \mid a*x = 1\} = \langle a \rangle$ because $a \in \mathcal{B}(X)$ and Proposition 7.2.8(2). □

Theorem 7.2.14 *Let X be a commutative BE-algebra and $a \in \mathcal{B}(X)$. Then $\theta_{\langle a \rangle} = \theta_a$.*

Proof Let $a \in \mathcal{B}(X)$. Assume that $(x, y) \in \theta_{\langle a \rangle}$. Then $x * y \in \langle a \rangle$ and $y * x \in \langle a \rangle$. Then by Proposition 7.2.8(2), we get $a * (x * y) = 1$ and $a * (y * x) = 1$. Since $a \in \mathcal{B}(X)$, we get $(a * x) * (a * y) = 1$ and $(a * y) * (a * x) = 1$. Since X is commutative, it yields that $a * x = a * y$. Hence $(x, y) \in \theta_a$. Therefore $\theta_{\langle a \rangle} \subseteq \theta_a$. Conversely, assume that $(x, y) \in \theta_a$. Then $a*x = a*y$. Since X is commutative, we get $a*(x*y) \geq (a*x)*(a*y) = 1$. Hence $x*y \in \theta_{\langle a \rangle}$. Similarly, we get $y*x \in \theta_{\langle a \rangle}$. Hence $(x, y) \in \theta_{\langle a \rangle}$. Therefore $\theta_a \subseteq \theta_{\langle a \rangle}$. □

Lemma 7.2.15 *Let $(X, *, 0, 1)$ be a bounded and commutative BE-algebra. If F is a filter of $\mathcal{B}(X)$, then $(x, y) \in \theta_F$ if and only if $a * x = a * y$ for some $a \in F$.*

Proof Let F be a filter of $\mathcal{B}(X)$. Since X is a distributive lattice, it can be easily seen that $F = \bigcup_{x \in F} \langle x \rangle$. Then by Proposition 7.2.8, the result follows immediately. □

Definition 7.2.16 Let $(X, *, 0, 1)$ be a bounded BE-algebra. For any $a \in X$, define the set Δ_a as $\Delta_a = \{x \in X \mid a * x = x\}$.

Obviously, we have $\Delta_0 = \{1\}$ and $\Delta_1 = X$. For any $a, b \in X$ with $a \leq b$, it can be observed that $\Delta_a \subseteq \Delta_b$. Some more properties of Δ_a, $a \in X$ are observed in the following lemmas.

Lemma 7.2.17 *Let $(X, *, 0, 1)$ be a bounded and transitive BE-algebra. Then Δ_a is a quasi-filter for any $a \in X$. Moreover, Δ_a is a subalgebra of X whenever $a \in \mathcal{B}(X)$.*

Proof Clearly $1 \in \Delta_a$. Let $x \in \Delta_a$. Then we get $a * x = x$. For any $t \in X$, we get $a * (t * x) = t * (a * x) = t * x$. Thus $t * x \in \Delta_a$. Therefore Δ_a is a quasi-filter of X. To prove the remaining, let $a \in \mathcal{B}(X)$. Let $x, y \in \Delta_a$. Then $a * x = x$ and $a * y = y$. Hence, by Proposition 7.2.8(3), we get $a * (x * y) = (a * x) * (a * y) = x * y$. Thus $x * y \in \Delta_a$. Therefore Δ_a is a subalgebra of X. □

Lemma 7.2.18 *Let X be a bounded BE-algebra and $a \in \mathcal{B}(X)$. For any $x, y \in X$, we have*

(1) $a^\diamond \in \Delta_a$,
(2) $a * x \in \Delta_a$,
(3) *If $(x, y) \in \theta_{\langle a \rangle}$ and $x, y \in \Delta_a$, then $x = y$.*

Proof (1) Since $a \in \mathcal{B}(X)$, we get that $a * a^\diamond = a^\diamond$. Therefore $a^\diamond \in \Delta_a$.
(2) Since $a \in \mathcal{B}(X)$, we get that $a * (a * x) = a * x$. Therefore $a * x \in \Delta_a$ for all $a \in X$.
(3) Let $(x, y) \in \theta_{\langle a \rangle}$ and $x, y \in \Delta_a$. Then it yields that $x = a * x = a * y = y$. □

Properties of multipliers are studied extensively in Chap. 5. An interconnection between multipliers and congruences is studied in Theorem 5.2.17 with the help of quasi-filters. In the following theorem, we now generalize this result with the help of the quasi-filter Δ_a.

Theorem 7.2.19 *Let $(X, *, 0, 1)$ be a bounded and commutative BE-algebra. For any $a \in \mathcal{B}(X)$, $\Delta_a \cap [x]_{\theta_{\langle a \rangle}}$ is a singleton set for all $x \in X$. Moreover, there exists an idempotent multiplier $f : X \to X$ such that $f(x * y) = x * f(y)$ for all $x, y \in X$.*

Proof Let $a \in \mathcal{B}(X)$. Then $\theta_{\langle a \rangle}$ is a congruence on X. Let x be an arbitrary element of X. Since $a \in \mathcal{B}(X)$, we get $a * (a * x) = a * x$. Hence by Lemma 7.2.18, we get $(a * x, x) \in \theta_{\langle a \rangle}$. Thus $a * x \in [x]_{\theta_{\langle a \rangle}}$. Also $a * x \in \Delta_a$ and so it implies that $a * x \in \Delta_a \cap [x]_{\theta_{\langle a \rangle}}$. Therefore $\Delta_a \cap [x]_{\theta_{\langle a \rangle}}$ is non-empty. Suppose c, d are two distinct elements of $\Delta_a \cap [x]_{\theta_{\langle a \rangle}}$. Then by above the lemma, we get that $c = d$. Therefore $\Delta_a \cap [x]_{\theta_{\langle a \rangle}}$ is a singleton set for all $x \in X$.

To prove the remaining, let us consider that x_0 be the single element of $\Delta_a \cap [x]_{\theta_{\langle a \rangle}}$ for $x \in X$. Define a self-map $f : X \to X$ such that $f(x) = x_0$ for all $x \in X$. Let

$b, c \in X$. By the definition of the mapping, we get $f^2(c) = f(f(c)) = x_0 = f(c)$. Hence f is an idempotent mapping. Also we have $f(c) = x_0 \in \Delta_a$. Since Δ_a is a quasi-filter, it yields $b * f(c) \in \Delta_a$. Now, we get

$$\begin{aligned} f(f(c)) = f(c) &\Rightarrow (f(c), c) \in \theta \\ &\Rightarrow (b * f(c), b * c) \in \theta \\ &\Rightarrow b * f(c) \in [b * c]_\theta \\ &\Rightarrow b * f(c) \in \Delta_a \cap [b * c]_\theta \qquad \text{since } b * f(c) \in \Delta_a \end{aligned}$$

Since $f(b * c) \in \Delta_a \cap [b * c]_\theta$ and $\Delta_a \cap [b * c]_\theta$ is a singleton set, we get $f(b * c) = b * f(c)$. Therefore f is an idempotent multiplier of X. Hence the theorem is proved. □

7.3 Closed Elements of BE-algebras

In this section, the notion of closed elements is introduced and their properties are studied. A set of equivalent conditions is established for every element of a BE-algebra to become a Boolean element. An equivalency is obtained between closed elements and closed congruences. Throughout this section X stands for a partially ordered BE-algebra unless otherwise mentioned.

Lemma 7.3.1 *Let $(X, *, 0, 1)$ be a bounded and commutative BE-algebra with condition **L**. Then, for any $x, y \in X$, the following conditions hold.*

(1) $x \wedge y = 0$ *implies* $x^\diamond \wedge y = y$,
(2) $(x \vee y)^\diamond = x^\diamond \wedge y^\diamond$,
(3) $(x^\diamond \wedge y^\diamond)^{\diamond\diamond} = x^\diamond \wedge y^\diamond$,
(4) $x \wedge y^\diamond = 0$ *implies* $x \leq y^{\diamond\diamond}$.

Proof (1) Assume that $x \wedge y = 0$ for all $x, y \in X$. Then by Proposition 2.1.35(2), we get that $x^\diamond \wedge y = (x*0) \wedge y = (x*(x \wedge y)) \wedge y = ((x*x) \wedge (x*y)) \wedge y = (x*y) \wedge y = y$ because $y \leq x * y$.
(2) Let $x, y \in X$. By Proposition 2.1.35(1), it yields $(x \vee y)^\diamond = (x \vee y) * 0 = (x * 0) \wedge (y * 0) = x^\diamond \wedge y^\diamond$.
(3) From the condition (2), we get that $(x^\diamond \wedge y^\diamond)^{\diamond\diamond} = (x \vee y)^{\diamond\diamond\diamond} = (x \vee y)^\diamond = x^\diamond \wedge y^\diamond$.
(4) Assume that $x \wedge y^\diamond = 0$. Then by (1), we get $y^{\diamond\diamond} \wedge x = x$. Hence $x \leq y^{\diamond\diamond}$. □

In the following, the notion of *closed elements* is introduced in BE-algebras.

Definition 7.3.2 An element a of a BE-algebra is said to be a *closed element* if $a^{\diamond\diamond} = a$.

We denote by $\mathcal{C}(X)$ the set of all closed elements of the BE-algebra X. Obviously $0, 1 \in \mathcal{C}(X)$. It is observed from Proposition 7.2.8(5) that every Boolean element of a BE-algebra is a closed element. It is evident from the following example that every closed element need not be Boolean.

Example 7.3.3 Let $X = \{1, a, b, c, d, 0\}$. Define a binary operation $*$ on X as follows:

$*$	1	a	b	c	d	0
1	1	a	b	c	d	0
a	1	1	a	c	c	d
b	1	1	1	c	c	c
c	1	a	b	1	a	b
d	1	1	a	1	1	a
0	1	1	1	1	1	1

Then it can be easily verified that $(X, *, 0, 1)$ is a bounded BE-algebra with smallest element 0. Observe that $a^{\diamond\diamond} = d^{\diamond} = a$; $b^{\diamond\diamond} = c^{\diamond} = b$; $c^{\diamond\diamond} = b^{\diamond} = c$; and $d^{\diamond\diamond} = a^{\diamond} = d$. Therefore $\mathcal{C}(X) = X$. But the elements a and d of the BE-algebra X are not Boolean because $a \vee a^{\diamond} = 1$ and $d \vee d^{\diamond} = 1$.

However, a set of equivalent conditions is derived for every element to become Boolean.

Theorem 7.3.4 *Let $(X, *, 0, 1)$ be a bounded and commutative BE-algebra with condition* **L**. *Then the following conditions are equivalent:*

(1) *X is a Boolean algebra,*
(2) *for all $x \in X$, $x \wedge x^{\diamond} = 0$,*
(3) *every element is Boolean,*
(4) *every closed element is Boolean.*

Proof (1) $\Rightarrow$ (2): It is clear.
(2) $\Rightarrow$ (3): Assume that $x \wedge x^{\diamond} = 0$ for all $\in X$. Since X is a commutative BE-algebra, we get that it is involutory and hence $X = \mathcal{C}(X)$. Let $a \in X$ be arbitrary. Then we get $(a \vee a^{\diamond})^{\diamond} = a^{\diamond} \wedge a^{\diamond\diamond} = a^{\diamond} \wedge a = 0$. Hence $a \vee a^{\diamond} = 1$. Thus $a \in \mathcal{B}(X)$. Therefore $X = \mathcal{B}(X)$.
(3) $\Leftrightarrow$ (4): Since X is an involutory BE-algebra, by Theorem 7.1.7, it is an easy consequence.
(3) $\Rightarrow$ (1): Since X is commutative with condition **L**, it is already observed that $(X, \vee, \wedge)$ is a distributive lattice. Then by Remark 7.2.7, it concludes that X is a Boolean algebra. □

Remark 7.3.5 For any bounded BE-algebra $(X, *, 0, 1)$, define the set $X^* = \{x \in X \mid x = a^{\diamond}$ for some $a \in X\}$. Since $x^{\diamond\diamond\diamond} = x^{\diamond}$ for all $x \in X$, it can be easily obtained that $X^* = \mathcal{C}(X)$.

Theorem 7.3.6 *Let $(X, *, 0, 1)$ be an ordered BE-algebra. Then $\mathcal{C}(X)$ is closed under $*$.*

Proof Let $x, y \in \mathcal{C}(X)$. Then $x = x^{\diamond\diamond}$ and $y = y^{\diamond\diamond}$. Hence by Proposition 7.1.4(5), we get $x * y = x * y^{\diamond\diamond} = (x * y^{\diamond\diamond})^{\diamond\diamond} = ((x * y^{\diamond\diamond})^{\diamond})^{\diamond}$. Thus by the above remark, it yields $x * y \in X^* = \mathcal{C}(X)$. □

In the following, the notion of spanning subalgebra of a BE-algebra is introduced.

Definition 7.3.7 Let $(X, *, 0, 1)$ be a bounded BE-algebra. A non-empty subset S of X is said to be a *spanning subalgebra* of X if it satisfies the following properties:

(1) $(S, \wedge)$ is a semilattice,
(2) for any $a, b \in S$, $a * b \in S$,
(3) $0, 1 \in S$.

Theorem 7.3.8 *Let $(X, *, 0, 1)$ be a bounded and transitive BE-algebra. Then $\mathcal{C}(X)$ is a bounded spanning subalgebra of X as well as a quasi-filter of X.*

Proof Clearly $(\mathcal{C}(X), \leq)$ is a poset. Clearly $0, 1 \in \mathcal{C}(X)$. For any $x, y \in \mathcal{C}(X)$, by the above two theorems, it yields that $x \wedge y \in \mathcal{C}(X)$ and $x * y \in \mathcal{C}(X)$. Therefore $\mathcal{C}(X)$ is a bounded spanning subalgebra of X. Let $x \in \mathcal{C}(X)$ and $a \in X$. Then, we get that $a * x = a * x^{\diamond\diamond} = (a * x^{\diamond\diamond})^{\diamond\diamond} \in X^* = \mathcal{C}(X)$. Therefore $\mathcal{C}(X)$ is a quasi-filter of X. Hence the theorem is proved. □

Theorem 7.3.9 *Let $(X, *, 0, 1)$ and $(Y, *, 0', 1')$ be two bounded BE-algebras and $\alpha : X \to Y$ an onto semilattice BE-morphism. Then we have the following conditions:*

(1) *α is isotone,*
(2) *if a is a closed element of X, then $\alpha(a)$ is a closed element of Y,*
(3) *if S is a bounded spanning subalgebra of X, then $\alpha(S)$ is a bounded spanning subalgebra of Y.*

Proof (1) Let $x, y \in X$ be such that $x \leq y$. Then we get that $x \wedge y = x$. Hence, it implies $\alpha(x) = \alpha(x \wedge y) = \alpha(x) \wedge \alpha(y)$. Thus $\alpha(x) \leq \alpha(y)$. Therefore α is isotone.
(2) Let a be a closed element of X. Then, we get that $a^{\diamond\diamond} = a$. Hence, it yields $\alpha(a)^{\diamond\diamond} = \alpha(a^{\diamond\diamond}) = \alpha(a)$. Therefore $\alpha(a)$ is a closed element of Y.
(3) Assume that S is a bounded spanning subalgebra of X. Since $(S, \leq)$ is a poset and $\alpha : X \to Y$ is an isotone, it yields that $(\alpha(S), \leq)$ is a poset. Let $a, b \in \alpha(S)$. Then, we get $a = \alpha(x)$ and $b = \alpha(y)$ for some $x, y \in S$. Since S is a bounded spanning subalgebra of X, we get that $x \wedge y \in S$; $x * y \in S$; $0 \in S$; and $1 \in S$. Now, for $a, b \in S$, we get $a \wedge b = \alpha(x) \wedge \alpha(y) = \alpha(x \wedge y) \in \alpha(S)$. Therefore $(\alpha(S), \wedge)$ is a semilattice of Y. Also we have $a * b = \alpha(x) * \alpha(y) = \alpha(x * y) \in \alpha(S)$. Therefore $\alpha(S)$ is closed under $*$. It is easy to see that $0' = \alpha(0) \leq \alpha(x) \leq \alpha(1) = 1'$ for all $x \in X$. Thus $\alpha(S), \leq)$ is bounded. Therefore $\alpha(S)$ is a bounded spanning subalgebra of Y. □

Theorem 7.3.10 *Let $(X, *, 0, 1)$ and $(Y, *, 0', 1')$ be two bounded BE-algebras and $\alpha : X \to Y$ a bijective semilattice BE-morphism. Then we have the following conditions.*

(1) *α is isotone,*

(2) *S is a spanning subalgebra of X if and only if $\alpha(S)$ is a spanning subalgebra of Y,*

(3) *S is a quasi-filter of X if and only if $\alpha(S)$ is a quasi-filter of Y.*

Proof It is an immediate consequence of the above theorem. □

Definition 7.3.11 Let X be a BE-algebra and $\mathfrak{L}$ a non-empty family of subsets of X. Then $\mathfrak{L}$ is called an *intersection structure* (or $\cap$-structure) on X if it satisfies the following condition:

$$\bigcap_{i\in I} A_i \in \mathfrak{L} \text{ for every non-empty family } \{A_i\}_{i\in I} \subseteq \mathfrak{L}.$$

Moreover, if $X \in \mathfrak{L}$ then it is called a topped $\cap$-structure on X.

Theorem 7.3.12 *Let X be a transitive BE-algebra. Then the family $\mathfrak{L} = \{A \subseteq X \mid C(A) = A\}$ of closed subsets of X is a $\cap$-structure on X. Moreover $\mathfrak{L}$ is topped $\cap$-structure whenever X is involuntary, and in this case it forms a complete lattice, when ordered by inclusion, in which*

$$\bigwedge_{i\in I} A_i = \bigcap_{i\in I} A_i \text{ and } \bigvee_{i\in I} A_i = C(\bigcup_{i\in I} A_i)$$

Proof Let $\{A_i\}_{i\in I} \subseteq \mathfrak{L}$ be a non-empty family in $\mathfrak{L}$. Then $C(A_i) = A_i$ for all $i \in I$, which means $x^{\circ\circ} = x$ for all $x \in X$ and $i \in I$. It is enough to show that $C(\bigcap A_i) = \bigcap A_i$. Now

$$\begin{aligned} x \in \bigcap_{i\in I} A_i &\Leftrightarrow x \in A_i \text{ for each } i \in I \\ &\Leftrightarrow x \in C(A_i) \text{ for each } i \in I \\ &\Leftrightarrow x^{\circ\circ} = x \text{ for each } i \in I \\ &\Leftrightarrow x \in C(A_i) \text{ for each } i \in I \\ &\Leftrightarrow x \in \bigcap_{i\in I} C(A_i). \end{aligned}$$

Hence $\bigcap_{i\in I} A_i \in \mathfrak{L}$. Therefore $\mathfrak{L}$ is a $\cap$-structure on X. If X is involuntary, then $C(X) = X$. Hence $X \in \mathfrak{L}$. Therefore $\mathfrak{L}$ is topped $\cap$-structure on X. □

For any congruence θ on a bounded BE-algebra X, it is already observed that the quotient algebra $X_{/\theta}$ is also a BE-algebra with respect to the induced operations given by

$$[x]_\theta * [y]_\theta = [x * y]_\theta$$

for all $x, y \in X$. Moreover, the quotient algebra $X_{/\theta}$ is also bounded with least element $[0]_\theta$. If X is pseudo-complemented BE-algebra, then the quotient algebra $X_{/\theta}$ is also pseudo-complemented in which the pseudo-complement of any congruence class $[x]_\theta$ is observed to be the congruence class $[x^\diamond]_\theta$. For any congruence θ on a bounded BE-algebra X, a congruence class $[x]_\theta, x \in X$ is said to be closed if $[x]_\theta^{\diamond\diamond} = [x]_\theta$.

Definition 7.3.13 A congruence θ on a bounded BE-algebra X is said to be a *closed congruence* if $(x, x^{\diamond\diamond}) \in \theta$ for all $x \in X$.

Theorem 7.3.14 *Let X be a partially ordered, transitive and bounded BE-algebra. Then the following conditions are equivalent:*

(1) *Every element of X is closed,*
(2) *For any congruence θ on X, every element of $X_{/\theta}$ is closed,*
(3) *every congruence is closed.*

Proof (1) $\Rightarrow$ (2): Assume that every element of X is closed. Then $x^{\diamond\diamond} = x$ for all $x \in X$. Let $[x]_\theta \in X_{/\theta}$. Then $[x]_\theta^{\diamond\diamond} = [x^{\diamond\diamond}]_\theta = [x]_\theta$. Therfore $[x]_\theta \in X_{/\theta}$ is a closed element in $X_{/\theta}$.
(2) $\Rightarrow$ (3): Let θ be a congruence on X. Then by the assumption, we get that every element of $X_{/\theta}$ is closed. Let $x \in X$. Then $[x]_\theta \in X_{/\theta}$. Then by the assumption, we get

$$\begin{aligned}[x]_\theta^{\diamond\diamond} = [x]_\theta &\Rightarrow [x^{\diamond\diamond}]_\theta = [x]_\theta \\ &\Rightarrow (x^{\diamond\diamond}, x) \in \theta \\ &\Rightarrow \theta \text{ is closed}\end{aligned}$$

Therefore, every congruence on X is closed.
(3) $\Rightarrow$ (1): Assume that every congruence is closed. Let $x \in X$. Clearly $x \leq x^{\diamond\diamond}$. Since $\langle 1 \rangle$ is a filter, we get that $\theta_{\langle 1 \rangle}$ is a congruence on X. Hence by the assumption, we get $\theta_{\langle 1 \rangle}$ is closed. Thus $(x^{\diamond\diamond}, x) \in \theta_{\langle 1 \rangle}$. Hence $x^{\diamond\diamond} * x \in \langle 1 \rangle$. Thus $x^{\diamond\diamond} \leq x$. Therefore $x^{\diamond\diamond} = x$. □

The following corollary is a direct consequence of the above theorem.

Corollary 7.3.15 *A partially ordered BE-algebra is involuntary if and only if every congruence defined on X is closed.*

In the following definition, the notion of *closed segments* of BE-algebras is introduced.

Definition 7.3.16 Let $(X, *, 0, 1)$ be a bounded BE-algebra. A final segment S of X is said to be a *closed segment* if $x^{\diamond\diamond} \in S$ implies $x \in S$ for all $x \in X$.

If every element of a bounded BE-algebra is closed, then it is obvious that every final segment is closed. However, in the following theorem, closed segments of BE-algebras are characterized.

Theorem 7.3.17 *A final segment S of a partially ordered bounded BE-algebra X is closed if and only if for any $x, y \in X$, $x^\diamond = y^\diamond$, and $x \in S$ imply $y \in S$.*

Proof Let S be a final segment of X. Then $S = [a, 1]$ for some $a \in X$. Assume that $[a, 1]$ is closed. Let $x, y \in X$ be such that $x^\diamond = y^\diamond$ and $x \in S$. Since $x \in S = [a, 1]$, we get $a \leq x \leq x^{\diamond\diamond} = y^{\diamond\diamond}$. Hence $y^{\diamond\diamond} \in [a, 1]$. Since $[a, 1]$ is closed, it yields that $y \in [a, 1] = S$. Conversely, assume that the final segment $S = [a, 1]$ satisfies the given condition. Let $x^{\diamond\diamond} \in [a, 1]$ for $x \in X$. Since $x^{\diamond\diamond\diamond} = x^\diamond$, by the assumed condition, we get $x \in [a, 1]$. Therefore $S = [a, 1]$ is a closed segment. □

In the following theorem, properties of homomorphic inverse images of closed segments of bounded BE-algebras are studied.

Theorem 7.3.18 *Let X and Y be two bounded BE-algebras and $\psi : X \to Y$ a bounded BE-morphism. If S is a closed segment of Y, then $\psi^{-1}(S)$ is a closed segment of X.*

Proof Let $S = [a, 1]$ be a closed segment of Y where $a \in Y$. Clearly $[\psi^{-1}(a), 1] = \psi^{-1}([a, 1])$ is a final segment of X. Let $x^{\diamond\diamond} \in \psi^{-1}([a, 1])$. Then we get $\psi(x)^{\diamond\diamond} = \psi(x^{\diamond\diamond}) \in [a, 1]$. Since $[a, 1]$ is a closed segment of Y, we get that $\psi(x) \in [a, 1]$. Hence $x \in \psi^{-1}([a, 1])$. Therefore $\psi^{-1}([a, 1]) = \psi^{-1}(S)$ is a closed segment of X. □

Theorem 7.3.19 *Let X be a partially ordered bounded BE-algebra. Then the following conditions are equivalent:*

(1) *X is involutory,*
(2) *every element is closed,*
(3) *every final segment is closed.*

Proof (1) $\Leftrightarrow$ (2), (2) $\Rightarrow$ (3) are obvious.
(3) $\Rightarrow$ (1): Assume that every final segment is closed. Let $x \in X$. Clearly $x^{\diamond\diamond} \in [x^{\diamond\diamond}, 1]$. Since $[x^{\diamond\diamond}, 1]$ is a closed segment, we get $x \in [x^{\diamond\diamond}, 1]$. Hence $x^{\diamond\diamond} \leq x$. Since $x \leq x^{\diamond\diamond}$, we get $x = x^{\diamond\diamond}$. Hence every element of X is closed. Therefore X is involutory. □

7.4 Dense Elements of BE-algebras

In this section, the properties of dense elements of BE-algebras are investigated. The concepts of dense BE-algebra and D-filters are introduced and characterized. Throughout this section X stands for a partially ordered BE-algebra unless otherwise mentioned.

Definition 7.4.1 An element x of a bounded BE-algebra $(X, *, 0, 1)$ is called *dense* if $x^\diamond = 0$.

It is obvious that 1 is a dense element of X. Let us denote the class of all dense elements of a bounded BE-algebra X by $\mathcal{D}(X)$. Then it can be observed that $\mathcal{D}(X)$ is a quasi-filter of X.

Example 7.4.2 Let $X = \{1, a, b, 0\}$ be a set and $*$ a binary operation defined on X as follows:

$*$	1	a	b	0
1	1	a	b	0
a	1	1	1	a
b	1	a	1	0
0	1	1	1	1

It can be easily verified that $(X, *, 0, 1)$ is a bounded BE-algebra with smallest element 0. Observe that $a^\diamond = a$; $b^\diamond = 0$. Therefore b and 1 are the dense elements but a is not a dense element.

Proposition 7.4.3 *For any transitive BE-algebra X, $\mathcal{D}(X)$ is a closed filter of X.*

Proof Clearly $1 \in \mathcal{D}(X)$. Let $x, x * y \in \mathcal{D}(X)$. Then we get $x^\diamond = 0$ and $(x * y)^\diamond = 0$. Since X is transitive, by Proposition 7.1.4(7), it yields that $1 = 0^\diamond = (x * y)^{\diamond\diamond} \leq x^{\diamond\diamond} * y^{\diamond\diamond} = 0^\diamond * y^{\diamond\diamond} = y^{\diamond\diamond}$. Thus $y^\diamond = 0$, which yields $y \in \mathcal{D}(X)$. Therefore $\mathcal{D}(X)$ is a filter of X. Let $x^{\diamond\diamond} \in \mathcal{D}(X)$. Then $x^\diamond = x^{\diamond\diamond\diamond} = 0$, which yields $x \in \mathcal{D}(X)$. Therefore $\mathcal{D}(X)$ is a closed filter of X. □

Theorem 7.4.4 *Let $(X, *, 0, 1)$ be a bounded and transitive BE-algebra. Then*

(1) $\mathcal{C}(X) \cap \mathcal{D}(X) = \{1\}$,
(2) *$a \in \mathcal{D}(X)$ implies $a * b^\diamond = b^\diamond$ for any $b \in X$,*
(3) *$a \in \mathcal{D}(X)$ implies $a * b^\diamond \in \mathcal{C}(X)$,*
(4) $\mathcal{B}(X) \cap \mathcal{D}(X) = \{1\}$,
(5) *$\mathcal{D}(X)$ is a quasi-filter of X,*
(6) *$a \in \mathcal{D}(X)$ implies $x^n * a \in \mathcal{D}(X)$ for any $x \in X$ and $n \in \mathbb{Z}^+$.*

Proof (1) Let $x \in \mathcal{C}(X) \cap \mathcal{D}(X)$. Then, we get $x = x^{\diamond\diamond} = 0^\diamond = 1$. Hence $\mathcal{C}(X) \cap \mathcal{D}(X) = \{1\}$.
(2) Let $a \in \mathcal{D}(X)$. Then $a^\diamond = 0$. For any $b \in X$, we get $a * b^\diamond = a * (b * 0) = b * (a * 0) = b * 0 = b^\diamond$.
(3) Let $a \in \mathcal{D}(X)$. Then by (2), we get $a * b^\diamond = b^\diamond \in \mathcal{C}(X)$.
(4) Let $x \in \mathcal{B}(X) \cap \mathcal{D}(X)$. Then $\{1\} = \langle x\rangle \cap \langle x^\diamond\rangle = \langle x\rangle \cap \langle 0\rangle = \langle x\rangle \cap X = \langle x\rangle$. Hence $x = 1$.
(5) Clearly $1 \in \mathcal{D}(X)$. Let $a \in \mathcal{D}(X)$. Then $a^\diamond = 0$. Clearly $a \leq x * a$ for any $x \in X$. Hence $(x * a)^\diamond \leq a^\diamond = 0$. Thus $x * a \in \mathcal{D}(X)$. Therefore $\mathcal{D}(X)$ is a quasi-filter of X.

(6) Let $a \in \mathcal{D}(X)$. From (5), we get that $x * a \in \mathcal{D}(X)$ for any $x \in X$. Continuing in this way, we get $x^n * a \in \mathcal{D}(X)$ for any $n \in \mathbb{Z}^+$. □

In the following, the notion of dense BE-algebra is introduced.

Definition 7.4.5 A bounded BE-algebra $(X, *, 0, 1)$ is said to be a dense BE-algebra if every nonzero element of X is dense (*i.e.*, $x^\diamond = 0$ for all $0 \neq x \in X$). A filter F of a bounded BE-algebra X is said to be a *dense filter* if every element of F is dense.

Clearly the two-element bounded BE-algebra $\{0, 1\}$ is a dense BE-algebra. The bounded BE-algebra given in the Example 7.4.2 is not a dense BE-algebra.

Example 7.4.6 Let $X = \{1, a, b, c, 0\}$. Define a binary operation $*$ on X as follows:

$*$	1	a	b	c	0
1	1	a	b	c	0
a	1	1	b	b	0
b	1	a	1	a	0
c	1	1	1	1	0
0	1	1	1	1	1

Then it can be easily verified that $(X, *, 0, 1)$ is a bounded BE-algebra with smallest element 0. Observe that $a^\diamond = b^\diamond = c^\diamond = 0$. Hence $\mathcal{D}(X) = X - \{0\}$. Therefore X is a dense BE-algebra.

Proposition 7.4.7 *Let* $(X, *, 0, 1)$ *be a dense BE-algebra. Then the following hold:*

(1) $\mathcal{C}(X) = \mathcal{B}(X) = \{1\}$,
(2) $(a^{\diamond\diamond} * a)^{\diamond\diamond} = 1$ *for all* $a \in X$,
(3) $\mathcal{D}(F) = \mathcal{D}(X) \cap F = \{X - \{0\}\} \cap F = F$ *for any filter* F *of* X.

Proof (1) From (1) and (4) of Theorem 7.4.4, it immediately follows.
(2) Let $a \in X$ be an arbitrary element. If $a = 0$, then the result is through. Suppose $a \neq 0$. Since X is dense, we get $a^\diamond = 0$. Hence $(a^{\diamond\diamond} * a)^{\diamond\diamond} = (0^\diamond * a)^{\diamond\diamond} = a^{\diamond\diamond} = 0^\diamond = 1$.
(3) Since $\mathcal{D}(F) = \{x \in F \mid x^\diamond = 0\}$ and X is a dense BE-algebra, it follows immediately. □

Proposition 7.4.8 *Let* X *and* Y *be two bounded BE-algebras and* $f : X \to Y$ *a bounded homomorphism. If* X *is dense, the homomorphic image* $f(X)$ *is dense. Moreover, the inverse image (if exists) of a dense element under* f *is dense whenever* $\{x \in X \mid f(x) = 0\} = \{0\}$ *and thus* $f^{-1}(F)$ *is dense in* X *whenever* F *is dense in* Y.

Proof Assume that X is dense. Let $x \in f(X)$. Then $x = f(a)$ for some $a \in X$. Since X is dense, we get $a^\diamond = 0$. Hence $x^\diamond = f(a)^\diamond = f(a^\diamond) = f(0) = 0$. Hence $x^\diamond = 0$ for all $x \in f(X)$. Therefore $f(X)$ is dense. Suppose $\{x \in X \mid f(x) = 0\} = \{0\}$. Let x be a dense element in Y. Suppose $a = f^{-1}(x)$. Then $f(a^\diamond) = f(a)^\diamond = x^\diamond = 0$. Hence $a^\diamond \in \{x \in X \mid f(x) = 0\} = \{0\}$. Thus $a^\diamond = 0$. Therefore $f^{-1}(x) = a$ is dense. To prove the remaining, let F be a dense filter of Y. Then clearly $f^{-1}(F)$ is a filter of X. Let $x \in f^{-1}(F)$. Then $f(x) \in F$. Since F is dense, we get $f(x^\diamond) = f(x)^\diamond = 0$. Then by hypothesis, we get $x^\diamond \in \{0\}$. Hence $x \in f^{-1}(F)$ is dense. Therefore $f^{-1}(F)$ is a dense filter of X. □

Theorem 7.4.9 *Let* $(X, *, 0, 1)$ *be a partially ordered and bounded BE-algebra. Then the following conditions are equivalent:*

(1) *X is dense,*
(2) *every filter is dense,*
(3) *every principal filter is dense,*
(4) *every final segment is dense.*

Proof (1) $\Rightarrow$ (2) and (2) $\Rightarrow$ (3) are obvious.
(3) $\Rightarrow$ (4): Let $a \in X$. Then $[a, 1]$ is a final segment in X. Clearly $[a, 1] \subseteq \langle a \rangle$. Since $\langle a \rangle$ is dense, we get that $[a, 1]$ is dense.
(4) $\Rightarrow$ (1): Let $x \in X$ be arbitrary. Since $[x, 1]$ is a final segment and $x \in [x, 1]$, we get $x^\diamond = 0$. Therefore X is dense. □

For any congruence θ on a bounded BE-algebra X, we know that the quotient algebra $X_{/\theta}$ containing the congruence classes is also a bounded BE-algebra. Using this fact, we now characterize dense BE-algebras in terms of congruence classes.

Theorem 7.4.10 *Let* $(X, *, 0, 1)$ *be a partially ordered and bounded BE-algebra. Then the following conditions are equivalent:*

(1) *X is dense,*
(2) *for any congruence θ, $X_{/\theta}$ is dense,*
(3) *for any congruence θ, $x^\diamond \in [0]_\theta$ for all $x \in X$.*

Proof (1) $\Rightarrow$ (2): Assume that X is a dense BE-algebra. Let θ be a congruence on X. Let $[x]_\theta \in X_{/\theta}$ where $x \in X$. Then by the hypothesis (1), we get $x^\diamond = 0$. Hence $[x]_\theta^\diamond = [x^\diamond]_\theta = [0]_\theta$. Thus every element of $X_{/\theta}$ is dense. Therefore $X_{/\theta}$ is a dense BE-algebra.
(2) $\Rightarrow$ (3): Assume that $X_{/\theta}$ is dense. Let θ be a congruence on X. Let $x \in X$. Then $[x]_\theta \in X_{/\theta}$. By the condition (2), we get that $[x]_\theta$ is dense in $X_{/\theta}$. Hence

$$
\begin{aligned}
[x]_\theta^\diamond = [0]_\theta &\Rightarrow [x^\diamond]_\theta = [0]_\theta \\
&\Rightarrow (x^\diamond, 0) \in \theta \\
&\Rightarrow x^\diamond \in [0]_\theta.
\end{aligned}
$$

Hence the condition (3) is proved.
(3) $\Rightarrow$ (1): Assume that condition (3) holds in X. Let $x \in X$ be an arbitrary element.

Since $\langle 1\rangle$ is a filter, we get that $\theta_{\langle 1\rangle}$ is a filter congruence on X. By the condition (3), we get $x^\diamond \in [0]_{\langle 1\rangle}$. By the definition of the congruence $\theta_{\langle 1\rangle}$, we get $x^{\diamond\diamond} = x^\diamond * 0 \in \langle 1\rangle$. Hence $x^\diamond = (x^{\diamond\diamond})^\diamond = 0$ for all $x \in X$. Therefore X is dense. □

In [187], Najafi introduced the notion of pseudo-commutators in BCK-algebras and investigated their properties. We now introduce the notion of pseudo-commutators in commutative BE-algebras. Later, Najafi, and Saeid [188] characterized solvable BCK-algebras in terms of commutators. We now generalize these concepts in BE-algebras with the help of pseudo-complements.

Definition 7.4.11 Let $(X, *, 0, 1)$ be a bounded BE-algebra. For any $x, y \in X$, the pseudo-commutator of the elements x and y is defined

$$[x, y] = (x^\diamond \vee y^\diamond) * (y^\diamond \vee x^\diamond)$$

where the operation $\vee$ is defined as $a \vee b = (b * a) * a$ for any $a, b \in X$.

In general, the element $[x_1, x_2, \ldots, x_n] = [[x_1, \ldots, x_{n1}], x_n]$ is a pseudo-commutator of weight $n \geq 2$, whereby convention $[x_1] = x_1$. A useful shorthand notation is $[x, y]_n = [x, \underbrace{y, \ldots, y}_{\text{n times}}]$.

Definition 7.4.12 Let $X_1, X_2, \ldots, X_n$ be non-empty subsets of a bounded BE-algebra X. Define the pseudo-commutator of subsets of X_1 and X_2 as

$$[X_1, X_2] = \{[x_1, x_2] \mid x_1 \in X_1 \text{ and } x_2 \in X_2\}$$

More generally, $[X_1, \ldots, X_n] = [[X_1, \ldots, X_{n1}], X_n]$ where $n \geq 2$. Furthermore, the subset $[X, X] = \{[a, b] \mid a, b \in X\}$ of X is called the derived subset of X, and we simply denote it by X'.

Lemma 7.4.13 *Let $(X, *, 0, 1)$ be a bounded BE-algebra. For any $x, y \in X$, we have*

(1) $x \leq y$ *implies* $[x, y] = 1$,
(2) $[x, 1] = [1, x] = [x, x] = 1$,
(3) $[0, x] = [x, 0] = 1$,
(4) $y^\diamond * [x, y] = 1$,
(5) $x^\diamond * [x, y] = 1$.

Proof (1) Let $x, y \in X$ be such that $x \leq y$. Then $y^\diamond \leq x^\diamond$. Hence $[x, y] = (x^\diamond \vee y^\diamond) * (y^\diamond \vee x^\diamond) = ((y^\diamond * x^\diamond) * x^\diamond) * ((x^\diamond * y^\diamond) * y^\diamond) = (1 * x^\diamond) * ((x^\diamond * y^\diamond) * y^\diamond) = x^\diamond * ((x^\diamond * y^\diamond) * y^\diamond) = (x^\diamond * y^\diamond) * (x^\diamond * y^\diamond) = 1$.
(2) Let $x \in X$. Then $[x, 1] = (x^\diamond \vee 1^\diamond) * (1\diamond \vee x^\diamond) = ((1^\diamond * x^\diamond) * x^\diamond) * ((x^\diamond * 1^\diamond) * 1^\diamond) = ((0^\diamond * x^\diamond) * x^\diamond) * ((x^\diamond * 0^\diamond) * 0^\diamond) = (1 * x^\diamond) * x^{\diamond\diamond\diamond} = x^\diamond * x^\diamond = 1$. Similarly, we can prove that $[1, x] = [x, x] = 1$. Therefore $[x, 1] = [1, x] = [x, x] = 1$.
(3) Similarly, it can be obtained.

(4) Let $x, y \in X$. Then $y^\diamond * [x, y] = y^\diamond * ((x^\diamond \vee y^\diamond) * (y^\diamond \vee x^\diamond)) = (x^\diamond \vee y^\diamond) * (y^\diamond * ((y^\diamond \vee x^\diamond)) = (x^\diamond \vee y^\diamond) * (y^\diamond * ((x^\diamond * y^\diamond) * y^\diamond) = (x^\diamond \vee y^\diamond) * ((x^\diamond * y^\diamond) * (y^\diamond * y^\diamond)) = (x^\diamond \vee y^\diamond) * 1 = 1$.
(5) Let $x, y \in X$. $x^\diamond * [x, y] = x^\diamond * ((x^\diamond \vee y^\diamond) * (y^\diamond \vee x^\diamond)) = (x^\diamond \vee y^\diamond) * (x^\diamond * ((y^\diamond \vee x^\diamond)) = (x^\diamond \vee y^\diamond) * (x^\diamond * ((x^\diamond * y^\diamond) * y^\diamond)) = (x^\diamond \vee y^\diamond) * ((x^\diamond * y^\diamond) * (x^\diamond * y^\diamond)) = (x^\diamond \vee y^\diamond) * 1 = 1$. □

Theorem 7.4.14 *If a bounded BE-algebra X is dense, then we have the following:*

(1) $X' = \{1\}$,
(2) $[A, B] = [B, A] = \{1\}$ *for any two subsets A, B of X.*

Proof (1) Assume that X is dense. Let $x, y \in X$. Then $x^\diamond = 0$ for all $x \in X$. Hence $[x, y] = (x^\diamond \vee y^\diamond) * (y^\diamond \vee x^\diamond) = (0 \vee 0) * (0 \vee 0) = 0 * 0 = 1$ for all $x, y \in X$. Hence $X' = \{[x, y] \mid x, y \in X\} = \{1\}$.
(2) Let A, B be two subsets of X. Then from (1), $[A, B] = \{1\}$ and $[B, A] = \{1\}$. □

Theorem 7.4.15 *For any partially ordered bounded BE-algebra X, $X' = \{1\}$ if and only if X satisfies the following condition for all $x, y \in X$:*

$$(x^\diamond * y^\diamond) * y^\diamond = (y^\diamond * x^\diamond) * x^\diamond$$

Proof Assume that $X' = \{1\}$. Let $x, y \in X$. Since $X' = \{1\}$, we get $[x, y] = 1$. Hence $(x^\diamond \vee y^\diamond) * (y^\diamond * x^\diamond) = 1$. Hence $(y^\diamond * x^\diamond) * x^\diamond = (x^\diamond \vee y^\diamond) \leq (y^\diamond * x^\diamond) = (x^\diamond * y^\diamond) * y^\diamond$. Since $[y, x] = 1$, similarly we get $(x^\diamond * y^\diamond) * y^\diamond \leq (y^\diamond * x^\diamond) * x^\diamond$. Therefore $(x^\diamond * y^\diamond) * y^\diamond = (y^\diamond * x^\diamond) * x^\diamond$.

Conversely, assume that X satisfies the given condition. Let $x, y \in X$. Then $[x, y] = (x^\diamond \vee y^\diamond) * (y^\diamond * x^\diamond) = ((x^\diamond * y^\diamond) * y^\diamond) * ((y^\diamond * x^\diamond) * x^\diamond) = 1$. Thus $[x, y] = 1$ for all $x, y \in X$. Hence $X' = \{[x, y] \mid x, y \in X\} = \{1\}$. □

The notion of D-filters is now introduced in the following:

Definition 7.4.16 A proper filter F of a bounded BE-algebra X is called *D-filter* if $\mathcal{D}(X) \subseteq F$.

Clearly $\mathcal{D}(X)$ is a D-filter, and every closed filter is a D-filter. Since every maximal filter of an ordered BE-algebra is closed it is also a D-filter. In the following theorem, some equivalent conditions are derived for every filter of a transitive BE-algebra to become D-filter.

Proposition 7.4.17 *Every maximal filter of an ordered BE-algebra is a D-filter.*

Proof Let M be a maximal filter of an ordered BE-algebra X. Suppose M is not a D-filter. Then $\mathcal{D}(X) \nsubseteq M$. Hence there exists $x \in \mathcal{D}(X)$ such the $x \notin M$. Since $x \in \mathcal{D}(X)$, we get $x * 0 = x^\diamond = 0$. Since M is maximal and $x \notin M$, we get

$\langle M \cup \{x\}\rangle = X$. Hence $0 \in \langle M \cup \{x\}\rangle$, which implies that $x^n * 0 \in M$ for some positive integer n. Hence

$$\begin{aligned}
x^n * 0 \in M &\Rightarrow \underbrace{x * (x * (\cdots (x}_{\text{n times}} *0) \cdots)) \in M \\
&\Rightarrow \underbrace{x * (x * (\cdots (x}_{\text{n-1 times}} *(x * 0)) \cdots)) \in M \\
&\Rightarrow \underbrace{x * (x * (\cdots (x}_{\text{n-1 times}} *0) \cdots)) \in M \\
&\Rightarrow \underbrace{x * (x * (\cdots (x}_{\text{n-2 times}} *(x * 0))) \cdots) \in F \\
&\qquad \cdots \\
&\Rightarrow x * 0 \in M \\
&\Rightarrow 0 \in M
\end{aligned}$$

which is a contradiction. Hence $\mathcal{D}(X) \subseteq M$. Therefore M is D-filter. □

Theorem 7.4.18 *Let X be a bounded BE-algebra. Then the set of all D-filters of X satisfies the following properties:*

(1) *If D is a D-filter and F a filter of X such that $D \subseteq F$, then F is also a D-filter of X,*

(2) *If F_1 and F_2 are two D-filters of X, then $F_1 \cap F_2$ is also a D-filter of X,*

(3) *If $\{F_\alpha\}_{\alpha\in\Delta}$ be an indexed set of D-filters, then $\prod\limits_{\alpha\in\Delta} F_\alpha$ is also a D-filter of X.*

Proof Routinely obtained from the definition of D-filters. □

Properties of multipliers of BE-algebras are studies in Chap. 5. As the filters of a BE-algebra X are the subalgebras of X, we restrict the multipliers to D-filters. By a bounded multiplier of a bounded BE-algebra, we mean a multiplier f that satisfies $f(0) = 0$.

Theorem 7.4.19 *Let X be a bounded and self-distributive BE-algebra and F be a D-filter of X. Then $Mult_X(F, X)$ is a bounded and self-distributive BE-algebra, where $Mult_X(F, X)$ is the set of all bounded multipliers restricted to F.*

Proof Let $f, g \in Mult_X(F, X)$. Define a binary operation $\circ$ and an order relation $\leq'$ on $Mult_X(F, X)$ as $(f \circ g)(x) = f(x) * g(x)$ for all $x \in F$, where $*$ represents the binary operation in X and $f \leq' g \Leftrightarrow f(x) \leq g(x)$ for all $x \in \mathcal{D}(X)$, where $\leq$ denotes the order relation on X. Then for any $f, g \in Mult_X(F, X)$ and $x, y \in F$, we get

$$\begin{aligned}
(f \circ g)(x * y) &= f(x * y) * g(x * y) \\
&= (x * f(y)) * (x * g(y)) \\
&= x * (f(y) * g(y)) \\
&= x * (f \circ g)(y).
\end{aligned}$$

Hence $f \circ g \in Mult_X(F, X)$. It can be easily verified that $Mult_X(F, X)$ satisfies all the axioms of a self-distributive BE-algebra. It can be noted that the unit map of X restricted to F plays the role of the largest element in $Mult_X(F, X)$. That is, for any $f \in Mult_X(F, X)$, the unit map $\overline{1}_F$ (defined as $\overline{1}_D(x) = 1$ for all $x \in F$) of X restricted to F is such that

$$(f \circ \overline{1})(x) = f(x) * \overline{1}(x) = f(x) * 1 = 1.$$

Clearly $\overline{1} \circ f = f$ and $f \circ f = \overline{1}$. It can also be noted that the identity map of X restricted to F plays the role of the smallest element. That is, for any $f \in Mult_X(F, X)$, the identity map $\overline{0}_F$ (defined as $\overline{0}(x) = x$ for all $x \in F$) of X restricted to F is such that $(\overline{0} \circ f)(x) = \overline{0}(x) * f(x) = x * f(x) = 1 = \overline{1}(x)$. Therefore $Mult_X(F, X)$ is a bounded and self-distributive BE-algebra. □

For any bounded and self-distributive BE-algebra X, we consider

$$M_X = \bigcup \{Mult_X(F, X) \mid F \text{ is a } D\text{-filter of } X\}.$$

It can be shown that M_X is itself a bounded and self-distributive BE-algebra. To see this, suppose $f, g \in M_X$ and let D_f and D_g denote the domain of f and g, respectively. Since D_f and D_g are D-filters of X, we get $D_f \cap D_g = \mathcal{D}(X)$ is also a D-filter of X.

Let us now define a binary operation $\circ$ and an order relation $\leq'$ on X_M using the above notation once again: For $f, g \in X_M$, $(f \circ g)(x) = f(x) * g(x)$, ($*$ represents the binary operation on X), for all $x \in \mathcal{D}(X) = D_f \cap D_g$; and $f \leq' g \Leftrightarrow f(x) \leq g(x)$ for all $x \in \mathcal{D}(X) = D_f \cap D_g$, where $\leq$ denotes the order relation on X. With these definitions, one can easily show that the system $\langle X_M, \circ, \leq' \rangle$ is a BE-algebra, which is bounded and self-distributive.

7.5 Ideals and Pseudo-complements

In this section, the notion of ideals is introduced in bounded BE-algebras as a generalization of dual ideals of BCK-algebras. Some properties of ideals are investigated. The notion of lower sets is introduced and observed the relation between ideals and lower sets. Throughout this section X stands for a partially ordered BE-algebra unless otherwise mentioned.

Definition 7.5.1 A non-empty subset I of a bounded BE-algebra X is called an *ideal* of X if it satisfies the following properties:

(1) $0 \in I$,
(2) $(x^\diamond * y^\diamond)^\diamond \in I$ and $x \in I$ imply that $y \in I$.

Obviously, the sets $\{0\}$ and the BE-algebra X are ideals of X.

Example 7.5.2 Let $X = \{1, a, b, c, d, 0\}$. Define a binary operation $*$ on X as follows:

$*$	1	a	b	c	d	0
1	1	a	b	c	d	0
a	1	1	a	c	c	d
b	1	1	1	c	c	c
c	1	a	b	1	a	b
d	1	1	a	1	1	a
0	1	1	1	1	1	1

Then clearly $(X, *, 0, 1)$ is a bounded BE-algebra with smallest element 0. It can be easily verified that the set $I = \{0, c, d\}$ is an ideal of X. Also the set $J = \{0, a, b, d\}$ is not an ideal of the BE-algebra X, because $a \in J$ and $(a^\diamond * c^\diamond)^\diamond = (d * b)^\diamond = a^\diamond = d \in J$ but $c \notin J$.

In [5], S.S. Ahn and K.S. So defined the ideal of a BE-algebra X as a non-empty subset I of X satisfying the properties (i) $a \in I$ implies $x * a \in I$ and (ii) $a, b \in I$ implies $(a * (b * x)) * x \in I$ for all $x \in X$. Theorem 4.1.7 is showing that definition of an ideal of S.S. Ahn and the definition of a filter are equivalent. Usually the notion of ideals is the dual concept to filters in any algebra, and they are the specific down sets containing the least element. According to the definition of an ideal, we are going to observe that $x \in I$ and $y \leq x$ imply that $y \in I$. This observation is supporting the Definition 7.5.1, and hence the theory of ideals is developed based on this definition.

Proposition 7.5.3 *Any ideal I of a transitive BE-algebra X satisfies the following properties:*

(1) *For any $x, y \in X, x \in I$ and $y \leq x$ imply $y \in I$,*
(2) *for any $x, y \in X, x^\diamond = y^\diamond, x \in I$ imply $y \in I$,*
(3) *for any $x \in X, x \in I$ if and only if $x^{\diamond\diamond} \in I$,*
(4) *for all $x \in X,\ y^\diamond \leq x^\diamond$ and $y \in I$ imply $x \in I$,*
(5) *$I \cap \mathcal{D}(X) \neq \emptyset$ if and only if $I = X$.*

Proof (1) Let $x \in I$ and suppose that $y \leq x$. Then $x^\diamond \leq y^\diamond$, which implies that $x^\diamond * y^\diamond = 1$. Hence $(x^\diamond * y^\diamond)^\diamond = 0 \in I$. Since $x \in I$ and I is an ideal, it concludes that $y \in I$.
(2) Let $x^\diamond = y^\diamond$ and $x \in I$. Then we get $(x^\diamond * y^\diamond)^\diamond = 1^\diamond = 0 \in I$. Since $x \in I$, it implies $y \in I$.
(3) Let $x \in I$. Then we get that $(x^\diamond * x^{\diamond\diamond\diamond})^\diamond = (x^\diamond * x^\diamond)^\diamond = 1^\diamond = 0 \in I$. Since $x \in I$, it yields $x^{\diamond\diamond} \in I$. Conversely, let $x^{\diamond\diamond} \in I$. Since $x \leq x^{\diamond\diamond}$, by condition (1), it concludes that $x \in I$.
(4) Let $y^\diamond \leq x^\diamond$ and $y \in I$. Then we get $y^\diamond * x^\diamond = 1$. Hence $(y^\diamond * x^\diamond)^\diamond = 1^\diamond = 0 \in I$. Since $y \in I$ and I an ideal of X, we get $x \in I$.
(5) Assume that $I \cap \mathcal{D}(X) \neq \emptyset$. Let $x \in I \cap \mathcal{D}(X)$. Then by (2), we get $x^\diamond = 0$ and $1 = x^{\diamond\diamond} \in I$. Hence by (1), we get $I = X$. Conversely, assume that $I = X$. Then $1 \in I$ and so $I \cap \mathcal{D}(X) \neq \emptyset$. □

Let $X_1, X_2, ..., X_n$ be a family of bounded and pseudo-complemented BE-algebras. Then their Cartesian product $X_1 \times X_2 \times ..... \times X_n$ is also a bounded and pseudo-complemented BE-algebra with respect to the point-wise operations in which the pseudo-complement of any element $(a_1, a_2, \ldots, a_n)$ is given by

$$(a_1, a_2, ..., a_n)^\diamond = (a_1^\diamond, a_2^\diamond, \ldots, a_n^\diamond)$$

where $a_i \in X_i$ for $1 \leq i \leq n$ and $n \in N$. The least element of this Cartesian product is nothing but $(0, 0, \ldots, 0)$. In the following, we observe the property of Cartesian product of ideals.

Theorem 7.5.4 *If I_1 and I_2 are ideals of bounded and transitive BE-algebras X_1 and X_2, respectively, then $I_1 \times I_2$ is an ideal of the product algebra $X_1 \times X_2$. Conversely, every ideal I of $X_1 \times X_2$ can be expressed as $I = I_1 \times I_2$ where I_i is an ideal of X_i for $i = 1, 2$.*

Proof Let I_1 and I_2 be the ideals of X_1 and X_2, respectively. Since $0 \in I_1$ and $0 \in I_2$, we get $(0, 0) \in I_1 \times I_2$. Let $(x_1, x_2) \in I_1 \times I_2$ and $((x_1^\diamond * y_1^\diamond)^\diamond, (x_2^\diamond * y_2^\diamond)^\diamond) = ((x_1 * y_1)^\diamond, (x_2 * y_2)^\diamond)^\diamond \in I_1 \times I_2$ where $x_1, y_1 \in X_1$ and $x_2, y_2 \in X_2$. Hence $(x_1^\diamond * y_1^\diamond)^\diamond \in I_1$ and $(x_2^\diamond * y_2^\diamond)^\diamond \in I_2$. Since $x_1 \in I_1$, $(x_1^\diamond * y_1^\diamond)^\diamond \in I_1$ and I_1 is an ideal of X_1, we get that $y_1 \in I_1$. Similarly, we get $y_2 \in I_2$. Hence $(y_1, y_2) \in I_1 \times I_2$. Therefore $I_1 \times I_2$ is an ideal of $X_1 \times X_2$.

Conversely, let I be any ideal of $X_1 \times X_2$. Consider the projections $\Pi_i : X_1 \times X_2 \longrightarrow X_i$ for $i = 1, 2$. Let I_1 and I_2 be the projections of I on X_1 and X_2, respectively. That is $\Pi_i(I) = I_i$ for $i = 1, 2$. We now prove that I_1 and I_2 are ideals of X_1 and X_2, respectively. Since $(0, 0) \in I$, we get $0 = \Pi_1(0, 0) \in I_1$. Let $x_1 \in I_1$ and $(x_1^\diamond * y_1^\diamond)^\diamond \in I_1$. Then $(x_1, 0) \in I$ and $((x_1, 0)^\diamond * (y_1, 0)^\diamond)^\diamond = ((x_1^\diamond * y_1^\diamond)^\diamond, (0^\diamond * 0^\diamond)^\diamond \in I$. Since I is an ideal, we get $(y_1, 0) \in I$. Thus $y_1 = \Pi_1(y_1, 0) \in \Pi_1(I) = I_1$. Therefore I_1 is an ideal of X_1. Similarly, we can get I_2 is an ideal of X_2.

We now prove that $I = I_1 \times I_2$. For any $a \in I$, we can express that $a = (\Pi_1(a), \Pi_2(a))$ and hence $I \subseteq I_1 \times I_2$. Let $(x, y) \in I_1 \times I_2$. Then $x \in I_1$ and $y \in I_2$. Since I_2 is an ideal of X_2, we get $y^{\diamond\diamond} \in I_2$. Thus $x \in I_1 = \Pi_1(I)$ and $y^{\diamond\diamond} \in I_2 = \Pi_2(I)$. Hence $(x, 0) \in I$ and $(0, y^{\diamond\diamond}) \in I$. Then

$$\begin{aligned}
((x, 0)^\diamond * (x, y)^\diamond)^\diamond &= (x^\diamond, 0^\diamond) * (x^\diamond, y^\diamond))^\diamond \\
&= (x^\diamond * x^\diamond, 0^\diamond * y^\diamond)^\diamond) \\
&= (1, y^\diamond)^\diamond \\
&= (0, y^{\diamond\diamond}) \in I
\end{aligned}$$

Since $(x, 0) \in I$ and I is an ideal, we get $(x, y) \in I$. Thus $I_1 \times I_2 \subseteq I$. Therefore $I = I_1 \times I_2$. □

The following corollary is a generalization of the above theorem.

Corollary 7.5.5 *Let $X = X_1 \times X_2 \times \cdots \times X_n$ be the product of the BE-algebras $X_1, X_2, \ldots, X_n$. Then I is an ideal of X if and only if I is of the form $I_1 \times I_2 \times \cdots \times I_n$, where I_i is an ideal of X_i, $1 \leq i \leq n$ and $n \in \mathbb{N}$.*

In 2008, S.S. Ahn and K.S. So [5, 6] introduced the notions of upper sets and generalized upper sets of BE-algebras. They presented several descriptions of ideals of transitive BE-algebras in terms of upper sets and generalized upper sets. In the following, we generalized these concepts and developed the theory of lower sets in terms of ideals and pseudo-complements.

Definition 7.5.6 Let X be a bounded BE-algebra. For any $a, b \in X$, we define

$$[a; b] := \{x \in X \mid a^\diamond * (b^\diamond * x^\diamond) = 1\}$$

We call $[a; b]$ a lower set of a and b in the BE-algebra X. In the following lemma, we observe some of the properties of lower sets of transitive BE-algebra. We first recall the significant property of pseudo-complements of transitive BE-algebras.

It is obvious that $0, a, b \in [a; b]$ for all $a, b \in X$.

Lemma 7.5.7 *Let X be a bounded and transitive BE-algebra. For any $a, b \in X$, we have*

(1) *for any $x, y \in X$, $x \in [a; b]$ implies $(y^\diamond * x^\diamond)^\diamond \in [a; b]$,*
(2) *for any $x, y \in X$, $x \in [a; b]$ and $y \leq x$ imply that $y \in [a; b]$,*
(3) *for any $x, y \in X$, $x^\diamond = y^\diamond$ and $x \in [a; b]$ imply that $y \in [a; b]$,*
(4) *for any $x \in X$, $x \in [a; b]$ if and only if $x^{\diamond\diamond} \in [a; b]$.*

Proof (1) Let $x \in [a; b]$. Then $a^\diamond * (b^\diamond * x^\diamond) = 1$. Since $x^\diamond \leq y^\diamond * x^\diamond$, we get $b^\diamond * x^\diamond \leq b^\diamond * (y^\diamond * x^\diamond) \leq b^\diamond * (y^\diamond * x^\diamond)^{\diamond\diamond}$. Hence $1 = a^\diamond * (b^\diamond * x^\diamond) \leq a^\diamond * (b^\diamond * (y^\diamond * x^\diamond)^{\diamond\diamond})$. Therefore $(y^\diamond * x^\diamond)^\diamond \in [a; b]$.
(2) Let $x \in [a; b]$ and $y \leq x$. Then we get $a^\diamond * (b^\diamond * x^\diamond) = 1$. Since $y \leq x$, we get $x^\diamond \leq y^\diamond$. Hence $1 = a^\diamond * (b^\diamond * x^\diamond) \leq a^\diamond * (b^\diamond * y^\diamond)$. Therefore $y \in [a; b]$.
(3) Suppose $x^\diamond = y^\diamond$ and $x \in [a; b]$. Then $a^\diamond * (b^\diamond * y^\diamond) = a^\diamond * (b^\diamond * x^\diamond) = 1$. Hence $y \in [a; b]$.
(4) Let $x \in [a; b]$. Then we get $a^\diamond * (b^\diamond * x^\diamond) = 1$. Since $x^\diamond = x^{\diamond\diamond\diamond}$, we get $1 = a^\diamond * (b^\diamond * x^\diamond) = a^\diamond * (b^\diamond * x^\diamond)^{\diamond\diamond}$. Hence $x^{\diamond\diamond} \in [a; b]$. Conversely, let $x^{\diamond\diamond} \in [a; b]$. Then we get $a^\diamond * (b^\diamond * x^\diamond) = a^\diamond * (b^\diamond * x^\diamond)^{\diamond\diamond} = 1$. Therefore $x \in [a; b]$. □

Proposition 7.5.8 *Let X be a bounded and transitive BE-algebra and $b \in X$ satisfy the equality $(b^\diamond * x^\diamond)^\diamond = 0$ for all $x \in X$. Then $[a; b] = X = [b; a]$ for all $a \in X$.*

Proof Let $b \in X$ satisfy the equality $(b^\diamond * x^\diamond)^\diamond = 0$ for all $x \in X$. Let $a \in X$. For any $x \in X$, we get $a^\diamond * (b^\diamond * x^\diamond) = a^\diamond * (b^\diamond * x^\diamond)^{\diamond\diamond} = a^\diamond * 1 = 1$. Hence $x \in [a; b]$. Thus $X \subseteq [a; b]$. Therefore $[a; b] = X$. Again, for any $x \in X$, $b^\diamond * (a^\diamond * x^\diamond) = a^\diamond * (b^\diamond * x^\diamond) = a^\diamond * (b^\diamond * x^\diamond)^{\diamond\diamond} = a^\diamond * 1 = 1$. Hence $x \in [b; a]$. Thus $[b; a] = X$. Therefore $[a; b] = X = [b; a]$ for all $a \in X$. □

In general, we can observe that the lower set $[a; b]$ of a bounded BE-algebra is not an ideal of X. It can be seen in the following example:

Example 7.5.9 Let $X = \{1, a, b, c, d, 0\}$. Define a binary operation $*$ on X as follows:

$*$	1	a	b	c	d	0
1	1	a	b	c	d	0
a	1	1	b	a	d	d
b	1	a	1	c	d	c
c	1	1	b	1	d	b
d	1	1	1	a	1	a
0	1	1	1	1	1	1

Then clearly $(X, *, 0, 1)$ is a bounded BE-algebra with smallest element 0. It can be easily verified that the lower set $[a; b] = \{0, a, b, d\}$ is not an ideal of the BE-algebra X, because $a \in [a; b]$ and $(a^\diamond * c^\diamond)^\diamond = (d * b)^\diamond = a^\diamond = d \in [a; b]$ but $c \notin [a; b]$.

Though every lower set $[a; b]$; $a, b \in X$ of a bounded BE-algebra need not be an ideal, in the following theorem, we derive a sufficient condition for every lower set of a bounded BE-algebra to become an ideal.

Proposition 7.5.10 *If X is self-distributive, then $[a; b]$ is an ideal for any $a, b \in X$.*

Proof Let $a, b \in X$ be arbitrary. Clearly $0 \in [a; b]$. Let $x \in [a; b]$ and $(x^\diamond * y^\diamond)^\diamond \in [a; b]$. Then we get $a^\diamond * (b^\diamond * ((x^\diamond * y^\diamond)^{\diamond\diamond})) = 1$. Hence

$$
\begin{aligned}
1 &= a^\diamond * (b^\diamond * ((x^\diamond * y^\diamond)^{\diamond\diamond})) \\
&= a^\diamond * (b^\diamond * (x^\diamond * y^\diamond)) \\
&= a^\diamond * ((b^\diamond * x^\diamond) * (b^\diamond * y^\diamond)) \\
&= (a^\diamond * (b^\diamond * x^\diamond)) * (a^\diamond * (b^\diamond * y^\diamond)) \\
&= 1 * (a^\diamond * (b^\diamond * y^\diamond)) \\
&= a^\diamond * (b^\diamond * y^\diamond)
\end{aligned}
$$

Hence $y \in [a; b]$. Therefore $[a; b]$ is an ideal of X. □

Theorem 7.5.11 *Let I be a non-empty subset of a transitive BE-algebra X. Then I is an ideal of X if and only if $[a; b] \subseteq I$ for every $a, b \in I$.*

Proof Assume that I is an ideal of X. Let $a, b \in I$. Suppose $x \in [a; b]$. Then $a^\diamond * (b^\diamond * x^\diamond) = 1$. Hence $(a^\diamond * (b^\diamond * x^\diamond)^{\diamond\diamond})^\diamond = (a^\diamond * (b^\diamond * x^\diamond))^\diamond = 0 \in I$. Since $a \in I$ and I is an ideal, we get $(b^\diamond * x^\diamond)^\diamond \in I$. Since $b \in I$ and I is an ideal, we get $x \in I$. Hence $[a; b] \subseteq I$.

Conversely, suppose that $[a; b] \subseteq I$ for all $a, b \in I$. Obviously $0 \in [a; b] \subseteq I$. Let $x, y \in X$ be such that $x \in I$ and $(x^\diamond * y^\diamond)^\diamond \in I$. Then $x^\diamond * ((x^\diamond * y^\diamond)^{\diamond\diamond} * y^\diamond) = x^\diamond * ((x^\diamond * y^\diamond) * y^\diamond) = (x^\diamond * y^\diamond) * (x^\diamond * y^\diamond) = 1$. Hence $y \in [x, (x^\diamond * y^\diamond)^\diamond] \subseteq I$. Therefore I is an ideal of X. □

Since every self-distributive BE-algebra is transitive, the following is an easy consequence:

Corollary 7.5.12 *Let X be a bounded and self-distributive BE-algebra. Every non-empty subset I of X containing $[a; b]$ for all $a, b \in I$ is an ideal of X.*

Since every ideal of a BE-algebra is a subalgebra, the following are clear.

Corollary 7.5.13 *Let X be a bounded and transitive BE-algebra. Every non-empty subset I of X containing $[a; b]$ for all $a, b \in I$ is a subalgebra of X.*

Corollary 7.5.14 *Let X be a bounded and self-distributive BE-algebra. Every non-empty subset I of X containing $[a; b]$ for all $a, b \in I$ is a subalgebra of X.*

Theorem 7.5.15 *If I is an ideal of a bounded and transitive BE-algebra X, then $I = \bigcup\limits_{a,b\in I} [a; b]$.*

Proof Assume that I is an ideal of X. Let $a, b \in I$. Clearly $[a; b] \subseteq I$. Hence $\bigcup\limits_{a,b\in I} [a; b] \subseteq I$. Again, let $x \in I$. Since $x \in [1; x]$, it follows that $I \subseteq \bigcup\limits_{x\in I}[1; x] \subseteq \bigcup\limits_{a,b\in I} [a; b]$. Hence $I = \bigcup\limits_{a,b\in I} [a; b]$. □

Corollary 7.5.16 *Let X be a bounded and self-distributive BE-algebra and I be an ideal of X. Then $I = \bigcup\limits_{a,b\in I} [a; b]$.*

The notion of quasi-ideals is introduced through the following definition.

Definition 7.5.17 Let $(X, *, 0, 1)$ be a bounded BE-algebra. A non-empty subset I of X is called a *quasi-ideal* if it satisfies the following conditions for all $x, y \in X$:

(1) $0 \in I$,
(2) $x \in X$ and $y \in I$ imply that $(x^\diamond * y^\diamond)^\diamond \in I$.

It is obvious that $\{0\}$ is a quasi-ideal of a BE-algebra X. Suppose that $\mathcal{QI}(X)$ denote the set of all quasi-ideal of a BE-algebra X. It is obvious that $\bigcap\limits_{I\in\mathcal{QI}(X)}$ is also a quasi-ideal of X.

Example 7.5.18 Let $X = \{0, a, b, c, 1\}$. Define a binary operation $*$ on X as follows:

$*$	1	a	b	c	0
1	1	a	b	c	0
a	1	1	a	a	0
b	1	1	1	a	0
c	1	1	a	1	0
0	1	1	1	1	1

Then it can be easily verified that $(X, *, 0, 1)$ is a BE-algebra. Consider $I = \{0, c\}$. Clearly I is a quasi-ideal of X. It is clear that I is not an ideal because c, $(c^\diamond * b^\diamond)^\diamond = 0 \in I$ but $b \notin I$.

In the following, we provide some more examples of quasi-ideals of BE-algebras.

Proposition 7.5.19 *For any $a \in X$, the set $\downarrow a = \{x \in X \mid a^\diamond * x^\diamond = 1\}$ is a quasi-ideal of X.*

Proof Clearly $0 \in\downarrow a$. Let $x \in X, y \in\downarrow a$. Then $a^\diamond * y^\diamond = 1$. Now $a^\diamond * (x^\diamond * y^\diamond)^{\diamond\diamond} = a^\diamond * (x^\diamond * y^\diamond) = x^\diamond * (a^\diamond * y^\diamond) = x^\diamond * 1 = 1$. Hence $(x^\diamond * y^\diamond)^\diamond \in\downarrow a$. Thus $\downarrow a$ is a quasi-ideal of X. □

Proposition 7.5.20 *Let X be a BE-algebra and $a \in X$ be a fixed element. Then the set $S_a = \{(a^\diamond * x^\diamond)^\diamond \mid x \in X\}$ is a quasi-ideal of X.*

Proof Clearly $0 = (a^\diamond * a^\diamond)^\diamond \in S_a$. Let $r \in X$ and $b \in S_a$. Then we get $b = (a^\diamond * x^\diamond)^\diamond$ for some $x \in X$. Hence $(r^\diamond * b^\diamond)^\diamond = (r^\diamond * (a^\diamond * x^\diamond)^{\diamond\diamond})^\diamond = (r^\diamond * (a^\diamond * x^\diamond))^\diamond = (a^\diamond * (r^\diamond * x^\diamond))^\diamond = (a^\diamond * (r^\diamond * x^\diamond)^{\diamond\diamond})^\diamond \in S_a$. Thus $(r^\diamond * b^\diamond)^\diamond$. Therefore S_a is a quasi-ideal of X. □

Theorem 7.5.21 *Let I and J be two quasi-ideals of a BE-algebra X. Then $I \cap J$ and $I \cup J$ are also quasi-ideal of X.*

Proof Let $x \in X$ and $a \in I \cap J$. Since $x \in X, a \in I$, and J is a quasi-ideal of X, it yields that $(x^\diamond * a^\diamond)^\diamond \in I$. Similarly, we obtain $(x^\diamond * a^\diamond)^\diamond \in J$. Hence $(x^\diamond * a^\diamond)^\diamond \in I \cap J$. Therefore $I \cap J$ is a quasi-ideal of X. Again, let $x \in X$ and $a \in I \cup J$. Then $a \in I$ or $a \in J$. If $a \in I$, then $(x^\diamond * a^\diamond)^\diamond \in I \subseteq I \cup J$ because I is a quasi-ideal of X. Therefore $I \cup J$ is a quasi-ideal of X. If $a \in J$, then similarly we can obtain that $I \cup J$ is a quasi-ideal of X. □

Corollary 7.5.22 *If $\{I_\alpha \mid \alpha \in \Delta\}$ is an indexed family of quasi-ideals of a BE-algebra X, then $\bigcap_{\alpha\in\Delta} I_\alpha$ and $\bigcup_{\alpha\in\Delta} I_\alpha$ are also quasi-ideals of X.*

Corollary 7.5.23 *The union of two ideals of a BE-algebra X is a quasi-ideal of X.*

Theorem 7.5.24 *Let X be a bounded and transitive BE-algebra. If I is a quasi-ideal and J a bounded subalgebra ($x \in J$ implies $x^\diamond \in J$ for all $x \in X$) of X, then $I \cap J$ is a quasi-ideal of J.*

Proof Let $x \in J$ and $a \in I \cap J$. Then $a \in I$ and $a \in J$. Since I is a quasi-ideal and $a \in I$, we get that $(x^\diamond * a^\diamond)^\diamond \in I$. Since $x, a \in J$ and J is a bounded subalgebra, we get $x^\diamond, a^\diamond \in J$. Hence $x^\diamond * a^\diamond \in J$. Thus $(x^\diamond * a^\diamond)^\diamond \in I \cap J$. Therefore $I \cap J$ is a quasi-ideal of J. □

Proposition 7.5.25 *Every ideal of a bounded and transitive BE-algebra is a quasi-ideal.*

Proof Let X be a bounded and transitive BE-algebra and I is an ideal of X. Suppose $x \in I$. Then $(x^\diamond * (y^\diamond * x^\diamond)^{\diamond\diamond})^\diamond = (x^\diamond * (y^\diamond * x^\diamond))^\diamond = (y^\diamond * (x^\diamond * x^\diamond))^\diamond = (y^\diamond * 1)^\diamond = 1^\diamond = 0 \in I$. Since $x \in I$ and I is an ideal of X, we get $(y^\diamond * x^\diamond)^\diamond \in I$. Therefore I is quasi-ideal of X. □

In general, the converse of the above proposition is not true; i.e., every quasi-ideal of a BE-algebra need not be an ideal. However, in the following theorem, we derive a necessary and sufficient condition for every quasi-ideal of a bounded BE-algebra to become an ideal.

Theorem 7.5.26 *A quasi-ideal I of a bounded and transitive BE-algebra X is an ideal if and only if the following conditions are satisfied:*

(1) *for all $a \in X$, $a^{\diamond\diamond} \in I$ implies $a \in I$,*
(2) *for all $x \in X$ and $a, b \in I$, $((a^{\diamond} * (b^{\diamond} * x^{\diamond})) * x^{\diamond})^{\diamond} \in I$.*

Proof Let I be a quasi-ideal of X. Assume that I is an ideal of X. Let $a \in X$. Suppose $a^{\diamond\diamond} \in I$. Since $a \leq a^{\diamond\diamond}$ and I is an ideal, we get $a \in I$. Let $a, b \in I$. Then by putting $b^{\diamond} * x^{\diamond} = t$, we get the following consequence:

$$\begin{aligned}(a^{\diamond} * ((a^{\diamond} * (b^{\diamond} * x^{\diamond})^{\diamond\diamond})^{\diamond\diamond} * (b^{\diamond} * x^{\diamond})^{\diamond\diamond})^{\diamond\diamond})^{\diamond} &= (a^{\diamond} * ((a^{\diamond} * t^{\diamond\diamond})^{\diamond\diamond} * t^{\diamond\diamond})^{\diamond\diamond})^{\diamond}\\ &= (a^{\diamond} * ((a^{\diamond} * t^{\diamond\diamond})^{\diamond\diamond} * t^{\diamond\diamond}))^{\diamond}\\ &= (a^{\diamond} * ((a^{\diamond} * t^{\diamond\diamond}) * t^{\diamond\diamond}))^{\diamond}\\ &= ((a^{\diamond} * t^{\diamond\diamond}) * (a^{\diamond} * t^{\diamond\diamond}))^{\diamond}\\ &= 1^{\diamond} \in I\end{aligned}$$

Since $a \in I$ and I is an ideal, it gives that $((a^{\diamond} * (b^{\diamond} * x^{\diamond})^{\diamond\diamond})^{\diamond\diamond} * (b^{\diamond} * x^{\diamond})^{\diamond\diamond})^{\diamond} \in I$. Now

$$\begin{aligned}(b^{\diamond} * ((a^{\diamond} * (b^{\diamond} * x^{\diamond})^{\diamond\diamond})^{\diamond\diamond} * x^{\diamond})^{\diamond\diamond})^{\diamond} &= (b^{\diamond} * ((a^{\diamond} * (b^{\diamond} * x^{\diamond})^{\diamond\diamond})^{\diamond\diamond} * x^{\diamond}))^{\diamond}\\ &= ((a^{\diamond} * (b^{\diamond} * x^{\diamond})^{\diamond\diamond})^{\diamond\diamond} * (b^{\diamond} * x^{\diamond}))^{\diamond}\\ &= ((a^{\diamond} * (b^{\diamond} * x^{\diamond})^{\diamond\diamond})^{\diamond\diamond} * (b^{\diamond} * x^{\diamond})^{\diamond\diamond})^{\diamond} \in I\end{aligned}$$

Since $b \in I$, it implies that $((a^{\diamond} * (b^{\diamond} * x^{\diamond})) * x^{\diamond})^{\diamond} = ((a^{\diamond} * (b^{\diamond} * x^{\diamond})^{\diamond\diamond})^{\diamond\diamond} * x^{\diamond})^{\diamond} \in I$.

Conversely, assume that I satisfies the given conditions. Let $x, y \in X$. Suppose that $x, (x^{\diamond} * y^{\diamond})^{\diamond} \in I$. Then we have the following:

$$\begin{aligned}((x^{\diamond} * ((x^{\diamond} * y^{\diamond})^{\diamond\diamond} * y^{\diamond})) * y^{\diamond})^{\diamond} &= ((x^{\diamond} * ((x^{\diamond} * y^{\diamond}) * y^{\diamond})) * y^{\diamond})^{\diamond}\\ &= (((x^{\diamond} * y^{\diamond}) * (x^{\diamond} * y^{\diamond})) * y^{\diamond})^{\diamond}\\ &= (1 * y^{\diamond})^{\diamond}\\ &= (0^{\diamond} * y^{\diamond})^{\diamond}\\ &= y^{\diamond\diamond}\end{aligned}$$

Since $x, (x^{\diamond} * y^{\diamond})^{\diamond} \in I$, by the condition (2), we get that $y^{\diamond\diamond} = ((x^{\diamond} * ((x^{\diamond} * y^{\diamond})^{\diamond\diamond} * y^{\diamond})) * y^{\diamond})^{\diamond} \in I$. By (1), we get $y \in I$. Therefore I is an ideal of X. □

Theorem 7.5.27 *Let $\phi : X_1 \to X_2$ be an epimorphism from a bounded BE-algebra X_1 onto a bounded BE-algebra X_2. If I is a quasi-ideal of X_1, then $\phi(I)$ is a quasi-ideal of X_2.*

Proof Clearly $1 = \phi(1) \in \phi(I)$. Let $x \in X_2$ and $\phi(a) \in \phi(I)$. Since ϕ is onto, there exists $t \in X_1$ such that $\phi(t) = x$. Since $a \in I$ and I is a quasi-ideal of X_1, we get $(t^\diamond * a^\diamond)^\diamond \in I$. Hence $(x^\diamond * \phi(a)^\diamond)^\diamond = (\phi(t)^\diamond * \phi(a)^\diamond)^\diamond = \phi((t^\diamond * a^\diamond)^\diamond) \in \phi(I)$. Therefore $\phi(I)$ is a quasi-ideal of X_2. □

For any congruence θ on a bounded BE-algebra X, it is already observed that the quotient algebra $X_{/\theta}$ is a BE-algebra with respect to the operation defined by $[x]_\theta * [y]_\theta = [x * y]_\theta$ for all $x, y \in X$, where $X_{/\theta} = \{[x]_\theta \mid x \in X\}$. Then clearly $\phi : X \to X_{/\theta}$ is a natural homomorphism.

Theorem 7.5.28 *Let θ be a congruence on a bounded and transitive BE-algebra X and $\emptyset \neq I \subseteq X$. Then I is a quasi-ideal of X if and only if $\phi(I) = I_{/\theta}$ is a quasi-ideal of $X_{/\theta}$.*

Proof Assume that I is a quasi-ideal of X. Since $0 \in I$, we get that $[0]_\theta \in I_{/\theta}$. Let $[x]_\theta \in X_{/\theta}$ and $[a]_\theta \in I_{/\theta}$. Then, it yields that $a \in I$. Since I is a quasi-ideal of X, it implies that $(x^\diamond * a^\diamond)^\diamond \in I$. Hence $([x]_\theta^\diamond * [a]_\theta^\diamond)^\diamond = [(x^\diamond * a^\diamond)^\diamond]_\theta \in I_{/\theta}$. Therefore $I_{/\theta}$ is a quasi-ideal of $X_{/\theta}$.

Conversely, assume that $I_{/\theta}$ is a quasi-ideal of $X_{/\theta}$. Since $[0]_\theta \in I_{/\theta}$, we get that $0 \in I$. Let $x \in X$ and $a \in I$. Then $[x]_\theta \in X_{/\theta}$ and $[a]_\theta \in I_{/\theta}$. Since $I_{/\theta}$ is a quasi-ideal of $X_{/\theta}$, we get $[(x^\diamond * a^\diamond)^\diamond]_\theta = ([x]_\theta^\diamond * [a]_\theta^\diamond)^\diamond \in I_{/\theta}$. Hence $(x^\diamond * a^\diamond)^\diamond \in I$. Therefore I is a quasi-ideal of X. □

The following results can be proved similarly.

Theorem 7.5.29 *Let $\phi : X_1 \to X_2$ be an epimorphism from a bounded BE-algebra X_1 into a bounded BE-algebra X_2. If J is a quasi-ideal of X_2, then $\phi^{-1}(J)$ is a quasi-ideal of X_1.*

Corollary 7.5.30 *Let θ be a congruence on a bounded BE-algebra X and $\phi : X \to X_{/\theta}$ be a natural homomorphism. If J is a quasi-ideal of $X_{/\theta}$, then $\phi^{-1}(J)$ is a quasi-ideal of X.*

Theorem 7.5.31 *Let I be a quasi-ideal of a bounded BE-algebra X. If S is a subalgebra of X which does not meet I, then there exists a subalgebra of X containing S which is maximal with respect to the property of not meeting I.*

Proof Consider the collection $\Im = \{T \subseteq X \mid T \text{ is a subalgebra of } X \text{ and } T \cap I = \emptyset\}$. Clearly $S \in \Im$ and hence $\Im \neq \emptyset$. Let $\{T_\alpha\}_{\alpha\in\Delta} \subseteq \Im$ be a chain. Let $x, y \in \bigcup_{\alpha\in\Delta} T_\alpha$. Then there exist $\alpha, \beta \in \Delta$ such that $x \in T_\alpha$ and $y \in T_\beta$. Since $\{T_\alpha\}_{\alpha\in\Delta}$ is a chain, we get either $x, y \in T_\alpha$ or $x, y \in T_\beta$. Since each T_α is a subalgebra, we get either $x*y \in T_\alpha$ or $x*y \in T_\beta$. Hence $x*y \in \bigcup_{\alpha\in\Delta} T_\alpha$. Therefore $\bigcup_{\alpha\in\Delta} T_\alpha$ is a subalgebra of X. Suppose $\big(\bigcup_{\alpha\in\Delta} T_\alpha\big) \cap I \neq \emptyset$. Choose $x \in \big(\bigcup_{\alpha\in\Delta} T_\alpha\big) \cap I$. Then there exists $\alpha \in \Delta$ such that $x \in T_\alpha$ and $x \in I$. Hence $x \in T_\alpha \cap I$, which is a contradiction to that $T_\alpha \in \Im$. Thus $\big(\bigcup_{\alpha\in\Delta} T_\alpha\big) \cap I = \emptyset$. Therefore $\bigcup_{\alpha\in\Delta} T_\alpha$ is an upper bound for the chain $\{T_\alpha\}_{\alpha\in\Delta}$. Thus by Zorn's lemma, $\Im$ has a maximal element, say M. Clearly M is a subalgebra of X such that $S \subseteq M$ and $M \cap I = \emptyset$. Hence the theorem is proved. □

The notion of initial segments is now introduced in BE-algebras, and the relations among initial segments, lower sets, and ideals are investigated.

Definition 7.5.32 Let $(X, *, 0, 1)$ be a bounded BE-algebra and $a \in X$. Then the set $[0, a] = \{x \in X | \, x \leq a\} = \{x \in X \mid x * a = 1\}$ is said to be an *initial segment* of X.

Example 7.5.33 Let $X = \{1, a, b, c, d, 0\}$. Define a binary operation $*$ on X as follows:

$*$	1	a	b	c	d	0
1	1	a	b	c	d	0
a	1	1	a	c	c	d
b	1	1	1	c	c	c
c	1	a	b	1	a	b
d	1	1	a	1	1	a
0	1	1	1	1	1	1

Clearly X is a BE-algebra. Its final segments have the form $[0, a] = \{0, a, b, d\}$, $[0, b] = \{0, b\}$, $[0, c] = \{0, c, d\}$ and $[0, d] = \{0, d\}$. All the above initial segments are *BE*-chains.

Definition 7.5.34 A non-empty subset S of a bounded BE-algebra X is said to be *pseudo subalgebra* of X if $(x^\diamond * y^\diamond)^\diamond \in S$ whenever $x, y \in S$.

Clearly every bounded subalgebra S (i.e., $x \in S$ implies $x^\diamond \in S$) of a BE-algebra is a pseudo subalgebra.

Proposition 7.5.35 *Every initial segment of a transitive and involutory BE-algebra X is a pseudo subalgebra.*

Proof Let $a \in X$. Clearly $0 \in [0, a]$. Let $x, y \in [0, a]$. Then $x \leq a$ and $y \leq a$. Hence $a^\diamond \leq x^\diamond$ and $a^\diamond \leq y^\diamond$. Thus $a^\diamond \leq x^\diamond * a^\diamond \leq x^\diamond * y^\diamond$. Hence $(x^\diamond * y^\diamond)^\diamond \leq a^{\diamond\diamond} = a$. Thus $(x^\diamond * y^\diamond)^\diamond \in [0, a]$. Therefore $[0, a]$ is a pseudo subalgebra of X for each $a \in X$. □

Proposition 7.5.36 *The set-theoretic union of any two initial segments of a transitive and involutory BE-algebra X is a pseudo subalgebra of X.*

Proof Let $a, b \in X$. Clearly $0 \in [0, a] \cup [0, b]$. Let $x, y \in [0, a] \cup [0, b]$. Suppose $x, y \in [0, a]$. Then $(x^\diamond * y^\diamond)^\diamond \in [0, a]$ and so $(x^\diamond * y^\diamond)^\diamond \in [0, a] \cup [0, b]$. If $x, y \in [0, b]$, then also $(x^\diamond * y^\diamond)^\diamond \in [0, b] \subseteq [0, a] \cup [0, b]$. Suppose $x \in [0, a]$ and $y \in [0, b]$. Then $x \leq a$ and $y \leq b$. Hence $b^\diamond \leq y^\diamond \leq x^\diamond * y^\diamond$. Hence $(x^\diamond * y^\diamond)^\diamond \leq b^{\diamond\diamond} = b$. Thus $(x^\diamond * y^\diamond)^\diamond \in [0, b] \subseteq [0, a] \cup [0, b]$. Hence $[0, a] \cup [0, b]$ is a pseudo subalgebra of X. □

In the following example, we observe a fact that the initial segments are not ideals of BE-algebras in general.

Example 7.5.37 Let $X = \{1, a, b, c, 0\}$. Define a binary operation $*$ on X as follows:

$*$	1	a	b	c	0
1	1	a	b	c	0
a	1	1	a	a	c
b	1	1	1	a	b
c	1	1	a	1	a
0	1	1	1	1	1

Then $(X, *, 0, 1)$ is a bounded BE-algebra. Consider the initial segment $[0, b] = \{0, b\}$. The initial segment $[0, b]$ is not an ideal of X, because $b \in [0, b]$ and $(b^\diamond * c^\diamond)^\diamond = (b * a)^\diamond = 1^\diamond = 0 \in [0, b]$ but $c \notin [0, b]$.

Theorem 7.5.38 *Let I be an ideal of a bounded and transitive BE-algebra X. Then $I = \bigcup_{a \in I} [0, a]$.*

Proof Let $a \in I$. Choose $x \in X$ such that $x \in [0, a]$. Then $x \leq a$. Since I is an ideal, we get $x \in I$. Hence $[0, a] \subseteq I$. Therefore $\bigcup_{a \in I} [0, a] \subseteq I$. Conversely, let $x \in I$. Since $x * x = 1$, we get $x \leq x$. Hence $x \in [0, x]$. Therefore $I \subseteq [0, x] \subseteq \bigcup_{a \in I} [0, a]$. □

In general, the converse of the above theorem is not true. Consider the following example.

Example 7.5.39 Let $X = \{1, a, b, c, d, 0\}$. Define a binary operation $*$ on X as follows:

$*$	1	a	b	c	d	0
1	1	a	b	c	d	0
a	1	1	b	c	d	c
b	1	a	1	c	c	b
c	1	1	b	1	b	a
d	1	1	1	1	1	d
0	1	1	1	1	1	1

Then $(X, *, 0, 1)$ is a bounded BE-algebra. Consider the set $A = \{0, b, c, d\}$. It is observed that $[0, b] = \{0, b, d\}$; $[0, c] = \{0, c, d\}$; and $[0, d] = \{0, d\}$. Clearly $A = [0, b] \cup [0, c] \cup [0, d] \cup \{0\}$. It can also be observed that A is not an ideal of X, because $d \in A$ and $(d^\diamond * a^\diamond)^\diamond = (d * c)^\diamond = 0 \in A$ but $a \neq A$.

Theorem 7.5.40 *Every ideal I of a bounded and transitive BE-algebra X satisfies the following properties:*

(1) $I = \bigcup_{a \in I} [0, a]$,

(2) *for all $a, b \in X$, $[0, a] = [0, b]$ and $a \in I$ imply that $b \in I$,*

(3) *$a \in I$ implies $[0, a] \subseteq I$ for all $a \in X$.*

Proof (1) It is proved in the above theorem.
(2) Let $a, b \in X$ be such that $[0, a] = [0, b]$. Suppose that $a \in I$. Then by (1), we get $[0, b] = [0, a] \subseteq \bigcup_{a \in I}[0, a] = I$. Since $b \in [0, b]$, it yields that $b \in I$.
(3) It is clear from (1). □

Proposition 7.5.41 *The Cartesian product of two initial segments of a bounded BE-algebra is also an initial segment.*

Proof Let X_1 and X_2 be two bounded BE-algebras. For any $a \in X_1$ and $b \in X_2$, we intend to prove that $[(0, 0), (a, b)] = [0, a] \times [0, b]$. For $a \in X_1$ and $b \in X_2$, we get

$$\begin{aligned} [(0,0),(a,b)] &= \{(x,y) \in X_1 \times X_2 \mid (x,y) \leq (a,b)\} \\ &= \{(x,y) \in X_1 \times X_2 \mid x \leq a, y \leq b\} \\ &= \{(x,y) \in X_1 \times X_2 \mid x \in [0,a] \text{ and } y \in [0,b]\} \\ &= [0,a] \times [0,b] \end{aligned}$$

Therefore $[(0, 0), (a, b)]$ is an initial segment in the product algebra $X_1 \times X_2$. □

We now establish a relation between initial segments and lower sets of BE-algebras.

Lemma 7.5.42 *Let X be a bounded and transitive BE-algebra and $a \in X$. Then $[0, a] \subseteq [0; a]$.*

Proof Let $a \in X$ be an arbitrary element. Suppose $x \in [0, a]$. Then we get $x \leq a$ and hence $a^\diamond \leq x^\diamond$. Thus $a^\diamond * x^\diamond = 1$. Hence $x \in [0; a]$. Therefore $[0, a] \subseteq [0; a]$. □

Theorem 7.5.43 *Let X be a bounded and transitive BE-algebra and $a \in X$. If X is involutory, then $[0, a] = [0; a]$.*

Proof Let $a \in X$. Clearly $[0, a] \subseteq [0; a]$. Conversely, let $x \in [0; a]$. Then $a^\diamond * x^\diamond = 1$. Hence $x = x^{\diamond\diamond} \leq a^{\diamond\diamond} = a$. Thus $x \in [0, a]$. Therefore $[0; a] \subseteq [0, a]$. □

Theorem 7.5.44 *An initial segment $[0, a]$ of a transitive and involutory BE-algebra X is an ideal if and only if for all $z \in X$*

$$a^\diamond \leq a^\diamond * z^\diamond \textit{ implies } a^\diamond \leq z^\diamond.$$

Proof Assume that the above implication holds. Let $x, y \in X$ be such that $x \in [0, a]$ and $(x^\diamond * y^\diamond)^\diamond \in [0, a]$. Then $x \leq a$ and $(x^\diamond * y^\diamond)^\diamond \leq a$. Hence $a^\diamond \leq x^\diamond$ and $a^\diamond \leq x^\diamond * y^\diamond$. Hence

$$\begin{aligned} a^\diamond \leq x^\diamond &\Rightarrow x^\diamond * y^\diamond \leq a^\diamond * y^\diamond \\ &\Rightarrow a^\diamond \leq x^\diamond * y^\diamond \leq a^\diamond * y^\diamond \\ &\Rightarrow a^\diamond \leq y^\diamond \qquad \text{by the given condition} \\ &\Rightarrow y = y^{\diamond\diamond} \leq a^{\diamond\diamond} = a \end{aligned}$$

Hence $y \in [0, a]$. Therefore the initial segment $[0, a]$ is an ideal of X.

Conversely, assume that the initial segment $[0, a]$ is an ideal of X. Suppose $a^\diamond \leq a^\diamond * z^\diamond$ for all $z \in X$. Then $(a^\diamond * z^\diamond)^\diamond \leq a^{\diamond\diamond} = a$. Hence $(a^\diamond * z^\diamond)^\diamond \in [0, a]$. Since $a \in [0, a]$ and $[0, a]$ is an ideal, it yields that $z \in [0, a]$. Therefore $a^\diamond \leq z^\diamond$. □

Theorem 7.5.45 *In a self-distributive and involutory BE-algebra, every initial segment is an ideal.*

Proof Let $a \in X$. Suppose $x, y \in X$ such that $x \in [0, a]$ and $(x^\diamond * y^\diamond)^\diamond \in [0, a]$. Then $x \leq a$ and $(x^\diamond * y^\diamond)^\diamond \leq a$. Hence $a^\diamond \leq x^\diamond$ and $a^\diamond \leq x^\diamond * y^\diamond$. Then

$$\begin{aligned} a^\diamond \leq x^\diamond &\Rightarrow x^\diamond * y^\diamond \leq a^\diamond * y^\diamond \\ &\Rightarrow a^\diamond \leq x^\diamond * y^\diamond \leq a^\diamond * y^\diamond \end{aligned}$$

Thus $a^\diamond*(a^\diamond*y^\diamond) = 1$. Since X is self-distributive, we get that $a^\diamond*y^\diamond = a^\diamond*(a^\diamond*y^\diamond) = 1$. Hence $a^\diamond \leq y^\diamond$. Since X is involutory, we get $y = y^{\diamond\diamond} \leq a^{\diamond\diamond} = a$. Hence $y \in [0, a]$. Therefore $[0, a]$ is an ideal of X. □

Definition 7.5.46 Let $(X, *, 0, 1)$ be a bounded BE-algebra. For any $a \in X$, define a relation θ_a on X by

$$(x, y) \in \theta_a \text{ if and only if } (x * y)^\diamond \leq a \text{ and } (y * x)^\diamond \leq a$$

for all $x, y \in X$.

Proposition 7.5.47 *If X is a self-distributive and involutory BE-algebra and $a \in X$, then the above relation θ_a is a congruence on X.*

Proof Clearly θ_a is reflexive and symmetric. To prove the transitivity, let $(x, y) \in \theta_a$ and $(y, z) \in \theta_a$. Then $(x*y)^\diamond \leq a$, $(y*x)^\diamond \leq a$ and $(y*z)^\diamond \leq a$, $(z*y)^\diamond \leq a$. Hence

$$y * z \leq (x * y) * (x * z) \leq (x * y)^{\diamond\diamond} * (x * z)^{\diamond\diamond}$$

Hence $((x * y)^{\diamond\diamond} * (x * z)^{\diamond\diamond})^\diamond \leq (y * z)^\diamond \leq a$. Thus$((x * y)^{\diamond\diamond} * (x * z)^{\diamond\diamond})^\diamond \in [0, a]$. Since $(x * y)^\diamond \in [0, a]$ and $[0, a]$ is an ideal, we get $(x * z)^\diamond \in [0, a]$. Therefore $(x * z)^\diamond \leq a$ Similarly, we can obtain $(z * x)^\diamond \leq a$. Hence $(x, z) \in \theta_a$. Therefore θ_a is an equivalence relation on X. Let $(x, y) \in \theta_a$ and $(u, v) \in \theta_a$. Then $(x * y)^\diamond \leq a$, $(y * x)^\diamond \leq a$, $(u * v)^\diamond \leq a$, and $(v * u)^\diamond \leq a$. Since X is transitive, we get $x*y \leq (u*x)*(u*y)$ and so $((u*x)*(u*y))^\diamond \leq (x*y)^\diamond \leq a$. Similarly, we can get $((u * y) * (u * x))^\diamond \leq a$ because $(y * x)^\diamond \leq a$. Hence, both together provide us $(u*x, u*y) \in \theta_a$. Again, since X is transitive, we get $v*y \leq (u*v)*(u*y)$. Hence

$$u * v \leq (v * y) * (u * y) \leq ((v * y) * (u * y))^{\diamond\diamond}$$

Hence $((v*y)*(u*y))^\diamond \leq (u*v)^\diamond \leq a$. Similarly, we can obtain $((u*y)*(v*y))^\diamond \leq a$ because $(v * u)^\diamond \leq a$. Thus $(u * y, v * y) \in \theta_a$. Therefore θ_a is a congruence on X. □

Definition 7.5.48 If θ_a is a congruence on a BE-algebra X for each $a \in X$, then the dual kernel of θ_a is defined by $DKer(\theta_a) = \{x \in X \mid (x, 0) \in \theta_a\}$.

Proposition 7.5.49 *Let X be an involutory BE-algebra and $a \in X$. If θ_a is a congruence on X, then $DKer(\theta_a)$ is an ideal of X such that $DKer(\theta_a) = [0, a]$.*

Proof Clearly $0 \in DKer(\theta_a)$. Let $x \in DKer(\theta_a)$ and $(x^\diamond * y^\diamond)^\diamond \in DKer(\theta_a)$. Then $(0, x) \in \theta_a$ and $(0, (x^\diamond * y^\diamond)^\diamond) \in \theta_a$. Since $(0, x) \in \theta_a$, we get $(0^\diamond, x^\diamond) \in \theta_a$. Hence $(y, (x^\diamond * y^\diamond)^\diamond) = (y^{\diamond\diamond}, (x^\diamond * y^\diamond)^\diamond) = ((0^\diamond * y^\diamond)^\diamond, (x^\diamond * y^\diamond)^\diamond) \in \theta_a$. Since $(0, (x^\diamond * y^\diamond)^\diamond) \in \theta_a$, it yields $(0, y) \in \theta_a$. Hence $y \in DKer(\theta_a)$. Therefore $DKer(\theta_a)$ is an ideal of X. Now, for any $x \in X$, it is clear that $(0, x) \in \theta_a$ if and only if $x \in [0, a]$. Hence $DKer(\theta_a) = [0, a]$. □

The following corollary is a direct consequence of the above proposition.

Corollary 7.5.50 *Let X be an involutory BE-algebra and $a \in X$. If θ_a is a congruence on X, then the initial segment $[0, a]$ is an ideal of X.*

The summation of all the above results yields the following:

Theorem 7.5.51 *Let X be a self-distributive and involutory BE-algebra and $a \in X$. Then θ_a is a congruence on X such that $DKer(\theta_a) = [0, a]$.*

7.6 Congruences and Pseudo-complements

In this section, a special type of congruence is introduced on a bounded and transitive BE-algebras with the help of ideals. Some properties of ideals are studied with the help of congruences and quotient algebras. A homomorphism theorem is proved with the help of congruences. Throughout this section, X stands for a partially ordered BE-algebra unless otherwise mentioned.

Definition 7.6.1 Let I be an ideal of a BE-algebra X. Then define a relation θ_I on X by

$$(x, y) \in \theta_I \text{ if and only if } (x * y)^\diamond \in I \text{ and } (y * x)^\diamond \in I \quad \text{for any } x, y \in X.$$

Proposition 7.6.2 *If X is a bounded and transitive BE-algebra and I an ideal of X, then the above relation θ_I is a congruence on X.*

Proof Clearly θ_I is reflexive and symmetric. To prove the transitivity, let (x, y), $(y, z) \in \theta_I$. Then $(x * y)^\diamond \in I$, $(y * x)^\diamond \in I$ and $(y * z)^\diamond \in I$, $(z * y)^\diamond \in I$. By Proposition 7.1.4(3), we get

$$y * z \leq (x * y) * (x * z) \leq (x * y)^{\diamond\diamond} * (x * z)^{\diamond\diamond}$$

Hence $((x * y)^{\diamond\diamond} * (x * z)^{\diamond\diamond})^{\diamond} \leq (y * z)^{\diamond}$. Since $(y * z)^{\diamond} \in I$, we get that $((x * y)^{\diamond\diamond} * (x * z)^{\diamond\diamond})^{\diamond} \in I$. Since $(x * y)^{\diamond} \in I$ and I is an ideal, we get $(x * z)^{\diamond} \in I$. Similarly, we can obtain $(z * x)^{\diamond} \in I$. Hence $(x, z) \in \theta_I$. Therefore θ_I is an equivalence relation on X. Let $(x, y) \in \theta_I$ and $(u, v) \in \theta_I$. Then $(x * y)^{\diamond} \in I$, $(y * x)^{\diamond} \in I$, $(u * v)^{\diamond} \in I$, and $(v * u)^{\diamond} \in I$. Since X is transitive, we get $x * y \leq (u * x) * (u * y)$ and so $((u * x) * (u * y))^{\diamond} \leq (x * y)^{\diamond}$. Since $(x * y)^{\diamond} \in I$, we get $((u * x) * (u * y))^{\diamond} \in I$. Similarly, we can get $((u*y)*(u*x))^{\diamond} \in I$ because $(y*x)^{\diamond} \in I$. Hence, both together provide us $(u*x, u*y) \in \theta_I$. Again, since X is transitive, we get $v*y \leq (u*v)*(u*y)$. Hence

$$u * v \leq (v * y) * (u * y) \leq ((v * y) * (u * y))^{\diamond\diamond}$$

Hence $((v * y) * (u * y))^{\diamond} \leq (u * v)^{\diamond}$. Since $(u * v)^{\diamond} \in I$ and I is an ideal, we get $((v * y) * (u * y))^{\diamond} \in I$. Similarly, we can obtain $((u * y) * (v * y))^{\diamond} \in I$ because $(v * u)^{\diamond} \in I$. Thus $(u * y, v * y) \in \theta_I$. Therefore θ_I is a congruence on X. □

From the above result, it is easy to see that the quotient algebra $X/I = \{I_x \mid x \in X\}$(where I_x is the congruence class of x modulo θ_I) is a bounded BE-algebra (with the least element I_0) in which the binary operation $*$ is defined as $I_x * I_y = I_{x*y}$ for $x, y \in X$. For any $x \in X$, the pseudo-complement of I_x in X/I is given by $I_x^{\diamond} = I_{x^{\diamond}}$. For any ideal I of a transitive BE-algebra X, it is natural to obtain the epimorphism $\nu : X \to X/I$ given by $\nu(x) = I_x$. Let X and Y be two bounded BE-algebras. Then for any bounded homomorphism $f : X \to Y$, a *dual kernel* of the morphism f is defined as $Dker(f) = \{x \in X \mid f(x) = 0\}$.

Lemma 7.6.3 *For any bounded homomorphism $f : X \to Y$, the dual kernel is an ideal of X.*

Proof Since $f(0) = 0$, it is clear that $0 \in Dker(f)$. Let $x \in Dker(f)$ and $(x^{\diamond} * y^{\diamond})^{\diamond} \in Dker(f)$. Then $f(x) = 0$ and $(f(x)^{\diamond} * f(y)^{\diamond})^{\diamond} = f((x^{\diamond} * y^{\diamond}))^{\diamond} = 0$. Thus $f(y)^{\diamond\diamond} = 0$. Since $y \leq y^{\diamond\diamond}$, we get $f(y) \leq f(y)^{\diamond\diamond} = 0$. Hence $y \in Dker(f)$. Therefore $Dker(f)$ is an ideal of X. □

Theorem 7.6.4 (Homomorphism theorem) *Let X and Y be two bounded BE-algebras where X is transitive. If $f : X \to Y$ is an epimorphism, then the following conditions are equivalent:*

(1) *$X/Dker(f)$ is isomorphic to Y,*
(2) *Y has a unique dense element,*
(3) *for any $x, y \in X$, $(x*y)^{\diamond} \in Dker(f)$, and $(y*x)^{\diamond} \in Dker(f)$ imply $f(x) = f(y)$.*

Proof (1) ⇒ (2): Assume that $X/Dker(f)$ is isomorphic to Y. Since $Dker(f)$ is an ideal of X, it is clear that $X/Dker(f)$ is a bounded and transitive BE-algebra. It is enough to show that $X/Dker(f)$ contains a unique dense element. Clearly I_1 is a dense element of $X/Dker(f)$. For $1 \neq x \in X$, suppose $I_{x^{\diamond}} = I_x^{\diamond} = I_0$. Then we get $(x^{\diamond}, 0) \in \theta_I$. Hence $x^{\diamond} = x^{\diamond\diamond\diamond} = (x^{\diamond} * 0)^{\diamond} \in I$. Thus $(1 * x)^{\diamond} \in I$ and $(x * 1)^{\diamond} \in I$.

Hence $(1, x) \in \theta_I$, which implies that $I_x = I_1$. Therefore I_1 is the unique dense element of $X/Dker(f)$.
(2) $\Rightarrow$ (3): Assume that Y has a unique dense element. Let $x, y \in X$. Suppose $(x * y)^\diamond \in Dker(f)$ and $(y * x)^\diamond \in Dker(f)$. Then $(f(x) * f(y))^\diamond = f(x * y)^\diamond = 0$ and $(f(y) * f(x))^\diamond = f(y * x)^\diamond = 0$. Hence $f(x) * f(y)$ and $f(y) * f(x)$ are two dense elements in Y. Since Y has a unique dense element, we get $f(x) = f(y)$.
(3) $\Rightarrow$ (1): Assume that the condition (3) holds. Put $I = Dker(f)$. Let $\nu : X/I \to Y$ be such that $\nu(I_x) = f(x)$ for all $x \in X$. Since f is onto, we get ν is onto. It is enough to show that ν is one-to-one. Let $I_x, I_y \in X/I$ be such that $\nu(I_x) = \nu(I_y)$ where $x, y \in X$. Then $f(x) = f(y)$. Hence $f((x * y)^\diamond) = (f(x) * f(y))^\diamond = 1^\diamond = 0$. Thus $(x * y)^\diamond \in Dker(f) = I$. Similarly, we can get $(y * x)^\diamond \in I$. Thus $(x, y) \in \theta_I$, which implies that $I_x = I_y$. Therefore ν is one-to-one. For any $I_x, I_y \in X/DKer(f)$, we have the following:

$$\nu(I_x * I_y) = \nu(I_{x*y}) = f(x * y) = f(x) * f(y) = \nu(I_x) * \nu(I_y).$$

Hence ν is a homomorphism, and therefore it is an isomorphism. □

Theorem 7.6.5 *Let X, X' be two bounded and transitive BE-algebras and $f : X \to X'$ be an epimorphism. If I' is an ideal of X' and $I = f^{-1}(I')$, then X/I is isomorphic to X'/I'.*

Proof It is clear that there is natural homomorphism ν from X' into X'/I'. Then $u = \nu \circ f$ is an epimorphism from X into X'/I'. We first show that $Dker(u) = f^{-1}(I')$.

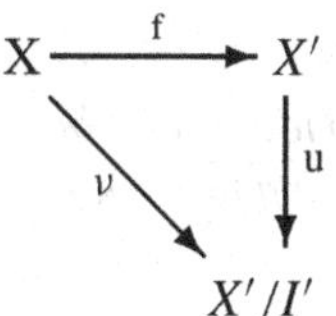

For $x \in X$, we get $x \in Dker(u) \Leftrightarrow u(x) = I'_{0'} \Leftrightarrow (\nu \circ f)(x) = I'_{0'} \Leftrightarrow \nu(f(x)) = I'_{0'} \Leftrightarrow I'_{f(x)} = I'_{0'} \Leftrightarrow f(x)^{\diamond\diamond} = (f(x) * 0')^\diamond \in I' \Leftrightarrow f(x) \in I' \Leftrightarrow x \in f^{-1}(I')$. Therefore $f^{-1}(I') = Dker(u)$. Since I' is an ideal of X', we get that X'/I' has a unique dense element. Then by the above theorem, $X/Dker(f)$ is isomorphic to X'/I'; that is, X/I is isomorphic to X/I'. □

Theorem 7.6.6 *Let X and Y be two bounded and transitive BE-algebras. If I and J are two ideals of X such that $J \subseteq I$, then X/I is isomorphic to $(X/J)/(I/J)$.*

Proof Let $u : X/J \to X/I$ be defined by $u(J_x) = I_x$. Suppose $J_x = J_y$. Then $(x, y) \in \theta_J$. Hence $(x * y)^\diamond \in J \subseteq I$ and $(y * x)^\diamond \in J \subseteq I$. Thus we get $u(J_x) = I_x = I_y = u(J_y)$. Therefore u is well-defined. Clearly u is onto. Also $u(J_x * J_y) = u(J_{x*y}) = I_{x*y} = I_x * I_y = u(J_x) * u(J_y)$. Therefore u is an epimorphism. We now show that

$Dker(u) = I/J$. For, this $J_x \in Dker(u) \Leftrightarrow u(J_x) = I_0 \Leftrightarrow I_x = I_0 \Leftrightarrow (x * 0)^\diamond \in I, (0 * x)^\diamond \in I \Leftrightarrow x^{\diamond\diamond} \in I \Leftrightarrow x \in I \Leftrightarrow J_x \in I/J$. Therefore $Dker(u) = I/J$. Since I is an ideal, we get X/I has a unique dense element. Then by the above theorem, we get $X/Dker\ u$ is isomorphic to X/I; that is, $(X/J)/(I/J)$ is isomorphic to X/I. □

Theorem 7.6.7 *Let X and Y be two bounded and transitive BE-algebras where $\mathcal{D}(Y) = \{1\}$. Suppose $f : X \to Y$ is an epimorphism such that $I = DKer(f) = I_0$. Then $\alpha : X/I \to Y$ defined by $\alpha(I_x) = f(x)$ is an isomorphism.*

Proof Clearly α is well-defined. Let $I_x, I_y \in X/I$ be such that $I_x = I_y$. Then

$$\begin{aligned} I_x = I_y &\Rightarrow (x, y) \in \theta_I \\ &\Rightarrow (x * y)^\diamond \in I = I_0 \text{ and } (y * x)^\diamond \in I = I_0 \\ &\Rightarrow (f(x) * f(y))^\diamond = f((x * y)^\diamond) = 0 \text{ and } (f(y) * f(x))^\diamond = f((y * x)^\diamond) = 0 \\ &\Rightarrow f(x) * f(y) \in \mathcal{D}(Y) = \{1\} \text{ and } f(y) * f(x) \in \mathcal{D}(Y) = \{1\} \\ &\Rightarrow f(x) \le f(y) \text{ and } f(y) \le f(x) \\ &\Rightarrow f(x) = f(y) \\ &\Rightarrow \alpha(I_x) = \alpha(I_y) \end{aligned}$$

Therefore α is one-to-one. Let $y \in Y$. Since f is onto, there exists $x \in X$ such that $f(x) = y$. For this $x \in X$, we get that $I_x \in X/I$ and $\alpha(I_x) = f(x) = y$. Therefore α is onto.

Let $I_x, I_y \in X/I$ where $x, y \in X$. Then $\alpha(I_x * I_y) = \alpha(I_{x*y}) = f(x * y) = f(x) * f(y) = \alpha(I_x) * \alpha(I_y)$. Hence α is a homomorphism. Therefore α is an isomorphism. □

Theorem 7.6.8 *Let I and J be two ideals of a bounded and transitive BE-algebra X such that $I \subseteq J$. Let $f : X \to X/I$ be a natural homomorphism. Denote $J/I = \{I_x \in X/I \mid x \in J\}$. Then*

(1) *$x \in J$ if and only if $I_x \in J/I$ for all $x \in X$,*
(2) *J/I is an ideal in X/I,*
(3) *Let I be an ideal of X. If S and T are the sets of all ideals of X and X/I, respectively, then the mapping $g : S \to T$ defined by $g(J) = J/I$ is a bijective map.*

Proof (1) The necessary part is clear. Conversely, assume that $I_x \in J/I$ for all $x \in X$. Then $I_x = I_y$ for some $y \in J$. Then $(x, y) \in \theta_I$. Hence $(x*y)^\diamond \in I \subseteq J$ and $(y*x)^\diamond \in I$. Since X is transitive, we get $x * y \le y^\diamond * x^\diamond$ and hence $(y^\diamond * x^\diamond)^\diamond \le (x * y)^\diamond$. Since J is an ideal, we get $(y^\diamond * x^\diamond)^\diamond \in J$. Since $y \in J$ and J is an ideal, we get that $x \in J$.
(2) Since $0 \in J$, we get $[0]_{\theta_I} = I_0 \in J/I$. Let $I_x, (I_x^\diamond * I_y^\diamond)^\diamond \in J/I$. Then $I_x \in J/I$ and $I_{(x^\diamond * y^\diamond)^\diamond} \in J/I$. Hence $x \in J$ and $(x^\diamond * y^\diamond)^\diamond \in J$. Since J is an ideal, we get $y \in J$. It implies that $I_y \in J/I$. Therefore J/I is an ideal of X/I.
(3) Let J and K be two ideals of X such that $g(J) = g(K)$. Then $J/I = K/I$. Let $x \in J$. Then it yields that $I_x \in J/I = K/I$. Hence $x \in K$. Thus $J \subseteq K$. Similarly, we can obtain $J \subseteq K$. Therefore $J = K$. Therefore g is one-to-one.

Let J/I be an ideal of X/I. We claim that J is an ideal of X. Since J/I is an ideal of X/I, we get $I_0 \in J/I$. Hence $0 \in J$. Let $x, (x^\diamond * y^\diamond)^\diamond \in J$. Then $I_x \in J/I$ and $(I_x^\diamond * I_y^\diamond)^\diamond = I_{(x^\diamond * y^\diamond)^\diamond} \in J/I$. Since J/I is an ideal of X/I, we get $I_y \in J/I$. Hence $y \in J$. Thus J is an ideal of X. Clearly $f(J) = J/I$. Thus g is onto. Therefore $g : S \to T$ is a bijection. If $f(J) = \cup\{I_x \mid I_x \in J/I\}$. Clearly $(g \circ f)(J/I) = (f \circ g)(J/I) = I_0$. Therefore f is the inverse of the mapping g. □

7.7 Congruences and Ideals

In this section, the notion of θ-ideals is introduced and characterized in BE-algebras. If θ is a congruence on a transitive BE-algebra X, then the quotient algebra X/θ is also a bounded BE-algebra in which $[x]_\theta * [y]_\theta = [x * y]_\theta$ and $[x]_\theta^\diamond = [x^\diamond]_\theta$ for all $x, y \in X$. Clearly $(x^\diamond, y^\diamond) \in \theta$ whenever $(x, y) \in \theta$ for all $x, y \in X$. The smallest congruence θ_0 is given by $\theta_0 = \{(a, a) \mid a \in X\}$. Throughout this section, X stands for a partially ordered BE-algebra unless otherwise mentioned.

Definition 7.7.1 Let θ be a congruence on a BE-algebra X. Define operations α and β by

(1) For any ideal I of X, define $\alpha(I) = \{\, [x]_\theta \mid (x^\diamond, y^\diamond) \in \theta$ for some $y \in I\,\}$.
(2) For any ideal $\widehat{I}$ of $X_{/\theta}$, define $\beta(\widehat{I}) = \{x \in X \mid (x^\diamond, y^\diamond) \in \theta$ for some $[y]_\theta \in \widehat{I}\}$.

In the following, we derive a characterization theorem of ideals of BE-algebras which has a significant role in the investigation of further properties of ideals with the help of congruences.

Theorem 7.7.2 *A non-empty subset I of a bounded and transitive BE-algebra X is an ideal of X if and only if it satisfies the following properties for all $x, y \in X$:*

(1) $x^{\diamond\diamond} \in I$ *implies* $x \in I$,
(2) $x \in I$ *implies* $(y^\diamond * x^\diamond)^\diamond \in I$,
(3) $a, b \in I$ *implies* $((a^\diamond * (b^\diamond * x^\diamond)) * x^\diamond)^\diamond \in I$.

Proof Assume that I is an ideal of X. Since $x \le x^{\diamond\diamond}$, condition (1) is clear. Let $x \in I$. By Proposition 7.1.4(5), we get

$$\begin{aligned}(x^\diamond * (y^\diamond * x^\diamond)^{\diamond\diamond})^\diamond &= (x^\diamond * (y^\diamond * x^\diamond))^\diamond \\ &= (y^\diamond * (x^\diamond * x^\diamond))^\diamond \\ &= (y^\diamond * 1)^\diamond \\ &= 1^\diamond \\ &= 0 \in I.\end{aligned}$$

Since $x \in I$, we get $(y^\diamond * x^\diamond)^\diamond \in I$ which proves the condition (2). To prove the condition (3), let $a, b \in I$. Then by putting $b^\diamond * x^\diamond = t$, we get the following:

$$\begin{aligned}(a^\diamond * ((a^\diamond * (b^\diamond * x^\diamond)^{\diamond\diamond})^{\diamond\diamond} * (b^\diamond * x^\diamond)^{\diamond\diamond})^{\diamond\diamond})^\diamond &= (a^\diamond * ((a^\diamond * t^{\diamond\diamond})^{\diamond\diamond} * t^{\diamond\diamond})^{\diamond\diamond})^\diamond \\ &= (a^\diamond * ((a^\diamond * t^{\diamond\diamond})^{\diamond\diamond} * t^{\diamond\diamond}))^\diamond \\ &= (a^\diamond * ((a^\diamond * t^{\diamond\diamond}) * t^{\diamond\diamond}))^\diamond \\ &= ((a^\diamond * t^{\diamond\diamond}) * (a^\diamond * t^{\diamond\diamond}))^\diamond \\ &= 1^\diamond \in I\end{aligned}$$

Since $a \in I$ and I is an ideal, it gives that $((a^\diamond * (b^\diamond * x^\diamond)^{\diamond\diamond})^{\diamond\diamond} * (b^\diamond * x^\diamond)^{\diamond\diamond})^\diamond \in I$. Again by using Proposition 7.1.4(5), we get

$$\begin{aligned}(b^\diamond * ((a^\diamond * (b^\diamond * x^\diamond)^{\diamond\diamond})^{\diamond\diamond} * x^\diamond)^{\diamond\diamond})^\diamond &= (b^\diamond * ((a^\diamond * (b^\diamond * x^\diamond)^{\diamond\diamond})^{\diamond\diamond} * x^\diamond))^\diamond \\ &= ((a^\diamond * (b^\diamond * x^\diamond)^{\diamond\diamond})^{\diamond\diamond} * (b^\diamond * x^\diamond))^\diamond \\ &= ((a^\diamond * (b^\diamond * x^\diamond)^{\diamond\diamond})^{\diamond\diamond} * (b^\diamond * x^\diamond)^{\diamond\diamond})^\diamond \in I\end{aligned}$$

Since $b \in I$, we get that $((a^\diamond * (b^\diamond * x^\diamond)) * x^\diamond)^\diamond = ((a^\diamond * (b^\diamond * x^\diamond)^{\diamond\diamond})^{\diamond\diamond} * x^\diamond)^\diamond \in I$.

Conversely, assume that I satisfies the given conditions. By taking $x = y$ in the condition (2), it can be seen that $0 \in I$. Let $x, y \in X$. Suppose $x, (x^\diamond * y^\diamond)^\diamond \in I$. Then

$$\begin{aligned}((x^\diamond * ((x^\diamond * y^\diamond)^{\diamond\diamond} * y^\diamond)) * y^\diamond)^\diamond &= ((x^\diamond * ((x^\diamond * y^\diamond) * y^\diamond)) * y^\diamond)^\diamond \\ &= (((x^\diamond * y^\diamond) * (x^\diamond * y^\diamond)) * y^\diamond)^\diamond \\ &= (1 * y^\diamond)^\diamond \\ &= (0^\diamond * y^\diamond)^\diamond \\ &= y^{\diamond\diamond}\end{aligned}$$

Since $x, (x^\diamond * y^\diamond)^\diamond \in I$, from (3), it infers that $y^{\diamond\diamond} \in I$. By condition (1), we get that $y \in I$. Therefore I is an ideal of X. □

Some basic properties of the above two operations α and β are observed.

Lemma 7.7.3 *Let θ be a congruence on a bounded BE-algebra X. Then*

(1) *For any ideal I of X, $\alpha(I)$ is an ideal in $X_{/\theta}$,*
(2) *For any ideal $\widehat{I}$ of $X_{/\theta}$, $\beta(\widehat{I})$ is an ideal in X,*
(3) *α and β are isotone,*
(4) *For any ideal I of X, $x \in I$ implies $[x]_\theta \in \alpha(I)$,*
(5) *For any ideal $\widehat{I}$ of $X_{/\theta}$, $[x]_\theta \in \widehat{I}$ implies $x \in \beta(\widehat{I})$.*

Proof (1) Let $[x]_\theta^{\diamond\diamond} \in X_{/\theta}$. Since $x^{\diamond\diamond\diamond} = x^\diamond$, we get $[x]_\theta \in X_{/\theta}$. Thus condition (1) of Theorem 7.7.2. is clear. Let $[x]_\theta \in X_{/\theta}$ and $[a]_\theta \in \alpha(I)$. Then $(a^\diamond, y^\diamond) \in \theta$ for some $y \in I$. Hence $((x^\diamond * a^\diamond)^\diamond, (x^\diamond * y^\diamond)^\diamond) \in \theta$ and $(x^\diamond * y^\diamond)^\diamond \in I$ because I is an ideal by Theorem 7.7.2. Thus $([x]_\theta^\diamond * [a]_\theta^\diamond)^\diamond = ([x^\diamond]_\theta * [a^\diamond]_\theta)^\diamond = [(x^\diamond * a^\diamond)^\diamond]_\theta \in \alpha(I)$. Let $[a]_\theta, [b]_\theta \in \alpha(I)$. Then $(a^\diamond, x^\diamond) \in \theta$ and $(b^\diamond, y^\diamond) \in \theta$ for some $x, y \in I$. Now, for any $t \in X$, we get

$$\begin{aligned}(b^\diamond, y^\diamond) \in \theta &\Rightarrow (b^\diamond * t^\diamond, y^\diamond * t^\diamond) \in \theta \\ &\Rightarrow (a^\diamond * (b^\diamond * t^\diamond), x^\diamond * (y^\diamond * t^\diamond)) \in \theta \qquad \text{since } (a^\diamond, x^\diamond) \in \theta \\ &\Rightarrow (((a^\diamond * (b^\diamond * t^\diamond)) * t^\diamond)^\diamond, ((x^\diamond * (y^\diamond * t^\diamond)) * t^\diamond)^\diamond) \in \theta\end{aligned}$$

Since $x, y \in I$, we get by Theorem 7.7.2 that $(x^\diamond * (y^\diamond * t^\diamond)) * t^\diamond \in I$. Hence

$$\begin{aligned}(([a]_\theta^\diamond * ([b]_\theta^\diamond * [t]_\theta^\diamond)) * [t]_\theta^\diamond)^\diamond &= [((a^\diamond * (b^\diamond * t^\diamond)) * t^\diamond)^\diamond]_\theta \\ &= [((x^\diamond * (y^\diamond * t^\diamond)) * t^\diamond)^\diamond]_\theta \in \alpha(I).\end{aligned}$$

Therefore by Theorem 7.7.2, it concludes that $\alpha(I)$ is an ideal of $X_{/\theta}$.
(2) Since $x^{\diamond\diamond\diamond} = x^\diamond$, condition (1) of Theorem 7.7.2 is clear. Let $x \in X$ and $a \in \beta(\widehat{I})$. Then $(a^\diamond, y^\diamond) \in \theta$ for some $[y]_\theta \in \widehat{I}$. Hence $((x^\diamond * a^\diamond)^\diamond, (x^\diamond * y^\diamond)^\diamond) \in \theta$. Since $\widehat{I}$ is an ideal, we get $[(x^\diamond * a^\diamond)^\diamond]_\theta = [(x^\diamond * y^\diamond)^\diamond]_\theta = ([x]_\theta^\diamond * [y]_\theta^\diamond)^\diamond \in \widehat{I}$. Thus $(x^\diamond * a^\diamond)^\diamond \in \beta(\widehat{I})$. Again, let $a, b \in \beta(\widehat{I})$ and $t \in X$. Then it implies that $(a^\diamond, x^\diamond) \in \theta$ and $(b^\diamond, y^\diamond) \in \theta$ for some $[x]_\theta \in \widehat{I}$ and $[y]_\theta \in \widehat{I}$. Since $\widehat{I}$ is an ideal of X/θ, we get the following:

$$[((x^\diamond * (y^\diamond * t^\diamond)) * t^\diamond)^\diamond]_\theta = (([x]_\theta^\diamond * ([y]_\theta^\diamond * [t]_\theta^\diamond)) * [t]_\theta^\diamond \in \widehat{I}.$$

Since $(a^\diamond, x^\diamond) \in \theta$ and $(b^\diamond, y^\diamond) \in \theta$, it is clear that $(((a^\diamond * (b^\diamond * t^\diamond)) * t^\diamond), ((x^\diamond * (y^\diamond * t^\diamond)) * t^\diamond)) \in \theta$. Since $[((x^\diamond * (y^\diamond * t^\diamond)) * t^\diamond)]_\theta \in \widehat{I}$, it infers that $((a^\diamond * (b^\diamond * t^\diamond)) * t^\diamond \in \beta(\widehat{I})$. Therefore by Theorem 7.7.2, it concludes that $\beta(\widehat{I})$ is an ideal of X.
(3) Let I_1, I_2 be two ideals in X such that $I_1 \subseteq I_2$. Let $[x]_\theta \in \alpha(I_1)$. Then, we get $(x^\diamond, y^\diamond) \in \theta$ for some $y \in I_1 \subseteq I_2$. Consequently, we get that $[x]_\theta \in \alpha(I_2)$. Therefore $\alpha(I_1) \subseteq \alpha(I_2)$. Again, let $\widehat{I_1}, \widehat{I_2}$ be two ideals of $X_{/\theta}$ such that $\widehat{I_1} \subseteq \widehat{I_2}$. Suppose $x \in \beta(\widehat{F_1})$. Then, it infers that $(x^\diamond, y^\diamond) \in \theta$ for some $[y]_\theta \in \widehat{F_1} \subseteq \widehat{F_2}$. Hence $y \in \beta(\widehat{I_2})$. Therefore $\beta(\widehat{I_1}) \subseteq \beta(\widehat{I_2})$.
(4) Let I be an ideal of X. For any $x \in I$, we have $(x^\diamond, x^\diamond) \in \theta$. Hence $[x]_\theta \in \alpha(I)$.
(5) Let $\widehat{I}$ be an ideal of X/θ. For any $[x]_\theta \in \widehat{I}$, we have $(x^\diamond, x^\diamond) \in \theta$. Hence we get $x \in \beta(\widehat{I})$. □

Lemma 7.7.4 *Let θ be a congruence on a BE-algebra X. For any ideal I of X, $\alpha\beta\alpha(I) = \alpha(I)$.*

Proof Let $[x]_\theta \in \alpha(I)$. Then we get $(x^\diamond, y^\diamond) \in \theta$ for some $y \in I$. Since $y \in I$, by Lemma 7.7.3(4), we get that $[y]_\theta \in \alpha(I)$. Since $(x, y) \in \theta$ and $[y]_\theta \in \alpha(I)$, we get that $x \in \beta\alpha(I)$. Hence $[x]_\theta \in \alpha\beta\alpha(I)$. Therefore $\alpha(I) \subseteq \alpha\beta\alpha(I)$.

Conversely, let $[x]_\theta \in \alpha\beta\alpha(I)$. Then $(x^\diamond, y^\diamond) \in \theta$ for some $y \in \beta\alpha(I)$. Since $y \in \beta\alpha(I)$, there exists $[a]_\theta \in \alpha(I)$ such that $(y^\diamond, a^\diamond) \in \theta$. Hence $[x]_\theta = [a]_\theta \in \alpha(I)$. Therefore $\alpha\beta\alpha(I) \subseteq \alpha(I)$. □

We now intend to show that the composition $\beta\alpha$ of the two operations is a closure operator on the set $\mathcal{I}(X)$ of all ideals of a BE-algebra X.

Proposition 7.7.5 *For any ideal I of a BE-algebra X, the map $I \longrightarrow \beta\alpha(I)$ is a closure operator on $\mathcal{I}(X)$. That is, for any two ideals I, J of X, we have the following:*

(a) $I \subseteq \beta\alpha(I)$,
(b) $\beta\alpha\beta\alpha(I) = \beta\alpha(I)$,
(c) $I \subseteq J \Rightarrow \beta\alpha(I) \subseteq \beta\alpha(J)$.

Proof (a) Let $x \in I$. Then by Lemma 7.7.3(4), we get that $[x]_\theta \in \alpha(I)$. Since $(x^\diamond, x^\diamond) \in \theta$ and $\alpha(I)$ is an ideal in $X_{/\theta}$, it results in $x \in \beta\alpha(I)$. Therefore $I \subseteq \beta\alpha(I)$.
(b) Since $\beta\alpha(I)$ is an ideal in X, by above condition (a), we get that $\beta\alpha(I) \subseteq \beta\alpha[\beta\alpha(I)]$. Conversely, let $x \in \beta\alpha[\beta\alpha(I)]$. Then we obtain $(x^\diamond, y^\diamond) \in \theta$ for some $[y]_\theta \in \alpha\beta\alpha(I)$. Thus by above Lemma 7.7.3, we get that $[y]_\theta \in \alpha(I)$. Hence $x \in \beta\alpha(I)$. Therefore $\beta\alpha[\beta\alpha(I)] \subseteq \beta\alpha(I)$.
(c) Suppose I and J are two ideals of X such that $I \subseteq J$. Let $x \in \beta\alpha(I)$. Then $[x]_\theta \in \alpha(I)$. Hence, it gives that $[x]_\theta = [y]_\theta$ for some $y \in I \subseteq J$. Since $y \in J$, we get that $[x]_\theta = [y]_\theta \in \alpha(J)$. Hence $x \in \beta\alpha(J)$. Therefore $\beta\alpha(I) \subseteq \beta\alpha(J)$. □

For any BE-algebra X, denote by $\mathcal{I}(X_{/\theta})$ the set of all ideals of $X_{/\theta} = \{[x]_\theta \mid x \in X\}$, we can therefore define a mapping $\alpha : \mathcal{I}(X) \longrightarrow \mathcal{I}(X_{/\theta})$ by $I \mapsto \alpha(I)$ and also another mapping $\beta : \mathcal{I}(X_{/\theta}) \longrightarrow \mathcal{I}(X)$ by $I \mapsto \beta(I)$. Then we have the following proposition.

Proposition 7.7.6 *Let θ be congruence on a bounded BE-algebra X. Then α is a residuated map with residual map β.*

Proof For every $I \in \mathcal{I}(L)$, by Proposition 7.7.5(a), we have $I \subseteq \beta\alpha(I)$. Let $I \in \mathcal{I}(X_{/\theta})$. Suppose $[x]_\theta \in I$. Then $x \in \beta(I)$. Since $\beta(I)$ is a ideal in X, we get $[x]_\theta \in \alpha\beta(I)$. Hence $I \subseteq \alpha\beta(I)$. Conversely, let $[x]_\theta \in \alpha\beta(I)$. Then $[x]_\theta = [y]_\theta$ for some $y \in \beta(I)$. Since $y \in \beta(I)$, we get $[x]_\theta = [y]_\theta \in I$. Hence $\alpha\beta(I) \subseteq J$. Therefore for every $I \in \mathcal{I}(X_{/\theta})$, we get $\alpha\beta(I) = I$. Since α and β are isotone, it follows that α is residuated and that the residual of α is nothing but β. □

We now introduce the notion of θ-ideals in a bounded BE-algebra.

Definition 7.7.7 Let θ be a congruence on X. An ideal I of X is said to be a *θ-ideal* if $\beta\alpha(I) = I$.

For any congruence θ on a bounded BE-algebra X, it can be easily observed that the ideal $\{1\}$ is a θ-ideal if and only if $[1]_\theta = \{1\}$. Moreover, we have the following:

Lemma 7.7.8 *Let θ be a congruence on a bounded BE-algebra X. For any ideal I of X, the following conditions hold:*

(1) *If I is a θ-ideal, then $[0]_\theta \subseteq I$,*
(2) *If I is a proper θ-ideal, then $I \cap [1]_\theta = \emptyset$.*

Proof (1) Let I be a θ-ideal of X. Suppose $x \in [0]_\theta$. Then $(x, 0) \in \theta$. Hence $(x^\diamond, 0^\diamond) \in \theta$ and $0 \in I$. Then $[x]_\theta \in \alpha(I)$. Hence $[x]_\theta = [y]_\theta$ for some $y \in I$. Thus $(x^\diamond, y^\diamond) \in \theta$ where $[y]_\theta \in \alpha(I)$. Hence $x \in \beta\alpha(I) = I$. Therefore $[0]_\theta \subseteq I$.
(2) Suppose $I \cap [1]_\theta \neq \emptyset$. Choose $x \in I \cap [1]_\theta = \emptyset$. Then $x \in I$ and $(x, 1) \in \theta$. Then $(x^\diamond, 1^\diamond) \in \theta$ and $x \in I$. Hence $[1]_\theta \in \alpha(I)$. Thus $[1]_\theta = [y]_\theta$ for some $y \in I$. Thus $(1, y) \in \theta$, which implies that $(1^\diamond, y^\diamond) \in \theta$. Since $[y]_\theta \in \alpha(I)$, we get $1 \in \beta\alpha(I)$ which is a contradiction. □

We now characterize θ-ideals BE-algebra in the following:

Theorem 7.7.9 *Let θ be a congruence on a BE-algebra X. For any ideal I of X, the following conditions are equivalent:*

(1) *I is a θ-ideal,*
(2) *For any $x, y \in X$, $[x]_\theta = [y]_\theta$ and $x \in I$ imply that $y \in I$,*
(3) $I = \bigcup\limits_{x \in I} [x]_\theta$,
(4) *$x \in I$ implies $[x]_\theta \subseteq I$.*

Proof (1) $\Rightarrow$ (2): Assume that I is a θ-ideal of X. Let $x, y \in X$ be such that $[x]_\theta = [y]_\theta$. Then we get $(x, y) \in \theta$. Suppose $x \in I = \beta\alpha(I)$. Then, it yields that $(x^\diamond, a^\diamond) \in \theta$ for some $[a]_\theta \in \alpha(I)$. Thus $(a^\diamond, y^\diamond) \in \theta$ and $[a]_\theta \in \alpha(I)$. Therefore $y \in \beta\alpha(I) = I$.
(2) $\Rightarrow$ (3): Assume the condition (2). Let $x \in I$. Since $x \in [x]_\theta$, we get $I \subseteq \bigcup\limits_{x \in I} [x]_\theta$. Conversely, let $a \in \bigcup\limits_{x \in I} [x]_\theta$. Then it gives that $(a, x) \in \theta$ for some $x \in I$. Hence $[a]_\theta = [x]_\theta$. By the condition (2), we get $a \in I$. Therefore $I = \bigcup\limits_{x \in I} [x]_\theta$.
(3) $\Rightarrow$ (4): Assume the condition (3). Let $a \in I$. Then $(x, a) \in \theta$ for some $x \in I$. Let $t \in [a]_\theta$. Then $(t, a) \in \theta$. Hence $(x, t) \in \theta$. Thus $t \in [x]_\theta \subseteq I$. Therefore $[a]_\theta \subseteq I$.
(4) $\Rightarrow$ (1): Assume the condition (4). Clearly $I \subseteq \beta\alpha(I)$. Conversely, let $x \in \beta\alpha(I)$. Then $(x^\diamond, y^\diamond) \in \theta$ for some $[y]_\theta \in \alpha(I)$. Hence $[x^{\diamond\diamond}]_\theta = [y^{\diamond\diamond}]_\theta$ and $[y^\diamond]_\theta = [a^\diamond]_\theta$ for some $a \in I$. Since $a \in I$, by Proposition 7.5.3(3), we get $a^{\diamond\diamond} \in I$. Then by condition (4), we get that $x^{\diamond\diamond} \in [y^{\diamond\diamond}]_\theta = [a^{\diamond\diamond}]_\theta \subseteq I$. Hence $x \leq x^{\diamond\diamond} \in I$. Thus $\beta\alpha(I) \subseteq I$. Therefore I is a θ-ideal of X. □

Theorem 7.7.10 *Let θ be a congruence on X. Then there exists a bijection between the set $\mathcal{I}_\theta(X)$ of all θ-ideals of X and the set of all ideals of the BE-algebra $X_{/\theta}$ of all congruence classes.*

Proof Define a mapping $\psi : \mathcal{I}_\theta(X) \mapsto \mathcal{I}_\theta(X_{/\theta})$ by $\psi(I) = \alpha(I)$ for all $I \in \mathcal{I}_\theta(X)$. Let $I, J \in \mathcal{I}_\theta(X)$. Then we get the following consequence:

$$\begin{aligned} \psi(I) = \psi(J) &\Rightarrow \alpha(I) = \alpha(J) \\ &\Rightarrow \beta\alpha(I) = \beta\alpha(J) \\ &\Rightarrow I = J \qquad \text{since } I, J \in \mathcal{I}_\theta(X) \end{aligned}$$

Hence ψ is one-to-one. We now show that ψ is onto. Let $\widehat{I}$ be an ideal of $\mathcal{I}(X_{/\theta})$. Then $\beta(\widehat{I})$ is an ideal in X. We now show that $\beta(\widehat{I})$ is a θ-ideal in X. We have always $\beta(\widehat{I}) \subseteq \beta\alpha\beta(\widehat{I})$. Let $x \in \beta\alpha\beta(\widehat{I})$. Then we get $(x^\diamond, y^\diamond) \in \theta$ for some $[y]_\theta \in \alpha\beta(\widehat{I}) = \widehat{I}$. Hence $x \in \beta(\widehat{I})$. Therefore $\beta(\widehat{I}) = \beta\alpha\beta(\widehat{I})$. Now for this $\beta(\widehat{I}) \in X$, we get $\psi[\beta(\widehat{I})] = \alpha\beta(\widehat{I}) = \widehat{I}$. Therefore ψ is onto. Therefore ψ is a bijection between $\mathcal{I}_\theta(X)$ and $\mathcal{I}(X_{/\theta})$. □

7.8 Irreducible Ideals of BE-algebras

In this section, the properties of irreducible ideals of BE-algebras are studied extensively. Some necessary conditions of irreducible ideals and sufficient conditions for an ideal to become an irreducible ideal are derived. An equivalent condition is derived for every proper ideal of a BE-algebra to become an irreducible ideal. Throughout this section, X stands for a partially ordered BE-algebra unless otherwise mentioned.

Definition 7.8.1 A proper ideal I of a BE-algebra is said to be *irreducible* if for any proper ideals A and B of X, $I = A \cap B$ implies $I = A$ or $I = B$.

Example 7.8.2 Let $X = \{1, a, b, c, 0\}$. Define a binary operation $*$ on X as follows:

$*$	1	a	b	c	0
1	1	a	b	c	0
a	1	1	b	b	c
b	1	a	1	a	b
c	1	1	1	1	a
0	1	1	1	1	1

It can be routinely verified that $(X, *, 0, 1)$ is a bounded BE-algebra. All the ideals of the BE-algebra are $\{0\}$, $\{0, b\}$, $\{0, c\}$ and X. It can be easily seen that the ideals $\{0, b\}$ and $\{0, c\}$ are irreducible ideals but the ideal $\{0\}$ is not irreducible, because $\{0\} = \{0, b\} \cap \{0, c\}$ and also neither $\{0\} = \{0, b\}$ nor $\{0\} = \{0, c\}$.

From the above example, it can also be observed that the intersection of irreducible ideals of a BE-algebra need not necessarily be an irreducible ideal. For, consider the irreducible ideals $I_1 = \{0, a\}$ and $I_2 = \{0, b\}$ of the BE-algebra X. Now, their intersection $I_1 \cap I_2 = \{0\}$ is not an irreducible ideal of F.

Theorem 7.8.3 *Let I be a proper ideal of a transitive BE-algebra X. Then I is an irreducible ideal of X if it satisfies the following condition for all $x, y \in X$:*

For $x, y \notin I$, there exists $z \notin I$ such that $z \leq x$ and $z \leq y$.

Proof Let I be a proper ideal of a transitive BE-algebra X. Assume that I satisfies the given condition. We prove this by contradiction. Suppose I is not an irreducible ideal of X. Then there exist two proper ideals A and B of X such that $I = A \cap B$, $I \neq A$, and $I \neq B$. Then clearly $I \subset A$ and $I \subset B$. Let $x \in A - F$ and $y \in B - F$. Then by the assumed condition, there exists $z \in X - I$ such that $z \leq x$ and $z \leq y$.

Since $x \in A, z \leq x$, and A is an ideal of the transitive BE-algebra X, we get $z \in A$. Since $y \in B$ and $z \leq y$, we get $z \in B$. Hence $z \in A \cap B = I$, which is a contradiction. Therefore I is an irreducible ideal of X. □

Theorem 7.8.4 *A proper ideal I of a transitive BE-algebra X is irreducible if and only if for every $x, y \notin I$, there exists $z \notin I$ such that $(x^\diamond * z^\diamond)^\diamond \in I$ and $(y^\diamond * z^\diamond)^\diamond \in I$.*

Proof The necessary condition follows from the above theorem. To prove the sufficient condition, assume that I satisfies the condition. Suppose I is not an irreducible ideal of X. Then there exist two proper ideals A and B of X such that $I = A \cap B$, $I \neq A$, and $I \neq B$. Choose $x \in A - I$ and $y \in B - I$. Then by the assumed condition, there exists $z \in X - I$ such that $(x^\diamond * z^\diamond)^\diamond \in I$ and $(y^\diamond * z^\diamond)^\diamond \in I$. Since $I \subset A$, we get $x, (x^\diamond * z^\diamond)^\diamond \in A$. Since A is an ideal, we get $z \in A$. Since $I \subset B$, we get $y, (y^\diamond * z^\diamond)^\diamond \in B$. Since B is an ideal, we get $z \in B$. Hence $z \in A \cap B = I$, which is a contradiction. Therefore I is an irreducible ideal of X. □

Theorem 7.8.5 *Let I be a proper ideal of a transitive BE-algebra X and $x \in X$. Suppose $x \notin I$. Then there exists an irreducible ideal M of X such that $I \subseteq M$ and $x \notin M$.*

Proof Consider the following collection of ideals of X:

$$\mathcal{IR} = \{J \mid J \text{ is an ideal of } X, I \subseteq J, x \notin J\}.$$

Clearly $I \in \mathcal{IR}$ and hence $\mathcal{IR} \neq \emptyset$. Let $\{J_\alpha\}_{\alpha \in \Delta}$ be a collection of elements of $\mathcal{IR}$ which is forming a chain. Clearly $I \subseteq J_\alpha$ and $x \notin J_\alpha$ for each $\alpha \in \Delta$. Hence $I \subseteq \bigcup_{\alpha \in \Delta} J_\alpha$ and $x \notin \bigcup_{\alpha \in \Delta} J_\alpha$. Thus $\bigcup_{\alpha \in \Delta} J_\alpha$ is an upper bound for $\{J_\alpha\}_{\alpha \in \Delta}$. Therefore, by Zorn's lemma, there exists a maximal element M in $\mathcal{IR}$. Then $I \subseteq M$ and $x \notin M$. We now prove that M is an irreducible ideal of X. Let A and B be two proper ideals of X such that $M = A \cap B$. Suppose $M \neq A$ and $M \neq B$. Then clearly $M \subset A$ and $M \subset B$. Also $I \subseteq M \subset A$ and $I \subseteq M \subset B$. By the maximality of M, we get $x \in A$ and $x \in B$. Hence $x \in A \cap B = M$, which is a contradiction. Therefore M is an irreducible ideal of X such that $I \subseteq M$ and $x \notin M$. □

Corollary 7.8.6 *Let X be a transitive BE-algebra and $0 \neq x \in X$. Then there exists an irreducible ideal M such that $x \notin M$.*

Proof Let $0 \neq x \in X$ and $I = \{0\}$. Clearly I is a proper ideal and $x \notin I$. Then by the main theorem, there exists an irreducible ideal M such that $x \notin M$. □

Theorem 7.8.7 *Let I be a proper ideal of a transitive BE-algebra X. Then*

$$I = \cap\{M \mid M \textit{ is an irreducible ideal of } X \textit{ such that } I \subseteq M\}$$

Proof Clearly $I \subseteq \cap\{M \mid M$ is an irreducible ideal of X such that $I \subseteq M\}$. Conversely, let $x \notin I$. Then by the above theorem, there exists an irreducible ideal M_x such that $I \subseteq M_x$ and $x \notin M_x$. Therefore

$$x \notin \cap\{M \mid M \text{ is an irreducible ideal of } X \text{ such that } I \subseteq M\}.$$

Therefore $\cap\{M \mid M$ is an irreducible ideal of X such that $I \subseteq M\} \subseteq I$. □

The following corollary is a direct consequence of the above results.

Corollary 7.8.8 *If* $\{0\}$ *is an irreducible ideal of a transitive BE-algebra* X*, then the intersection of all irreducible ideals of a BE-algebra is equal to* $\{0\}$.

Theorem 7.8.9 *Let* X *and* Y *be two BE-algebras where* X *is commutative and* $f : X \to Y$ *a homomorphism which maps ideals to ideals. If* I *is an irreducible ideal of* Y *and* $f^{-1}(I) \neq X$*, then* $f^{-1}(I)$ *is an irreducible ideal of* X.

Proof Let I be an irreducible ideal of Y. Then clearly $f^{-1}(I)$ is an ideal of X. Let A and B be two proper ideals of X such that $f^{-1}(I) = A \cap B$. Then $I = f(A \cap B)$. We first prove that $f(A \cap B) = f(A) \cap f(B)$. Clearly $f(A \cap B) \subseteq f(A) \cap f(B)$. Conversely, let $x \in f(A) \cap f(B)$. Then $x = f(a)$ and $x = f(b)$ for some $a \in A$ and $b \in B$. Since A and B are filters of X, we get $(b * a) * a \in A$ and $(a * b) * b \in B$. Since X is commutative, we get $(a * b) * b = (b * a) * a \in A \cap B$. Hence

$$\begin{aligned} x &= (x * x) * x \\ &= (f(a) * f(b)) * f(b) \\ &= f((a * b) * b) \in f(A \cap B). \end{aligned}$$

Hence $f(A) \cap f(B) \subseteq f(A \cap B)$. Thus $I = f(A \cap B) = f(A) \cap f(B)$. Since $f(A)$ and $f(B)$ are proper ideals of Y and I is irreducible in Y, we get $I = f(A)$ or $I = f(B)$. Hence $f^{-1}(I) = A$ or $f^{-1}(I) = B$. Thus $f^{-1}(I)$ is an irreducible ideal of X. □

In the following theorem, a necessary and sufficient condition is derived for the class $\mathcal{I}(X)$ of all ideals of a transitive BE-algebra X to become a totally ordered set.

Theorem 7.8.10 *Let* X *be a transitive BE-algebra. Then* $\mathcal{I}(X)$ *is a totally ordered set or a chain if and only if every proper ideal of* X *is an irreducible ideal.*

Proof Assume that $\mathcal{I}(X)$ is a totally ordered set. Let I be a proper ideal of X. Let A and B be two proper ideals of X such that $I = A \cap B$. Since $\mathcal{I}(X)$ is totally ordered and $A, B \in \mathcal{I}(X)$, we get that either $A \subseteq B$ or $B \subseteq A$. Hence $I = A \cap B = A$ or $I = A \cap B = B$. Therefore I is an irreducible ideal of X.

Conversely assume that every proper ideal of X is an irreducible ideal. Let I and J be two proper ideals of X. Clearly $I \cap J$ is a proper ideal of X and hence by the assumed condition $I \cap J$ is irreducible. Since $I \cap J = I \cap J$, we get

$$I = I \cap J \text{ or } J = I \cap J$$

Hence $I \subseteq J$ or $J \subseteq I$. Therefore $\mathcal{A}(X)$ is a totally ordered set. □

Exercise

1. Let X be a bounded and commutative BE-algebra. If there exists a complement of any element of X, then prove that it is unique.
2. A BE-algebra X is called i-invariant if $x \wedge (x * y) = x \wedge y$ for all $x, y \in X$. Prove that a bounded, transitive, and implicative BE-algebra X is i-invariant.
3. A BE-algebra X is called m-invariant if $x * (x \wedge y) = x * y$ for all $x, y \in X$. Prove that a bounded, transitive, and implicative BE-algebra X is m-invariant.
4. Let $(X, *, 0, 1)$ be a bounded and commutative BE-algebra with condition **L**. Then prove that $X = \mathcal{B}(X)$ and hence X is a Boolean algebra.
5. Let $(X, *, 0, 1)$ be a bounded and commutative BE-algebra. For every $a \in \mathcal{B}(X)$ and $x, y \in X$, prove that $a \vee (x * y) = (a \vee x) * (a \vee y)$.
6. Let $(X, *, 0, 1)$ be a bounded and commutative BE-algebra. For every $a, b \in \mathcal{B}(X)$ and $x \in X$, prove that $(a * b) \vee x = (a \vee x) * (b \vee x)$.
7. For any bounded BE-morphism f on a bounded BE-algebra X, prove that the image of a Boolean element(closed element) of X under f is again a Boolean element (closed element).
8. For any two bounded BE-algebras, X, Y and $f : X \to Y$ a bounded BE-morphism, prove that $f^{-1}(a)$ is a closed element of X whenever a is a closed element of Y. Moreover the inverse image of a spanning subalgebra of Y is a again a spanning subalgebra of X.
9. Let F_1 and F_2 be two filters of bounded BE-algebras X_1 and X_2, respectively. Prove that F_1 and F_2 be two closed filters of the BE-algebras X_1 and X_2, respectively, if and only if $F_1 \times F_2$ is a closed filter of $X_1 \times X_2$.
10. For any proper filter F of a bounded BE-algebra, prove that X is dense if and only if $F \cap \mathcal{D}(X) = F$ if and only if $\{F_x \mid F_{x^{\diamond\diamond}} = F_1\} = \{1\}$.
11. Let A and B be two ideals of a self-distributive BE-algebra X such that $B \subseteq A$, then prove that $x \in A$ if and only if $B_x \in A/B$ where $A/B = \{B_x \mid x \in A\}$.
12. Let X, X' be two bounded BE-algebras and $f : X \to X'$ a epimorphism. If $Dker(f) \subseteq I$, then $f^{-1}(f(I)) = I$.
13. Let I be an ideal and A a non-empty subset of a transitive BE-algebra, then prove that $\langle I \cup A\rangle = \{x \in X \mid a_1^{\diamond} * (\cdots * (a_1^{\diamond} * x^{\diamond})\cdot) \in I$ for some $a_1, a_2, \ldots, a_n \in A\}$.
14. Suppose that $I_x^{\diamond} * I_y^{\diamond} = I_0$ implies $I_x = I_y$ for any ideal I of a self-distributive BE-algebra X. Then prove that $(I \vee J)/J \cong I/(I \cap J)$ for any two ideals I and J of X.

Chapter 8
Stabilizers of BE-algebras

A stabilizer is a part of an algebra acting on a set. Specifically, let X be any algebra operating on a set X and let A be a subset of X. The stabilizer of A, sometimes denoted $St(A)$, is the set of elements a of A for which $a(S) \subseteq S$. The strict stabilizer is the set of $a \in A$ for which $a(A) = A$. In the other words, the stabilizer of A is the transporter of A to itself. The concept of stabilizers is introduced in Hilbert algebras by I. Chajda and R. Halaš [37]. In this paper, the authors studied the properties of stabilizers and relative stabilizers of a given subset of a Hilbert algebra. They proved that every stabilizer of a deductive system C of $\mathcal{H}$ is also a deductive system which is a pseudo-complement of C in the lattice of all deductive systems of $\mathcal{H}$. In [15], A. Borumand Saeid and N. Mohtashamnia constructed quotient of residuated lattices via stabilizer and studied its properties. L. Torkzadeh [232] introduced dual right and dual left stabilizers in bounded BCK-algebras and investigated the relationship between of them.

In this chapter, the notions of stabilizers and relative stabilizers of a subset are generalized in BE-algebras. Some properties of relative stabilizers are studied in BE-algebras. It is proved that every stabilizer of a filter of a commutative BE-algebra is also a filter which is the pseudo-complement of it. The notions of dual stabilizers and relative dual stabilizers of a given subset of a BE-algebra are introduced in terms of pseudo-complements of BE-algebras. It is proved that the dual stabilizer of any ideal of a transitive BE-algebra is again an ideal.

Properties of the stabilizers induced by a given filter are studied. For any given filter, the set of all stabilizers induced by the filters is observed to form a complete distributive lattice. The celebrated Stone's theorem of prime filters is generalized to the case of associative filters of commutative BE-algebras with the help of stabilizers. Topological properties of closure operators are studied in BE-algebras. A base for a topology is observed via left and right stabilizers. It is proved that the above topological space is a Bair space, locally connected and separable space.

S. R. Mukkamala, *A Course in BE-algebras*,
https://doi.org/10.1007/978-981-10-6838-6_8

8.1 Properties of Stabilizers

In this section, the concepts of a stabilizer and a relative stabilizer of a given subset are introduced in BE-algebras. We prove that every stabilizer of a filter of a BE-algebra is also a filter. Some properties of these stabilizers and relative stabilizers of BE-algebras are studied.

Definition 8.1.1 Let A a non-empty subset of a BE-algebra X. Define

$$A_r^{\sim} = \{x \in X \mid x * a \leq a \text{ for all } a \in A\};$$
$$A_l^{\sim} = \{x \in X \mid a * x \leq x \text{ for all } a \in A\}.$$

$A_r^{\sim}$ and $A_l^{\sim}$ are called *right and left stabilizers* of X and denote the stabilizer of X as

$$A^{\sim} = A_r^{\sim} \cap A_l^{\sim}.$$

In the following, we introduce the concept of relative stabilizers of BE-algebras.

Definition 8.1.2 Let A, B be two non-empty subsets of a BE-algebra X. Define

$$(A, B)_r^{\sim} = \{x \in X \mid (x * a) * a \in B \text{ for all } a \in A\};$$
$$(A, B)_l^{\sim} = \{x \in X \mid (a * x) * x \in B \text{ for all } a \in A\}.$$

$(A, B)_r^{\sim}$ and $(A, B)_l^{\sim}$ are called right and left relative stabilizers of A with respect to B and denote the relative stabilizer of A with respect to B as

$$(A, B)^{\sim} = (A, B)_r^{\sim} \cap (A, B)_l^{\sim}.$$

Proposition 8.1.3 *Let X be a commutative BE-algebra and A and B be two non-empty subsets of X. Then the following conditions hold.*

(1) $(A, B)^{\sim} = (A, B)_r^{\sim} = (A, B)_l^{\sim}$,
(2) *If* $B = \{1\}$, *then* $(A, B)_r^{\sim} = A_r^{\sim}$ *and* $(A, B)_l^{\sim} = A_l^{\sim}$,
(3) *If* $B = \{1\}$, *then* $A^{\sim} = A_r^{\sim} = A_l^{\sim}$.

Proof (1) Let $x \in (A, B)_r^{\sim}$. Then we get $(x * a) * a \in B$ for all $a \in A$. Since X is commutative, we get $(a * x) * x \in B$ for all $a \in A$. Hence $x \in (A, B)_l^{\sim}$. Thus $(A, B)_r^{\sim} \subseteq (A, B)_l^{\sim}$. The converse can be obtained similarly. Therefore $(A, B)^{\sim} = (A, B)_r^{\sim} \cap (A, B)_l^{\sim} = (A, B)_r^{\sim} = (A, B)_l^{\sim}$.
(2) Suppose that $B = \{1\}$. Then we get that $(A, B)_r^{\sim} = (A, \{1\})_r^{\sim} = \{x \in X \mid (x * a) * a = 1 \text{ for all } a \in A\} = \{x \in X \mid x * a \leq a \text{ for all } a \in A\} = A_r^{\sim}$. Similarly, we get that $(A, B)_l^{\sim} = A_l^{\sim}$.
(3) From the above two, it is clear. □

In the following lemma, some of the properties of relative stabilizers can be observed.

Lemma 8.1.4 *Let X be a BE-algebra and $\emptyset \neq A, B, C \subseteq X$. Then*

(1) *If $(A, B)_r^{\sim} = X$, then $A \subseteq B$,*
(2) *If $(A, B)_l^{\sim} = X$, then $A \subseteq B$,*
(3) *If $B \subseteq C$, then $(A, B)_r^{\sim} \subseteq (A, C)_r^{\sim}$ and $(A, B)_l^{\sim} \subseteq (A, C)_l^{\sim}$,*
(4) *If $A \subseteq B$, then $(B, C)_r^{\sim} \subseteq (A, C)_r^{\sim}$ and $(B, C)_l^{\sim} \subseteq (A, C)_l^{\sim}$,*
(5) *$(A, B \cap C)_r^{\sim} = (A, B)_r^{\sim} \cap (A, C)_r^{\sim}$ and $(A, B \cap C)_l^{\sim} = (A, B)_l^{\sim} \cap (A, C)_l^{\sim}$,*
(6) *If X is commutative, then $(A, \{1\})_r^{\sim} = A_r^{\sim}$ and $(A, \{1\})_l^{\sim} = A_l^{\sim}$ and hence $(A, \{1\})^{\sim} = A^{\sim}$.*

Proof (1) Let $\emptyset \neq A, B \subseteq X$. Suppose that $(A, B)_r^{\sim} = X$. Then $A \subseteq X = (A, B)_r^{\sim}$. Let $t \in A$ be an arbitrary element. Then we get $t \in (A, B)_r^{\sim}$, which implies that $(t * a) * a \in B$ for all $a \in A$. In particular for $a = t$, it yields that $a = 1 * a = (a * a) * a \in B$. Therefore $A \subseteq B$.
(2) Let $\emptyset \neq A, B \subseteq X$. Suppose that $(A, B)_l^{\sim} = X$. Since $A \subseteq X = (A, B)_l^{\sim}$, we get that $a = (a * a) * a \in B$ for all $a \in A$. Therefore $A \subseteq B$.
(3) Let $\emptyset \neq A, B \subseteq X$. Assume that $B \subseteq C$. Let $x \in (A, B)_r^{\sim}$. Then $(x*a)*a \in B \subseteq C$ for all $a \in A$. Hence $x \in (A, C)_r^{\sim}$. Therefore $(A, B)_r^{\sim} \subseteq (A, C)_r^{\sim}$. Similarly, we get $(A, B)_l^{\sim} \subseteq (A, C)_l^{\sim}$.
(4) Let $\emptyset \neq A, B, C \subseteq X$. Suppose $A \subseteq B$. Let $x \in (B, C)_r^{\sim}$. Then we get that $(x * a) * a \in C$ for all $a \in B$. Since $A \subseteq B$, it yields that $(x * a) * a \in C$ for all $a \in A$. Thus $x \in (A, C)_r^{\sim}$. Therefore $(B, C)_r^{\sim} \subseteq (A, C)_r^{\sim}$. Similarly, we get $(B, C)_l^{\sim} \subseteq (A, C)_l^{\sim}$.
(5) Since $B \cap C \subseteq B, C$, from (3), we get $(A, B \cap C)_r^{\sim} \subseteq (A, B)_r^{\sim}$ and $(A, B \cap C)_r^{\sim} \subseteq (A, C)_r^{\sim}$. Hence $(A, B \cap C)_r^{\sim} \subseteq (A, B)_r^{\sim} \cap (A, C)_r^{\sim}$. Conversely, let $x \in (A, B)_r^{\sim} \cap (A, C)_r^{\sim}$. Then $(x * a) * a \in B$ for all $a \in A$ and $(x * a) * a \in C$ for all $a \in A$. Hence $(x * a) * a \in B \cap C$ for all $a \in A$. Hence $x \in (A, B \cap C)_r^{\sim}$. Therefore $(A, B \cap C)_r^{\sim} = (A, B)_r^{\sim} \cap (A, C)_r^{\sim}$. Similarly, we get $(A, B \cap C)_l^{\sim} = (A, B)_l^{\sim} \cap (A, C)_l^{\sim}$.
(6) From the above observation, it is clear. □

In case of filters, we have some more conditions.

Lemma 8.1.5 *Let F be a filter of a BE-algebra X and $\emptyset \neq A \subseteq X$. Then we have*

(1) *If $(F, A)_r^{\sim} = (F, A)_l^{\sim} = X$ if and only if $F \subseteq A$,*
(2) *$(F, F)_r^{\sim} = X$,*
(3) *$A_r^{\sim} \subseteq (A, F)_r^{\sim}$ and $A_l^{\sim} \subseteq (A, F)_l^{\sim}$,*
(4) *$(\{1\}, F)_r^{\sim} = X$ and $(\{1\}, F)_l^{\sim} = X$, hence $(\{1\}, F)^{\sim} = X$,*
(5) *$F \subseteq \bigcup_{a \in F} \{a\}_r^{\sim} = X$ and $F \subseteq \bigcup_{a \in F} \{a\}_l^{\sim} = X$.*

Proof (1) Assume that $F \subseteq A$. Let $x \in X$ and $a \in F$ be arbitrary elements. Since $a \in F, a \leq (x * a) * a$ and F is a filter, it yields that $(x * a) * a \in F \subseteq A$ for all $a \in F$. Hence $x \in (F, A)_r^{\sim}$, which is true for all $x \in X$. Thus, we get $X \subseteq (F, A)_r^{\sim}$. Hence $(F, A)_r^{\sim} = X$. Similarly, we get $(F, A)_l^{\sim} = X$. The converse is clear by above Lemma 8.1.4(1).
(2) It is clear by (1), since F is a filter and $F \subseteq F$.
(3) Let $x \in A_r^{\sim}$. Then $x * a \leq a$ for all $a \in A$. Since F is a filter, we get $(x * a) * a =$

$1 \in F$ for all $a \in A$. Hence $x \in (A, F)_r^{\sim}$. Therefore $A_r^{\sim} \subseteq (A, F)_r^{\sim}$. Similarly, we can obtain $A_l^{\sim} \subseteq (A, F)_l^{\sim}$.
(4) Let $x \in X$. Since F is a filter, we get $(x * 1) * 1 = 1 \in F$. Hence $x \in (\{1\}, F)_r^{\sim}$, which implies $X \subseteq (\{1\}, F)_r^{\sim}$. Therefore $(\{1\}, F)_r^{\sim} = X$. Again, let $x \in X$. Since F is a filter, we get $(1 * x) * x = 1 \in F$. Hence $x \in (\{1\}, F)_l^{\sim}$, which implies $X \subseteq (\{1\}, F)_l^{\sim}$. Therefore $(\{1\}, F)_l^{\sim} = X$. By combining the above two results, we get that $(\{1\}, F)^{\sim} = (\{1\}, F)_r^{\sim} \cap (\{1\}, F)_l^{\sim} = X$.
(5) Let $x \in F$. Since $x * 1 = 1$ and $1 \in F$, we get that $x \in \{1\}_r^{\sim} \subseteq \bigcup\limits_{a \in F} \{a\}_r^{\sim}$. Hence $F \subseteq \bigcup\limits_{a \in F} \{a\}_r^{\sim}$. Let $x \in X$. Since $x * 1 = 1$, it yields that $x \in \{1\}_r^{\sim} \subseteq \bigcup\limits_{a \in F} \{a\}_r^{\sim}$. Hence $\bigcup\limits_{a \in F} \{a\}_r^{\sim} = X$. Therefore $F \subseteq \bigcup\limits_{a \in F} \{a\}_r^{\sim} = X$. Similarly, we obtain $F \subseteq \bigcup\limits_{a \in F} \{a\}_l^{\sim} = X$. □

Lemma 8.1.6 *Let X be a BE-algebra and $\{A_i\}_{i \in I}$ and $\{B_i\}_{i \in I}$ be two family of subsets of X such that $\bigcap\limits_{i \in I} A_i \neq \emptyset$ and $\bigcap\limits_{i \in I} B_i \neq \emptyset$. Then the following conditions hold.*

(1) $\bigcap\limits_{i \in I} (A_i, B_i)_r^{\sim} = (\bigcap\limits_{i \in I} A_i, \bigcap\limits_{i \in I} B_i)_r^{\sim}$,
(2) $\bigcup\limits_{i \in I} (A_i, B_i)_r^{\sim} \subseteq (\bigcap\limits_{i \in I} A_i, \bigcap\limits_{i \in I} B_i)_r^{\sim}$,
(3) $(\bigcup\limits_{i \in I} A_i, \bigcup\limits_{i \in I} B_i)_r^{\sim} \subseteq (\bigcap\limits_{i \in I} A_i, \bigcup\limits_{i \in I} B_i)_r^{\sim}$.

Proof (1) Let $\{A_i\}_{i \in I}$ and $\{B_i\}_{i \in I}$ be as in the statement. Then, we get $x \in \bigcap\limits_{i \in I} (X_i, Y_i)_r^{\sim} \Leftrightarrow x \in (X_i, Y_i)_r^{\sim}$ for all $i \in I \Leftrightarrow (x * a) * a \in Y_i$ for all $a \in X_i$ and $i \in I \Leftrightarrow (x * a) * a \in \bigcap\limits_{i \in I} Y_i$ for all $y \in \bigcap\limits_{i \in I} X_i \Leftrightarrow x \in (\bigcap\limits_{i \in I} X_i, \bigcap\limits_{i \in I} Y_i)_r^{\sim}$. Therefore $\bigcap\limits_{i \in I} (A_i, B_i)_r^{\sim} = (\bigcap\limits_{i \in I} A_i, \bigcap\limits_{i \in I} B_i)_r^{\sim}$.
(2) Let $x \in \bigcup\limits_{i \in I} (X_i, Y_i)_r^{\sim}$. Then, we get that $x \in (X_i, Y_i)_r^{\sim}$ for some $i \in I$. Hence there exists $i \in I$ such that $(x * a) * a \in Y_i$ for all $a \in X_i$. Therefore $(x * a) * a \in \bigcup\limits_{i \in} Y_i$, for all $a \in \bigcap\limits_{i \in I} X_i$. Hence $x \in (\bigcap\limits_{i \in I} X_i, \bigcup\limits_{i \in I} Y_i)_r^{\sim}$ for all $i \in I$. Therefore $\bigcup\limits_{i \in I} (A_i, B_i)_r^{\sim} \subseteq (\bigcap\limits_{i \in I} A_i, \bigcap\limits_{i \in I} B_i)_r^{\sim}$.
(3) Let $x \in (\bigcup\limits_{i \in I} X_i, \bigcup\limits_{i \in I} Y_i)_r^{\sim}$. Then, for all $a \in \bigcup\limits_{i \in I} X_i$, we get that $(x * a) * a \in \bigcup\limits_{i \in I} Y_i$. Hence, for all $a \in \bigcap\limits_{i \in I} X_i$, we get that $(x * a) * a \in \bigcup\limits_{i \in I} Y_i$. Therefore $x \in (\bigcap\limits_{i \in I} A_i, \bigcup\limits_{i \in I} B_i)_r^{\sim}$. □

Lemma 8.1.7 *Let X be a BE-algebra and $\emptyset \neq A, B \subseteq X$. Then the following hold.*

(1) *If $A \subseteq B$, then $B^{\sim} \subseteq A^{\sim}$,*
(2) $(A \cup B)^{\sim} = A^{\sim} \cap B^{\sim}$,
(3) $X^{\sim} = \{1\}$ *and* $\{1\}^{\sim} = X$,
(4) $A^{\sim} = \bigcap\limits_{x \in A} \{x\}^{\sim}$,
(5) *If $h : X \to X$ be a homomorphism and $a \in X$, then $h(\{a\}^{\sim}) \subseteq \{h(a)\}^{\sim}$.*

Proof (1) Suppose $A \subseteq B$. Let $x \in B^{\sim} = B_r^{\sim} \cap B_l^{\sim}$. Then we get $x * a \leq a$ and $a * x \leq x$ for all $a \in B$. Hence, we get $x * a \leq a$ and $a * x \leq x$ for all $a \in A$. Therefore $x \in A_r^{\sim} \cap A_l^{\sim} = A^{\sim}$.
(2) Since $A, B \subseteq A \cup B$, we get from (1) that $(A \cup B)^{\sim} \subseteq A^{\sim} \cap B^{\sim}$. Conversely, let $x \in A^{\sim} \cap B^{\sim}$. Then $x \in A_r^{\sim} \cap A_l^{\sim}$ and $x \in B_r^{\sim} \cap B_l^{\sim}$. Hence $x * a \leq a$ and $a * x \leq x$ for all $a \in A$. Also $x * a \leq a$ and $a * x \leq x$ for all $a \in B$. Thus it yields $x \in (A \cup B)^{\sim}$. Therefore $A_r^{\sim} \cap B_r^{\sim} \subseteq (A \cup B)^{\sim}$.
(3) Let $x \in X^{\sim}$. Then $x \in X_r^{\sim} \cap X_l^{\sim}$. Hence $x * a \leq a$ and $a * x \leq x$ for all $a \in X$. Since $x \in X$, we get $1 = x * x \leq x$. Thus $x = 1$, which implies $X^{\sim} \subseteq \{1\}$. For any $x \in X$, we have $1 * x = x$ and $x * 1 = 1$. Hence $1 \in X_r^{\sim} \cap X_l^{\sim} = X^{\sim}$. Therefore $X^{\sim} = \{1\}$.

Let $x \in X$ be an arbitrary element. Since $1 * x = x$ and $x * 1 = 1$, it yields $x \in \{1\}_r^{\sim} \cap \{1\}_l^{\sim} = \{1\}^{\sim}$. Hence, we get that $X \subseteq \{1\}^{\sim}$. On the other hand, clearly $\{1\}^{\sim} \subseteq X$. Therefore $\{1\}^{\sim} = X$.
(4) Let $x \in A^{\sim} = A_r^{\sim} \cap A_l^{\sim}$. Then, we get the following consequence:

$$\begin{aligned} x \in A_r^{\sim} \cap A_l^{\sim} &\Leftrightarrow x * a \leq a \text{ and } a * x \leq x \text{ for all } a \in A \\ &\Leftrightarrow x \in \{a\}_r^{\sim} \text{ and } x \in \{a\}_l^{\sim} \text{ for all } a \in A \\ &\Leftrightarrow x \in \bigcap_{a \in A} \{a\}_r^{\sim} \text{ and } x \in \bigcap_{a \in A} \{a\}_l^{\sim} \\ &\Leftrightarrow x \in \Big(\bigcap_{a \in A} \{a\}_r^{\sim}\Big) \cap \Big(\bigcap_{a \in A} \{a\}_l^{\sim}\Big) \\ &\Leftrightarrow x \in \bigcap_{a \in A} \{a\}^{\sim} \end{aligned}$$

It concludes from the above observation that $A^{\sim} = \bigcap_{x \in A} \{x\}^{\sim}$.
(5) Let $a \in X, h : X \to X$ be a homomorphism and $y \in h(\{a\}^{\sim})$. Then there exists $x \in \{a\}^{\sim}$ such that $y = h(x)$. Hence $x * a \leq a$ and $a * x \leq x$. Since h is a homomorphism, we get $h(x) * h(a) \leq h(a)$ and $h(a) * h(x) \leq h(x)$. Hence $y = h(x) \in \{h(a)\}^{\sim}$. Therefore $h(\{a\}^{\sim}) \subseteq \{h(a)\}^{\sim}$. □

Theorem 8.1.8 *Let F and G be two filters of a self-distributive BE-algebra X. Then $(F, G)_r^{\sim}$ is a filter of X such that $F_r^{\sim} \subseteq (F, G)_r^{\sim}$.*

Proof Let F and G be two filters of a self-distributive BE-algebra X. Since $(1 * x) * x = 1 \in G$ for all $x \in F$, we get that $1 \in (F, G)_r^{\sim}$. Let $x, x * y \in (F, G)_r^{\sim}$. Let $a \in F$ be an arbitrary element. Since F is a filter, we get $x * a \in F$ for all $x \in X$. Since $x * y \in (F, G)_r^{\sim}$, we get $((x * y) * (x * a)) * (x * a) \in G$. Since X is self-distributive, we get the following consequence.

$$
\begin{aligned}
(y * a) * (x * a) &= x * ((y * a) * a) \\
&= (x * (y * a)) * (x * a) \\
&= ((x * y) * (x * a)) * (x * a) \in G
\end{aligned} \quad (*)
$$

Since $x \in (F, G)_r^{\sim}$ and $a \in F$, we get $(x * a) * a \in G$. Now, we have the following:

$$
\begin{aligned}
(x * a) * a &\leq (y * a) * ((x * a) * a) \\
&= ((y * a) * (x * a)) * ((y * a) * a)
\end{aligned}
$$

Since $(x*a)*a \in G$ and G is a filter of X, we get that $((y*a)*(x*a))*((y*a)*a) \in G$. Since $(y * a) * (x * a) \in G$, because of the above condition $(*)$, it yields that $(y*a)*a \in G$. Hence $y \in (F, G)_r^{\sim}$. Therefore $(F, G)_r^{\sim}$ is a filter of X. Let $a \in F_r^{\sim}$. Then $a * x \leq x$ for all $x \in F$. Since G is a filter, we get $(a * x) * x = 1 \in G$. Hence $a \in (F, G)_r^{\sim}$. Therefore $F_r^{\sim} \subseteq (F, G)_r^{\sim}$. □

8.2 Filters and Stabilizers

In this section, some properties of stabilizers are studied in the lattice of filters. It is proved that every stabilizer of a filter of a commutative BE-algebra is also a filter which is a pseudo-complement of the filter. We first observe the following result which is required in the later text.

Proposition 8.2.1 *Let X be a transitive BE-algebra. Then, for any $\emptyset \neq A \subseteq X$, the stabilizer $A_r^{\sim}$ is a filter of X. Moreover, $A \cap A_r^{\sim} = \{1\}$ for any $\emptyset \neq A \subseteq X$.*

Proof Since $1 * a = a$ for all $a \in A$, we get that $1 \in A_r^{\sim}$. Now, let $x, x * y \in A_r^{\sim}$. Then $x * a \leq a$ and $(x * y) * a \leq a$ for all $a \in A$. Since X is a transitive BE-algebra, we get

$$
\begin{aligned}
y * a &\leq (x * y) * (x * a) \\
&= x * ((x * y) * a) \\
&\leq x * a \\
&\leq a \qquad \text{for all } a \in A.
\end{aligned}
$$

Thus $y \in A_r^{\sim}$. Therefore $A_r^{\sim}$ is a filter of X. Let $x \in A \cap A_r^{\sim}$. Then $x \in A$ and $x \in A_r^{\sim}$. Hence $1 = x * x \leq x$. Therefore $A \cap A_r^{\sim} = \{1\}$. □

Proposition 8.2.2 *Let X be a transitive BE-algebra and $\emptyset \neq A \subseteq X$. Then*

(1) *If $1 \in A$, then $A \cap A_r^{\sim} = \{1\}$, otherwise, $A \cap A_r^{\sim} = \emptyset$,*
(2) *$A \subseteq (A_r^{\sim})_l^{\sim}$ and $A \subseteq (A_l^{\sim})_r^{\sim}$,*
(3) *$A_r^{\sim} = ((A_r^{\sim})_l^{\sim})_r^{\sim}$ and $A_l^{\sim} = ((A_l^{\sim})_r^{\sim})_{lr}^{\sim}$.*

Proof (1) Suppose $1 \in A$. Since $A_r^{\sim}$ is a filter, we get that $1 \in A_r^{\sim}$. Hence, it yields that $1 \in A \cap A_r^{\sim}$. Now, let $x \in A \cap A_r^{\sim}$. Then we get that $1 = x * x \leq x$. Hence $A \cap A_r^{\sim} = \{1\}$. Otherwise, suppose that $A \cap A_r^{\sim} \neq \emptyset$. Choose $x \in A \cap A_r^{\sim} = \emptyset$. Then, we have just observed that $x = 1$. Hence $1 = x \in A$, which is a contradiction. Therefore $A \cap A_r^{\sim} = \emptyset$.
(2) Since $1 \in A_r^{\sim}$, it is obvious that $A_r^{\sim} \neq \emptyset$. Let $x \in A$. Then we get that $a * x \leq x$ for any $a \in A_r^{\sim}$. Hence $x \in (A_r^{\sim})_l^{\sim}$. Therefore $A \subseteq (A_r^{\sim})_l^{\sim}$. The other result is proved similarly.
(3) Since $(A_r^{\sim})_l^{\sim}$ is non-empty, it is clear that $((A_r^{\sim})_l^{\sim})_r^{\sim}$ is well-defined. By the condition (2), we get that $((A_r^{\sim})_l^{\sim})_r^{\sim} \subseteq A_r^{\sim}$. On the other hand, if $x \in A_r^{\sim}$, then we get that $x * a \leq a$ for all $a \in (A_r^{\sim})_l^{\sim}$. Hence $x \in ((A_r^{\sim})_l^{\sim})_r^{\sim}$. Thus $A_r^{\sim} \subseteq ((A_r^{\sim})_l^{\sim})_r^{\sim}$. Therefore $A_r^{\sim} = ((A_r^{\sim})_l^{\sim})_r^{\sim}$. The other is proved similarly. □

In the following example, it can be observed that $A_l^{\sim}$ is not a filter for every $\emptyset \neq A \subseteq X$.

Example 8.2.3 Let $X = \{1, a, b, c\}$. Define an operation $*$ on X as follows:

$*$	1	a	b	c
1	1	a	b	c
a	1	1	1	1
b	1	a	1	1
c	1	a	b	1

Then $(X, *, 1)$ is a BE-algebra. Consider $F = \{1, c\}$. Clearly F is a filter of X and $F_l^{\sim} = \{1, a, b\}$. Observed that $F_l^{\sim}$ is not a filter of X, because of $a \in F_l^{\sim}$ and $a * c = 1 \in F_l^{\sim}$ but $c \neq F_l^{\sim}$.

However, in the following theorem, a sufficient condition is derived for $A_l^{\sim}$ of a BE-algebra to become a filter for every non-empty subset A of X. If X is a commutative BE-algebra, then $A_l^{\sim} = \{x \in X \mid a * x = x \text{ for all } a \in A\}$ and $A_r^{\sim} = \{x \in X \mid x * a = a \text{ for all } a \in A\}$.

Theorem 8.2.4 *Let X be a commutative BE-algebra. Then $A_l^{\sim}$ is a filter for every $\emptyset \neq A \subseteq X$.*

Proof Since $x * 1 = 1$ for all $x \in A$, we get $1 \in A_l^{\sim}$. Let $x \in A_l^{\sim}$ and $x * y \in A_l^{\sim}$. Then $a * x = x$ and $a * (x * y) = x * y$ for all $a \in A$. Then $a * y \leq x * (a * y) = a * (x * y) = x * y$. Hence

$$\begin{aligned}(y * a) * (x * a) &= x * ((y * a) * a) \\ &= x * ((a * y) * y) \\ &= (a * y) * (x * y) \\ &= 1.\end{aligned}$$

Thus $y * a \leq x * a = a$. Hence $a * y \leq y$. Thus $y \in A_l^{\sim}$, which implies $A_l^{\sim}$ is a filter of X. □

Theorem 8.2.5 *Let F be a filter of a transitive BE-algebra $(X, *, 1)$. Then $F^{\sim}$ is a filter of X.*

Proof Let F be a filter of X. By Proposition 8.2.1, we get that $F_r^{\sim}$ is a filter of X. We first prove that $F_r^{\sim} \subseteq F_l^{\sim}$. Let $x \in F_r^{\sim}$. Let $a \in F$ be an arbitrary element. Since $a * ((a * x) * x) = (a * x) * (a * x) = 1$, we get $a \leq (a * x) * x$. Since F is a filter and $a \in F$, it yields $(a * x) * x \in F$. Since $x \in F_r^{\sim}$ and $(a * x) * x \in F$, we get that

$$x * ((a * x) * x) = (a * x) * x$$

As the left side of the above equation is 1, it implies $(a*x)*x = 1$. Hence $(a*x) \leq x$ for all $a \in F$. Thus $x \in F_l^{\sim}$. Therefore $F_r^{\sim} \subseteq F_l^{\sim}$. Now $F^{\sim} = F_r^{\sim} \cap F_l^{\sim} = F_r^{\sim}$, which is a filter of X. □

Proposition 8.2.6 *Let $(X, *, 1)$ be a commutative BE-algebra. If FandG are two filters of X, then*

(1) $F \cap G = \{(b * a) * a \mid a \in F$ *and* $b \in G\}$,
(2) $F \cap G = \{1\}$ *if and only if* $b * a = a$ *for all* $a \in F$ *and* $b \in G$,
(3) $F \cap G = \{1\}$ *implies* $F \subseteq G_l^{\sim}$.

Proof (1) Consider $A = \{(b*a)*a \mid a \in F$ and $b \in G\}$. Let $x \in A$. Then $x = (b*a)*a$ for some $a \in F$ and $b \in G$. Since F is a filter and $a \in F$, it is obvious that $x \in F$. Since $x = (b * a) * a = (a * b) * a \in G$ because of $b \in G$. Hence $x \in F \cap G$. Therefore $A \subseteq F \cap G$. Conversely, let $x \in F \cap G$. Take $a = x = y$. Then, it implies $x = 1 * x = (x * x) * x$. Hence, it gives $x \in A$. Thus $F \cap G \subseteq A$. Therefore $F \cap G = \{(b * a) * a \mid a \in F$ and $b \in G\}$.
(2) Suppose $F \cap G = \{1\}$. Let $a \in F$ and $b \in G$. Then from (1), we get $(b * a) * a \in F \cap G = \{1\}$. Hence $b * a \leq a$. Since $a \leq b * a$ and X is commutative, it yields that $b * a = a$ for all $a \in F$ and $b \in G$. Conversely, let $b * a = a$ for all $a \in F$ and $b \in G$. Let $x \in F \cap G$. Then, we get $x = (b * a) * a = 1$ for all $a \in F$ and $b \in G$. Therefore $x = 1$, which yields that $F \cap G = \{1\}$.
(3) Suppose that $F \cap G = \{1\}$. Let $x \in F$. Then for any $a \in G$, it is clear that $x \leq (a * x) * x$. Since $x \in F$, we get that $(a * x) * x \in F$. Since $F \cap G = \{1\}$, from condition (2), it yields $a * x = x$. Hence $(a * x) * x = 1 \in G$. Thus we get that $(a * x) * x \in F \cap G = \{1\}$. Hence $a * x \leq x$, which implies that $x \in G_l^{\sim}$. Therefore $F \subseteq G_l^{\sim}$. □

It was already observed in Chap. 2 that the class $\mathcal{F}(X)$ of all filters of a transitive BE-algebra X forms a complete lattice in which, for any two filters F and G, the infimum is $F \cap G$ and suprimum is $F \vee G = \langle F \cup G \rangle$. Now, we prove the following theorem.

Theorem 8.2.7 *Let F be a filter of a commutative BE-algebra $(X, *, 1)$. Then the stabilizer $F_r^{\sim}$ is also a filter of X, and it is a pseudo-complement of F in the lattice $\mathcal{F}(X)$.*

Proof Since every commutative BE-algebra is transitive, we get that $\mathcal{F}(X)$ is a lattice. By using Proposition 8.2.1, we get immediately $F_r^{\sim}$ is a filter of X. Clearly $F \cap F_r^{\sim} = \{1\}$. Suppose G is a filter of X such that $F \cap G = \{1\}$. Let $x \in G$. Then by above proposition, we get $x * a = a$ for each $a \in F$. Hence $x \in F_r^{\sim}$, which implies that $G \subseteq F_r^{\sim}$. Thus $F_r^{\sim}$ is the smallest filter of X such that $F \cap F_r^{\sim} = \{1\}$. Therefore $F_r^{\sim}$ is the pseudo-complement of F in the lattice $\mathcal{F}(X)$ of filters. □

Theorem 8.2.8 *Let F and G be two filters of a commutative BE-algebra $(X, *, 1)$. Then the relative stabilizer $(F, G)_r^{\sim}$ is also a filter of X, and it is the relative pseudo-complement of F with respect to G in the lattice $\mathcal{F}(X)$ of all filters of X.*

Proof Since every commutative BE-algebra is a transitive BE-algebra, we get that $\mathcal{F}(X)$ is a lattice. By using Theorem 8.1.8, we immediately infer that $(F, G)_r^{\sim}$ is a filter of X. To show that $(F, G)_r^{\sim}$ is the relative pseudo-complement of F with respect to G, it is enough to show that $(F, G)_r^{\sim}$ is the greatest element of $\mathcal{F}(X)$ such that $F \cap (F, G)_r^{\sim} \subseteq G$. Let $x \in F \cap (F, G)_r^{\sim}$. Then, it gives that $(x * a) * a \in G$ for all $a \in F$. In particular for $x \in F$, we get $x = 1 * x = (x * x) * x \in G$. Hence $F \cap (F, G)_r^{\sim} \subseteq G$. Let P be a filter of X such that $F \cap P \subseteq G$. Let $x \in P$. Then $(x * a) * a \in F \cap P \subseteq G$ for all $a \in F$. Hence $(x * a) * a \in G$ for all $a \in F$. Therefore $x \in (F, G)_r^{\sim}$, which yields that $(F, G)_r^{\sim}$ is the largest filter of X such that $F \cap (F, G)_r^{\sim} \subseteq G$. □

The intension of the following result is to show that a set and the filter generated by this set have the same left stabilizers in BE-algebras. For any non-empty subset A of a transitive BE-algebra $(X, *, 1)$, properties of the filter $\langle A \rangle = \{x \in X \mid a_n * (\cdots * (a_1 * x) \cdots) = 1$ for some $a_1, \ldots, a_n \in A\}$ generated by the set A are extensively studied in Sect. 2.1. For $A = \{a\}$, we simply denoted $\langle \{a\} \rangle = \{x \in X \mid a^n * x = 1$ for some $n \in \mathbb{N}\}$ by $\langle a \rangle$ and we called this filter a principal filter. Keeping in view of these facts, the following theorem is constructed.

Theorem 8.2.9 *Let A be a non-empty subset of a transitive BE-algebra X. Then $A_l^{\sim} = \langle A \rangle_l^{\sim}$ and $A_l^{\sim} \cap \langle A \rangle = \{1\}$.*

Proof Since $A \subseteq \langle A \rangle$, it is clear that $\langle A \rangle_l^{\sim} \subseteq A_l^{\sim}$. Conversely, let $x \in A_l^{\sim}$. Then $a * x \leq x$ for all $a \in A$. Let $b \in \langle A \rangle$. Then there exists $a_1, a_2, \ldots, a_n \in A$ for some $n \in \mathbb{N}$ such that

$$a_n * (a_{n-1} * (\cdots * (a_1 * b) \cdots)) = 1 \tag{8.1}$$

Since $a_i \in A$ for $i = 1, 2, \ldots, n$, we get that $a_i * x \leq x$ for $i = 1, 2, \ldots, n$. Hence

$$\begin{aligned} a_1 * x \leq x &\Rightarrow a_2 * (a_1 * x) \leq a_2 * x \leq x \\ &\Rightarrow a_3 * (a_2 * (a_1 * x)) \leq a_3 * x \leq x \\ &\ldots \\ &\ldots \\ &\Rightarrow a_n * (a_{n-1} * (\cdots * (a_1 * x) \cdots)) \leq x \end{aligned} \tag{8.2}$$

From the above (8.1) and (8.2), we immediately infer the following:

$$\begin{aligned}(a_n * (\cdots * (a_1 * b) \cdots)) * (a_n * (\cdots * (a_1 * x) \cdots)) &= 1 * (a_n * (\cdots * (a_1 * x) \cdots)) \\ &\leq 1 * x \\ &= x \end{aligned} \tag{8.3}$$

Since X is a transitive BE-algebra, we get the following consequence.

$$\begin{aligned} b * x &\leq (a_1 * b) * (a_1 * x) \\ &\leq (a_2 * (a_1 * b)) * (a_2 * (a_1 * x)) \\ &\ldots \\ &\ldots \\ &\leq (a_n * (\cdots * (a_1 * b) \cdots)) * (a_n * (\cdots * (a_1 * x) \cdots)) \\ &\leq x \end{aligned}$$

Hence $y \in \langle A \rangle_l^{\sim}$. Thus $A_l^{\sim} \subseteq \langle A \rangle_l^{\sim}$. Therefore $A_l^{\sim} = \langle A \rangle_l^{\sim}$. □

In the following, it is observed that the above fact is not true in the case of right stabilizers.

Example 8.2.10 Let $X = \{1, a, b, c, d\}$. Define an operation $*$ on X as follows:

$*$	1	a	b	c	d
1	1	a	b	c	d
a	1	1	b	c	b
b	1	a	1	b	a
c	1	a	1	1	a
d	1	1	1	b	1

Then clearly $(X, *, 1)$ is a BE-algebra. Consider $A = \{a, c\}$. Then clearly $\langle A \rangle = \{1, a, b, c\}$. It is easy to check that $A_r^{\sim} = \{1, b\}$ and $\langle A \rangle_r^{\sim} = X_r^{\sim} = \{1\}$. Therefore $A_r^{\sim} \neq \langle A \rangle_r^{\sim}$.

However, in the following theorem, an equivalent condition is derived in a BE-algebra under which a set and the filter generated by this set have the same right stabilizers.

Theorem 8.2.11 *Let $(X, *, 1)$ be a transitive BE-algebra. Then for any non-empty subset A of X, $A_r^{\sim} = \langle A \rangle_r^{\sim}$ if and only if $x * y \leq y \Leftrightarrow y * x \leq x$ for each $x, y \in X$.*

Proof Assume that $A_r^{\sim} = \langle A \rangle_r^{\sim}$ for each $\emptyset \neq A \subseteq X$. Let $x, y \in X$. Suppose $x * y \leq y$. Hence we get $x \in \{y\}_r^{\sim}$. Then by the assumed condition, we get that $x \in \langle y \rangle_r^{\sim}$ where $\langle y \rangle = \{t \in X \mid y^n * t = 1 \text{ for some } n \in \mathbb{N}\}$. Since $y^n * ((y * x) * x) = y^{n-1} * ((y * x) * (y * x)) = y^{n-1} * 1 = 1$ for any $n \in \mathbb{N}$, we get that $(y * x) * x \in \langle y \rangle$. Since $x \in \langle y \rangle_r^{\sim}$ and $(y * x) * x \in \langle y \rangle$, it yields

$$1 = (y * x) * 1 = (y * x) * (x * x) = x * ((y * x) * x) = (y * x) * x.$$

Thus $y * x \leq x$. The converse can be obtained by interchanging x and y in the above proof.

Conversely assume that $x * y \leq y \Leftrightarrow y * x \leq x$ for each $x, y \in X$. Let $x, y \in X$. Then by the definition of the right stabilizer, we get the following observation.

$$x \in \{y\}_r^{\sim} \text{ if and only if } y \in \{x\}_r^{\sim} \tag{8.4}$$

Let A be a non-empty subset of X. Suppose A is a singleton set, say $A = \{c\}$. Since $\{c\} \subseteq \langle c\rangle$, we get $\langle c\rangle_r^{\sim} \subseteq \{c\}_r^{\sim}$. Conversely, let $x \in \{c\}_r^{\sim}$. Then by the above observation, we get $c \in \{x\}_r^{\sim}$. Hence $\langle c\rangle \subseteq \{x\}_r^{\sim}$. Suppose $t \in \langle c\rangle$. Then $t \in \{x\}_r^{\sim}$. Hence $x \in \{t\}_r^{\sim}$ for all $t \in \langle c\rangle$. Thus

$$x \in \bigcap_{t \in \langle c\rangle} \{t\}_r^{\sim} = \langle c\rangle_r^{\sim}$$

Hence $\{c\}_r^{\sim} \subseteq \langle c\rangle_r^{\sim}$. Therefore $A_r^{\sim} = \langle A\rangle_r^{\sim}$ for every singleton subset A of X. Let A be any non-empty subset of X. Clearly $\langle A\rangle_r^{\sim} \subseteq A_r^{\sim}$. Then it is obvious that

$$A_r^{\sim} = \bigcap_{a \in A} \{a\}_r^{\sim} = \bigcap_{a \in A} \langle a\rangle_r^{\sim}.$$

If $x \in \langle a\rangle_r^{\sim}$ for each $a \in A$, then by the assumed condition, it yields that $a \in \{x\}_r^{\sim}$. Hence $\langle A\rangle \subseteq \{x\}_r^{\sim}$. Hence $x \in (\{x\}_r^{\sim})_r^{\sim} \subseteq \langle A\rangle_r^{\sim}$. Therefore it concludes that $A_r^{\sim} \subseteq \langle A\rangle_r^{\sim}$. □

Theorem 8.2.12 *Let* $(X, *, 1)$ *be a self-distributive BE-algebra which satisfies the condition of the above theorem, then* $\{a\}_l^{\sim}$ *is a filter of* X *for each* $a \in X$.

Proof Assume that X satisfies the condition of the above theorem. Let $a \in X$ be an arbitrary element. Clearly $1 \in \{a\}_l^{\sim}$. Let $x, x * y \in \{a\}_l^{\sim}$. Then $a * x \leq x$ and $a*(x*y) \leq x*y$. Then by the assumed condition, we get $x*a \leq a$ and $(x*y)*a \leq a$. Hence

$$\begin{aligned} y * a &\leq x * (y * a) \\ &= (x * y) * (x * a) && \text{since } X \text{ is self-distributive} \\ &\leq (x * y) * a && \text{since } X \text{ is transitive} \\ &\leq a \end{aligned}$$

By the assumed condition, we get $a * y \leq y$. Hence $y \in \{a\}_l^{\sim}$. Therefore $\{a\}_l^{\sim}$ is a filter of X. □

Theorem 8.2.13 *Let* $(X, *, 1)$ *be a self-distributive BE-algebra. Then* $\langle a\rangle = \{(a*x) * x \mid x \in X\}$.

Proof Put $\Im_a = \{(a * x) * x \mid x \in X\}$. Let $x \in \langle a \rangle$. Then $a * x = 1$. Hence $x = 1 * x = (a * x) * x$. Thus $x \in \Im_a$. Hence $\langle a \rangle \subseteq \Im_a$. Conversely, let $t \in \Im_a$. Then $t = (a * x) * x$ for some $x \in X$. Then

$$a * t = a * ((a * x) * x) = (a * x) * (a * x) = 1.$$

Hence $t \in \langle a \rangle$, which implies that $\Im_a \subseteq \langle a \rangle$. Therefore $\Im_a = \langle a \rangle$. □

8.3 Stabilizers Induced by Filters

In this section, properties of the stabilizers induced by a filter of a BE-algebra are studied. It is proved that the class of all stabilizers induced by a filter forms a complete distributive lattice. A sufficient condition is derived, with the help of stabilizers, for a filter to become an intersection of all prime filters. The celebrated Stone's theorem of prime filters is generalized to the case of associative filters of BE-algebras with the help of stabilizers.

Definition 8.3.1 Let $(X, *, 1)$ be a BE-algebra and $a \in X$. For any non-empty subset A of X, the *stabilizer* of a induced by A is defined as follows:

$$\langle a, A \rangle = \{x \in X \mid a * x \in A\}$$

Some of the basic properties of these stabilizers are studied in the following.

Lemma 8.3.2 *Let A, B be two non-empty subsets of a transitive BE-algebra $(X, *, 1)$. Then for any $a, b \in X$, we have the following:*

(1) $\langle a, X \rangle = X$,
(2) $a \leq b$ *implies* $\langle b, A \rangle \subseteq \langle a, A \rangle$, *provided X is order reversing,*
(3) $A \subseteq B$ *implies* $\langle a, A \rangle \subseteq \langle a, B \rangle$,
(4) $\langle a, A \rangle \cap \langle a, B \rangle = \langle a, A \cap B \rangle$,
(5) $a * b \in \langle c, A \rangle$ *if and only if* $c * b \in \langle a, A \rangle$ *for any* $c \in X$.

Proof (1) It is clear.
(2) Suppose $a \leq b$. Let $x \in \langle b, F \rangle$. Then, we get $b * x \in F$. Since $a \leq b$ and X is order reversing, it yields that $b * x \leq a * x$. Hence $a * x \in F$. Thus $x \in \langle a, A \rangle$. Therefore $\langle b, A \rangle \subseteq \langle a, A \rangle$.
(3) Let A and B be two non-empty subsets of X such that $A \subseteq B$. Let $x \in \langle a, A \rangle$. Then we get that $a * x \in A \subseteq B$. Hence $x \in \langle a, B \rangle$. Therefore $\langle a, A \rangle \subseteq \langle a, B \rangle$.
(4) From the result (3), we can obtain that $\langle a, F \cap G \rangle \subseteq \langle a, F \rangle \cap \langle a, G \rangle$. Conversely let $x \in \langle a, F \rangle \cap \langle a, G \rangle$. Then we get $a * x \in F \cap G$. Therefore $x \in \langle a, F \cap G \rangle$.
(5) For any $c \in X$, we have $a * b \in \langle c, F \rangle \Leftrightarrow c * (a * b) \in A \Leftrightarrow a * (c * b) \in A \Leftrightarrow c * b \in \langle a, F \rangle$. □

Proposition 8.3.3 *Let $(X, *, 1)$ be a transitive BE-algebra and $a \in X$. For any quasi-filter F of X, the set $\langle a, F \rangle$ is a quasi-filter of X.*

Proof Let $a \in X$. Since $a * 1 = 1 \in F$, it gives that $1 \in \langle a, F\rangle$. Let $x \in \langle a, F\rangle$ and $t \in X$. Then, we get $a * x \in F$. Since F is a quasi-filter and $a * x \in F$, we get that $a * (t * x) = t * (a * x) \in F$. Hence $t * x \in \langle a, F\rangle$. Therefore $\langle a, F\rangle$ is a quasi-filter of X. □

Proposition 8.3.4 *Let $(X, *, 1)$ be a self-distributive BE-algebra and $a \in X$. For any simple filter F of X, the set $\langle a, F\rangle$ is a simple filter of X.*

Proof Let $a \in X$. Since $a * 1 = 1 \in F$, it gives that $1 \in \langle a, F\rangle$. Let $x * y \in \langle a, F\rangle$ where $x, y \in X$. Then, we get $(a * x) * (a * y) = a * (x * y) \in F$. Since F is a simple filter, by the self-distributivity of X, we get that $a*(x*(x*y)) = (a*x)*(a*(x*y)) = (a * x) * ((a * x) * (a * y)) \in F$. Hence $x * (x * y) \in \langle a, F\rangle$. Therefore $\langle a, F\rangle$ is a simple filter of X. □

Proposition 8.3.5 *Let $(X, *, 1)$ be a self-distributive BE-algebra and $a \in X$. For any filter F of X, the set $\langle a, F\rangle$ is a filter of X.*

Proof Let $a \in X$. Since $a * 1 = 1 \in F$, we get that $1 \in \langle a, F\rangle$. Let $x \in \langle a, F\rangle$ and $x * y \in \langle a, F\rangle$. Then, we get that $a * x \in F$ and $(a * x) * (a * y) = a * (x * y) \in F$. Since F is a filter and $a * x \in F$, we get that $a * y \in F$. Thus $y \in \langle a, F\rangle$. Therefore $\langle a, F\rangle$ is a filter of X. □

Some properties of the above stabilizers can be observed in the following theorem.

Theorem 8.3.6 *Let F, G be two filters of a transitive BE-algebra X. Then for any $a, b \in X$, we have*

(1) *$a \in F$ if and only if $\langle a, F\rangle = F$,*
(2) *$\langle a, \langle b, F\rangle\rangle = \langle b, \langle a, F\rangle\rangle$,*
(3) *$\langle 1, F\rangle = F$.*

Proof (1) Assume that $a \in F$. Let $x \in F$. Since $x \le a * x$, we can get $a * x \in F$. Hence $x \in \langle a, F\rangle$, which yields that $F \subseteq \langle a, F\rangle$. On the other hand, let $x \in \langle a, F\rangle$. Then we get that $a * x \in F$. Since $a \in F, a * x \in F$ and F is a filter, we get $x \in F$. Hence, it concludes $\langle a, F\rangle \subseteq F$. Therefore $\langle a, F\rangle = F$. Conversely assume that $\langle a, F\rangle = F$. Then, it concludes $a \in \langle a, F\rangle = F$.
(2) For any $a, b \in X$, we have the following

$$\begin{aligned} x \in \langle b, \langle a, F\rangle\rangle &\Leftrightarrow b * x \in \langle a, F\rangle \\ &\Leftrightarrow a * (b * x) \in F \\ &\Leftrightarrow b * (a * x) \in F \\ &\Leftrightarrow a * x \in \langle b, F\rangle \\ &\Leftrightarrow x \in \langle a, \langle b, F\rangle\rangle \end{aligned}$$

Therefore $\langle a, \langle b, F\rangle\rangle = \langle b, \langle a, F\rangle\rangle$.
(3) It is clear by (1). □

In Sect. 2.1, the condition **L** was introduced on a commutative BE-algebra and then proved that the BE-algebra $(X, *, 1)$ was converted into a distributive lattice where $x \wedge y = \inf\{x, y\}$ and the supremum was defined as $x \vee y = (y * x) * x$ for all $x, y \in X$. For any filter F of a commutative BE-algebra with condition **L**, $x \wedge y \in F$ for all $x, y \in F$. For, let $x, y \in F$. Clearly $y \leq x * y = x * (x \wedge y) \in F$. Since $x \in F$ and F is a filter, it yields that $x \wedge y \in F$. Using this fact, we obtain the following result. First we observe the following lemma which is useful in the later section.

Lemma 8.3.7 *Let $(X, *, 1)$ be a self-distributive and commutative BE-algebra with condition* **L**. *Then for any $x, y, z \in X$, the following condition holds:*

$$(x \wedge y) * z = x * (y * z).$$

Proof Let $x, y, z \in X$. Since X is self-distributive, it is negatively ordered. Hence $((y * z) * z) * z = y * z$. Then by the Proposition 2.1.35(5), we get the following consequence.

$$\begin{aligned}
(x \wedge y) * z &= (x * z) \vee (y * z) \\
&= ((y * z) * (x * z)) * (x * z) \\
&= (x * ((y * z) * z)) * (x * z) \\
&= x * (((y * z) * z) * z) \\
&= x * (y * z)
\end{aligned}$$

□

Theorem 8.3.8 *Let $(X, *, 1)$ be a self-distributive and commutative BE-algebra with condition* **L**. *Then the following conditions hold for any filter F of X.*

(1) $\langle a, F\rangle \cap \langle b, F\rangle = \langle a \vee b, F\rangle$,
(2) $\langle a \wedge b, F\rangle = \langle a, \langle b, F\rangle\rangle = \langle b, \langle a, F\rangle\rangle$,
(3) $\langle a, \langle a, F\rangle\rangle = \langle a, F\rangle$,
(4) $\langle a, F\rangle = \langle b, F\rangle$ *implies* $\langle a \wedge c, F\rangle = \langle b \wedge c, F\rangle$ *for any* $c \in X$,
(5) $\langle a, F\rangle = \langle b, F\rangle$ *implies* $\langle a \vee c, F\rangle = \langle b \vee c, F\rangle$ *for any* $c \in X$.

Proof (1) For any $a, b \in X$, it is clear that $\langle a \vee b, F\rangle \subseteq \langle a, F\rangle \cap \langle b, F\rangle$. Conversely, let $x \in \langle a, F\rangle \cap \langle b, F\rangle$. Then, we get that $a * x, b * x \in F$. Since F is a filter, it implies that $(a \vee b) * x = (a * x) \wedge (b * x) \in F$. Hence $x \in \langle a \vee b, F\rangle$. Therefore $\langle a, F\rangle \cap \langle b, F\rangle = \langle a \vee b, F\rangle$.
(2) Let $a, b \in X$. Then by the above lemma, we get $x \in \langle a \wedge b, F\rangle \Leftrightarrow (a \wedge b) * x \in F \Leftrightarrow a*(b*x) \in F \Leftrightarrow b*(a*x) \in F \Leftrightarrow a*x \in \langle b, F\rangle \Leftrightarrow x \in \langle a, \langle b, F\rangle\rangle$. Therefore $\langle a \wedge b, F\rangle = \langle a, \langle b, F\rangle\rangle$. It is clear that $\langle a, \langle b, F\rangle\rangle = \langle a \wedge b, F\rangle = \langle b \wedge a, F\rangle = \langle b, \langle a, F\rangle\rangle$.
(3) For $a, b \in X$, it is observed from (2) that $\langle a, \langle a, F\rangle\rangle = \langle a \wedge a, F\rangle = \langle a, F\rangle$.
(4) Suppose $\langle a, F\rangle = \langle b, F\rangle$ for $a, b \in X$. Let $x \in \langle a \wedge c, F\rangle$. Then from condition (2), we get $a * (c * x) = (a \wedge c) * x \in F$. Hence $c * x \in \langle a, F\rangle = \langle b, F\rangle$. Thus $(b \wedge c) * x = b * (c * x) \in F$. Hence $x \in \langle b \wedge c, F\rangle$. Therefore $\langle a \wedge c, F\rangle \subseteq \langle b \wedge c, F\rangle$.

Similarly, we can get $\langle a \wedge c, F\rangle \subseteq \langle b \wedge c, F\rangle$.
(5) From (1), we get $\langle a \vee c, F\rangle = \langle a, F\rangle \cap \langle c, F\rangle = \langle b, F\rangle \cap \langle c, F\rangle = \langle b \vee c, F\rangle$. □

The following is a direct consequence of the property (2) of above theorem.

Corollary 8.3.9 *Let $(X, *, 1)$ be a self-distributive and commutative BE-algebra with condition* **L**. *If F is a filter of X, then for any $a_1, a_2, \ldots\ldots, a_n \in X$. Then*

$$\langle a_1 \wedge a_2 \wedge \ldots\ldots \wedge a_n, F\rangle = \langle a_1 \langle a_2 \ldots . \langle a_n, F\rangle \ldots .\rangle\rangle$$

Theorem 8.3.10 *Let $(X, *, 1)$ be a commutative BE-algebra with condition* **L**. *For any filter F of X, define a relation θ_F on X as follows:*

$$(a, b) \in \theta_F \text{ if and only if } \langle a, F\rangle = \langle b, F\rangle.$$

Then θ_F is a congruence on X having F as a congruence class.

Proof Clearly θ_F is an equivalence relation on X. Form the implications (4) and (5) of Theorem 8.3.8, it is clear that θ_F is a congruence on X. Now, let $x, y \in F$. Then by Theorem 8.3.6(1), we get that $\langle x, F\rangle = F = \langle y, F\rangle$. Hence $(x, y) \in \theta_F$. Therefore F is a congruence class *w.r.t.* θ_F. □

Corollary 8.3.11 *Let $(X, *, 1)$ be a commutative BE-algebra with condition* **L**. *For any filter F of X, the Co-kernel of θ_F which is defined as Coker $\theta_F = \{x \in X \mid (x, 1) \in \theta_F\}$ is a lattice filter of X such that Coker $\theta_F = F$.*

Let $\mathcal{S}_F(X)$ denotes the class of all stabilizers induced by a filter F of a bounded self-distributive and commutative BE-algebra X with condition **L**. Then the following result shows that $\mathcal{S}_F(X)$ is a complete distributive lattice on its own.

Theorem 8.3.12 *Let F be a filter of a bounded self-distributive and commutative BE-algebra $(X, *, 0, 1)$ with condition* **L**. *Then the class $\mathcal{S}_F(X)$ of all stabilizers induced by F forms a complete distributive lattice with smallest element F and the greatest element $\langle 0, F\rangle$.*

Proof For any $a, b \in X$, define the operations $\sqcap$ and $\sqcup$ as follows:

$$\langle a, F\rangle \sqcap \langle b, F\rangle = \langle a \vee b, F\rangle \quad \text{and} \quad \langle a, F\rangle \sqcup \langle b, F\rangle = \langle a \wedge b, F\rangle$$

Then clearly $\langle a \vee b, F\rangle$ is the infimum of $\langle a, F\rangle$ and $\langle b, F\rangle$ in $\mathcal{S}_F(X)$. We now show that $\langle a \wedge b, F\rangle$ is the supremum of $\langle a, F\rangle$ and $\langle b, F\rangle$ in $\mathcal{S}_F(X)$. Clearly $\langle a \wedge b, F\rangle$ is an upper bound for both $\langle a, F\rangle$ and $\langle b, F\rangle$ in $\mathcal{S}_F(X)$. Let $\langle c, F\rangle$ be an upper bound for both $\langle a, F\rangle$ and $\langle b, F\rangle$. Let $x \in \langle a \wedge b, F\rangle$. Then we get $a * (b * x) = (a \wedge b) * x \in F$. Hence $b * x \in \langle a, F\rangle \subseteq \langle c, F\rangle$. Thus $b * (c * x) = c * (b * x) \in F$. Hence $c * x \in \langle b, F\rangle \subseteq \langle c, F\rangle$. Thus we can see that

$$\begin{aligned} c * x &= (c \wedge c) * x \\ &= c * (c * x) \in F. \end{aligned}$$

Thus $x \in \langle c, F\rangle$. Therefore $\langle a \wedge b, F\rangle$ is the supremum of $\langle a, F\rangle$ and $\langle b, F\rangle$. Now it can be routinely observed that $(\mathcal{S}_F(X), \sqcap, \sqcup, F, X)$ is a bounded distributive lattice. For any $F, G \in \mathcal{S}_F(X)$, define $F \leq G$ if and only if $F \subseteq G$. Then clearly $(\mathcal{S}_F(X), \leq)$ is a partially ordered set. Now by the extension of the property (1) of Theorem 8.3.8, we can get the infimum (which is nothing but the set intersection of stabilizers induced by a filter) for any subset of $\mathcal{S}_F(X)$. Therefore $(\mathcal{S}_F(X), \subseteq)$ is a complete distributive lattice. □

Proposition 8.3.13 *Let F be a filter of a self-distributive and commutative BE-algebra $(X, *, 1)$ with condition **L**. For any $a, b \in X$, defined an operation $\to$ as follows:*

$$\langle a, F\rangle \to \langle b, F\rangle = \{\, x \in X \mid x \vee t \in \langle b, F\rangle \textit{ for all } t \in \langle a, F\rangle\}$$

Then $\sqcap$ and $\to$ form an adjoint pair(residuation) on $\mathcal{S}_F(X)$.

Proof Suppose X is self-distributive and commutative with condition L. Let $a, b, c \in X$. Assume that $\langle a, F\rangle \sqcap \langle b, F\rangle \subseteq \langle c, F\rangle$. Let $x \in \langle a, F\rangle$. For any $t \in \langle b, F\rangle$, we get that $x \vee t \in \langle a, F\rangle \sqcap \langle b, F\rangle \subseteq \langle c, F\rangle$. Hence $x \in \langle b, F\rangle \to \langle c, F\rangle$. Therefore $\langle a, F\rangle \subseteq \langle b, F\rangle \to \langle c, F\rangle$. Conversely assume that $\langle a, F\rangle \subseteq \langle b, F\rangle \to \langle c, F\rangle$. Let $x \in \langle a, F\rangle \sqcap \langle b, F\rangle$. Then, we get $x \in \langle a, F\rangle \subseteq \langle b, F\rangle \to \langle c, F\rangle$. Hence $x \vee t \in \langle c, F\rangle$ for all $t \in \langle b, F\rangle$. Since $x \in \langle b, F\rangle$, we get that $x = x \vee x \in \langle c, F\rangle$. Hence $\langle a, F\rangle \sqcap \langle b, F\rangle \subseteq \langle c, F\rangle$. Therefore $\sqcap$ and $\to$ formed an adjoint pair (residuation) on $\mathcal{S}_F(X)$. □

Proposition 8.3.14 *If F is a filter of a self-distributive and commutative BE-algebra $(X, *, 1)$ with condition **L** and $a \in X$, then $\langle a, F\rangle$ is a filter of X containing F. Moreover $\langle a, F\rangle$ is the smallest filter containing both a and F.*

Proof In the Proposition 8.3.5, it is already observed that $\langle a, F\rangle$ is a filter of X. We now observe the following fact. Clearly $a \in \langle a, F\rangle$ and $F \subseteq \langle a, F\rangle$. Let G be a filter of L such that $a \in G$ and $F \subseteq G$. Let $x \in \langle a, F\rangle$. Then $a * x \in F \subseteq G$. Since $a \in G$, we get that $a \wedge (a * x) \in G$. Since G is a filter, it implies that $a * x \in G$. Since $a \in G$ and G is a filter, it yields that $x \in G$. Thus $\langle a, F\rangle \subseteq G$. Therefore $\langle a, F\rangle$ is the smallest filter containing both a, F. □

In the following, we present a new version of the proof for the prime filter theorem of self-distributive and commutative BE-algebras with the help of stabilizers induced by filters.

Theorem 8.3.15 *Let F be a filter of a self-distributive and commutative BE-algebra $(X, *, 1)$ with condition **L**. Then we have the following*

$$F = \bigcap \{P \mid P \textit{ is a prime filter such that } F \subseteq P\}$$

Proof Let $F_0 = \bigcap\{P \mid P$ is a prime filter such that $F \subseteq P\}$. Then clearly $F \subseteq F_0$. Conversely, let $a \notin F$. Consider $\sum = \{G \mid G$ is a filter , $F \subseteq G$ and $a \notin G\}$. Clearly

$\sum$ satisfies the hypothesis of Zorn's lemma. Let M be the maximal element of $\sum$. Choose $x, y \in L$ such that $x \notin M$ and $y \notin M$. Then we have $M \subset \langle x, M\rangle$ and $M \subset \langle y, M\rangle$. By the maximality of M, we can get $a \in \langle x, M\rangle$ and $a \in \langle y, M\rangle$. Therefore

$$\begin{aligned} a &\in \langle x, M\rangle \cap \langle y, M\rangle \\ &= \langle x \vee y, M\rangle \end{aligned}$$

If $x \vee y \in M$, then $a \in \langle x \vee y, M\rangle = M$, which is a contradiction. Thus M is a prime filter such that $a \notin M$. It can be concluded that $a \notin F_0$. Therefore $F_0 \subseteq F$, which concludes that $F = F_0$. □

The following corollary is a direct consequence of the above result.

Corollary 8.3.16 *Let F be a filter of a self-distributive and commutative BE-algebra $(X, *, 1)$ with condition* **L** *and $a \in X$ such that $a \notin F$. Then there exists a prime filter P of X such that $F \subseteq P$ and $a \notin P$.*

Proposition 8.3.17 *If F is an associative filter of a self-distributive BE-algebra X and $a \in X$, then $\langle a, F\rangle$ is the smallest associative filter of X containing both a and F.*

Proof Since X is self-distributive, it is observed from Proposition 8.3.5 that $\langle a, F\rangle$ is a filter containing both a and F. Let $x, y, z \in X$ such that $x * (y * z) \in \langle a, F\rangle$. Since F is an associative filter, we get the following consequence.

$$\begin{aligned} x * (y * z) \in \langle a, F\rangle &\Rightarrow a * (x * (y * z)) \in F \\ &\Rightarrow x * (a * (y * z)) \in F \\ &\Rightarrow x * (y * (a * z)) \in F \\ &\Rightarrow (x * y) * (a * z) \in F \\ &\Rightarrow a * ((x * y) * z) \in F \\ &\Rightarrow (x * y) * z \in \langle a, F\rangle. \end{aligned}$$

Therefore $\langle a, F\rangle$ is an associative filter of X. By applying the Proposition 8.3.5, it can be concluded that $\langle a, F\rangle$ is the smallest associative filter containing both a and F. □

In the following, the famous Stone's theorem of prime filters is generalized to the case of associative filters of BE-algebras and proved with the help of their stabilizers.

Theorem 8.3.18 *Let F be an associative filter and S a $\vee$-closed subset of X such that $F \cap S = \emptyset$. Then there exists a prime associative filter P of X such that $F \subseteq P$ and $S \cap P = \emptyset$.*

Proof Since X is self-distributive, clearly it is an ordered BE-algebra. Suppose F is an associative filter and S a $\vee$-closed subset of X such that $F \cap S = \emptyset$. Consider the following set:

$$\sum = \{G \mid G \text{ is an associative filter, } F \subseteq G \text{ and } G \cap S = \emptyset\}.$$

Clearly $F \in \sum$. Let $\{G_\alpha \mid \alpha \in \Delta\}$ be a chain in $\sum$. Then G_α is an associative filter such that $F \subseteq G_\alpha$ and $G_\alpha \cap S = \emptyset$ for each $\alpha \in \Delta$. Clearly $\bigcup\limits_{\alpha\in\Delta} G_\alpha$ is an associative filter of X such that $F \subseteq \bigcup\limits_{\alpha\in\Delta} G_\alpha$ and $(\bigcup\limits_{\alpha\in\Delta} G_\alpha) \cap S = \emptyset$. Hence $\bigcup\limits_{\alpha\in\Delta} G_\alpha$ is an upper bound for the chain $\{G_\alpha \mid \alpha \in \Delta\}$. Then by Zorn's lemma, $\sum$ has a maximal element, say P. Let $a, b \in X$ be such that $a \notin P$ and $b \notin P$. Then we get $P \subset \langle a, P\rangle$ and $P \subset \langle b, P\rangle$. By the maximality of P, we can get that $\langle a, P\rangle \cap S \neq \emptyset$ and $\langle b, P\rangle \cap S \neq \emptyset$. Choose $x \in \langle a, P\rangle \cap S$ and $y \in \langle b, P\rangle \cap S$. Since $x \in \langle a, P\rangle$, we get that $a*x \in P$. Since $x \leq x \vee y$ and X is an ordered BE-algebra, it yields that $a * x \leq a * (x \vee y)$ and hence $a * (x \vee y) \in P$. By using the similar argument, we can get $b * (x \vee y) \in P$. Since P is an associative filter, it implies that $(a \vee b) * (x \vee y) = (a * (x \vee y)) \wedge (b * (x \vee y)) \in P$. Hence $x \vee y \in \langle a \vee b, P\rangle$. If $a \vee b \in P$, then $x \vee y \in P$ and hence $x \vee y \in P \cap S$, which is a contradiction. Therefore P is a prime associative filter of X such that $F \subseteq P$ and $P \cap S = \emptyset$. □

8.4 Topology with Stabilizers

In this section, a base for a topology is observed by using left and right stabilizers. It is proved that the above topological space is a Bair space and locally connected and separable space. Some other topological properties are studied by using left and right stabilizers.

Definition 8.4.1 Let A be a non-empty set. Then the mapping $C : \mathcal{P}(A) \to \mathcal{P}(A)$ is called a *closure operator* on A, if for all $X, Y \in \mathcal{P}(A)$, the following conditions hold:

(1) $X \subseteq C(X)$,
(2) $X \subseteq T$ implies $C(X) \subseteq C(Y)$,
(3) $C^2(X) = C(X)$.

Theorem 8.4.2 *For any BE-algebra X, define a function $\alpha : \mathcal{P}(X) \to \mathcal{P}(X)$ by $\alpha(A) = (A_l^{\sim})_r^{\sim}$ for $A \in \mathcal{P}(X)$, where $\mathcal{P}(X)$ is the set of all subsets of X. Then α is a closure operator on X.*

Proof By Proposition 8.2.2(2), it is evident that $A \subseteq \alpha(A)$ for all $A \in \mathcal{P}(X)$. Let A, B be two subsets of X such that $A \subseteq B$. Let $x \in \alpha(A) = (A_l^{\sim})_r^{\sim}$. Then by Lemma 8.1.7, we get that $x \in (B_l^{\sim})_r^{\sim} = \alpha(B)$. Hence $\alpha(A) \subseteq \alpha(B)$. Again by Proposition 8.2.2(3), it is clear $\alpha^2(A) = (((A_l^{\sim})_r^{\sim})_l^{\sim})_r^{\sim} = (A_l^{\sim})_r^{\sim} = \alpha(A)$. Therefore α is a closure operator on X. □

The following result can be obtained similarly.

Theorem 8.4.3 *Let X be a BE-algebra. Then the function $\beta : \mathcal{P}(X) \longrightarrow \mathcal{P}(X)$, where $\beta(A) = (A_r^{\sim})_l^{\sim}$ is a closure operator on X.*

Theorem 8.4.4 *For any two subsets A, B of a BE-algebra X, the following conditions hold:*

(1) *$A \subseteq B$ implies $\alpha(A) \subseteq \alpha(B)$,*
(2) *$\alpha(A) \cap \alpha(B) = \alpha(A \cap B)$,*
(3) *$\alpha(A)$ is a filter whenever X is transitive.*

Proof (1) It is clear.
(2) Since α is a closure operator, we get $A \cap B \subseteq \alpha(A \cap B)$. Since $A \cap B \subseteq A, B$, we get $A_l^{\sim}, B_l^{\sim} \subseteq (A \cap B)_l^{\sim}$. Hence $\alpha(A \cap B) = ((A \cap B)_l^{\sim})_r^{\sim} \subseteq ((A)_l^{\sim})_r^{\sim}, ((B)_l^{\sim})_r^{\sim} = \alpha(A), \alpha(B)$. Therefore $\alpha(A \cap B) \subseteq \alpha(A) \cap \alpha(B)$. Thus $\alpha(A \cap B) = \alpha(A) \cap \alpha(B)$.
(3) It is clear by Proposition 8.2.1. □

Theorem 8.4.5 *Let $(X, *, 1)$ be a BE-algebra. Consider the function α given in Theorem 8.4.2. Then the set $\mu_\alpha = \{A \in \mathcal{P}(X) \mid \alpha(A) = A\}$ is a basis for a topology on X.*

Proof It is clear that $X_l^{\sim} = \{1\}$ and $\{0\}_r^{\sim} = X$. Hence $\alpha(X) = X$ and so it yields that $X \in \mu_\alpha$. Thus, for each $x \in X$, there exists an element of μ_α containing x. Let $x \in A \cap B$ for some $A, B \in \mu_\alpha$. Since α is a closure operator, from the above lemma, we get $\alpha(A \cap B) = \alpha(A) \cap \alpha(B) = A \cap B$. Hence $A \cap B \in \mu_\alpha$ such that $x \in A \cap B$. Therefore μ_α is a basis for a topology on X. □

The following result can be obtained similarly.

Theorem 8.4.6 *Let X be a BE-algebra. Consider the function β given in Theorem 8.4.3. Then $\mu_\beta = \{A \in \mathcal{P}(X) \mid \beta(A) = A\}$ is a basis for a topology on X.*

Define the topologies τ_α and τ_β generated by basis μ_α and μ_β respectively observed in the above theorems. Then, in the following theorem, we show that (X, τ_α) is a Hausdorff space.

Theorem 8.4.7 *Let $(X, *, 1)$ be a transitive BE-algebra. Then (X, τ_α) $((X, \tau_\beta))$ is a Hausdorff space if and only if $X = \{1\}$.*

Proof Suppose that $X \neq \{1\}$. Clearly $1 \in \alpha(A)$ for any $\emptyset \neq A \subseteq X$. Hence $1 \in U$ for any $U \in \tau_\alpha$. Let $U, V \in \tau_\alpha$. Then we get $U \cap V \neq \emptyset$. Hence, we get (X, τ_α) is a not Housdorff space. Conversely, assume that $X = \{1\}$. Then $\tau_\alpha = \{\emptyset, X\}$. Therefore (X, τ_α) is a Housdorff space. □

Theorem 8.4.8 *(X, τ_α) $((X, \tau_\beta))$ is connected.*

Proof Since $1 \in U$ for any non-empty open set U of X, then there are no non-empty open subsets U and V of X such that $X = U \cup V$ and $U \cap V = \emptyset$. Therefore (X, τ_α) is a connected space. □

The following corollary is a direct consequence of the above.

Corollary 8.4.9 *Let U be a non-empty and non-connected subset of (X, τ_α) $((X, \tau_\beta))$. Then $1 \in U$.*

Theorem 8.4.10 *Let $(X, *, 1)$ be a BE-algebra and $\emptyset \neq A \neq X$ a closed subset of $(X, \tau_\alpha)((X, \tau_\beta))$. Then A is a connected set of X.*

Proof Since $\emptyset \neq A \neq X$ is a closed set of $(X, \tau_\alpha)((X, \tau_\beta))$, it yields that $\emptyset \neq X - A$ is an open set of the topological space $(X, \tau_\alpha)((X, \tau_\beta))$. Clearly $1 \in X - A$ and hence $1 \notin A$. Therefore by Corollary 8.4.9, we get that A is a connected set of X. □

Theorem 8.4.11 *Let $(X, *, 1)$ be a transitive BE-algebra and A a subset of the topological space $(X, \tau_\alpha)((X, \tau_\beta))$ and $1 \in A$. Then $\overline{A} = X$.*

Proof Let $x \in X$. If $x = 1$, then $1 \in \overline{A}$. Let $x \neq 0$. Since $1 \in U$ for any non-empty open subset U of X, we get that $U \cap A \neq \emptyset$ for any open set containing x. Hence $x \in \overline{A}$. Therefore $X \subseteq \overline{A}$. □

The following corollary is a direct consequence of the above result.

Corollary 8.4.12 *Let X be a BE-algebra and U a non-empty open subset of X. Then $\overline{U} = X$.*

Theorem 8.4.13 *Let $(X, *1)$ be a BE-algebra and A a non-empty subset of the topological space $(X, \tau_\alpha)((X, \tau_\beta))$. Then $1 \in \overline{A}$ if and only if $\overline{A} = X$.*

Proof Let A be a non-empty subset of X and $1 \in \overline{A}$. Then $1 \in C$ for all closed subset C of X containing A. Since 1 is in any non-empty open subset of the topological space X, then the only closed subset of X containing 1 and A is X. So $\overline{A} = X$. The proof of the converse is clear. □

Lemma 8.4.14 *$\{1\}$ is an open subset of the topological space $(X, \tau_\alpha)((X, \tau_\beta))$.*

Proof From the preliminary results of stabilizers, it is clear that $\{1\}_r^\sim = X$ and $X_r^\sim = \{1\}$. Hence, it yields $\{0\} \in \mu_\alpha$. Thus $\{1\}$ is an open set of the topological space $(X, \tau_\alpha)((X, \tau_\beta))$. □

Theorem 8.4.15 *For any BE-algebra X, $(X, \tau_\alpha)((X, \tau_\beta))$ are separable.*

Proof By Lemma 8.4.14 and Theorem 8.4.13 we get $\{1\} = X$. Then (X, τ_α) is separable. □

Theorem 8.4.16 *For any BE-algebra X, $(X, \tau\alpha)((X, \tau_\beta)$ is a locally connected space.*

Proof Let x be an arbitrary element of X and U be an open set containing x. By Theorem 8.4.8, we get that U is connected and also containing x. Therefore (X, τ_α) is locally connected. □

8.5 Pseudo-complements and Stabilizers

In this section, the notion of dual stabilizers is introduced with the help of pseudo-complements. Some properties of dual stabilizers are then studied. It is proved that the stabilizer of a non-empty set of a commutative BE-algebra is an ideal. Throughout this section, X stands for a bounded and partially ordered set.

Definition 8.5.1 Let A a non-empty subset of a bounded BE-algebra X. Define

$$DA_r^{\sim} = \{x \in X \mid x^{\diamond} * a^{\diamond} \leq a^{\diamond} \text{ for all } a \in A\};$$
$$DA_l^{\sim} = \{x \in X \mid a^{\diamond} * x^{\diamond} \leq x^{\diamond} \text{ for all } a \in A\}.$$

For any non-empty subset A of X, the above sets $DA_r^{\sim}$ and $DA_l^{\sim}$ are called right and left dual stabilizers of A and denote the dual stabilizer as $DA^{\sim} = DA_r^{\sim} \cap DA_l^{\sim}$. For a singleton set $A = \{a\}$, the left and right dual stabilizers are denoted by $D\{a\}_l^{\sim}$ and $D\{a\}_r^{\sim}$, respectively. In what follows, $A^{\diamond}$ denotes the set $\{x^{\diamond} \mid x \in A\}$ for all $\emptyset \neq A \subseteq X$, unless otherwise mentioned.

Lemma 8.5.2 *Let X be a bounded BE-algebra and $\emptyset \neq A, B \subseteq X$. Then the following hold:*

(1) *If $A \subseteq B$, then $DB^{\sim} \subseteq DA^{\sim}$,*
(2) *$D(A \cup B)^{\sim} = DA^{\sim} \cap DB^{\sim}$,*
(3) *$DX^{\sim} = \{0\}$ and $D\{0\}^{\sim} = X$,*
(4) *$DA^{\sim} = \bigcap_{x \in A} D\{x\}^{\sim}$,*
(5) *$1 \in A$ implies $DA^{\sim} = DA_l^{\sim} = DA_r^{\sim} = \{0\}$,*
(6) *$DA_l^{\sim} = X$ implies $A = \{0\}$ and $DA_r^{\sim} = X$ implies $A = \{0\}$,*
(7) *If $h : X \to X$ be a bounded homomorphism and $a \in X$, then $h(D\{a\}^{\sim}) \subseteq D\{h(a)\}^{\sim}$.*

Proof (1) Suppose $A \subseteq B$. Let $x \in DB^{\sim} = DB_r^{\sim} \cap DB_l^{\sim}$. Then $x^{\diamond} * a^{\diamond} \leq a^{\diamond}$ and $a^{\diamond} * x^{\diamond} \leq x^{\diamond}$ for all $a \in B$. Hence $x^{\diamond} * a^{\diamond} \leq a^{\diamond}$ and $a^{\diamond} * x^{\diamond} \leq x^{\diamond}$ for all $a \in A$. Therefore $x \in DA_r^{\sim} \cap DA_l^{\sim} = DA^{\sim}$.
(2) Since $A, B \subseteq A \cup B$, we get from (1) that $D(A \cup B)^{\sim} \subseteq DA^{\sim} \cap DB^{\sim}$. Conversely, let $x \in DA^{\sim} \cap DB^{\sim}$. Then we get $x \in DA_r^{\sim} \cap DA_l^{\sim}$ and $x \in DB_r^{\sim} \cap DB_l^{\sim}$. Hence $x^{\diamond} * a^{\diamond} \leq a^{\diamond}$ and $a^{\diamond} * x^{\diamond} \leq x^{\diamond}$ for all $a \in A$. Also we can obtain $x^{\diamond} * a^{\diamond} \leq a^{\diamond}$ and $a^{\diamond} * x^{\diamond} \leq x^{\diamond}$ for all $a \in B$. Thus $x \in D(A \cup B)^{\sim}$. Therefore $DA_r^{\sim} \cap DB_r^{\sim} \subseteq D(A \cup B)^{\sim}$.
(3) Let $x \in DX^{\sim}$. Then $x \in DX_r^{\sim} \cap DX_l^{\sim}$. Hence $x^{\diamond} * a^{\diamond} \leq a^{\diamond}$ and $a^{\diamond} * x^{\diamond} \leq x^{\diamond}$ for all $a \in X$. Since $x \in X$, we get that $1 = x^{\diamond} * x^{\diamond} \leq x^{\diamond}$. Thus $x^{\diamond} = 1$, which implies $x \leq x^{\diamond\diamond} = 0$. Therefore $DX^{\sim} = \{0\}$. For any $x \in X$, we have $0^{\diamond} * x^{\diamond} = 1 * x^{\diamond} = x^{\diamond}$ and $x^{\diamond} * 0^{\diamond} = x^{\diamond} * 1 = 1 = 0^{\diamond}$. Hence $0 \in DX_r^{\sim} \cap DX_l^{\sim} = DX^{\sim}$. Therefore $DX^{\sim} = \{0\}$.
Let $x \in X$, be arbitrary. Since $0^{\diamond} * x^{\diamond} = x^{\diamond}$ and $x^{\diamond} * 0^{\diamond} = 0^{\diamond}$, it yields $x \in D\{0\}_r^{\sim} \cap D\{0\}_l^{\sim} = D\{0\}^{\sim}$. Hence $X \subseteq D\{0\}^{\sim}$. On the other hand, clearly $D\{0\}^{\sim} \subseteq X$. Therefore $D\{0\}^{\sim} = X$.
(4) Let $x \in DA^{\sim} = DA_r^{\sim} \cap DA_l^{\sim}$. Then, we get the following consequence:

$$
\begin{aligned}
x \in DA_r^{\sim} \cap DA_l^{\sim} &\Leftrightarrow x^{\diamond} * a^{\diamond} \leq a^{\diamond} \text{ and } a^{\diamond} * x^{\diamond} \leq x^{\diamond} \text{ for all } a \in A \\
&\Leftrightarrow x \in D\{a\}_r^{\sim} \text{ and } x \in D\{a\}_l^{\sim} \text{ for all } a \in A \\
&\Leftrightarrow x \in \bigcap_{a \in A} D\{a\}_r^{\sim} \text{ and } x \in \bigcap_{a \in A} D\{a\}_l^{\sim} \\
&\Leftrightarrow x \in \Big(\bigcap_{a \in A} D\{a\}_r^{\sim}\Big) \cap \Big(\bigcap_{a \in A} D\{a\}_l^{\sim}\Big) \\
&\Leftrightarrow x \in \bigcap_{a \in A} D\{a\}^{\sim}
\end{aligned}
$$

Therefore that $DA^{\sim} = \bigcap\limits_{x \in A} D\{x\}^{\sim}$.
(5) Suppose $1 \in A$. Clearly $0 \in DA_l^{\sim} \cap DA_r^{\sim}$. Let $x \in DA_l^{\sim}$. Then $a^{\diamond} * x^{\diamond} \leq x^{\diamond}$ for all $a \in A$. Since $1 \in A$, we get $1 = 0 * x^{\diamond} = 1^{\diamond} * x^{\diamond} \leq x^{\diamond}$. Hence $x \leq x^{\diamond\diamond} = 0$. Therefore $DA_l^{\sim} = \{0\}$. Again, let $x \in DA_r^{\sim}$. Then $x^{\diamond} * a^{\diamond} \leq a^{\diamond}$ for all $a \in A$. Since $1 \in A$, it yields $x \leq x^{\diamond\diamond} = x^{\diamond} * 0 = x^{\diamond} * 1^{\diamond} \leq 1^{\diamond} = 0$. Hence $DA_r^{\sim} = \{0\}$. Therefore $DA^{\sim} = DA_l^{\sim} \cap DA_r^{\sim} = \{0\}$.
(6) Suppose $DA_l^{\sim} = X$. Let $x \in A \subseteq X = DA_l^{\sim}$. Then $a^{\diamond} * x^{\diamond} \leq x^{\diamond}$ for all $a \in A$. Hence $1 = x^{\diamond} * x^{\diamond} = x^{\diamond}$. Thus $x = 0$, which implies $DA_l^{\sim} = \{0\}$. Similarly $DA_r^{\sim} = X$ implies $A = \{0\}$.
(7) Let $a \in X$. Suppose $h : X \to X$ is a bounded BE-morphism and $y \in h(D\{a\}^{\sim})$. Then there exists $x \in D\{a\}^{\sim}$ such that $y = h(x)$. Hence $x^{\diamond} * a^{\diamond} \leq a^{\diamond}$ and $a^{\diamond} * x^{\diamond} \leq x^{\diamond}$. Since h is a bounded BE-morphism, we get $h(x)^{\diamond} * h(a)^{\diamond} = h(x^{\diamond} * a^{\diamond}) \leq h(a^{\diamond}) = h(a)^{\diamond}$ and $h(a)^{\diamond} * h(x)^{\diamond} = h(a^{\diamond} * x^{\diamond}) \leq h(x^{\diamond}) = h(x)^{\diamond}$. Hence $y = h(x) \in D\{h(a)\}^{\sim}$. Therefore $Dh(\{a\}^{\sim}) \subseteq D\{h(a)\}^{\sim}$. □

Proposition 8.5.3 *Let A be a non-empty subset of a BE-algebra X. Then $A \cap DA_l^{\sim} = \{0\}$.*

Proof Let $x \in A \cap DA_l^{\sim}$. Then $x \in A$ and $x \in DA_l^{\sim}$. Hence $a^{\diamond} *^{\diamond} \leq x^{\diamond}$ for all $a \in A$. Hence $1 = x^{\diamond} * x^{\diamond} \leq x^{\diamond}$. Thus $x \leq x^{\diamond\diamond} = 0$. Therefore $x = 0$, which implies $A \cap DA_l^{\sim} = \{0\}$. □

Proposition 8.5.4 *Let A be a non-empty subset of a bounded BE-algebra X. If A has a dense element, then $DA_r^{\sim} = DA_l^{\sim} = DA^{\sim} = \{0\}$.*

Proof Suppose A has a dense element, say $a \in A$. Let $x \in DA_l^{\sim}$. Then we get $a^{\diamond} * x^{\diamond} \leq x^{\diamond}$. Since a is dense, we get $1 = 0 * x^{\diamond} \leq x^{\diamond}$. Hence $x^{\diamond} = 1$, which implies that $x \leq x^{\diamond\diamond} = 0$. Thus $x = 0$. Therefore $DA_l^{\sim}$. Similarly, we can obtain $DA_r^{\sim} = \{0\}$. Therefore $DA^{\sim} = \{0\}$. □

Proposition 8.5.5 *Let X be a transitive BE-algebra. Then, for any $\emptyset \neq A \subseteq X$, the dual stabilizer $DA_r^{\sim}$ is an ideal of X. Moreover, $A \cap DA_r^{\sim} = \{0\}$ for any $\emptyset \neq A \subseteq X$.*

Proof Clearly $0 \in DA_r^{\sim}$. Let $x \in DA_r^{\sim}$ and $(x^{\diamond} * y^{\diamond})^{\diamond} \in DA_r^{\sim}$. For any $a \in A$, we have

$$\begin{aligned}
y^\diamond * a^\diamond &\leq (x^\diamond * y^\diamond) * (x^\diamond * a^\diamond) \\
&= x^\diamond * ((x^\diamond * y^\diamond) * a^\diamond) \\
&= x^\diamond * (a * (x^\diamond * y^\diamond)^\diamond) \\
&= x^\diamond * (a * (x^\diamond * y^\diamond)^{\diamond\diamond\diamond}) \\
&= x^\diamond * ((x^\diamond * y^\diamond)^{\diamond\diamond} * a^\diamond) \\
&= (x^\diamond * y^\diamond)^{\diamond\diamond} * (x^\diamond * a^\diamond) \\
&\leq (x^\diamond * y^\diamond)^{\diamond\diamond} * a^\diamond && \text{since } x^\diamond * a^\diamond \leq a^\diamond \\
&\leq a^\diamond && \text{since } (x^\diamond * y^\diamond)^\diamond \in DA_r^{\sim}
\end{aligned}$$

Since $a \in A$ is arbitrary, we get $y \in DA_r^{\sim}$. Therefore $DA_r^{\sim}$ is an ideal of X. Let $x \in A \cap DA_r^{\sim}$. Then we get $x \in A$ and $x \in DA_r^{\sim}$. Hence $1 = x^\diamond * x^\diamond \leq x^\diamond$, which implies that $x^\diamond = 1$. Hence $x \leq x^{\diamond\diamond} = 0$. Hence $x = 0$, which immediately infers that $A \cap DA_r^{\sim} = \{0\}$. □

If A is a non-empty subset of a transitive BE-algebra X, then the following example shows that $DA_l^{\sim}$ need not be an ideal of X in general.

Example 8.5.6 Let $X = \{1, a, b, c, 0\}$. Define an operation $*$ on X as follows:

$*$	1	a	b	c	0
1	1	a	b	c	0
a	1	1	a	c	c
b	1	1	1	c	c
c	1	1	1	1	1
0	1	1	1	1	0

Then $(X, *, 1)$ is a bounded and transitive BE-algebra. Clearly $A = \{0, c\}$ is an ideal of X. Now $DA_l^{\sim} = \{0, a, b\}$ is not an ideal of X, because of $(b^\diamond * c^\diamond)^\diamond \in DA_l^{\sim}$ and $b \in DA_l^{\sim}$ but $c \notin DA_l^\diamond$.

However, in the following theorem, a sufficient condition is derived for every ideal A of a BE-algebra X, the left dual stabilizer $DA_l^{\sim}$ to become an ideal of X.

Theorem 8.5.7 *If X is a commutative BE-algebra, then $DA_l^{\sim}$ is an ideal for every non-empty subset A of X. Moreover $DA_r^{\sim} = DA_l^{\sim} = DA^{\sim}$.*

Proof Let $x \in DA_l^{\sim}$ and $(x^\diamond * y^\diamond)^\diamond \in DA_l^{\sim}$. For any $a \in A$, we get $a^\diamond * x^\diamond \leq x^\diamond$ and $a^\diamond * (x^\diamond * y^\diamond)^{\diamond\diamond} \leq (x^\diamond * y^\diamond)^{\diamond\diamond}$. Hence $(a^\diamond * x^\diamond) * x^\diamond = 1$ and $(a^\diamond * (x^\diamond * y^\diamond)^{\diamond\diamond}) * (x^\diamond * y^\diamond)^{\diamond\diamond} = 1$. Since X is commutative, it yields $(x^\diamond * a^\diamond) * a^\diamond = 1$ and $((x^\diamond * y^\diamond)^{\diamond\diamond} * a^\diamond) * a^\diamond = 1$. Therefore $x^\diamond * a^\diamond \leq a^\diamond$ and $(x^\diamond * y^\diamond)^{\diamond\diamond} * a^\diamond \leq a^\diamond$, which provides $x \in DA_r^{\sim}$ and $(x^\diamond * y^\diamond)^\diamond \in DA_r^{\sim}$. Since X is commutative, it is transitive. Hence by Proposition 8.5.5, we get $DA_r^{\sim}$ is an ideal. Hence $y \in DA_r^{\sim}$, which yields $y^\diamond * a^\diamond \leq a^\diamond$. Since X is commutative, we get $a^\diamond * y^\diamond \leq y^\diamond$. Hence $y \in DA_l^{\sim}$. Therefore $DA_l^{\sim}$ is an ideal of X. Let $x \in X$. Since X is commutative, we get

$$\begin{aligned} x \in DA_l^{\sim} &\Leftrightarrow a^{\diamond} * x^{\diamond} \leq x^{\diamond} \text{ for all } a \in A \\ &\Leftrightarrow (a^{\diamond} * x^{\diamond}) * x^{\diamond} = 1 \text{ for all } a \in A \\ &\Leftrightarrow (x^{\diamond} * a^{\diamond}) * a^{\diamond} = 1 \text{ for all } a \in A \\ &\Leftrightarrow x^{\diamond} * a^{\diamond} \leq a^{\diamond} \text{ for all } a \in A \\ &\Leftrightarrow x \in DA_r^{\sim}. \end{aligned}$$

Thus $DA_r^{\sim} = DA_l^{\sim}$. Therefore $DA_r^{\sim} = DA_l^{\sim} = DA^{\sim}$. □

Lemma 8.5.8 *Let X be a bounded and transitive BE-algebra and $x, a_1, a_2, \cdots a_n \in X$. Then*

$$(a_1^{\diamond} * (a_2^{\diamond} * (\cdots (a_n^{\diamond} * x^{\diamond}) \cdots))^{\diamond\diamond} \leq a_1^{\diamond} * (a_2^{\diamond} * (\cdots (a_n^{\diamond} * x^{\diamond}) \cdots).$$

Proof For any $x, a_1, a_2, \cdots a_n \in X$, put $t = a_1^{\diamond} * (\cdots (a_n^{\diamond} * x^{\diamond}) \cdots)$. Then by Lemma 7.1.3(3),

$$\begin{aligned} t^{\diamond\diamond} * (a_1^{\diamond} * (\cdots (a_n^{\diamond} * x^{\diamond}) \cdots)) &= a_1^{\diamond} * (\cdots (a_n^{\diamond} * (t^{\diamond\diamond} * x^{\diamond})) \cdots) \\ &= a_1^{\diamond} * (\cdots (a_n^{\diamond} * (x * t^{\diamond\diamond\diamond})) \cdots) \\ &= a_1^{\diamond} * (\cdots (a_n^{\diamond} * (x * t^{\diamond})) \cdots) \\ &= a_1^{\diamond} * (\cdots (a_n^{\diamond} * (t * x^{\diamond})) \cdots) \\ &= t * (a_1^{\diamond} * (\cdots (a_n^{\diamond} * x^{\diamond}) \cdots)) \\ &= 1 \end{aligned}$$

Therefore $(a_1^{\diamond} * (a_2^{\diamond} * (\cdots (a_n^{\diamond} * x^{\diamond}) \cdots))^{\diamond\diamond} \leq a_1^{\diamond} * (a_2^{\diamond} * (\cdots (a_n^{\diamond} * x^{\diamond}) \cdots)$. □

The following corollaries are direct consequences of the above lemma.

Corollary 8.5.9 *Let X be a bounded BE-algebra and $x, a_1, a_2, \cdots a_n \in X$. Then*

$$(a_1 * (a_2 * (\cdots (a_n * x^{\diamond}) \cdots))^{\diamond\diamond} \leq a_1 * (a_2 * (\cdots (a_n * x^{\diamond}) \cdots).$$

Corollary 8.5.10 *Let X be a bounded and transitive BE-algebra and $x, a \in X$. Then $(a^n * x^{\diamond})^{\diamond\diamond} \leq a^n * x^{\diamond}$ for any positive integer $n \in \mathbb{N}$.*

Corollary 8.5.11 *Let X be a bounded and transitive BE-algebra and $a, b \in X$. Then $(a * b^{\diamond})^{\diamond\diamond} \leq a * b^{\diamond}$.*

Theorem 8.5.12 *If A is an ideal of a transitive BE-algebra X, then $DA^{\sim}$ is an ideal of X.*

Proof Let $(x^{\diamond} * y^{\diamond})^{\diamond} \in DA^{\sim}$ and $x \in DA^{\sim}$. Then $(x^{\diamond} * y^{\diamond})^{\diamond} \in DA_r^{\sim} \cap DA_l^{\sim}$ and $x \in DA_r^{\sim} \cap DA_l^{\sim}$. Since $DA_r^{\sim}$ is an ideal (by Proposition 8.5.4), we get $y \in DA_r^{\sim}$. Hence $y^{\diamond\diamond} \in DA_r^{\sim}$. Since A is an ideal, we get $a^{\diamond\diamond} \in A$ for all $a \in A$. Since X is transitive, we get $a^{\diamond} = 1 * a^{\diamond} \leq (a^{\diamond} * y^{\diamond}) * (1 * y^{\diamond}) = (a^{\diamond} * y^{\diamond}) * y^{\diamond}$, which implies $((a^{\diamond} * y^{\diamond}) * y^{\diamond})^{\diamond} \leq a^{\diamond\diamond} \in A$. Also $y^{\diamond} \leq (a^{\diamond} * y^{\diamond}) * y^{\diamond}$ gives that

$((a^\diamond * y^\diamond) * y^\diamond)^\diamond \le y^{\diamond\diamond} \in DA_r^\sim$. Hence by the above two observations, we get $((a^\diamond * y^\diamond) * y^\diamond)^\diamond \in A \cap DA_r^\sim = \{0\}$. Hence $(a^\diamond * y^\diamond) * y^\diamond = 1$, which implies $a^\diamond * y^\diamond \le y^\diamond$ for all $a \in A$. Thus $y \in DA_l^\sim$. Hence $y \in DA_r^\sim \cap DA_l^\sim = DA^\sim$. Therefore $DA^\sim$ is an ideal of X. □

Theorem 8.5.13 *Let A and B be two ideals of a transitive BE-algebra X. Then $A \cap B = \{0\}$ if and only if $A \subseteq DB_r^\sim$.*

Proof Assume that $A \cap B = \{0\}$. Let $x \in A$. As A and B are ideals of X, it results in $x^{\diamond\diamond} \in A$ and $y^{\diamond\diamond} \in B$ for all $x \in A$ and $y \in B$. Since $((x^\diamond * y^\diamond) * y^\diamond)^\diamond \le x^{\diamond\diamond}$ and $((x^\diamond * y^\diamond) * y^\diamond)^\diamond \le y^{\diamond\diamond}$, we get that $((x^\diamond * y^\diamond) * y^\diamond)^\diamond \in A$ and $((x^\diamond * y^\diamond) * y^\diamond)^\diamond \in B$. Hence $((x^\diamond * y^\diamond) * y^\diamond)^\diamond \in A \cap B = \{0\}$. Thus $((x^\diamond * y^\diamond) * y^\diamond)^{\diamond\diamond} = 1$. Then by Corollary 8.5.9, we get $(x^\diamond * y^\diamond) * y^\diamond = 1$. Hence $x^\diamond * y^\diamond \le y^\diamond$ for all $y \in B$. Thus $x \in DB_r^\sim$. Therefore $A \subseteq DB_r^\sim$.

Conversely, assume that $A \subseteq DB_r^\sim$. Let $x \in A \cap B$. Then $x \in A \subseteq DB_r^\sim$ and $x \in B$. Then $1 = x^\diamond * x^\diamond \le x^\diamond$. Hence $x^\diamond = 1$, which implies $x \le x^{\diamond\diamond} = 0$. Therefore $A \cap B = \{0\}$. □

Theorem 8.5.14 *Let A and B be two ideals of a transitive BE-algebra X. Then $A \cap B = \{0\}$ if and only if $A \subseteq DB_l^\sim$.*

Proof It can be obtained similar to the above proof. □

Theorem 8.5.15 *Let A and B be two ideals of a transitive BE-algebra X. Then $A \cap B = \{0\}$ if and only if $A \subseteq DB^\sim$.*

Proof It can be concluded by the above two results. □

Theorem 8.5.16 *Let A be a non-empty subset of an involutory BE-algebra X. Then*

(1) $(DA_l^\sim)^\diamond = (A^\diamond)_l^\sim$, $(DA_l^\sim)^\diamond = (DA^\diamond)_l^\sim$ *and* $(DA^\sim)^\diamond = (DA^\diamond)^\sim$,
(2) $(A_l^\sim)^\diamond = D(A^\diamond)_l^\sim$, $D(A_r^\sim)^\diamond = D(A^\diamond)_r^\sim$ *and* $D(A^\sim)^\diamond = D(A^\diamond)^\sim$.

Proof (1) Let $x \in (DA_l^\sim)^\diamond$. Then $x = a^\diamond$ for some $a \in DA_l^\sim$. Hence $t^\diamond * a^\diamond \le a^\diamond$ for all $t \in A$. Thus $t^\diamond * x \le x$. Since $t^\diamond \in A^\diamond$, we get $x \in (A^\diamond)_l^\sim$. Therefore $(DA_l^\sim)^\diamond \subseteq (A^\diamond)_l^\sim$. Conversely, let $x \in (A^\diamond)_l^\sim$. Then $a * x \le x$ for all $a \in A^\diamond$. Since $a \in A^\diamond$, we get $t^\diamond * x \le x$ for all $t \in A$. Since X is involutory, we get $t^\diamond * x^{\diamond\diamond} \le x^{\diamond\diamond}$ for all $t \in A$. Thus $x^\diamond \in DA_l^\sim$, which implies that $x = x^{\diamond\diamond} \in (DA_l^\sim)^\diamond$. Therefore $(A^\diamond)_l^\sim \subseteq (DA_l^\sim)^\diamond$. Similarly, we can obtain $(DA_l^\sim)^\diamond = (DA^\diamond)_l^\sim$. Therefore $(DA^\sim)^\diamond = (DA_l^\sim \cap (DA_r^\sim))^\diamond = (DA_l^\sim)^\diamond \cap (DA_l^\sim)^\diamond = (A^\diamond)_l^\sim \cap (A^\diamond)_r^\sim = (A^\diamond)^\sim$.
(2) Similarly it can be obtained. □

Lemma 8.5.17 *For any non-empty subset A of a transitive BE-algebra, the set $[A] = \{x \in X \mid a_n^\diamond * (a_2^\diamond * (\cdots * (a_1^\diamond * x^\diamond) \cdots)) = 1$ for some $a_1, a_2, \ldots, a_n \in A\}$ is a quasi-ideal of X.*

Proof Clearly $0 \in [A]$. Let $x \in X$ and $y \in A$. Then there exists $a_1, a_2, \ldots, a_n \in A$ such that $a_n^\diamond * (a_2^\diamond * (\cdots * (a_1^\diamond * y^\diamond)\cdots)) = 1$. Clearly $y^\diamond \leq x^\diamond * y^\diamond$. Hence $y^\diamond = y^{\diamond\diamond\diamond} \leq (x^\diamond * y^\diamond)^{\diamond\diamond}$. Since X is transitive, we get

$$1 = a_n^\diamond * (a_2^\diamond * (\cdots * (a_1^\diamond * y^\diamond)\cdots)) \leq a_n^\diamond * (a_2^\diamond * (\cdots * (a_1^\diamond * (x^\diamond * y^\diamond)^{\diamond\diamond})\cdots))$$

Hence $(x^\diamond * y^\diamond)^\diamond \in [A]$. Therefore $[A]$ is a quasi-ideal of X. □

Theorem 8.5.18 *Let A be a non-empty subset of a transitive BE-algebra X. Then we have*

(1) $[A] \cap DA_l^\sim = \{0\}$,
(2) $DA_l^\sim = D[A]_l^\sim$.

Proof (1) Let $x \in [A] \cap DA_l^\sim$. Since $x \in [A]$, there exists $a_1, a_2, \ldots, a_n \in A$ such that $a_n^\diamond * (a_2^\diamond * (\cdots * (a_1^\diamond * x^\diamond)\cdots)) = 1$. Since $x \in DA_l^\sim$ and $a_1 \in A$, we get $a_1^\diamond * x^\diamond \leq x^\diamond$. Since $a_2 \in A$, it implies $a_2^\diamond * (a_1^\diamond * x^\diamond) \leq a_2^\diamond * x^\diamond \leq x^\diamond$. Continuing in this way, we get $1 = a_n^\diamond * (\cdots * (a_1^\diamond * x^\diamond)\cdots) \leq x^\diamond$. Hence $x^\diamond = 1$, which implies $x \leq x^{\diamond\diamond} = 0$. Therefore $[A] \cap DA_l^\sim = \{0\}$.
(2) Since $A \subseteq [A]$, we get $D[A]_l^\sim \subseteq DA_l^\sim$. Conversely, let $x \in DA_l^\sim$. Then $a^\diamond * x^\diamond \leq x^\diamond$ for all $a \in A$. Let $a \in [A]$. Then there exists $a_1, a_2, \ldots, a_n \in A$ such that $a_n^\diamond * (\cdots * (a_1^\diamond * x^\diamond)\cdots) = 1$. Since $a_1, a_2, \ldots, a_n \in A$, we get $a_n^\diamond * (\cdots * (a_1^\diamond * x^\diamond)\cdots) \leq a_{n-1}^\diamond * (\cdots * (a_1^\diamond * x^\diamond)\cdots) \leq \ldots \leq a_1^\diamond * x^\diamond \leq x^\diamond$. Hence $(a^\diamond * x^\diamond) * (a_n^\diamond * (\cdots * (a_1^\diamond * x^\diamond)\cdots)) \leq (a^\diamond * x^\diamond) * x^\diamond$.

$$\begin{aligned}(a^\diamond * x^\diamond) * x^\diamond &\geq (a^\diamond * x^\diamond) * (a_n^\diamond * (\cdots * (a_1^\diamond * x^\diamond)\cdots))\\ &= a_n^\diamond * (\cdots * (a_1^\diamond * ((a^\diamond * x^\diamond) * x^\diamond)\cdots))\\ &\geq a_n^\diamond * (\cdots * (a_1^\diamond * a^\diamond)\cdots) \qquad \text{since } a^\diamond \leq (a^\diamond * x^\diamond) * x^\diamond\\ &= 1\end{aligned}$$

Thus $a^\diamond * x^\diamond \leq x^\diamond$. Hence $x \in D[A]_l^\sim$. Therefore $DA_l^\sim \subseteq D[A]_l^\sim$. □

Theorem 8.5.19 *Let A be an ideal of a transitive BE-algebra X. If $DA_l^\sim$ is an ideal of X, then $DA^\sim = DA_l^\sim = DA_r^\sim$.*

Proof Clearly $DA^\sim = DA_l^\sim \cap DA_r^\sim \subseteq DA_r^\sim$. Since A and $DA_r^\sim$ are ideals such that $A \cap DA_r^\sim = \{0\}$, we get $DA_r^\sim \subseteq DA^\sim$. Hence $DA_r^\sim = DA^\sim \subseteq DA_l^\sim$. If $DA_l^\sim$ is an ideal, then we get $DA^\sim = DA_l^\sim$. Hence $DA^\sim = DA_r^\sim = DA_l^\sim$. □

Theorem 8.5.20 *The following conditions are equivalent in a transitive BE-algebra X:*

(1) *For each $a \in X$, $D\{a\}_l^\sim$ is an ideal of X,*
(2) *for each $a \in X$, $D\{a\}_l^\sim \subseteq D\{a\}_r^\sim$,*
(3) *for each $a \in X$, $D\{a\}_l^\sim = D\{a\}_r^\sim$,*
(4) *for each $x, y \in X$, $x^\diamond * y^\diamond \leq y^\diamond$ implies $y^\diamond * x^\diamond \leq x^\diamond$.*

Proof (1) $\Rightarrow$ (2): Assume that $D\{a\}_l^\sim$ is an ideal for each $a \in X$. Then by Lemma 8.5.17, we get $D\{a\}_l^\sim = D[a]_l^\sim$. Again by Theorem 8.5.18, we get that $D\{a\}_l^\sim = D[a]_l^\sim = D[a]_r^\sim \subseteq D\{a\}_r^\sim$.
(2) $\Rightarrow$ (3): Assume that $D\{a\}_l^\sim \subseteq D\{a\}_r^\sim$ for each $a \in X$. Let $a \in X$ and $x \in D\{a\}_r^\sim$. Then $x^\diamond * a^\diamond \leq a^\diamond$, which implies $a \in D\{x\}_l^\sim$. Then by the condition (2), it yields $a \in D\{x\}_r^\sim$. Therefore $a^\diamond * x^\diamond \leq x^\diamond$ and so $x \in D\{a\}_l^\sim$. Thus $D\{a\}_r^\sim \subseteq D\{a\}_l^\sim$. Therefore $D\{a\}_l^\sim = D\{a\}_r^\sim$.
(3) $\Rightarrow$ (4): Assume that $D\{a\}_l^\sim = D\{a\}_r^\sim$ for each $a \in X$. Suppose $x, y \in X$ and $x^\diamond * y^\diamond \leq y^\diamond$. Then we get $x \in D\{y\}_r^\sim = D\{y\}_l^\sim$. Therefore $y^\diamond * x^\diamond \leq x^\diamond$.
(4) $\Rightarrow$ (1): Assume the condition (4). Let $a \in X$. Then $D\{a\}_l^\sim = \{x \in X \mid a^\diamond * x^\diamond \leq x^\diamond\} = \{x \in X \mid x^\diamond * a^\diamond \leq a^\diamond\} = D\{a\}_r^\sim$. Since $D\{a\}_r^\sim$ is an ideal, we get $D\{a\}_l^\sim$ is also an ideal of X. □

Exercise

1. If X is a commutative BE-algebra and A be a non-empty subset of X, then prove that $A_l^\sim = A_r^\sim = A^\sim$ and hence A_l is a filter of X.
2. A filter F of a transitive BE-algebra X is called a direct factor of X if there exists a filter G of X such that $F \cap G = \{1\}$ and $X = \langle F \cup G \rangle$. If every filter of X is a direct factor of X, then prove that the stabilizer $\{a\}_l^\sim$ is a filter of X for each $a \in X$.
3. If each principal filter of a transitive BE-algebra X is a direct factor of X, then prove that the stabilizer $\{a\}_l^\sim$, for each $a \in X$, is a direct factor of X.
4. For any BE-algebra $(X, *, 1)$, prove that all proper subsets of $(X, \tau_\alpha)((X, \tau_\beta))$ are connected, whenever they are closed or open.
5. Let X be a BE-algebra. Is there any $A \subseteq (X, \tau_\alpha)((X, \tau_\beta))$ such that $A = X$ and $1 \notin A$.
6. Let X be a bounded and transitive BE-algebra. For any non-empty subset A of a X, prove that $DA_l^\sim$ is a subalgebra of X if and only if $A = \{0\}$.
7. Let X be a bounded BE-algebra, $a \in X - \{0, 1\}$ and $x^\diamond = 0$, for all $x \in X - \{a, 1\}$. Then prove that $DA_r^\sim = DA_l^\sim = DA^\sim = \{1\}$, for any non-empty subset $A \neq \{1\}$ of X.
8. Let A be a non-empty subset of a transitive BE-algebra X. If $DA_l^\sim$ is an ideal of X, then prove that $DA^\sim = DA_l^\sim$.

Chapter 9
States on BE-algebras

Many new fields of science require a probability theory based on non-classical logics. We know multiple-valued logics are non-classical logics and became popular in computer science since it was understood that they play a fundamental role in fuzzy logics. In analogous to probability measure, the states on multiple-valued algebras proved to be the most suitable models for averaging the truth-value in their corresponding logics. Mundici introduced states (an analog of probability measures) on MV-algebras in 1995, as averaging of the truth-value in [184]. Since middle 1990s, mainly after Mundici's paper [184], on probability theory on MV-algebras, there has been an increasing amount of study on generalizations of probability theoretical concepts, most notable states, on various logic origin algebraic structures. In [24], R. Borzooei et al. studied the states on BE-algebras. Bosbach state was introduced by R. Bosbach in [25] and [26]. The notion of a Bosbach state has been studied for other algebras of fuzzy structures such as pseudo BL-algebras [99], bounded non-commutative $R\ell$-monoids [90, 91], residuated lattices [55], pseudo BCK-semilattices and pseudo BCK-algebras [163]. In [31], C. Busneag developed the theory of state-morphisms on Hilbert algebras and get some results relative to the theory of Bosbach states on bounded and non-bounded Hilbert algebras.

In this chapter, the notions of bounded states and pseudo states are introduced on bounded BE-algebras. A set of equivalent conditions is derived for a self-map of BE-algebra to become a pseudo states. An equivalency is obtained between pseudo states and states of bounded commutative BE-algebras. Properties of pseudo states are studied with the help of congruences. The existence and uniqueness of pseudo state on a domain (co-domain) of a bounded homomorphism corresponding to a given pseudo state on the co-domain (domain) are derived. Properties of pseudo state-morphisms of bounded BE-algebras are studied. Properties of extremal pseudo states are studied.

S. R. Mukkamala, *A Course in BE-algebras*,
https://doi.org/10.1007/978-981-10-6838-6_9

9.1 States on Bounded BE-algebras

In this section, the notion of bounded states is introduced on bounded BE-algebra. Some properties of these states are derived in bounded BE-algebras. For any commutative BE-algebra X, a set of equivalent conditions is derived for a mapping $s : X \to [0, 1]$ to become a state.

Definition 9.1.1 A *BE-semilattice* is an algebra $(X, \vee, *, 1)$ where $(X, *, 1)$ is a BE-algebra whose induced poset $(x, \leq)$ is a join-semilattice with the associated operation $\vee$. A *bounded BE-semilattice* $(X, \vee, *, 0, 1)$ is a BE-semilattice containing the least element 0 w.r.t $\leq$.

Definition 9.1.2 Let $(X, *, 0, 1)$ be a bounded BE-algebra. A mapping $s : X \to [0, 1]$ is called a *bounded Bosbach state* on $(X, \vee, *, 1)$ if it satisfies the following axioms for all $a, b \in X$:

(1) $s(0) = 0$ and $s(1) = 1$,
(2) $s(a) + s(a * b) = s(b) + s(b * a)$.

Example 9.1.3 Let $(X, *, 0, 1)$ be a bounded BE-algebra. Then the function $s : X \to [0, 1]$ defined by $s(0) = 0$ and $s(x) = 1$ for every $0 \neq x \in X$ is a bounded Bosbach state on X.

Lemma 9.1.4 *Let $s : X \to [0, 1]$ be a bounded Bosbach state on a bounded BE-algebra $(X, *, 0, 1)$. Then the following conditions hold for all $a, b \in X$:*

(1) $s(a^\diamond) = 1 - s(a)$,
(2) $s(a^{\diamond\diamond}) = s(a)$,
(3) if $a \leq b$, then $s(b * a^{\diamond\diamond}) = 1 + s(a) - s(b)$.

Proof (1) For any $a \in X$, we get $s(a^\diamond) = s(a*0) = 1-s(a)+s(0) = 1-s(a)+0 = 1 - s(a)$.
(2) Let $a, b \in X$. From (1), we get $s(a^{\diamond\diamond}) = 1 - s(a^\diamond) = 1 - (1 - s(a)) = s(a)$.
(3) Let $a, b \in X$ be such that $a \leq b$. Then $s(b * a^{\diamond\diamond}) = 1 + s(a^{\diamond\diamond}) - s(b) = 1 + s(a) - s(b)$. □

Lemma 9.1.5 *Let $s : X \to [0, 1]$ be a bounded Bosbach state on a bounded and commutative BE-semilattice $(X, \vee, *, 0, 1)$. Then the following conditions hold for all $a, b \in X$:*

(1) $a \leq b$ implies $s(a) \leq s(b)$ and $s(b * a) = 1 - s(b) + s(a)$,
(2) $s(a * b) = 1 - s(a \vee b) + s(b)$,
(3) $s((a \vee b) * a^{\diamond\diamond}) = 1 - s(a^{\diamond\diamond} \vee b) + s(a)$,
(4) $s((a \vee b) * b^{\diamond\diamond}) = 1 - s(a \vee b^{\diamond\diamond}) + s(b)$.

Proof (1) Let $a \leq b$. Then $a * b = 1$. Hence $s(a) + 1 = s(a) + s(a * b) = s(b) + s(b * a)$. Hence $s(a) - s(b) = s(b * a) - 1 \leq 0$. Therefore $s(a) \leq s(b)$. Also $s(b * a) = 1 - s(b) + s(a)$.

(2) Since $b \leq a \vee b$, by (1), we get $s(a * b) = s(((a * b) * b) * b) = s((a \vee b) * b) = 1 - s(a \vee b) + s(b)$.
(3) From (2), we get $s((a \vee b) * a^{\circ\circ}) = 1 - s(a \vee b \vee a^{\circ\circ}) + s(a^{\circ\circ}) = 1 - s(a^{\circ\circ} \vee b) + s(a)$.
(4) From (3), it is clear. □

Definition 9.1.6 Let $s : Y \to [0, 1]$ be a map on a bounded BE-algebra X and $f : X \to Y$ a bounded homomorphism. Define a map $s^f(x) = (s \circ f)(x)$ for all $x \in X$.

Theorem 9.1.7 *Let $s : Y \to [0, 1]$ be a map and $f : X \to Y$ a bounded homomorphism. Then s is a bounded Bosbach state on Y if and only if s^f is a bounded Bosbach state on X.*

Proof Assume that s is a bounded Bosbach state on Y. Clearly $s^f : X \to [0, 1]$ is well-defined. First observe that $s^f(1) = s(f(1)) = s(1) = 1$. Similarly, we get $s^f(0) = 0$. Let $x, y \in X$. Then clearly $f(x), f(y) \in Y$. Since s is a bounded Bosbach state on Y, we get

$$\begin{aligned} s^f(x) + s^f(x * y) &= s(f(x)) + s(f(x * y)) \\ &= s(f(x)) + s(f(x) * f(y)) \\ &= s(f(y)) + s(f(y) * f(x)) \\ &= s(f(y)) + s(f(y * x)) \\ &= s^f(y) + s^f(y * x) \end{aligned}$$

Therefore s^f is a bounded Bosbach state on X. Conversely, assume that s^f is a bounded Bosbach state on X. Then $s(1) = s(f(1)) = s^f(1) = 1$. Similarly, we get $s(0) = 0$. Let $x, y \in Y$. Since f is onto, there exist $a, b \in X$ such that $f(a) = x$ and $f(b) = y$. Now

$$\begin{aligned} s(x) + s(x * y) &= s(f(a)) + s(f(a) * f(b)) \\ &= s^f(a) + s(f(a * b)) \\ &= s^f(a) + s^f(a * b) \\ &= s^f(b) + s^f(b * a) \\ &= s(f(b)) + s(f(b * a)) \\ &= s(f(b)) + s(f(b) * f(a)) \\ &= s(y) + s(y * x) \end{aligned}$$

Therefore s is a bounded Bosbach state on Y. □

Definition 9.1.8 Let $s_1 : X \to [0, 1]$ and $s_2 : Y \to [0, 1]$ be two bounded Bosbach states on bounded BE-algebras X and Y, respectively. Then their *direct product*

$s_1 \times s_2 : X \times Y \to [0, 1] \times [0, 1]$ is defined for all $a \in X$ and $b \in Y$ as follows:

$$(s_1 \times s_2)(a, b) = (s_1(a), s_2(b))$$

Theorem 9.1.9 *Let $s_1 : X \to [0, 1]$ and $s_2 : Y \to [0, 1]$ be two bounded Bosbach states on bounded BE-algebras X and Y, respectively. Then their direct product $s_1 \times s_2 : X \times Y \to [0, 1] \times [0, 1]$ is a bounded Bosbach state on $X \times Y$.*

Proof Since s_1, s_2 are bounded Bosbach states, we get $(s_1 \times s_2)(0, 0) = (s_1(0), s_2(0)) = (0, 0)$. Also $(s_1 \times s_2)(1, 1) = (s_1(1), s_2(1)) = (1, 1)$. Let $(a, b), (c, d) \in X \times Y$. Then we get

$$\begin{aligned}
(s_1 \times s_2)(a, b) + (s_1 \times s_2)((a, b) * (c, d)) &= (s_1(a), s_2(b)) + (s_1 \times s_2)(a * c, b * d) \\
&= (s_1(a), s_2(b)) + (s_1(a * c), s_2(b * d)) \\
&= (s_1(a) + s_1(a * c), s_2(b) + s_2(b * d)) \\
&= (s_1(c) + s_1(c * a), s_2(d) + s_2(d * b)) \\
&= (s_1(c), s_2(d)) + (s_1(c * a), s_2(d * b)) \\
&= (s_1 \times s_2)(c, d) + (s_1 \times s_2)(c * a, d * b) \\
&= (s_1 \times s_2)(c, d) + (s_1 \times s_2)((c, d) * (a, b))
\end{aligned}$$

Therefore $s_1 \times s_2$ is a bounded Bosbach state on $X \times Y$. □

Definition 9.1.10 Let X be a bounded BE-algebra. A *state relation* on X is a mapping $s : X \times X \to [0, 1] \times [0, 1]$.

Definition 9.1.11 Let $\nu : X \to [0, 1]$ be a mapping. Then the *strongest state relation* s_ν is a state relation on $X \times X$ defined by $s_\nu(x, y) = (\nu(x), \nu(y))$ for all $x, y \in X$.

Theorem 9.1.12 *Let $\nu : X \to [0, 1]$ be a mapping and s_ν the strongest state relation on $X \times X$. If ν is a bounded Bosbach state on X, then s_ν is a bounded Bosbach state on $X \times X$.*

Proof Assume that ν is a bounded Bosbach state on X. Then $s_\nu(0, 0) = (\nu(0), \nu(0)) = (0, 0)$. Similarly, we get $s_\nu(1, 1) = (1, 1)$. Let $(a, b), (c, d) \in X \times X$. Then we get

$$\begin{aligned}
s_\nu(a, b) + s_\nu((a, b) * (c, d)) &= s_\nu(a, b) + s_\nu(a * c, b * d) \\
&= (\nu(a), \nu(b)) + (\nu(a * c), \nu(b * d)) \\
&= (\nu(a) + \nu(a * c), \nu(b) + \nu(b * d)) \\
&= (\nu(c) + \nu(c * a), \nu(d) + \nu(d * b)) \\
&= (\nu(c), \nu(d)) + (\nu(c * a), \nu(d * b)) \\
&= s_\nu(c, d) + s_\nu(c * a, d * b) \\
&= s_\nu(c, d) + s_\nu((c, d) * (a, b))
\end{aligned}$$

Therefore s_ν is a bounded Bosbach state on $X \times X$. □

Definition 9.1.13 Let s be a state on a bounded BE-algebra $(X, *, 0, 1)$. Then define the *kernel* of the state s as $Ker(s) = \{x \in X \mid s(x) = 1\}$.

Lemma 9.1.14 *For any Bosbach state s on a bounded BE-algebra $(X, *, 0, 1)$, $Ker(s)$ is a filter of X. If s is a bounded state, then $Ker(s)$ is a closed filter of X.*

Proof Suppose s is a Bosbach state on X. Since $s(1) = 1$, we get $1 \in Ker(s)$. Let $a, a * b \in Ker(s)$. Then $s(a) = 1$ and $s(a * b) = 1$. Since s is a state, we get $s(b * a) + s(b) = s(a) + s(a * b) = 1 + 1 = 2$. Therefore $s(b)$ and $s(b * a)$ both must be equal to 1. Hence $s(b) = 1$, which implies $b \in Ker(s)$. Therefore $Ker(s)$ is a filter of X. Suppose s is a bounded Bosbach state on X. Let $x^{\diamond\diamond} \in Ker(s)$. Then we get $s(x^{\diamond\diamond}) = 1$. Since $x \leq x^{\diamond\diamond}$, we get $1 = s(x^{\diamond\diamond}) \leq s(x)$. Hence $x \in Ker(s)$. Conversely, let $x \in Ker(s)$. Then $s(x^{\diamond\diamond}) = 1 - s(x^{\diamond}) = 1 - (1 - s(x)) = s(x) = 1$. Hence $x^{\diamond\diamond} \in Ker(s)$. Therefore $Ker(s)$ is a closed filter of X. □

9.2 Pseudo-states on BE-algebras

In this section, the notion of pseudo-states is introduced on a bounded BE-algebra. Some important properties of these pseudo-states are derived in bounded BE-algebras. For any bounded BE-algebra X, a set of equivalent conditions is derived for a mapping $s : X \to [0, 1]$ to become a pseudo-state. Properties of dual kernels of pseudo-states of bounded BE-algebras are studied.

Definition 9.2.1 Let X be a bounded BE-algebra. A mapping $s : X \to [0, 1]$ is said to be a 0-*state* on X if it satisfies the following conditions for all $a \in X$:

(1) $s(0) = 0$,
(2) $s(a) + s(a * 0) = s(1)$.

Clearly every state on a BE-algebra X is a 0-state on X.

Theorem 9.2.2 *For any map $s : X \to [0, 1]$ with $s(1) = 1$, the following are equivalent:*

(1) *s is a 0-state,*
(2) $s(a^{\diamond}) = 1 - s(a)$,
(3) $s(a) + s(a^{\diamond}) = 1$.

Proof (1) ⇒ (2): Assume that s is a 0-state on X. Let $a \in X$, Since $s(0) = 0$, we get that $s(a^{\diamond}) = s(a * 0) = s(1) - s(a) = 1 - s(a)$.
(2) ⇒ (3): Let $a \in X$. Then by (1), we get $s(a) + s(a^{\diamond}) = s(a) + 1 - s(a) = 1$.
(3) ⇒ (1): Putting $a = 0$ in (2), we get $1 = s(0) + s(0^{\diamond}) = s(0) + s(1) = s(0) + 1$. Hence, we must have $s(0) = 0$. Again $s(a) + s(a * 0) = s(a) + s(a^{\diamond}) = 1 = s(1)$. Therefore s is a 0-state on X. □

The notion of a pseudo-state is now introduced on bounded BE-algebras.

Definition 9.2.3 Let $(X, *, 0, 1)$ be a bounded BE-algebra. A mapping $s : X \to [0, 1]$ is called a *pseudo-state* on X if it satisfies the following axioms for all $a, b \in X$:

(1) $s(0) = 0$ and $s(1) = 1$,
(2) $s(a) + s(b^\diamond * a^\diamond) = s(b) + s(a^\diamond * b^\diamond)$.

It can be easily observed that every pseudo-state is a 0-state.

Example 9.2.4 Let $(X, *, 0, 1)$ be a bounded BE-algebra. Then the function $s : X \to [0, 1]$ defined by $s(1) = 1$ and $s(x) = 0$ for every $1 \neq x \in X$ is a pseudo-state on X.

Lemma 9.2.5 *Let $s : X \to [0, 1]$ be a pseudo-state on a bounded BE-algebra $(X, *, 0, 1)$. Then the following conditions hold for all $a, b \in X$:*

(1) $s(a^\diamond) = 1 - s(a)$,
(2) $a \leq b$ *implies* $s(a) \leq s(b)$,
(3) $s(a^{\diamond\diamond}) = s(a)$,
(4) $a \leq b$ *implies* $s(a^\diamond * b^\diamond) = 1 + s(a) - s(b)$.

Proof (1) Let $a \in X$. Then, we get $s(a) + s(a^\diamond) = s(a) + s(1 * a^\diamond) = s(a) + s(0^\diamond * a^\diamond) = s(0) + s(a^\diamond * 0^\diamond) = 0 + s(a^\diamond * 1) = 0 + s(1) = 1$. Hence $s(a^\diamond) = 1 - s(a)$.
(2) Let $a \leq b$. Then $b^\diamond * a^\diamond = 1$. Hence, we get $s(a) + 1 = s(a) + s(b^\diamond * a^\diamond) = s(b) + s(a^\diamond * b^\diamond)$. Thus $s(a) - s(b) = s(a^\diamond * b^\diamond) - 1 \leq 0$. Therefore $s(a) \leq s(b)$.
(3) Let $a \in X$. From (1), we get $s(a^{\diamond\diamond}) = 1 - s(a^\diamond) = 1 - (1 - s(a)) = s(a)$.
(4) Since $a \leq b$. Then $b^\diamond \leq a^\diamond$, which implies $b^\diamond * a^\diamond = 1$. Hence $s(a) + 1 = s(a) + s(1) = s(a) + s(b^\diamond * a^\diamond) = s(b) + s(a^\diamond * b^\diamond)$. Thus $s(a^\diamond * b^\diamond) = 1 + s(a) - s(b)$. □

Proposition 9.2.6 *Every bounded Bosbach state s on a bounded BE-algebra X is a pseudo-state.*

Proof Let s be a bounded Bosbach state on X. Then clearly $s(0) = 0$. Let $x \in X$. Hence, we get $s(x^\diamond) = s(x * 0) = 1 + s(0) - s(x) = 1 - s(x)$. Let $a, b \in X$. Since s is a state, we get

$$\begin{aligned} 1 - s(a) + s(a^\diamond * b^\diamond) &= s(a^\diamond) + s(a^\diamond * b^\diamond) \\ &= s(b^\diamond) + s(b^\diamond * a^\diamond) \\ &= 1 - s(b) + s(b^\diamond * a^\diamond). \end{aligned}$$

Hence $s(a) + s(b^\diamond * a^\diamond) = s(b) + s(a^\diamond * b^\diamond)$. Therefore s is a pseudo-state on X. □

The converse of the above proposition is not true. However, in the following theorem, a sufficient condition is derived for every pseudo-state on a bounded BE-algebra to become a state.

Theorem 9.2.7 *A pseudo-state s on a bounded BE-algebra X is a bounded state if it satisfies the following condition for all $x, y \in X$:*

$$s(y^\diamond * (x * y)^\diamond) = s(x^\diamond * (y * x)^\diamond)$$

Proof Let $x, y \in X$. Since $y \leq x*y$, we get that $s(y^\diamond*(x*y)^\diamond) = 1+s(y)-s(x*y)$. Since $x \leq y*x$, similarly we get $s(x^\diamond * (y*x)^\diamond) = 1+s(x)-s(y*x)$. Then by the given condition, we get that $1+s(y)-s(x*y) = s(y^\diamond*(x*y)^\diamond) = s(x^\diamond*(y*x)^\diamond) = 1 + s(x) - s(y * x)$. Hence $s(x) + s(x * y) = s(y) + s(y * x)$. Therefore s is a bounded Bosbach state on X. □

Let $(X, *, 0, 1)$ be a bounded and commutative BE-algebra. For any $x, y \in X$, define $x \wedge y = (x^\diamond \vee y^\diamond)^\diamond$, where $x \vee y = (y * x) * x$ for all $x, y \in X$. In the following results, the equivalency between states and pseudo-states is observed. For this, we need the following facts:

Theorem 9.2.8 *Let $(X, *, 0, 1)$ be a bounded and commutative BE-algebra. For any $x, y \in X$, the following conditions hold:*

(1) $(x \vee y)^\diamond = x^\diamond \wedge y^\diamond$,
(2) $(x \wedge y)^\diamond = x^\diamond \vee y^\diamond$,
(3) $x^{\diamond\diamond} = x$,
(4) $x^\diamond * y^\diamond = y * x$.

Proof (1) Since $x, y \leq x \vee y$, we get $(x \vee y)^\diamond \leq x^\diamond, y^\diamond$. Hence $(x \vee y)^\diamond$ is a lower bound for $x^\diamond$ and $y^\diamond$. Again, let $t \in X$ such that $t \leq x^\diamond, y^\diamond$. Then we get $x \leq x^{\diamond\diamond} \leq t^\diamond$ and $y \leq y^{\diamond\diamond} \leq t^\diamond$, which implies that $x \vee y \leq t^\diamond$. Hence $t \leq t^{\diamond\diamond} \leq (x \vee y)^\diamond$. Therefore $(x \vee y)^\diamond = x^\diamond \wedge y^\diamond$.
(2) Let $x, y \in X$ be two arbitrary elements. Since X is a commutative BE-algebra, it is clear that $x^{\diamond\diamond} = x$. Then by the definition of $\wedge$, it implies that $(x \wedge y)^\diamond = (x^\diamond \vee y^\diamond)^{\diamond\diamond} = x^\diamond \vee y^\diamond$.
(3) Since X is commutative, we get that $x^{\diamond\diamond} = (x*0)*0 = (0*x)*x = 1*x = x$.
(4) Let $x, y \in X$. Then we get $x^\diamond * y^\diamond = (x * 0) * (y * 0) = y * ((x * 0) * 0) = y * x^{\diamond\diamond} = y * x$. □

Corollary 9.2.9 *Let $(X, *, 0, 1)$ be a bounded and commutative BE-algebra. Then X is a lattice with respect to the operations $\vee$ and $\wedge$ defined above.*

Theorem 9.2.10 *Let $(X, \vee, *, 1)$ be a BE-semilattice and $s : X \to [0, 1]$ a mapping such that $s(x^\diamond) = 1 - s(x)$ for all $x \in X$. Then the following are equivalent.*

(1) *s is a bounded state,*
(2) *s is a pseudo-state,*
(3) *$a \leq b$ implies $s(a^\diamond * b^\diamond) = 1 + s(a) - s(b)$,*
(4) *$s(a^\diamond * b^\diamond) = 1 + s(a \wedge b) - s(b)$,*
(5) *$s(a^\diamond) + s(a^\diamond * b^\diamond) = s(b^\diamond) + s(b^\diamond * a^\diamond)$.*

Proof (1) ⇒ (2) Assume that s is a state on X. Then $s(0) = s(1^\diamond) = 1 - s(1) = 1 - 1 = 0$. Then by the Proposition 9.2.6, s is pseudo-state on X.
(2) ⇒ (3): Assume that s is a pseudo-state on X. Let $a, b \in X$ such that $a \leq b$. Then we get $b^\diamond \leq a^\diamond$, which implies that $b^\diamond * a^\diamond = 1$. Hence $s(a)+1 = s(a)+s(b^\diamond * a^\diamond) = s(b) + s(a^\diamond * b^\diamond)$. Therefore $s(a^\diamond * b^\diamond) = 1 + s(a) - s(b)$.
(3) ⇒ (4): Let $a, b \in X$. Since $a \wedge b \leq a$, by the condition (3), we get

$$\begin{aligned} s(a^\diamond * b^\diamond) &= s(((a^\diamond * b^\diamond) * b^\diamond) * b^\diamond) \\ &= s((b^\diamond \vee a^\diamond) * b^\diamond) \\ &= s((a^\diamond \vee b^\diamond) * b^\diamond) \\ &= s((a \wedge b)^\diamond * b^\diamond) \\ &= 1 + s(a \wedge b) - s(b) \end{aligned}$$

Hence condition (4) is proved.
(4) ⇒ (5): Assume that the condition (4) holds in X. Let $a, b \in X$. Then we get the following:

$$\begin{aligned} s(a^\diamond) + s(a^\diamond * b^\diamond) &= 1 - s(a) + 1 + s(a \wedge b) - s(b) \\ &= [1 + s(a \wedge b) - s(a)] + 1 - s(b) \\ &= [1 + s(b \wedge a) - s(a)] + s(b^\diamond) \\ &= s(b^\diamond * a^\diamond) + s(b^\diamond) \end{aligned}$$

Hence condition (5) is proved.
(5) ⇒ (2): Assume the condition (5). For any $a \in X$, by the condition (5), we get $s(1) = s(a^\diamond * a^\diamond) = 1 + s(a) - s(a) = 1$. Again, we have $s(0) = s(1^\diamond) = 1 - s(1) = 0$. Let $a, b \in X$. Then, we get $1 - s(a) + s(a^\diamond * b^\diamond) = s(a^\diamond) + s(a^\diamond * b^\diamond) = s(b^\diamond) + s(b^\diamond * a^\diamond) = 1 - s(b) + s(b^\diamond * a^\diamond)$. Hence $s(a) + s(b^\diamond * a^\diamond) = s(b) + s(a^\diamond * b^\diamond)$. Therefore s is a pseudo-state on X.
(2) ⇒ (1): Let $x, y \in X$. Since X is commutative, by Theorem 9.2.8(4), we get $s(a) + s(a * b) = s(a) + s(b^\diamond * a^\diamond) = s(b) + s(a^\diamond * b^\diamond) = s(b) + s(b * a)$. Therefore s is a bounded state on X. □

Definition 9.2.11 Let s be a pseudo-state on a bounded BE-algebra $(X, *, 0, 1)$. Then define the *dual kernel* of the pseudo-state s as $Dker(s) = \{x \in X \mid s(x) = 0\}$.

Lemma 9.2.12 *For any pseudo-state s on a bounded BE-algebra X, the dual kernel $Dker(s)$ is an ideal of X.*

Proof Since $s(0) = 0$, we get that $0 \in Dker(s)$. Let $x \in Dker(s)$ and $(x^\diamond * y^\diamond)^\diamond \in Dker(s)$. Then we get $s(x) = 0$ and $s((x^\diamond * y^\diamond)^\diamond) = 0$. Hence $s(x^\diamond * y^\diamond) = 1 - s((x^\diamond * y^\diamond)^\diamond) = 1 - 0 = 1$. Since $x^\diamond \leq y^\diamond * x^\diamond$, we get $1 = s(x^\diamond) \leq s(y^\diamond * x^\diamond)$. Hence $s(y^\diamond * x^\diamond) = 1$. Now, we get $1 = 1 - 0 = s(x^\diamond * y^\diamond) - s(x) = s(y^\diamond * x^\diamond) - s(y) = 1 - s(y)$. Hence $s(y) = 0$, which implies $y \in Dker(s)$. Therefore $Dker(s)$ is an ideal of X. □

For any ideal I of a bounded and commutative BE-algebra X, we know that the relation θ_I defined, for all $x, y \in X$, by $(x, y) \in \theta_I$ if and only if $(x * y)^\diamond \in I$ and $(y * x)^\diamond \in I$ is a congruence on X. It is easily seen that the quotient algebra $(X/\theta_I, *, I_0, I_1)$ is also a bounded and transitive BE-algebra with respect to the operation given by $I_x * I_y = I_{x*y}$ and $I_x^\diamond = I_{x^\diamond}$ where $I_x = \{y \in X \mid (x, y) \in \theta_I\}$.

Proposition 9.2.13 *Let $s : X \to [0, 1]$ be a pseudo-state on a bounded and transitive BE-algebra X and put $I = Dker(s)$. Then we have the following statements:*

(1) $I_a^\diamond \subseteq I_b^\diamond \Leftrightarrow s(a^\diamond * b^\diamond) = 1 \Leftrightarrow s(a) + s(b^\diamond * a^\diamond) - s(b) = 1$,
(2) $I_a^\diamond = I_b^\diamond \Leftrightarrow s(a^\diamond * b^\diamond) = s(b^\diamond * a^\diamond) = 1 \Leftrightarrow s(a) = s(b) = s(a \wedge b)$.

Proof (1) Suppose $I_a^\diamond \subseteq I_b^\diamond \Leftrightarrow I_{a^\diamond * b^\diamond} = I_a^\diamond * I_b^\diamond = I_1 \Leftrightarrow (a^\diamond * b^\diamond, 1) \in \theta_I \Leftrightarrow (a^\diamond * b^\diamond)^\diamond = (1 * (a^\diamond * b^\diamond))^\diamond \in I = Dker(s) \Leftrightarrow s((a^\diamond * b^\diamond)^\diamond) = 0$. Hence $s(a^\diamond * b^\diamond) = 1$. From Lemma 8.2.5(4), it is clear that $s(a^\diamond * b^\diamond) = 1 \Leftrightarrow s(b) + s(b^\diamond * a^\diamond) - s(b) = 1$.
(2) Let $a, b \in X$. Since $I = Dker(s)$, we get

$$
\begin{aligned}
I_a^\diamond = I_b^\diamond &\Leftrightarrow I_{a^\diamond} = I_{b^\diamond} \\
&\Leftrightarrow (a^\diamond, b^\diamond) \in \theta_I \\
&\Leftrightarrow (a^\diamond * b^\diamond)^\diamond \in I \text{ and } (b^\diamond * a^\diamond)^\diamond \in I \\
&\Leftrightarrow s((a^\diamond * b^\diamond)^\diamond) = 0 \text{ and } s((b^\diamond * a^\diamond)^\diamond) = 0 \\
&\Leftrightarrow s(a^\diamond * b^\diamond) = 1 \text{ and } s(b^\diamond * a^\diamond) = 1 \\
&\Leftrightarrow 1 + s(a \wedge b) - s(b) = 1 \text{ and } 1 + s(b \wedge a) - s(a) = 1 \\
&\Leftrightarrow s(a \wedge b) = s(b) \text{ and } s(b \wedge a) = s(a) \\
&\Leftrightarrow s(a) = s(a \wedge b) = s(b)
\end{aligned}
$$

Hence all the three conditions of (2) are equivalent. □

If $(X, \vee, *, 1)$ is a bounded and commutative BE-semilattice, then for any ideal I of X the quotient algebra $(X/\theta_I, \vee, *, I_0, I_1)$ is also a bounded and commutative BE-semilattice in which $I_x \vee I_y = I_{x \vee y}$. In view of the condition (4) of the Theorem 9.2.10(4), the following corollary is a direct consequence of the above theorem.

Corollary 9.2.14 *Let $s : X \to [0, 1]$ be a pseudo-state on $(X, \vee, *, 0, 1)$ and put $I = Dker(s)$. Then we have the following statements.*

(1) $I_a^\diamond \subseteq I_b^\diamond \Leftrightarrow s(a^\diamond * b^\diamond) = 1 \Leftrightarrow s(a \wedge b) = s(b)$,
(2) $I_a^\diamond = I_b^\diamond \Leftrightarrow s(a^\diamond * b^\diamond) = s(b^\diamond * a^\diamond) = 1 \Leftrightarrow s(a) = s(b) = s(a \wedge b)$.

Definition 9.2.15 Let $s_1 : X \to [0, 1]$ and $s_2 : Y \to [0, 1]$ be two pseudo-states on bounded BE-algebras X and Y, respectively. Then their *direct product* $s_1 \times s_2 : X \times Y \to [0, 1] \times [0, 1]$ is defined for all $a \in X$ and $b \in Y$ as follows:

$$(s_1 \times s_2)(a, b) = (s_1(a), s_2(b))$$

Theorem 9.2.16 *Let $s_1 : X \to [0,1]$ and $s_2 : Y \to [0,1]$ be two pseudo-states on bounded BE-algebras X and Y, respectively. Then their direct product $s_1 \times s_2 : X \times Y \to [0,1] \times [0,1]$ is a pseudo-state on $X \times Y$.*

Proof Since s_1, s_2 are pseudo-states, we get $(s_1 \times s_2)(0,0) = (s_1(0), s_2(0)) = (0,0)$. Also $(s_1 \times s_2)(1,1) = (s_1(1), s_2(1)) = (1,1)$. Let $(a,b),(c,d) \in X \times Y$. Then we get

$$\begin{aligned}
(s_1 \times s_2)(a,b) + (s_1 \times s_2)((a,b)^\diamond * (c,d)^\diamond) &= (s_1 \times s_2)(a,b) + (s_1 \times s_2)((a^\diamond, b^\diamond) * (c^\diamond, d^\diamond)) \\
&= (s_1(a), s_2(b)) + (s_1 \times s_2)(a^\diamond * c^\diamond, b^\diamond * d^\diamond) \\
&= (s_1(a), s_2(b)) + (s_1(a^\diamond * c^\diamond), s_2(b^\diamond * d^\diamond)) \\
&= (s_1(a) + s_1(a^\diamond * c^\diamond), s_2(b) + s_2(b^\diamond * d^\diamond)) \\
&= (s_1(c) + s_1(c^\diamond * a^\diamond), s_2(d) + s_2(d^\diamond * b^\diamond)) \\
&= (s_1(c), s_2(d)) + (s_1(c^\diamond * a^\diamond), s_2(d^\diamond * b^\diamond)) \\
&= (s_1 \times s_2)(c,d) + (s_1 \times s_2)(c^\diamond * a^\diamond, d^\diamond * b^\diamond) \\
&= (s_1 \times s_2)(c,d) + (s_1 \times s_2)(c^\diamond * a^\diamond, d^\diamond * b^\diamond) \\
&= (s_1 \times s_2)(c,d) + (s_1 \times s_2)((c^\diamond, d^\diamond) * (a^\diamond, b^\diamond)) \\
&= (s_1 \times s_2)(c,d) + (s_1 \times s_2)((c,d)^\diamond * (a,b)^\diamond)
\end{aligned}$$

Therefore $s_1 \times s_2$ is a pseudo-state on $X \times Y$. □

Definition 9.2.17 Let X be a bounded BE-algebra. A *state relation* on X is a mapping $s : X \times X \to [0,1] \times [0,1]$.

Definition 9.2.18 Let $\nu : X \to [0,1]$ be a mapping. Then the strongest state relation s_ν is a state relation on $X \times X$ defined by $s_\nu(x,y) = (\nu(x), \nu(y))$ for all $x, y \in X$.

Theorem 9.2.19 *Let $\nu : X \to [0,1]$ be a mapping and s_ν the strongest state relation on $X \times X$. If ν is a pseudo-state on X, then s_ν is a pseudo-state on $X \times X$.*

Proof Assume that ν is a pseudo state on X. Then $s_\nu(0,0) = (\nu(0), \nu(0)) = (0,0)$. Similarly, we get $s_\nu(1,1) = (1,1)$. Let $(a,b),(c,d) \in X \times X$. Then we get

$$\begin{aligned}
s_\nu(a,b) + s_\nu((a,b)^\diamond * (c,d)^\diamond) &= s_\nu(a,b) + s_\nu(a^\diamond * c^\diamond, b^\diamond * d^\diamond) \\
&= (\nu(a), \nu(b)) + (\nu(a^\diamond * c^\diamond), \nu(b^\diamond * d^\diamond)) \\
&= (\nu(a) + \nu(a^\diamond * c^\diamond), \nu(b) + \nu(b^\diamond * d^\diamond)) \\
&= (\nu(c) + \nu(c^\diamond * a^\diamond), \nu(d) + \nu(d^\diamond * b^\diamond)) \\
&= (\nu(c), \nu(d)) + (\nu(c^\diamond * a^\diamond), \nu(d^\diamond * b^\diamond)) \\
&= s_\nu(c,d) + s_\nu(c^\diamond * a^\diamond, d^\diamond * b^\diamond) \\
&= s_\nu(c,d) + s_\nu((c^\diamond, d^\diamond) * (a^\diamond, b^\diamond)) \\
&= s_\nu(c,d) + s_\nu((c,d)^\diamond * (a,b)^\diamond)
\end{aligned}$$

Therefore s_ν is a pseudo-state on $X \times X$. □

9.3 State-morphisms on BE-algebras

In this section, the notion of state-morphisms is introduced in BE-algebras. Some properties of state-morphisms are obtained in BE-algebras. A necessary and sufficient condition is derived for a state on a BE-algebra to become a state-morphism. The concept of extremal states is also introduced and obtained that every non-trivial state-morphism is an extremal state.

Definition 9.3.1 Let $(X, *, 1)$ be a BE-algebra. A function $m : X \to [0, 1]$ is a *state-morphism* of X if it satisfies the following axiom:

(1) $m(1) = 1$,
(2) $m(a * b) = m(a) * m(b)$

for every $a, b \in X$, where the interval $[0, 1]$ is equipped with the implication $*$ defined by $a * b = \min\{1 - a + b, 1\}$ converted into a bounded BE-algebra $([0, 1], *, 0, 1)$.

Definition 9.3.2 Let $(X, *, 1)$ be a BE-algebra. A state $m : X \to [0, 1]$ is said to be *simple* if $\langle m(a)\rangle = [0, 1]$ for any $1 \neq m(a) \in [0, 1]$.

Proposition 9.3.3 [24] *Every state-morphism on a BE-algebra is a state.*

Theorem 9.3.4 [24] *Let $(X, \vee, *, 1)$ be a commutative BE-semilattice. A state $m : X \to [0, 1]$ on X is a state-morphism if and only if $m(a \vee b) = \max\{m(a), m(b)\}$ for all $a, b \in X$.*

Proposition 9.3.5 *Let $(X, \vee, *, 1)$ be a BE-semilattice with $(x * y) \vee (y * x) = 1$ for all $x, y \in X$. If $m : X \to [0, 1]$ is a state on X such that it is linearly ordered on $[0, 1]$ with respect to $\leq$, then m is a non-trivial state-morphism if and only if $ker(m)$ is a prime filter of X.*

Proof Assume that m is a non-trivial state-morphism. Hence $Ker(m)$ is a proper filter of X. Let $x, y \in X$ be such that $x \vee y \in Ker(m)$. Then $m(x \vee y) = 1$. Hence

$$\begin{aligned} m(x \vee y) = 1 &\Rightarrow m((y * x) * x) = 1 \\ &\Rightarrow m(y * x) * m(x) = 1 \\ &\Rightarrow (m(y) * m(x)) * m(x) = 1 \end{aligned}$$

Since m is linearly ordered $[0, 1]$, we get either $m(x) \leq m(y)$ or $m(y) \leq m(x)$. Suppose $m(y) \leq m(x)$. Then from the above, we get $m(x) = 1$. Hence $x \in Ker(m)$. Since X is commutative, we get $m(y \vee x) = 1 \Rightarrow (m(x) * m(y)) * m(y) = 1$. If $m(x) \leq m(y)$, then from the above, we get $m(y) = 1$. Hence, it implies that $y \in Ker(m)$. Therefore $Ker(m)$ is a prime filter in X.

Conversely, assume that $Ker(m)$ is a prime filter of X. Let $x, y \in X$. Since X satisfies the condition $(x*y) \vee (y*x) = 1$, we get $x*y \in Ker(m)$ or $y*x \in Ker(m)$.

If $x * y \in Ker(m)$. Then $m(x * y) = 1$. Since m is a state, we get the following consequence.

$$\begin{aligned} m(x) + m(x * y) = m(y * x) + m(y) &\Rightarrow m(x) + 1 = m(y * x) + m(y) \\ &\Rightarrow m(x) - m(y) = m(y * x) - 1 \leq 0 \\ &\Rightarrow m(x) \leq m(y) \\ &\Rightarrow m(x) * m(y) = 1 \end{aligned}$$

Hence $m(x * y) = 1 = m(x) * m(y)$. If $y * x \in Ker(m)$. Then $m(y * x) = 1$. Since m is a state, we get the following consequence.

$$\begin{aligned} m(x) + m(x * y) = m(y * x) + m(y) &\Rightarrow m(x) + m(x * y) = 1 + m(y) \\ &\Rightarrow m(y) - m(x) = m(x * y) - 1 \leq 0 \\ &\Rightarrow m(y) \leq m(x) \\ &\Rightarrow m(y) * m(x) = 1 \end{aligned}$$

Hence, it concludes that $m(y * x) = 1 = m(y) * m(x)$. Therefore m is a state-morphism on X. □

Proposition 9.3.6 *Let $(X, \vee, *, 1)$ be a commutative BE-semilattice with $(x * y) \vee (y * x) = 1$ for all $x, y \in X$. If $m : X \to [0, 1]$ is a simple state on X, then m is a non-trivial state-morphism if and only if its kernel $Ker(m)$ is a maximal filter of X.*

Proof Assume that m is a non-trivial state-morphism. Then clearly $Ker(m)$ is a proper filter in X. Let $a \notin Ker(m)$. We claim that $Ker(m) \vee \langle a \rangle = \langle Ker(m) \cup \{a\} \rangle = X$. Since $a \notin Ker(m)$, we get $m(a) \neq 1$. Since m is simple, we get that $\langle m(a) \rangle = [0, 1]$. Let $x \in X$. Hence $m(x) \in [0, 1] = \langle m(a) \rangle$. Hence $m(a * x) = m(a) * m(x) = 1$. Hence $a * x \in Ker(m)$. Thus $x \in Ker(m) \vee \langle a \rangle$. Hence $Ker(m) \vee \langle a \rangle = X$. Therefore $Ker(m)$ is a maximal filter in X.

Conversely, assume that $Ker(m)$ is a maximal filter. Clearly $Ker(m)$ is prime. Then by the argument of the above proposition, we get that m is a non-trivial state-morphism. □

Definition 9.3.7 Let $(X, *, 1)$ be a bounded BE-algebra. A state $s : X \to [0, 1]$ on X is said to be an *extremal state* if for any $0 < \lambda < 1$ and for any two states $s_1, s_2 : X \to [0, 1]$ such that $s = \lambda s_1 + (1 - \lambda)s_2$, then $s = s_1 = s_2$.

Proposition 9.3.8 *Let $(X, *, 1)$ be a BE-algebra and $s : X \to [0, 1]$ an extremal state on X, then $s(x) = 1$ for all $x \in X$.*

Proof Assume that $s : X \to [0, 1]$ an extremal state on X. Define $s_1, s_2 : X \to [0, 1]$ as follows:

$$s_1(x) = s(a * x) \text{ and } s_2(x) = s(x) - s(x * a) + s(a)$$

It is easy to observe that s_1 is a state on X. Let $x, y \in X$. Then

$$\begin{aligned} s_2(x * y) &= s(x * y) - s((x * y) * a) + s(a) \\ &= s(x * y) + s(x * y) - s(a * (x * y)) \\ &= 2s(x * y) - s((a * x) * (a * y)) \end{aligned}$$

Similarly, we can obtain that $s_2(y * x) = 2s(y * x) - s((a * y) * (a * x))$. Hence

$$\begin{aligned} s_2(x) + s_2(x * y) &= s(x) - s(x * a) + s(a) + 2s(x * y) - s((a * x) * (a * y)) \\ &= s(x) + s(x) - s(a * x) + 2s(x * y) - s((a * x) * (a * y)) \\ &= 2[s(x) + s(x * y)] - [s(a * x) + s((a * x) * (a * y))] \\ &= 2[s(y) + s(y * x)] - [s(a * y) + s((a * y) * (a * x))] \\ &= 2s(y) - s(a * y) + 2s(y * x) - s((a * y) * (a * x)) \\ &= 2s(y) - s(a * y) + s_2(y * x) \\ &= 2s(y) - [s(y) + s(y * a) - s(a)] + s_2(y * x) \\ &= s(y) - s(y * a) + s(a) + s_2(y * x) \\ &= s_2(y) + s_2(y * x) \end{aligned}$$

Therefore s_2 is a state on X. Clearly $s = \frac{1}{2}s_1 + \frac{1}{2}s_2$. Since s is extremal, we get that $s = s_1 = s_2$. Let $a \in X$. Then for any $x \in X$, we get $x \leq a * x$. Hence $s(x) \leq s(a * x)$. Now we have the following consequence:

$$\begin{aligned} s_1(x) = s_2(x) &\Rightarrow s(a * x) = s(x) - s(x * a) + s(a) \\ &\Rightarrow s(a * x) - s(x) = s(a) - s(x * a) \\ &\Rightarrow s(a * x) - s(x) = s(a) - s(x * a) \leq 0 \qquad \text{since } a \leq x * a \\ &\Rightarrow s(a * x) - s(x) \leq 0 \end{aligned}$$

Hence that $s(x) = s(a*x)$ for all $a, x \in X$. In particular for $a = x, s(x) = s(a*a) = s(1) = 1$. Therefore $s(x) = 1$ for all $x \in X$. □

Lemma 9.3.9 *Let $(X, *, \vee, 1)$ be a BE-semilattice and $s_1, s_2 : X \to [0, 1]$ two non-trivial simple states such that s_1 is a state-morphism. If $Ker(s_1) = Ker(s_2)$, then $s_1 = s_2$.*

Proof Since s_1 is simple, we get $Ker(s_1)$ is a maximal filter in X. Since $Ker(s_1) = Ker(s_2)$, we get $Ker(s_2)$ is also a maximal filter in X. Hence s_2 is also a state-morphism on X. Choose $x \in X$. If $x \in Ker(s_1) = Ker(s_2)$. Clearly $s_1(x) = s_2(x) = 1$. Hence $s_1 = s_2$. Suppose $x \notin Ker(s_1) = Ker(s_2)$. Hence $\langle Ker(s_1) \cup \{x\}\rangle = X$. Hence $1 \in \langle Ker(s_1) \cup \{x\}\rangle$. Thus $x = 1 * x \in Ker(s_1)$. Therefore $s_1(x) = 1$. Similarly, we get $s_2(x) = 1$. Therefore $s_1 = s_2$. □

Lemma 9.3.10 *Let $(X, *, 1)$ be a BE-algebra and $m, m_1, m_2 : X \to [0, 1]$ be three states on X such that $m = \lambda m_1 + (1 - \lambda m_2)m_2$ where $0 < \lambda < 1$. Then*

$$Ker(m) = Ker(m_1) \cap Ker(m_2)$$

Proof Suppose m, m_1, and m_2 are three states on X such that $m = \lambda m_1 + (1 - \lambda m_2)m_2$ where $0 < \lambda < 1$. Let $x \in Ker(m_1 \cap ker(m_2)$. Then, we get $m_1(x) = 1$ and $m_2(x) = 1$. Hence $m(x) = \lambda m_1(x) + (1 - \lambda)m_2(x) = \lambda + (1 - \lambda) = 1$. Hence $x \in Ker(m)$. Therefore $Ker(m_1) \cap Ker(m_2) \subseteq Ker(m)$. Conversely, let $x \in Ker(m)$. Then $m(x) = 1$. Now

$$\begin{aligned}
m(x) = 1 &\Rightarrow \lambda m_1(x) + (1 - \lambda)m_2(x) = 1 \\
&\Rightarrow 1 - \lambda m_1(x) - (1 - \lambda)m_2(x) = 0 \\
&\Rightarrow \lambda + (1 - \lambda) - \lambda m_1(x) - (1 - \lambda)m_2(x) = 0 \\
&\Rightarrow \lambda(1 - m_1(x)) + (1 - \lambda)(1 - m_2(x)) = 0 \\
&\Rightarrow 1 - m_1(x) = 0 \text{ and } 1 - m_2(x) = 0 \\
&\Rightarrow m_1(x) = 1 \text{ and } m_2(x) = 1 \\
&\Rightarrow x \in Ker(m_1) \cap Ker(m_2)
\end{aligned}$$

Hence $Ker(m) \subseteq Ker(m_1) \cap Ker(m_2)$. Therefore $Ker(m) = Ker(m_1) \cap Ker(m_2)$. □

Proposition 9.3.11 *Let $(X, \vee, *, 1)$ be a commutative BE-semilattice. Then every non-trivial state-morphism on X is an extremal state.*

Proof Assume that $m : X \to [0, 1]$ is a non-trivial state-morphism. Then $Ker(m)$ is a maximal filter of X. Suppose m_1 and m_2 be two states on X such that $m = \lambda m_1 + (1 - \lambda)m_2$ where $0 < \lambda < 1$. Then by above lemma, we get

$$Ker(m) \subseteq Ker(m_1) \cap Ker(m_2)$$

Since $Ker(m)$ is maximal, we get that $Ker(m) = Ker(m_1) = Ker(m_2)$. Hence $Ker(m_1)$ and $Ker(m_2)$ are both maximal and so m_1 and m_2 are state-morphisms on X. Hence by Lemma 8.3.9, it concludes that $m_1 = m_2$. Therefore m is an extremal state on X. □

9.4 Pseudo State-morphisms on BE-algebras

In this section, the notion of pseudo state-morphisms is introduced in BE-algebras and their properties are studied. A necessary and sufficient condition is derived for a pseudo-state on a BE-algebra to become a pseudo state-morphism. The concept of extremal states is also introduced and obtained that every non-trivial pseudo state-morphism is an extremal state.

Definition 9.4.1 Let X be a bounded BE-algebra. A function $m : X \to [0, 1]$ is said to be a *pseudo state-morphism* of X if it satisfies the following axioms:

(1) $m(0) = 0$,
(2) $m(a^\diamond * b^\diamond) = m(b) * m(a)$

for every $a, b \in X$, where the interval $[0, 1]$ is equipped with the implication $*$ defined by $a * b = \min\{1 - a + b, 1\}$ converted into a bounded BE-algebra $([0, 1], *, 0, 1)$. Moreover, $m(1) = m(0^\diamond * 0^\diamond) = m(0) * m(0) = 0 * 0 = 1$.

Definition 9.4.2 Let $(X, *, 0, 1)$ be a bounded BE-algebra. A pseudo-state $m : X \to [0, 1]$ is said to be *simple* if the lower set $[0; m(a)] = [0, 1]$ for any $1 \neq m(a) \in [0, 1]$.

Proposition 9.4.3 *Let $(X, *, 0, 1)$ be a bounded BE-algebra. Then every pseudo state-morphism on X is a pseudo-state on X.*

Proof Let m be a pseudo state-morphism on X. Clearly $m(0) = 0$ and $m(1) = 1$. Let $a, b \in X$. Then we can obtain the following consequence:

$$\begin{aligned} m(a) + m(b^\diamond * a^\diamond) &= m(a) + m(a) * m(b) \\ &= m(a) + \min\{1 - m(a) + m(b), 1\} \\ &= \min\{m(a) + 1 - m(a) + m(b), 1 + m(a)\} \\ &= \min\{1 + m(b), 1 + m(a)\} \\ &= m(b) + \min\{1, 1 - m(b) + m(a)\} \\ &= m(b) + m(b) * m(a) \\ &= m(b) + m(a^\diamond * b^\diamond). \end{aligned}$$

Therefore m is a pseudo-state on $(X, *, 0, 1)$. □

In general, the converse of the above proposition is not true. However, in the following theorem, a necessary and sufficient condition is derived for every pseudo-state on a bounded BE-algebra to become a pseudo state-morphism.

Proposition 9.4.4 *A pseudo-state m on a bounded BE-algebra is a pseudo-state-morphism if it satisfies the following condition for all $x, y \in X$:*

$$x^\diamond * y^\diamond \in Ker(m) \text{ or } y^\diamond * x^\diamond \in Ker(m)$$

Proof Let $x, y \in X$. Suppose $x^\diamond * y^\diamond \in Ker(m)$. Then $m(x^\diamond * y^\diamond) = 1$. Since m is a pseudo-state, we get the following consequence.

$$\begin{aligned} m(y) + m(x^\diamond * y^\diamond) = m(y^\diamond * x^\diamond) + m(x) &\Rightarrow m(y) + 1 = m(y^\diamond * x^\diamond) + m(x) \\ &\Rightarrow m(y) - m(x) = m(y^\diamond * x^\diamond) - 1 \leq 0 \\ &\Rightarrow m(y) \leq m(x) \\ &\Rightarrow m(y) * m(x) = 1. \end{aligned}$$

Hence $m(x^\diamond * y^\diamond) = 1 = m(y) * m(x)$. If $y^\diamond * x^\diamond \in Ker(m)$. Then $m(y^\diamond * x^\diamond) = 1$. Since m is a pseudo state, we get the following consequence.

$$\begin{aligned} m(x) + m(y^\diamond * x^\diamond) = m(x^\diamond * y^\diamond) + m(y) &\Rightarrow m(x) + 1 = m(x^\diamond * y^\diamond) + m(y) \\ &\Rightarrow m(x) - m(y) = m(x^\diamond * y^\diamond) - 1 \le 0 \\ &\Rightarrow m(x) \le m(y) \\ &\Rightarrow m(x) * m(y) = 1. \end{aligned}$$

Hence $m(y^\diamond * x^\diamond) = 1 = m(x) * m(y)$. Thus m is a pseudo state-morphism. □

Theorem 9.4.5 *A pseudo-state m on a bounded and commutative BE-semilattice $(X, \vee, *, 0, 1)$ is a pseudo state-morphism if and only if for all $a, b \in X$, $m(a \wedge b) = \min\{m(a), m(b)\}$.*

Proof Let m be a pseudo state-morphism on X. Then we have the following:

$$\begin{aligned} m(a \wedge b) &= -1 + m(a^\diamond * b^\diamond) + m(b) \\ &= -1 + (m(b) * m(a)) + m(b) \\ &= -1 + \min\{1 - m(b) + m(a), 1\} + m(b) \\ &= \min\{m(a), m(b)\}. \end{aligned}$$

Conversely, assume that the pseudo-state m satisfies the given condition. Hence

$$\begin{aligned} m(a^\diamond * b^\diamond) &= 1 + m(a \wedge b) - m(b) \\ &= 1 + \min\{m(a), m(b)\} - m(b) \\ &= \min\{1 - m(b) + m(a), 1\} \\ &= m(b) * m(a). \end{aligned}$$

Therefore m is a pseudo state-morphism on $(X, \vee, *, 0, 1)$. □

Proposition 9.4.6 *If $m : X \to [0, 1]$ is a simple and non-trivial pseudo state-morphism on a bounded BE-algebra X such that $m(x^\diamond * y^\diamond)^\diamond = m(x^\diamond) * m(y^\diamond)$ for all $x, y \in X$, then $DKer(m)$ is a proper ideal of X such that $a^\diamond * x^\diamond \in Dker(m)$ for all $a \notin Dker(m)$.*

Proof Assume that m is a simple and non-trivial pseudo state-morphism on X. Then clearly $Dker(m)$ is a proper ideal in X. Let $a \notin Dker(m)$. We claim that $[Dker(m) \cup \{a\}] = X$. Since $a \notin Dker(m)$, we get that $m(a) \neq 0$. Since m is simple, we get that $[0; m(a)] = [0, 1]$. Let $x \in X$. Hence $m(x) \in [0, 1] = [0; m(a)]$. Thus $m(a^\diamond * x^\diamond)^\diamond = m(a^\diamond) * m(x^\diamond) = m(a)^\diamond * m(x)^\diamond = 1$. Then $m(a^\diamond * x^\diamond) = 0$. Hence $a^\diamond * x^\diamond \in Dker(m)$. □

Theorem 9.4.7 *Let $(X, *, 0, 1)$ be a bounded BE-algebra and $m, m_1, m_2 : X \to [0, 1]$ be three pseudo-states on X such that $m = \lambda m_1 + (1 - \lambda m_1)m_2$ where $0 < \lambda < 1$. Then $Dker(m) = Dker(m_1) \cap Dker(m_2)$.*

Proof Let m, m_1, and m_2 be three pseudo-states on X such that $m = \lambda m_1 + (1 - \lambda m_1)m_2$ where $0 < \lambda < 1$. Let $x \in Dker(m_1) \cap Dker(m_2)$. Then we get $m_1(x) = 0$

and $m_2(x) = 0$. Hence $m(x) = \lambda m_1(x) + (1-\lambda)m_2(x) = 0$. Thus $x \in Dker(m)$. Therefore $Dker(m_1) \cap Dker(m_2) \subseteq Dker(m)$. Conversely, let $x \in Dker(m)$. Then we get $m(x) = 0$. Hence $\lambda m_1(x) + (1-\lambda)m_2(x) = 0$. Thus $m_1(x) = 0$ and $m_2(x) = 0$. Hence $x \in Dker(m_1) \cap Dker(m_2)$. Thus $Dker(m) \subseteq Dker(m_1) \cap Dker(m_2)$. Therefore $Dker(m) = Dker(m_1) \cap Dker(m_2)$ □

Theorem 9.4.8 *Let m be a state-morphism on a bounded and transitive BE-algebra X. Define a relation θ on X by $(x, y) \in \theta$ if and only if $(x * y)^\diamond \in Dker(m)$ and $(y * x)^\diamond \in Dker(m)$ for all $x, y \in Ker(m)$. Then θ is a congruence on X.*

Proof Since $Dker(m)$ is an ideal of X, by Proposition 7.6.2, the proof is clear. □

Theorem 9.4.9 *Let m be a state-morphism on a bounded and transitive BE-algebra X. Then the structure $(X_{/Dker(m)}, \circledast, \bar{0}, \bar{1})$ is a bounded and transitive BE-algebra where $\bar{x} = [x]_{Dker(m)}$ and $\bar{x} \circledast \bar{y} = \overline{x * y}$ for all $x, y \in X$.*

Proof By Theorem 2.3.3, we get $(X_{/Dker(m)}, \circledast, \bar{1})$ is a BE-algebra in which the ordering $\leq$ is given by $\bar{x} \leq \bar{y}$ if and only if $\bar{x} \circledast \bar{y} = \bar{1}$. For any $x \in X$, we have $\bar{0} \circledast \bar{x} = \overline{0 * x} = \bar{1}$. Hence $\bar{0}$ is the smallest element of $X_{/Dker(m)}$ such that $\bar{0} \leq \bar{x}$. Hence $(X_{/Dker(m)}, \circledast, \bar{0}, \bar{1})$ is a bounded BE-algebra. Since X is transitive, it can be routinely obtained that $(X_{/Dker(m)}, \circledast, \bar{0}, \bar{1})$ is transitive. For any $x \in X$, the pseudo-complement of $\bar{x}^\diamond$ is given by $\overline{x^\diamond}$. □

Theorem 9.4.10 *Let m be a state-morphism on a bounded and transitive BE-algebra X. Define a map $\widehat{m} : X_{/Dker(m)} \to [0, 1]$ by $\widehat{m}(\bar{x}) = m(x)$ for all $\bar{x} \in X_{/Dker(m)}$. Then*

(1) $\bar{x} \leq \bar{y} \Leftrightarrow m(y^\diamond * x^\diamond)^\diamond = 0 \Leftrightarrow \widehat{m}(\bar{x}) \leq \widehat{m}(\bar{y})$,
(2) $\bar{x} = \bar{y} \Leftrightarrow m(x^\diamond * y^\diamond)^\diamond = m(y^\diamond * x^\diamond)^\diamond = 0 \Leftrightarrow \widehat{m}(\bar{x}) = \widehat{m}(\bar{y})$,
(3) *$\widehat{m}$ is a pseudo state-morphism on $X_{/Dker(m)}$.*

Proof (1) Let $x, y \in X$. Then $\bar{x} \leq \bar{y} \Leftrightarrow \bar{y}^\diamond * \bar{x}^\diamond = \bar{1} \Leftrightarrow (\bar{y}^\diamond * \bar{x}^\diamond)^\diamond = (\bar{1})^\diamond \Leftrightarrow \overline{(y^\diamond * x^\diamond)^\diamond} = \bar{0} \Leftrightarrow ((y^\diamond * x^\diamond)^\diamond, 0) \in \theta \Leftrightarrow (y^\diamond * x^\diamond)^\diamond = ((y^\diamond * x^\diamond)^\diamond * 0)^\diamond \in Dker(m)$ $\Leftrightarrow (y^\diamond * x^\diamond)^\diamond \in Dker(m) \Leftrightarrow m((y^\diamond * x^\diamond)^\diamond) = 0 \Leftrightarrow 1 - m(y^\diamond * x^\diamond) = 0 \Leftrightarrow 1 - \min\{1 - m(y^\diamond) + m(x^\diamond), 1\} \Leftrightarrow 1 - \min\{1 - (1 - m(y)) + (1 - m(x)), 1\} = 0 \Leftrightarrow 1 - \min\{1 - m(x) + m(y), 1\} = 0 \Leftrightarrow 1 - m(x * y) = 0 \Leftrightarrow m(x * y) = 1 \Leftrightarrow m(x) * m(y) = 1 \Leftrightarrow m(x) \leq m(y) \Leftrightarrow \widehat{m}(\bar{x}) \leq \widehat{m}(\bar{y})$.

(2) Similar to (1), it can be proved.

(3) Let $\bar{x}, \bar{y} \in X_{/Dker(m)}$. By (2), we get $\bar{x} = \bar{y}$ if and only if $m(x) = m(y)$ if and only if $\widehat{m}(\bar{x}) = \widehat{m}(\bar{y})$. Therefore $\widehat{m}$ is well-defined. Clearly $\widehat{m}(\bar{0}) = m(0) = 0$. Also $\widehat{m}(\bar{1}) = 1$. Now $\widehat{m}(\bar{x}^\diamond * \bar{y}^\diamond) = \widehat{m}(\overline{x^\diamond * y^\diamond}) = m(x^\diamond * y^\diamond) = \min\{1 - m(x^\diamond) + m(y^\diamond), 1\} = \{1 - (1 - m(x)) + (1 - m(y)), 1\} = \{1 - m(y) + m(x), 1\} = \min\{1 - \widehat{m}(\bar{y}) + \widehat{m}(\bar{x}), 1\} = \widehat{m}(\bar{y}) * \widehat{m}(\bar{x})$. Therefore $\widehat{m}$ is a pseudo state-morphism on $X_{/Dker(m)}$. □

9.5 Homomorphisms and Pseudo-states

In this section, some more properties of pseudo-states of bounded BE-algebras are studied. Properties of the composition of pseudo-states and homomorphisms are studied. Let us first recall the definition of a homomorphism of BE-algebras.

Definition 9.5.1 Let X and Y be two bounded BE-algebras. A mapping $f : X \to Y$ is said to be a *bounded homomorphism* if $f(0) = 0$ and $f(x * y) = f(x) * f(y)$ for all $x, y \in X$.

If $f : X \to Y$ is a bounded homomorphism, then it can be easily observed that $f(0) = 0$ if and only if $f(x^\diamond) = f(x)^\diamond$ for all $x \in X$. Moreover, a bounded homomorphism f is injective if and only if $Dker(f) = \{0\}$.

Definition 9.5.2 Let $s : Y \to [0, 1]$ be a map on a bounded BE-algebra X and $f : X \to Y$ a bounded homomorphism. Define a map $s^f(x) = (s \circ f)(x)$ for all $x \in X$.

Theorem 9.5.3 *Let $s : Y \to [0, 1]$ be a map and $f : X \to Y$ a bounded surjective homomorphism. Then s is a pseudo-state on Y if and only if s^f is a pseudo-state on X.*

Proof Assume that s is a pseudo-state on Y. Clearly $s^f : X \to [0, 1]$ is well-defined. First observe that $s^f(1) = s(f(1)) = s(1) = 1$. Similarly, we get $s^f(0) = 0$. Let $x, y \in X$. Since s is a pseudo-state on X, we get

$$\begin{aligned} s^f(x) + s^f(y^\diamond * x^\diamond) &= s(f(x)) + s(f(y^\diamond * x^\diamond)) \\ &= s(f(x)) + s(f(y)^\diamond * f(x)^\diamond) \\ &= s(f(y)) + s(f(x)^\diamond * f(y)^\diamond) \\ &= s(f(y)) + s(f(x^\diamond * y^\diamond)) \\ &= s^f(y) + s^f(x^\diamond * y^\diamond). \end{aligned}$$

Therefore s^f is a pseudo-state on X. Conversely, assume that s^f is a pseudo-state on X. Then $s(1) = s(f(1)) = s^f(1) = 1$. Similarly, we get $s(0) = 0$. Let $x, y \in Y$. Since f is surjective, there exist $a, b \in X$ such that $f(a) = x$ and $f(b) = y$. Now

$$\begin{aligned} s(x) + s(y^\diamond * x^\diamond) &= s(f(a)) + s(f(b)^\diamond * f(a)^\diamond) \\ &= s^f(a) + s(f(b^\diamond * a^\diamond)) \\ &= s^f(a) + s^f(b^\diamond * a^\diamond) \\ &= s^f(b) + s^f(a^\diamond * b^\diamond) \\ &= s(f(b)) + s(f(a^\diamond * b^\diamond)) \\ &= s(f(b)) + s(f(a)^\diamond * f(b)^\diamond) \\ &= s(y) + s(x^\diamond * y^\diamond). \end{aligned}$$

Therefore s is a pseudo-state on Y. □

Theorem 9.5.4 *Let X and Y be two bounded BE-algebras and $f : X \to Y$ a bounded homomorphism. If $s : Y \to [0, 1]$ is a pseudo-state on Y, then there exists a unique pseudo-state $t : X \to [0, 1]$ such that $t = s \circ f$.*

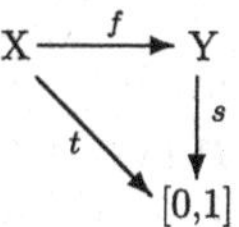

Proof Let $t = s \circ f$. Then clearly $t : X \to [0, 1]$ is a function which is well-defined. We now show that t is a pseudo-state on X. We have $t(1) = (s \circ f)(1) = s(f(1)) = s(1) = 1$. Also $t(0) = (s \circ f)(0) = s(f(0)) = s(0) = 0$. Let $x, y \in X$. Then we have the following consequence:

$$\begin{aligned}
t(x) + t(y^\diamond * x^\diamond) &= (s \circ f)(x) + (s \circ f)(y^\diamond * x^\diamond)\\
&= s(f(x)) + s(f(y^\diamond * x^\diamond))\\
&= s(f(x)) + s(f(y^\diamond) * f(x^\diamond))\\
&= s(f(x)) + s(f(y)^\diamond * f(x)^\diamond)\\
&= s(f(y)) + s(f(x)^\diamond * f(y)^\diamond)\\
&= s(f(y)) + s(f(x^\diamond) * f(y^\diamond))\\
&= s(f(y)) + s(f(x^\diamond * y^\diamond))\\
&= (s \circ f)(y) + (s \circ f)(x^\diamond * y^\diamond)\\
&= t(y) + t(x^\diamond * y^\diamond).
\end{aligned}$$

Therefore t is a pseudo-state on X. It is remaining to show that t is unique. Suppose there exists a pseudo-state $r : X \to [0, 1]$ such that $r = s \circ f$. Then we get $r(x) = (s \circ f)(x) = t(x)$ for all $x \in X$. Hence $r = t$. Therefore t is a unique pseudo-state on X such that $t = s \circ f$. □

Theorem 9.5.5 *Let X and Y be two bounded BE-algebras and $f : X \to Y$ a bounded isomorphism. If $s : X \to [0, 1]$ is a pseudo-state on X, then there exists a unique pseudo-state $t : Y \to [0, 1]$ such that $s = t \circ f$.*

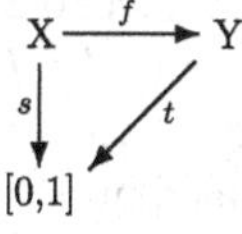

Proof Let $y \in Y$. Since f is an epimorphism, there exists $x \in X$ such that $f(x) = y$. Suppose there exists $x_1 \in X$ such that $f(x_1) = y$. Hence $f(x) = f(x_1)$. Since f is injective, $x = x_1$. Thus to each $y \in Y$, there exists a unique $s(x) \in [0, 1]$ such that y maps to $s(x)$. Hence there exists a unique function $t : Y \to [0, 1]$ such that $s = t \circ f$. We now show that t is a pseudo state on Y. Since f is bounded, we get $t(0) = t(f(0)) = (t \circ f)(0) = s(0) = 0$. Since s is a pseudo-state on X, we get $t(1) = t(f(1)) = (t \circ f)(1) = s(1) = 1$. Let $y_1, y_2 \in Y$. Since f is an epimorphism, there exist $x_1, x_2 \in X$ such that $f(x_1) = y_1$ and $f(x_2) = y_2$. Then we get the following consequence:

$$\begin{aligned}
t(y_1) + t(y_2^\diamond * y_1^\diamond) &= t(f(x_1)) + t(f(x_2)^\diamond * f(x_1)^\diamond) \\
&= t(f(x_1)) + t(f(x_2^\diamond) * f(x_1^\diamond)) \\
&= t(f(x_1)) + t(f(x_2^\diamond * x_1^\diamond)) \\
&= (t \circ f)(x_1) + (t \circ f)(x_2^\diamond * x_1^\diamond) \\
&= s(x_1) + s(x_2^\diamond * x_1^\diamond) \\
&= s(x_2) + s(x_1^\diamond * x_2^\diamond) \\
&= (t \circ f)(x_2) + (t \circ f)(x_1^\diamond * x_2^\diamond) \\
&= t(f(x_2)) + t(f(x_1^\diamond * x_2^\diamond)) \\
&= t(f(x_2)) + t(f(x_1^\diamond) * f(x_2^\diamond)) \\
&= t(f(x_2)) + t(f(x_1)^\diamond * f(x_2)^\diamond) \\
&= t(y_2) + t(y_1^\diamond * y_2^\diamond).
\end{aligned}$$

Hence t is a pseudo-state on Y. Suppose that there exists another pseudo-state $r : Y \to [0, 1]$ such that $s = r \circ f$. Let $y \in Y$. Then there exists $x \in X$ such that $f(x) = y$. Hence we get

$$\begin{aligned}
r(y) &= r(f(x)) \\
&= (r \circ f)(x) \\
&= s(x) \\
&= (t \circ f)(x) \\
&= t(f(x)) \\
&= t(y).
\end{aligned}$$

Hence $r = t$. Therefore t is a unique pseudo-state on Y such that $s = t \circ f$. □

Let X be a bounded BE-algebra and $f : X \to X$ a bounded endomorphism. Define a binary relation θ_f on X by $(x, y) \in \theta_f$ if and only if $f(x) = f(y)$ for all $x, y \in X$. Then θ_f is a congruence on X. Denote the congruence class of any $x \in X$ by $[x]_{\theta_f} = \{y \in X \mid (x, y) \in \theta_f\} = \{y \in X \mid f(x) = f(y)\}$. Then the class of all congruence classes of X denoted by $X_{/\theta_f}$, *i.e.*, $X_{/\theta_f} = \{[x]_{\theta_f} \mid x \in X\}$ is a bounded BE-algebra w.r.t. the operation $\circledast$ defined by

$$[x]_{\theta_f} \circledast [y]_{\theta_f} = [x * y]_{\theta_f},$$

and the pseudo-complement of any $[x]_{\theta_f} \in X_{/\theta_f}$ is observed as $[x]_{\theta_f}^\diamond = [x^\diamond]_{\theta_f}$. Also we denote $p_\theta : X \to X_{/\theta_f}$ by $p_\theta(x) = [x]_{\theta_f}$ for all $x \in X$. It can be easily observed that p_θ is an epimorphism. Now introduce a relation "$\leq$" on $X_{/\theta_f}$ by $[x]_{\theta_f} \leq [y]_{\theta_f}$ if and only if $[x]_{\theta_f} \circledast [y]_{\theta_f} = [1]_{\theta_f}$.

Proposition 9.5.6 *Let f be a bounded endomorphism on a bounded BE-algebra X. If $s : X \to [0,1]$ is a pseudo state on X such that $Ker(s) = Ker(f)$. Then*

(1) $[a]^{\diamond}_{\theta_f} \leq [b]^{\diamond}_{\theta_f}$ *if and only if* $s(a^{\diamond} * b^{\diamond}) = 1$,
(2) $[a]^{\diamond}_{\theta_f} = [b]^{\diamond}_{\theta_f}$ *if and only if* $s(a^{\diamond} * b^{\diamond}) = s(b^{\diamond} * a^{\diamond}) = 1$.

Proof (1) Let $a, b \in X$. Then $[a]^{\diamond}_{\theta_f} \subseteq [b]^{\diamond}_{\theta_f} \Leftrightarrow [a^{\diamond}]_{\theta_f} \subseteq [b^{\diamond}]_{\theta_f} \Leftrightarrow [a^{\diamond} * b^{\diamond}]_{\theta_f} = [a^{\diamond}]_{\theta_f} \circledast [b^{\diamond}]_{\theta_f} = [1]_{\theta_f} \Leftrightarrow (a^{\diamond} * b^{\diamond}, 1) \in \theta_f \Leftrightarrow f(a^{\diamond} * b^{\diamond}) = f(1) = 1 \Leftrightarrow a^{\diamond} * b^{\diamond} \in Ker(f) = Ker(s) \Leftrightarrow s(a^{\diamond} * b^{\diamond}) = 1$.
(2) For any $a, b \in X$, we have

$$\begin{aligned}
[a]^{\diamond}_{\theta_f} = [b]^{\diamond}_{\theta_f} &\Leftrightarrow [a^{\diamond}]_{\theta_f} = [b^{\diamond}]_{\theta_f} \\
&\Leftrightarrow (a^{\diamond}, b^{\diamond}) \in \theta_f \\
&\Leftrightarrow (a^{\diamond} * b^{\diamond}, b^{\diamond} * b^{\diamond}) \in \theta_f \text{ and } (a^{\diamond} * a^{\diamond}, b^{\diamond} * a^{\diamond}) \in \theta_f \\
&\Leftrightarrow (a^{\diamond} * b^{\diamond}, 1) \in \theta_f \text{ and } (1, b^{\diamond} * a^{\diamond}) \in \theta_f \\
&\Leftrightarrow f(a^{\diamond} * b^{\diamond}) = 1 \text{ and } f(b^{\diamond} * a^{\diamond}) = 1 \\
&\Leftrightarrow s(a^{\diamond} * b^{\diamond}) = 1 \text{ and } s(b^{\diamond} * a^{\diamond}) = 1
\end{aligned}$$

□

The following two theorems are direct consequences of Theorem 9.5.4 and Theorem 9.5.5.

Theorem 9.5.7 *Let f be a bounded endomorphism on a bounded BE-algebra X. If $s : X_{/\theta_f} \to [0,1]$ is a pseudo state on $X_{/\theta_f}$, then there exists a unique pseudo state $t : X \to [0,1]$ such that $t = s \circ p_{\theta}$.*

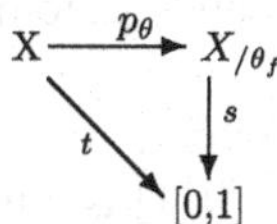

Theorem 9.5.8 *Let f be a bounded endomorphism on a transitive BE-algebra X such that $Ker(f) = \{1\}$. If $s : X \to [0,1]$ is a pseudo state on X, then there exists a unique pseudo state $t : X_{/\theta_f} \to [0,1]$ such that $s = t \circ p_{\theta}$.*

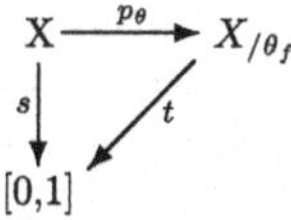

Exercise

1. If F is a maximal filter of a bounded and commutative BE-algebra X, then prove that there exists a bounded state $s : X \to [0,1]$ such that $Ker(s) = F$.
2. For any bounded Bosbach state $s : X \to [0,1]$ on a bounded BE-algebra X, prove that $s((x * y) * y) = s((y * x) * x)$ for all $x, y \in X$.

3. Let X and Y be two bounded BE-algebras and $f : X \to Y$ be a bounded homomorphism. If $s : Y \to [0, 1]$ be a bounded Bosbach state on Y, then prove that there exists a unique bounded Bosbach state $t : X \to [0, 1]$ such that $t = s \circ f$.
4. Let X and Y be two bounded BE-algebras where X is self-distributive and $f : X \to Y$ be a bounded and epimorphism such that $Ker(f) = \{1\}$. If $s : X \to [0, 1]$ is a bounded state on X, then prove that there exists a unique bounded state $t : Y \to [0, 1]$ such that $s = t \circ f$.
5. Let F be a filter of a commutative and self-distributive BE-algebra. Consider the quotient algebra $X_{/\theta}$ where θ is the known congruence on X defined by $(x, y) \in \theta$ such that $x * y \in F$ and $y * x \in F$. If $s : X \to [0, 1]$ is a state on X, then there exists a unique state $t : X_{/\theta} \to [0, 1]$.
6. Let X be a bounded and self-distributive BE-algebra and $s : X \to [0, 1]$ a bounded Bosbach state on X. Then prove that $Ker(s)$ is a maximal filter of X.
7. Let $s : X \to [0, 1]$ be a bounded Bosbach state on X and $f : Y \to X$ a bounded homomorphism. Then prove that $s \circ f$ is a bounded state-morphism on X.
8. Let $s : X \to [0, 1]$ be a bounded Bosbach state and $f : Y \to X$ a bounded homomorphism. Then $s \circ f$ is a bounded Bosbach state on Y.
9. Let $s : X \to [0, 1]$ be a pseudo state-morphism and $f : Y \to X$ a bounded homomorphism. Then prove that $s \circ f$ is a pseudo state-morphism on Y.
10. Let X and Y be two bounded BE-algebras. If $m_1 : X \to [0, 1]$ and $m_2 : Y \to [0, 1]$ are two pseudo state-morphisms on X and Y respectively, then prove that their Cartesian product $m_1 \times m_2$ is also a pseudo state-morphism on $X \times Y$.
11. Let X and Y be two bounded BE-algebras and $f : X \to Y$ be a bounded homomorphism. If $s : Y \to [0, 1]$ be a pseudo state-morphism on Y, then prove that there exists a unique pseudo state-morphism $t : X \to [0, 1]$ such that $t = s \circ f$.
12. Let $f : X \to Y$ be a bounded epimorphism, where X is self-distributive, such that $Ker(f) = \{1\}$. If $s : X \to [0, 1]$ is a pseudo state-morphism on X, then prove that there exists a unique pseudo state-morphism $t : Y \to [0, 1]$ such that $s = t \circ f$.

Chapter 10
State BE-algebras

Flaminio and Montagna were the first to present a unified approach to states and probabilistic many-valued logic in a logical and algebraic setting [95]. They added a unary operation, called internal state or state operator to the language of MV-algebras which preserves the usual properties of states. A more powerful type of logic can be given by algebraic structures with internal states, and they are also very interesting varieties of universal algebras. Di Nola and Dvurecenskij introduced the notion of a state-morphism MV-algebra which is a stronger variation of a state MV-algebra [65]. The notion of a state operator was extended by Rachunek and Salounova in [196] for the case of GMV-algebras (pseudo MV-algebras). State operators and state-morphism operators on BL-algebras were introduced and investigated in [57] and subdirectly irreducible state-morphism BL-algebras were studied in [89]. Recently, the state BCK-algebras and state-morphism BCK-algebras were defined and studied in [22].

In this chapter, the notion of state operators is introduced in BE-algebras. Properties of state operators are studied in terms of Cartesian products and strongest self-maps. Some interconnections are studied among the structures generated by the state operators. The concept of state filters is introduced, and the elements of the smallest state filter generated by a given subset are characterized. Characterization theorems of prime state filters and maximal state filters are derived, and then the celebrated Stone's theorem is generalized to the case of state filters.

The concept of state congruences is introduced in BE-algebras. State filters are characterized in terms of state congruences. A bijection is obtained between the set of all state congruences and the set of all state filters. The notion of injective filters is introduced in terms of state operators and obtained a one-to-one correspondence between the set of all injective filters of a BE-algebra and the set of all injective filters of the quotient algebra corresponding to an injective congruence. Properties of injective congruences are observed.

S. R. Mukkamala, *A Course in BE-algebras*,
https://doi.org/10.1007/978-981-10-6838-6_10

10.1 State Operators on BE-algebras

In this section, the notion of state operators is introduced in BE-algebras. Properties of state operators with respect to Cartesian products and state relations are studied in BE-algebras. A state operator is constructed on quotient structure with respect to a congruence.

Definition 10.1.1 Let $(X, *, 1)$ be a BE-algebra. A mapping $\mu : X \to X$ is called a *state operator* on X if it satisfies the following properties for all $x, y \in X$.

(SO1) $x \le y$ implies $\mu(x) \le \mu(y)$,
(SO2) $\mu(x * y) = \mu((x * y) * y) * \mu(y)$,
(SO3) $\mu(\mu(x) * \mu(y)) = \mu(x) * \mu(y)$.

In this case, the pair (X, μ) is called a state BE-algebra. The state BE-algebra (X, μ) is called self-distributive (commutative) if the underline BE-algebra X is self-distributive (commutative).

Example 10.1.2 Let $X = \{1, a, b, c\}$. Define a binary operation $*$ on X as follows:

$*$	1	a	b	c
1	1	a	b	c
a	1	1	a	a
b	1	1	1	b
c	1	1	1	1

Then $(X, *, 1)$ is a BE-algebra. Define $\mu : X \to X$ by

$$\mu(1) = \mu(a) = 1 \text{ and } \mu(b) = \mu(c) = b$$

It is easy to check that μ is a state operator on X.

Example 10.1.3 Let $(X, *, 1)$ be a BE-algebra. Then the mapping $Id_X : X \to X$ defined by $Id_X(x) = x$ for all $x \in X$ is a state operator on X.

Properties of state operators of BE-algebras are observed in the following lemma:

Lemma 10.1.4 *Let (X, μ) be a state BE-algebra. Then the following conditions hold:*

(1) $\mu(1) = 1$,
(2) $\mu(\mu(x)) = \mu(x)$,
(3) $\mu(x) * \mu(y) \ge \mu(x * y)$,
(4) $y \le x$ *implies* $\mu(x) * \mu(y) = \mu(x * y)$

for all $x, y \in X$.

Proof (1) $\mu(1) = \mu(1 * 1) = \mu((1 * 1) * 1) * \mu(1) = \mu(1) * \mu(1) = 1$.
(2) For any $x \in X$, $\mu(\mu(x)) = \mu(1 * \mu(x)) = \mu(\mu(1) * \mu(x)) = \mu(1) * \mu(x) = 1 * \mu(x) = \mu(x)$.

(3) Let $x, y \in X$. Since $x \leq (x * y) * y$, we get that $\mu(x) \leq \mu((x * y) * y)$. Therefore $\mu(x * y) = \mu((x * y) * y) * \mu(y) \leq \mu(x) * \mu(y)$.
(4) Let $x, y \in X$ be such that $y \leq x$. Then we get $y * x = 1$. Therefore $\mu(x * y) = \mu((y * x) * x) * \mu(y) = \mu(1 * x) * \mu(y) = \mu(x) * \mu(y)$. □

Theorem 10.1.5 *Let (X, μ_1) and (X, μ_2) be two state BE-algebras. Then the mapping $\mu_1 \times \mu_2$ defined on $X \times Y$, for all $a \in X$ and $b \in Y$, as follows:*

$$(\mu_1 \times \mu_2)(a, b) = (\mu_1(a), \mu_2(b))$$

is a state operator on $X \times Y$ and hence $(X \times Y, \mu_1 \times \mu_2)$ is a state BE-algebra.

Proof Clearly $X \times Y$ is a BE-algebra with respect to point-wise operations where the greatest element, say $(1, 1)$. Then we observe the following:
(SO1). Let $(a_1, b_1), (a_2, b_2) \in X \times Y$ where $a_1, a_2 \in X$ and $b_1, b_2 \in Y$. Suppose $(a_1, b_1) * (a_2, b_2) = (1, 1)$. Then $(a_1 * a_2, b_1 * b_2) = (1, 1)$. Hence $a_1 * a_2 = 1$ and $b_1 * b_2 = 1$. Since μ_1 and μ_2 are state operators on X and Y respectively, we get $\mu_1(a_1) * \mu_1(a_2) = 1$ and $\mu_2(b_1) * \mu_2(b_2) = 1$. Hence

$$\begin{aligned}(\mu_1 \times \mu_2)(a_1, b_1) * (\mu_1 \times \mu_2)(a_2, b_2) &= (\mu_1(a_1), \mu_2(b_1)) * (\mu_1(a_2), \mu_2(b_2)) \\ &= (\mu_1(a_1) * \mu_1(a_2), \mu_2(b_1) * \mu_2(b_2)) \\ &= (1, 1)\end{aligned}$$

(SO2). Let $t_1 = (a_1, b_1), t_2 = (a_2, b_2) \in X \times Y$. Since μ_1 and μ_2 are state operators on X and Y respectively, we get the following consequence:

$$\begin{aligned}(\mu_1 \times \mu_2)((t_1 * t_2) * t_2) * (\mu_1 \times \mu_2)(t_2) &= (\mu_1 \times \mu_2)((a_1 * a_2) * a_2, (b_1 * b_2) * b_2) * (\mu_1 \times \mu_2)(a_2, b_2) \\ &= (\mu_1((a_1 * a_2) * a_2), \mu_2((b_1 * b_2) * b_2)) * (\mu_1(a_2), \mu_2(b_2)) \\ &= (\mu_1((a_1 * a_2) * a_2) * \mu_1(a_2), \mu_2((b_1 * b_2) * b_2) * \mu_2(b_2)) \\ &= (\mu_1(a_1 * a_2), \mu_1(b_1 * b_2)) \\ &= (\mu_1 \times \mu_2)(a_1 * a_2, b_1 * b_2) \\ &= (\mu_1 \times \mu_2)((a_1, b_1) * (a_2, b_2)) \\ &= (\mu_1 \times \mu_2)(t_1 * t_2)\end{aligned}$$

(SO3). Let $t_1 = (a_1, b_1), t_2 = (a_2, b_2) \in X \times Y$. Since μ_1 and μ_2 are state operators on X and Y respectively, we get the following consequence:

$$\begin{aligned}(\mu_1 \times \mu_2)((\mu_1 \times \mu_2)(t_1) * (\mu_1 \times \mu_2)(t_2)) &= (\mu_1 \times \mu_2)((\mu_1(a_1), \mu_2(b_1)) * (\mu_1(a_2), \mu_2(b_2))) \\ &= (\mu_1 \times \mu_2)(\mu_1(a_1) * \mu_1(a_2), \mu_2(b_1) * \mu_2(b_2)) \\ &= (\mu_1(\mu_1(a_1) * \mu_1(a_2)), \mu_2(\mu_2(b_1) * \mu_2(b_2))) \\ &= (\mu_1(a_1) * \mu_1(a_2), \mu_2(b_1) * \mu_2(b_2)) \\ &= (\mu_1(a_1), \mu_2(b_1)) * (\mu_1(a_2), \mu_2(b_2)) \\ &= (\mu_1 \times \mu_2)(a_1, b_1) * (\mu_1 \times \mu_2)(a_2, b_2)\end{aligned}$$

Hence $\mu_1 \times \mu_2$ is a state operator on $X \times Y$. Therefore $(X \times Y, \mu_1 \times \mu_2)$ is a state BE-algebra. □

The following corollary is an immediate consequence of the above theorem.

Corollary 10.1.6 *Let μ_1 and μ_2 be two state operators on a BE-algebra X. Then $\mu_1 \times \mu_2$ is a state operator on X^2 and hence $(X^2, \mu_1 \times \mu_2)$ is a state BE-algebra.*

Definition 10.1.7 Let ν be a self-map on a BE-algebra X. Then the *strongest self-map* μ_ν is a self-map on X^2 defined by

$$\mu_\nu(x, y) = (\nu(x), \nu(y)) \quad \text{for all } x, y \in X.$$

Theorem 10.1.8 *Let ν be a self-map on a BE-algebra X and μ_ν the strongest self-map on X^2. Then ν is state operator on X if and only if μ_ν is a state operator on X^2.*

Proof Assume that ν is a state operator on X. Let $(a_1, b_1), (a_2, b_2) \in X \times X$. Suppose $(a_1, b_1) * (a_2, b_2) = (1, 1)$. Then $(a_1 * a_2, b_1 * b_2) = (1, 1)$. Hence $a_1 * a_2 = 1$ and $b_1 * b_2 = 1$. Since ν is a state operators on X, we get $\nu(a_1) * \nu(a_2) = 1$ and $\nu(b_1) * \nu(b_2) = 1$. Hence

$$\begin{aligned}
\mu_\nu(a_1, b_1) * \mu_\nu(a_2, b_2) &= (\nu(a_1), \nu(b_1)) * (\nu(a_2), \nu(b_2)) \\
&= (\nu(a_1) * \nu(a_2), \nu(b_1) * \nu(b_2)) \\
&= (1, 1).
\end{aligned}$$

Again, let $t_1 = (a_1, b_1), t_2 = (a_2, b_2) \in X \times X$. Since ν is a state operators on X, we get

$$\begin{aligned}
\mu_\nu((t_1 * t_2) * t_2) * \mu_\nu(t_2) &= \mu_\nu((a_1 * a_2) * a_2, (b_1 * b_2) * b_2) * \mu_\nu(a_2, b_2) \\
&= (\nu((a_1 * a_2) * a_2), \nu((b_1 * b_2) * b_2)) * (\nu(a_2), \nu(b_2)) \\
&= (\nu((a_1 * a_2) * a_2) * \nu(a_2), \nu((b_1 * b_2) * b_2) * \nu(b_2)) \\
&= (\nu(a_1 * a_2), \nu(a_1 * a_2)) \\
&= \mu_\nu(a_1 * a_2, b_1 * b_2) \\
&= \mu_\nu((a_1, b_1) * (a_2, b_2)) \\
&= \mu_\nu(t_1 * t_2)
\end{aligned}$$

and also

$$\begin{aligned}
\mu_\nu(\mu_\nu(t_1) * \mu_\nu(t_2)) &= \mu_\nu((\nu(a_1), \nu(b_1)) * (\nu(a_2), \nu(b_2))) \\
&= \mu_\nu(\nu(a_1) * \nu(a_2), \nu(b_1) * \nu(b_2)) \\
&= (\nu(\nu(a_1) * \nu(a_2)), \nu(\nu(b_1) * \nu(b_2))) \\
&= (\nu(a_1) * \nu(a_2), \nu(b_1) * \nu(b_2)) \\
&= (\nu(a_1), \nu(b_1)) * (\nu(a_2), \nu(b_2)) \\
&= \mu_\nu(a_1, b_1) * \mu_\nu(a_2, b_2).
\end{aligned}$$

Hence μ_ν is a state operator on X^2. Therefore (X^2, μ_ν) is a state BE-algebra.

Conversely, assume that μ_ν is a state operator on X^2. To show that ν is state operator on X, we need to obtain the following:
(SO1). Let $a, b \in X$ be such that $a*b = 1$. Then we get $(1, a)*(1, b) = (1*1, a*b) = (1, 1)$. Since μ_ν is a state operator on X^2, we get that $\mu_\nu(1, a) * \mu_\nu(1, b) = (1, 1)$. Hence $(\nu(1) * \nu(1), \nu(a) * \nu(b)) = (\nu(1), \nu(a)) * (\nu(1), \nu(b)) = (1, 1)$. Therefore $\nu(a) * \nu(b) = 1$.
(SO2). For $a, b \in X$, we have $(1, a), (1, b) \in X^2$. Since μ_ν is a state operator on X^2, we get

$$\begin{aligned}
(1, \nu((a * b) * b) * \nu(b)) &= (\nu(1) * \nu(1), \nu((a * b) * b) * \nu(b)) \\
&= (\nu(1), \nu((a * b) * b)) * (\nu(1), \nu(b)) \\
&= \mu_\nu(1, (a * b) * b) * \mu_\nu(1, b) \\
&= \mu_\nu((1 * 1) * 1, (a * b) * b) * \mu_\nu(1, b) \\
&= \mu_\nu(((1, a) * (1, b)) * (1, b)) * \mu_\nu(1, b) \\
&= \mu_\nu(1, a) * \mu_\nu(1, b) \\
&= (\nu(1), \nu(a)) * (\nu(1), \nu(b)) \\
&= (\nu(1) * \nu(1), \nu(a) * \nu(b)) \\
&= (1, \nu(a) * \nu(b)).
\end{aligned}$$

Therefore $\nu((a * b) * b) * \nu(b) = \nu(a) * \nu(b)$.
(SO3). Since μ_ν is a state operator on X^2, we get $\mu_\nu(1, 1) = (1, 1)$. Hence $(\nu(1), \nu(1)) = (1, 1)$, which implies that $\nu(1) = 1$. For $a, b \in X$, we have $(a, a), (a, b) \in X^2$. Now

$$\begin{aligned}
(1, \nu(\nu(a) * \nu(b))) &= (\nu(1), \nu(\nu(a) * \nu(b))) \\
&= \mu_\nu(1, \nu(a) * \nu(b)) \\
&= \mu_\nu(\nu(a) * \nu(a), \nu(a) * \nu(b)) \\
&= \mu_\nu((\nu(a), \nu(a)) * (\nu(a), \nu(b))) \\
&= \mu_\nu(\mu_\nu(a, a) * \mu_\nu(a, b)) \\
&= \mu_\nu(a, a) * \mu_\nu(a, b) \\
&= (\nu(a), \nu(a)) * (\nu(a), \nu(b)) \\
&= (\nu(a) * \nu(a), \nu(a) * \nu(b)) \\
&= (1, \nu(a) * \nu(b)).
\end{aligned}$$

Hence $\nu(\nu(a) * \nu(b)) = \nu(a) * \nu(b)$. Therefore ν is a state operator on X. □

Corollary 10.1.9 *Let $(X, *, 1)$ be a BE-algebra. Then (X, ν) is a state BE-algebra if and only if (X^2, μ_ν) is a state BE-algebra.*

For any state BE-algebra (X, μ), define the kernel of the state operator μ as $Ker(\mu) = \{x \in X \mid \mu(x) = 1\}$. The image of the BE-algebra X is defined as $\mu(X) = \{\mu(x) \mid x \in X\}$. Also the identity set of X is defined as $I_m(\mu) = \{x \in X \mid \mu(x) = x\}$.

Proposition 10.1.10 *Let (X, μ) be a state BE-algebra. Then the following conditions hold:*

(1) *$Ker(\mu)$ is a filter in X,*
(2) *$\mu(X)$ is a subalgebra of X,*
(3) *$Ker(\mu) \cap I_m(X) = \{1\}$.*

Proof (1) Clearly $1 \in Ker(\mu)$. Let $x, x * y \in Ker(\mu)$. Then we get that $\mu(x) = \mu(x * y) = 1$. Hence $\mu(y) = 1 * \mu(y) = \mu(x) * \mu(y) \geq \mu(x * y) = 1$. Hence $\mu(y) = 1$ and thus we get $y \in Ker(\mu)$. Therefore $Ker(\mu)$ is a filter of X.
(2) Let $\mu(x), \mu(y) \in \mu(X)$ where $x, y \in X$. Since $\mu(x) * \mu(y) = \mu(\mu(x) * \mu(y)) \in \mu(X)$, it implies that $\mu(X)$ is a subalgebra of X.
(3) Let $x \in Ker(\mu) \cap I_m(\mu)$. Then we get that $\mu(x) = 1$ and $\mu(x) = x$. Hence $x = \mu(x) = 1$. Therefore $Ker(\mu) \cap I_m(X) = \{1\}$. □

For any state BE-algebra (X, μ), consider a relation $\theta_{Ker(\mu)}$ defined for all $x, y \in X$ as

$$(x, y) \in \theta_{Ker(\mu)} \text{ if and only if } x * y \in Ker(\mu) \text{ and } y * x \in Ker(\mu)$$

It was already observed that $\theta_{Ker(\mu)}$ is a congruence on X as $Ker(\mu)$ a filter in X. Now consider $X_{/Ker(\mu)} = \{[x]_{/Ker(\mu)} \mid x \in X\}$ the set of all congruence classes where $[x]_{Ker(\mu)} = \{y \in X \mid (x, y) \in \theta_{Ker(\mu)}\}$. Then clearly $\langle X_{/Ker(\mu)}, *, 1_{/Ker(\mu)}\rangle$ is a BE-algebra where the operation $*$ defined as follows:

$$[x]_{Ker(\mu)} * [y]_{Ker(\mu)} = [x * y]_{Ker(\mu)}$$

Note that the largest element of the BE-algebra $X_{/Ker(\mu)}$ is the unit congruence class $[1]_{Ker(\mu)}$.

Theorem 10.1.11 *Let (X, μ) be a state BE-algebra. Then the map $\overline{\mu} : X_{/Ker(\mu)} \to X_{/Ker(\mu)}$ defined by*

$$\overline{\mu}([x]_{Ker(\mu)}) = [\mu(x)]_{Ker(\mu)} \quad \textit{for all } x \in X$$

is a state operator on $X_{/Ker(\mu)}$.

Proof We first observe that $\overline{\mu}$ is well-defined. Suppose $[x]_{Ker(\mu)} = [y]_{Ker(\mu)}$ where $x, y \in X$. Then we get $(x, y) \in \theta_{Ker(\mu)}$. Hence $x * y \in Ker(\mu)$ and $y * x \in Ker(\mu)$. Thus $\mu(x * y) = 1 = \mu(y * x)$. Hence $\mu(x) * \mu(y) \geq \mu(x * y) = 1$. Therefore $\mu(x) * \mu(y) = 1$. Similarly, we can obtain $\mu(y) * \mu(x) = 1$. Hence $\mu(x) = \mu(y)$. Therefore

$$\overline{\mu}([x]_{Ker(\mu)}) = [\mu(x)]_{Ker(\mu)} = [\mu(y)]_{Ker(\mu)} = \overline{\mu}([y]_{Ker(\mu)}).$$

Therefore the map $\overline{\mu}$ is well-defined. We now prove the postulates of a state operator for $\overline{\mu}$.

(SO1). Suppose $[x]_{Ker(\mu)} * [y]_{Ker(\mu)} = [1]_{Ker(\mu)}$ where $x, y \in X$. Then $[x * y]_{Ker(\mu)} = [1]_{Ker(\mu)}$. Hence $\mu(x * y) = 1$. Therefore we get the following consequence:

$$\begin{aligned}\overline{\mu}([x]_{Ker(\mu)}) * \overline{\mu}([y]_{Ker(\mu)}) &= [\mu(x)]_{Ker(\mu)} * [\mu(y)]_{Ker(\mu)} \\ &= [\mu(x) * \mu(y)]_{Ker(\mu)} \\ &\supseteq [\mu(x * y)]_{Ker(\mu)} \qquad \text{since } \mu(x) * \mu(y) \geq \mu(x * y) \\ &= [1]_{Ker(\mu)}\end{aligned}$$

Therefore $\overline{\mu}([x]_{Ker(\mu)}) * \overline{\mu}([y]_{Ker(\mu)}) = [1]_{Ker(\mu)}$.
(SO2). For any $[x]_{Ker(\mu)}, [y]_{Ker(\mu)} \in X_{/Ker(\mu)}$ where $x, y \in X$, we get the following:

$$\begin{aligned}\overline{\mu}([x]_{Ker(\mu)}) * \overline{\mu}([y]_{Ker(\mu)}) &= \overline{\mu}([x * y]_{Ker(\mu)}) \\ &= [\mu(x * y)]_{Ker(\mu)} \\ &= [\mu((x * y) * y) * \mu(y)]_{Ker(\mu)} \\ &= [\mu((x * y) * y)]_{Ker(\mu)} * [\mu(y)]_{Ker(\mu)} \\ &= \overline{\mu}([(x * y) * y)]_{Ker(\mu)} * \overline{\mu}[y]_{Ker(\mu)} \\ &= \overline{\mu}(([x]_{Ker(\mu)} * [y]_{Ker(\mu)}) * [y]_{Ker(\mu)}) * \overline{\mu}[y]_{Ker(\mu)}\end{aligned}$$

(SO3). For any $[x]_{Ker(\mu)}, [y]_{Ker(\mu)} \in X_{/Ker(\mu)}$ where $x, y \in X$, we get the following:

$$\begin{aligned}\overline{\mu}\{\overline{\mu}([x]_{Ker(\mu)}) * \overline{\mu}([y]_{Ker(\mu)})\} &= \overline{\mu}\{[\mu(x)]_{Ker(\mu)} * [\mu(y)]_{Ker(\mu)}\} \\ &= \overline{\mu}\{[\mu(x) * \mu(y)]_{Ker(\mu)}\} \\ &= [\mu(\mu(x) * \mu(y))]_{Ker(\mu)} \\ &= [\mu(x) * \mu(y)]_{Ker(\mu)} \\ &= [\mu(x)]_{Ker(\mu)} * [\mu(y)]_{Ker(\mu)} \\ &= \overline{\mu}([x]_{Ker(\mu)}) * \overline{\mu}([y]_{Ker(\mu)})\end{aligned}$$

Hence $\overline{\mu}$ is a state operator on $X_{/Ker(\mu)}$ and thus $(X_{/Ker(\mu)}, *, [1]_{Ker(\mu)})$ is a state BE-algebra. □

10.2 Structures Generated by State Operators

In this section, some structures are introduced in BE-algebras with the help of state operators. Some properties of these structures are then studied. Some interconnections among these structures are established. For any state BE-algebra (X, μ), it is proved that $\mu(X)$ is lattice.

Definition 10.2.1 A state BE-algebra (X, μ) is called *μ-distributive* if $\mu(x)*(\mu(y)*\mu(z)) = (\mu(x)*\mu(y))*(\mu(x)*\mu(z))$ for all $x, y, z \in X$.

Definition 10.2.2 A state BE-algebra (X, μ) is called *μ-commutative* if $(\mu(x)*\mu(y))*\mu(y) = (\mu(y)*\mu(x))*\mu(x)$ for all $x, y \in X$.

Definition 10.2.3 A state BE-algebra (X, μ) is called *μ-transitive* if $\mu(y)*\mu(z) \leq (\mu(x)*\mu(y))*(\mu(x)*\mu(z))$ for all $x, y \in X$.

In the following, the notion of ordered BE-algebras with respect to μ is introduced.

Definition 10.2.4 A state BE-algebra (X, μ) is called positively ordered *w.r.t.* μ if $x \leq y$ implies $\mu(z)*\mu(x) \leq \mu(z)*\mu(y)$ for all $x, y, z \in X$. It is called negatively ordered *w.r.t.* μ if $x \leq y$ implies $\mu(y)*\mu(z) \leq \mu(x)*\mu(z)$ for all $x, y, z \in X$. X is called ordered BE-algebra *w.r.t.* μ, if it is both positively ordered and negatively ordered *w.r.t.* μ.

Observation: Let X be a BE-algebra and μ a state operator on X. If X is self-distributive (commutative), then (X, μ) is μ-distributive (μ-commutative). Though the converse is not true, it is easily observed that X is self-distributive (commutative) whenever the state operator μ is a homomorphism or at least surjective.

Theorem 10.2.5 *Every μ-distributive BE-algebra is positively ordered w.r.t. μ.*

Proof Let (X, μ) be a μ-distributive BE-algebra. Let $x, y \in X$ be such that $x \leq y$. Then we get $\mu(x) \leq \mu(y)$ and thus $\mu(x)*\mu(y) = 1$. Since X is μ-distributive, we get that $(\mu(z)*\mu(x))*(\mu(z)*\mu(y)) = \mu(z)*(\mu(x)*\mu(y)) = \mu(z)*1 = 1$. Hence $\mu(z)*\mu(x) \leq \mu(z)*\mu(y)$ for all $x, y, z \in X$. Therefore X is positively ordered *w.r.t.* μ. □

Theorem 10.2.6 *Every μ-commutative BE-algebra is negatively ordered w.r.t. μ.*

Proof Let (X, μ) be a μ-commutative BE-algebra. If $x \leq y$, then we get

$$\begin{aligned}(\mu(y)*\mu(z))*(\mu(x)*\mu(z)) &= \mu(x)*((\mu(y)*\mu(z))*\mu(z)) \\ &= \mu(x)*((\mu(z)*\mu(y))*\mu(y)) \\ &= (\mu(z)*\mu(y))*(\mu(x)*\mu(y)) \\ &= (\mu(z)*\mu(y))*1 \\ &= 1.\end{aligned}$$

Hence $\mu(y)*\mu(z) \leq \mu(x)*\mu(z)$. Therefore (X, μ) is negatively ordered *w.r.t.* μ. □

Proposition 10.2.7 *Let (X, μ) be a μ-transitive BE-algebra. If $x \leq y$, then*

(1) $\mu(z)*\mu(x) \leq \mu(z)*\mu(y)$ *and* $\mu(y)*\mu(z) \leq \mu(x)*\mu(z)$,
(2) $\mu(y)*\mu(z) \leq (\mu(z)*\mu(x))*(\mu(y)*\mu(x))$ *for all* $x, y, z \in X$.

Proposition 10.2.8 *Let* (X, μ) *be a negatively ordered state BE-algebra* $w.r.t.$ μ. *Then*

$$(\mu(x) * \mu(y)) * (\mu(x) * \mu(z)) \leq \mu(x) * (\mu(y) * \mu(z)) \text{ for any } x, y, z \in X.$$

Proof Let $x, y, z \in X$. Then we get that

$$\begin{aligned}\mu(y) \leq \mu(x) * \mu(y) &\Rightarrow (\mu(x) * \mu(y)) * (\mu(x) * \mu(z)) \leq \mu(y) * (\mu(x) * \mu(z)) \\ &\Rightarrow (\mu(x) * \mu(y)) * (\mu(x) * \mu(z)) \leq \mu(x) * (\mu(y) * \mu(z))\end{aligned}$$

Therefore $(\mu(x)*\mu(y))*(\mu(x)*\mu(z)) \leq \mu(x)*(\mu(y)*\mu(z))$ for all $x, y, z \in X$. □

Proposition 10.2.9 *Let* (X, μ) *be a commutative* μ*-transitive state BE-algebra. Then for any* $x, y \in X$,

$$((\mu(x) * \mu(y)) * \mu(y)) * \mu(y) = \mu(x) * \mu(y)$$

Proof For any $x, y \in X$, consider $\mu(x) * ((\mu(x) * \mu(y)) * \mu(y)) = (\mu(x) * \mu(y)) * (\mu(x) * \mu(y)) = 1$. Hence $\mu(x) \leq (\mu(x) * \mu(y)) * \mu(y)$. By Theorem 10.2.6, X is negatively ordered $w.r.t.$ μ. Hence $((\mu(x) * \mu(y)) * \mu(y)) * \mu(y) \leq \mu(x) * \mu(y)$. Again, consider

$$(\mu(x) * \mu(y)) * (((\mu(x) * \mu(y)) * \mu(y)) * \mu(y)) = ((\mu(x) * \mu(y)) * \mu(y)) * ((\mu(x) * \mu(y)) * \mu(y)) = 1$$

Hence $\mu(x) * \mu(y) \leq ((\mu(x) * \mu(y)) * \mu(y)) * \mu(y)$. Therefore $((\mu(x) * \mu(y)) * \mu(y)) * \mu(y) = \mu(x) * \mu(y)$ for all $x, y \in X$. Hence the proof is completed. □

Theorem 10.2.10 *A* μ*-commutative BE-algebra is* μ*-distributive if and only if it satisfies the following condition for all* $x, y \in X$:

$$\mu(x) * (\mu(x) * \mu(y)) = \mu(x) * \mu(y)$$

Proof Let (X, μ) be a μ-commutative state BE-algebra. Let $x, y, z \in X$. Since (X, μ) is μ-commutative, we get $\mu(y) * \mu(z) \leq (\mu(x) * \mu(y)) * (\mu(x) * \mu(z))$. Hence, we get

$$\begin{aligned}\mu(x) * (\mu(y) * \mu(z)) &\leq \mu(x) * ((\mu(x) * \mu(y)) * (\mu(x) * \mu(z))) \\ &= (\mu(x) * \mu(y)) * (\mu(x) * (\mu(x) * \mu(z))) \\ &= (\mu(x) * \mu(y)) * (\mu(x) * \mu(z))\end{aligned}$$

On the other hand we also have by treating $\mu(x) = a$; $\mu(y) = b$ and $\mu(z) = c$

$$\begin{aligned}((a*b)*(a*c))*(a*(b*c)) &= ((a*b)*(a*c)*(b*(a*c))\\ &= b*(((a*b)*(a*c))*(a*c))\\ &= b*(((a*c)*(a*b))*(a*b))\\ &= ((a*c)*(a*b))*(b*(a*b))\\ &= ((a*c)*(a*b))*1\\ &= 1\end{aligned}$$

It follows that $(\mu(x)*\mu(y))*(\mu(x)*\mu(z)) \leq \mu(x)*(\mu(y)*\mu(z))$ and hence that $\mu(x)*(\mu(y)*\mu(z)) = (\mu(x)*\mu(y))*(\mu(x)*\mu(z))$. Therefore (X, μ) is μ-distributive. Conversely, assume that (X, μ) is μ-distributive. Hence $\mu(x)*(\mu(y)*\mu(z)) = (\mu(x)*\mu(y))*(\mu(x)*\mu(z))$ for all $x, y, z \in X$. Putting $x = y$, it yields $\mu(x)*(\mu(x)*\mu(y)) = (\mu(x)*\mu(x))*(\mu(x)*\mu(y)) = 1*(\mu(x)*\mu(y)) = \mu(x)*\mu(y)$. Hence the theorem is proved. □

Definition 10.2.11 A state BE-algebra (X, μ) is said to be μ-*implicative* if it satisfies the implicative condition $\mu(x) = (\mu(x)*\mu(y))*\mu(x)$ for all $x, y \in X$.

Proposition 10.2.12 *Every μ-distributive and μ-commutative state BE-algebra is μ-implicative.*

Proof Let (X, μ) be a μ-distributive and μ-commutative state BE-algebra. Let $x, y \in X$. Clearly $\mu(x) \leq (\mu(x)*\mu(y))*\mu(x)$. Again, we get the following:

$$\begin{aligned}((\mu(x)*\mu(y))*\mu(x))*\mu(x) &= (\mu(x)*(\mu(x)*\mu(y)))*(\mu(x)*\mu(y))\\ &= ((\mu(x)*\mu(x))*(\mu(x)*\mu(y)))*(\mu(x)*\mu(y))\\ &= (1*(\mu(x)*\mu(y)))*(\mu(x)*\mu(y))\\ &= (\mu(x)*\mu(y))*(\mu(x)*\mu(y))\\ &= 1\end{aligned}$$

Hence $(\mu(x)*\mu(y))*\mu(x) \leq \mu(x)$. Thus $(\mu(x)*\mu(y))*\mu(x) = \mu(x)$. Therefore (X, μ) is a μ-implicative state BE-algebra. □

In the following proposition, a sufficient condition is derived for a state BE-algebra to become a μ-commutative BE-algebra.

Proposition 10.2.13 *Every positively ordered and μ-implicative state BE-algebra is μ-commutative.*

Proof Let (X, μ) be a positively ordered $w.r.t.$ μ and μ-implicative state BE-algebra. Let $x, y \in X$. Since (X, μ) is μ-implicative, we get that $(\mu(y)*\mu(x))*\mu(y) = \mu(y)$. Hence

$$\mu(x)*((\mu(x)*\mu(y))*\mu(y)) = (\mu(x)*\mu(y))*(\mu(x)*\mu(y)) = 1$$

which implies that $\mu(x) \leq ((\mu(x)*\mu(y))*\mu(y))$. Since (X, μ) is positively ordered state BE-algebra $w.r.t.$ μ, we get the following consequence:

$$\begin{aligned}(\mu(y) * \mu(x)) * \mu(x) &\leq (\mu(y) * \mu(x)) * ((\mu(x) * \mu(y)) * \mu(y)) \\ &= (\mu(x) * \mu(y)) * ((\mu(y) * \mu(x)) * \mu(y)) \\ &= (\mu(x) * \mu(y)) * \mu(y)\end{aligned}$$

Hence $(\mu(y) * \mu(x)) * \mu(x) \leq (\mu(x) * \mu(y)) * \mu(y)$. Interchanging x and y, we get $(\mu(x) * \mu(y)) * \mu(y) \leq (\mu(y) * \mu(x)) * \mu(x)$. Thus $(\mu(x) * \mu(y)) * \mu(y) = (\mu(y) * \mu(x)) * \mu(x)$ for all $x, y \in X$. Therefore (X, μ) is a μ-commutative BE-algebra. □

Definition 10.2.14 A state BE-algebra (X, μ) is said to be *μ-transitive* if it satisfies the condition: $\mu(y) * \mu(z) \leq (\mu(x) * \mu(y)) * (\mu(x) * \mu(z))$ for all $x, y, z \in X$.

Proposition 10.2.15 *Every μ-commutative state BE-algebra is μ-transitive.*

Proof Let (X, μ) be a μ-commutative state BE-algebra. Let $x, y, z \in X$. Then by treating $\mu(x) = a$; $\mu(y) = b$ and $\mu(z) = c$, we get the following consequence:

$$\begin{aligned}(b * c) * ((a * b) * (a * c)) &= (a * b) * ((b * c) * (a * c)) \\ &= (a * b) * (a * ((b * c) * c)) \\ &= (a * b) * (a * ((c * b) * b)) \\ &= (a * b) * ((c * b) * (a * b)) \\ &= (c * b) * ((a * b) * (a * b) \\ &= (c * b) * 1 \\ &= 1.\end{aligned}$$

Hence $\mu(y) * \mu(z) \leq (\mu(x) * \mu(y)) * (\mu(x) * \mu(z))$. Therefore (X, μ) is μ-transitive. □

Proposition 10.2.16 *Every μ-distributive state BE-algebra is μ-transitive.*

Proof Let (X, μ) be a μ-distributive state BE-algebra. Let $x, y, z \in X$. Since (X, μ) is μ-distributive, we get that $\mu(y) * \mu(z) \leq \mu(x) * (\mu(y) * \mu(z)) = (\mu(x) * \mu(y)) * (\mu(x) * \mu(z))$. Therefore (X, μ) is μ-transitive. □

Proposition 10.2.17 *A state BE-algebra (X, μ) is μ-commutative if and only if $(\mu(x) * \mu(y)) * \mu(y) \leq (\mu(y) * \mu(x)) * \mu(x)$ for all $x, y \in X$.*

Proof The condition for necessity is clear. Conversely, assume the condition. Let $x, y \in X$. Then by the condition, we have $(\mu(x)*\mu(y))*\mu(y) \leq (\mu(y)*\mu(x))*\mu(x)$. By interchanging x and y in the given condition, we get $(\mu(y) * \mu(x)) * \mu(x) \leq (\mu(x) * \mu(y)) * \mu(y)$. Hence $(\mu(x) * \mu(y)) * \mu(y) = (\mu(y) * \mu(x)) * \mu(x)$ for all $x, y \in X$. Therefore (X, μ) is a μ-commutative state BE-algebra. □

Theorem 10.2.18 *Suppose that (X, μ) is a negatively ordered state BE-algebra w.r.t. μ. Then the following conditions are equivalent, for $x, y \in X$:*

(1) (X, μ) *is* μ*-commutative,*
(2) $x \leq y$ *implies* $\mu(y) = (\mu(y) * \mu(x)) * \mu(x)$,
(3) $(\mu(y) * \mu(x)) * \mu(x) = (((\mu(y) * \mu(x)) * \mu(x))) * \mu(y)) * \mu(y)$.

Proof (1) $\Rightarrow$ (2): Assume that (X, μ) is μ-commutative. Suppose $x \leq y$. Then $\mu(x) * \mu(y) = 1$. Hence $\mu(y) = 1 * \mu(y) = (\mu(x) * \mu(y)) * \mu(y) = (\mu(y) * \mu(x)) * \mu(x)$.
(2) $\Rightarrow$ (3): Assume the condition (2). Since $\mu(y) \leq (\mu(y)*\mu(x))*\mu(x)$, by condition (2), we get $(\mu(y) * \mu(x)) * \mu(x) = (((\mu(y) * \mu(x)) * \mu(x)) * \mu(y)) * \mu(y)$.
(3) $\Rightarrow$ (1): Assume the condition (3). Let $x, y \in X$. From the condition (3), we have $(((\mu(y) * \mu(x)) * \mu(x)) * \mu(y)) * \mu(y) = (\mu(y) * \mu(x)) * \mu(x)$. Since $\mu(x) \leq (\mu(y) * \mu(x)) * \mu(x)$, we get $((\mu(y) * \mu(x)) * \mu(x)) * \mu(y) \leq \mu(x) * \mu(y)$. Hence $(\mu(x)*\mu(y))*\mu(y) \leq (((\mu(y)*\mu(x))*\mu(x))*\mu(y))*\mu(y) = (\mu(y)*\mu(x))*\mu(x)$. Thus by the above proposition, we get (X, μ) is μ-commutative. □

Theorem 10.2.19 *A state BE-algebra* (X, μ) *is* μ*-commutative if and only if* $(\mu(z)* \mu(x)) * (\mu(y) * \mu(x)) = (\mu(x) * \mu(z)) * (\mu(y) * \mu(z))$ *for all* $x, y, z \in X$.

Proof Assume that (X, μ) is μ-commutative. Let $x, y, z \in X$. Then

$$\begin{aligned}(\mu(z) * \mu(x)) * (\mu(y) * \mu(x)) &= \mu(y) * ((\mu(z) * \mu(x)) * \mu(x)) \\ &= \mu(y) * ((\mu(x) * \mu(z)) * \mu(z)) \\ &= (\mu(x) * \mu(z)) * (\mu(y) * \mu(z)).\end{aligned}$$

Conversely, assume that $(\mu(z) * \mu(x)) * (\mu(y) * \mu(x)) = (\mu(x) * \mu(z)) * (\mu(y) * \mu(z))$ for all $x, y, z \in X$. Putting $y = 1$, it yields that $(\mu(z) * \mu(x)) * (\mu(1) * \mu(x)) = (\mu(x) * \mu(z)) * (\mu(1) * \mu(z))$. Hence, we get that $(\mu(z) * \mu(x)) * (1 * \mu(x)) = (\mu(x) * \mu(z)) * (1 * \mu(z))$. Thus $(\mu(z) * \mu(x)) * \mu(x) = (\mu(x) * \mu(z)) * \mu(z)$ for all $x, z \in X$. Therefore (X, μ) is μ-commutative. □

Theorem 10.2.20 *A negatively ordered state BE-algebra* (X, μ) *is* μ*-commutative if and only if* $(\mu(X), \leq)$ *is an upper semilattice with* $\mu(x) \vee \mu(y) = (\mu(y)*\mu(x))*\mu(x)$ *for any* $x, y \in X$.

Proof Since $\mu(y) \leq (\mu(y) * \mu(x)) * \mu(x)$ and $\mu(x) \leq (\mu(y) * \mu(x)) * \mu(x)$, we have $(\mu(y) * \mu(x)) * \mu(x)$ is an upper bound of $\mu(x)$ and $\mu(y)$ for any $x, y \in X$. Let $\mu(z)$ be any upper bound of $\mu(x)$ and $\mu(y)$. Since $\mu(x) \leq \mu(z)$, we get $\mu(z) = 1 * \mu(z) = \mu(1) * \mu(z) = (\mu(x) * \mu(z)) * \mu(z) = (\mu(z) * \mu(x)) * \mu(x)$. Also, since (X, μ) is negatively ordered *w.r.t.* μ and $\mu(y) \leq \mu(z)$, we obtain $(\mu(y) * \mu(x)) * \mu(x) \leq (\mu(z) * \mu(x)) * \mu(x) = \mu(z)$. Hence $(\mu(y) * \mu(x)) * \mu(x) \leq \mu(z)$ and $(\mu(y) * \mu(x)) * \mu(x)$ must be least upper bound of $\mu(x)$ and $\mu(y)$.

Conversely, assume that $(\mu(X), \leq)$ is an upper semilattice with $\mu(x) \vee \mu(y) = (\mu(y) * \mu(x)) * \mu(x)$ for any $x, y \in X$. Hence $(\mu(y) * \mu(x)) * \mu(x) = \mu(x) \vee \mu(y) = \mu(y) \vee \mu(x) = (\mu(x) * \mu(y)) * \mu(y)$. Therefore (X, μ) is μ-commutative. □

Proposition 10.2.21 *Let* (X, μ) *be a* μ*-commutative state BE-algebra and* $x, y, z \in X$. *Then the following conditions hold.*

(1) $\mu(x) * (\mu(y) \vee \mu(z)) = (\mu(z) * \mu(y)) * (\mu(x) * \mu(y))$,
(2) $x \leq y$ *implies* $\mu(x) \vee \mu(y) = \mu(y)$,
(3) $z \leq x$ *and* $\mu(x) * \mu(z) \leq \mu(y) * \mu(z)$ *imply* $y \leq x$.

Proof (1) Let $x, y, z \in X$ be three arbitrary elements. Then, we get that $\mu(x)*(\mu(y) \vee \mu(z)) = \mu(x) * ((\mu(z) * \mu(y)) * \mu(y)) = (\mu(z) * \mu(y)) * (\mu(x) * \mu(y))$.
(2) Let $x, y \in X$ be such that $x \leq y$. Then, we get $\mu(x) * \mu(y) = 1$. Hence $\mu(y) = 1 * \mu(y) = (\mu(x) * \mu(y)) * \mu(y) = (\mu(y) * \mu(x)) * \mu(x) = \mu(x) \vee \mu(y)$.
(3) Let $x, y, z \in X$ be such that $z \leq x$ and $\mu(x) * \mu(z) \leq \mu(y) * \mu(z)$. Then, we get $\mu(z) * \mu(x) = 1$ and $(\mu(x) * \mu(z)) * (\mu(y) * \mu(z)) = 1$. Hence

$$\begin{aligned}\mu(y) * \mu(x) &= \mu(y) * (1 * \mu(x)) \\ &= \mu(y) * ((\mu(z) * \mu(x)) * \mu(x)) \\ &= \mu(y) * ((\mu(x) * \mu(z)) * \mu(z)) \\ &= (\mu(x) * \mu(z)) * (\mu(y) * \mu(z)) \\ &= 1\end{aligned}$$

Therefore $\mu(y) \leq \mu(x)$. Hence the proof is completed. □

Theorem 10.2.22 *Let* (X, μ) *be a* μ*-commutative state BE-algebra. If there is a lower bound* $\mu(a)$ *of* $\mu(x)$ *and* $\mu(y)$*, then the greatest lower bound* $\mu(x) \wedge \mu(y)$ *of* $\mu(x)$ *and* $\mu(y)$ *exists and* $\mu(x) \wedge \mu(y) = ((\mu(x) * \mu(a)) \vee (\mu(y) * \mu(a))) * \mu(a)$.

Proof Assume that $\mu(a) \leq \mu(x), \mu(y)$. Clearly $\mu(a) \leq \mu(x) \wedge \mu(y)$. Since $\mu(x) * \mu(a) \leq (\mu(x) * \mu(a)) \vee (\mu(y) * \mu(a))$, we get

$$\begin{aligned}[(\mu(x) * \mu(a)) \vee (\mu(y) * \mu(a))] * \mu(a) &\leq (\mu(x) * \mu(a)) * \mu(a) \\ &= \mu(a) \vee \mu(x) \\ &= \mu(x)\end{aligned}$$

Similarly, we can obtain that $[(\mu(x) * \mu(a)) \vee (\mu(y) * \mu(a))] * \mu(a) \leq \mu(y)$. Hence $[(\mu(x) * \mu(a)) \vee (\mu(y) * \mu(a))] * \mu(a)$ is a lower bound of $\mu(x)$ and $\mu(y)$. Suppose $\mu(b)$ is another lower bound for $\mu(x)$ and $\mu(y)$, i.e., $\mu(b) \leq \mu(x), \mu(y)$. Hence $\mu(x) * \mu(a) \leq \mu(b) * \mu(a)$ and $\mu(y) * \mu(a) \leq \mu(b) * \mu(a)$. Hence $(\mu(x) * \mu(a)) \vee (\mu(y) * \mu(a)) \leq \mu(b) * \mu(a)$. Therefore we get

$$\begin{aligned}\mu(b) &\leq \mu(b) \vee \mu(a) \\ &= \mu(a) \vee \mu(b) \\ &= (\mu(b) * \mu(a)) * \mu(a) \\ &\leq [(\mu(x) * \mu(a)) \vee (\mu(y) * \mu(a))] * \mu(a)\end{aligned}$$

Hence $[(\mu(x) * \mu(a)) \vee (\mu(y) * \mu(a))] * \mu(a)$ is the greatest lower bound of $\mu(x)$ and $\mu(y)$. Therefore $\mu(x) \wedge \mu(y) = [(\mu(x) * \mu(a)) \vee (\mu(y) * \mu(a))] * \mu(a)$. □

Corollary 10.2.23 *Let (X, μ) is a μ-commutative state BE-algebra. If the lower bound exists for every two elements of $\mu(X)$, then $(\mu(X), \vee, \wedge)$ is a lattice.*

Proposition 10.2.24 *Let (X, μ) be a μ-commutative state BE-algebra and $x, y, z \in X$. Then the following conditions hold.*

(1) $(\mu(x) \vee \mu(y)) * \mu(z) = (\mu(x) * \mu(z)) \wedge (\mu(y) * \mu(z))$,
(2) $(\mu(x) \wedge \mu(y)) * \mu(z) = (\mu(x) * \mu(z)) \vee (\mu(y) * \mu(z))$.

Proof (1). Since $\mu(x), \mu(y) \leq \mu(x) \vee \mu(y)$, we get that $(\mu(x) \vee \mu(y)) * \mu(z) \leq \mu(x) * \mu(z)$ and $(\mu(x) \vee \mu(y)) * \mu(z) \leq \mu(y) * \mu(z)$. Hence $(\mu(x) \vee \mu(y)) * \mu(z)$ is a lower bound for $\mu(x) * \mu(z)$and $\mu(y) * \mu(z)$. Let $\mu(u)$ be a lower bound for $\mu(x)*\mu(z)$ and $\mu(y)*\mu(z)$. Hence $\mu(u) \leq \mu(x)*\mu(z)$ and $\mu(u) \leq \mu(y)*\mu(z)$ and so $\mu(x) \leq \mu(u) * \mu(z)$ and $\mu(y) \leq \mu(u) * \mu(z)$. Therefore $\mu(x) \vee \mu(y) \leq \mu(u) * \mu(z)$ and thus $\mu(u) \leq (\mu(x) \vee \mu(y)) * \mu(z)$. Therefore $(\mu(x) \vee \mu(y)) * \mu(z)$ is the greatest lower bound for $\mu(x) * \mu(z)$ and $\mu(y) * \mu(z)$. Hence $(\mu(x) \vee \mu(y)) * \mu(z) = (\mu(x) * \mu(z)) \wedge (\mu(y) * \mu(z))$.
(2). By using the dual argument, it can be followed by (1). □

Theorem 10.2.25 *Let (X, μ) be a μ-commutative BE-algebra in which $\mu(X)$ is closed under $\wedge$. Then $\mu(X)$ is a distributive lattice with respect to the operations $\vee$ and $\wedge$ defined in the above theorems.*

Proof By (1) and (2) of the above proposition, we get

$$\begin{aligned}\mu(x) \vee (\mu(y) \wedge \mu(z)) &= ((\mu(y) \wedge \mu(z)) * \mu(x)) * \mu(x)\\ &= ((\mu(y) * \mu(x)) \vee (\mu(z) * \mu(x))) * \mu(x)\\ &= ((\mu(y) * \mu(x)) * \mu(x)) \wedge ((\mu(z) * \mu(x)) * \mu(x))\\ &= (\mu(x) \vee \mu(y)) \wedge (\mu(x) \vee \mu(z))\end{aligned}$$

Therefore $(\mu(X), \vee, \wedge)$ is a distributive lattice. □

10.3 State Filters of State BE-algebras

In this section, the notion of state filters is introduced in state BE-algebras. Properties of state filters are studied with respect to homomorphic images and Cartesian products. The elements of the smallest state filters generated by a non-empty set are characterized.

Definition 10.3.1 Let (X, μ) be a state BE-algebra. A non-empty subset A of X is said to be a *state subset* of (X, μ) if $\mu(A) \subseteq A$. Moreover, a filter F of X is called a *state filter* of (X, μ) if $\mu(F) \subseteq F$.

Obviously, the filters $\{1\}$ and X are state filters of the state BE-algebra (X, μ). For any state BE-algebra (X, μ), it is clear that the filter $Ker(\mu)$ is a state filter in X.

It can be easily seen that the intersection of any two state filters of a state BE-algebra is again a state filter. Moreover, we have $\bigcap_{F \in \mathcal{F}_s(X)} F = \{1\}$, where $\mathcal{F}_s(X)$ is the set of all state filters of a state BE-algebra (X, μ).

Example 10.3.2 Let $(X, *, 1)$ be a BE-algebra and F a filter of X. For any $a \in X - F$, define a self-map $\mu : X \to X$ as follows:

$$\mu_a(x) = \begin{cases} 1 \text{ if } x \in F \\ a \text{ if } x \notin F \end{cases}$$

for all $x \in X$. Then clearly (X, μ) is a state BE-algebra. For any $x \in X$ with $x \in F$, we get $\mu(x) = 1 \in F$. Hence $\mu(F) \subseteq F$. Therefore F is a state filter of (X, μ).

Proposition 10.3.3 *Let (X, μ) be a self-distributive state BE-algebra. Then for any $a \in X$, the set μ_a defined by $\mu_a = \{x \in X \mid a * \mu(x) = 1\}$ is a state filter of (X, μ).*

Proof Clearly $1 \in \mu_a$. Let $x, x*y \in \mu_a$. Then we get $a*\mu(x) = 1$ and $a*\mu(x*y) = 1$. Hence $a * \mu(y) = 1 * (a * \mu(y)) = (a * \mu(x)) * (a * \mu(y)) = a * (\mu(x) * \mu(y)) \geq a * \mu(x * y) = 1$. Thus $y \in \mu_a$, which implies that μ_a is a filter of X. Now, let $x \in \mu_a$. Then $a * \mu(x) = 1$, which gives $a * \mu^2(x) = 1$. Hence $\mu(x) \in \mu_a$. Therefore μ_a is a state filter of (X, μ). □

In the following theorem, a set of equivalent conditions is derived for every filter of a self-distributive state BE-algebra to become a state filter.

Theorem 10.3.4 *Let (X, μ) be a self-distributive state BE-algebra. Then the following conditions are equivalent:*

(1) *every filter is a state filter,*
(2) *for each $x \in X$, $\langle x \rangle$ is a state filter,*
(3) *for each $x \in X$, $x * \mu(x) = 1$.*

Proof (1) $\Rightarrow$ (2): It is clear.
(2) $\Rightarrow$ (3): Assume that $\langle x \rangle$ is a state filter for each $x \in X$. Since $x \in \langle x \rangle$ and $\langle x \rangle$ is a state filter of X, we get that $\mu(x) \in \langle x \rangle$. Since X is self-distributive, it yields that $x * \mu(x) = 1$.
(3) $\Rightarrow$ (1): Let F be a filter of X. Let $x \in F$. Then we get $x * \mu(x) = 1$, which implies $x \leq \mu(x)$. Hence $\mu(x) \in F$. Therefore F is a state filter of X. □

In the following proposition, we observe that the images and inverse images of a state filter under a state operator on a state BE-algebra are again state filters.

Proposition 10.3.5 *Let (X, μ) be a state BE-algebra. Then for any state filter F of (X, μ), we have the following conditions.*

(1) *$\mu(F)$ is a state filter of X provided $\mu(F)$ is a filter of X.*
(2) *$\mu^{-1}(F)$ is a state filter of X provided μ^{-1} exists.*

Proof (1) Let $x \in \mu(F)$. Then $x = \mu(a)$ for some $a \in F$. Since F is a state filter, we get $\mu(a) \in F$. Hence $\mu(x) = \mu^2(a) \in \mu(F)$. Therefore $\mu(F)$ is a state filter of X.
(2) Let F be a state filter of X and $x \in \mu^{-1}(F)$. Then $\mu(x) \in F$. Since F is a state filter, we get that $\mu^2(x) = \mu(x) \in F$. Hence $\mu(x) \in \mu^{-1}(F)$. Therefore $\mu^{-1}(F)$ is a state filter of X. □

Theorem 10.3.6 *If F_1 and F_2 are two state filters of state BE-algebras (X_1, μ) and (X_2, μ) respectively, then $F_1 \times F_2$ is a state filter of the product algebra $X_1 \times X_2$. Conversely, every state filter F of $X_1 \times X_2$ can be expressed as $F = F_1 \times F_2$ where F_i is a state filter of X_i for $i = 1, 2$.*

Proof Let F_1 and F_2 be two state filters of (X_1, μ) and (X_2, μ), respectively. By Theorem 4.1.9, $F_1 \times F_2$ is a filter of $X_1 \times X_2$. Define $\overline{\mu} : X_1 \times X_2 \to X_1 \times X_2$ by $\overline{\mu}(a, b) = (\mu(a), \mu(b))$ for all $a \in X_1$ and $b \in X_2$. It can be easily observed that $\overline{\mu}$ is a state operator on $X_1 \times X_2$. Hence $(X_1 \times X_2, \overline{\mu})$ is a state BE-algebra. We now intend to show that $F_1 \times F_2$ is a state filter of $X_1 \times X_2$. Let $(a, b) \in F_1 \times F_2$. Then $a \in F_1$ and $b \in F_2$. Since F_1 and F_2 are state filters, we get that $\overline{\mu}(a, b) = (\mu(a), \mu(b)) \in F_1 \times F_2$. Hence $F_1 \times F_2$ is a state filter of $(X_1 \times X_2, \overline{\mu})$.

Conversely, let F be any state filter of $X_1 \times X_2$. Consider the projections $\Pi_i : X_1 \times X_2 \to X_i$ for $i = 1, 2$. Let F_1 and F_2 be the projections of F on X_1 and X_2, respectively. That is $\Pi_i(F) = F_i$ for $i = 1, 2$. By Theorem 4.1.9, we get F_1 and F_2 are filters of X_1 and X_2, respectively. Let $x \in F_1$. Then $(x, 1) \in F$. Since F is a state filter, we get $(\mu(x), \mu(1)) = \overline{\mu}(x, 1) \in F$. Thus $\mu(x) = \Pi_1(x, 1) \in \Pi_1(F) = F_1$. Therefore F_1 is a state filter of X_1. Similarly, we can get F_2 is a state filter of X_2. Moreover $F = F_1 \times F_2$ due to Theorem 4.1.9. □

Corollary 10.3.7 *Let $X = X_1 \times X_2 \times \cdots \times X_n$ be the Cartesian product of the BE-algebras $X_1, X_2, \ldots, X_n$ and F is a subset of X. Then F is a state filter of X if and only if F is of the form $F_1 \times F_2 \times \cdots \times F_n$ where F_i is a state filter of X_i for $1 \le i \le n$ and $n \in \mathbb{N}$.*

For any subset A of a BE-algebra X, $\langle A\rangle_s$ is the smallest state filter containing A. In the following theorem, we now characterize the elements of $\langle A\rangle_s$.

Theorem 10.3.8 *Let A be a non-empty subset of a transitive state BE-algebra (X, μ). Then*

$$\langle A\rangle_s = \{x \in X \mid \mu(a_1) * (\cdots * (\mu(a_n) * (b_1 * (\cdots * (b_m * x) \cdots))) \cdots)) = 1 \textit{ for some } m, n \in \mathbb{N} \textit{ and } a_1, \ldots, a_n, b_1, \ldots, b_m \in A\}$$

is the smallest state filter containing A.

Proof Clearly $1 \in \langle A\rangle_s$. Let $x \in \langle A\rangle_s$ and $x * y \in \langle A\rangle_s$. Then there exist $a_1, \ldots, a_n, b_1, \ldots, b_m \in A$ and $c_1, \ldots, c_s, d_1, \ldots, d_t \in A$ such that

$$\mu(a_1) * (\cdots * (\mu(a_n) * (b_1 * (\cdots * (b_m * x) \cdots))) \cdots)) = 1 \text{ and}$$
$$\mu(c_1) * (\cdots * (\mu(c_s) * (d_1 * (\cdots * (d_t * (x * y)) \cdots))) \cdots)) = 1$$

Hence

$$\begin{aligned}1 &= \mu(c_1) * (\cdots * (\mu(c_s) * (d_1 * (\cdots * (d_t * (x * y)) \cdots))) \cdots)) \\ &= \mu(c_1) * (\cdots * (\mu(c_s) * (d_1 * (\cdots * (x * (d_t * y)) \cdots))) \cdots)) \\ &\ldots\ldots \\ &\ldots\ldots \\ &= x * (\mu(c_1) * (\cdots * (\mu(c_s) * (d_1 * (\cdots * (d_t * y) \cdots))) \cdots)))\end{aligned}$$

Hence $x \leq \mu(c_1) * (\cdots * (\mu(c_s) * (d_1 * (\cdots * (d_t * y) \cdots))) \cdots))$. By the sequential application of BE-ordering $\leq$ with the elements $b_1, \ldots, b_m$, we get

$$b_1 * (\cdots * (b_m * x) \cdots) \leq b_1 * (\cdots * (b_m * (\mu(c_1) * (\cdots * (\mu(c_s) * (d_1 * (\cdots * (d_t * y) \cdots))) \cdots)))) \cdots))$$

Again, by the sequential application of BE-ordering $\leq$ with the elements $\mu(b_1)$, $\mu(b_2), \ldots, \mu(b_m)$ we get the following consequence:

$$\begin{aligned}1 &= \mu(a_1) * (\cdots * (\mu(a_n) * (b_1 * (\cdots * (b_m * x) \cdots))) \cdots)) \\ &\leq \mu(a_1) * (\cdots * (\mu(a_n) * (b_1 * (\cdots * (b_m * (\mu(c_1) * (\cdots * (\mu(c_s) * (d_1 * (\cdots * (d_t * y) \cdots))) \cdots)))) \cdots))) \\ &= \mu(a_1) * (\cdots * (\mu(a_n) * (\mu(c_1) * (\cdots * (\mu(c_s) * (b_1 * (\cdots * (b_m * (d_1 * (\cdots * (d_t * y) \cdots))) \cdots)))) \cdots)))\end{aligned}$$

Hence $y \in \langle A \rangle_s$. Therefore $\langle A \rangle_s$ is a filter of X. For any $x \in A$, we get $\mu(x) * (\cdots * (\mu(x) * (x * (\cdots * (x * x) \cdots))) \cdots) = 1$. Hence $x \in \langle A \rangle_s$. Therefore $A \subseteq \langle A \rangle_s$. Let $x \in \langle A \rangle_s$. Then there exist $a_1, \ldots, a_n, b_1, \ldots, b_m \in A$ such that

$$\mu(a_1) * (\cdots * (\mu(a_n) * (b_1 * (\cdots * (b_m * x) \cdots))) \cdots)) = 1.$$

Hence $\mu(\mu(a_1) * (\cdots * (\mu(a_n) * (b_1 * (\cdots * (b_m * x) \cdots))) \cdots))) = \mu(1) = 1$. Since μ is a state operator, by (SO2), we get that $\mu(a_1) * (\cdots * (\mu(a_n) * (\mu(b_1) * (\cdots * (\mu(b_m) * \mu(x)) \cdots))) \cdots)) = 1$. Hence $\mu(x) \in \langle A \rangle_s$. Therefore $\langle A \rangle_s$ is a state filter in X such that $A \subseteq \langle A \rangle_s$.

We now prove that $\langle A \rangle_s$ is the smallest state filter containing A. Let F be a state filter of X containing A. Let $x \in \langle A \rangle_s$. Then there exist $a_1, \ldots, a_n, b_1, \ldots, b_m \in A$ such that

$$\mu(a_1) * (\cdots * (\mu(a_n) * (b_1 * (\cdots * (b_m * x) \cdots))) \cdots)) = 1 \in F$$

Since $A \subseteq F$, we get $a_1, \ldots, a_n, b_1, \ldots, b_m \in F$. Since F is a state filter, we get $\mu(a_1), \ldots, \mu(a_n) \in F$. Since $\mu(a_1) \in F$ and $\mu(a_1) * (\cdots * (\mu(a_n) * (b_1 * (\cdots * (b_m * x) \cdots))) \cdots)) \in F$, we get $\mu(a_2) * (\cdots * (\mu(a_n) * (b_1 * (\cdots * (b_m * x) \cdots))) \cdots)) \in F$. Continuing in this way finally we get $x \in F$. Hence $\langle A \rangle_s \subseteq F$. Therefore $\langle A \rangle_s$ is the smallest state filter of X containing A. □

Corollary 10.3.9 *Let (X, μ) be a transitive state BE-algebra and F a state filter of X. If $\emptyset \neq B \subseteq X$, then*

$$\langle F \cup B\rangle_s = \{x \in X \mid \mu(a_1) * (\cdots * (\mu(a_n) * (b_1 * (\cdots * (b_m * x) \cdots))) \cdots)) \in F \text{ for some } m, n \in \mathbb{N} \text{ and } a_1, \ldots, a_n, b_1, \ldots, b_m \in B\}$$

Corollary 10.3.10 *Let (X, μ) be a transitive state BE-algebra and F a state filter of X. For any $a \in X$,*

$$\langle F \cup \{a\}\rangle_s = \{x \in X \mid \mu(a)^m * (a^n * x) \in F \text{ for some } m, n \in \mathbb{N}\}$$

Since $\{1\}$ and X are state filters of (X, μ), the following is a direct consequence of the above.

Corollary 10.3.11 *Let (X, μ) be a transitive state BE-algebra and $a \in X$. Then $G = \{x \in X \mid \mu(a)^m * (a^n * x) = 1$ for some $m, n \in \mathbb{N}\}$ is a state filter of X.*

For any self-distributive state BE-algebra (X, μ), the above state filter G in X is the principal state filter generated by a and is denoted by $\langle a\rangle_s$.

10.4 Prime State Filters of State BE-algebras

In this section, the concepts of prime and maximal state filters are introduced in BE-algebras. Some properties of prime filters are studied. The celebrated Stone's theorem is generalized to the case of prime state filters of BE-algebras. An equivalent condition is derived for the set of all prime state filters of a BE-algebra to become a chain.

Definition 10.4.1 Let (X, μ) be a state BE-algebra. A state filter F of X is said to be *proper* if $F \neq X$.

Definition 10.4.2 Let (X, μ) be a state BE-algebra. A proper state filter P of X is said to be *prime* if $F \cap G \subseteq P$ implies $F \subseteq P$ or $G \subseteq P$ for all state filters F and G of X.

In the following, we characterize the prime state filters of a state BE-algebra. For this, we need the following proposition.

Proposition 10.4.3 *Let (X, μ) be a state BE-algebra. A proper state filter P is prime if and only if $\langle x\rangle_s \cap \langle y\rangle_s \subseteq P$ implies $x \in P$ or $y \in P$ for all $x, y \in X$.*

Proof Assume that P is a prime state filter of X. Let $x, y \in X$ be such that $\langle x\rangle_s \cap \langle y\rangle_s \subseteq P$. Since P is prime, it implies that $x \in \langle x\rangle_s \subseteq P$ or $y \in \langle y\rangle_s \subseteq P$. Conversely, assume the condition. Let F and G be two state filters of X such that $F \cap G \subseteq P$. Let $x \in F$ and $y \in G$ be the arbitrary elements. Then $\langle x\rangle_s \subseteq F$ and $\langle y\rangle_s \subseteq G$. Hence $\langle x\rangle_s \cap \langle y\rangle_s \subseteq F \cap G \subseteq P$. Then by the assumed condition, we get $x \in P$ or $y \in P$. Thus $F \subseteq P$ or $G \subseteq P$. Therefore P is prime. □

Definition 10.4.4 Let (X, μ) be a state BE-algebra. A proper state filter M of (X, μ) is said to be a *maximal state filter* if $\langle M \cup \{x\}\rangle_s = X$ for any $x \in X - M$.

In the following theorem, we derive a necessary and sufficient condition for every proper state filter of a state BE-algebra to become a maximal state filter.

Theorem 10.4.5 *Let (X, μ) be a state BE-algebra. A proper state filter M of X is maximal if and only if $M \subseteq F \subseteq X$ implies $M = F$ or $F = X$ for any state filter F of X.*

Proof Assume that M is a maximal state filter of X. Suppose there is a state filter F of X such that $M \subseteq F \subseteq X$. Let $M \neq F$. Then $M \subset F$. Choose $x \in F$ such that $x \notin M$. Since M is a maximal state filter, we get that $\langle M \cup \{x\}\rangle_s = X$. Let $a \in X$ be arbitrary. Hence $a \in \langle M \cup \{x\}\rangle_s$. Then there exist $m, n \in \mathbb{N}$ such that

$$(\mu(x))^m * (x^n * a) \in M \subseteq F$$

Since $x \in F$ and F is a state filter, we get $\mu(x) \in F$. Since $x \in F$ and $\mu(x) \in F$, it implies that $a \in F$. Hence $X \subseteq F$. Therefore $F = X$.

Conversely, assume that the condition holds. Let M be a proper filter of X. Let $x \in X - M$. Suppose $\langle M \cup \{x\}\rangle_s \neq X$. Choose $a \in X$ such that $a \notin \langle M \cup \{x\}\rangle_s$. Hence $M \subseteq \langle M \cup \{x\}\rangle_s \subset X$. Then by the assumed condition, we get that $M = \langle M \cup \{x\}\rangle_s$. Hence $x \in M$, which is a contradiction. Therefore M is a maximal state filter of X. □

In the following, we prove that the class of all prime state filters properly contains the class of all maximal state filters of a state BE-algebra. To prove this, we need the following result.

Proposition 10.4.6 *Let F be a state of a state BE-algebra (X, μ). Then for any $a, b \in X$*

$$\langle a\rangle_s \cap \langle b\rangle_s \subseteq F \text{ if and only if } \langle F \cup \{a\}\rangle_s \cap \langle F \cup \{b\}\rangle_s = F$$

Proof Assume that $\langle F \cup \{a\}\rangle_s \cap \langle F \cup \{b\}\rangle_s = F$. Since $a \in \langle F \cup \{a\}\rangle_s$, we get that $\langle a\rangle_s \subseteq \langle F \cup \{a\}\rangle_s$. Similarly, we get $\langle b\rangle_s \subseteq \langle F \cup \{b\}\rangle_s$. Hence $\langle a\rangle_s \cap \langle b\rangle_s \subseteq \langle F \cup \{a\}\rangle_s \cap \langle F \cup \{b\}\rangle_s = F$.

Conversely, assume that $\langle a\rangle_s \cap \langle b\rangle_s \subseteq F$ for all $a, b \in X$. Clearly $F \subseteq \langle F \cup \{a\}\rangle_s \cap \langle F \cup \{b\}\rangle_s$. Let $x \in \langle F \cup \{a\}\rangle_s \cap \langle F \cup \{b\}\rangle_s$. Then there exist $m, n, r, s \in \mathbb{N}$ such that

$$\mu(a)^m * (a^n * x) \in F \text{ and } \mu(b)^r * (b^s * x) \in F$$

Hence there exists $u, v \in F$ such that

$$\mu(a)^m * (a^n * x) = u \text{ and } \mu(b)^r * (b^s * x) = v$$

Hence

$$\mu(a)^m * (a^n * (u * x)) = u * (\mu(a)^m * (a^n * x)) = u * u = 1 \in \langle a\rangle_s$$

Since $\langle a \rangle_s$ is a state filter and $a \in \langle a \rangle_s$, we get $u * x \in \langle a \rangle_s$. By the similar argument, we get that $v * x \in \langle b \rangle_s$. Also $u * x \leq v * (u * x) = u * (v * x)$ and $v * x \leq u * (v * x)$. Since $u * x \in \langle a \rangle_s$, we get that $u * (v * x) \in \langle a \rangle_s$. Since $v * x \in \langle b \rangle_s$, we get that $u * (v * x) \in \langle b \rangle_s$. Hence

$$u * (v * x) \in \langle a \rangle_s \cap \langle b \rangle_s \subseteq F.$$

Since $u, v \in F$ and F is a filter, we get that $x \in F$. Therefore $\langle F \cup \{a\} \rangle_s \cap \langle F \cup \{b\} \rangle_s \subseteq F$. □

Theorem 10.4.7 *Every maximal state filter of a state BE-algebra (X, μ) is a prime state filter.*

Proof Let M be a maximal state filter of (X, μ). Let $\langle x \rangle_s \cap \langle y \rangle_s \subseteq M$ for some $x, y \in X$. Suppose $x \notin M$ and $y \notin M$. Then $\langle M \cup \{x\} \rangle_s = X$ and $\langle M \cup \{y\} \rangle_s = X$. Hence

$$\langle M \cup \{x\} \rangle_s \cap \langle M \cup \{y\} \rangle_s = X$$

Hence, by Proposition 10.4.6, it yields that $\langle x \rangle_s \cap \langle y \rangle_s \nsubseteq M$, which is a contradiction. Hence $x \in M$ or $y \in M$. Therefore M is a prime state filter of X. □

Corollary 10.4.8 *Let (X, μ) be a state BE-algebra. If $M_1, M_2, ..., M_n$ and M be maximal state filters of X such that $\bigcap_{i=1}^{n} M_i \subseteq M$. Then there exists $j \in \{1, 2, ..., n\}$ such that $M_j = M$.*

Proof Let $M_1, M_2, ..., M_n$ and M be maximal state filters X such that $\bigcap_{i=1}^{n} M_i \subseteq M$. By the above theorem, M is prime. Hence there exists $j \in \{1, 2, ..., n\}$ such that $M_j \subseteq M$. Since M_j is maximal, it gives $M_j = M$. □

Theorem 10.4.9 *Let (X, μ) ba a state BE-algebra and $a \in X$. If F is a state filter of X such that $a \notin F$, then there exists a prime state filter P such that $a \notin P$ and $F \subseteq P$.*

Proof Let F be a state filter of X such that $a \notin F$. Consider $\Im = \{G \in \mathcal{F}_s(X) \mid a \notin G \text{ and } F \subseteq G\}$, where $\mathcal{F}_s(X)$ is the class of all state filters of X. Clearly $F \in \Im$. Then by Zorn's lemma, $\Im$ has a maximal element, say M. Clearly $a \notin M$. We now prove that M is prime. Let $x, y \in X$ be such that $\langle x \rangle_s \cap \langle y \rangle_s \subseteq M$. Then by Proposition 10.4.6, we get

$$\langle M \cup \{x\} \rangle_s \cap \langle M \cup \{y\} \rangle_s = M$$

Since $a \notin M$, we can obtain that $a \notin \langle M \cup \{x\} \rangle_s$ or $a \notin \langle M \cup \{y\} \rangle_s$. By the maximality of M, we get that $\langle M \cup \{x\} \rangle_s = M$ or $\langle M \cup \{y\} \rangle_s = M$. Hence $x \in M$ or $y \in M$. Therefore M is prime. □

Corollary 10.4.10 *Let (X, μ) be a state BE-algebra and $1 \neq x \in X$. Then there exists a prime state filter P of (X, μ) such that $x \notin P$.*

Proof Let $1 \neq x \in X$ and $F = \{1\}$. Then F is a state filter of X such that $x \notin F$. Hence by the main theorem, there exists a prime state filter P such that $x \notin P$. □

The following corollary is a direct consequence of the above results.

Corollary 10.4.11 *Let (X, μ) be a state BE-algebra. Then the intersection of all prime state filters of X is equal to $\{1\}$.*

Corollary 10.4.12 *Let (X, μ) be a state BE-algebra and F a proper state filter of X. Then*

$$F = \cap\{P \mid P \text{ is a prime state filter of } X \text{ such that } F \subseteq P\}$$

Proof Clearly $F \subseteq \cap\{P \mid P$ is a prime state filter of X such that $F \subseteq P\}$. Conversely, let $x \notin F$. Then by the main theorem, there exists a prime state filter P_x such that $x \notin P_x$ and $F \subseteq P$. Therefore $x \notin \cap\{P \mid P$ is a prime state filter of X such that $F \subseteq P\}$. Therefore $\cap\{P \mid P$ is a prime state filter of X such that $F \subseteq P\} \subseteq F$. □

Theorem 10.4.13 *Let (X, μ) be a state BE-algebra such that $\mu(X)$ is a state filter and $\mu(x)^{m+n} * \mu(a) \leq \mu(x)^m * (x^n * a)$ for all $m, n \in \mathbb{N}$ and $x, a \in X$. If F is a prime state filter of X and $\mu^{-1}(F) \neq X$, then $\mu^{-1}(F)$ is a prime state filter.*

Proof Since $\mu(1) = 1 \in F$, we get $1 \in \mu^{-1}(F)$. Let $x, x*y \in \mu^{-1}(F)$. Hence $\mu(x) \in F$ and $\mu(x*y) \in F$. Since $\mu(x*y) \leq \mu(x) * \mu(y)$, it yields $\mu(x) * \mu(y) \in F$. Since $\mu(x) \in F$, we get $\mu(y) \in F$ and hence $y \in \mu^{-1}(F)$. Therefore $\mu^{-1}(F)$ is a filter of X. Let $x \in \mu^{-1}(F)$. Then $\mu(x) \in F$. Since F is a state filter, we get $\mu(\mu(x)) \in F$. Hence $\mu(x) \in \mu^{-1}(F)$. Therefore $\mu^{-1}(F)$ is a state filter of X. Let $x, y \in X$ be such that $\langle x\rangle_s \cap \langle y\rangle_s \subseteq \mu^{-1}(F)$. Let $u \in \langle\mu(x)\rangle_s \cap \langle\mu(y)\rangle_s$. Then there exists $m, n, s, t \in \mathbb{N}$ such that $\mu(x)^{m+n} * u = \mu(x)^m * (\mu(x)^n * u) = \mu(\mu(x))^m * (\mu(x)^n * u) = 1 = \mu(1) \in \mu(X)$ and $\mu(y)^{s+t} * u = \mu(y)^s * (\mu(y)^t * u) = \mu(\mu(y))^s * (\mu(y)^t * u) = 1 = \mu(1) \in \mu(X)$. Since $\mu(X)$ is a filter and $\mu(x) \in \mu(X)$, we get $u \in \mu(X)$. Thus $u = \mu(a)$ for some $a \in X$. By the given condition, we get

$$\begin{aligned} 1 &= \mu(x)^{m+n} * u \\ &= \mu(x)^{m+n} * \mu(a) \\ &\leq \mu(x)^m * (x^n * a) \end{aligned}$$

Hence $\mu(\mu(x)^m * (x^n * a)) = \mu(1) = 1$. Thus $\mu(x)^m * (x^n * a) \in \mu^{-1}(F)$. Therefore $a \in \langle\mu^{-1}(F) \cup \{x\}\rangle_s$. Similarly, we can get $a \in \langle\mu^{-1}(F) \cup \{y\}\rangle_s$. Hence

$$a \in \langle\mu^{-1}(F) \cup \{x\}\rangle_s \cap \langle\mu^{-1}(F) \cup \{y\}\rangle_s$$

Since $\langle x\rangle_s \cap \langle y\rangle_s \subseteq \mu^{-1}(F)$, by Proposition 10.4.6, we get $a \in \langle\mu^{-1}(F) \cup \{x\}\rangle_s \cap \langle\mu^{-1}(F) \cup \{y\}\rangle_s = \mu^{-1}(F)$. Hence $u = \mu(a) \in F$. Thus $\langle f(x)\rangle_s \cap \langle f(y)\rangle_s \subseteq F$. Since F is prime, we get $\mu(x) \in F$ or $\mu(y) \in F$. Hence $x \in \mu^{-1}(F)$ or $y \in \mu^{-1}(F)$. Thus $\mu^{-1}(F)$ is a prime state filter of X. □

In the following theorem, a necessary and sufficient condition is derived, in terms of primeness of state filters of a state BE-algebra (X, μ), for the class $\mathcal{F}_s(X, \mu)$ to become a chain.

Theorem 10.4.14 *Let (X, μ) be a state BE-algebra. Then $\mathcal{F}_s(X, \mu)$ is a totally ordered set (chain) if and only if every proper state filter of (X, μ) is prime.*

Proof Assume that $\mathcal{F}_s(X, \mu)$ is a totally ordered set. Let F be a proper state filter of (X, μ). Let $a, b \in X$ be such that $\langle a\rangle_s \cap \langle b\rangle_s \subseteq F$. Since $\langle a\rangle_s$ and $\langle b\rangle_s$ are state filters of (X, μ), we get that either $\langle a\rangle_s \subseteq \langle b\rangle_s$ or $\langle b\rangle_s \subseteq \langle a\rangle_s$. Hence $a \in F$ or $b \in F$. Therefore F is prime.

Conversely assume that every proper state filter of (X, μ) is prime. Let F and G be two proper state filters of (X, μ). Since $F \cap G$ is a proper state filter of (X, μ), we get that $F \subseteq F \cap G$ or $G \subseteq F \cap G$. Hence $F \subseteq G$ or $G \subseteq F$. Therefore $\mathcal{F}_s(X, \mu)$ is a totally ordered set. □

10.5 State Congruences of State BE-algebras

In this section, the notion of state congruences is introduced in state BE-algebras. An equivalent condition is derived for a congruence of a BE-algebra to become a state congruence on the state BE-algebra. A set of equivalent conditions is derived for a filter of a state BE-algebra to become a state filter in terms of a state congruence.

Definition 10.5.1 Let (X, μ) be a state BE-algebra. A congruence θ on X is called a *state congruence* on (X, μ) if it satisfies the following property for all $x, y \in X$:

$$(x, y) \in \theta \text{ implies } (\mu(x), \mu(y)) \in \theta$$

Proposition 10.5.2 *Let (X, μ) be a self-distributive state BE-algebra. If F is a state filter of (X, μ), then $\theta_F = \{(x, y) \in X \times X \mid x * y \in F$ and $y * x \in F\}$ is a state congruence on (X, μ).*

Proof Since X is self-distributive, it was already obtained that θ_F is a congruence on X. Let $(x, y) \in \theta_F$. Then we get $x * y \in F$ and $y * x \in F$. Since F is a state filter, we get $\mu(x) * \mu(y) \geq \mu(x * y) \in F$ and $\mu(y) * \mu(x) \geq \mu(y * x) \in F$. Thus $\mu(x) * \mu(y) \in F$ and $\mu(y) * \mu(x) \in F$. Hence $(\mu(x), \mu(y)) \in \theta_F$. Therefore θ_F is a state congruence on (X, μ). □

Theorem 10.5.3 *Let (X, μ) be a commutative state BE-algebra. A congruence θ on X is a state congruence on (X, μ) if and only if $(1, x) \in \theta$ implies $(1, \mu(x)) \in \theta$ for all $x \in X$.*

Proof Let θ be a congruence on X. Then the condition for necessity is clear. Conversely, assume the condition. Let $x, y \in X$. Let $(x, y) \in \theta$. Then we get the following consequence:

$$\begin{aligned}(x, y) \in \theta &\Rightarrow (x * x, y * x) \in \theta\\ &\Rightarrow (1, y * x) \in \theta\\ &\Rightarrow (1, \mu(y * x)) \in \theta\\ &\Rightarrow (1 * \mu(x), \mu(y * x) * \mu(x)) \in \theta\\ &\Rightarrow (\mu(x), \mu((y * x) * x)) \in \theta \qquad \text{since } x \leq y * x.\end{aligned}$$

Again

$$\begin{aligned}(x, y) \in \theta &\Rightarrow (x * y, y * y) \in \theta\\ &\Rightarrow (x * y, 1) \in \theta\\ &\Rightarrow (\mu(x * y), 1) \in \theta\\ &\Rightarrow (\mu(x * y) * \mu(y), 1 * \mu(y)) \in \theta\\ &\Rightarrow (\mu((x * y) * y), \mu(y)) \in \theta \qquad \text{since } y \leq x * y.\end{aligned}$$

Since X is commutative, we get that $\mu((y * x) * x) = \mu((x * y) * y)$. Hence, it concludes from the above two observations that $(\mu(x), \mu(y)) \in \theta$. Therefore θ is a state congruence on (X, μ). □

Let θ be a state congruence on a state BE-algebra (X, μ). Then clearly $(X_{/\theta}, *, [1]_\theta)$ is a BE-algebra with respect to the operation $*$ defined as $[x]_\theta * [y]_\theta = [x * y]_\theta$ for all $x, y \in X$.

Theorem 10.5.4 *Let θ be a state congruence on a state BE-algebra (X, μ). Then the map $\overline{\mu} : X_{/\theta} \to X_{/\theta}$ defined by*

$$\overline{\mu}([x]_\theta) = [\mu(x)]_\theta \quad \textit{for all } x \in X$$

is a state operator on $X_{/\theta}$.

Proof We first observe that $\overline{\mu}$ is well-defined. Suppose $[x]_\theta = [y]_\theta$ where $x, y \in X$. Then $(x, y) \in \theta$. Since θ is a state congruence, we get that $(\mu(x), \mu(y)) \in \theta$. Therefore

$$\overline{\mu}([x]_\theta) = [\mu(x)]_\theta = [\mu(y)]_\theta = \overline{\mu}([y]_\theta).$$

Therefore the map $\overline{\mu}$ is well-defined. We now prove the postulates of a state operator for $\overline{\mu}$.
(SO1). Suppose $[x]_\theta * [y]_\theta = [1]_\theta$ where $x, y \in X$. Then $[x * y]_\theta = [1]_\theta$, which implies that $(x * y, 1) \in \theta$. Since θ is a state congruence, we get $(\mu(x * y), 1) \in \theta$. Therefore we get the following:

$$\begin{aligned}\overline{\mu}([x]_\theta) * \overline{\mu}([y]_\theta) &= [\mu(x)]_\theta * [\mu(y)]_\theta\\ &= [\mu(x) * \mu(y)]_\theta\\ &\supseteq [\mu(x * y)]_\theta \qquad \text{since } \mu(x) * \mu(y) \geq \mu(x * y)\\ &= [1]_\theta\end{aligned}$$

Therefore $\overline{\mu}([x]_\theta) * \overline{\mu}([y]_\theta) = [1]_\theta$.
(SO2). For any $[x]_\theta, [y]_\theta \in X_{/\theta}$ where $x, y \in X$, we get the following:

$$\begin{aligned}
\overline{\mu}([x]_\theta) * \overline{\mu}([y]_\theta) &= \overline{\mu}([x * y]_\theta) \\
&= [\mu(x * y)]_\theta \\
&= [\mu((x * y) * y) * \mu(y)]_\theta \\
&= [\mu((x * y) * y)]_\theta * [\mu(y)]_\theta \\
&= \overline{\mu}([(x * y) * y)]_\theta * \overline{\mu}[y]_\theta \\
&= \overline{\mu}(([x]_\theta * [y]_\theta) * [y]_\theta) * \overline{\mu}[y]_\theta.
\end{aligned}$$

(SO3). For any $[x]_\theta, [y]_\theta \in X_{/\theta}$ where $x, y \in X$, we get the following:

$$\begin{aligned}
\overline{\mu}\{\overline{\mu}([x]_\theta) * \overline{\mu}([y]_\theta)\} &= \overline{\mu}\{[\mu(x)]_\theta * [\mu(y)]_\theta\} \\
&= \overline{\mu}\{[\mu(x) * \mu(y)]_\theta\} \\
&= [\mu(\mu(x) * \mu(y))]_\theta \\
&= [\mu(x) * \mu(y)]_\theta \\
&= [\mu(x)]_\theta * [\mu(y)]_\theta \\
&= \overline{\mu}([x]_\theta) * \overline{\mu}([y]_\theta).
\end{aligned}$$

Therefore $\overline{\mu}$ is a state operator on $X_{/\theta}$. □

Corollary 10.5.5 *Let (X, μ) be a state BE-algebra and θ a state congruence on (X, μ). Then $(X_{/\theta}, \overline{\mu})$ is a state BE-algebra.*

Using the above theorem, we now obtain a one-to-one correspondence between the set of all state filter of state BE-algebra (X, μ) and the set of all state filters of the state BE-algebra $(X_{/\theta}, \overline{\mu})$. For any subset S of a BE-algebra X, let us denote $\overline{S} = \{[x]_\theta \mid x \in S\}$.

Theorem 10.5.6 *Let θ a state congruence on a state BE-algebra (X, μ) and F a non-empty subset of (X, μ). Then F is a state filter of X if and only if $\overline{F}$ is a state filter of $X_{/\theta}$.*

Proof Assume that F is a state filter of (X, μ). Since $1 \in F$, we get that $[1]_\theta \in \overline{F}$. Let $[x]_\theta$ and $[x * y]_\theta = [x]_\theta * [y]_\theta \in \overline{F}$. Then $x, x * y \in F$. Since F is a filter, we get that $y \in F$. Hence $[y]_\theta \in \overline{F}$. Therefore $\overline{F}$ is a filter of $X_{/\theta}$. Let $[x]_\theta \in \overline{F}$. Then $x \in F$. Since F is a state filter, we get that $\mu(x) \in F$. Hence $\overline{\mu}([x]_\theta) = [\mu(x)]_\theta \in \overline{F}$. Therefore $\overline{F}$ is a state filter of $(X_{/\theta}, \overline{\mu})$.

Conversely, assume that $\overline{F}$ is a state filter of $X_{/\theta}$. Then $[1]_\theta \in \overline{F}$, which implies that $1 \in F$. Let $x, x * y \in F$. Then $[x]_\theta, [x]_\theta * [y]_\theta = [x * y]_\theta \in \overline{F}$. Since $\overline{F}$ is a filter of $X_{/\theta}$, we get $[y]_\theta \in \overline{F}$. Hence $y \in F$. Therefore F is a filter of X. Let $x \in F$. Then we get $[x]_{/\theta} \in \overline{F}$. Since $\overline{F}$ is a state filter of $[\mu(x)]_{/\theta} = \overline{\mu}([x]_\theta \in F_{/\theta}$. Hence $\mu(x) \in F$. Therefore F is a state filter of (X, μ). □

In the following, we derive a characterization theorem for state filters of a state BE-algebra.

Theorem 10.5.7 *Let (X, μ) be a self-distributive state BE-algebra. If F is a filter of X, then the following conditions are equivalent:*

(1) *F is a state filter,*
(2) *there exists a state congruence on (X, μ) whose kernel is F,*
(3) *$x, y \in F$ implies $\mu(x) * \mu(y) \in F$.*

Proof (1) $\Rightarrow$ (2): Assume that F is a state filter of (X, μ). Then by Proposition 10.5.2, θ_F is a state congruence on (X, μ). It is enough to show that $Ker(\theta_F) = F$. Let $x \in Ker(\theta_F)$. Then $(1, x) \in \theta_F$. Hence $x = 1 * x \in F$. Therefore $Ker(\theta_F) \subseteq F$. Again, let $x \in F$. Then $1 * x = x \in F$ and $x * 1 = 1 \in F$. Hence $(1, x) \in \theta_F$. Therefore $F \subseteq Ker(\theta_F)$.
(2) $\Rightarrow$ (3): Assume that the condition (2) holds. Let $x, y \in F = Ker(\theta)$. Then $(1, x) \in \theta$ and $(1, y) \in \theta$. Since θ is a state congruence, we get that $(1, \mu(x)) \in \theta$ and $(1, \mu(y)) \in \theta$. Hence $(\mu(x) * \mu(y), 1) = (\mu(x) * \mu(y), 1 * 1) \in \theta$. Therefore $\mu(x) * \mu(y) \in Ker(\theta) = F$.
(3) $\Rightarrow$ (1): Let $x \in F$. Since $1, x \in F$, by the condition (3), it yields that $\mu(x) = 1 * \mu(x) = \mu(1) * \mu(x) \in F$. Therefore F is a state filter of (X, μ). □

Corollary 10.5.8 *Let (X, μ) be a self-distributive and commutative state BE-algebra where X. If F is a state filter of (X, μ), then θ_F is the smallest state congruence whose kernel is F.*

Proof Since F is a state filter of (X, μ), by the main theorem θ_F is a state congruence on (X, μ) such that $Ker(\theta_F) = F$. It is enough to prove that θ_F is the smallest congruence on X. Suppose R is a congruence on X such that $F = Ker(R)$.

$$\begin{aligned}
(x, y) \in \theta_F &\Rightarrow x * y \in F = Ker(R) \text{ and } y * x \in F = Ker(R)\\
&\Rightarrow (x * y, 1) \in R \text{ and } (y * x, 1) \in R\\
&\Rightarrow ((x * y) * y, 1 * y) \in R \text{ and } ((y * x) * x, 1 * x) \in R\\
&\Rightarrow ((x * y) * y, y) \in R \text{ and } ((y * x) * x, x) \in R\\
&\Rightarrow (x, y) \in R. \qquad\qquad \text{since } X \text{ is commutative}
\end{aligned}$$

Therefore θ_F is the smallest state congruence on (X, μ) whose kernel if F. □

Theorem 10.5.9 *Let (X, μ) be a commutative state BE-algebra. Then there exists an isotone bijection between the set $C_s(X, \mu)$ of all state congruences of (X, μ) and the set $\mathcal{F}_s(X, \mu)$ of all state filters of the state BE-algebra (X, μ).*

Proof Define $\Psi : C_s(X, \mu) \to \mathcal{F}_s(X, \mu)$ by $\Psi(\theta) = Ker(\theta)$. Since $Ker(\theta)$ is a state filter of (X, μ), it is clear that the map Ψ is well-defined. Let $\theta_1, \theta_2 \in C_s(X, \mu)$ be such that $\Psi(\theta_1) = \Psi(\theta_2)$. Hence $Ker(\theta_1) = Ker(\theta_2)$. We now prove that $\theta_1 = \theta_2$. Let $(x, y) \in \theta_1$. Then we get the following:

$$\begin{aligned}(x, y) \in \theta_1 &\Rightarrow (x * y, y * y) \in \theta_1 \\ &\Rightarrow (x * y, 1) \in \theta_1 \\ &\Rightarrow x * y \in Ker(\theta_1) = Ker(\theta_2) \\ &\Rightarrow (x * y, 1) \in \theta_2 \\ &\Rightarrow ((x * y) * y, 1 * y) \in \theta_2 \\ &\Rightarrow ((x * y) * y, y) \in \theta_2\end{aligned}$$

Similarly, we get $((y*x)*x, x) \in \theta_2$. Hence, it implies $(x, y) \in \theta_2$. Therefore $\theta_1 \subseteq \theta_2$. Similarly, we can obtain that $\theta_2 \subseteq \theta_1$. Therefore Ψ is one-to-one. We now claim that Ψ is onto. Let F be a state filter of (X, μ). Then by Theorem 9.5.2, θ_F is a state congruence on (X, μ) such that $Ker(\theta_F) = F$. For this state congruence θ_F, we get

$$\Psi(\theta_F) = Ker(\theta_F) = F$$

Hence, it yields that Ψ is onto. Therefore Ψ is a bijection. To show that Ψ is an isotone, let us assume that $\theta_1 \subseteq \theta_2$ for $\theta_1, \theta_2 \in \mathcal{C}_s(X, \mu)$. Let $x \in \Psi(\theta_1) = Ker(\theta_1)$. Then we get $(x, 1) \in \theta_1 \subseteq \theta_2$. Hence $x \in Ker\ \theta_2 = \Psi(\theta_2)$. Thus $\Psi(\theta_1) \subseteq \Psi(\theta_2)$. Therefore Ψ is isotone. □

10.6 Injective Filters of State BE-algebras

In this section, the notion of injective filters is introduced in BE-algebras. It is derived that every filter is an injective filter if and only if the respective state operator is injective. It is observed that every injective filter is a state filter but not the converse. However we derive a necessary and sufficient condition for every state filter to become an injective filter.

Definition 10.6.1 Let (X, μ) be a state BE-algebra. A filter F of X is said to be an *injective filter* if for all $x, y \in X$, $\mu(x) = \mu(y)$ and $x \in F$ imply that $y \in F$.

For any state BE-algebra (X, μ), it is clear that the filter $Ker(\mu)$ is an injective filter of (X, μ) and it is the smallest state filter of (X, μ). It can be easily seen that the intersection of any two injective filters of a state BE-algebra is again an injective filter. Moreover, we have that $\bigcap_{F \in \mathcal{F}^i(X)} F = Ker(\mu)$, where $\mathcal{F}^i(X)$ is the set of all injective filters of a state BE-algebra (X, μ). Though the unit filter $\{1\}$ is a state filter of a state BE-algebra (X, μ), it is not sure that $\{1\}$ is an injective filter with respect to μ. However a set of equivalent conditions is established for the filter $\{1\}$ of a state BE-algebra to become an injective filter.

Theorem 10.6.2 *Let (X, μ) be a state BE-algebra. Then the following conditions are equivalent.*

(1) $\{1\}$ *is injective w.r.t.* μ,
(2) $Ker(\mu) = \{1\}$,
(3) *for all* $x \in X$, $\mu(x) = 1$ *implies* $x = 1$.

Proof (1) $\Rightarrow$ (2): Assume that $\{1\}$ is an injective filter with respect to μ. Since $\mu(1) = 1$, we get that $\{1\} \subseteq Ker(\mu)$. Conversely, let $x \in Ker(\mu)$. Then we get $\mu(x) = 1 = \mu(1)$. Since $\{1\}$ is injective, we get that $x \in \{1\}$. Hence $Ker(\mu) \subseteq \{1\}$. Therefore $Ker(\mu) = \{1\}$.
(2) $\Rightarrow$ (3): It is clear.
(3) $\Rightarrow$ (1): Assume the condition (3). Let $x, y \in X$ and $\mu(x) = \mu(y)$. Suppose $x \in \{1\}$. Then we get that $\mu(y) = \mu(x) = 1 = \mu(1)$. Thus by condition (3), we get that $y = 1$. Hence $y \in \{1\}$. Therefore $\{1\}$ is an injective filter of (X, μ). □

Theorem 10.6.3 *Let* (X, μ) *be a self-distributive and commutative state BE-algebra. Then the following conditions are equivalent:*

(1) μ *is injective,*
(2) *every filter is injective,*
(3) *every principal filter is injective.*

Proof (1) $\Rightarrow$ (2) and (2) $\Rightarrow$ (3) are clear.
(3) $\Rightarrow$ (1): Assume that every principal filter of X is injective. Let $x, y \in X$ be such that $\mu(x) = \mu(y)$. Since $\langle x \rangle$ and $\langle y \rangle$ are injective, $x \in \langle x \rangle$ and $y \in \langle y \rangle$, we get that $y \in \langle x \rangle$ and $x \in \langle y \rangle$. Hence $x * y = 1$ and $y * x = 1$. Since X is commutative, we get that $x = y$. Therefore μ is an injective map. □

Theorem 10.6.4 *Let* (X, μ) *be a state BE-algebra. Then every injective filter is a state filter.*

Proof Let F be an injective filter of (X, μ). Let $x \in F$. Then $\mu(x) \in \mu(F)$. Hence $\mu(\mu(x)) = \mu(x) = \mu(a)$ for some $a \in F$. Since F is injective, we get that $\mu(x) \in F$. Hence F is a state filter of (X, μ). □

The converse of the above theorem is not true. Hence every state filter of a state BE-algebra need not be an injective filter with respect to μ, which can be seen in the following example.

Example 10.6.5 Let $X = \{1, a, b, c, d\}$. Define a binary operation $*$ on X as follows:

$*$	1	a	b	c	d
1	1	a	b	c	d
a	1	1	b	c	b
b	1	a	1	b	a
c	1	a	1	1	a
d	1	1	1	b	1

Then it can be easily verified that $(X, *, 1)$ is a BE-algebra. It is easy to check that $F = \{b, c, 1\}$ is a filter in X. Define a mapping $\mu : X \to X$ as follows:

$$\mu(a) = \mu(1) = 1;\ \mu(b) = \mu(d) = b \text{ and } \mu(c) = c$$

It can be easily seen that F is a state filter of (X, μ). But F is not an injective filter of (X, μ) because of $\mu(b) = \mu(d)$, $b \in F$ and $d \in F$.

However, in the following theorem, we derive a necessary and sufficient condition for every state filter of a state BE-algebra to become an injective filter.

Theorem 10.6.6 *Let (X, μ) be a state BE-algebra and F a state filter of (X, μ). Then the following conditions are equivalent:*

(1) *F is injective,*
(2) *for all $x \in X$, $\mu(x) \in F \Leftrightarrow x \in F$.*

Proof (1) $\Rightarrow$ (2): Let F be a state filter of (X, μ). Assume that F is injective. Let $x \in X$. Clearly $x \in F$ implies $\mu(x) \in F$. Suppose $\mu(x) \in F$. Clearly $\mu(\mu(x)) = \mu(x)$. Since F is injective and $\mu(x) \in F$, we get $x \in F$.

Conversely, assume that $\mu(x) \in F$ if and only if $x \in F$ for all $x \in X$. Let $x, y \in X$ be such that $\mu(x) = \mu(y)$. Now

$$\begin{aligned} x \in F &\Rightarrow \mu(x) \in \mu(F) \subseteq F \\ &\Rightarrow \mu(y) = \mu(x) \in F \\ &\Rightarrow y \in F \qquad \text{by the assumed condition} \end{aligned}$$

Therefore F is injective. $\square$

For any state operator μ on a transitive BE-algebra X, let us denote by $\langle S\rangle_i$ the smallest injective filter containing S, where S is a non-empty subset of X. It is called the injective filter generated by S. Instead of $\langle\{a\}\rangle_i$, we will write rather $\langle a\rangle_i$ where $a \in X$.

Theorem 10.6.7 *Let S be a subset of a transitive state BE-algebra (X, μ). Then*

$$\langle S\rangle_i = \{x \in X \mid \mu(a_1) * (\cdots * (\mu(a_n) * \mu(x))\cdots) = 1 \text{ for some } a_1, a_2, \cdots, a_n \in S \text{ and } n \in \mathbb{N}\}.$$

Proof Let $A = \{x \in X \mid \mu(a_1) * (\cdots * (\mu(a_n) * \mu(x))\cdots) = 1$ for some $n \in \mathbb{N}$ and $a_1, a_2, \ldots, a_n \in S\}$. Clearly $1 \in A$. Let $x, x * y \in A$. Then there exist $a_1, a_2, \ldots, a_n, b_1, b_2, \ldots, b_m \in S$ where $m, n \in \mathbb{N}$ such that $\mu(a_1) * (\cdots * (\mu(a_n) * \mu(x))\cdots) = 1$ and $\mu(b_1) * (\cdots * (\mu(b_m) * \mu(x * y))\cdots) = 1$. Hence

$$\begin{aligned} 1 &= \mu(b_1) * (\cdots * (\mu(b_m) * \mu(x * y))\cdots) \\ &\leq \mu(b_1) * (\cdots * (\mu(b_m) * (\mu(x) * \mu(y)))\cdots) \\ &= \mu(x) * (\mu(b_1) * (\cdots * (\mu(b_m) * \mu(y))\cdots)) \end{aligned}$$

which implies

$$\mu(x) \leq \mu(b_1) * (\cdots * (\mu(b_m) * \mu(y))\cdots)$$

Hence by the sequential application of BE-ordering $\leq$ with $\mu(a_1), \mu(a_2), \ldots, \mu(a_n)$, we get

$$1 = \mu(a_1)*(\cdots*(\mu(a_n)*\mu(x))\cdots) \leq \mu(a_1)*(\cdots*(\mu(a_n)*(\mu(b_1)*(\cdots*(\mu(b_m)*\mu(y))\cdots)).$$

Thus we get $y \in A$. Therefore A is a filter of X. Let $x \in S$. Since $\mu(x) * \mu(x) = 1$, we get $x \in A$. Therefore $S \subseteq A$. Let $x, y \in X$ such that $\mu(x) = \mu(y)$ and $x \in A$. Then there exist $a_1, a_2, \ldots, a_n \in S$ and $n \in \mathbb{N}$ such that $\mu(a_1) * (\cdots * (\mu(a_n) * \mu(x))\cdots) = 1$. Hence $\mu(a_1) * (\cdots * (\mu(a_n) * \mu(y))\cdots) = 1$, which implies that $y \in A$. Therefore A is an injective filter of (X, μ) such that $S \subseteq A$. Suppose G is an injective filter of (X, μ) such that $S \subseteq G$. Let $x \in A$. Then there exists $a_1, a_2, \ldots, a_n \in S \subseteq G$ and $n \in \mathbb{N}$ such that

$$\mu(a_1) * (\cdots * (\mu(a_n) * \mu(x))\cdots) = 1$$

Since G is a state filter of (X, μ), we get $\mu(a_1), \mu(a_2), \ldots, \mu(a_n) \in G$. Hence $\mu(x) \in G$. Since G is injective, by the above theorem, we get $x \in G$. Hence $A \subseteq G$. Hence A is the smallest injective filter of (X, μ) that is containing S. □

Theorem 10.6.8 *Let F be a state filter of a transitive state BE-algebra (X, μ). Then for any $a \in X$, $\langle F \cup \{a\}\rangle_i = \{x \in X \mid \mu(a)^n * \mu(x) \in F$ for some $n \in \mathbb{N}\}$.*

Proof Let $A = \{x \in X \mid \mu(a)^n * \mu(x) \in F$ for some $n \in \mathbb{N}\}$. Clearly $1 \in A$. Let $x, x * y \in A$. Then there exists $m, n \in \mathbb{N}$ such that $\mu(a)^m * \mu(x) \in F$ and $\mu(a)^n * \mu(x * y) \in F$. Since $\mu(a)^n * \mu(x * y) \leq \mu(a)^n * (\mu(x) * \mu(y))$ and F is a filter of X, we get that $\mu(a)^n * (\mu(x) * \mu(y)) \in F$. Hence

$$\begin{aligned}
(\mu(a)^m * \mu(x))* &((\mu(a)^n * (\mu(x) * \mu(y))) * (\mu(a)^{m+n} * \mu(y))) \\
&= (\mu(a)^m * \mu(x)) * (\mu(x) * ((\mu(a)^n * \mu(y))) * (\mu(a)^m * (\mu(a)^n * \mu(y))) \\
&\geq (\mu(a)^m * \mu(x)) * (\mu(a)^m * \mu(x)) \\
&= 1 \in F
\end{aligned}$$

Since $\mu(a)^m * \mu(x) \in F$ and $\mu(a)^n * (\mu(x) * \mu(y)) \in F$, we get that $\mu(a)^{m+n} * \mu(y) \in F$. Hence $y \in A$. Therefore A is a filter of X. Let $x, y \in X$ be such that $\mu(x) = \mu(y)$ and $x \in A$. Then $\mu(a)^n * \mu(y) = \mu(a)^n * \mu(x) \in F$ for some $n \in \mathbb{N}$. Hence we get $y \in A$, which implies that A is an injective filter of (X, μ). Clearly $a \in A$. Let $x \in F$. Since F is a state filter, we get $\mu(x) \in F$. Since F is a filter, we get $\mu(a) * \mu(x) \in F$ for any $a \in X$. Hence $x \in A$, which implies $F \subseteq A$. Hence A is an injective filter of (X, μ) containing $F \cup \{a\}$. We now show that A is the smallest injective filter of (X, μ) that containing $F \cup \{a\}$. Let G be an injective filter of (X, μ) such that $F \cup \{a\} \subseteq G$. Let $x \in A$. Then there exists $n \in \mathbb{N}$ such that $\mu(a)^n * \mu(x) \in F \subseteq G$. Since $a \in G$ and G is a state filter, we get $\mu(a) \in G$. Hence $\mu(x) \in G$. Since G is an injective filter of (X, μ), by the characterization theorem of injective filters, we get that $x \in G$. Hence A is the smallest injective filter of (X, μ) that is containing $F \cup \{a\}$. □

Proposition 10.6.9 *Let F be an injective filter of a transitive state BE-algebra (X, μ). Then for any $a, b \in X$*

$$\langle a\rangle_i \cap \langle b\rangle_i \subseteq F \text{ if and only if } \langle F \cup \{a\}\rangle_i \cap \langle F \cup \{b\}\rangle_i = F$$

Proof Assume that $\langle F \cup \{a\}\rangle_i \cap \langle F \cup \{b\}\rangle_i = F$. Since $a \in \langle F \cup \{a\}\rangle_i$, we get that $\langle a\rangle_i \subseteq \langle F \cup \{a\}\rangle_i$. Similarly, we get $\langle b\rangle_i \subseteq \langle F \cup \{b\}\rangle_i$. Hence $\langle a\rangle_i \cap \langle b\rangle_i \subseteq \langle F \cup \{a\}\rangle_i \cap \langle F \cup \{b\}\rangle_i = F$.

Conversely, assume that $\langle a\rangle_i \cap \langle b\rangle_i \subseteq F$ for all $a, b \in X$. Clearly $F \subseteq \langle F \cup \{a\}\rangle_i \cap \langle F \cup \{b\}\rangle_i$. Let $x \in \langle F \cup \{a\}\rangle_i \cap \langle F \cup \{b\}\rangle_i$. Then there exist $m, n \in \mathbb{N}$ such that $\mu(a)^m * \mu(x) \in F$ and $\mu(b)^n * \mu(x) \in F$. Hence there exists $u, v \in F$ such that $\mu(a)^m * \mu(x) = u$ and $\mu(b)^n * \mu(x) = v$. Hence

$$\mu(a)^m * (u * \mu(x)) = u * (\mu(a)^m * \mu(x)) = u * u = 1 \in \langle a\rangle_i$$

Since $\langle a\rangle_i$ is an injective filter and $a \in \langle a\rangle_i$, we get $u * \mu(x) \in \langle a\rangle_i$. By the similar argument, we get that $v*\mu(x) \in \langle b\rangle_i$. Also $u*\mu(x) \leq v*(u*\mu(x)) = u*(v*\mu(x))$ and $v*\mu(x) \leq u*(v*\mu(x))$. Since $u*\mu(x) \in \langle a\rangle_i$, we get that $u*(v*\mu(x)) \in \langle a\rangle_i$. Since $v*\mu(x) \in \langle b\rangle_i$, we get that $u*(v*\mu(x)) \in \langle b\rangle_i$. Hence $u*(v*\mu(x)) \in \langle a\rangle_i \cap \langle b\rangle_i \subseteq F$. Since $u, v \in F$ and F is a filter, we get $\mu(x) \in F$. Since F is injective, we get $x \in F$. Therefore $\langle F \cup \{a\}\rangle_i \cap \langle F \cup \{b\}\rangle_i \subseteq F$. □

Theorem 10.6.10 *Let F, G be two injective filters of a transitive BE-algebra X. Then $F \vee G = \{x \in X \mid \mu(a) * (\mu(b) * \mu(x)) = 1 \text{ for some } a \in F \text{ and } b \in G\}$ is the smallest injective filter of X which is containing both F and G.*

Proof Clearly $1 \in F \vee G$. Let $x, x * y \in F \vee G$. Then there exists $a, c \in F$ and $b, d \in G$ such that $\mu(a) * (\mu(b) * \mu(x)) = 1$ and $\mu(c) * (\mu(d) * \mu(x * y)) = 1$. Then

$$1 = \mu(c)*(\mu(d)*\mu(x*y)) \leq \mu(c)*(\mu(d)*(\mu(x)*\mu(y))) = \mu(x)*(\mu(c)*(\mu(d)*\mu(y))).$$

Hence $\mu(x) \leq \mu(c) * (\mu(d) * \mu(y))$. Since X is transitive, we get

$$\begin{aligned}
1 &= \mu(a) * (\mu(b) * \mu(x)) \\
&\leq \mu(a) * (\mu(b) * (\mu(c) * (\mu(d) * \mu(y)))) \\
&= \mu(a) * (\mu(c) * (\mu(b) * (\mu(d) * (y))))
\end{aligned}$$

Hence $\mu(a)*(\mu(c)*(\mu(b)*(\mu(d)*\mu(y)))) = 1 \in F$ such that $a, c \in F$ and $b, d \in G$. Since F is a state filter, we get $\mu(a), \mu(c) \in F$. Hence $\mu(b) * (\mu(d) * \mu(y)) \in F$. Then there exists $f \in F$ such that $\mu(b) * (\mu(d) * \mu(y)) = f$. Hence by (SO3), we get

$$\begin{aligned}\mu(f) &= \mu(\mu(b) * (\mu(d) * \mu(y))) \\ &= \mu(\mu(b) * \mu(\mu(d) * \mu(y))) \\ &= \mu(b) * \mu(\mu(d) * \mu(y)) \\ &= \mu(b) * (\mu(d) * \mu(y)).\end{aligned}$$

Thus

$$\begin{aligned}\mu(b) * (\mu(d) * (\mu(f) * \mu(y))) &= \mu(f) * (\mu(b) * (\mu(d) * \mu(y))) \\ &= (\mu(b) * (\mu(d) * \mu(y))) * (\mu(b) * (\mu(d) * \mu(y))) \\ &= 1 \in G\end{aligned}$$

Since $\mu(b), \mu(d) \in G$, it yields $\mu(f) * \mu(y) \in G$. Then there exists $g \in G$ such that $\mu(f) * \mu(y) = g$. Then $\mu(g) = \mu(\mu(f) * \mu(y)) = \mu(f) * \mu(y)$. Hence

$$\mu(f) * (\mu(g) * \mu(y)) = \mu(g) * (\mu(f) * \mu(y)) = (\mu(f) * \mu(y)) * (\mu(f) * \mu(y)) = 1$$

Since $f \in F, g \in G$, we get $y \in F \vee G$. Therefore $F \vee G$ is a filter of X. Let $x \in F$. Then $\mu(x) * (\mu(1) * \mu(x)) = \mu(x) * \mu(x) = 1$. Since $1 \in G$, we get $x \in F \vee G$. Hence $F \subseteq F \vee G$. Similarly, we get $G \subseteq F \vee G$. Let H be an injective filter of X such that $F \subseteq H$ and $G \subseteq H$. Let $x \in F \vee G$. Then there exists $a \in F \subseteq H$ and $b \in G \subseteq H$ such that $\mu(a) * (\mu(b) * \mu(x)) = 1 \in H$. Since $\mu(a), \mu(b) \in H$, we get $\mu(x) \in H$. Since H is injective, we get $x \in H$. Hence $F \vee G \subseteq H$. Therefore $F \vee G$ is the smallest injective filter which contains both F and G. □

For any transitive state BE-algebra (X, μ), denote the set of all injective filters of X by $\mathcal{F}_I(X, \mu)$. Then $\mathcal{F}_I(X, \mu)$ is a partially ordered set with respect to set inclusion. Then following is a direct consequence of the above theorem.

Corollary 10.6.11 *Let (X, μ) be a transitive state BE-algebra. Then $(\mathcal{F}_I(X, \mu), \cap, \vee)$ forms a complete lattice with respect to the set inclusion.*

10.7 Congruences and Injective Filters

In this section, the notion of injective congruences is introduced. An equivalent condition is derived for every congruence on a BE-algebra to become an injective congruence. A bijection is obtained between the set of all injective filters of a state BE-algebra (X, μ) and the set of all injective congruences on X. Let θ be a state congruence on a state BE-algebra (X, μ). Then clearly $(X_{/\theta}, *, [1]_\theta)$ is a BE-algebra with respect to point-wise operation.

Definition 10.7.1 A congruence θ on a state BE-algebra (X, μ) is said to be an *injective congruence* if $(\mu(x), \mu(y)) \in \theta$ implies $(x, y) \in \theta$ for all $x, y \in X$.

Proposition 10.7.2 *Let (X, μ) be a transitive state BE-algebra. For any injective filter F of (X, μ), define*

$$(x, y) \in \theta_F \text{ if and only if } \mu(x) * \mu(y) \in F \text{ and } \mu(y) * \mu(x) \in F$$

for all $x, y \in X$. Then θ_F is an injective congruence on (X, μ).

Proof Clearly θ_F is reflexive and symmetric. Let $x, y, z \in X$ be such that $(x, y) \in \theta_F$ and $(y, z) \in \theta_F$. Then $\mu(x) * \mu(y) \in F, \mu(y) * \mu(x) \in F, \mu(y) * \mu(z) \in F$ and $\mu(z) * \mu(y) \in F$. $\mu(x) * (\mu(y) * \mu(z)) \in F$. Since X is transitive, we get $\mu(y) * \mu(z) \leq (\mu(x) * \mu(y)) * (\mu(x) * \mu(z))$. Since $\mu(y) * \mu(z) \in F$, we get $(\mu(x) * \mu(y)) * (\mu(x) * \mu(z)) \in F$. Since $\mu(x) * \mu(y) \in F$, we get $\mu(x) * \mu(z) \in F$. Similarly, we get $\mu(z) * \mu(x) \in F$. Thus $(x, z) \in \theta$. Therefore θ_F is an equivalence relation on X. Let $(x, y) \in \theta_F$ and $(u, v) \in \theta_F$. Then $\mu(x) * \mu(y) \in F, \mu(y) * \mu(x) \in F, \mu(u) * \mu(v) \in F$ and $\mu(v) * \mu(u) \in F$. Since $\mu(x) * \mu(y) \in F$, we get $(\mu(u) * \mu(x)) * (\mu(u) * \mu(y)) = \mu(u) * (\mu(x) * \mu(y)) \in F$. Since $\mu(y) * \mu(x) \in F$, we get $(\mu(u) * \mu(y)) * (\mu(u) * \mu(x)) = \mu(u) * (\mu(y) * \mu(x)) \in F$. Hence $(\mu(u) * \mu(x), \mu(u) * \mu(y)) \in \theta_F$. Again,

$$\begin{aligned}(\mu(v) * \mu(y)) * (\mu(u) * \mu(y)) &= \mu(u) * ((\mu(v) * \mu(y)) * \mu(y)) \\ &= (\mu(u) * (\mu(v) * \mu(y))) * (\mu(u) * \mu(y)) \\ &= ((\mu(u) * \mu(v)) * (\mu(u) * \mu(y))) * (\mu(u) * \mu(y))\end{aligned}$$

Hence by considering $a = \mu(u)$; $b = \mu(v)$ and $\mu(y) = c$, we get

$$\begin{aligned}(a * b) * ((b * c) * (a * c)) &= (a * b) * (((a * b) * (a * c)) * (a * c)) \\ &= ((a * b) * (a * c)) * ((a * b) * (a * c)) \\ &= 1 \in F\end{aligned}$$

Hence $(\mu(u)*\mu(v))*((\mu(v)*\mu(y))*(\mu(u)*\mu(y))) \in F$. Since $\mu(u)*\mu(v) \in F$, we get $(\mu(v)*\mu(y))*(\mu(u)*\mu(y)) \in F$. Similarly, we can obtain $(\mu(u)*\mu(y))*(\mu(v)*\mu(y)) \in F$. Hence $(\mu(u)*\mu(y), \mu(v)*\mu(y)) \in \theta_F$. Thus $(\mu(u)*\mu(x), \mu(v)*\mu(y)) \in \theta_F$. Therefore θ_F is a congruence on X. Since $\mu(\mu(x)) = \mu(x)$ for all $x \in X$, it is an easy consequence that θ_F is an injective congruence on (X, μ). □

In general, every congruence of a state BE-algebra is not an injective congruence. However, a necessary and sufficient condition is derived for a congruence to become an injective congruence.

Theorem 10.7.3 *Let (X, μ) be a commutative state BE-algebra. A congruence θ on X is an injective congruence on (X, μ) if and only if $(1, \mu(x)) \in \theta$ implies $(1, x) \in \theta$ for all $x \in X$.*

Proof Let θ be a congruence on X. Then the condition for necessity is clear. Conversely, assume the condition. Let $x, y \in X$. Let $(\mu(x), \mu(y)) \in \theta$. Then

$$
\begin{aligned}
(\mu(x), \mu(y)) \in \theta &\Rightarrow (\mu(x) * \mu(x), \mu(y) * \mu(x)) \in \theta \\
&\Rightarrow (1, \mu(y) * \mu(x)) \in \theta \\
&\Rightarrow (1 * \mu(x), (\mu(y) * \mu(x)) * \mu(x)) \in \theta \\
&\Rightarrow (\mu(x), (\mu(\mu(y) * \mu(x)) * \mu(x))) \in \theta \\
&\Rightarrow (\mu(x), \mu(\mu(\mu(y) * \mu(x)) * \mu(x))) \in \theta \\
&\Rightarrow (x, \mu(\mu(y) * \mu(x)) * \mu(x)) \in \theta \\
&\Rightarrow (x, (\mu(y) * \mu(x)) * \mu(x)) \in \theta.
\end{aligned}
$$

Again

$$
\begin{aligned}
(\mu(x), \mu(y)) \in \theta &\Rightarrow (\mu(x) * \mu(y), \mu(y) * \mu(y)) \in \theta \\
&\Rightarrow (\mu(x) * \mu(y), 1) \in \theta \\
&\Rightarrow (\mu(\mu(x) * \mu(y)), \mu(1)) \in \theta \\
&\Rightarrow (\mu(\mu(x) * \mu(y)) * \mu(y), 1 * \mu(y)) \in \theta \\
&\Rightarrow (\mu(\mu(x) * \mu(y)) * \mu(y), \mu(y)) \in \theta \\
&\Rightarrow (\mu(\mu(\mu(x) * \mu(y)) * \mu(y)), \mu(y)) \in \theta \\
&\Rightarrow (\mu(\mu(x) * \mu(y)) * \mu(y), y) \in \theta \\
&\Rightarrow ((\mu(x) * \mu(y)) * \mu(y), y) \in \theta.
\end{aligned}
$$

Since X is commutative, we get that $(\mu(y) * \mu(x)) * \mu(x) = (\mu(x) * \mu(y)) * \mu(y)$. From the above two observations, we get $(x, y) \in \theta$. Therefore θ is an injective congruence on (X, μ). □

In the following, we derive a characterization theorem for injective filters.

Theorem 10.7.4 *Let (X, μ) be a state BE-algebra where X is transitive. If F is a state filter of X, then the following conditions are equivalent:*

(1) F is an injective filter,
(2) there exists an injective congruence on (X, μ) whose kernel is F,
(3) for all $x, y \in X$, $\mu(x), \mu(y) \in F$ implies $x * y \in F$.

Proof (1) $\Rightarrow$ (2): Assume that F is an injective filter of (X, μ). Then by Proposition 10.7.2, θ_F is an injective congruence on (X, μ). It is enough to show that $Ker(\theta_F) = F$. Let $x \in Ker(\theta_F)$. Then we get that $(1, x) \in \theta_F$. Hence $\mu(x) = 1 * \mu(x) = \mu(1) * \mu(x) \in F$. Since F is injective, we get $x \in F$. Therefore $Ker(\theta_F) \subseteq F$. Conversely, let $x \in F$. Then we get $\mu(1) * \mu(x) = \mu(x) \in F$ and $\mu(x) * \mu(1) = 1 \in F$. Hence $(\mu(1), \mu(x)) \in \theta_F$. Since θ_F is injective, it immediately infers that $(1, x) \in \theta_F$. Hence $x \in Ker(\theta_F)$.
(2) $\Rightarrow$ (3): Assume that the condition (2) holds. Let θ be an injective congruence on X such that $F = Ker(\theta)$. Let $\mu(x), \mu(y) \in F = Ker(\theta)$. Then $(\mu(1), \mu(x)) = (1, \mu(x)) \in \theta_F$ and $(\mu(1), \mu(y)) = (1, \mu(y)) \in \theta_F$. Since θ_F is an injective congruence, we get that $(1, x) \in \theta_F$ and $(1, y) \in \theta_F$. Hence $(x * y, 1) = (x * y, 1 * 1) \in \theta_F$. Therefore $x * y \in Ker(\theta_F) = F$.

(3) ⇒ (1): Let $\mu(x) \in F$. Since $\mu(1), \mu(x) \in F$, by the condition (3), it yields that $x = 1 * x \in F$. Therefore F is an injective filter of (X, μ). □

Corollary 10.7.5 *Let (X, μ) be a commutative state BE-algebra. If F is an injective filter of (X, μ), then θ_F is the smallest injective congruence whose kernel is F.*

Proof Since F is an injective filter of (X, μ), by the main theorem θ_F is an injective congruence on (X, μ) such that $Ker(\theta_F) = F$. It is enough to prove that θ_F is the smallest injective congruence on X. Suppose R is an injective congruence on X such that $F = Ker(R)$.

$$\begin{aligned}
(x, y) \in \theta_F &\Rightarrow \mu(x) * \mu(y) \in F = Ker(R) \text{ and } \mu(y) * \mu(x) \in F = Ker(R)\\
&\Rightarrow (\mu(x) * \mu(y), \mu(1)) \in R \text{ and } (\mu(y) * \mu(x), \mu(1)) \in R\\
&\Rightarrow ((\mu(x) * \mu(y)) * \mu(y), \mu(1) * \mu(y)) \in R \text{ and}\\
&\qquad\qquad ((\mu(y) * \mu(x)) * \mu(x), \mu(1) * \mu(x)) \in R\\
&\Rightarrow ((\mu(x) * \mu(y)) * \mu(y), \mu(y)) \in R \text{ and}\\
&\qquad\qquad ((\mu(y) * \mu(x)) * \mu(x), \mu(x)) \in R\\
&\Rightarrow (\mu(x), \mu(y)) \in R \qquad \text{since } X \text{ is commutative}\\
&\Rightarrow (x, y) \in R. \qquad \text{since } R \text{ is injective}
\end{aligned}$$

Therefore θ_F is the smallest injective congruence on (X, μ) whose kernel is F. □

Theorem 10.7.6 *Let (X, μ) be a subdirectly irreducible state BE-algebra. Then $Ker(\mu)$ is a prime injective filter of (X, μ).*

Proof Let F and G be two injective filters of (X, μ) such that $F \cap G \subseteq Ker(\mu)$. For simplicity, represent $Ker(\mu)$ by K. Define $\phi : X_{/K} \to \mu(X)_{/F} \times \mu(X)_{/G}$, by $\phi(K_x) = (F_x, G_x)$ for all $x \in X$. Let $x, y \in X$ be such that $K_x = K_y$. Then $(x, y) \in \theta_K$, which implies $\mu(x)*\mu(y) \in K$ and $\mu(y)*\mu(x) \in K$. Hence $\mu(x)*\mu(y) \geq \mu(x*y) = 1 = \mu(y * x) \leq \mu(y) * \mu(x)$. Thus we get $\mu(x) = \mu(y)$. Therefore ϕ is a well-defined. Clearly ϕ is a homomorphism. Let $x, y \in X$ be such that $\phi(K_x) = \phi(K_y)$. Then $(F_{\mu(x)}, G_{\mu(x)}) = (F_{\mu(y)}, G_{\mu(y)})$, so that $\mu(x) * \mu(y), \mu(y) * \mu(x) \in F \cap G$. Hence $\mu(x) * \mu(y), \mu(y) * \mu(x) \in K$. It follows that $K_x = K_y$, which implies that ϕ is one-to-one. Clearly $\pi_1 \circ \phi(X_{/K}) = \mu(X)_{/F}$ and $\pi_2 \circ \phi(X_{/K}) = \mu(X)_{/G}$, where $\pi_1 : \mu(X)_{/F} \times \mu(X)_{/G} \to \mu(X)_{/F}$ and $\pi_2 : \mu(X)_{/F} \times \mu(X)_{/G} \to \mu(X)_{/G}$ are natural projection maps. Since $X_{/Ker(\mu)}$ and $\mu(X)$ are isomorphic, we get $X_{/Ker(\mu)}$ is a subdirectly irreducible BE-algebra and so $\pi_1 \circ \phi : X_{/K} \to \mu(X)_{/F}$ or $\pi_2 \circ \phi : X_{/K} \to \mu(X)_{/G}$ is an isomorphism. Without lost of generality, we can assume that $\pi_1 \circ \phi$ is an isomorphism. For any $x \in F$, $\pi_1(\phi(K_x)) = \pi_1(F_{\mu(x)}, G_{\mu(x)}) = \mu(x)_{/F}$. Since F is an injective filter, we get $\mu(x) \in F$ and hence $F_{\mu(x)} = F_1$. It follows that $K_x = K_1$ (since $\pi_1 \circ \phi$ is an isomorphism) and $x \in K$. Therefore $F \subseteq K = Ker(\mu)$ and so $K = Ker(\mu)$ is prime. □

Theorem 10.7.7 *Let (X, μ) be a commutative state BE-algebra. Then there exists an isotone bijection between the set $\mathcal{C}_I(X, \mu)$ of all injective congruences of (X, μ) and the set $\mathcal{F}_I(X, \mu)$ of all injective filters of (X, μ).*

Proof Define $\Psi : \mathcal{C}_I(X, \mu) \to \mathcal{F}_I(X, \mu)$ by $\Psi(\theta) = Ker(\theta)$. Since $Ker(\theta)$ is an injective filter of (X, μ), it is clear that the map Ψ is well-defined. Let $\theta_1, \theta_2 \in \mathcal{C}_I(X, \mu)$ be such that $\Psi(\theta_1) = \Psi(\theta_2)$. Hence $Ker(\theta_1) = Ker(\theta_2)$. We now prove that $\theta_1 = \theta_2$. Let $(x, y) \in \theta_1$. Then we get the following:

$$\begin{aligned}
(x, y) \in \theta_1 &\Rightarrow (x * y, y * y) \in \theta_1 \\
&\Rightarrow (x * y, 1) \in \theta_1 \\
&\Rightarrow x * y \in Ker(\theta_1) = Ker(\theta_2) \\
&\Rightarrow (x * y, 1) \in \theta_2 \\
&\Rightarrow ((x * y) * y, 1 * y) \in \theta_2 \\
&\Rightarrow ((x * y) * y, y) \in \theta_2
\end{aligned}$$

Similarly, we get $((y*x)*x, x) \in \theta_2$. Hence $(x, y) \in \theta_2$. Therefore $\theta_1 \subseteq \theta_2$. Similarly, we get $\theta_2 \subseteq \theta_1$. Therefore Ψ is one-to-one. We now claim that Ψ is onto. Let F be an injective filter of (X, μ). Since X is self-distributive, by Proposition 10.7.2, we get θ_F is an injective congruence on (X, μ) such that $Ker\ \theta_F = F$. For this injective congruence θ_F, we get

$$\Psi(\theta_F) = Ker(\theta_F) = F$$

Hence Ψ is onto. To show that Ψ is an isotone, let us assume that $\theta_1 \subseteq \theta_2$ for $\theta_1, \theta_2 \in \mathcal{C}_I(X)$. Let $x \in \Psi(\theta_1) = Ker(\theta_1)$. Then $(x, 1) \in \theta_1 \subseteq \theta_2$. Hence $x \in Ker(\theta_2) = \Psi(\theta_2)$. Thus $\Psi(\theta_1) \subseteq \Psi(\theta_2)$. Therefore Ψ is isotone. □

Exercise

1. Prove that every idempotent endomorphism is a state operator on a BE-algebra. Show that the converse of this statement is not true. Give an example of a state operator which is not an endomorphism.
2. Let μ be a state operator on a commutative BE-algebra X. If X is linearly ordered, then prove that μ is a homomorphism.
3. Let μ be a state operator on a BE-algebra X such that $Ker(\mu) = \{1\}$. Then prove that μ is the identity mapping on X.
4. Let $a \in [0, 1], X = ((a, 1], *_{\mathbb{R}}, 1)$ and (X, μ) be a state BE-algebra. Then prove that $\mu : X \to [0, 1]$ is a measure. In addition, if $X = ([0, 1], *_{\mathbb{R}}, 1)$ and (X, μ) is a state BE-algebra such that $\mu(1) = 1$, then prove that $\mu : X \to [0, 1]$ is a state-morphism.
5. Let θ be a state congruence on a BE-algebra X and $\overline{\mu}$ a self-map on $X_{/\theta}$ by $[x]_\theta \mapsto [\mu(x)]_\theta$. Then prove that (X, μ) is a state BE-algebra if and only if $(X_{/\theta}, \overline{\mu})$ is a state BE-algebra.
6. Let F be a filter of a state BE-algebra (X, μ). If F is a prime state filter of (X, μ), then prove that $\overline{F}$ is a prime state filter of the quotient state algebra $(X_{/\theta}, \overline{\mu})$ where

θ is the state congruence on X defined by $(x, y) \in \theta$ such that $x * y \in F$ and $y * x \in F$.

7. Let θ be a state congruence and F a filter of a state BE-algebra (X, μ) such that $[1]_\theta = \{1\}$. Then $\overline{F}$ is a prime state filter of $(X_{/\theta}, \overline{\mu})$ whenever F is a prime state filter of (X, μ).
8. Prove that a state BE-algebra (X, μ) is μ-commutative if and only if $(\mu(z) * \mu(x)) * (\mu(y) * \mu(x)) = (\mu(x) * \mu(z)) * (\mu(y) * \mu(z))$ for all $x, y, z \in X$.
9. Let (X, μ) is a μ-distributive state BE-algebra. Then prove that (X, μ) is μ-commutative if and only if it is μ-implicative.
10. Let (X, μ) be a negatively ordered state BE-algebra. Then prove that $(\mu(x) * \mu(y)) * (\mu(x) * \mu(z)) \leq \mu(x) * (\mu(y) * \mu(z))$ for any $x, y, z \in X$.
11. For any subdirectly irreducible state BE-algebra (X, μ), prove that $Ker(\mu)$ is a prime state filter of the state BE-algebra.
12. Let θ a state congruence on a state BE-algebra (X, μ) and F a non-empty subset of (X, μ). Then prove that F is a prime injective filter of X if and only if $\overline{F}$ is a prime injective filter of $X_{/\theta}$, where $\overline{F} = \{[x]_\theta \mid x \in F\}$.
13. In any transitive state BE-algebra (X, μ), prove that every maximal injective filter is prime. Moreover, if $M_1, M_2, ..., M_n$ and M are maximal injective filters of X such that $\bigcap_{i=1}^{n} M_i \subseteq M$, then prove that there exists $j \in \{1, 2, ..., n\}$ such that $M_j = M$.
14. Consider a state BE-algebra (X, μ) and $a \in X$. If F is an injective filter of X such that $a \notin F$, then prove that there exists a prime injective filter P such that $a \notin P$ and $F \subseteq P$.
15. Let (X, μ) be a state BE-algebra. If F is a prime injective filter of X and $\mu^{-1}(F) \neq X$, then prove that $\mu^{-1}(F)$ is a prime injective filter of X.
16. Let (X, μ) be a self-distributive and commutative state BE-algebra. If F is an injective filter of (X, μ), then prove that θ_F is the smallest injective congruence whose kernel is F.

Chapter 11
Self-maps of BE-algebras

Self-mappings of special kinds are the subjects of many important theories: see, for instance, Lie group, mapping class group, permutation group. In category theory, "map" is often used as a synonym for morphism or arrow, thus for something more general than a function. In formal logic, the term map is sometimes used for a functional predicate, whereas a function is a model of such a predicate in set theory. During the past twenty years, various aspects of BE-algebras were studied by a number of researchers, but so far, the notions like self-maps of BE-algebras and categorical aspects of BE-algebras had not been investigated intensively. In 1991, M. Kondo [158] studied extensively the properties of left mappings in BCK-algebras. Later in 1999, Z. Chen and Y. Huang [43] extended the study of Kondo and derived a set of equivalent conditions for a left map to become an endomorphism. Recently, S.S. Ahn, H.S. Kim, and H.D. Lee [7] studied the relationship between the left and right mappings of a Q-algebra.

The purpose of this chapter is to study the properties of self-maps of BE-algebras. This has divided into four sections. Section 11.1 gives a brief survey of the properties of homomorphisms of BE-algebras which are used in the subsequent sections. In Sect. 11.2, an equivalency between the two left and right self-maps of BE-algebras is observed. A set of equivalent conditions is derived for a right map on a BE-algebra to become a surjective map. The kernel of the right map of a BE-algebra is observed to become a filter. Section 11.3 is devoted to the self-maps of bounded BE-algebras in terms of pseudo-complements. Some sufficient conditions are derived for a bounded self-map of a bounded BE-algebra to become identical.

In Sect. 11.4, some characterization theorems are derived to characterize self-distributive and commutative BE-algebras in terms of bounded right maps. It is observed that the set of all bounded right maps of a bounded BE-algebra forms a self-distributive BE-algebra. Some properties of bounded multipliers of the BE-algebra of bounded self-maps are studied.

S. R. Mukkamala, *A Course in BE-algebras*,
https://doi.org/10.1007/978-981-10-6838-6_11

11.1 Homomorphisms of BE-algebras

In this section, some important properties of homomorphism of BE-algebras are studied. It is proved that a homomorphism f of a BE-algebra is one-to-one if and only if its kernel is the set containing the greatest element. Compositions of homomorphisms of BE-algebras are studied.

Definition 11.1.1 Let $(X, *, 1)$ and $(Y, \circ, 1')$ be two BE-algebras. A mapping $f : X \longrightarrow Y$ is called a *homomorphism* of X into Y if it satisfies the following property:

$$f(x * y) = f(x) * f(y)$$

for all $x, y \in X$. A homomorphism of X into itself is called an *endomorphism* on X.

Example 11.1.2 Let $X = \{1, a, b, c\}$. Define a binary operation $*$ on X as follows:

$*$	1	a	b	c
1	1	a	b	c
a	1	1	a	a
b	1	1	1	a
c	1	1	a	1

Then it can be easily verified that $(X, *, 1)$ is a BE-algebra. Now define a mapping $f : X \to X$ as follows:

$$f(x) = \begin{cases} 1 \text{ if } x = 1, a, c \\ a \quad \text{x} = \text{b} \end{cases}$$

Then it can be easily check that f is a homomorphism on X.

Example 11.1.3 Let $(X, *, 1)$ and $(Y, \circ, 1')$ be two BE-algebras. Define a mapping $\tau : X \to Y$ by $\tau(x) = 1'$ for all $x \in X$. Then τ is a homomorphism of X into Y. For, let $x, y \in X$, we get $\tau(x * y) = 1' = 1' \circ 1' = \tau(x) \circ \tau(y)$. Hence τ is a homomorphism of X into Y.

Example 11.1.4 Let $(X, *, 1)$ be a BE-algebra. Define a mapping $\omega : X \to X$ by $\omega(x) = x$ for all $x \in X$. Then ω is an endomorphism on X. For, let $x, y \in X$, we get $\omega(x * y) = x * y = \omega(x) * \omega(y)$. Therefore ω is an endomorphism on X.

Example 11.1.5 Let $(X, *, 1)$ be a self-distributive BE-algebra. For any fixed $a \in X$, define a mapping $\psi_a : X \to X$ by $\psi_a(x) = a * x$ for all $x \in X$. Then ψ_a is an endomorphism on X. For, let $x, y \in X$, we get that $\psi_a(x * y) = a * (x * y) = (a * x) * (a * y) = \psi_a(x) * \psi_a(y)$.

For any two BE-algebras $(X_1, *, 1)$ and $(X_2, *, 1)$, it can be routinely verified that their direct product $X_1 \times X_2$ is also a BE-algebra with respect to point-wise operations. In the following proposition, we provide two more homomorphisms on product algebras.

Proposition 11.1.6 *Suppose $f_1 : X_1 \times X_2 \to X_1$ and $f_2 : X_1 \times X_2 \to X_2$ are two mappings defined as $f_1(x, y) = x$ and $f_2(x, y) = y$ for all $(x, y) \in X_1 \times X_2$. Then f_1 and f_2 are homomorphisms.*

Proof For any $(x_1, y_1), (x_2, y_2) \in X_1 \times X_2$ where $x_1, x_2 \in X_1$ and $y_1, y_2 \in X_2$. Hence

$$\begin{aligned} f_1((x_1, y_1) * (x_2, y_2)) &= f_1(x_1 * x_2, y_1 * y_2) \\ &= x_1 * x_2 \\ &= f_1(x_1, y_1) * f_2(x_2, y_2). \end{aligned}$$

Hence f_1 is a homomorphism. Similarly, it can be proved that f_2 is a homomorphism. □

In the following proposition, we observe some properties of homomorphisms.

Proposition 11.1.7 *Let $(X, *, 1)$ and $(Y, \circ, 1')$ be two BE-algebras and $f : X \to Y$ a homomorphism. Then the following conditions hold for all $x, y \in X$.*

(1) $f(1) = 1'$,
(2) $x * y = 1$ *implies* $f(x) \circ f(y) = 1'$,
(3) $x \leq y$ *implies* $f(x) \leq f(y)$, *where $\leq$ is the same BE-ordering on X and Y.*

Proof (1) For any $x \in X, f(1) = f(x * x) = f(x) \circ f(x) = 1'$.
(2) Assume that $x * y = 1$. Then $1' = f(1) = f(x * y) = f(x) \circ f(y)$.
(3) Let $x \leq y$. Then $x * y = 1$. By (2), we get $f(x) \circ f(y) = 1'$. Hence $f(x) \leq f(y)$. □

Definition 11.1.8 $(X, *, 1)$, $(Y, \circ, 1')$, and $(Z, \diamond, 1^0)$ be three BE-algebras. If $f_1 : Y \to Z$ and $f_2 : X \to Y$ are two homomorphisms, then define their composition $f_1 \bullet f_2$ as follows:

$$(f_1 \bullet f_2)(x) = f_1(f_2(x)) \quad \text{for all } x \in X.$$

Proposition 11.1.9 *Let $(X, *, 1)$, $(Y, *, 1')$ and $(Z, \diamond, 1^0)$ be there BE-algebras. If $f_1 : Y \to Z$ and $f_2 : X \to Y$ are two homomorphisms, then $f_1 \bullet f_2$ is a homomorphism from X to Z.*

Proof Let $x, y \in X$. Then $(f_1 \bullet f_2)(x * y) = f_1(f_2(x * y)) = f_1(f_2(x) \circ f_2(y)) = f_1(f_2(x)) \diamond f_1(f_2(y)) = (f_1 \bullet f_2)(x) \diamond (f_1 \bullet f_2)(y)$. Therefore $f_1 \bullet f_2$ is a homomorphism from X to Z. □

Corollary 11.1.10 *Let $(X, *, 1)$ be a BE-algebra. If f and g are two endomorphisms on X, then $f \bullet g$ and $g \bullet f$ are also endomorphisms on X.*

Definition 11.1.11 Let $(X, *, 1)$ be a BE-algebra and f an endomorphism on X. Then define the *fixed set* $Fix_f(X)$ of the endomorphism f as follows:

$$Fix_f(X) = \{x \in X \mid f(x) = x\}$$

Proposition 11.1.12 *Let $(X, *, 1)$ be a BE-algebra and f an endomorphism of X. Then $Fix_f(X)$ is a subalgebra of X.*

Proof Let $x, y \in Fix_f(X)$. Then we get $f(x) = x$ and $f(y) = y$. Hence $f(x * y) = f(x) * f(y) = x * y$. Thus it implies that $x * y \in Fix_f(X)$. Therefore $Fix_f(X)$ is a subalgebra of X. □

Definition 11.1.13 Let $(X, *, 1)$ and $(Y, \circ, 1')$ be two BE-algebras and $f : X \to Y$ a homomorphism. For any subset A of X, the *image set* $Im_f(A)$ is defined as follows:

$$Im_f(A) = f(A) = \{f(x) \mid x \in A\}.$$

Theorem 11.1.14 *Let $(X, *, 1)$ and $(Y, \circ, 1')$ be two BE-algebras and $f : X \to Y$ a homomorphism. If A is a subalgebra of X, then $Im_f(A)$ is a subalgebra of Y with respect to the induced operation $\circ$ of Y.*

Proof Let A be a subalgebra of X. Let $x, y \in Im_f(A)$. Then there exist $a, b \in A$ such that $x = f(a)$ and $y = f(b)$. Now $x \circ y = f(a) \circ f(b) = f(a * b) \in Im_f(A)$ because of $a * b \in A$. Therefore $Im_f(A)$ is a subalgebra of Y. □

Since $1' = f(1) \in Im_f(X)$, the following corollaries are direct consequences.

Corollary 11.1.15 *Let $(X, *, 1)$ and $(Y, \circ, 1')$ be two BE-algebras. If $f : X \to Y$ is a homomorphism, then $Im_f(X)$ is a BE-algebra with respect to the induced operation $\circ$ of Y.*

Corollary 11.1.16 *Let $(X, *, 1)$ and $(Y, \circ, 1')$ be two BE-algebras and $f : X \to Y$ a homomorphism. If X is a commutative BE-algebra, then $Im_f(X)$ is also commutative with respect to the induced operation $\circ$ of Y.*

Definition 11.1.17 Let $(X, *, 1)$ and $(Y, \circ, 1')$ be two BE-algebras and $f : X \to Y$ a homomorphism. Then the *kernel* of the homomorphism f is defined as follows:

$$Ker(f) = \{x \in X \mid f(x) = 1\}.$$

It is obvious that $1' \in Ker(f)$. It can be easily seen that $Ker(f)$ is a subalgebra of X.

Proposition 11.1.18 *Let $(X, *, 1)$ and $(Y, \circ, 1')$ be two BE-algebras and $f : X \to Y$ a homomorphism. If $x \in Ker(f)$ and $x \leq y$, then $y \in yer(f)$ for all $x, y \in X$.*

Proof Let $x \in Ker(f)$ and $x \leq y$. Then we get that $f(x) = 1'$ and $f(x) \leq f(y)$. Hence $1' = f(x) \leq f(y)$, which implies that $f(y) = 1'$. Therefore $y \in Ker(f)$. □

Theorem 11.1.19 *Let $(X, *, 1)$ is a is commutative BE-algebra and $(Y, \circ, 1')$ a BE-algebra. Let $f : X \to Y$ be a homomorphism. Then f is one-to-one if and only if $Ker(f) = \{1\}$.*

Proof Assume that f is one-to-one. Let $x \in Ker(f)$. Then we get $f(x) = 1' = f(1)$. Since f is one-to-one, we get that $x = 1$. Hence, it concludes that $Ker(f) = \{1\}$. Conversely, assume that $Ker(f) = \{1\}$. Let $x, y \in X$ be such that $f(x) = f(y)$. Then we get that $f(x * y) = f(x) \circ f(y) = f(x) \circ f(x) = 1'$. Hence $x * y \in Ker(f) = \{1\}$. Thus $x \leq y$. Similarly, we get that $y \leq x$. Since X is commutative, we get that $x = y$. Therefore f is one-to-one. □

Theorem 11.1.20 *Let X, Y, Z be three BE-algebras where X is commutative and let $h : X \to Y$ be an epimorphism. If $g : X \to Z$ is a homomorphism such that $Ker(h) \subseteq Ker(g)$. Then there exists a unique homomorphism $f : Y \to Z$ such that $f \circ h = g$.*

Proof Let $y \in Y$. Since h is an epimorphism, there exists $x \in X$ such that $h(x) = y$. Suppose there exists $x_1 \in X$ such that $h(x_1) = y$. Then $h(x * x_1) = h(x) * h(x_1) = y * y = 1$. Hence $x * x_1 \in Ker(h) \subseteq Ker(g)$. Hence $g(x) * g(x_1) = g(x * x_1) = 1$. Hence $g(x) \leq g(x_1)$. Similarly, we obtain, $g(x_1) \leq g(x)$. Since X is commutative, we get $g(x) = g(x_1)$. Thus for each $y \in Y$, there exists a unique $g(x) \in Z$ such that y maps into $g(x)$. Therefore there exists a function $f : Y \to Z$ such that $f \circ h = g$. We now show that f is a homomorphism. Let $y_1, y_2 \in Y$. Since h an epimorphism, there exists $x_1, x_2 \in X$ such that $h(x_1) = y_1$ and $h(x_2) = y_2$. Then

$$\begin{aligned} f(y_1) * f(y_2) &= f(h(x_1)) * f(h(x_2)) \\ &= g(x_1) * g(x_2) \\ &= g(x_1 * x_2) \\ &= f(h(x_1 * x_2)) \\ &= f(h(x_1) * h(x_2)) \\ &= f(y_1 * y_2). \end{aligned}$$

Therefore f is a homomorphism from Y into Z. We now finally prove the uniqueness of f. Suppose $f' : Y \to Z$ is another homomorphism such that $f' \circ h = g$. Let $y \in Y$. Since h is an epimorphism, there exists $x \in X$ such that $h(x) = y$. Then we get the following consequence:

$$\begin{aligned} f(y) &= f(h(x)) \\ &= g(x) \\ &= f'(h(x)) \\ &= f'(y). \end{aligned}$$

Therefore f is the unique homomorphism such that $f \circ h = g$. □

Corollary 11.1.21 *Let X, Y, Z be three BE-algebras where X is self-distributive and let $h : X \to Y$ be an epimorphism. If $g : X \to Z$ is a homomorphism such that $Ker(h) \subseteq Ker(g)$. Then the following conditions hold.*

(1) $Ker(f) = h(Ker(g))$,

(2) *$Im(f) = Im(g)$,*
(3) *f is an epimorphism if and only if g is an epimorphism.*

Proof (1) Let $y \in Ker(f)$. Since $y \in Ker(f) \subseteq Y$ and h is onto, there exists $x \in X$ such that $h(x) = y$. Then $1 = f(y) = f(h(x)) = g(x)$. Hence we get $x \in Ker(g)$. Therefore $y = h(x) \in h(Ker(g))$. Thus it yields that $Ker(f) \subseteq h(Ker(g))$. Conversely, let $x \in h(Ker(g))$. Then there exists an element $y \in Ker(g)$ such that $h(y) = x$. Then $f(x) = f(h(y)) = g(y) = 1$. Hence $x \in Ker(f)$. Therefore $h(Ker(g)) \subseteq Ker(f)$.
(2) Let $x \in Im(f)$. Then $x = f(a)$ for some $a \in Y$. Since h surjective, there exists $y \in X$ such that $h(y) = a$. Hence $x = f(a) = f(h(y)) = g(y) \in Im(g)$. Hence, we get that $Im(f) \subseteq Im(g)$. Conversely, let $x \in Im(g)$. Then $x = g(a)$ for some $a \in X$. Hence $x = g(a) = f(h(a)) \in Im(f)$. Thus $Im(g) \subseteq Im(f)$. Therefore $Im(f) = Im(g)$.
(3) Since $f \circ h = g$, it is obvious. □

Theorem 11.1.22 *Let X, Y, Z be three BE-algebras and let $g : X \to Z$ be a homomorphism. If $h : Y \to Z$ be a monomorphism such that $Im(g) \subseteq Im(h)$, then there exists a unique homomorphism $f : X \to Y$ such that $g = h \circ f$.*

Proof Let $x \in X$ be arbitrary. Then $g(x) \in Im(g) \subseteq Im(h)$. Since h is a monomorphism, there exists a unique $y \in Y$ such that $g(x) = h(y)$. Therefore for each $x \in X$, there is a unique $y \in Y$ such that x maps to y, i.e., $f(x) = y$. Hence there exists a function $f : X \to Y$ such that $h(f(x)) = h(y) = g(x)$, i.e., $h \circ f = g$. Let $x_1, x_2 \in X$. Then we get the following consequence:

$$\begin{aligned} h(f(x_1) * f(x_2)) &= h(f(x_1)) * h(f(x_2)) \\ &= g(x_1) * g(x_2) \\ &= g(x_1 * x_2) \\ &= h(f(x_1 * x_2)) \end{aligned}$$

Since h is a monomorphism, we get $f(x_1 * x_2) = f(x_1) * f(x_2)$. Therefore f is a homomorphism such that $h \circ f = g$. Suppose $f' : X \to Y$ is a homomorphism such that $h \circ f' = g$. Let $x \in X$. Suppose $f(x) \neq f'(x)$. Since $f(x), f'(x) \in Y$, and h is a monomorphism, we get that $g(x) = h(f(x)) \neq h(f'(x)) = g(x)$, which is a contradiction. $f(x) = f'(x)$ for all $x \in X$. Therefore f is the unique homomorphism such that $g = h \circ f$. □

11.2 Self-maps of BE-algebras

In this section, some properties of self-maps are studied. A set of equivalent conditions is derived for two self-maps of a BE-algebras to become equal in the sense of mappings. The notion of idempotent endomorphisms is introduced on BE-algebras. Some equivalent conditions are derived for a self-map of a BE-algebra to become a surjective map.

Definition 11.2.1 Let $(X, *, 1)$ be a BE-algebra. A mapping $f : X \to X$ is said to be a *self-map* on X. A self-map f is called *idempotent* if $f^2(x) = f(x)$ for all $x \in X$ where $f^2 = f \circ f$.

For any self-map f on X, the image of f is defined as $Im(f) = f(X) = \{f(x) \mid x \in X\}$. The set of all fixed elements of the mapping f is denoted by $Fix_f(X) = \{x \in X \mid f(x) = x\}$. If f is an endomorphism on X, then it can be easily observed that both $Im(f)$ and $Fix_f(X)$ are subalgebras of X. If f is an idempotent self-map on X, then it can be easily observed that $f(x) \in Fix_f(X)$ for all $x \in X$. Since the composition of two self-maps f and g on a BE-algebra X is also a self-map on X where $(f \bullet g)(x) = f(g(x))$ for all $x \in X$, we have the following result.

Theorem 11.2.2 *Let f and g be two idempotent self-maps on a BE-algebra X such that $f \bullet g = g \bullet f$. Then the following conditions are equivalent.*

(1) $f = g$,
(2) $Im(f) = Im(g)$,
(3) $Fix_f(X) = Fix_g(X)$.

Proof (1) $\Rightarrow$ (2): It is obvious.

(2) $\Rightarrow$ (3): Assume that $Im(f) = Im(g)$. Let $x \in Fix_f(X)$. Then we get $x = f(x) \in Im(f) = Im(g)$. Hence $x = g(y)$ for some $y \in X$. Now $g(x) = g(g(y)) = g^2(y) = g(y) = x$. Thus $x \in Fix_g(X)$. Therefore $Fix_f(X) \subseteq Fix_g(X)$. Similarly, we can obtain that $Fix_g(X) \subseteq Fix_f(X)$. Therefore $Fix_f(X) = Fix_g(X)$.

(3) $\Rightarrow$ (1): Assume that $Fix_f(X) = Fix_g(X)$. Let $x \in X$ be an arbitrary element. Since $f(x) \in Fix_f(X) = Fix_g(X)$, we can get $g(f(x)) = f(x)$. Also we have $g(x) \in Fix_g(X) = Fix_f(X)$. Hence $f(g(x)) = g(x)$. Thus we have

$$f(x) = g(f(x)) = (g \bullet f)(x) = (f \bullet g)(x) = f(g(x)) = g(x).$$

Therefore f and g are equal in the sense of mappings. □

In the following, we introduce the notions of right mappings and left mappings in a BE-algebra.

Definition 11.2.3 Let $(X, *, 1)$ be a BE-algebra. For any $a \in X$, define a self-mapping $R_a : X \to X$ by $R_a(x) = x * a$ for all $x \in X$.

Definition 11.2.4 Let $(X, *, 1)$ be a BE-algebra. For any $a \in X$, define a self-mapping $L_a : X \to X$ by $L_a(x) = a * x$ for all $x \in X$.

Let us denote the set of all right maps of a BE-algebra X by $\mathbb{R}(X)$ and the set of all left maps of X by $\mathbb{L}(X)$. For any $a, b \in X$, it can be easily observed that $R_a(b) = L_b(a)$. Hence there is one-to-one correspondence between $\mathbb{R}(X)$ and $\mathbb{L}(X)$ of a BE-algebra. Let Id_X be the identity map, *i.e.*, $Id_X(x) = x$ for all $x \in X$. If $\overline{1}$ is the unit map defined by $\overline{x} = 1$ for all $x \in X$.

Proposition 11.2.5 *If X is commutative, then for any $a \in X$, $R_a^3 = R_a$.*

Proof Let $a \in X$. Since X is commutative, by Proposition 2.1.12, we get, for any $x \in X$, $R_a^3(x) = R_a^2(x * a) = R_a((x * a) * a) = ((x * a) * a) * a = x * a = R_a(x)$. Therefore $R_a^3 = R_a$. □

Corollary 11.2.6 *If X is a simple BE-algebra, then every left map is idempotent.*

Proposition 11.2.7 *Let $(X, *, 1)$ be a commutative BE-algebra. For any fixed $a \in X$, the following conditions hold for all $x, y \in X$.*

(1) *R_a^2 is idempotent,*
(2) *R_a^2 is isotone,*
(3) $R_a^2(x) * R_a(y) = R_a^2(y) * R_a(x)$,
(4) $x * R_a^2(y) = R_a(y) * R_a(x) = R_a^2(x) * R_a^2(y)$.

Proof (1) By Proposition 11.2.5, it is clear.
(2) Let $x, y \in X$ be such that $x \leq y$. Then for any $a \in X$, we get $y * a \leq x * a$ and hence $(x * a) * a \leq (y * a) * a$. Thus $R_a^2(x) \leq R_a^2(y)$. Therefore R_a^2 is isotone.
(3) Let $x, y \in X$. Since X is commutative, we get the following consequence:

$$\begin{aligned} R_a^2(x) * R_a(y) &= ((x * a) * a) * (y * a) \\ &= y * (((x * a) * a) * a) \\ &= y * (x * a) \\ &= x * (y * a) \\ &= x * (((y * a) * a) * a) \\ &= ((y * a) * a) * (x * a) \\ &= R_a^2(y) * R_a(x). \end{aligned}$$

(4) For $x, y \in X$, $x * R_a^2(y) = x * ((y * a) * a) = (y * a) * (x * a) = R_a(y) * R_a(x)$. Again we get

$$\begin{aligned} R_a^2(x) * R_a^2(y) &= ((x * a) * a) * ((y * a) * a) \\ &= (y * a) * (((x * a) * a) * a) \\ &= (y * a) * (x * a) \\ &= R_a(y) * R_a(x) \end{aligned}$$

Therefore $R_a^2(x) * R_a(y) = R_a^2(y) * R_a(x)$. □

In the following definition, we introduce the notion of kernel of the map R_a^2.

Definition 11.2.8 Let $(X, *, 1)$ be a BE-algebra. For any fixed $a \in X$, the *kernel* of the right map R_a^2 is defined as $Ker(R_a^2) = \{x \in X \mid R_a^2(x) = 1\}$.

Proposition 11.2.9 *Let X be a transitive BE-algebra and $a \in X$. Then $Ker(R_a^2)$ is a filter of X.*

Proof Since $R_a^2(1) = 1$, we get that $1 \in Ker(R_a^2)$. Let $x, x * y \in Ker(R_a^2)$. Then $R_a^2(x) = 1$ and $R_a^2(x * y) = 1$. Then by Preposition 11.1.7(4), we get the following:

$$\begin{aligned} R_a^2(y) &= 1 * R_a^2(y) \\ &= R_a^2(x * y) * R_a^2(x) \\ &= (x * y) * R_a^2(y) \\ &= (x * y) * (1 * R_a^2(y)) \\ &= (x * y) * (R_a^2(x) * R_a^2(y)) \\ &= (x * y) * (x * R_a^2(y)) \\ &\geq y * R_a^2(y) \\ &= R_a^2(y) * R_a^2(y) \\ &= 1. \end{aligned}$$

Hence $y \in Ker(R_a^2)$. Therefore $Ker(R_a^2)$ is a filter of X. □

Theorem 11.2.10 *Let $(X, *, 1)$ be a commutative BE-algebra. For any fixed $a \in X$, the following conditions are equivalent:*

(1) *R_a is surjective,*
(2) *R_a is injective,*
(3) *R_a^2 is identical.*

Proof (1) ⇒ (2): Assume that R_a is surjective. Let $x, y \in X$ be such that $R_a(x) = R_a(y)$. Since R_a is surjective, there exists $x_1, y_1 \in X$ such that $R_a(x_1) = x$ and $R_a(y_1) = y$. Hence

$$\begin{aligned} R_a(x) = R_a(y) &\Rightarrow R_a^2(x_1) = R_a^2(y_1) \\ &\Rightarrow R_a^3(x_1) = R_a^3(y_1) \\ &\Rightarrow R_a(x_1) = R_a(y_1) \\ &\Rightarrow x = y. \end{aligned}$$

Therefore R_a is injective.
(2) ⇒ (3): Assume that R_a is injective. Let $x \in X$. Since $R_a^3(x) = R_a(x)$ and R_a is injective, we get that $R_a^2(x) = x$. Therefore R_a is identical.
(3) ⇒ (1): Assume that R_a is an identical mapping. Let $x \in X$. Since R_a is identical, we get $R_a(R_a(x)) = R_a^2(x) = x$. Thus to each $x \in X$, there exists $R_a(x)$ such that $R_a(R_a(x)) = x$.
Hence R_a is surjective. □

In a commutative BE-algebra, it was already observed that R_a^2 is an idempotent, *i.e.*, $R_a^4 = R_a^2 \circ R_a^2 = R_a^2$. Then the following theorem is a direct consequence of the above theorem.

Theorem 11.2.11 *Let* $(X, *, 1)$ *be a commutative BE-algebra. For any fixed* $a \in X$, *the following conditions are equivalent:*

(1) R_a^2 *is surjective,*
(2) R_a^2 *is injective,*
(3) R_a^2 *is identical.*

Definition 11.2.12 A BE-algebra X is said to be 3-*potent* with respect to the right map R_a for any $a \in X$ if $R_a^3 = R_a$.

Example 11.2.13 Let $X = \{1, a, b\}$. Define a binary operation $*$ on X as follows:

$*$	1	a	b
1	1	a	b
a	1	1	1
b	1	1	1

It can be easily seen that $(X, *, 1)$ is a 3-potent BE-algebra with respect to R.

Proposition 11.2.14 *Every commutative BE-algebra is a* 3-*potent BE-algebra.*

Proof Let X be a commutative BE-algebra. Let $a \in X$. Since X is commutative, by Theorem 1.1.12, we get that for any $x \in X$, $R_a^3(x) = R_a^2(x * a) = R_a((x * a) * a) = ((x * a) * a) * a = x * a = R_a(x)$. Thus $R_a^3 = R_a$. Therefore X is a 3-potent BE-algebra. □

Proposition 11.2.15 *Let* F *and* G *be two filters of a commutative BE-algebra. Then* $F \cap G = \{1\}$ *if and only if* $R_a^2(b) = R_b^2(a) = 1$ *for any* $a \in F$ *and* $b \in G$.

Proof Assume that $F \cap G = \{1\}$. Let $a \in F$ and $b \in G$. Since $a \leq (b * a) * a$, we get $(b * a) * a \in F$. Similarly $(a * b) * b \in G$. Since $a \in F$, we get $(b * a) * a \in F$. Similarly $(a * b) * b \in G$. Since X is commutative, we get $(a * b) * b = (b * a) * a \in F \cap G = \{1\}$. Therefore $R_a^2(b) = 1$ and $R_b^2(a) = 1$. Conversely, assume that $R_a^2(b) = R_b^2(a) = 1$ for any $a \in F$ and $b \in G$. Let $x \in F \cap G$. Then by the assumed condition, we get $R_x^2(x) = 1$. Hence $x = (x * x) * x = 1$. Therefore $F \cap G = \{1\}$. □

Theorem 11.2.16 *Let* X *be a commutative BE-algebra and* $a \in X$. *If* $Fix(L_a)$ *is a filter of* X, *then* $Fix(L_a) \subseteq Ker(R_a^2)$.

Proof Let $a \in X$ and suppose $Fix(L_a)$ is a filter of X. We first claim that $\langle a \rangle \cap Fix(L_a) = \{1\}$. Let $x \in \langle a \rangle \cap Fix(L_a)$. Then we get that $a^n * x = 1$ and $a * x = L_a(x) = x$ for some positive integer $n \in \mathbb{N}$. Since $a * x = x$ and $a^n * x = 1$, we get that $a * x = 1$. Hence $x = a * x = 1$. Therefore $\langle a \rangle \cap Fix(L_a) = \{1\}$. Let $t \in Fix(L_a)$. Since $a \in \langle a \rangle$ and $t \in Fix(L_a)$, by the above proposition, $R_a^2(t) = 1$. Hence $t \in Ker(R_a^2)$. Therefore $Fix(L_a) \subseteq Ker(R_a^2)$. □

Corollary 11.2.17 *Let X be a commutative BE-algebra and $a \in X$. If $Fix(L_a)$ is a filter of X, then $Fix(L_a) \subseteq Ker(R_a)$.*

11.3 Self-maps and Pseudo-complements

In this section, some properties of self-maps are studied with respect to pseudo-complements. The notion of bounded self-maps is introduced on BE-algebras. Some sufficient conditions are derived for a bounded self-map of a BE-algebra to become an endomorphism.

Definition 11.3.1 Let $(X, *, 0, 1)$ be a bounded BE-algebra. For any $a \in X$, define a self-mapping $\dot{R}_a : X \to X$ by $\dot{R}_a(x) = x * a^\diamond$ for all $x \in X$.

Definition 11.3.2 Let $(X, *, 0, 1)$ be a bounded BE-algebra. For any $a \in X$, define a self-mapping $\dot{L}_a : X \to X$ by $\dot{L}_a(x) = a^\diamond * x$ for all $x \in X$.

Let us denote the set of all right maps of a BE-algebra X by $\dot{\mathbb{R}}(X)$ and the set of all left maps of X by $\dot{\mathbb{L}}(X)$. For any $a, b \in X$, it can be easily observed that $\dot{R}_a(b) = \dot{R}_b(a)$ and $\dot{L}_a(b^\diamond) = \dot{L}_b(a^\diamond)$. Let Id_X be the identity map, *i.e.*, $Id_X(x) = x$ for all $x \in X$. If $\overline{1}$ is the unit map defined by $\overline{x} = 1$ for all $x \in X$. Then the following lemma requires a routine verification.

Lemma 11.3.3 *Let $(X, *, 0, 1)$ be a bounded and transitive BE-algebra. For any $a, b \in X$,*

(1) $\dot{L}_1 = \overline{1}$,
(2) $\dot{L}_0 = Id_X$,
(3) $\dot{L}_1 \circ \dot{R}_a = \dot{L}_1$,
(4) $\dot{L}_0 \circ \dot{R}_a = \dot{R}_1$,
(5) $\dot{L}_a \circ \dot{L}_b = \dot{L}_b \circ \dot{L}_a$,
(6) $\dot{L}_a \circ (\dot{L}_b \circ \dot{L}_c) = \dot{L}_b \circ (\dot{L}_a \circ \dot{L}_c)$.

Proposition 11.3.4 *Let $(X, *, 0, 1)$ be a bounded BE-algebra and $a \in X$. Then $\dot{R}_a$ is an endomorphism on X if and only if $a = 0$.*

Proof Assume that $\dot{R}_a$ is an endomorphism for $a \in X$. Let $x, y \in X$. Then we get $a^\diamond = 1 * a^\diamond = \dot{R}_a(1) = \dot{R}_a(1 * 1) = \dot{R}_a(1) * \dot{R}_a(1) = 1$. Hence $a \leq a^{\diamond\diamond} = 1^\diamond = 0$, which means $a = 0$. Conversely, assume that $a = 0$. Let $x, y \in X$. Then $R_a(x * y) = R_0(x * y) = (x * y) * 0^\diamond = (x * y) * 1 = 1 = (x * 0^\diamond) * (x * 0^\diamond) = \dot{R}_0(x) * \dot{R}_0(y)$. Therefore $\dot{R}_0$ is an endomorphism on X. □

Theorem 11.3.5 *Let $(X, *, 0, 1)$ be a bounded BE-algebra. Then $\dot{L}_0$ is an endomorphism on X.*

Proof Let $x, y \in X$. Then we get $\dot{L}_0(x * y) = 0^\diamond * (x * y) = 1 * (x * y) = x * y = (1 * x) * (1 * y) = (0^\diamond * x) * (0^\diamond * y) = \dot{L}_0(x) * \dot{L}_0(y)$. Therefore $\dot{L}_0$ is an endomorphism on X. □

Theorem 11.3.6 *Let $(X, *, 0, 1)$ be a bounded BE-algebra. Each $\dot{L}_a$; $a \in X$ is an endomorphism if and only if it satisfies the following condition for all $x, y, z \in X$:*

$$x^\diamond * (y * z) = (x^\diamond * y) * (x^\diamond * z).$$

Proof Let $a \in X$. Assume that X satisfies the given condition. Let $x, y \in X$. Then $\dot{L}_a(x * y) = a^\diamond * (x * y) = (a^\diamond * x) * (a^\diamond * y) = \dot{L}_a(x) * \dot{L}_a(y)$. Hence $\dot{L}_a$ is an endomorphism. Conversely, assume that $\dot{L}_a$ is an endomorphism for each $a \in X$. Let $x, y, z \in X$. Then

$$\begin{aligned} x^\diamond * (y * z) &= \dot{L}_x(y * z) \\ &= \dot{L}_x(y) * \dot{L}_x(z) \\ &= (x^\diamond * y) * (x^\diamond * z) \end{aligned}$$

Therefore X is satisfying the condition. □

Proposition 11.3.7 *If X is commutative, then for any $a \in X$, $R_a^3 = R_a$.*

Proof Let $a \in X$. Since X is commutative, by Proposition 2.1.12, we get that for any $x \in X$, $\dot{R}_a^3(x) = \dot{R}_a^2(x * a^\diamond) = \dot{R}_a((x * a^\diamond) * a^\diamond) = ((x * a^\diamond) * a^\diamond) * a^\diamond = x * a^\diamond = \dot{R}_a(x)$. □

Corollary 11.3.8 *If X is a simple BE-algebra, then every left map $\dot{L}_a$; $a \in X$ is idempotent.*

Proposition 11.3.9 *Let $(X, *, 0, 1)$ be a bounded and commutative BE-algebra. For any fixed $a \in X$, the following conditions hold for all $x, y \in X$:*

(1) *$\dot{R}_a^2$ is idempotent,*
(2) *$\dot{R}_a^2$ is isotone,*
(3) *$\dot{R}_a^2(x) * \dot{R}_a(y) = \dot{R}_a^2(y) * \dot{R}_a(x)$,*
(4) *$x * \dot{R}_a^2(y) = \dot{R}_a(y) * \dot{R}_a(x) = \dot{R}_a^2(x) * \dot{R}_a^2(y)$.*

Proof (1) By Proposition 11.3.7, it is clear.
(2) Let $x, y \in X$ be such that $x \leq y$. Then for any $a \in X$, we get $y * a^\diamond \leq x * a^\diamond$ and hence $(x * a^\diamond) * a^\diamond \leq (y * a^\diamond) * a^\diamond$. Thus $\dot{R}_a^2(x) \leq \dot{R}_a^2(y)$. Therefore $\dot{R}_a^2$ is isotone.
(3) For any $x, y \in X$, we get the following consequence:

$$\begin{aligned}\dot{R}_a^2(x) * \dot{R}_a(y) &= ((x * a^\diamond) * a^\diamond) * (y * a^\diamond)\\ &= y * (((x * a^\diamond) * a^\diamond) * a^\diamond)\\ &= y * (x * a^\diamond)\\ &= x * (y * a^\diamond)\\ &= x * (((y * a^\diamond) * a^\diamond) * a^\diamond)\\ &= ((y * a^\diamond) * a^\diamond) * (x * a^\diamond)\\ &= \dot{R}_a^2(y) * \dot{R}_a(x).\end{aligned}$$

(4) For $x, y \in X$, $x * \dot{R}_a^2(y) = x * ((y * a^\diamond) * a^\diamond) = (y * a^\diamond) * (x * a^\diamond) = \dot{R}_a(y) * \dot{R}_a(x)$. Again

$$\begin{aligned}\dot{R}_a^2(x) * \dot{R}_a^2(y) &= ((x * a^\diamond) * a^\diamond) * ((y * a^\diamond) * a^\diamond)\\ &= (y * a^\diamond) * (((x * a^\diamond) * a^\diamond) * a^\diamond)\\ &= (y * a^\diamond) * (x * a^\diamond)\\ &= \dot{R}_a(y) * \dot{R}_a(x).\end{aligned}$$

Therefore $x * \dot{R}_a^2(y) = \dot{R}_a(y) * \dot{R}_a(x) = \dot{R}_a^2(x) * \dot{R}_a^2(y)$. □

Form (4) of the above theorem, the following result is clear.

Theorem 11.3.10 *Let* $(X, *, 0, 1)$ *be a bounded and commutative BE-algebra. For any fixed* $a \in X$*, the following conditions are equivalent for all* $x, y \in X$*:*

(1) $\dot{R}_a^2$ *is endomorphism,*
(2) $\dot{R}_a^2(x * y) = x * \dot{R}_a^2(y)$,
(3) $\dot{R}_a^2(x * y) = \dot{R}_a(y) * \dot{R}_a(x)$.

11.4 Characterization Theorems

In this section, self-maps are used to characterize some algebraic structures like commutative and self-distributive BE-algebras. It is also proved that the class of all bounded right maps is a self-distributive BE-algebra. Some properties of bounded right maps are studied with the help of congruences.

Lemma 11.4.1 *Let* $(X, *, 1)$ *be a BE-algebra and* $a \in X$*, If* $a = 1$*, then* L_a *is an endomorphism.*

Proof Let $x, y \in X$. Suppose $a = 1$. Then $L_a(x * y) = L_1(x * y) = 1 * (x * y) = x * y = (1 * x) * (1 * y) = L_1(x) * L_1(y)$. Therefore L_a is a homomorphism. □

Theorem 11.4.2 *A BE-algebra* $(X, *, 1)$ *is self-distributive if and only if each* $L_a, a \in X$ *is an endomorphism of* X.

Proof Assume that X is self-distributive. Let $a \in X$. Since X is self-distributive, we get $L_a(x * y) = a * (x * y) = (a * x) * (a * y) = L_a(x) * L_a(y)$. Hence L_a is an endomorphism. Conversely, assume that each $L_a, a \in X$ is an endomorphism. Let $x, y, z \in X$. Then

$$\begin{aligned} x * (y * z) &= L_x(y * z) \\ &= L_x(y) * L_x(z) \\ &= (x * y) * (x * z). \end{aligned}$$

Therefore X is self-distributive. □

Theorem 11.4.3 *Let $(X, *, 1)$ be a BE-algebra. Then the following conditions are equivalent:*

(1) *X is commutative,*
(2) *for all $x, y \in X$, $L_{R_x(y)}(x) = L_{R_y(x)}(y)$,*
(3) *for all $x, y \in X$, $L_{L_y(x)}(x) = L_{L_x(y)}(y)$.*

Proof (1) $\Rightarrow$ (2): Assume that X is commutative. Let $x, y \in X$. Since X is commutative, we get $L_{R_x(y)}(x) = R_x(y) * x = (y * x) * x = (x * y) * y = R_y(x) * y = L_{R_y(x)}(y)$.
(2) $\Rightarrow$ (3): Assume the condition (2). Let $x, y \in X$. Since $R_x(y) = L_y(x)$, it is clear.
(3) $\Rightarrow$ (1): Assume the condition (3). Let $x, y \in X$. Then we get

$$(y * x) * x = L_y(x) * x = L_{L_y(x)}(x) = L_{L_x(y)}(y) = L_x(y) * y = (x * y) * y.$$

Therefore X is a commutative BE-algebra. □

Theorem 11.4.4 *Let $(X, *, 0, 1)$ be a bounded BE-algebra. Each $\dot{R}_a$; $a \in X$ is an endomorphism if and only if it satisfies the following condition for all $x, y, z \in X$:*

$$(x * y) * z^\diamond = (x * z^\diamond) * (y * z^\diamond).$$

Proof Let $a \in X$. Assume that X satisfies the given condition. Let $x, y \in X$. Then $\dot{R}_a(x * y) = (x * y) * a^\diamond = (x * a^\diamond) * (y * a^\diamond) = \dot{R}_a(x) * \dot{R}_a(y)$. Therefore $\dot{R}_a$ is an endomorphism. Conversely, assume that $\dot{R}_a$ is an endomorphism. Let $x, y, z \in X$. Then

$$\begin{aligned} (x * y) * z^\diamond &= \dot{R}_z(x * y) \\ &= \dot{R}_z(x) * \dot{R}_z(y) \\ &= (x * z^\diamond) * (y * z^\diamond). \end{aligned}$$

Therefore X is satisfying the condition of the theorem. □

Here after, we consider a bounded BE-algebra $(X, *, 0, 1)$ as a bounded BE-algebra which satisfies the condition of the above theorem, unless otherwise mentioned. For any $\dot{R}_a, \dot{R}_b \in \dot{\mathbb{R}}(X)$, define an operation $\circledast$ on $\dot{\mathbb{R}}(X)$ as follows:

$$(\dot{R}_a \circledast \dot{R}_b)(x) = \dot{R}_a(x) * \dot{R}_b(x) \quad \text{for all } x \in X.$$

It can be easily observe that $(\dot{R}_a \circledast \dot{R}_b)(x) = \dot{R}_a(x) * \dot{R}_b(x) = (x * a^\diamond) * (x * b^\diamond) = (a * x^\diamond) * (b * x^\diamond) = (a * b) * x^\diamond = x * (a * b)^\diamond = \dot{R}_{a*b}(x)$. Hence $\dot{R}_a \circledast \dot{R}_b = \dot{R}_{a*b} \in \dot{\mathbb{R}}(X)$. Therefore $\circledast$ is well-defined.

Theorem 11.4.5 *If $(X, *, 0, 1)$ is a bounded and self-distributive BE-algebra, then the algebra $(\dot{\mathbb{R}}(X), \circledast, \dot{R}_0)$ is a self-distributive BE-algebra.*

Proof For any $a \in X$, it is clear that $\dot{R}_0 \circledast \dot{R}_a = \dot{R}_a$, $\dot{R}_a \circledast \dot{R}_0 = \dot{R}_0$, and $\dot{R}_a \circledast \dot{R}_a = \dot{R}_0$. Now, let $a, b, c \in X$. Then for any $x \in X$, we get $\dot{R}_a \circledast (\dot{R}_b \circledast \dot{R}_c)(x) = \dot{R}_a(x) * (\dot{R}_b(x) * \dot{R}_c(x)) = \dot{R}_b(x) * (\dot{R}_a(x) * \dot{R}_c(x)) = \dot{R}_b \circledast (\dot{R}_a \circledast \dot{R}_c)(x)$. Hence $\dot{R}_a \circledast (\dot{R}_b \circledast \dot{R}_c) = \dot{R}_b \circledast (\dot{R}_a \circledast \dot{R}_c)$. Therefore $(\dot{\mathbb{R}}(X), \circledast, \dot{R}_0)$ is a BE-algebra. Again let $\dot{R}_a, \dot{R}_b, \dot{R}_c \in \dot{\mathbb{R}}(X)$. Then for any $x \in X$,

$$\begin{aligned}(\dot{R}_a \circledast (\dot{R}_b \circledast \dot{R}_c))(x) &= \dot{R}_a(x) * (\dot{R}_b \circledast \dot{R}_c)(x)\\ &= \dot{R}_a(x) * (\dot{R}_b(x) * \dot{R}_c(x))\\ &= (\dot{R}_a(x) * \dot{R}_b(x)) * (\dot{R}_a(x) * \dot{R}_c(x))\\ &= (\dot{R}_a \circledast \dot{R}_b)(x) * (\dot{R}_a \circledast \dot{R}_c)(x)\\ &= ((\dot{R}_a \circledast \dot{R}_b) \circledast (\dot{R}_a \circledast \dot{R}_c))(x).\end{aligned}$$

Therefore $(\dot{\mathbb{R}}(X), \circledast, \dot{R}_0)$ is a self-distributive BE-algebra. □

Corollary 11.4.6 *Let $(X, *, 0, 1)$ be a bounded BE-algebra. Then X is epimorphic to $\dot{\mathbb{R}}(X)$.*

Proof Define a mapping $f : X \to \dot{\mathbb{R}}(X)$ by $f(a) = \dot{R}_a$ for all $a \in X$. Let $a, b \in X$. Then clearly $f(a * b) = \dot{R}_{a*b} = \dot{R}_a \circledast \dot{R}_b = f(a) \circledast f(b)$. Therefore f is a homomorphism. Let $\dot{R}_a \in \dot{\mathbb{R}}(X)$. For this $a \in X$, it is obvious that $f(a) = \dot{R}_a$. Therefore f is onto and hence an epimorphism. □

Corollary 11.4.7 *Let X be a bounded BE-algebra. If X is commutative, then so is $(\dot{\mathbb{R}}(X), \circledast, \dot{R}_0)$.*

Corollary 11.4.8 *Let X be a bounded BE-algebra. If X is implicative, then so is $(\dot{\mathbb{R}}(X), \circledast, \dot{R}_0)$.*

Let us recall that a self-map f on a BE-algebra X is called a multiplier on X if $f(x * y) = x * f(y)$ for all $x, y \in X$. We mean a bounded multiplier on a bounded BE-algebra is a multiplier which satisfies $f(0) = 0$. Then clearly $f(x^\diamond) = f(x)^\diamond$ for all $x \in X$.

Theorem 11.4.9 *Let f be a bounded multiplier on a bounded BE-algebra X. Define $\overline{f} : \dot{\mathbb{R}}(X) \to \dot{\mathbb{R}}(X)$ by $\overline{f}(\dot{R}_x) = \dot{R}_{f(x)}$ for all $x \in X$. Then $\overline{f}$ is a multiplier on $\dot{\mathbb{R}}(X)$.*

Proof We first prove that $\overline{f}$ is well-defined. Choose $\dot{R}_a, \dot{R}_b \in \dot{\mathbb{R}}(X)$ such that $\dot{R}_a = \dot{R}_b$. Then $t * a^\diamond = \dot{R}_a(t) = \dot{R}_b(t) = t * b^\diamond$ for all $t \in X$. Since f is a bounded multiplier on X, we get $t * f(a)^\diamond = t * f(a^\diamond) = f(t * a^\diamond) = f(t * b^\diamond) = t * f(b^\diamond) = t * f(b)^\diamond$ for all $t \in X$. Let $x \in X$. Then

$$\dot{R}_{f(a)}(x) = x * f(a)^\diamond = x * f(b)^\diamond = \dot{R}_{f(b)}(x)$$

Hence $\dot{R}_{f(a)} = \dot{R}_{f(b)}$. Thus $\overline{f}(\dot{R}_a) = \overline{f}(\dot{R}_b)$. Hence $\overline{f}$ is well-defined. Let $\dot{R}_a, \dot{R}_b \in \dot{\mathbb{R}}(X)$. Then

$$\overline{f}(\dot{R}_a \circledast \dot{R}_b) = \overline{f}(\dot{R}_{a*b}) = \dot{R}_{f(a*b)} = \dot{R}_{a*f(b)} = \dot{R}_a \circledast \dot{R}_{f(b)} = \dot{R}_a \circledast \overline{f}(\dot{R}_b).$$

Therefore $\overline{f}$ is a multiplier on $\dot{\mathbb{R}}(X)$. □

Definition 11.4.10 Let f be a bounded multiplier on a bounded BE-algebra X. Define the *fixed set* of $\dot{\mathbb{R}}(X)$ with respect to the multiplier $\overline{f}$ as $Fix_{\overline{f}}(\dot{\mathbb{R}}(X)) = \{\dot{R}_x | \overline{f}(\dot{R}_x) = \dot{R}_x\}$.

Theorem 11.4.11 *Let f be a bounded multiplier on a bounded BE-algebra X. Then the fixed set $Fix_{\overline{f}}(\dot{\mathbb{R}}(X))$ is a quasi-filter of $\dot{\mathbb{R}}(X)$.*

Proof Since $f(0) = 0$, we get that $\overline{f}(\dot{R}_0) = \dot{R}_{f(0)} = \dot{R}_0$. Hence $\dot{R}_0 \in Fix_{\overline{f}}(\dot{\mathbb{R}}(X))$. Let $\dot{R}_x \in \dot{\mathbb{R}}(X)$ and $\dot{R}_a \in Fix_{\overline{f}}(\dot{\mathbb{R}}(X))$. Hence $\dot{R}_{f(a)} = \overline{f}(\dot{R}_a) = \dot{R}_a$. Now, we get the following consequence:

$$\overline{f}(\dot{R}_x \circledast \dot{R}_a) = \overline{f}(\dot{R}_{x*a}) = \dot{R}_{f(x*a)} = \dot{R}_{x*f(a)} = \dot{R}_x \circledast \dot{R}_{f(a)} = \dot{R}_x \circledast \dot{R}_a.$$

Hence $\dot{R}_x \circledast \dot{R}_a \in Fix_{\overline{f}}(\dot{\mathbb{R}}(X))$. Therefore $Fix_{\overline{f}}(\dot{\mathbb{R}}(X))$ is a quasi-filter of $\dot{\mathbb{R}}(X)$. □

Let X, Y be two BE-algebras and $f : X \to Y$ be a homomorphism. For any $x, y \in X$, the binary relation θ_f defined by $(x, y) \in \theta_f$ if and only if $f(x) = f(y)$ is a congruence on X. The kernel of the congruence θ_f is defined by $Ker(f) = \{x \in X \mid f(x) = 1\}$. The congruence class of any $x \in X$ with respect to the congruence θ_f is given by $[x]_{\theta_f} = \{y \in X \mid (x, y) \in \theta_f\}$. Then it can be easily observed that $[1]_{\theta_f}$ is a filter such that $[1]_{\theta_f} = Ker(f)$. Define an operation $\circledast$ on $X_{/\theta}$ as follows:

$$[x]_{\theta_f} \otimes [y]_{\theta_f} = [x * y]_{\theta_f}.$$

Since θ_f is a congruence relation on X, it is clear that the operation $\otimes$ is well-defined. Hence it is clear that $(X_{/\theta_f}, \otimes, [1]_{\theta_f})$ is a BE-algebra.

Theorem 11.4.12 *Let* $(X, *, 0, 1)$ *be a bounded BE-algebra and* $f : X \to \dot{\mathbb{R}}(X)$ *be a homomorphism. If* θ_f *is the congruence on* X *defined by* $(x, y) \in \theta_f$ *if and only if* $f(x) = f(y)$ *(i.e.,* $\dot{R}_x = \dot{R}_y$*) for all* $x, y \in X$*, then the image of* X *is isomorphic to the quotient algebra* $X_{/\theta f}$.

Proof Define $\psi : X_{/\theta_f} \to f(X)$ by $\psi([x]_{\theta_f}) = f(x) = \dot{R}_x$. Let $[x]_{\theta_f} = [y]_{\theta_f}$ for $x, y \in X$. Then $(x, y) \in \theta_f$ and hence $\dot{R}_x = \dot{R}_y$. Thus $\psi([x]_{\theta_f}) = \psi([y]_{\theta_f})$. Therefore ψ is well-defined. To show that ψ is one-to-one, let us consider $\psi([x]_{\theta_f}) = \psi([y]_{\theta_f})$ where $x, y \in X$. Then it yields $\dot{R}_x = \dot{R}_y$. Hence $(x, y) \in \theta_f$, which implies $[x]_{\theta_f} = [y]_{\theta_f}$. Therefore ψ is one-to-one. Let $y \in f(X)$. Then there exists $x \in X$ such that $y = f(x)$. For this $x \in X$, we can seen that $\psi([x]_{\theta_f}) = \dot{R}_x = f(x) = y$. Hence ψ is onto. Let $[x]_{\theta_f}, [y]_{\theta_f} \in X_{/\theta_f}$ where $x, y \in X$. Now

$$\begin{aligned}\psi([x]_{\theta_f} * [y]_{\theta_f}) &= \psi([x * y]_{\theta_f})\\ &= \dot{R}_{x*y}\\ &= \dot{R}_x \otimes \dot{R}_y\\ &= \psi([x]_{\theta_f}) \otimes \psi([y]_{\theta_f}).\end{aligned}$$

Thus ψ is a homomorphism. Therefore $X_{/\theta_f}$ is isomorphic to $f(X)$. □

Exercise

1. Let X, Y, Z be three BE-algebras and $g : X \to Z$ a homomorphism. If $h : Y \to Z$ is a monomorphism such that $Im(g) \subseteq Im(h)$, then prove that f is an epimorphism if and only if $Im(h) = Im(g)$.
2. Let X, Y, Z be three BE-algebras where X is self-distributive and $h : X \to Y$ an epimorphism. If $g : X \to Z$ is a homomorphism such that $Ker(h) \subseteq Ker(g)$, then prove that f is a monomorphism if and only if $Ker(h) = Ker(g)$.
3. Let $(X, *, 1)$ be a self-distributive BE-algebra. For any fixed $a \in X$, prove that $Fix(R_a^2) = Im(R_a)$ where $Fix(R_a^2) = \{x \in X \mid R_a^a(x) = x\}$ and $Im(R_a) = \{R_a(x) \mid x \in X\}$.
4. Let X, Y, Z be three BE-algebras and let $g : X \to Z$ be a homomorphism. If $h : Y \to Z$ be a monomorphism such that $Im(g) \subseteq Im(h)$, then prove that $Ker(f) = Ker(g)$.
5. Let X, Y, Z be three BE-algebras and let $g : X \to Z$ be a homomorphism. If $h : Y \to Z$ be a monomorphism such that $Im(g) \subseteq Im(h)$, then prove that $Im(f) = h^{-1}(Im(g))$. Moreover, f is a monomorphism if and only if g is a monomorphism.
6. For any BE-algebra X, consider the self-map $f : X \to X$ defined by $f(x) = x^{\diamond\diamond}$ for all $x \in X$ (where $x^{\diamond} = x * 0$). Then prove that $Fix_X(f) = Im(f)$. Moreover, prove that X is involutory if and only if $Fix_X(f) = Im(X) = X$.
7. For any fixed element a of a BE-algebra X, consider the right map R_a on X. Then prove that $Fix(R_a) \cap Ker(R_a) = \{1\}$.
8. Let F and G be two filters of a 3-potent BE-algebra X. Then prove that $F \cap G = \{1\}$ if and only if $R_a(b) = a$ and $R_b(a) = b$ for any $a \in F$ and $b \in G$.

9. Prove that a non-empty subset F of a self-distributive BE-algebra is a simple filter of X if and only if R_F is a simple filter of $\mathbb{R}(X)$.
10. Let X and Y be two BE-algebras. If $f : X \to Y$ is a homomorphism of BE-algebras and θ_f is the equivalence relation given by $f(x) = f(y)$, then prove that the image of f is isomorphic to the quotient BE-algebra $X_{/\theta_f}$, i.e., $X_{/\theta_f} \cong f(X)$.

Chapter 12
Endomorphisms of BE-algebras

The term endomorphism derives from the Greek adverb endon ("inside") and morphosis ("to form" or "to shape"). In an algebra, an endomorphism of a group, module, ring, vector space, etc. is a homomorphism from the algebra to itself (with surjectivity not required). In 2001, Sergio Celani [34] gave a representation theorem for Hilbert algebras by means of ordered sets and characterize the homomorphisms of Hilbert algebras in terms of applications defined between the sets of all irreducible deductive systems of the associated algebras. In [11], Chul Kon Bae investigated some properties on homomorphisms in BCK-algebras. In his paper, he mainly studied the properties of the compositions of homomorphisms of BCK-algebras. In [46], Z. Chen, Y. Huang and E.H. Roh considered the centralizer $C(S)$ of a given set with respect to the semigroup $End(X)$ of all endomorphisms of an implicative BCK-algebras X with the condition (S). They obtained a series of interesting results those indicated the embedding of X into the centralizer $C(S)$.

The main aim of this chapter is to make an extensive study of some interesting properties of endomorphisms of BE-algebras. The relation between endomorphisms and state operators of BE-algebras is observed. Some structure theorems related to the class of all endomorphisms of BE-algebras are proved. The notions of semi-endomorphic BE-algebras and semi-endomorphic ideals are introduced, and their properties are studied. A set of equivalent conditions is established for every ideal of a BE-algebra to become a semi-endomorphic ideal.

The notion of semi-endomorphic congruences is introduced in BE-algebras. Properties of semi-endomorphic congruences of BE-algebras are studied. The concept of endomorphic congruences is introduced in BE-algebras and studied their properties. Properties of a congruence generated by the image set of an idempotent endomorphism are derived. A sufficient condition is derived for an endomorphic BE-algebra to become subdirectly irreducible. Divisibility properties of endomorphic BE-algebra are studied in terms of endomorphisms.

S. R. Mukkamala, *A Course in BE-algebras*,
https://doi.org/10.1007/978-981-10-6838-6_12

12.1 Properties of Endomorphisms

In this section, properties of endomorphisms of BE-algebras are studied. A relation is established between the endomorphisms and the state operators. Properties of endomorphisms are studied with respect to Cartesian products, strong self-mappings, and compositions of mappings.

Definition 12.1.1 Let $(X, *, 1)$ be a BE-algebra. Then a self-map $f : X \to X$ is said to be an *endomorphism* if $f(x * y) = f(x) * f(y)$ for all $x, y \in X$.

For any endomorphism on X, it is observed that $f(1) = 1$ and for all $x, y \in X$, $f(x) \leq f(y)$ whenever $x \leq y$. The kernel of the endomorphism is defined as $Ker(f) = \{x \in X \mid f(x) = 1\}$. Clearly $Ker(f)$ is a filter of X. The image of X is defined as $Im(f) = \{f(x) | x \in X\}$. It is obvious that $Im(f)$ is a subalgebra of X with $1 \in Im(f)$.

Definition 12.1.2 An endomorphism f on a BE-algebra X is called idempotent if $f^2(x) = f(x)$ for all $x \in X$, where f^2 is the usual representation of $f \circ f$.

Example 12.1.3 Let $(X, *, 1)$ be a BE-algebra. Then the identity map $Id_X : X \to X$ and the unit map $1_X(x) = 1$ for all $x \in X$ are idempotent endomorphisms on X.

Example 12.1.4 Let $(X, *, 1)$ be a self-distributive BE-algebra. For any fixed $a \in X$, the map $\alpha_a : X \to X$ defined by $\alpha_a(x) = a * x$ for all $x \in X$ is state-morphism operator on X. First, we show that α_a is a homomorphism. For $x, y \in X$, $\alpha_a(x * y) = a * (x * y) = (a * x) * (a * y) = \alpha_a(x) * \alpha_a(y)$. Again, for any $x \in X$, $\alpha_a(\alpha_a(x)) = a * (a * x) = (a * a) * (a * x) = 1 * (a * x) = a * x = \alpha_a(x)$. Therefore α_a is an idempotent endomorphism on X.

In the following lemma, some basic properties of idempotent endomorphisms on BE-algebras are derived. The proof of the following is routine and so it is omitted.

Lemma 12.1.5 *Let f be an endomorphism on a BE-algebra X. Then the following hold:*

(1) *$f(X)$ is a subalgebra X,*
(2) *$Ker(f) = \{x \in X \mid f(x) = 1\}$ is a filter of X,*
(3) *$Ker(f) \cap I_m(X) = \{1\}$.*

Proposition 12.1.6 *Let $(X, *, 1)$ be a commutative BE-algebra. Then every idempotent endomorphism on X is a state operator on X.*

Proof Since X is a commutative BE-algebra, we get $((x * y) * y) * y = x * y$ for all $x, y \in X$. Let f be an idempotent endomorphism on X. Let $x, y \in X$. Suppose that $x \leq y$. Then $x * y = 1$ and hence $f(x * y) = f(1) = 1$. Thus $f(x) * f(y) = 1$, which implies that $f(x) \leq f(y)$. Since X is commutative, we get $f((x * y) * y) * f(y) = f(((x * y) * y) * y) = f(x * y)$. Finally, $f(f(x) * f(y)) = f^2(x) * f^2(y) = f(x) * f(y)$. Therefore f is a state operator on X. □

Example 12.1.7 Let $X = \{1, a, b, c\}$. Define a binary operation $*$ on X as follows:

$*$	1	a	b	c
1	1	a	b	c
a	1	1	a	a
b	1	1	1	b
c	1	1	1	1

Then $(X, *, 1)$ is a BE-algebra. Define $\mu : X \to X$ by

$$f(1) = f(a) = 1 \quad \text{and} \quad f(b) = f(c) = b$$

It is easy to check that f is a state operator on X. But f is not a homomorphism. Indeed, we have $f(a * b) = f(a) = 1 \neq b = 1 * b = f(a) * f(b)$. Thus f is not an endomorphism on X.

In the following, we derive a sufficient condition for a state operator on a commutative BE-algebra X to become an idempotent endomorphism on X.

Theorem 12.1.8 *Every state operator on a linearly ordered commutative BE-algebra is an idempotent endomorphism.*

Proof Let $(X, *, 1)$ be a linearly ordered commutative BE-algebra and f a state operator on X. Let $x, y \in X$. Then either $x \leq y$ or $y \leq x$. Suppose $x \leq y$. Then we get $x * y = 1$. Hence $f(x) * f(y) \geq f(x * y) = f(1) = 1$. Also $f(x * y) = f(1) = 1$. Hence $f(x * y) = 1 = f(x) * f(y)$. Hence f is an idempotent endomorphism on X. Suppose $y \leq x$. Then $y * x = 1$. Since X is commutative, we get $f(x * y) = f((x * y) * y) * f(y) = f((y * x) * x) * f(y) = f(1 * x) * f(y) = f(x) * f(y)$. Hence f is a state-morphism operator on X. Therefore f is a state operator on X. □

For any two idempotent endomorphisms f_1 and f_2 on a BE-algebra $(X, *, 1)$, define their composition $f_1 \circ f_2$ on X as $(f_1 \circ f_2)(x) = f_1(f_2(x))$ for all $x \in X$. Clearly $f_1 \circ f_2$ is an idempotent endomorphism on X. In the following result, we observe the direct products of the idempotent endomorphisms.

Theorem 12.1.9 *Let f_1 and f_2 be two idempotent endomorphisms on a BE-algebra X. For any $x, y \in X$, define their direct product $f_1 \times f_2$ on X^2 as follows:*

$$(f_1 \times f_2)(x, y) = (f_1(x), f_2(y))$$

Then $f_1 \times f_2$ is an idempotent endomorphism on X^2.

Definition 12.1.10 Let ν be a self-map on a BE-algebra X. Then the *strongest self-map* f_ν is a self-map on X^2 defined, for all $x, y \in X$, as follows:

$$f_\nu(x, y) = (\nu(x), \nu(y)).$$

Theorem 12.1.11 *Let ν be a self-map on X and f_ν the strongest self-map on X^2. Then ν is an idempotent endomorphism on X if and only if f_ν is an idempotent endomorphism on X^2.*

Proof Assume that ν is an idempotent endomorphism on X. Let $(a_1, b_1), (a_2, b_2) \in X \times X$. Then

$$\begin{aligned} f_\nu(a_1, b_1) * f_\nu(a_2, b_2) &= (\nu(a_1), \nu(b_1)) * (\nu(a_2), \nu(b_2)) \\ &= (\nu(a_1) * \nu(a_2), \nu(b_1) * \nu(b_2)) \\ &= (\nu(a_1 * a_2), \nu(b_1 * b_2)) \\ &= f_\nu(a_1 * a_2, b_1 * b_2) \\ &= f_\nu((a_1, b_1) * (a_2, b_2)). \end{aligned}$$

Hence f_ν is a homomorphism on X^2. Again, $f_\nu^2(a_1, b_1) = f_\nu(f_\nu(a_1, b_1)) = f_\nu(\nu(a_1), \nu(b_1)) = (\nu^2(a_1), \nu^2(b_1)) = (\nu(a_1), \nu(b_1)) = f_\nu(a_1, b_1)$. Thus f_ν is an idempotent endomorphism on X^2.

Conversely, assume that f_ν is an idempotent endomorphism on X^2. Since f_ν is an idempotent endomorphism on X^2, we get $f_\nu(1, 1) = (1, 1)$. Hence $(\nu(1), \nu(1)) = (1, 1)$, which implies that $\nu(1) = 1$. Since f_ν is an idempotent endomorphism on X^2, for $a, b \in X$, we get the following:

$$\begin{aligned} (1, \nu(a) * \nu(b)) &= (\nu(1) * \nu(1), \nu(a) * \nu(b)) \\ &= ((\nu(1), \nu(a)) * (\nu(1), \nu(b))) \\ &= f_\nu(1, a) * f_\nu(1, b) \\ &= f_\nu((1, a) * (1, b)) \\ &= f_\nu(1 * 1, a * b) \\ &= (\nu(1), \nu(a * b)) \\ &= (1, \nu(a * b)). \end{aligned}$$

Hence it implies that $\nu(a) * \nu(b) = \nu(a * b)$. Therefore ν is an idempotent endomorphism on X. □

Theorem 12.1.12 *Let X be a partially ordered BE-algebra and f an endomorphism on X. Then f is a monomorphism (injective endomorphism) on X if and only if $Ker(f) = \{1\}$.*

Proof Assume that f is a monomorphism. Clearly $\{1\} \subseteq Ker(f)$. Let $x \in Ker(f)$. Then we get $f(x) = 1 = f(1)$. Since f is a monomorphism, we get $x = 1$. Conversely, assume that $Ker(f) = \{1\}$. Let $x, y \in X$ be such that $f(x) = f(y)$. Then $f(x * y) = f(x) * f(y) = f(x) * f(x) = 1$. Hence $x * y \in Ker(f) = \{1\}$, which implies that $x * y = 1$. Hence $x \leq y$. Similarly, we obtain $y \leq x$. Since X is partially ordered, we get $x = y$. Therefore f is a monomorphism. □

It is already observed that if $g \in$ **End(X)** and F a filter of X, then $g^{-1}(F)$ is also a filter of X. In the following result, the same fact is observed for maximal filters of X.

Theorem 12.1.13 *If Q is a maximal filter of a BE-algebra X and $g \in$ **End(X)** is such that $g^{-1}(Q) \neq X$ then $g^{-1}(Q)$ is a maximal filter of X.*

Proof Suppose that $g^{-1}(Q)$ is not a maximal filter of X. Then there exists $a \notin g^{-1}(Q)$ such that $a * b \notin g^{-1}(Q)$ for some $b \in X$. Hence $g(a) \notin Q$ is such that $g(a) * g(b) \notin Q$ and this is against the hypothesis that Q is a maximal filter of X. □

Theorem 12.1.14 *Let X be a transitive BE-algebra. If g is an idempotent endomorphism and h is a surjective endomorphism in X such that $Ker(h) \subseteq Ker(g)$. Then there exists a unique idempotent endomorphism f on X such that $f \circ h = g$.*

Proof Let $x \in X$. Since h is an epimorphism, there exists $x \in X$ such that $h(x) = y$. Suppose there exists $x_1 \in X$ such that $h(x_1) = y$. Then $h(x * x_1) = h(x) * h(x_1) = y * y = 1$. Hence $x * x_1 \in Ker(h) \subseteq Ker(g)$. Hence $g(x) * g(x_1) = g(x * x_1) = 1$. Hence $g(x) \leq g(x_1)$. Similarly, we obtain, $g(x_1) \leq g(x)$. Since X is transitive, we get $g(x) = g(x_1)$. Thus for each $y \in X$, there exists a unique $g(x) \in X$ such that y maps into $g(x)$. Therefore there exists a function $f : X \to X$ such that $f \circ h = g$. We now show that f is an endomorphism. Let $y_1, y_2 \in X$. Since h an epimorphism, there exists $x_1, x_2 \in X$ such that $h(x_1) = y_1$ and $h(x_2) = y_2$. Then

$$\begin{aligned} f(y_1) * f(y_2) &= f(h(x_1)) * f(h(x_2)) \\ &= g(x_1) * g(x_2) \\ &= g(x_1 * x_2) \\ &= f(h(x_1 * x_2)) \\ &= f(h(x_1) * h(x_2)) \\ &= f(y_1 * y_2). \end{aligned}$$

Therefore f is an endomorphism on X. Let $x \in X$. Since h is surjective, there exists $y \in X$ such that $h(y) = x$. Now $f^2(x) = f(f(x)) = f(h(y)) = g(y) = f(h(y)) = f(x)$. Hence f is idempotent. We now prove the uniqueness of f. Suppose f' is another endomorphism such that $f' \circ h = g$. Let $y \in X$. Since h is an epimorphism, there exists $x \in X$ such that $h(x) = y$. Then $f(y) = f(h(x)) = g(x) = f'(h(x)) = f'(y)$. Hence f is the unique endomorphism. □

12.2 Structure Theorems of Endomorphisms

In this section, some structures are constructed on the class of all idempotent endomorphisms of a BE-algebra. It is observed in a self-distributive BE-algebra that the class of all idempotent endomorphisms forms a semilattice.

Definition 12.2.1 For any BE-algebra X, denote by **End(X)** the set of all endomorphisms on X. Then define a binary operation $\odot$ on **End(X)** by $(f \odot g)(x) = f(x) * g(x)$ for all $x \in X$.

In the following theorem, it is observed that the class **End(X)** forms a BE-algebra.

Theorem 12.2.2 (First structure theorem) *For any BE-algebra* X*,* $(\textbf{End(X)}, \odot, \overline{1}_X)$ *is a BE-algebra where* $\overline{1}_X$ *is the identity map defined by* $\overline{1}_X(x) = x$ *for all* $x \in X$*.*

Proof It is a routine verification. □

Theorem 12.2.3 *Let* X *be a self-distributive BE-algebra. For any* $a \in X$*, the map* L_a *is an idempotent endomorphism on* X*.*

Proof Let $a \in X$. Then for any $x, y \in X$, we get $L_a(x * y) = a * (x * y) = (a * x) * (a * y) = L_a(x) * L_a(y)$. Hence L_a is an endomorphism. Since X is self-distributive, we get $L_a^2(x) = a * (a * x) = (a * a) * (a * x) = 1 * (a * x) = a * x = L_a(x)$. Therefore L_a is idempotent. □

Corollary 12.2.4 *If* X *is a self-distributive BE-algebra, then* **L(X)** *is a non-empty closed subset of* **End(X)***, where* **L(X)** *is the set of all left maps on the BE-algebra* X*.*

Proof Since $L_1(x) = 1 * x = x$, it is clear that $L_1 \in$ **End(X)**. Then by the above theorem, we get that $\emptyset \neq$ **L(X)** $\subseteq$ **End(X)**. Let $L_a, L_b \in$ **L(X)** where $a, b \in X$. For any $x, y \in X$, we get $(L_a \circ L_b)(x * y) = L_a(L_b(x * y)) = L_a(L_b(x) * L_b(y)) = L_a(L_b(x)) * L_a(L_b(y)) = (L_a \circ L_b)(x) * (L_a \circ L_b)(y)$. Therefore $L_a \circ L_b$ is an endomorphism on X. Also

$$\begin{aligned}
(L_a \circ L_b)^2(x) &= (L_a \circ L_b)(L_a \circ L_b)(x) \\
&= (L_a \circ L_b)(L_a(L_b(x)) \\
&= L_a(L_b(L_a(L_b(x)))) \\
&= L_a(L_a(L_b(L_b(x)))) \\
&= L_a^2(L_b^2(x)) \\
&= L_a(L_b(x)) \\
&= (L_a \circ L_b)(x)
\end{aligned}$$

Thus $L_a \circ L_b$ is idempotent. Hence $L_a \circ L_b \in$ **End(X)**. Thus the proof is completed. □

Given a BE-algebra X, the algebraic system (**End(X)**; $\circ, \overline{1}_X$) forms a monoid with $\overline{1}_X$ as the unit element, where the underlying set **End(X)** consists of all endomorphisms of X, the operation $\circ$ is the composition of mappings such that $(f \circ g)(x) = f(g(x))$ for all $x \in X$. Let $L_a(X)$ denote the set of all left maps on a BE-algebra X, i.e., **L(X)** $= \{L_a \mid a \in X$ and $L_a(x) = a * x$ for all $x \in X\}$. In the following, it is observed that **L(X)** is a non-empty subset of **End(X)**.

Theorem 12.2.5 (Second structure theorem) *If* X *is a self-distributive BE-algebra, then* **L(X)** *is a commutative submonoid of* **End(X)** *with unit map* $\overline{1}_X$*.*

Proof By above Theorem 12.2.3, we get that $\emptyset \neq$ **L(X)** $\subseteq$ **End(X)**. Let $L_a, L_b \in$ **L(X)**. Then for any $x \in X$, $(L_a \circ L_b)(x) = L_a(L_b(x)) = L_a(b * x) = a * (b * x) = b * (a * x) = b * (L_a(x)) = L_b(L_a(x)) = (L_b \circ L_a)(x)$. Therefore $L_a \circ L_b = L_b \circ L_a$, which implies that **L(X)** is commutative. □

Definition 12.2.6 Let $(X, *, 1)$ be a BE-algebra. For any $f, g \in$ **End(X)**, we define the binary operation "+" on **End(X)** as the following:

$$(f+g)(x) = (g(x) * f(x)) * f(x) \text{ for all } x \in X.$$

If X is a commutative BE-algebra, then it is easily seen that $f+g = g+f$ for all $f, g \in$ **End(X)**. Define an ordering $\leq$ on **End(X)** as $f \leq g$ or equivalently $f * g = \overline{1}_X$ if and only if $f(x) \leq g(x)$ or equivalently $f(x) * g(x) = 1$ for all $f, g \in$ **End(X)** and for all $x \in X$. Then it is obvious that (**End(X)**, $\leq$) is a partially ordered set.

Lemma 12.2.7 *Let X be a commutative BE-algebra. Then for any $f, g, h \in$ **End(X)***

(a) $f * (f+g) = \overline{1}_X$,
(b) $f * g = g * h = \overline{1}_X$ *implies* $f * h = \overline{1}_X$,
(c) $f * g = \overline{1}_X$ *implies* $(f+h) * (g+h) = \overline{1}_X$,
(d) $f * h = g * h = \overline{1}_X$ *implies* $(f+g) * h = \overline{1}_X$.

Proof (a) For any $x \in X$, we have $(f * (f+g))(x) = f(x) * (f+g)(x) = f(x) * ((g(x) * f(x)) * f(x)) = f(x) * ((f(x) * g(x)) * g(x)) = (f(x) * g(x)) * (f(x) * g(x)) = 1 = \overline{1}_X(x)$. Therefore $f * (f+g) = \overline{1}_X$.
(b) Suppose that $f * g = g * f = \overline{1}_X$. Then for any $x \in X$, we get that $\overline{1}_X(x) = (g * h)(x) = g(x) * h(x) \leq (f(x) * g(x)) * (f(x) * h(x)) = (f * g)(x) * (f * h)(x) = \overline{1}_X(x) * (f * h)(x) = 1 * (f * h)(x) = (f * h)(x)$. Hence $(f * h)(x) = \overline{1}_X(x)$. Therefore $f * h = \overline{1}_X$.
(c) Let $f * g = \overline{1}_X$. Then $\overline{1}_X(x) = (f * g)(x) = f(x) * g(x) \leq (g(x) * h(x)) * (f(x) * h(x)) = (g * h)(x) * (f * h)(x) = ((g * h) * (f * h))(x)$. Hence $(g * h) * (f * h) = \overline{1}_X$, which implies that $g * h \leq f * h$. Thus $f + h = (f * h) * h \leq (g * h) * h = g + h$. Therefore $(f+h) * (g+h) = \overline{1}_X$.
(d) Suppose that $f * h = g * h = \overline{1}_X$. From (c), we obtain that $(f+g) * (g+h) = \overline{1}_X$ and $(g+h) * (h+h) = \overline{1}_X$. Then from (b), it yields $(f+g) * (h+h) = \overline{1}_X$. Therefore $(f+g) * h = \overline{1}_X$. □

Theorem 12.2.8 (Third structure theorem) *If $(X, *, 1)$ is a commutative BE-algebra, then (**End(X)**, +) is a semilattice with the unit element $\overline{1}_X$.*

Proof Obviously $f+f = f$ and $f+g = g+f$ for all $f, g \in$ **End(X)**. We will now prove that + is associative on **End(X)**. Let $f, g, h \in$ **End(X)**. Then from (a) the above lemma, we get $f * (f+g) = \overline{1}_X$ and $(f+g) * [(f+g)+h] = \overline{1}_X$. Therefore

$$f * [(f+g)+h] = \overline{1}_X \tag{12.1}$$

Since $g * (f+g) = \overline{1}_X$, we get from (c) of the above lemma that

$$(g+h) * [(f+g)+h] = \overline{1}_X \tag{12.2}$$

Again from (d) of the above lemma and from (12.1) and (12.2), we obtain

$$[f + (g + h)] * [(f + g) + h] = \overline{1}_X \tag{12.3}$$

Similarly,

$$[(f + g) + h] * [(f + g) + h] = \overline{1}_X \tag{12.4}$$

From (12.3) and (12.4), we conclude that $(f + g) + h = f + (g + h)$. □

Corollary 12.2.9 *If X is self-distributive, then* **L(X)** *is a sub-semilattice of* **End(X)**.

Definition 12.2.10 For any BE-algebra X, define the centralizer of **L(X)** w.r.t. **End(X)** as

$$\mathbf{C(L(X))} = \{f \in \mathbf{End(X)} \mid f \circ L_a = L_a \circ f \text{ for all } L_a \in \mathbf{L(X)}\}.$$

It is clear that **C(L(X))** is a submonoid of **End(X)** such that **L(X)** ⊂ **C(L(X))**.

Proposition 12.2.11 *Let X be a partially ordered and transitive BE-algebra and $f \in$ **End(X)**. Then the following are equivalent:*

(1) $f \in$ ***C(L(X))***,
(2) $f(a * x) = a * f(x)$ *for any* $a \in X$,
(3) $a \leq f(a)$ *for any* $a \in X$.

Proof (1) ⇒ (2): Assume that $f \in$ **C(L(X))**. Let $a \in X$ be an arbitrary element. Then, we get that $a * f(x) = L_a(f(x)) = (L_a \circ f)(x) = (f \circ L_a)(x) = f(L_a(x)) = f(a * x)$.
(2) ⇒ (3): It is obtained by putting $x = a$ in (1).
(3) ⇒ (1): Assume that condition (3) holds. Then for any $x \in X$, $(L_a \circ f)(x) = L_a(f(x)) = a * f(x) \leq f(a) * f(x) = f(a * x) = f(L_a(x)) = (f \circ L_a)(x)$.
On the other hand,

$$\begin{aligned}
(f \circ L_a)(x) * (L_a \circ f)(x) &= f(a * x) * (a * f(x)) \\
&= a * (f(a * x) * f(x)) \\
&= a * ((f(a) * f(x)) * f(x)) \\
&= a * ((f(x) * f(a)) * f(a)) \\
&\geq a * f(a) \qquad \text{since } f(a) \leq (f(x) * f(a)) * f(a) \\
&= 1 \qquad \text{since } a \leq f(a)
\end{aligned}$$

Hence $(L_a \circ f)(x) * (f \circ L_a)(x)$. Thus $L_a \circ f = f \circ L_a$ for any $a \in X$. Therefore $f \in$ **C(L(X))**. □

Proposition 12.2.12 *Let X be a partially ordered and transitive BE-algebra and $f \in$ **C(L(X))**. Then $Ker(f) = \{f(x) * x \mid x \in X\}$.*

Proof Put $K = \{f(x) * x \mid x \in X\}$. Let $x \in Ker(f)$. Then we get $f(x) = 1$. Hence $x = 1 * x = f(x) * x$. Therefore $Ker(f) \subseteq K$. On the other hand, let $a \in K$. Then $a = f(x) * x$ for some $x \in X$. Hence $f(f(x) * x) = f(L_{f(x)}(x)) = (f \circ L_{f(x)})(x) = (L_{f(x)} \circ f)(x) = f(x) * f(x) = 1$. Hence $a = f(x) * x \in Ker(f)$. Thus $K \subseteq Ker(f)$. Therefore $Ker(f) = \{f(x) * x \mid x \in X\}$. □

Proposition 12.2.13 (Fourth structure theorem) *Let* $(X, *, 1)$ *be a self-distributive and commutative BE-algebra. Then* **C(L(X))** *is a sub-semilattice of* **End(X)**.

Proof Let $f_1, f_2 \in$ **C(L(X))**. Then $f_i \circ L_a = L_a \circ f_i$ for all $a \in X$ and $i = 1, 2$. For any $a \in X$, $(f_1 \circ f_2) \circ L_a(x) = f_1(f_2(L_a(x)) = f_1(L_a(f_2(x))) = L_a(f_1(f_2(x)))$.
Hence $(f_1 f_2)L_a = L_a(f_1 f_2)$. Thus $f_1 \circ f_2 \in$ **C(L(X))**. Therefore **C(L(X))** is closed under the operation $\circ$ of **End(X)**, which implies that **C(L(X))** is a sub-semilattice of **End(X)**. Let $f_1, f_2 \in$ **C(L(X))**. Let $x \in X$. Then for any $a \in X$, we get

$$
\begin{aligned}
(f_1 + f_2)L_a(x) &= (f_1 + f_2)(a * x) \\
&= (f_2(a * x) * f_1(a * x)) * f_1(a * x) \\
&= ((f_2(L_a(x))) * (f_1(L_a(x)))) * (f_1(L_a(x))) \\
&= ((L_a(f_2(x))) * (L_a(f_1(x)))) * (L_a(f_1(x))) \\
&= L_a((f_2(x) * f_1(x)) * f_1(x)) \\
&= L_a(f_1 + f_2)(x)
\end{aligned}
$$

Hence $(f_1 + f_2)L_a = L_a(f_1 + f_2)$. Therefore $f_1 + f_2 \in$ **C(L(X))**. □

12.3 Endomorphic BE-algebras

In this section, the notion of bounded semi-endomorphic BE-algebras is introduced. Some properties of semi-endomorphic BE-algebras are studied. The concepts of semi-endomorphic ideals and semi-endomorphic filters are introduced in a bounded BE-algebra and derived a set of equivalent conditions for every lower set (final segment) of a self-distributive BE-algebra to become a semi-endomorphic ideal (filter).

Definition 12.3.1 Let X be a BE-algebra. A self-mapping $f : X \to X$ is said to be a *semi-endomorphism* if it satisfies the following conditions for all $x, y \in X$:

(1) $f(1) = 1$,
(2) $f(x * y) \leq f(x) * f(y)$.

It is obvious that every endomorphism of a BE-algebra is a semi-endomorphism but the converse is not true in general. In the following example, it is observed that a semi-endomorphism of a BE-algebra need not be an endomorphism.

Example 12.3.2 Let $X = \{a, b, c, 1\}$ be a finite set. Define a binary operation $*$ on X as

$*$	1	a	b	c
1	1	a	b	c
a	1	1	b	c
b	1	1	1	c
c	1	1	1	1

Clearly $(X, *, 1)$ is a BE-algebra. Define a self-map $f : X \to X$ as $f(1) = f(a) = 1$ and $f(b) = f(c) = b$. It can be easily observed that f is a bounded semi-endomorphism. However, f is not an endomorphism because of $f(b * c) = f(c) = b \neq 1 = b * b = f(b) * f(c)$.

Lemma 12.3.3 *Let f be a semi-endomorphism on a transitive BE-algebra X. Then $x \leq y$ implies $f(x) \leq f(y)$ for all $x, y \in X$.*

Proof Suppose $x \leq y$. Then $x * y = 1$. Hence $1 = f(x * y) \leq f(x) * f(y)$. Thus $f(x) \leq f(y)$. □

For any semi-endomorphism f of a transitive BE-algebra X, define the kernel of f by $Ker(f) = \{x \in X \mid f(x) = 1\}$. Properties of $Ker(f)$ are now observed in the following:

Lemma 12.3.4 *Let f be a semi-endomorphism on a transitive BE-algebra X. Then*

(1) *$Ker(f)$ is a filter of X,*
(2) *Inverse image of a filter under f is a filter.*

Proof (1) Since $f(1) = 1$, we get $1 \in Ker(f)$. Let $x, y \in X$. Suppose $x \in Ker(f)$ and $x * y \in Ker(f)$. Then we get $f(x) = 1$ and $f(x * y) = 1$. Hence $1 = f(x * y) \leq f(x) * f(y) = 1 * f(y) = f(y)$. Thus $f(y) = 1$, which means $y \in Ker(f)$. Therefore $Ker(f)$ is a filter of X.
(2) Let F be a filter of X. Since $f(1) = 1 \in F$, we get $1 \in f^{-1}(F)$. Let $x, y \in X$. Suppose $x \in f^{-1}(F)$ and $x * y \in f^{-1}(F)$. Then we get $f(x) \in F$ and $f(x * y) \in F$. Since $f(x * y) \leq f(x) * f(y)$ and F is a filter, we get $f(x) * f(y) \in F$. Since $f(x) \in F$ and F is a filter, we get $f(y) \in F$. Hence $y \in f^{-1}(F)$, which concludes that $f^{-1}(F)$ is a filter of X. □

Definition 12.3.5 Let $(X, *, 1)$ be a BE-algebra and f an idempotent semi-endomorphism on X. Then the pair (X, f) is said to be a *semi-endomorphic BE-algebra*. If f is an endomorphism, then (X, f) is called an *endomorphic BE-algebra.*

Theorem 12.3.6 *Let $(X, *, 0, 1)$ be a transitive BE-algebra and ν a self-map on X. Define the self-map f^* on $X \times X$ as $f^*(x, y) = (\nu(x), \nu(y))$. Then (X, ν) is a semi-endomorphic BE-algebra if and only if $(X \times X, f^*)$ is a semi-endomorphic BE-algebra.*

Proof It is an easy consequence of Theorem 12.1.11. □

Definition 12.3.7 Let (X, f) be a semi-endomorphic BE-algebra. A filter F of X is said to be a *semi-endomorphic filter* of (X, f) if $f(F) \subseteq F$.

For any semi-endomorphic BE-algebra (X, f), it is clear that the filters $\{1\}$ and $ker(f)$ are semi-endomorphic filters of (X, f). It can be seen that the intersection of any two semi-endomorphic filters of a semi-endomorphic BE-algebra is again a semi-endomorphic filter. Moreover, the intersection of all semi-endomorphic filters of a semi-endomorphic BE-algebra is $\{1\}$.

Theorem 12.3.8 *Let (X, f) be a semi-endomorphic BE-algebra and X is self-distributive. Then the following conditions are equivalent:*

(1) *Every filter of X is a semi-endomorphic filter,*
(2) *for each $x \in X$, the final segment $[x, 1]$ is a semi-endomorphic filter,*
(3) *for each $x \in X$, $x * f(x) = 1$.*

Proof (1) $\Rightarrow$ (2): Assume that every filter of X is semi-endomorphic. Let $x \in X$. Since X is self-distributive, by Theorem 4.6.15, $[x, 1]$ is a filter of X. Therefore $[x, 1]$ is semi-endomorphic.
(2) $\Rightarrow$ (3): Assume that $[x, 1]$ is a semi-endomorphic filter for each $x \in X$. Since $x \in [x, 1]$ and $[x, 1]$ is a semi-endomorphic filter of X, we get $f(x) \in [x, 1]$. Hence $x * f(x) = 1$.
(3) $\Rightarrow$ (1): Let F be a filter of X and $x \in F$. Then $x * f(x) = 1$. Hence $x \leq f(x)$. Since F is a filter, we get $f(x) \in F$. Therefore F is a semi-endomorphic filter of X. □

Definition 12.3.9 Let X be a bounded BE-algebra. A self-mapping $f : X \to X$ is said to be a *lower semi-endomorphism* if it satisfies the following conditions for all $x, y \in X$:

(1) $f(0) = 0$,
(2) $f(x) * f(y) \leq f(x * y)$.

Lemma 12.3.10 *Let f be a lower semi-endomorphism on a bounded and transitive BE-algebra X. Then the following properties hold in X:*

(1) $f(x)^{\diamond} \leq f(x^{\diamond})$ *for all* $x \in X$,
(2) $x \leq y$ *implies* $f(y^{\diamond} * x^{\diamond})^{\diamond} = 0$ *for all* $x, y \in X$.

Proof (1) Let $x \in X$. Then $f(x)^{\diamond} = f(x) * 0 = f(x) * f(0) \leq f(x * 0) = f(x^{\diamond})$.
(2) Let $x, y \in X$ be such that $x \leq y$. Then $y^{\diamond} \leq x^{\diamond}$ and thus $y^{\diamond} * x^{\diamond} = 1$. Hence $f(y^{\diamond} * x^{\diamond})^{\diamond} \leq f((y^{\diamond} * x^{\diamond})^{\diamond}) = f(1^{\diamond}) = f(0) = 0$. Therefore $f(y^{\diamond} * x^{\diamond})^{\diamond} = 0$. □

Definition 12.3.11 Let $(X, *, 0, 1)$ be a bounded BE-algebra and f an idempotent lower semi-endomorphism on X. Then the pair (X, f) is said to be a *lower semi-endomorphic BE-algebra.* If f is an endomorphism, then (X, f) is called an *lower endomorphic BE-algebra.*

Definition 12.3.12 Let (X, f) be a lower semi-endomorphic BE-algebra. An ideal I of a bounded BE-algebra X is called a *semi-endomorphic ideal* of (X, f) if $f(I) \subseteq I$.

Example 12.3.13 Let $(X, *, 0, 1)$ be a bounded BE-algebra and I an ideal of X. For any $a \in X - I$, define a self-map $f : X \to X$ as follows:

$$f_a(x) = \begin{cases} 0 & \text{if } x \in I \\ a & \text{if } x \notin I \end{cases}$$

for all $x \in X$. Then clearly (X, f) is a lower semi-endomorphic BE-algebra. For any $x \in X$ with $x \in I$, we get $f(x) = 0 \in I$. Hence $f(I) \subseteq I$. Therefore I is a semi-endomorphic ideal of (X, f).

In the following theorem, a set of equivalent conditions is derived for every ideal of a self-distributive lower semi-endomorphic BE-algebra to become a semi-endomorphic ideal.

Theorem 12.3.14 *Let X be a self-distributive and involutory BE-algebra. If (X, f) be a lower semi-endomorphic BE-algebra with $f(x^\diamond)^\diamond = f(x)$, then the following conditions are equivalent:*

(1) *Every ideal of X is a semi-endomorphic ideal,*
(2) *for each $x \in X$, the lower set $[0; x]$ is a semi-endomorphic ideal,*
(3) *for each $x \in X$, $x^\diamond * f(x^\diamond) = 1$.*

Proof (1) $\Rightarrow$ (2): Assume that every ideal of X is semi-endomorphic. Let $x \in X$. Since X is self-distributive, by Proposition 7.5.10, $[0; x]$ is an ideal of X. Therefore $[0; x]$ is semi-endomorphic.
(2) $\Rightarrow$ (3): Assume that $[0; x]$ is a semi-endomorphic ideal for each $x \in X$. Since $x \in [0; x]$ and $[0; x]$ is a semi-endomorphic ideal of X, we get $f(x) \in [0; x]$. Hence $x^\diamond * f(x^\diamond) = 0^\diamond * (x^\diamond * f(x^\diamond)) = 1$.
(3) $\Rightarrow$ (1): Let I be an ideal of X. Let $x \in I$ be arbitrary. Then we get $x^\diamond * f(x^\diamond) = 1$. Hence $(x^\diamond * f(x^\diamond)^{\diamond\diamond})^\diamond = (x^\diamond * f(x^\diamond))^\diamond = 1^\diamond = 0 \in I$. Since I is an ideal and $x \in I$, we get $f(x^\diamond)^\diamond \in I$. By the given condition, $f(x) = f(x^\diamond)^\diamond \in I$. Therefore I is a semi-endomorphic ideal of X. □

For any filter F of a commutative BE-algebra X, it can be easily observed that the set $End_X(F)$ of all endomorphisms defined on F forms a commutative BE-algebra with respect to the point-wise operations given by:

$$(f * g)(x) = f(x) * g(x) \quad \text{and} \quad (f \vee g)(x) = f(x) \vee g(x)$$

for all $x \in F$ and $f, g \in Hom_X(F)$ in which the unit element is given by $\bar{1}(x) = 1$ for all $x \in F$.

Lemma 12.3.15 *Let F be a filter of a self-distributive and commutative BE-algebra X. For any $a \in X$, define $\phi_a : F \to F$ by $\phi_a(x) = a * x$ for all $x \in F$. Then we have the following:*

(1) *ϕ_a is an endomorphism,*
(2) *for $a, b \in X$, $\phi_{a\vee b} \leq \phi_a \vee \phi_b$,*
(3) *$\phi_1 = I_d$, where I_d is the identity map of $End_X(F)$.*

Proof (1) Since F is a filter of X, we get that $a * x \in F$ for any $x \in F$ and $a \in X$. Thus ϕ_a is well-defined for all $a \in X$. Let $x, y \in F$. Then $\phi_a(x * y) = a * (x * y) = (a * x) * (a * y) = \phi_a(x) * \phi_a(y)$. Again $\phi_a(1) = a * 1 = 1$. Therefore ϕ_a is an endomorphism on F. Hence $\phi_a \in End_X(F)$.
(2) Let $a, b \in X$. Since $a \leq a \vee b$, we get $(a \vee b) * x \leq a * x \leq (a * x) \vee (b * x)$. For any $x \in F$,

$$\begin{aligned}\phi_{a\vee b}(x) &= (a \vee b) * x\\ &\leq (a * x) \vee (b * x)\\ &= \phi_a(x) \vee \phi_b(x).\end{aligned}$$

Therefore $\phi_{a\vee b} \leq \phi_a \vee \phi_b$.
(3) For $x \in F$, we get $\phi_1(x) = 1 * x = x = I_d(x)$. Therefore $\phi_1 = I_d$. □

From the above result, we have seen that $(End_X(F), \vee)$ is a semilattice. It is now derived that for any filter F of a commutative BE-algebra X, there exists a semihomomorphism (i.e., $f(x \vee y) \leq f(x) \vee f(y)$ for all $x, y \in X$) which characterizes the dual annihilator of F.

Theorem 12.3.16 *Let $(X, *, 1)$ be a self-distributive and commutative BE-algebra. Then for any filter F of X, there exists a semihomomorphism f of semilattices from X into $End_X(F)$ such that $F^+ = \{x \in X \mid f(x) = I_d\}$.*

Proof Let F be a filter of X and $a \in X$. Then by above lemma, we get $\phi_a \in End_X(F)$. Define $f : X \longrightarrow End_X(F)$ by $f(a) = \phi_a$ for all $a \in X$. Clearly f is well-defined. By (2) of above lemma, we get f is a semihomomorphism. Let $a \in \{x \in X \mid f(x) = I_d\}$. Then $f(a) = I_d$ and hence

$$\begin{aligned}\phi_a = f(a) = I_d &\Rightarrow \phi_a(x) = I_d(x) && \text{for all } x \in F\\ &\Rightarrow a * x = x && \text{for all } x \in F\\ &\Rightarrow (a * x) * x = x * x && \text{for all } x \in F\\ &\Rightarrow a \vee x = 1 && \text{for all } x \in F\\ &\Rightarrow a \in F^+.\end{aligned}$$

Hence $\{x \in X \mid f(x) = I_d\} \subseteq F^+$. Conversely, let $a \in F^+$. Then $a \vee x = 1$ for all $x \in F$. Hence

$$\begin{aligned}(a*x)*x = a \vee x = 1 &\Rightarrow a*x = ((a*x)*x)*x = 1*x\\ &\Rightarrow a*x = x \qquad \text{for all } x \in F\\ &\Rightarrow \phi_a(x) = I_d(x) \quad \text{for all } x \in F\\ &\Rightarrow f(a) = \phi_a = I_d\\ &\Rightarrow a \in \{x \in X \mid f(x) = I_d\}\end{aligned}$$

which implies that $F^+ \subseteq \{x \in X \mid f(x) = I_d\}$. Therefore $F^+ = \{x \in X \mid f(x) = I_d\}$. □

12.4 Endomorphic Congruences of BE-algebras

In this section, the notion of semi-endomorphic congruences is introduced in BE-algebras. Semi-endomorphic congruences generated by filter and ideals are observed. Properties of congruences generated by idempotent endomorphisms are studied. A sufficient condition is derived for an endomorphic BE-algebra to become subdirectly irreducible.

Definition 12.4.1 Let (X, f) be a semi-endomorphic BE-algebra. A congruence θ on X is said to be a *semi-endomorphic congruence* on (X, f) if it satisfies

$$(x, y) \in \theta \text{ implies } (f(x), f(y)) \in \theta \text{ for all } x, y \in X.$$

Proposition 12.4.2 *Let (X, f) be a semi-endomorphic BE-algebra where X is transitive. If F is a semi-endomorphic filter of (X, f), then $\theta_F = \{(x, y) \in X \times X \mid x * y \in F$ and $y * x \in F\}$ is a semi-endomorphic congruence on (X, f).*

Proof Since X is transitive, we know that θ_F is a congruence on X. Let $(x, y) \in \theta_F$. Then $x * y \in F$ and $y * x \in F$. Since F is a semi-endomorphic filter, we get $f(x * y) \in F$ and $f(y * x) \in F$. Since $f(x * y) \leq f(x) * f(y)$ and $f(y * x) \leq f(y) * f(x)$, we get $f(x) * f(y) \in F$ and $f(y) * f(x) \in F$. Hence $(f(x), f(y)) \in \theta_F$. Therefore θ_F is a semi-endomorphic congruence on (X, f). □

Definition 12.4.3 If f is an idempotent endomorphism on a BE-algebra X, then the pair (X, f) is called an *endomorphic BE-algebra*.

For any endomorphic BE-algebra (X, f), the congruence given in Definition 12.4.1 is called an *endomorphic congruence* on (X, f). Let us denote by Con X and Con (X, f), the set of all congruences on X and (X, f), respectively. Then clearly Con $(X, f) \subseteq$ Con X.

Proposition 12.4.4 *For any endomorphic BE-algebra (X, f), define a relation Ψ_f on (X, f) by*

$$(x, y) \in \Psi_f \text{ if and only if } f(x) = f(y)$$

for all $x, y \in X$. *Then* Ψ_f *is an endomorphic congruence on* (X, f)*, i.e.,* $\Psi_f \in Con\,(X, f)$.

Proof Clearly Ψ_f is an equivalence relation on X. Since f is an endomorphism, clearly Ψ_f is a congruence on X. Let $(x, y) \in \Psi_f$. Since f is idempotent, we get $f(f(x)) = f(x) = f(y) = f(f(y))$. Hence $(f(x), f(y)) \in \Psi_f$. Therefore Ψ_f is an endomorphic congruence on (X, f). □

For any endomorphism f on a BE-algebra X, the image set $\text{Img}_X(f)$ is a subalgebra of X. Hence $\langle \text{Img}_X(f), *, f(1)\rangle$ is a BE-algebra in which $f(1)$ is the greatest element. Therefore $(\text{Img}_X(f), \text{Id}_f)$ is an endomorphic BE-algebra where Id_f is the identity map on $\text{Img}_X(f)$. For any endomorphic BE-algebra (X, f) and $\theta \subseteq X \times X$, we mean $\Omega(\theta)$ the congruence on X generated by θ and $\Omega_f(\theta)$ the congruence on (X, f) generated by θ. Then clearly $\Omega(\theta) \subseteq \Omega_f(\theta)$.

Theorem 12.4.5 *Let* (X, f) *be an endomorphic BE-algebra. For any congruence* θ *on* $Img_X(f)$*, define a relation* Φ_θ *on* X *by* $(x, y) \in \Phi_\theta$ *if and only if* $(f(x), f(y)) \in \theta$ *for all* $x, y \in X$. *Then* Φ_θ *is an endomorphic congruence on* (X, f). *Moreover*

(1) $\theta = \Delta(f)$ *implies* $\Phi_\theta \subseteq \psi_f$ *where* $\Delta(f) = \{(f(x), f(y)) \mid f(x) = f(y)\}$.
(2) $\Phi_\theta \cap Img_X(f)^2 = \theta$.
(3) $\theta \subseteq \Phi_\theta$ *and* $\Omega_f(\theta) \subseteq \Omega(\theta)$.

Proof Clearly Φ_θ is reflexive and symmetric. Let $(x, y) \in \Phi_\theta$; $(y, z) \in \Phi_\theta$. Then $(f(x), f(y)) \in \theta$ and $(f(y), f(z)) \in \theta$. Since θ is transitive on $\text{Img}_X(f)$, we get $(f(x), f(z)) \in \theta$. Hence $(x, z) \in \Phi_\theta$, which yields that Φ_θ is an equivalence relation on X. Let $(x, y) \in \Phi_\theta$ and $(z, w) \in \Phi_\theta$. Then $(f(x), f(y)) \in \theta$ and $(f(z), f(w)) \in \theta$. Since θ is a congruence on $\text{Img}_X(f)$, we get $(f(x * z), f(y * w)) = (f(x) * f(z), f(y) * f(w)) \in \theta$. Hence $(x * z, y * w) \in \Phi_\theta$. Thus Φ_θ is a congruence on X. Let $(x, y) \in \Phi_\theta$. Then $(f(x), f(y)) \in \theta$. Since f is idempotent, we get $(f(f(x)), f(f(y))) = (f(x), f(y)) \in \theta$. Hence $(f(x), f(y)) \in \Phi_\theta$. Therefore Φ_θ is an endomorphic congruence on X.
(1) Suppose $\theta = \Delta_{Img_X(f)}$. Let $(x, y) \in \Phi_\theta$. Then we get $(f(x), f(y)) \in \theta = \Delta(f)$. Hence $f(x) = f(y)$, which implies that $(x, y) \in \Psi_f$. Therefore $\Phi_\theta \subseteq \Psi_f$.
(2) Since $\theta \subseteq \text{Img}_X(f)^2$ and $\theta \subseteq \Phi_\theta$, we get that $\theta \subseteq \Phi_\theta \cap \text{Img}_X(f)^2$. Conversely, let $(x, y) \in \Phi_\theta \cap \text{Img}_X(f)^2$. Then $(x, y) \in \Phi_\theta$ and $(x, y) \in \text{Img}_X(f)^2$. Since $(x, y) \in \text{Img}_X(f)^2$, we get $x, y \in \text{Img}_X(f)$. Hence $x = f(a)$ and $y = f(b)$ for some $a, b \in X$. Thus $f(x) = f(f(a)) = f(a) = x$. Similarly, we get $f(y) = y$. Hence $x = f(x) \in \text{Img}_X(f)$ and $y = f(y) \in \text{Img}_X(f)$. Thus $(x, y) = (f(x), f(y)) \in \theta$. Therefore $\Phi_\theta \cap \text{Img}_X(f)^2 \subseteq \theta$.
(3) Let $(x, y) \in \theta$. Then $x, y \in \text{Img}_X(f)$. Then $f(x) = x$ and $f(y) = y$. Hence $(f(x), f(y)) = (x, y) \in \theta$, which gives that $(x, y) \in \Phi_\theta$. Thus $\theta \subseteq \Phi_\theta$, which yields that $\Omega_f(\theta) \subseteq \Omega(\theta)$. □

Lemma 12.4.6 *Let* θ *be a congruence on* X *such that* $\theta \subseteq \Omega(f)$. *Then* θ *is an endomorphic congruence on* (X, f). *Moreover* $\Omega(x, y) = \Omega_f(x, y)$ *for all* $x, y \in X$ *with* $(x, y) \in \Phi_\theta$*, where* $\Omega(x, y)$ *and* $\Omega_f(x, y)$ *are congruences generated by* (x, y) *on* X *and* (X, f)*, respectively.*

Proof Let $(x, y) \in \theta$. Since $\theta \subseteq \Psi_f$, we get $(x, y) \in \Psi_f$. Hence $f(x) = f(y)$, which in turn gives $(f(x), f(y)) = (f(x), f(x)) \in \theta$. Therefore θ is an endomorphic congruence on (X, f).

Moreover, let $x, y \in X$ be such that $(x, y) \in \Psi_f$. Clearly $\Omega(x, y) \subseteq \Omega_f(x, y)$. Since $(x, y) \in \Psi_f$, we get $f(x) = f(y)$. Hence $\Omega(x, y) \subseteq \Psi_f$. By the first part, we get $\Omega(x, y)$ is an endomorphic congruence on (X, f). Therefore $\Omega_f(x, y) \subseteq \Omega(x, y)$. □

Theorem 12.4.7 *Let X be a BE-algebra and f an idempotent endomorphism on X. Then the endomorphic BE-algebra (X, f) is subdirectly irreducible whenever X is subdirectly irreducible.*

Proof Assume that X is subdirectly irreducible. Suppose f is an identity map on X. Then clearly Con X = Con (X, f). Hence (X, f) is subdirectly irreducible. Suppose f is non-identity and endomorphism on X. Then Ψ_f is non-trivial congruence on X. Hence $\theta_{\min} \subseteq \Psi_f$, where $\theta_{\min} \in$ Con X is the least non-trivial congruence. Thus by the above lemma, we get $\theta_{\min} \in$ Con (X, f). Hence Con $(X, f) \subseteq$ Con X. Therefore (X, f) is subdirectly irreducible. □

12.5 Divisibility and Endomorphisms in BE-algebras

In this section, the notion of divisibility is introduced with the help of endomorphisms. The notions of associates, units, prime elements, and irreducible elements are introduced in terms of endomorphisms, and their properties are investigated. Images and pre-images of filters are studied in terms of endomorphisms.

Definition 12.5.1 Let (X, f) be an endomorphic BE-algebra. For any $a \in X$, define the *cluster set* (a, f) of the element a with respect to f as follows:

$$(a, f) = \{x \in X \mid f(x) = a * f(s) \text{ for some } s \in X\}$$

If $(X, *, 0, 1)$ is a bounded BE-algebra, then it can be easily seen that $(1, f) = X$ and $(0, f) = Ker(f)$. Let us denote X' the set $\{x \in X \mid f(x) \neq 1\}$.

Proposition 12.5.2 *Let (X, f) be an endomorphic BE-algebra $a, b \in X$. Then*

(1) $1 \in (a, f)$,
(2) $(a, f) \subseteq (f(a), f)$,
(3) $f(a) = f(b)$ *implies* $(a, f) \subseteq (f(b), f)$.

Proof (1) Let $a \in X$. Since $f(1) = 1 = a * 1 = a * f(1)$, it gives that $1 \in (a, f)$.
(2) Let $a \in X$ be arbitrary. Suppose $x \in (a, f)$. Then we get $f(x) = a * f(s)$ for some $s \in X$. Since f is idempotent, we get $f(x) = f(f(x)) = f(a) * f(f(s)) = f(a) * f(s)$. Hence $x \in (f(a), f)$. Therefore $(a, f) \subseteq (f(a), f)$.
(3) Let $a, b \in X$. Suppose $f(a) = f(b)$. Let $x \in (a, f)$. Then $f(x) = a * f(s)$ for some $s \in X$. Since f is idempotent, we get $f(x) = f(f(x)) = f(a) * f(f(s)) = f(a) * f(s) = f(b) * f(s)$. Hence $x \in (f(b), f)$. Therefore $(a, f) \subseteq (f(b), f)$. □

Proposition 12.5.3 *Let (X, f) be a self-distributive endomorphic BE-algebra and $a \in X$. Then the cluster set (a, f) is a weak filter of X.*

Proof Since $f(1) = a * f(1)$, we get $1 \in (a, f)$. Let $x, x * y \in (a, f)$. Then $f(x) = a * f(s)$ and $f(x * y) = a * f(t)$ for some $s, t \in X$. Then

$$\begin{aligned} f(x * (x * y)) &= f(x) * f(x * y) \\ &= (a * f(s)) * (a * f(t)) \\ &= a * (f(s) * f(t)) \\ &= a * f(s * t). \end{aligned}$$

Hence $x * (x * y) \in (a, f)$. Therefore (a, f) is a weak filter of X. □

Definition 12.5.4 An endomorphic BE-algebra (X, f) is said to be *associative* with respect to f (or f-associative), for all $x, y, z \in X$, if it satisfies the following condition:

$$(x * f(y)) * f(z) = x * f(y * z)$$

Proposition 12.5.5 *Let (X, f) is an endomorphic BE-algebra. If X is associative w.r.t. f, then we get the following:*

(1) *for any $a \in X$, $f(a) = a * f(a)$,*
(2) *for any $a \in X$, $a \in (a, f)$,*
(3) *for any $a, b \in X$, $a \in (b, f)$ implies $(a, f) \subseteq (b, f)$,*
(4) *for any $a, b \in X$, $f(a) = f(b)$ implies $(a, f) = (b, f)$,*
(5) *for any $a, b \in X$, $f(a) = f(b)$ implies $(a/b)_f$ and $(b/a)_f$.*

Proof (1) Let $a \in X$ be an arbitrary element. Then we get $f(a) = 1 * f(a) = (a * 1) * f(a) = (a * f(1)) * f(a) = a * f(1 * a) = a * f(a)$. Hence $f(a) = a * f(a)$.
(2) From (1), it is clear.
(3) Let $a, b \in X$ and $a \in (b, f)$. Then $f(a) = b * f(s)$ for some $s \in X$. Let $x \in (a, f)$. Then $f(x) = a * f(t)$ for some $t \in X$. Hence

$$\begin{aligned} f(x) = a * f(t) &\Rightarrow f(f(x)) = f(a) * f(f(t)) \\ &\Rightarrow f(x) = f(a) * f(t) \\ &\Rightarrow f(x) = (b * f(s)) * f(t) \\ &\Rightarrow f(x) = b * f(s * t) \end{aligned}$$

Hence $x \in (b, f)$. Therefore $(a, f) \subseteq (b, f)$.
(4) Let $a, b \in X$ and suppose that $f(a) = f(b)$. Since X is f-associative, we get $a * f(a) = f(a) = f(b)$. Hence $b \in (a, f)$. From (3), we get $(b, f) \subseteq (a, f)$. Similarly, we can obtain $b * f(b) = f(b) = f(a)$, we get $a \in (b, f)$. Hence $(a, f) \subseteq (b, f)$. Therefore $(a, f) = (b, f)$.
(5) From (4), it can be easily obtained. □

Definition 12.5.6 Let (X, f) be an endomorphic BE-algebra. An element $a \in X'$ is said to be a *unit divisor* if there exists $b \in X'$ such that $a * f(b) = b * f(a) = 1$.

Theorem 12.5.7 *Let (X, f) be an endomorphic BE-algebra and $a \in X$. If X is f-associative and a is a unit divisor in X, then $(a, f) = X$.*

Proof Since a is a unit divisor in X, there exists $b \in X'$ such that $a * f(b) = b * f(a) = 1$. Let $x \in X$ be arbitrary. Since X is f-associative, we get

$$\begin{aligned} f(x) &= f(1 * x) \\ &= f(1) * f(x) \\ &= 1 * f(x) \\ &= (a * f(b)) * f(x) \\ &= a * f(b * x) \end{aligned}$$

Hence $x \in (a, f)$. Therefore $(a, f) = X$. Similarly, we get $(b, f) = X$. □

Definition 12.5.8 Let (X, f) be an endomorphic BE-algebra and $a, b \in X'$. We say that a divides b with respect to f, if there exists $c \in X'$ such that

$$f(b) = f((a * c) * c) = f((c * a) * a).$$

In this case, we represent it by $(a/b)_f$. Clearly $(a/a)_f$. We call the element b, the multiple of a with respect to f. From the above definition, it can be observed that b is also a multiple of c.

Definition 12.5.9 Let (X, f) be an endomorphic BE-algebra and $a, b \in X'$. Then we say that a and b are f-associates if and only if $(a/b)_f$ and $(b/a)_f$.

Theorem 12.5.10 *Let (X, f) be an endomorphic BE-algebra and $a, b \in X'$. Then a and b are f-associates if and only if $(a, f) = (b, f)$.*

Proof Let $a, b \in X'$. Assume that a and b are f-associates. Then we get $(a/b)_f$ and $(b/a)_f$. Hence $f(b) = a * f(s)$ and $f(a) = b * f(t)$ for some $s, t \in X$. Thus $b \in (a, f)$ and $a \in (b, f)$. Since $b \in (a, f)$ and X is f-associative, we get $(b, f) \subseteq (a, f)$. Since $a \in (b, f)$ and X is f-associative, we get $(a, f) \subseteq (b, f)$. Therefore $(a, f) = (b, f)$.

Conversely, assume that $(a, f) = (b, f)$. Since $a \in (a, f) = (b, f)$, we get $f(a) = b * f(s)$ for some $s \in X$. Hence $(b/a)_f$. Since $b \in (b, f) = (a, f)$, we get $f(b) = a * f(t)$ for some $t \in X$. Hence $(a/b)_f$. Therefore a and b are f-associates. □

Corollary 12.5.11 *Let (X, f) be an endomorphic BE-algebra. Then the relation of being f-associate is an equivalence relation on X'.*

Definition 12.5.12 Let (X, f) be an endomorphic and commutative BE-algebra, $a \in X'$. Then a is said to be an *f-prime element* if $(a/b \vee c)_f$ implies $(a/b)_f$ or $(a/c)_f$ for all $b, c \in X'$.

Theorem 12.5.13 *Let (X, f) be an endomorphic BE-algebra where X is commutative and f-associative. Suppose $a \in X'$ such that (a, f) is a proper filter of X. Then a is f-prime if and only if (a, f) is a prime filter of X such that $(a, f) \neq \{1\}$.*

Proof Let (a, f) be a proper filter of X. Assume that a is f-prime. Then $f(a) \neq 1$, which implies that $a \neq 1$. Since $1 \neq a \in (a, f)$, we get $(a, f) \neq \{1\}$. Let $x, y \in X$ and $x \vee y \in (a, f)$. Then $f(x \vee y) = a * f(s)$ for some $s \in X$. Thus $(a/x \vee y)_f$. Since a is f-prime, we get $(a/x)_f$ or $(a/y)_f$. If $(a/x)_f$, then $f(x) = a * f(c)$ for some $c \in X$. Thus $x \in (a, f)$. If $(a/y)_f$, then $f(y) = a * f(d)$ for some $d \in X$. Thus $y \in (a, f)$. Hence (a, f) is a prime filter of X such that $(a, f) \neq \{1\}$.

Conversely, assume that (a, f) is a prime filter of X such that $(a, f) \neq \{1\}$. Suppose $f(a) = 1$. Then $f(x) = 1 * f(x) = f(a) * f(x) = (a * f(a)) * f(x) = a * f(a * x)$ for all $x \in X$. Hence $x \in (a, f)$, which is a contradiction to that (a, f) is proper. Therefore $f(a) \neq 1$. Suppose $(a/b \vee c)$ for some $b, c \in X$. Then $f(b \vee c) = a * f(t)$ for some $t \in X$. Hence $b \vee c \in (a, f)$. Since (a, f) is a prime filter, we get $b \in (a, f)$ or $c \in (a, f)$. Since $b \in (a, f)$, we get $f(b) = a * f(s)$ for some $s \in X$. Hence $(a/b)_f$. Since $c \in (a, f)$, we get $f(c) = a * f(t)$ for some $t \in X$. Hence $(a/c)_f$. □

Theorem 12.5.14 *Let (X, f) be an f-associative endomorphic BE-algebra. Then $f((a, f)) = (f(a), f)$ for all $a \in X'$.*

Proof Let $x \in f((a, f))$. Then $x = f(t)$ for some $t \in (a, f)$. Since $t \in (a, f)$, we get $f(t) = a * f(s)$ for some $s \in X$. Hence $x = f(t) = f(f(t)) = f(a) * f(f(s)) = f(a) * f(s)$. Hence $x = f(t) \in (f(a), f)$. Therefore $(a, f) \subseteq (f(a), f)$.

Conversely, let $a \in X'$ and $x \in (f(a), f)$. Then we get $f(x) = f(a) * f(s)$ for some $s \in X$. Since X is f-associative, we get

$$\begin{aligned} f(x) &= (a * f(a)) * f(s) \\ &= a * f(a * s) \end{aligned}$$

Hence $x \in (a, f)$. Thus $(f(a), f) \subseteq (a, f)$. Therefore $(a, f) = (f(a), f)$. □

Theorem 12.5.15 *Let (X, f) be an isomorphic BE-algebra. Then $f^{-1}((a, f)) = (f^{-1}(a), f)$ for all $a \in X'$.*

Proof Let $x \in (f^{-1}(a), f)$. Then $f(x) = f^{-1}(a) * f(s)$ for some $s \in X$. Hence $f(f(x)) = f(f^{-1}(a)) * f(f(s)) = a * f(s)$. Thus we get $f(x) \in (a, f)$, which means $x \in f^{-1}((a, f))$. Therefore $(f^{-1}(a), f) \subseteq f^{-1}((a, f))$.

Conversely, let $x \in f^{-1}((a, f))$. Then $f(x) \in (a, f)$. Hence $f(x) = f(f(x)) = a * f(s)$ for some $s \in X$. Since f is isomorphism, we get

$$\begin{aligned} f(f(x)) = a * f(s) &= f(f(x)) = a * f(f(s)) \\ &= f^{-1}(f(f(x))) = f^{-1}(a * f(f(s))) \\ &= f(x) = f^{-1}(a) * f^{-1}(f(f(s))) \\ &= f(x) = f^{-1}(a) * f(s) \end{aligned}$$

Hence $x \in (f^{-1}(a), f)$. Therefore $f^{-1}((a, f)) \subseteq (f^{-1}(a), f)$. □

Exercise

1. Let (X, μ) be a state BE-algebra. Then the mapping $\overline{\mu} : X \times X \to X \times X$ defined by $\overline{\mu}(a, b) = (\mu(a), \mu(b))$ is an idempotent endomorphism on $X \times X$.
2. Let X be a transitive BE-algebra. If g is an idempotent endomorphism and h is a surjective endomorphism in X such that $Ker(h) \subseteq Ker(g)$. Then prove that $Ker(f) = h(Ker(g))$ and $Im(f) = Im(g)$. Moreover, f is injective if and only if $Ker(h) = Ker(g)$.
3. Let X be a transitive BE-algebra. If g is an idempotent endomorphism and h is a surjective endomorphism in X such that $Ker(h) \subseteq Ker(g)$. Then prove that f is surjective if and only g is surjective.
4. Let X be a BE-algebra. If g be an endomorphism and h an injective endomorphisms on X such that $Im(g) \subseteq Im(h)$. Then prove that there exists a unique endomorphism on X such that $g = h \circ f$.
5. For any commutative BE-algebra X and $f \in$ **End(X)**, prove that $f \in$ **C(L(X))** if and only if every principal filter $\langle a \rangle$ is an endomorphic filter, i.e., $\langle a \rangle \subseteq f(\langle a \rangle)$.
6. If X is a commutative BE-algebra and $f \in$ **End(X)**, then prove that $f \in$ **C(L(X))** if and only if $f(Im(L_a)) = L_a(Im(f))$ for all $a \in X$.
7. For any BE-algebra X, prove that $Ker(f) = \{f(x) * x \mid x \in X\}$ whenever $f \in$ **C(L(X))**.
8. If I_1 and I_2 are two endomorphic ideals of an endomorphic BE-algebras (X_1, f) and (X_2, f), respectively, then $I_1 \times I_2$ is an endomorphic ideal of the product algebra $X_1 \times X_2$. Conversely, every endomorphic ideal I of $X_1 \times X_2$ can be expressed as $I = I_1 \times I_2$ where I_i is an endomorphic ideals of X_i for $i = 1, 2$.
9. Let θ an endomorphic congruence on an endomorphic BE-algebra (X, f) and I a non-empty subset of (X, f). Then prove that I is an endomorphic ideal of X if and only if $\overline{I}$ is an endomorphic ideal of $X_{/\theta}$, where $\overline{I} = \{[x]_\theta \mid x \in I\}$.
10. Let f be a bounded endomorphism and (X, f) an endomorphic BE-algebra where X is self-distributive and commutative. If I is an endomorphic ideal of (X, f), then prove that θ_I is the smallest endomorphic congruence whose kernel is I.

Chapter 13
Fuzzification of Filters

It is well-known that an important task of the artificial intelligence is to make computer simulate human being in dealing with certainty and uncertainty in information. Logic gives a technique for laying the foundations of this task. Information processing dealing with certain information is based on the classical logic. Non-classical logic includes many valued logic and fuzzy logic which takes the advantage of the classical logic to handle information with various facets of uncertainty [253], such as fuzziness and randomness. Therefore, non-classical logic has become a formal and useful tool for computer science to deal with fuzzy information and uncertain information. Fuzziness and incomparability are two kinds of uncertainties often associated with humans' intelligent activities in the real word, and they exist not only in the processed object itself, but also in the course of the object being dealt with.

The concept of fuzzy sets was introduced by Zadeh [250]. At present, these ideals have been applied to other algebraic structures such as groups and rings. Liu et al. [165, 166] introduced the notions of fuzzy filters and fuzzy prime filters in BL-algebras and investigated some of their properties. In [3], S.S. Ahn and J.S. Han studied several degrees in defining a fuzzy positive implicative filter, which is a generalization of a fuzzy filter in BE-algebras.

In this chapter, fuzziness of filters and weak filters is investigated in BE-algebras. Some characterization theorems are derived for fuzzy weak filters of BE-algebras. Homomorphic images and inverse images of fuzzy weak filters of BE-algebras are investigated. The notions of fuzzy translations and fuzzy multiplications of BE-algebras are introduced and then some important properties of fuzzy weak filters in terms of fuzzy translations and fuzzy multiplications are investigated. The concept of anti fuzzy weak filters of BE-algebras is introduced, and then, an equivalent condition is derived for a fuzzy set of a BE-algebra to become an anti fuzzy weak filter. Properties of normal fuzzy weak filters of BE-algebras are investigated.

S. R. Mukkamala, *A Course in BE-algebras*,
https://doi.org/10.1007/978-981-10-6838-6_13

13.1 Fuzzy Filters/Weak Filters of BE-algebras

In this section, the notions of fuzzy filters and fuzzy weak filters are introduced in BE-algebras and their properties are studied. Some characterization theorems are derived for fuzzy filters of fuzzy weak filters. Level filters are generalized in BE-algebras and a relation between level filters and fuzzy weak filters of BE-algebras is obtained.

Definition 13.1.1 A fuzzy set μ in a BE-algebra X is called a *fuzzy filter* of X if it satisfies:

(FF1) $\mu(1) \geq \mu(x)$ for all $x \in X$,
(FF2) $\mu(y) \geq \min\{\mu(x), \mu(x * y)\}$ for all $x, y \in X$.

Example 13.1.2 Let $X = \{1, a, b, c\}$ be a set. Define a binary operation $*$ on X as follows:

$*$	1	a	b	c
1	1	a	b	c
a	1	1	a	a
b	1	1	1	a
c	1	1	a	1

Then $(X, *, 1)$ is a BE-algebra. Define a fuzzy set $\mu : X \longrightarrow [0, 1]$ as follows:

$$\mu(x) = \begin{cases} 1 & \text{if } x = 1 \\ 0 & \text{otherwise} \end{cases}$$

Then it can be easily verified that μ is a fuzzy filter of X.

Lemma 13.1.3 *Let μ be a fuzzy filter of a BE-algebra X. Then the following conditions hold:*
(1) *For any $x, y \in X$, $x \leq y$ implies $\mu(x) \leq \mu(y)$,*
(2) *For any $x, y, z \in X$, $z \leq x * y$ implies $\mu(y) \geq \min\{\mu(x), \mu(z)\}$.*

Proof (1) Let $x, y \in X$ be such that $x \leq y$. Then $x * y = 1$. Since $\mu(1) \geq \mu(x)$, we get $\mu(y) \geq \min\{\mu(x), \mu(x*y)\} = \min\{\mu(x), \mu(1)\} = \mu(x)$. Therefore $\mu(x) \leq \mu(y)$ whenever $x \leq y$.
(2) Let μ be a fuzzy filter of X. Let $x, y, z \in X$. Suppose $z \leq x * y$. By (1), we get that $\mu(z) \leq \mu(x * y)$. Since μ is a fuzzy filter, we get $\mu(y) \geq \min\{\mu(x), \mu(x * y)\} \geq \min\{\mu(x), \mu(z)\}$. □

Proposition 13.1.4 *Every fuzzy filter of a BE-algebra is a fuzzy subalgebra of X.*

Proof Let μ be a fuzzy filter of X. Let $x, y \in X$. Since $y \leq x * y$, we get $\mu(y) \leq \mu(x * y)$. Hence

$$\mu(x * y) \geq \mu(y) \geq \min\{\mu(x), \mu(x * y)\} \geq \min\{\mu(x), \mu(y)\}.$$

Therefore μ is a fuzzy subalgebra of the BE-algebra X. □

A fuzzy set μ of a BE-algebra X is called *order-preserving* if $x \leq y$ implies $\mu(x) \leq \mu(y)$ for all $x, y \in X$. From Lemma 13.1.3, it is clear that every fuzzy filter is order-preserving. Though the converse is not true, in the following couple of results, necessary and sufficient conditions are derived for an order-preserving fuzzy set of a BE-algebra to become a fuzzy filter.

Theorem 13.1.5 *An order-preserving fuzzy set μ of a BE-algebra X is a fuzzy filter if and only if it satisfies the following condition for all $x, y, z \in X$*

$$\mu(x * z) \geq \min\{\mu(x * (y * z)), \mu(y)\}.$$

Proof Let μ be an order-preserving fuzzy subset of X. Assume that μ is a fuzzy filter of X. Let $x, y, z \in X$. Since μ is a fuzzy filter, we get

$$\begin{aligned} \mu(x * z) &\geq \min\{\mu(y * (x * z)), \mu(y)\} \\ &= \min\{\mu(x * (y * z)), \mu(y)\}. \end{aligned}$$

Conversely, assume that μ satisfies the condition. Take $x = 1$ in the condition, we get $\mu(z) = \mu(1 * z) \geq \min\{\mu(1 * (y * z)), \mu(y)\} = \min\{\mu(y * z), \mu(y)\}$. Therefore μ is a fuzzy filter of X. □

Theorem 13.1.6 *An order-preserving fuzzy set μ of a BE-algebra X is a fuzzy filter of X if and only if it satisfies the following conditions for all $a, b, x, y \in X$:*

$$\mu((a * (b * x)) * x) \geq \min\{\mu(a), \mu(b)\}.$$

Proof Assume that μ is a fuzzy filter of X. Let $a, b, x \in X$. By Theorem 13.1.5, we get

$$\begin{aligned} \mu((a * (b * x)) * x) &\geq \min\{\mu((a * (b * x)) * (b * x)), \mu(b)\} \\ &\geq \min\{\mu(a), \mu(b)\} \qquad \text{since } a \leq (a * (b * x)) * (b * x) \end{aligned}$$

Conversely, assume the condition. Let $x, y \in X$. Then by the assumed condition, we get $\mu(y) = \mu(1 * y) = \mu(((x * y) * (x * y)) * y) \geq \min\{\mu(x * y), \mu(x)\}$. Hence μ is a fuzzy filter of X. □

Definition 13.1.7 Let μ be a fuzzy set in a BE-algebra X. For any $\alpha \in [0, 1]$, the set $\mu_\alpha = \{x \in X \mid \mu(x) \geq \alpha\}$ is called a *level subset* of μ.

In the following theorem, a necessary and sufficient condition is derived for a fuzzy set of a BE-algebra to become a fuzzy filter of X.

Theorem 13.1.8 *Let μ be a fuzzy set of a BE-algebra X. Then μ is a fuzzy filter of X if and only if for each $\alpha \in [0, 1]$, the level subset μ_α is a filter in X, when $\mu_\alpha \neq \emptyset$.*

Proof Let μ be a fuzzy set of a BE-algebra X. Assume that μ is a fuzzy filter of X. Then we get $\mu(1) \geq \mu(x)$ for all $x \in X$. In particular, $\mu(1) \geq \mu(x) \geq \alpha$ for all $x \in \mu_\alpha$. Hence $1 \in \mu_\alpha$. Let $x, x * y \in \mu_\alpha$. Then we get $\mu(x) \geq \alpha$ and $\mu(x * y) \geq \alpha$. Since μ is a fuzzy filter, we get that $\mu(y) \geq \min\{\mu(x), \mu(x * y)\} \geq \alpha$. Thus $y \in \mu_\alpha$. Therefore μ_α is a filter of X.

Conversely, assume that μ_α is a filter of X for each $\alpha \in [0, 1]$ with $\mu_\alpha \neq \emptyset$. Suppose there exists $x_0 \in X$ such that $\mu(1) < \mu(x_0)$. Let $\alpha_0 = \frac{1}{2}(\mu(1) + \mu(x_0))$. Then $\mu(1) < \alpha_0$ and $0 \leq \alpha_0 < \mu(x_0) \leq 1$. Hence $x_0 \in \mu_{\alpha_0}$ and $\mu_{\alpha_0} \neq \emptyset$. Since μ_{α_0} is a filter of X, we get $1 \in \mu_{\alpha_0}$ and hence $\mu(1) \geq \alpha_0$, which is a contradiction. Therefore $\mu(1) \geq \mu(x)$ for all $x \in X$. Let $x, y \in X$ be such that $\mu(x) = \alpha_1$ and $\mu(x * y) = \alpha_2$. Then $x \in \mu_{\alpha_1}$ and $x * y \in \mu_{\alpha_2}$. Without loss of generality, assume that $\alpha_1 \leq \alpha_2$. Clearly $\mu_{\alpha_2} \subseteq \mu_{\alpha_1}$. Hence $x * y \in \mu_{\alpha_1}$. Since μ_{α_1} is a filter, we get $y \in \mu_{\alpha_1}$. Hence $\mu(y) \geq \alpha_1 = \min\{\alpha_1, \alpha_2\} = \min\{\mu(x), \mu(x * y)\}$. Therefore μ is a fuzzy filter of X. □

For any fuzzy filter μ of a BE-algebra X, the filters μ_α, $\alpha \in [0, 1]$ are called *level filters* of X. The concept of fuzzy weak filters is introduced in BE-algebras.

Definition 13.1.9 A fuzzy subset μ of a BE-algebra X is said to be a *fuzzy weak filter* of X if it satisfies the following conditions:

(FWF1) $\mu(1) \geq \mu(x)$ for all $x \in X$,
(FWF2) $\mu(x * (x * y)) \geq \min\{\mu(x), \mu(x * y)\}$ for all $x, y \in X$.

Proposition 13.1.10 *A fuzzy weak filter μ of a BE-algebra X satisfies the following properties:*
(1) *$x \leq y$ implies $\mu(x) \leq \mu(x * (x * y))$ for all $x, y \in X$,*
(2) *$\mu(x * (y * (y * z))) \geq \min\{\mu(y), \mu(x * (y * z)))\}$ for all $x, y, z \in X$.*

Proof (1) Let $x, y \in X$. Suppose $x \leq y$. Then $x * y = 1$. Since μ is fuzzy weak filter,

$$\mu(x * (x * y)) \geq \min\{\mu(x), \mu(x * y)\} = \min\{\mu(x), \mu(1)\} = \mu(x).$$

Hence $\mu(x) \leq \mu(x * (x * y))$ for all $x, y \in X$.
(2) Let $x, y, z \in X$. Since $x * z \leq x * (x * z)$ and μ is a fuzzy weak filter, we get

$$\begin{aligned} \mu(x * y * (y * z))) &= \mu(y * (y * (x * z))) \\ &\geq \min\{\mu(y), \mu(y * (x * z))\} \\ &= \min\{\mu(y), \mu(x * (y * z))\}. \end{aligned}$$

Therefore $\mu(x * (y * (y * z))) \geq \min\{\mu(y), \mu(x * (y * z))\}$ for all $x, y, z \in X$. □

Proposition 13.1.11 *Every fuzzy filter of a BE-algebra is a fuzzy weak filter.*

Proof Let μ be a fuzzy filter of a BE-algebra X. Clearly $\mu(1) \geq \mu(x)$ for all $x \in X$. Let $x, y \in X$. Clearly $x * y \leq x * (x * y) \leq x * (x * (x * y))$. Since μ is a fuzzy filter, we get

$$\begin{aligned}\mu(x * (x * y)) &\geq \min\{\mu(x), \mu(x * (x * (x * y)))\}\\ &\geq \min\{\mu(x), \mu(x * (x * y))\}\\ &\geq \min\{\mu(x), \mu(x * y)\}\end{aligned}$$

Therefore μ is a fuzzy weak filter of X. □

Though the converse of the above proposition is not true in general, an equivalent condition is derived for every fuzzy weak filter of a BE-algebra to become a fuzzy filter.

Theorem 13.1.12 *A fuzzy weak filter μ of a BE-algebra X is a fuzzy filter of X if and only if it satisfies the following conditions:*

(1) $\mu(1) \geq \mu(x)$ *for all* $x \in X$,

(2) $\mu(y) \geq \min\{\mu(x), \mu(x * (x * y))\}$ *for all* $x, y \in X$.

Proof Assume that μ is a fuzzy filter of X. Clearly $\mu(1) \geq \mu(x)$ for all $x \in X$. Let $x, y \in X$. Since μ is a fuzzy filter of X, we get $\mu(x * y) \geq \min\{\mu(x), \mu(x * (x * y))\}$. Hence

$$\begin{aligned}\mu(y) &\geq \min\{\mu(x), \mu(x * y)\}\\ &\geq \min\{\mu(x), \min\{\mu(x), \mu(x * (x * y))\}\}\\ &= \min\{\mu(x), \mu(x * (x * y))\}\end{aligned}$$

Therefore μ is satisfying the condition.

Conversely, assume that μ satisfies the condition. Let $x, y \in X$. Since μ is a fuzzy weak filter, we get $\mu(x * (x * y)) \geq \min\{\mu(x), \mu(x * y)\}$. By the assumed condition, we get

$$\begin{aligned}\mu(y) &\geq \min\{\mu(x), \mu(x * (x * y))\}\\ &\geq \min\{\mu(x), \min\{\mu(x), \mu(x * y)\}\}\\ &\geq \min\{\mu(x), \mu(x * y)\}\end{aligned}$$

Therefore μ is a fuzzy filter of X. □

Theorem 13.1.13 *Let $\{\mu_\alpha\}_{\alpha\in\Delta}$ be an indexed family of fuzzy weak filters of a BE-algebra. Then the set intersection $\bigcap\limits_{\alpha\in\Delta} \mu_\alpha$, of the members of the family $\{\mu_\alpha\}_{\alpha\in\Delta}$, which is defined as $\big(\bigcap\limits_{\alpha\in\Delta} \mu_\alpha\big)(x) = \inf\{\mu_\alpha(x) \mid \alpha \in \Delta\}$ for all $x \in X$ is also a fuzzy weak filter of X.*

Proof Let $x \in X$. Since each μ_α is a fuzzy weak filter, we get $\mu_\alpha(1) \geq \mu_\alpha(x)$ for each $\alpha \in \Delta$. Then $\Big(\bigcap\limits_{\alpha\in\Delta} \mu_\alpha\Big)(1) = \inf\{\mu_\alpha(1) \mid \alpha \in \Delta\} \geq \inf\{\mu_\alpha(x) \mid \alpha \in \Delta\} = \Big(\bigcap\limits_{\alpha\in\Delta} \mu_\alpha\Big)(x)$. Let $x, y \in X$. Suppose $x, x * y \in \bigcap\limits_{\alpha\in\Delta} \mu_\alpha$. Since each μ_α is a fuzzy weak filter of X, we get

$$\begin{aligned}
\Big(\bigcap_{\alpha\in\Delta} \mu_\alpha\Big)(x * (x * y)) &= \inf\{\mu_\alpha(x * (x * y)) \mid \alpha \in \Delta\} \\
&\geq \inf\Big\{\min\{\mu_\alpha(x), \mu_\alpha(x * y)\} \mid \alpha \in \Delta\Big\} \\
&= \min\Big\{\inf\{\mu_\alpha(x) \mid \alpha \in \Delta\}, \inf\{\mu_\alpha(x * y) \mid \alpha \in \Delta\}\Big\} \\
&= \min\Big\{\Big(\bigcap_{\alpha\in\Delta} \mu_\alpha\Big)(x), \Big(\bigcap_{\alpha\in\Delta} \mu_\alpha\Big)(x * y)\Big\}
\end{aligned}$$

Therefore $\bigcap\limits_{\alpha\in\Delta} \mu_\alpha$ is a fuzzy weak filter of X. □

Definition 13.1.14 Let μ be a fuzzy subset of a BE-algebra X and $\lambda \in [0, 1]$. Define a fuzzy set $\mu_\lambda : X \to [0, 1]$ by $\mu_\lambda(x) = \mu(x) \cdot \lambda$ for all $x \in X$ where $(\cdot)$ indicates the multiplication.

Theorem 13.1.15 *Let μ be a fuzzy set of a BE-algebra X. Then μ is a fuzzy weak filter of X if and only if μ_λ is a fuzzy weak filter of X for each $\lambda \in [0, 1]$.*

Proof Assume that μ is a fuzzy weak filter of X. Let $\lambda \in [0, 1]$. Then for any $x \in X$, we get $\mu_\lambda(1) = \mu(1) \cdot \lambda \geq \mu(x) \cdot \lambda = \mu_\lambda(x)$. Let $x, y \in X$. Since μ is a fuzzy weak filter of X, we get

$$\begin{aligned}
\mu_\lambda(x * (x * y)) &= \mu(x * (x * y)) \cdot \lambda \\
&\geq \min\{\mu(x), \mu(x * y)\} \cdot \lambda \\
&= \min\{\mu(x) \cdot \lambda, \mu(x * y) \cdot \lambda\} \\
&= \min\{\mu_\lambda(x), \mu_\lambda(x * y)\}
\end{aligned}$$

Therefore μ_λ is a fuzzy weak filter of X. Conversely, assume that μ_λ is a fuzzy weak filter of X for each $\lambda \in [0, 1]$. Let $x \in X$. Then $\mu(1)\cdot\lambda = \mu_\lambda(1) \geq \mu_\lambda(x) = \mu(x)\cdot\lambda$. Hence $\mu(1) \geq \mu(x)$ for all $x \in X$. Let $x, y \in X$. Since μ_λ is a fuzzy weak filter of X, we get

$$\begin{aligned}
\mu(x * (x * y)) \cdot \lambda &= \mu_\lambda(x * (x * y)) \\
&\geq \min\{\mu_\lambda(x), \mu_\lambda(x * y)\} \cdot \lambda \\
&= \min\{\mu(x) \cdot \lambda, \mu(x * y) \cdot \lambda\} \\
&= \min\{\mu(x), \mu(x * y)\} \cdot \lambda
\end{aligned}$$

Hence $\mu(x * (x * y)) \geq \min\{\mu(x), \mu(x * y)\}$. Therefore μ is a fuzzy weak filter of X. □

The proof of the following lemma is routine.

Lemma 13.1.16 *Let μ and ν be two fuzzy sets in a BE-algebra X. Then the following hold.*

(1) *$\mu \times \nu$ is a fuzzy relation on X,*

(2) *$(\mu \times \nu)_\alpha = \mu_\alpha \times \nu_\alpha$ for all $\alpha \in [0, 1]$.*

Proposition 13.1.17 *Let μ and ν be two fuzzy weak filters of a BE-algebra X. Then $\mu \times \nu$ is a fuzzy weak filter of $X \times X$.*

Proof Let $(x, y) \in X \times X$. Then $(\mu \times \nu)(1, 1) = \min\{\mu(1), \nu(1)\} \geq \min\{\mu(x), \nu(y)\} = (\mu \times \nu)(x, y)$. Now let $(x, y), (z, w) \in X \times X$. Then

$$\begin{aligned}
(\mu \times \nu)((x, y) * ((x, y) * (z, w))) &= (\mu \times \nu)(x * (x * z), y * (y * w)) \\
&= \min\{\mu(x * (x * z)), \nu(y * (y * w))\} \\
&\geq \min\{\min\{\mu(x), \mu(x * z)\}, \min\{\nu(y), \nu(y * w)\}\} \\
&= \min\{\min\{\mu(x), \nu(y)\}, \min\{\mu(x * z), \nu(y * w)\}\} \\
&= \min\{(\mu \times \nu)(x, y), (\mu \times \nu)(x * z, y * w)\} \\
&= \min\{(\mu \times \nu)(x, y), (\mu \times \nu)((x, y) * (z, w))\}.
\end{aligned}$$

Therefore $\mu \times \nu$ is a fuzzy weak filter of $X \times X$. □

Definition 13.1.18 Let ν be a fuzzy set in a BE-algebra X. Then the *strongest fuzzy relation* μ_ν is a fuzzy relation on X defined by $\mu_\nu(x, y) = \min\{\nu(x), \nu(y)\}$ for all $x, y \in X$.

Theorem 13.1.19 *Let ν be a fuzzy set in a BE-algebra X and μ_ν the strongest fuzzy relation on X. If ν is a fuzzy weak filter of X, then μ_ν is a fuzzy weak filter of $X \times X$.*

Proof Assume that ν is a fuzzy weak filter of X. For any $(x, y) \in X \times X$, we get $\mu_\nu(1, 1) = \min\{\nu(1), \nu(1)\} \geq \min\{\nu(x), \nu(y)\} = \mu_\nu(x, y)$. Let $(x, y), (z, w) \in X \times X$. Then

$$\begin{aligned}
\mu_\nu((x, y) * ((x, y) * (z, w))) &= \mu_\nu(x * (x * z), y * (y * w)) \\
&= \min\{\nu(x * (x * z)), \nu(y * (y * w))\} \\
&\geq \min\{\min\{\nu(x), \nu(x * z)\}, \min\{\nu(y), \nu(y * w)\}\} \\
&= \min\{\min\{\nu(x), \nu(y)\}, \min\{\nu(x * z), \nu(y * w)\}\} \\
&= \min\{\mu_\nu(x, y), \mu_\nu(x * z, y * w)\} \\
&= \min\{\mu_\nu(x, y), \mu_\nu((x, y) * (z, w))\}.
\end{aligned}$$

Therefore μ_ν is a fuzzy weak filter of $X \times X$. □

13.2 Homomorphic Images of Fuzzy Weak Filters

In this section, the notion of invariant sets is introduced under homomorphisms of BE-algebras. Inverse homomorphic images of fuzzy sets are studied. The *sup property* of fuzzy sets is introduced, and then, homomorphic images of fuzzy weak filters are studied.

Definition 13.2.1 Let $f : X \longrightarrow Y$ be a homomorphism of BE-algebras, and μ is a fuzzy set in Y. Then define a mapping $\mu^f : X \longrightarrow [0, 1]$ such that $\mu^f(x) = \mu(f(x))$ for all $x \in X$.

Clearly the above mapping μ^f is well-defined and a fuzzy set in X.

Theorem 13.2.2 *Let $f : X \longrightarrow Y$ be an onto homomorphism of BE-algebras, and μ is a fuzzy set of Y. Then μ is a fuzzy weak filter in Y if and only if μ^f is a fuzzy weak filter of X.*

Proof Assume that μ is a fuzzy weak filter of Y. For any $x \in X$, we have $\mu^f(1) = \mu(f(1)) = \mu(1') \geq \mu(f(x)) = \mu^f(x)$. Let $x, y \in X$. Then

$$\begin{aligned}\mu^f(x * (x * y)) &= \mu(f(x * (x * y)) \\ &= \mu(f(x) * (f(x) * f(y))) \\ &\geq \min\{\mu(f(x)), \mu(f(x) * f(y))\} \\ &= \min\{\mu(f(x)), \mu(f(x * y))\} \\ &= \min\{\mu^f(x), \mu^f(x * y)\}\end{aligned}$$

Therefore μ^f is a fuzzy weak filter of X. Conversely, assume that μ^f is a fuzzy weak filter of X. Let $x \in Y$. Since f is onto, there exists $y \in X$ such that $f(y) = x$. Then $\mu(1') = \mu(f(1)) = \mu^f(1) \geq \mu^f(y) = \mu(f(y)) = \mu(x)$. Let $x, y \in Y$. Then there exists $a, b \in X$ such that $f(a) = x$ and $f(b) = y$. Hence we get

$$\begin{aligned}\mu(x * (x * y)) &= \mu(f(a) * (f(a) * f(b)) \\ &= \mu(f(a * (a * b)) \\ &= \mu^f(a * (a * b)) \\ &\geq \min\{\mu^f(a), \mu^f(a * b)\} \\ &= \min\{\mu(f(a)), \mu(f(a) * f(b))\} \\ &= \min\{\mu(x), \mu(x * y)\}\end{aligned}$$

Therefore μ is a fuzzy weak filter in Y. □

Definition 13.2.3 Let X and X' be any two sets and $f : X \longrightarrow X'$ be any function. If μ is a fuzzy set in X, then the fuzzy set ν in X' defined for all $x \in X'$ by

$$\nu(x) = \begin{cases} \sup\limits_{t \in f^{-1}(x)} \mu(t) & \text{if } f^{-1}(x) \neq \emptyset \\ 1 & \text{otherwise} \end{cases}$$

is called the image of μ under f and is denoted by $f(\mu)$.

Definition 13.2.4 Let X and X' be any two sets and $f : X \longrightarrow X'$ be a function. If ν is a fuzzy set in $f(X)$, then the fuzzy set μ in X defined for all $x \in X$ by $\mu(x) = \nu(f(x))$ is called the *pre-image* of ν under f and is denoted by $f^{-1}(\nu)$. Clearly $f^{-1}(\nu) = \nu \circ f$

Definition 13.2.5 Let X and X' be any two sets and $f : X \longrightarrow X'$ be a function. A fuzzy set μ in X is called f-*invariant* if $f(x) = f(y)$ implies $\mu(x) = \mu(y)$ for all $x, y \in X$

For any fuzzy sets μ and ν in a set X, we define $\mu \subseteq \nu$ if and only if $\mu(x) \leq \nu(x)$ for all $x \in X$ and also $(\mu \cap \nu)(x) = \min\{\mu(x), \nu(x)\}$ for all $x \in X$.

Lemma 13.2.6 *Let f be a function defined on a BE-algebra X. Then we have*

(1) *for any fuzzy set μ in X, $\mu \subseteq f^{-1}(f(\mu))$,*
(2) *if μ is an f-invariant fuzzy set in X, then $\mu = f^{-1}(f(\mu))$,*
(3) *for any fuzzy set ν in $f(X)$, $\nu \subseteq f(f^{-1}(\nu))$,*
(4) *for any fuzzy sets μ_1, μ_2 in X, $\mu_1 \subseteq \mu_2 \Rightarrow f(\mu_1) \subseteq f(\mu_2)$,*
(5) *for any fuzzy sets ν_1, ν_2 in $f(X)$, $\nu_1 \subseteq \nu_2 \Rightarrow f^{-1}(\nu_1) \subseteq f^{-1}(\nu_2)$.*

Proof (1) Let μ be a fuzzy set in X. For any $x \in X$, we get

$$\begin{aligned} f^{-1}(f(\mu))(x) &= f(\mu)f(x) \\ &= \sup_{t \in f^{-1}(f(x))} \mu(t) \\ &\geq \mu(t) \qquad \text{for all } t \in f^{-1}f(x) \end{aligned}$$

Since $x \in f^{-1}(f(x))$, it yields $f^{-1}(f(\mu))(x) \geq \mu(x)$. Therefore $\mu \subseteq f^{-1}(f(\mu))$.
(2) Suppose μ is f-invariant. Let $x \in X$. Then, for $t \in f^{-1}(f(x))$, we get $f(t) = f(x)$. Since μ is f-invariant, it yields $\mu(t) = \mu(x)$. Hence $\mu(x) = \sup_{t \in f^{-1}(f(x))} \mu(t) = f(\mu)f(x) = f^{-1}(f(\mu))(x)$.
(3) Let $y \in f(X)$. Then $f(x) = y$ for some $x \in X$. Hence we get

$$\begin{aligned} f(f^{-1}(\nu))(y) &= f(f^{-1}(\nu))f(x) \\ &= \sup_{t \in f^{-1}(f(x))} f^{-1}(\nu)(t) \\ &= \sup_{t \in f^{-1}(f(x))} \nu(f(t)) \\ &= \nu(f(x)) \qquad \text{since } t \in f^{-1}(f(x)), \text{ we get } f(t) = f(x) \\ &= \nu(y) \end{aligned}$$

(4) and (5) can be easily seen. □

Definition 13.2.7 A fuzzy set μ in a BE-algebra X has the *sup property* if, for any subset S of X, there exists $x_0 \in S$ such that $\mu(x_0) = \sup_{t\in S} \mu(t)$.

Theorem 13.2.8 *Let $f : X \longrightarrow Y$ be an onto homomorphism of BE-algebras. If ν is a fuzzy weak filter in Y, then $f^{-1}(\nu)$ is a fuzzy weak filter in X.*

Proof For any $x \in X$, $f^{-1}(\nu)(1) = \nu(f(1)) = \nu(1) \geq \nu(f(x)) = f^{-1}(\nu)(x)$. Let $x, y \in X$. Then it yields $f^{-1}(\nu)(x * (x * y)) = \nu(f(x * (x * y))) \geq \min\{\nu(a), \nu(a * f(y))\}$ for some $a \in Y$. Since f is onto, there exists $x_a \in X$ such that $f(x_a) = a$. The following is a consequence of the above:

$$\begin{aligned} f^{-1}(\nu)(x * (x * y)) &\geq \min\{\nu(a), \nu(a * f(y))\} \\ &= \min\{\nu(f(x_a)), \nu(f(x_a) * f(y))\} \\ &= \min\{\nu(f(x_a)), \nu(f(x_a * y))\} \\ &= \min\{f^{-1}(\upsilon)(x_a), f^{-1}(\nu)(x_a * y)\} \end{aligned}$$

Since a is arbitrary, it holds for all $x, y \in X$. Hence $f^{-1}(\nu)(x * (x * y)) \geq \min\{f^{-1}(\upsilon)(x), f^{-1}(\nu)(x * y)\}$. Therefore $f^{-1}(\nu)$ is a fuzzy weak filter in X. □

Theorem 13.2.9 *Let $f : X \longrightarrow Y$ be a homomorphism of X onto Y. If μ is a fuzzy weak filter of X which has the sup property, then the image of μ under f is a fuzzy weak filter of Y.*

Proof Since $1 \in f^{-1}(1)$, we get $f(\mu)(1) = \sup_{t\in f^{-1}(1)} \mu(t) = \mu(1) \geq \mu(x)$ for all $x \in X$. Hence $f(\mu)(1) \geq \sup_{t\in f^{-1}(a)} \mu(t) = f(\mu)(a)$ for all $a \in Y$. For any $a, b \in Y$, let $x_a \in f^{-1}(a)$ and $x_b \in f^{-1}(b)$ be such that $\mu(x_a) = \sup_{t\in f^{-1}(a)} \mu(t)$, $\mu(x_a * x_b) = \sup_{t\in f^{-1}(a*b)} \mu(t)$ and $\mu(x_a * (x_a * x_b)) = \sup_{t\in f^{-1}(a*(a*b))} \mu(t)$. Then we get the following consequence:

$$\begin{aligned} f(\mu)(a * (a * b)) &= \sup_{t\in f^{-1}(a*(a*b))} \mu(t) \\ &= \mu(x_a * (x_a * x_b)) \\ &\geq \min\{\mu(x_a), \mu(x_a * x_b)\} \\ &= \min\{\sup_{t\in f^{-1}(a)} \mu(t), \sup_{t\in f^{-1}(a*b)} \mu(t)\} \\ &= \min\{f(\mu)(a), f(\mu)(a * b)) \end{aligned}$$

Therefore $f(\mu)$ is a fuzzy weak filter in Y. □

Denote the set of all fuzzy weak filters of a BE-algebra X by $\mathcal{FW}_F(X)$ and the set of all fuzzy weak filters of X with sup property by $\mathcal{FW}^s_F(X)$. Then the following theorem is an easy consequence of the above theorems.

Theorem 13.2.10 *Let $f : X \to Y$ be a homomorphism from a BE-algebra X into a BE-algebra Y. Suppose every fuzzy weak filter is f-invariant. Then there exists a one-to-one correspondence between the set of all fuzzy weak filters of X with sup property and the set of all fuzzy weak filters of Y.*

Proof Define a mapping $\Psi : \mathcal{FW}^s_F(X) \to \mathcal{FW}_F(Y)$ as $\mu \mapsto f(\mu)$ for all $\mu \in \mathcal{FW}^s_F(X)$. Clearly Ψ is well-defined. Then by the above two theorems, the proof follows. □

13.3 Fuzzy Translations of Weak Filters

In this section, properties of extensions, translations, and multiplications of weak filters are investigated based on the set theory. Interconnections among fuzzy extensions, fuzzy translations, and fuzzy multiplications of BE-algebras are derived.

Let $(X, *, 1)$ be a BE-algebra, and μ is a fuzzy set on X. Define $\top = 1 - \sup\{\mu(x) \mid x \in X\}$. Since μ is a fuzzy set, we know that $\mu(x) \in [0, 1]$ and hence $\top \in [0, 1]$.

Definition 13.3.1 [164] Let μ be a fuzzy subset of a BE-algebra X and $\alpha \in [0, \top]$. A mapping $\mu^T_\alpha : X \to [0, 1]$ is called *fuzzy α-translation* of μ if it satisfies the following:

$$\mu^T_\alpha(x) = \mu(x) + \alpha \text{ for all } x \in X.$$

Theorem 13.3.2 *Let X be a BE-algebra, and μ is a fuzzy weak filter of X; then, the fuzzy α-translation μ^T_α of μ is a fuzzy weak filter of X for all $\alpha \in [0, \top]$.*

Proof Assume that μ is a fuzzy weak filter of X. Let $\alpha \in [0, \top]$. Since $\mu(1) \geq \mu(x)$ for all $x \in X$, we get $\mu^T_\alpha(1) = \mu(1) + \alpha \geq \mu(x) + \alpha = \mu^T_\alpha(x)$. Let $x, y \in X$. Since μ is a fuzzy weak filter, we get

$$\begin{aligned}\mu^T_\alpha(x * (x * y)) &= \mu(x * (x * y)) + \alpha \\ &\geq \min\{\mu(x), \mu(x * y)\} + \alpha \\ &= \min\{\mu(x) + \alpha, \mu(x * y) + \alpha\} \\ &= \min\{\mu^T_\alpha(x), \mu^T_\alpha(x * y)\}.\end{aligned}$$

Therefore μ^T_α is a fuzzy weak filter of X for all $\alpha \in [0, \top]$. □

Theorem 13.3.3 *Let μ be a fuzzy subset of a BE-algebra X such that the fuzzy α-translation μ^T_α of μ is a fuzzy weak filter of X for some $\alpha \in [0, \top]$. Then μ is a fuzzy weak filter of X.*

Proof Assume that μ_α^T is a fuzzy weak filter of X for some $\alpha \in [0, \top]$. Let $x \in X$. Then $\mu(1) + \alpha = \mu_\alpha^T(1) \geq \mu_\alpha^T(x) = \mu(x) + \alpha$. Hence $\mu(1) \geq \mu(x)$ for all $x \in X$. Let $x, y \in X$. Since μ_α^T is a fuzzy weak filter of X, we get

$$\begin{aligned}\mu(x * (x * y)) + \alpha &= \mu_\alpha^T(x * (x * y))\\ &\geq \min\{\mu_\alpha^T(x), \mu_\alpha^T(x * y)\}\\ &= \min\{\mu(x) + \alpha, \mu(x * y) + \alpha\}\\ &= \min\{\mu(x), \mu(x * y)\} + \alpha.\end{aligned}$$

Hence $\mu(x * (x * y)) \geq \min\{\mu(x), \mu(x * y)\}$. Therefore μ is a fuzzy weak filter of X. □

Theorem 13.3.4 *Let μ be a fuzzy subset of a BE-algebra X. Then the intersection and union of any two fuzzy translations of the fuzzy weak filter μ of X is also a fuzzy weak filter of X.*

Proof Assume that μ is a fuzzy weak filter of X. Let $\alpha, \beta \in [0, \top]$. Then μ_α^T and μ_β^T be the fuzzy α-translation and fuzzy β-translation of μ, respectively. Then μ_α^T and μ_β^T are fuzzy weak filters of X. Without loss of generality, assume that $\alpha \leq \beta$. For any $x \in X$, we get

$$\begin{aligned}(\mu_\alpha^T \cap \mu_\beta^T)(x) &= \min\{\mu_\alpha^T(x), \mu_\beta^T(x)\}\\ &= \min\{\mu(x) + \alpha, \mu(x) + \beta\}\\ &= \mu(x) + \alpha\\ &= \mu_\alpha^T(x)\end{aligned}$$

Hence $\mu_\alpha^T \cap \mu_\beta^T = \mu_\alpha^T$. Since μ_α^T is a fuzzy weak filter of X, we get $\mu_\alpha^T \cap \mu_\beta^T$ is also a fuzzy weak filter of X. Again for any $x \in X$, we get

$$\begin{aligned}(\mu_\alpha^T \cup \mu_\beta^T)(x) &= \max\{\mu_\alpha^T(x), \mu_\beta^T(x)\}\\ &= \max\{\mu(x) + \alpha, \mu(x) + \beta\}\\ &= \mu(x) + \beta\\ &= \mu_\beta^T(x)\end{aligned}$$

Hence $\mu_\alpha^T \cup \mu_\beta^T = \mu_\beta^T$. Since μ_β^T is a fuzzy weak filter of X, we get $\mu_\alpha^T \cup \mu_\beta^T$ is also a fuzzy weak filter of X. □

Theorem 13.3.5 *Let $\alpha \in [0, \top]$, and μ is fuzzy subalgebra of a BE-algebra X. Then the fuzzy α-translation μ_α^T of μ is a fuzzy weak filter of X.*

Proof Let $x \in X$. Since μ is a fuzzy subalgebra, we get $\mu_\alpha^T(1) = \mu(1) + \alpha = \mu(x * x) + \alpha \geq \min\{\mu(x), \mu(x)\} + \alpha = \mu(x) + \alpha = \mu_\alpha^T(x)$. Let $x, y \in X$. Then we get

$$\begin{aligned}\mu_\alpha^T(x*(x*y)) &= \mu(x*(x*y))+\alpha\\ &\geq \min\{\mu(x),\mu(x*y)\}+\alpha\\ &= \min\{\mu(x)+\alpha,\mu(x*y)+\alpha\}\\ &\geq \min\{\mu_\alpha^T(x),\mu_\alpha^T(x*y)\}.\end{aligned}$$

Therefore μ_α^T is a fuzzy weak filter of X. □

Since every fuzzy filter of a BE-algebra is a fuzzy subalgebra, the following corollary is a direct consequence of the above theorem.

Corollary 13.3.6 *Let $\alpha \in [0, \top]$, and μ is fuzzy filter of a BE-algebra X. Then the fuzzy α-translation μ_α^T of μ is a fuzzy weak filter of X.*

Theorem 13.3.7 *Let $\alpha \in [0, \top]$, and μ is fuzzy weak filter of a BE-algebra X. Then the fuzzy α-translation μ_α^T of μ is a fuzzy filter of X if and only if it satisfies the following condition:*

$$\mu(y) \geq \min\{\mu(x),\mu(x*(x*y))\} \textit{ for all } x, y \in X.$$

Proof Assume that μ_α^T is a fuzzy filter of X. Let $x, y \in X$. Since μ_α^T is a fuzzy filter of X, we get $\mu_\alpha^T(x*y) \geq \min\{\mu_\alpha^T(x), \mu_\alpha^T(x*(x*y))\}$. Hence

$$\begin{aligned}\mu(y)+\alpha &= \mu_\alpha^T(y)\\ &\geq \min\{\mu_\alpha^T(x),\mu_\alpha^T(x*y)\}\\ &\geq \min\{\mu_\alpha^T(x),\min\{\mu_\alpha^T(x),\mu_\alpha^T(x*(x*y))\}\}\\ &= \min\{\mu_\alpha^T(x),\mu_\alpha^T(x*(x*y))\}\\ &= \min\{\mu(x)+\alpha,\mu(x*(x*y))+\alpha\}\\ &= \min\{\mu(x),\mu(x*(x*y))\}+\alpha.\end{aligned}$$

Hence $\mu(y) \geq \min\{\mu(x), \mu(x*(x*y))\}$. Therefore μ is satisfying the condition.

Conversely, assume that μ satisfies the condition. Let $x, y \in X$. Since μ is a fuzzy weak filter, we get $\mu(x*(x*y)) \geq \min\{\mu(x), \mu(x*y)\}$. By the assumed condition, we get

$$\begin{aligned}\mu_\alpha^T(y) &= \mu(y)+\alpha\\ &\geq \min\{\mu(x),\mu(x*(x*y))\}+\alpha\\ &\geq \min\{\mu(x),\min\{\mu(x),\mu(x*y)\}\}+\alpha\\ &= \min\{\mu(x),\mu(x*y)\}+\alpha\\ &= \min\{\mu(x)+\alpha,\mu(x*y)+\alpha\}\\ &= \min\{\mu_\alpha^T(x),\mu_\alpha^T(x*y)\}.\end{aligned}$$

Therefore μ_α^T is a fuzzy filter of X. □

Definition 13.3.8 Let μ be a fuzzy subset of a BE-algebra X, For any $\alpha \in [0, \top]$ and $t \in [0, 1]$ with $t \geq \alpha$ define $U_\alpha(\mu, t) = \{x \in X \mid \mu(x) \geq t - \alpha\}$.

Clearly $U_\alpha(\mu, t)$ is a subset of X for all $t \in Im(\mu)$ with $t \geq \alpha$.

Lemma 13.3.9 *If μ is a fuzzy weak filter of a BE-algebra X, then $U_\alpha(\mu, t)$ is also a weak filter of X for all $t \in Im(\mu)$ with $t \geq \alpha$.*

Proof Let $t \in Im(\mu)$ with $t \geq \alpha$. Assume that μ is a fuzzy weak filter of X. Clearly $\mu(1) \geq \mu(x)$ for all $x \in X$, In particular, $\mu(1) \geq \mu(x) \geq t - \alpha$ for all $x \in U_\alpha(\mu, t)$. Hence $1 \in U_\alpha(\mu, t)$. Let $x, y \in X$. Suppose $x, x * y \in U_\alpha(\mu, t)$. Then $\mu(x) \geq t - \alpha$ and $\mu(x * y) \geq t - \alpha$. Since μ is a fuzzy weak filter, we get $\mu(x * (x * y)) \geq \min\{\mu(x), \mu(x * y)\} = \min\{t - \alpha, t - \alpha\} = t - \alpha$. Hence $x * (x * y) \in U_\alpha(\mu, t)$. Therefore $U_\alpha(\mu, t)$ is a weak filter of X. □

Proposition 13.3.10 *Let μ be a fuzzy subset of a BE-algebra X and $\alpha \in [0, \top]$. If μ^T_α is a fuzzy α-translation of μ, then μ^T_α is a fuzzy weak filter of X if and only if $U_\alpha(\mu, t)$ is a weak filter of X for all $t \in Im(\mu)$ with $t \geq \alpha$.*

Proof Assume that μ^T_α is a fuzzy weak filter of X. Let $t \in Im(\mu)$ be such that $t \geq \alpha$. Since μ^T_α is a fuzzy weak filter of X, we get $\mu^T_\alpha(1) \geq \mu^T_\alpha(x)$ for all $x \in X$. Hence $\mu(1) + \alpha = \mu^T_\alpha(1) \geq \mu^T_\alpha(x) = \mu(x) + \alpha \geq t$ for $x \in U_\alpha(\mu, t)$. Hence $x \in U_\alpha(\mu, t)$. Let $x, y \in X$ be such that $x \in U_\alpha(\mu, t)$ and $x * y \in U_\alpha(\mu, t)$. Then $\mu(x) \geq t - \alpha$ and $\mu(x * y) \geq t - \alpha$. Hence $\mu^T_\alpha(x) = \mu(x) + \alpha \geq t$ and $\mu^T_\alpha(x * y) = \mu(x * y) + \alpha \geq t$. Since μ^T_α is a fuzzy weak filter of X, we get

$$\begin{aligned}\mu(x * (x * y)) + \alpha &= \mu^T_\alpha(x * (x * y)) \\ &\geq \min\{\mu^T_\alpha(x), \mu^T_\alpha(x * y)\} \\ &= \min\{t, t\} \\ &= t\end{aligned}$$

Hence $\mu(x * (x * y)) \geq t - \alpha$. Thus $x * (x * y) \in U_\alpha(\mu, t)$. Hence $U_\alpha(\mu, t)$ is a weak filter of X.

Conversely, assume that $U_\alpha(\mu, t)$ is a weak filter of X for all $t \in Im(\mu)$ with $t \geq \alpha$. Suppose there exists $a \in X$ such that $\mu^T_\alpha(1) < \beta \leq \mu^T_\alpha(a)$. Then $\mu(a) \geq \beta - \alpha$ but $\mu(1) < \beta - \alpha$. Hence $a \in U_\alpha(\mu, t)$ and $1 \notin U_\alpha(\mu, t)$, which is a contradiction to that $U_\alpha(\mu, t)$ is a weak filter of X. Therefore $\mu^T_\alpha(1) \geq \mu^T_\alpha(x)$ for all $x \in X$. Let $x, y \in X$ be such that $\mu(x) = t_1 - \alpha$ and $\mu(x * y) = t_2 - \alpha$ for some $t_1, t_2 \geq \alpha$. Then $x \in U_\alpha(\mu, t_1)$ and $x * y \in U_\alpha(\mu, t_2)$. Without loss of generality, assume that $t_1 \leq t_2$. Then clearly $U_\alpha(\mu, t_2) \subseteq U_\alpha(\mu, t_1)$. Hence $x * y \in U_\alpha(\mu, t_1)$. Since $U_\alpha(\mu, t_1)$ is a weak filter, we get $x * (x * y) \in U_\alpha(\mu, t_1)$. Thus $\mu(x * (x * y)) \geq t_1 - \alpha$. Hence

$$\begin{aligned}\mu_\alpha^T(x*(x*y)) &= \mu(x*(x*y))+\alpha\\ &\ge t_1\\ &= \min\{t_1,t_2\}\\ &= \min\{\mu(x)+\alpha, \mu(x*y)+\alpha\}\\ &= \min\{\mu_\alpha^T(x), \mu_\alpha^T(x*y)\}\end{aligned}$$

Therefore μ_α^T is a fuzzy weak filter of X. □

Definition 13.3.11 [164] Let μ_1 and μ_2 be two fuzzy subsets of a BE-algebra X. If $\mu_1(x) \le \mu_2(x)$ for all $x \in X$, then we say that μ_2 is a fuzzy extension of μ_1.

Definition 13.3.12 Let μ_1 and μ_2 be two fuzzy subsets of a BE-algebra X. Then μ_2 is said to be a fuzzy weak filter extension of μ_1 if it satisfies the following:

(1) μ_2 is a fuzzy extension of μ_1,

(2) If μ_1 is a fuzzy weak filter of X, then so is μ_2.

Theorem 13.3.13 *Let μ be a fuzzy weak filter of a BE-algebra X and $\alpha, \beta \in [0, \top]$. If $\alpha \ge \beta$, then μ_α^T is a fuzzy weak filter extension of μ_β^T.*

Proof Let $\alpha, \beta \in [0, \top]$ be such that $\alpha \ge \beta$. For any $x \in X$, we get $\mu_\beta^T(x) = \mu(x)+\beta \le \mu(x)+\alpha = \mu_\alpha^T$. Hence μ_α^T is a fuzzy extension of μ_β^T. Suppose μ_β^T is a fuzzy weak filter of X. Then μ is a fuzzy weak filter X. Hence

$$\begin{aligned}\mu_\alpha^T(x*(x*y)) &= \mu(x*(x*y))+\alpha\\ &\ge \min\{\mu(x), \mu(x*y)\}+\alpha\\ &= \min\{\mu(x)+\alpha, \mu(x*y)+\alpha\}\\ &= \min\{\mu_\alpha^T(x), \mu_\alpha^T(x*y)\}.\end{aligned}$$

Therefore μ_α^T is a fuzzy weak filter and hence a fuzzy weak filter extension of μ_β^T. □

Theorem 13.3.14 *Let μ be a fuzzy weak filter of a BE-algebra X and $\alpha \in [0, \top]$. Then the fuzzy α-translation μ_α^T of μ is a fuzzy weak filter extension of μ.*

Proof Since $\alpha \ge 0$, we get $\mu_\alpha^T(x) = \mu(x)+\alpha \ge \mu(x)$. Hence μ_α^T is a fuzzy extension of μ. Suppose μ is a fuzzy weak filter of X. Clearly $\mu_\alpha^T(1) = \mu(1)+\alpha \ge \mu(x)+\alpha = \mu_\alpha^T(x)$ for all $x \in X$. Let $x, y \in X$. Since μ is a fuzzy weak filter of X, we get the following:

$$\begin{aligned}\mu_\alpha^T(x*(x*y)) &= \mu(x*(x*y))+\alpha\\ &\ge \min\{\mu(x), \mu(x*y)\}+\alpha\\ &= \min\{\mu(x)+\alpha, \mu(x*y)+\alpha\}\\ &= \min\{\mu_\alpha^T(x), \mu_\alpha^T(x*y)\}.\end{aligned}$$

Therefore μ_α^T is a fuzzy weak filter and hence a fuzzy weak filter extension of μ. □

Definition 13.3.15 Let μ be a fuzzy subset of a BE-algebra X and $\alpha \in [0, 1]$. A mapping $\mu_\alpha^m : X \to [0, 1]$ is called *fuzzy α-multiplication* of μ if it satisfies the following:

$$\mu_\alpha^m(x) = \mu(x) \cdot \alpha \text{ for all } x \in X.$$

Theorem 13.3.16 *Let μ be a fuzzy subset of a BE-algebra X. Then μ is a fuzzy weak filter of X if and only if μ_α^m is a fuzzy weak filter of X for all $\alpha \in (0, 1]$.*

Proof Assume that μ is a fuzzy weak filter of X. Let $\alpha \in (0, 1]$. Then $\mu_\alpha^m(1) = \mu(1) \cdot \alpha \geq \mu(x) \cdot \alpha = \mu_\alpha^m(x)$ for all $x \in X$. Let $x, y \in X$. Since μ is a fuzzy weak filter of X, we get

$$\begin{aligned}
\mu_\alpha^m(x * (x * y)) &= \mu(x * (x * y)) \cdot \alpha \\
&\geq \min\{\mu(x), \mu(x * y)\} \cdot \alpha \\
&= \min\{\mu(x) \cdot \alpha, \mu(x * y) \cdot \alpha\} \\
&= \min\{\mu_\alpha^m(x), \mu_\alpha^m(x * y)\}.
\end{aligned}$$

Therefore μ_α^m is a fuzzy weak filter of X.

Conversely, assume that μ_α^T is a fuzzy weak filter of X for all $\alpha \in (0, 1]$. Then $\mu(1) \cdot \alpha = \mu_\alpha^T(1) \geq \mu_\alpha^T(x) = \mu(x) \cdot \alpha$ for all $\alpha \in (0, 1]$. Hence $\mu(1) \geq \mu(x)$ for all $x \in X$. Let $x, y \in X$. Since μ_α^m is fuzzy weak filter of X, we get

$$\begin{aligned}
\mu(x * (x * y)) \cdot \alpha &= \mu_\alpha^m(x * (x * y)) \\
&\geq \min\{\mu_\alpha^m(x), \mu_\alpha^m(x * y)\} \\
&= \min\{\mu(x) \cdot \alpha, \mu(x * y) \cdot \alpha\} \\
&= \min\{\mu(x), \mu(x * y)\} \cdot \alpha.
\end{aligned}$$

Hence $\mu(x * (x * y)) \geq \min\{\mu(x), \mu(x * y)\}$. Therefore μ is a fuzzy weak filter of X. □

Theorem 13.3.17 *Let $\alpha \in (0, 1]$, and μ is fuzzy subalgebra of a BE-algebra X. Then the fuzzy α-multiplication μ_α^m of μ is a fuzzy weak filter of X.*

Proof Let $x \in X$. Since μ is a fuzzy subalgebra of X, we get $\mu_\alpha^m(1) = \mu(1) \cdot \alpha = \mu(x * x) \cdot \alpha \geq \min\{\mu(x), \mu(x)\} \cdot \alpha = \mu(x) \cdot \alpha = \mu_\alpha^m(x)$. Let $x, y \in X$. Then we get

$$\begin{aligned}
\mu_\alpha^m(x * (x * y)) &= \mu(x * (x * y)) \cdot \alpha \\
&\geq \min\{\mu(x), \mu(x * y)\} \cdot \alpha \\
&= \min\{\mu(x) \cdot \alpha, \mu(x * y) \cdot \alpha\} \\
&\geq \min\{\mu_\alpha^m(x), \mu_\alpha^m(x * y)\}.
\end{aligned}$$

Therefore μ_α^m is a fuzzy weak filter of X. □

Since every fuzzy filter of a BE-algebra is a fuzzy subalgebra, the following corollary is a direct consequence of the above theorem.

Corollary 13.3.18 *Let $\alpha \in (0, 1]$, and μ is fuzzy filter of a BE-algebra X. Then the fuzzy α-multiplication μ_α^m of μ is a fuzzy weak filter of X.*

Theorem 13.3.19 *Let $\alpha \in (0, 1]$, and μ is a fuzzy weak filter of a BE-algebra X. Then the fuzzy α-multiplication μ_α^m of μ is a fuzzy filter of X if and only if it satisfies*

$$\mu(y) \geq \min\{\mu(x), \mu(x * (x * y))\} \text{ for all } x, y \in X.$$

Proof Assume that μ_α^m is a fuzzy filter of X. Let $x, y \in X$. Since μ_α^m is a fuzzy filter of X, we get $\mu_\alpha^m(x * y) \geq \min\{\mu_\alpha^m(x), \mu_\alpha^m(x * (x * y))\}$. Since μ_α^m is a fuzzy filter of X, we get

$$\begin{aligned}
\mu(y) \cdot \alpha &= \mu_\alpha^m(y) \\
&\geq \min\{\mu_\alpha^m(x), \mu_\alpha^m(x * y)\} \\
&\geq \min\{\mu_\alpha^m(x), \min\{\mu_\alpha^m(x), \mu_\alpha^m(x * (x * y))\}\} \\
&= \min\{\mu_\alpha^m(x), \mu_\alpha^m(x * (x * y))\} \\
&= \min\{\mu(x) \cdot \alpha, \mu(x * (x * y)) \cdot \alpha\} \\
&= \min\{\mu(x), \mu(x * (x * y))\} \cdot \alpha.
\end{aligned}$$

Hence $\mu(y) \geq \min\{\mu(x), \mu(x * (x * y))\}$. Therefore μ is satisfying the condition.

Conversely, assume that μ satisfies the condition. Let $x, y \in X$. Since μ is a fuzzy weak filter, we get $\mu(x * (x * y)) \geq \min\{\mu(x), \mu(x * y)\}$. By the assumed condition, we get

$$\begin{aligned}
\mu_\alpha^m(y) &= \mu(y) \cdot \alpha \\
&\geq \min\{\mu(x), \mu(x * (x * y))\} \cdot \alpha \\
&\geq \min\{\mu(x), \min\{\mu(x), \mu(x * y)\}\} \cdot \alpha \\
&= \min\{\mu(x), \mu(x * y)\} \cdot \alpha \\
&= \min\{\mu(x) \cdot \alpha, \mu(x * y) \cdot \alpha\} \\
&= \min\{\mu_\alpha^m(x), \mu_\alpha^m(x * y)\}.
\end{aligned}$$

Therefore μ_α^m is a fuzzy filter of X. □

Theorem 13.3.20 *Let μ be a fuzzy subset of a BE-algebra X, $\alpha \in [0, \top]$, and $\beta \in (0, 1]$. Then every fuzzy α-translation μ_α^T is a fuzzy weak filter extension of the fuzzy β-multiplication μ_β.*

Proof Let $x \in X$. Let $\alpha \in [0, \top]$ and $\beta \in (0, 1]$. Since $\beta \in (0, 1]$, we get $\mu_\alpha^T(x) = \mu(x) + \alpha \geq \mu(x) \geq \mu(x) \cdot \beta = \mu_\beta^m(x)$ for all $x \in X$. Hence μ_α^T is a fuzzy extension of μ_α^m. Assume that μ_α^m is a fuzzy weak filter of X. Then μ is a fuzzy weak filter of X.

Hence μ_α^T is a fuzzy weak filter of X. Therefore μ_α^T is a fuzzy weak filter extension of μ_α^m. □

Definition 13.3.21 Let μ be a fuzzy subset of a BE-algebra $X, \alpha \in [0, \top]$, and $b \in (0, 1]$. A mapping $\mu_{\beta\alpha}^{mT} : X \to [0, 1]$ is said to be a *fuzzy magnified $\beta\alpha$-translation* of μ if it satisfies

$$\mu_{\beta\alpha}^{mT}(x) = \beta \cdot \mu(x) + \alpha \text{ for all } x \in X.$$

Theorem 13.3.22 *Let μ be a fuzzy subset of a BE-algebra $X, \alpha \in [0, \top]$, and $\beta \in (0, 1]$. Then μ is a fuzzy weak filter of X if and only if $\mu_{\beta\alpha}^{mT}$ is a fuzzy weak filter of X.*

Proof Assume that μ is a fuzzy weak filter of X. Let $\alpha \in [0, \top]$ and $b \in (0, 1]$. Let $x \in X$. Then clearly $\mu(1) \geq \mu(x)$. Hence $\mu_{\beta\alpha}^{mT}(1) = \beta \cdot \mu(1) + \alpha \geq \beta\mu(x) + \alpha = \mu_{\beta\alpha}^{mT}(x)$. Let $x, y \in X$. Since μ is a fuzzy weak filter of X, we get

$$\begin{aligned}\mu_{\beta\alpha}^{mT}(x * (x * y)) &= \beta \cdot \mu(x * (x * y)) + \alpha \\ &\geq \beta \cdot \left(\min\{\mu(x), \mu(x * y)\}\right) + \alpha \\ &= \min\{\beta \cdot \mu(x) + \alpha, \beta \cdot \mu(x * y) + \alpha\} \\ &= \min\{\mu_{\beta\alpha}^{mT}(x), \mu_{\beta\alpha}^{mT}(x * y)\}.\end{aligned}$$

Therefore $\mu_{\beta\alpha}^{mT}$ is a fuzzy weak filter of X.

Conversely, assume that $\mu_{\beta\alpha}^{mT}$ is a fuzzy weak filter of X for all $\alpha \in [0, \top]$ and $\beta \in (0, 1]$. Let $x \in X$. Then $\beta \cdot \mu(1) + \alpha = \mu_{\beta\alpha}^{mT}(1) \geq \mu_{\beta\alpha}^{mT}(x) = \beta \cdot \mu(x) + \alpha$. Hence $\mu(1) \geq \mu(x)$ for all $x \in X$. Let $x, y \in X$. Since $\mu_{\beta\alpha}^{mT}$ is a fuzzy weak filter of X, we get

$$\begin{aligned}\beta \cdot \mu(x * (x * y)) + \alpha &= \mu_{\beta\alpha}^{mT}(x * (x * y)) \\ &\geq \min\{\mu_{\beta\alpha}^{mT}(x), \mu_{\beta\alpha}^{mT}(x * y)\} \\ &= \min\{\beta \cdot \mu(x) + \alpha, \beta \cdot \mu(x * y) + \alpha\} \\ &= \beta \cdot \min\{\mu(x), \mu(x * y)\} + \alpha.\end{aligned}$$

Hence $\mu(x * (x * y)) \geq \min\{\mu(x), \mu(x * y)\}$. Therefore μ is a fuzzy weak filter of X. □

13.4 Anti Fuzzy Weak Filters of BE-algebras

In this section, the notion of anti fuzzy weak filters is introduced. An equivalent condition is derived for a fuzzy set to become an anti fuzzy weak filter. An equivalent condition is derived for an anti fuzzy weak filter to become an anti fuzzy filter.

Definition 13.4.1 A fuzzy subset μ of a BE-algebra X is called an *anti fuzzy filter* of X if for all $x, y \in X$ it satisfies the following conditions:

(AF1) $\mu(1) \leq \mu(x)$,
(AF2) $\mu(y) \leq \max\{\mu(x), \mu(x * y)\}$.

Definition 13.4.2 A fuzzy subset μ of a BE-algebra X is called an *anti fuzzy weak filter* of X if for all $x, y \in X$ it satisfies the following conditions:

(AFW1) $\mu(1) \leq \mu(x)$,
(AFW2) $\mu(x * (x * y)) \leq \max\{\mu(x), \mu(x * y)\}$.

Clearly every constant map $\mu_c : X \to [0, 1]$ is an anti fuzzy weak filter of X

Example 13.4.3 Let $X = \{1, a, b, c\}$. Define a binary operation $*$ on X as follows:

$*$	1	a	b	c
1	1	a	b	c
a	1	1	a	c
b	1	1	1	c
c	1	a	b	1

Then $(X, *, 1)$ is a BE-algebra. Define a fuzzy set $\mu : X \longrightarrow [0, 1]$ as $\mu(1) = \mu(a) = \mu(b) = \alpha$ and $\mu(c) = \beta$ where $\alpha, \beta \in [0, 1]$ such that $\alpha > \beta$. Obviously μ is an anti fuzzy weak filter of X.

Theorem 13.4.4 *Let μ be an anti fuzzy weak filter of a BE-algebra X. Then the set $X_\mu = \{x \in X \mid \mu(x) = \mu(1)\}$ is a weak filter of X.*

Proof Assume that μ is an anti fuzzy weak filter of X. Clearly $1 \in X_\mu$. Let $x, y \in X$. Suppose $x, x*y \in X_\mu$. Then $\mu(x) = \mu(x*y) = \mu(1)$. Clearly $\mu(1) \leq \mu(x*(x*y))$. Again, $\mu(x * (x * y)) \leq \max\{\mu(x), \mu(x * y)\} = \max\{\mu(1), \mu(1)\} = \mu(1)$. Then we get $\mu(x * (x * y)) = \mu(1)$. Thus $x * (x * y) \in X_\mu$. Therefore X_μ is a weak filter of X. □

Lemma 13.4.5 *Let μ be an anti fuzzy weak filter of a BE-algebra X and $x, y \in X$. Then $x \leq y$ implies $\mu(x * (x * y)) \leq \mu(x)$.*

Proof Let $x, y \in X$ be such that $x \leq y$. Then $x*y = 1$. Clearly $\mu(1) \leq \mu(x*(x*y))$. Since μ is a fuzzy weak filter of X, we get $\mu(x * (x * y)) \leq \max\{\mu(x), \mu(x * y)\} = \max\{\mu(x), \mu(1)\} = \mu(x)$. Therefore $\mu(x * (x * y)) \leq \mu(x)$ whenever $x \leq y$. □

Theorem 13.4.6 *A fuzzy set μ of a BE-algebra X is an anti fuzzy weak filter in X if it satisfies the following conditions for all $x, y, a, b \in X$:*

(AFW3) $\mu(1) \leq \mu(x)$ *for all* $x \in X$,
(AFW4) $\mu((x * a) * (y * b)) \leq \max\{\mu(x * y), \mu(a * b)\}$.

Proof Assume that μ satisfies the conditions (AFW3) and (AFW4). Let $x, y \in X$. Then

$$\begin{aligned}\mu(x * (x * y)) &= \mu((1 * x) * (x * y))\\ &\leq \max\{\mu(1 * x), \mu(x * y)\}\\ &= \max\{\mu(x), \mu(x * y)\}.\end{aligned}$$

Therefore μ is fuzzy weak filter of X. □

Theorem 13.4.7 *Every anti fuzzy filter of a BE-algebra is an anti fuzzy weak filter.*

Proof Let μ be an anti fuzzy filter of a BE-algebra X. Then μ is order reversing. Clearly $\mu(1) \leq \mu(x)$ for all $x \in X$. Let $x, y \in X$. Clearly $x * y \leq x \leq (x * y) \leq x * (x * (x * y))$. Since μ is an anti fuzzy filter, we get

$$\begin{aligned}\mu(x * (x * y)) &\leq \max\{\mu(x), \mu(x * (x * (x * y)))\}\\ &\leq \max\{\mu(x), \mu(x * (x * y))\}\\ &\leq \max\{\mu(x), \mu(x * y)\}\end{aligned}$$

Therefore μ is an anti fuzzy weak filter of X. □

In general, the converse of the above theorem is not true. However, in the following theorem, a set of equivalent conditions is derived for every anti fuzzy weak filter of a BE-algebra to become an anti fuzzy filter.

Theorem 13.4.8 *An anti fuzzy weak filter μ of a BE-algebra X is an anti fuzzy filter of X if and only if it satisfies the following conditions:*

(1) $\mu(1) \leq \mu(x)$ *for all* $x \in X$,
(2) $\mu(y) \leq \max\{\mu(x), \mu(x * (x * y))\}$ *for all* $x, y \in X$.

Proof Assume that μ is an anti fuzzy filter of X. Clearly $\mu(1) \leq \mu(x)$ for all $x \in X$. Let $x, y \in X$. Since μ is an anti fuzzy filter of X, we get $\mu(x * y) \leq \max\{\mu(x), \mu(x * (x * y))\}$. Hence

$$\begin{aligned}\mu(y) &\leq \max\{\mu(x), \mu(x * y)\}\\ &\leq \max\{\mu(x), \min\{\mu(x), \mu(x * (x * y))\}\}\\ &= \max\{\mu(x), \mu(x * (x * y))\}\end{aligned}$$

Therefore μ is satisfying the condition.

Conversely, assume that μ satisfies the condition. Let $x, y \in X$. Since μ is an anti fuzzy weak filter, we get $\mu(x * (x * y)) \leq \max\{\mu(x), \mu(x * y)\}$. By the assumed condition, we get

$$\begin{aligned}\mu(y) &\leq \max\{\mu(x), \mu(x * (x * y))\}\\ &\leq \max\{\mu(x), \max\{\mu(x), \mu(x * y)\}\}\\ &\leq \max\{\mu(x), \mu(x * y)\}\end{aligned}$$

Therefore μ is an anti fuzzy filter of X. □

Definition 13.4.9 Let μ be a fuzzy set in a BE-algebra X. For any $\alpha \in [0, 1]$, the set $\mu_\alpha = \{x \in X \mid \mu(x) \leq \alpha\}$ is said to be a *level subset* of μ.

Theorem 13.4.10 *Let μ be a fuzzy set of a BE-algebra X. Then μ is an anti fuzzy weak filter of X if and only if for each $\alpha \in [0, 1]$, the level subset μ_α is a weak filter of X, when $\mu_\alpha \neq \emptyset$.*

Proof Assume that μ is an anti fuzzy weak filter of X. Let $\alpha \in [0, 1]$. Then $\mu(1) \leq \mu(x) \leq \alpha$ for all $x \in \mu_\alpha$. Hence $1 \in \mu_\alpha$. Let $x \in \mu_\alpha$ and $x * y \in \mu_\alpha$. Then $\mu(x) \leq \alpha$ and $\mu(x * y) \leq \alpha$. Then $\mu(x * (x * y)) \leq \max\{\mu(x), \mu(x * y)\} \leq \max\{\alpha, \alpha\} = \alpha$. Hence $x * (x * y) \in \mu_\alpha$. Therefore μ_α is a weak filter of X.

Conversely, assume that μ_α is a weak filter of X for each $\alpha \in [0, 1]$ with $\mu_\alpha \neq \emptyset$. Suppose there exists $x_0 \in X$ such that $\mu(1) > \mu(x_0)$. Then $\mu(1) > \alpha_0$ and $0 \leq \mu(x_0) \leq \alpha_0 < 1$. Hence $x_0 \in \mu_{\alpha_0}$ and $\mu_{\alpha_0} \neq \emptyset$. Since μ_{α_0} is a weak filter of X, we get $1 \in \mu_{\alpha_0}$ and hence $\mu(1) \leq \alpha_0$, which is a contradiction. Therefore $\mu(1) \leq \mu(x)$ for all $x \in X$. Again, let $\alpha_0 = \frac{1}{2}(\mu(1)+\mu(x_0))$. Let $x, y \in X$ be such that $\mu(x) = \alpha_1$ and $\mu(x * y) = \alpha_2$. Then $x \in \mu_{\alpha_1}$ and $x * y \in \mu_{\alpha_2}$. Without loss of generality, assume that $\alpha_2 \leq \alpha_1$. Clearly $\mu_{\alpha_2} \subseteq \mu_{\alpha_1}$. Hence $x*y \in \mu_{\alpha_1}$. Since μ_{α_1} is a weak filter of X, we get $x*(x*y) \in \mu_{\alpha_1}$. Thus $\mu(x*(x*y)) \leq \alpha_1 = \max\{\alpha_1, \alpha_2\} = \max\{\mu(x), \mu(x*y)\}$. Therefore μ is an anti fuzzy weak filter of X. □

Definition 13.4.11 For any fuzzy subset μ of a BE-algebra X, define the *complement* μ^c of μ as the fuzzy subset of X given by $\mu^c(x) = 1 - \mu(x)$ for all $x \in X$.

Theorem 13.4.12 *A fuzzy subset μ of a BE-algebra X is a fuzzy weak filter of X if and only if its complement μ^c is an anti fuzzy weak filter of X.*

Proof Assume that μ is a fuzzy weak filter. For any $x \in X$, we get $\mu^c(1) = 1-\mu(1) \leq 1 - \mu(x) = \mu^c(x)$. Let $x, y \in X$. Then $\mu^c(x * (x * y)) = 1 - \mu(x * (x * y)) \leq 1-\min\{\mu(x), \mu(x*y)\} = 1-\min\{1-\mu^c(x), 1-\mu^c(x*y)\} = \max\{\mu^c(x), \mu^c(x*y)\}$. Hence μ^c is an anti fuzzy weak filter.

Conversely, assume that μ^c is an anti fuzzy weak filter of X. Then we get $\mu(1) = 1 - \mu^c(x) \geq 1 - \mu^c(x) = \mu(x)$. Again, let $x, y \in X$. Then $\mu(x * (x * y)) = 1-\mu^c(x*(x*y)) \geq 1-\max\{\mu^c(x), \mu^c(x*y)\} = 1-\max\{1-\mu(x), 1-\mu(x*y)\} = \min\{\mu(x), \mu(x * y)\}$. Therefore μ is a fuzzy weak filter. □

13.5 Normal Fuzzy Weak Filters of BE-algebras

In this section, the notion of normal fuzzy weak filters is introduced in BE-algebras. These classes of normal fuzzy weak filters are then characterized. Some properties of normal fuzzy weak filters are studied with respect to fuzzy relations and Cartesian products.

Definition 13.5.1 A fuzzy weak filter μ of a BE-algebra X is said to be a *normal fuzzy weak filter* if there exists $x \in X$ such that $\mu(x) = 1$.

For any normal fuzzy weak filter μ, we obviously have $\mu(1) = 1$.

Proposition 13.5.2 *For any fuzzy weak filter μ of X, define a fuzzy set μ^+ in X as $\mu^+(x) = \mu(x) + 1 - \mu(1)$ for all $x \in X$. Then μ^+ is a normal fuzzy weak filter of X such that $\mu \subseteq \mu^+$.*

Proof Let $x \in X$. Then we have $\mu^+(1) = \mu(1) + 1 - \mu(1) = 1 \geq \mu^+(x)$. Let $x, y \in X$. Then

$$\begin{aligned}\mu^+(x * (x * y)) &= \mu(x * (x * y)) + 1 - \mu(1)\\ &\geq \min\{\mu(x), \mu(x * y)\} + 1 - \mu(1)\\ &= \min\{\mu(x) + 1 - \mu(1), \mu(x * y) + 1 - \mu(1)\}\\ &= \min\{\mu^+(x), \mu^+(x * y)\}.\end{aligned}$$

Therefore μ^+ is a fuzzy weak filter of X. Clearly $\mu^+(1) = 1$. Hence μ^+ is a normal fuzzy weak filter of X. Now, for any $x \in X$, it is clear that $\mu(x) \leq \mu^+(x)$. Therefore $\mu \subseteq \mu^+$. □

Corollary 13.5.3 *If $\mu^+(x_0) = 0$ for some $x_0 \in X$, then so $\mu(x_0) = 0$.*

Proposition 13.5.4 *Let A be a weak filter of X and X_A a fuzzy set of X defined by*

$$X_A(x) = \begin{cases} 1 \text{ if } x \in A \\ 0 \text{ otherwise}\end{cases}$$

Then X_A is a normal fuzzy weak filter of X.

Proof Since A is a weak filter of X, we get $1 \in A$. Hence $X_A(1) = 1 \geq X_A(x)$ for all $x \in X$. Let $x, y \in X$. Suppose $x \in A$ and $x * y \in A$. Since A is a weak filter, we get $x * (x * y) \in A$. Then $X_A(x) = X_A(x * (x * y)) = X_A(x * y) = 1$. Hence $X_A(x * (x * y)) \geq 1 = \min\{X_A(x), X_A(x * y)\}$. Suppose $x \notin A$ and $x * y \notin A$. Then $X_A(x) = X_A(x * y) = 0$. Hence $X_A(x * (x * y)) \geq 0 = \min\{X_A(x), X_A(x * y)\}$. If exactly one of x and $x * y$ belongs to A, then exactly one of $X_A(x)$ and $X_A(x * y)$ is equal to 0. Hence $X_A(x*(x*y)) \geq \min\{X_A(x), X_A(x*y)\}$. Hence $X_A(x*(x*y)) \geq \min\{X_A(x), X_A(x * y)\}$ for all $x, y \in X$. Therefore X_A is a fuzzy weak filter of X. Since $1 \in A$, we get $X_A(1) = 1$. Therefore X_A is a normal fuzzy weak filter of X. □

Theorem 13.5.5 *Let μ be a fuzzy weak filter of X. Then the following are equivalent:*

(1) *μ is normal,*
(2) $\mu(1) = 1$,
(3) $\mu = \mu^+$.

Proof (1) ⇒ (2): Assume that μ is normal. Then there exists some $x \in X$ such that $\mu(x) = 1$. Since μ is a fuzzy weak filter, we get $\mu(1) \geq \mu(x) = 1$. Therefore $\mu(1) = 1$.

(2) $\Rightarrow$ (3): For any $x \in X$, we get that $\mu^+(x) = \mu(x) + 1 - \mu(1) = \mu(x)$. Therefore $\mu = \mu^+$.

(3) $\Rightarrow$ (1): Assume that $\mu = \mu^+$. Then for any $x \in X$, we get $\mu^+(x) = \mu(x) + 1 - \mu(1)$. Since $\mu^+(x) = \mu(x)$, we get that $\mu(1) = 1$. Therefore μ is a normal fuzzy weak filter of X. □

Definition 13.5.6 Let μ be a fuzzy set in a BE-algebra X. Define the sets X_μ and Δ_μ as $X_\mu = \{x \in X \mid \mu(x) = \mu(1)\}$ and $\Delta_\mu = \{x \in X \mid \mu(x) = 1\}$.

If μ is normal, then it can be easily observed that $X_\mu = \Delta_\mu$.

Proposition 13.5.7 *Let μ be a fuzzy weak filter of a BE-algebra X. Then we have*

(1) *If μ is normal, then X_μ is a weak filter of X,*

(2) *μ is normal if and only if Δ_μ is a weak filter of X.*

Proof (1) Clearly $1 \in X_\mu$. Let $x, y \in X$ be such that $x \in X_\mu$ and $x * y \in X_\mu$. Then $\mu(x) = \mu(x * y) = \mu(1)$. Since μ is a fuzzy weak filter, we get $\mu(x * (x * y)) \geq \min\{\mu(x), \mu(x * y)\} = \min\{\mu(1), \mu(1)\} = \mu(1)$. Since μ is normal, by Theorem 13.5.5, we get $\mu(x * (x * y)) \geq \mu(1) = 1$. Hence $\mu(x * (x * y)) = 1 = \mu(1)$. Thus $x * (x * y) \in X_\mu$. Hence X_μ is a weak filter of X.
(2) Assume that μ is normal. Then $\mu(1) = 1$. Hence $1 \in \Delta_\mu$. Let $x, x * y \in \Delta_\mu$. Then $\mu(x) = \mu(x * y) = 1$. Since μ is a fuzzy weak filter, we get $\mu(x * (x * y)) \geq \min\{\mu(x), \mu(x * (y)\} = 1$. Hence $\mu(x * (x * y)) = 1$. Thus $x * (x * y) \in \Delta_\mu$. Hence Δ_μ is a weak filter of X. Conversely, assume that Δ_μ is a weak filter of X. Hence $1 \in \Delta_\mu$. Thus $\mu(1) = 1$. Therefore, by Theorem 13.5.5, μ is a normal fuzzy weak filter of X. □

Proposition 13.5.8 *Let μ and ν be two fuzzy weak filters of a BE-algebra X such that $\mu \subseteq \nu$. Then $\Delta_\mu \subseteq \Delta_\nu$. Moreover, if μ and ν are normal and $\mu \subseteq \nu$, then $X_\mu \subseteq X_\nu$.*

Proof Let $x \in \Delta_\mu$. Then we get $\nu(x) \geq \mu(x) = 1$. Hence $\nu(x) = 1$, which implies that $x \in \Delta_\nu$. Therefore $\Delta_\mu \subseteq \Delta_\nu$. To prove the remaining, let $x \in X_\mu$. Then $\nu(x) \geq \mu(x) = \mu(1) = 1$. Hence $\nu(x) = 1 = \nu(1)$, which concludes that $x \in X_\nu$. Therefore $X_\mu \subseteq X_\nu$. □

Definition 13.5.9 A non-constant fuzzy weak filter μ of a BE-algebra X is said to be a *maximal fuzzy weak filter* if there exists non-constant fuzzy weak filter ν such that $\mu \subseteq \nu$.

Proposition 13.5.10 *Every maximal fuzzy weak filter is a normal fuzzy weak filter.*

Proof Let μ be a maximal fuzzy weak filter of a BE-algebra X. Then μ is non-constant, and hence, μ^+ is non-constant. Otherwise, suppose $\mu^+(x) = c$ for all $x \in X$, where c is a constant. Then for all $x \in X, c = \mu^+(x) = \mu(x) + 1 - \mu(1)$, which shows that μ is constant. Since $\mu \subseteq \mu^+$ and μ is maximal, we get that $\mu = \mu^+$. Therefore, by Theorem 13.5.5, we get that μ is normal. □

Theorem 13.5.11 *Every maximal fuzzy weak filter takes only the values* 0 *and* 1.

Proof Let μ be a maximal fuzzy weak filter of a BE-algebra X. Since μ is maximal, by above proposition, μ is normal and hence $\mu(1) = 1$. Let $x \in X$ be such that $\mu(x) \neq 0$. Suppose $\mu(x) \neq 1$. Then there exists some $x_0 \in X$ such that $0 < \mu(x_0) < 1$. Define a fuzzy set ν in X as $\nu(x) = \frac{1}{2}(\mu(x)+\mu(x_0))$ for all $x \in X$. Clearly ν is well-defined. Let $x \in X$. Then $\nu(1) = \frac{1}{2}(\mu(1)+\mu(x_0)) = \frac{1}{2}(1+\mu(x_0)) \geq \frac{1}{2}(\mu(x)+\mu(x_0)) = \nu(x)$. Let $x, y \in X$. Then

$$\begin{aligned}\nu(x*(x*y)) &= \frac{1}{2}\{\mu(x*(x*y)) + \mu(x_0)\}\\ &\geq \frac{1}{2}\{\min\{\mu(x), \mu(x*y)\} + \mu(x_0)\}\\ &= \frac{1}{2}\{\min\{\mu(x) + \mu(x_0), \mu(x*y) + \mu(x_0)\}\\ &= \min\{\frac{1}{2}(\mu(x) + \mu(x_0)), \frac{1}{2}(\mu(x*y) + \mu(x_0))\}\\ &= \min\{\nu(x), \nu(x*y)\}.\end{aligned}$$

Therefore ν is a fuzzy weak filter of X. Hence by Proposition 13.5.2, we get that ν^+ is a normal fuzzy weak filter of X. Clearly $\nu^+(x) \geq \mu(x)$ for all $x \in X$. Now

$$\begin{aligned}\nu^+(x_0) &= \nu(x_0) + 1 - \nu(1)\\ &= \frac{1}{2}\{\mu(x_0) + \mu(x_0)\} + 1 - \frac{1}{2}\{\mu(1) + \mu(x_0)\}\\ &= \frac{1}{2}\{\mu(x_0) + 1\}\\ &> \mu(x_0).\end{aligned}$$

and also $\nu^+(x_0) < 1 = \nu^+(1)$. Hence ν^+ is non-constant such that $\mu \subseteq \nu^+$. Therefore μ is not maximal, which is a contradiction. Hence $\mu(x) = 1$. □

Let us denote the set of all weak filters of a BE-algebra X by $\mathcal{WF}(X)$. Then it is obvious that $\mathcal{WF}(X)$ forms a partially ordered set with respect to set inclusion. By a maximal weak filter of X, we mean a maximal element in the poset $\langle \mathcal{WF}(X), \subseteq \rangle$ of weak filters of X.

Theorem 13.5.12 *Let μ be a non-constant fuzzy weak filter of a BE-algebra X. Then*

(1) *If μ is maximal, then X_μ is a maximal weak filter of X,*
(2) *μ is maximal if and only if Δ_μ is a maximal weak filter.*

Proof (1) Assume that μ is a maximal fuzzy weak filter of X. Then by Proposition 13.5.7, X_μ is a weak filter of X. Suppose $X_\mu = X$. Then we get $\mu(x) = \mu(1)$ for all $x \in X$. Thus μ is constant, which is a contradiction. Therefore X_μ is proper. Let F be a weak filter of X such that $X_\mu \subseteq F$. Then by Proposition 13.5.4, we get that $\mu = \mu_{X_\mu} \subseteq \mu_F$. Since μ is maximal, we get either $\mu = \mu_F$ or μ_F is constant.

Suppose μ_F is constant. Then $F = X$, which is a contradiction. Suppose $\mu = \mu_F$. Then we get $X_\mu = X_{\mu_F} = F$. Therefore X_μ is a maximal weak filter of X.
(2) Assume that μ is a maximal fuzzy weak filter of X. Then μ is normal, and hence, by (1), we get $X_\mu = \Delta_\mu$ is a maximal weak filter on X. Conversely, assume that Δ_μ is a maximal weak filter of X. Let ν be a non-constant fuzzy weak filter of X such that $\mu \subset \nu$. Then we get $\Delta_\mu \subseteq \Delta_{nu}$. Since Δ_μ is maximal, we get either $\Delta_\nu = X$. Hence for all $x \in X = \Delta_\nu$, we get $\nu(x) = 1$. Thus ν is constant, which is a contradiction. Hence μ is a maximal fuzzy weak filter of X. □

Theorem 13.5.13 *Let $f : X \longrightarrow Y$ be onto homomorphism. For a fuzzy set μ in Y, μ is a normal fuzzy weak filter of Y if and only if the mapping $\mu^f : X \longrightarrow [0, 1]$ such that $\mu^f(x) = \mu(f(x))$ for all $x \in X$ is a normal fuzzy weak filter of X.*

Proof Assume that μ is a normal fuzzy weak filter of Y. By Theorem 13.2.2, we get that μ^f is a fuzzy weak filter of X. We now show that μ^f is normal. Since μ is normal in Y, we get $\mu^f(1) = \mu(f(1)) = \mu(1') = 1$. Hence μ^f is a normal fuzzy weak filter of X.

Conversely, assume that μ^f is a normal fuzzy weak filter of X. By Theorem 13.2.2, μ is a fuzzy weak filter of Y. Since μ^f is normal, we get $\mu(1') = \mu(f(1)) = \mu^f(1) = 1$. Therefore μ is a normal fuzzy weak filter of Y. □

Proposition 13.5.14 *Let μ and ν be two normal fuzzy weak filters of a BE-algebra X. Then $\mu \times \nu$ is a normal fuzzy weak filter of $X \times X$.*

Proof Suppose μ and ν be two normal fuzzy weak filters of a BE-algebra X. Then by Proposition 13.1.17, $\mu \times \nu$ is a fuzzy weak filter of $X \times X$. Now $(\mu \times \nu)(1, 1) = \min\{\mu(1), \nu(1)\} = \min\{1, 1\} = 1$. Therefore $\mu \times \nu$ is a normal fuzzy weak filter of $X \times X$. □

Theorem 13.5.15 *Let ν be a fuzzy set in a BE-algebra X and μ_ν the strongest fuzzy relation on X. If ν is a normal fuzzy weak filter of X, then μ_ν is a normal fuzzy weak filter of $X \times X$.*

Proof Assume that ν is a normal fuzzy weak filter of X. By Theorem 13.1.19, we get that μ_ν is a fuzzy weak filter of $X \times X$. Again $\mu_\nu(1, 1) = \min\{\nu(1), \nu(1)\} = \min\{1, 1\} = 1$. Therefore μ_ν is a normal fuzzy weak filter of $X \times X$. □

Exercise

1. Let F be a non-empty subset of a BE-algebra X. Define a fuzzy set $\mu_F : X \to [0, 1]$ as $\mu_F(x) = \alpha$ if $x \in F$ and $\mu_F(x) = 0$ if $x \notin F$ where $0 < \alpha < 1$ is fixed. If μ_F is a fuzzy filter in X, then prove that F is a filter in X.
2. Let μ be a fuzzy filter of a BE-algebra X with $Im(\mu) = \{\alpha_i \mid i \in \Delta\}$ and $\mathcal{F} = \{\mu_{\alpha_i} \mid i \in \Delta\}$ where Δ is an arbitrary indexed set. Then prove that $X = \bigcup_{i \in \Delta} \mu_{\alpha_i}$.
3. Let F be a filter of a BE-algebra X and let $1 \neq \alpha < \beta$ be elements in $[0, 1]$. If $\mu : X \to [0, 1]$ is a fuzzy set defined by $\mu(x) = \alpha$ if $x \in F$ and $\mu(x) = \beta$ if $x \notin F$, then prove that μ is a fuzzy filter of X.

4. For any onto homomorphism $f : X \to Y$ and any fuzzy set μ in Y, define a mapping $\mu^f : X \to [0, 1]$ such that $\mu^f(x) = \mu(f(x))$ for all $x \in X$. Prove that μ is a fuzzy filter in Y if and only if μ^f is a fuzzy filter in X.
5. If μ and ν are two fuzzy weak filters of a BE-algebra X, then prove that the direct product of fuzzy sets $\mu \times \nu$ is a fuzzy weak filter of $X \times X$.
6. Let μ be a fuzzy subset of a BE-algebra X. Prove that the intersection and union of any two fuzzy multiplications of the fuzzy weak filter μ of X is also a fuzzy weak filter of X.
7. Let μ be a fuzzy subset of a BE-algebra X. For any $\alpha, t \in (0, 1]$ with $t \geq \alpha$, define $U_\alpha^m(\mu, t) = \{x \in X \mid \mu(x) \geq \frac{\alpha}{t}\}$. Prove that μ_α^m is a fuzzy weak filter of X if and only if $U_\alpha^m(\mu, t)$ is a weak filter of X for all $t \in Im(\mu)$ with $t \geq \alpha$.
8. For any onto homomorphism $f : X \to Y$ and any fuzzy set μ in Y, define a mapping $\mu^f : X \to [0, 1]$ such that $\mu^f(x) = \mu(f(x))$ for all $x \in X$. If μ is an anti fuzzy weak filter of Y, then prove that μ^f is an anti fuzzy weak filter of X.
9. Let μ and ν be two fuzzy subsets in a BE-algebra X. If μ and ν are anti fuzzy weak filters of X, then prove that their Cartesian product $\mu \times \nu : X \times X \to [0, 1]$ defined by $(\mu \times \nu)(x, y) = \max\{\mu(x), \nu(y)\}$ for all $x, y \in X$ is also an anti fuzzy weak filter in $X \times X$.
10. Let μ and ν be two fuzzy weak filters of a BE-algebra X such that $\mu \subseteq \nu$. If μ is normal, then ν is also normal.
11. Let $\mathcal{ND}(X)$ be the class of all normal fuzzy weak filters of a BE-algebra X. Then prove that $\mathcal{ND}(X)$ is a complete semilattice with respect to set-theoretic intersection.
12. If μ is a fuzzy weak filter of a BE-algebra X, then prove that $(\mu^+)^+ = \mu^+$. Moreover, if μ is a normal fuzzy weak filter, then prove that $(\mu^+)^+ = \mu$.
13. Let X and Y be two BE-algebras and $f : X \to Y$ an onto homomorphism. If ν is a normal fuzzy weak filter of Y, then prove that $f^{-1}(\nu)$ is a normal fuzzy weak filter of X.
14. Let X and Y be two BE-algebras and $f : X \to Y$ a homomorphism. If μ is a fuzzy normal weak filter of X, then prove that $f(\mu)$ is a normal fuzzy weak filter of Y.

Chapter 14
Implicative Filters

In 2001, Turunen [233] proposed the notions of implicative filters and Boolean filters (Boolean deductive systems) and proved that implicative filters are equivalent to Boolean filters in BL-algebras. Boolean filters are important filters, because the quotient algebras induced by Boolean filters are Boolean algebras. Jun and Ahn [132] provided several degrees in defining a fuzzy implicative filter. Meng [171] introduced the concept of implicative ideals in BCK-algebras and investigated the relationship of it with the concepts of positive implicative ideals and commutative ideals. Meng et al. [179] and Mostafa [182] fuzzified the concept of implicative ideals in BCK-algebras independently. Also, Xu and Qin [245] proposed the notions of implicative filters and fuzzy implicative filters lattice implication algebras.

In this chapter, the notion of implication filters is introduced in BE-algebras. Some characterization theorems are derived for implicative filters of BE-algebras including an extension property for implicative filters. The relation between implicative filters and associative filters is established. It is observed that an implicative filter induces a congruence with respect to which the respective quotient algebra turned to be a self-distributive BE-algebra. The notion of weak implicative filters is introduced in BE-algebras. Some relations are established among weak implicative filters, implicative filters, and associative filters of BE-algebras. AN extension property is also derived for weak implicative filters. Fuzzification is also considered for implicative filters. A set of equivalent conditions is derived for a fuzzy filter of a BE-algebra to become a fuzzy implicative filter. The properties of fuzzy implicative filters are discussed in terms of homomorphic images and Cartesian products. The notion of ε-generalized weak implicative filters is generalized in BE-algebras. Theory of vague sets is applied to weak implicative filters of BE-algebras, and some relations are established among implicative filters, weak implicative filters, and associative filters. Theory of falling shadows is applied to weak implicative filters of BE-algebras.

S. R. Mukkamala, *A Course in BE-algebras*,
https://doi.org/10.1007/978-981-10-6838-6_14

14.1 Definition and Properties

In this section, the notion of implicative filters is introduced in BE-algebras. A set of equivalent conditions is derived for every filter of a BE-algebra to become an implicative filter. An extension property is obtained for implicative filters of BE-algebras.

Definition 14.1.1 Let $(X, *, 1)$ be a BE-algebra. A subset F of X is called an *implicative filter* if it satisfies the following conditions for all $x, y, z \in X$;

(IF1) $1 \in F$,
(IF2) $x * (y * z) \in F$ and $x * y \in F$ imply $x * z \in F$.

It can be easily observed that every filter is an implicative filter in a self-distributive BE-algebra. For, consider $x = 1$ in the above definition, yields that $y * z \in F$ and $y \in F$ imply that $z \in F$. Hence every implicative filter is a filter of a BE-algebra. But, in general, the converse of this consequence is not true in a BE-algebra. For consider the following Example.

Example 14.1.2 Let $X = \{1, a, b, c, d\}$. Define a binary operation $*$ on X as follows:

$*$	1	a	b	c	d
1	1	a	b	c	d
a	1	1	b	c	b
b	1	a	1	b	a
c	1	a	1	1	a
d	1	1	1	b	1

It is easy to observe that $(X, *, 1)$ is a BE-algebra. Clearly $F = \{1, a\}$ is a filter but not an implicative filter of X, because of $b * (d * c) = b * b = 1 \in F$ and $b * d = a \in F$ but $b * c = b \notin F$.

However, in the following theorem, a sufficient condition is derived for every filter of a transitive BE-algebra to become an implicative filter.

Theorem 14.1.3 *Let F be a filter of a transitive BE-algebra X. Then F is an implicative filter of X if it satisfies the following condition for all $x, y, z \in F$;*

$$x * (x * z) \in F \text{ imply } x * z \in F.$$

Proof Assume that F is a filter of X. Let $x, y, z \in X$ be such that $x * (y * z) \in F$ and $x * y \in F$. Since X is transitive, we get the following:

$$x * (y * z) = y * (x * z) \leq (x * y) * (x * (x * z)).$$

Since F is a filter and $x * (y * z) \in F$, we get $(x * y) * (x * (x * z)) \in F$. Since $x * y \in F$, it yields $x * (x * z) \in F$. By the assumed condition, we get $x * z \in F$. Hence F is an implicative filter. □

In the following theorem, a set of equivalent conditions is derived for every filter of a transitive BE-algebra to become an implicative filter.

Theorem 14.1.4 *Let X be a transitive BE-algebra and F a filter of X. For all $x, y, z \in X$, the following conditions are equivalent:*

(1) *F is an implicative filter,*
(2) *Each stabilizer $\langle a, F\rangle$ is a filter,*
(3) *$x * (x * y) \in F$ implies $x * y \in F$,*
(4) *$x * (y * z) \in F$ implies $(x * y) * (x * z) \in F$.*

Proof The equivalency among (1), (2), and (3) is obtained in [209]. We now prove the remaining.

(1) $\Rightarrow$ (2) : Assume that F is an implicative filter of X. Let $x \in \langle a, F\rangle$ and $x * y \in \langle a, F\rangle$ for all $a \in X$. Then we get $a * x \in F$ and $a * (x * y) \in F$. Since F is an implicative filter, it implies that $a * y \in F$. Hence $y \in \langle a, F\rangle$. Therefore $\langle a, F\rangle$ is a filter of X.

(2) $\Rightarrow$ (3) : Assume that the condition (2) holds. Let $x * (x * y) \in F$ for all $x, y \in X$. Then $x * y \in \langle x, F\rangle$. Since $x * x = 1 \in F$, it yields that $x \in \langle x, F\rangle$. Since $\langle x, F\rangle$ is a filter, it implies $y \in \langle x, F\rangle$. Hence $x * y \in F$. It proves the condition (3). □

Corollary 14.1.5 *In a self-distributive BE-algebra, every filter is an implicative filter.*

Proposition 14.1.6 *Every associative filter of a BE-algebra is an implicative filter.*

Proof Let F be an associative filter of a BE-algebra X. Clearly F is a filter of X. Let $x, y, z \in X$ be such that $x * (y * z) \in F$ and $x * y \in F$. Since F is associative, we get that $z \in F$. Since F is a filter, it yields that $x * z \in F$. Therefore F is an implicative filter of X. □

The converse of the above proposition is not true. That is, every implicative filter of a BE-algebra need not be an associative filter. It can be observed in the following example.

Example 14.1.7 Let $X = \{1, a, b, c, d\}$. Define a binary operation $*$ on X as follows:

$*$	1	a	b	c	d
1	1	a	b	c	d
a	1	1	b	c	d
b	1	a	1	c	c
c	1	1	b	1	b
d	1	1	1	1	1

It is easy to observe that $(X, *, 1)$ is a BE-algebra. Clearly $F = \{1, a, b\}$ is an implicative filter but not an associative filter of X, because of $d * (b * c) = d * c = 1 \in F$ and $d * b = 1 \in F$ but $c \notin F$.

However, in the following theorem, a necessary and sufficient condition is derived for an implicative filter of a BE-algebra to become an associative filter.

Theorem 14.1.8 *Let F be an implicative of a BE-algebra X. Then F is associative if and only if it satisfies the condition: $x * y \in F$ implies $y \in F$ for all $x, y \in X$.*

Proof Let F be an implicative filter of a BE-algebra X. Then clearly F is a filter of X. Assume that F is an associative filter of X. Let $x, y \in X$. Suppose $x * y \in F$. It gives $x * (1 * y) = x * y \in F$. Since F is associative, we get $y = 1 * y = (x * 1) * y \in F$. Conversely, assume that F satisfies the condition. Let $x * (y * z) \in F$ and $x * y \in F$ for all $x, y, z \in X$. Since F is implicative, it yields $x * z \in F$. From the assumed condition, we get $z \in F$. Thus F is an associative filter of X. □

Let F be a filter of a self-distributive BE-algebra X. Then the relation θ_F defined by

$$(x, y) \in \theta_F \text{ if and only if } x * y \in F \text{ and } y * x \in F \text{ for all } x, y \in X$$

is a congruence [Chap. 2, Theorem 2.3.20] on X. Also the congruence class $[1]_\theta$ is the smallest congruence containing F. It can be easily observed that the quotient algebra $X_{/\theta} = \{[x]_\theta \mid x \in X\}$ is a self-distributive BE-algebra whenever X is a self-distributive BE-algebra. Now, in the following, a necessary and sufficient condition is derived for a filter of a self-distributive BE-algebra to become an implicative filter with the help of the above congruence. For this, we first need the following lemma whose proof is clear from Theorem 14.1.3.

Lemma 14.1.9 *Let F be a filter of a transitive BE-algebra X. Then F is an implicative filter if and only if $(x * (y * z)) * ((x * y) * (x * z)) \in F$ for all $x, y, z \in X$.*

Theorem 14.1.10 *Let F be a filter of a commutative BE-algebra X. Then F is an implicative filter if and only if $X_{/\theta}$ is self-distributive.*

Proof Assume that F is an implicative filter of X. Let $x, y, z \in X$. Then by the above lemma $(x * (y * z)) * ((x * y) * (x * z)) \in F$. Since X is commutative, we get $((x * y) * (x * z)) * (x * z) = ((x * z) * (x * y)) * (x * y)$. Hence, we get the following consequence:

$$\begin{aligned}
((x * y) * (x * z)) * (x * (y * z)) &= ((x * y) * (x * z)) * (y * (x * z)) \\
&= y * (((x * y) * (x * z)) * (x * z)) \\
&= y * (((x * z) * (x * y)) * (x * y)) \\
&= ((x * z) * (x * y)) * (y * (x * y)) \\
&= ((x * z) * (x * y)) * 1 \\
&= 1 \in F.
\end{aligned}$$

Hence $(x * (y * z), (x * y) * (x * z)) \in \theta_F$, which implies that $[x]_\theta * ([y]_\theta * [z]_\theta) = [x * (y * z)]_\theta = [(x * y) * (x * z)]_\theta = ([x]_\theta * [y]_\theta) * ([x]_\theta * [z]_\theta)$. Therefore $X_{/\theta}$ is self-distributive.

Conversely, assume that $X_{/\theta}$ is a self-distributive BE-algebra. Let $x, y, z \in X$ be such that $x * (y * z) \in F$ and $x * y \in F$. Then $x * (y * z) \in [1]_\theta$ and $x * y \in [1]_\theta$, which imply that $[x * (y * z)]_\theta = [1]_\theta$ and $[x * y]_\theta = [1]_\theta$. Hence, we get the following consequence:

$$\begin{aligned}[x * z]_\theta &= [x]_\theta * [z]_\theta \\ &= [1]_\theta * ([x]_\theta * [z]_\theta) \\ &= ([x]_\theta * [y]_\theta) * ([x]_\theta * [z]_\theta) \\ &= [x]_\theta * ([y]_\theta * [z]_\theta) \\ &= [x * (y * z)]_\theta \\ &= [1]_\theta.\end{aligned}$$

Hence $x * z = 1 * (x * z) \in F$. Therefore F is an implicative filter of X. □

14.2 Weak Implicative Filters of BE-algebras

In this section, the notion of weak implicative filters is introduced in BE-algebras. Some interconnections among weak implicative filters, associative filters, and implicative filters are established. Extension property for weak implicative filters of BE-algebra is also proved.

Definition 14.2.1 Let $(X, *, 1)$ be a BE-algebra. A subset F of X is called a *weak implicative filter* if it satisfies the following conditions for all $x, y, z \in F$.

(WIF1) $1 \in F$,
(WIF2) $x * (y * z) \in F$ and $x * y \in F$ imply $x * (x * z) \in F$.

Proposition 14.2.2 *Every weak implicative filter of a BE-algebra is a filter.*

Proof Let F be a weak implicative filter of a BE-algebra X. Let $x \in F$ and $x * y \in F$ for $x, y \in X$. Then, it implies $1 * (x * y) = x * y \in F$ and $1 * x = x \in F$. Since F is a weak implicative filter, it yields that $y = 1 * (1 * y) \in F$. Therefore F is a filter of X. □

Proposition 14.2.3 *Every implicative filter of a BE-algebra is a weak implicative filter.*

Proof Let $(X, *, 1)$ be a BE-algebra and F an implicative filter of a X. Then clearly F is a filter of X. Let $x, y, z \in X$ be such that $x * (y * z) \in F$ and $x * y \in F$. Since F is an implicative filter, we get that $x * z \in F$. Since F is a filter and $x * z \le x * (x * z)$, it immediately infers $x * (x * z) \in F$. Therefore F is a weak implicative filter of X. □

Since every associative filter of a BE-algebra is an implicative filter(by Proposition 14.1.6), the following corollary is an immediate consequence of the above proposition.

Corollary 14.2.4 *Every associative filter of a BE-algebra is a weak implicative filter.*

The converse of the above corollary is not true. It can be seen from Example 14.1.2. In the BE-algebra X of the example, the set $F = \{1, a, b\}$ is observed as a weak implicative filter but not an associative filter of X, because of $d * (b * c) = d * c = 1 \in F$ and $d * b = 1 \in F$ but $c \notin F$. However, in the following theorem, a necessary and sufficient condition is derived for every weak implicative filter of a BE-algebra to become an associative filter.

Theorem 14.2.5 *A weak implicative filter of a BE-algebra X is an associative filter if and only if it satisfies the condition: $x * (x * y) \in F$ implies $y \in F$ for all $x, y \in X$.*

Proof Let F be a weak implicative filter of X. Then clearly F is a filter. Assume that F is an associative filter. Let $x, y \in X$ be such that $x * (x * y) \in F$. Since F is associative, we get $y = 1 * y = (x * x) * y \in F$. Conversely, assume that F satisfies the condition. Let $x * (y * z) \in F$ and $x * y \in F$ for all $x, y, z \in X$. Since F is weak implicative, it yields $x * (x * z) \in F$. Hence, from the assumed condition, we get that $z \in F$. Therefore F is an associative filter of X. □

It is observed in Proposition 14.2.3 that every implicative filter is a weak implicative filter. But the converse is not true. For, consider the subset $F = \{1, a\}$ of the BE-algebra presented in Example 14.1.2. Clearly F is a weak implicative filter but not an implicative filter. However, in the following theorem, a set of equivalent conditions is derived for every weak implicative filter of a BE-algebra to become an implicative filter.

Theorem 14.2.6 *Let F be a weak implicative filter of a BE-algebra X. Then F is an implicative filter if and only if $x * (x * y) \in F \Leftrightarrow x * y \in F$ for all $x, y \in X$.*

Proof Let F be a weak implicative filter of X. Then clearly F is a filter. Assume that F is an implicative filter. Let $x * (x * y) \in F$. Clearly $x * x = 1 \in F$. Since F is an implicative filter, it yields that $x * y \in F$. Again, let $x * y \in F$. Since F is a filter, it is clear that $x * (x * y) \in F$. Conversely, assume that F satisfies the given condition. Let $x, y, z \in X$ be such that $x * (y * z) \in F$ and $x * y \in F$. Since F is a weak implicative filter, we get that $x * (x * z) \in F$. Then by the assumed condition, it yields that $x * z \in F$. Therefore F is an implicative filter of X. □

Corollary 14.2.7 *In a simple BE-algebra, every weak implicative filter is an implicative filter.*

Theorem 14.2.8 *In a transitive BE-algebra, every filter is a weak implicative filter.*

Proof Let $(X, *, 1)$ be a transitive BE-algebra and F a filter of X. Let $x, y, z \in X$ be such that $x * (y * z) \in F$ and $x * y \in F$. Since X is transitive, we get $y * z \leq (x * y) * (x * z)$. Hence, we get

$$\begin{aligned} x * (y * z) &\leq x * ((x * y) * (x * z)) \\ &= (x * y) * (x * (x * z)) \end{aligned}$$

Since F is a filter and $x * (y * z) \in F$, we get $(x * y) * (x * (x * z)) \in F$. Again, since F is a filter and $x * y \in F$, it yields that $x * (x * z) \in F$. Therefore F is a weak implicative filter of X. □

In the following theorem, a sufficient condition is derived for every filter of a BE-algebra to become a weak implicative filter.

Theorem 14.2.9 *Let X be a BE-algebra and F a filter of X. Then F is a weak implicative filter if it satisfies the following condition for all $x, y, z \in X$:*

$$x * (y * z) \in F \textit{ implies } x * ((x * y) * (x * z)) \in F.$$

Proof Let F be a filter of X. Let $x, y, z \in X$ and $x * (y * z) \in F$ implies $x * ((x * y) * (x * z)) \in F$. Let $x * (y * z) \in F$ and $x * y \in F$ for all $x, y, z \in X$. Then by the assumed condition, we get

$$(x * y) * (x * (x * z)) = x * ((x * y) * (x * z)) \in F.$$

Since F is a filter and $x * y \in F$, we get $x * (x * z) \in F$. Hence F is a weak implicative filter. □

Theorem 14.2.10 (Extension property for weak implicative filters) *Let F and G be two filters of a BE-algebra $(X, *, 1)$ such that $F \subseteq G$. If F is a weak implicative filter, then so is G.*

Proof Suppose F is a weak implicative filter of X. Let $x * (y * z) \in G$. Then $x * (y * ((x * (y * z)) * z)) = (x * (y * z)) * (x * (y * z)) = 1 \in F$. Since F is a weak implicative filter, by Theorem 14.1.3, it immediately infers that $x * ((x * y) * (x * ((x * (y * z)) * z))) \in F$. Hence

$$\begin{aligned}(x * (y * z)) * (x * ((x * y) * (x * z))) &= x * ((x * (y * z)) * ((x * y) * (x * z)))\\ &= x * ((x * y) * ((x * (y * z)) * (x * z)))\\ &= x * ((x * y) * (x * ((x * (y * z)) * z))) \in F \subseteq G.\end{aligned}$$

Since G is a filter and $x * (y * z) \in G$, we get that $x * ((x * y) * (x * z)) \in G$. Thus by Theorem 14.1.3, it concludes that G is a weak implicative filter of X. This completes the proof. □

For any congruence θ on a BE-algebra X, it is already observed that the quotient algebra $X_{/\theta}$ is a BE-algebra with respect to the operation defined by $[x]_\theta * [y]_\theta = [x * y]_\theta$ for all $x, y \in X$, where $X_{/\theta} = \{[x]_\theta \mid x \in X\}$. Then clearly $\phi : X \to X_{/\theta}$ is a natural homomorphism.

Theorem 14.2.11 *Let θ be a congruence on X and F a non-empty subset of X. Then F is a weak implicative filter of X if and only if $\phi(F) = F_{/\theta}$ is a weak implicative filter in $X_{/\theta}$.*

Proof Assume that F is a weak implicative filter of X. Since $1 \in F$, we get that $[1]_\theta \in F_{/\theta}$. Let $[x]_\theta, [y]_\theta, [z]_\theta \in X_\theta$ be such that $[x]_\theta * ([y]_\theta * [z]_\theta) \in F_{/\theta}$ and $[x]_\theta * [y]_\theta \in F_\theta$. Then $[x * (y * z)]_\theta \in F_\theta$ and $[x * y]_\theta \in F_\theta$. Hence $x * (y * z) \in F$ and $x * y \in F$. Since F is a weak implicative filter of X, it implies that $x * (x * z) \in F$. Hence $[x]_\theta * ([x]_\theta * [z]_\theta = [x * (x * z)]_\theta \in F_{/\theta}$. Therefore $F_{/\theta}$ is a weak implicative filter of $X_{/\theta}$.

Conversely, assume that $F_{/\theta}$ is a weak implicative filter of $X_{/\theta}$. Since $[1]_\theta \in F_{/\theta}$, we get that $1 \in F$. Let $x * (y * z) \in F$ and $x * y \in F$. Then $[x]_\theta * ([y]_\theta * [z]_\theta) = [x * (y * z)]_\theta \in F_{/\theta}$ and $[x]_\theta * [y]_\theta = [x * y]_\theta \in F_{/\theta}$. Since $F_{/\theta}$ is a weak implicative filter of $X_{/\theta}$, we get $[x * (x * z)]_\theta = [x]_\theta * ([x]_\theta * [z]_\theta) \in F_{/\theta}$. Hence $x * (x * z) \in F$. Therefore F is a weak implicative filter of X. □

14.3 Fuzzification of Implicative Filters

In this section, the fuzzification of implicative filters is discussed in BE-algebras. Some characterization theorems for fuzzy implicative filters of BE-algebras are derived. Extension property for fuzzy implicative filters of transitive BE-algebras is obtained.

Definition 14.3.1 A fuzzy set μ of a BE-algebra X is said to be a *fuzzy implicative filter* if it satisfies the following conditions:

(1) $\mu(1) \geq \mu(x)$,
(2) $\mu(x * z) \geq \min\{\mu(x * (y * z)), \mu(x * y)\}$ for all $x, y, z \in X$.

If we replace x of Definition 14.1.1 by 1, then it is easily observed that every implicative filter is a filter as well as every fuzzy implicative filter is a fuzzy filter. However, every fuzzy filter need not be a fuzzy implicative filter as shown in the following example.

Example 14.3.2 Let $X = \{1, a, b, c, d\}$. Define a binary operation $*$ on X as follows:

$*$	1	a	b	c	d
1	1	a	b	c	d
a	1	1	b	c	b
b	1	a	1	b	a
c	1	a	1	1	a
d	1	1	1	b	1

Then it can be easily verified that $(X, *, 1)$ is a BE-algebra. Define a fuzzy set μ on X as follows:

$$\mu(x) = \begin{cases} 0.9 & \text{if } x = a, 1 \\ 0.2 & \text{otherwise} \end{cases}$$

for all $x \in X$. Then clearly μ is a fuzzy filter of X, but μ is not a fuzzy implicative filter of X since $\mu(b * c) \ngeq min\{\mu(b * (d * c)), \mu(b * d)\}$.

Theorem 14.3.3 *Every fuzzy filter of a self-distributive BE-algebra is a fuzzy implicative filter.*

Proof Let F be a fuzzy filter of a self-distributive BE-algebra X. Clearly $\mu(1) \geq \mu(x)$ for all $x \in X$. Let $x, y, z \in X$. Since X is self-distributive, we get $\mu(x * z) \geq \min\{\mu((x * y) * (x * z)), \mu(x * y)\} = \min\{\mu(x * (y * z)), \mu(x * y)\}$. Therefore μ is a fuzzy implicative filter in X. □

However, in the following theorem, a sufficient condition is derived for every fuzzy filter of a transitive BE-algebra to become a fuzzy implicative filter.

Theorem 14.3.4 *A fuzzy filter μ of a transitive BE-algebra X is a fuzzy implicative filter if it satisfies $\mu(y * z) \geq \min\{\mu(x), \mu(x * (y * (y * z)))\}$ for all $x, y, z \in X$.*

Proof Assume that μ is a fuzzy filter of X satisfying the given condition. Let $x, y, z \in X$. Since X is a transitive BE-algebra, we get $\mu(x * (y * z)) = \mu(y * (x * z)) \leq \mu((x * y) * (x * (x * z)))$. Hence by assumed condition, we get $\mu(x * z) \geq \min\{\mu(x * y), \mu((x * y) * (x * (x * z)))\} \geq \min\{\mu(x * y), \mu(x * (y * z))\}$. Therefore μ is a fuzzy implicative filter in X. □

In the following theorem, we derive a set of equivalent conditions for every fuzzy filter of a transitive BE-algebra to become a fuzzy implicative filter.

Theorem 14.3.5 *Let μ be a fuzzy filter of a transitive BE-algebra X. Then the following conditions are equivalent.*

(1) *μ is a fuzzy implicative filter in X;*
(2) *$\mu(x * y) = \mu(x * (x * y))$ for all $x, y \in X$;*
(3) *$\mu((x * y) * (x * z)) \geq \mu(x * (y * z))$ for all $x, y, z \in X$.*

Proof (1) $\Rightarrow$ (2) : Assume that μ is a fuzzy implicative filter in X. Let $x, y \in X$. Since $x * y \leq x * (x * y)$, we get $\mu(x * y) \leq \mu(x * (x * y))$. Since μ is a fuzzy filter in X, we get that

$$\begin{aligned}\mu(x * y) &\geq \min\{\mu(x * (x * y)), \mu(x * x)\}\\ &= \min\{\mu(x * (x * y)), \mu(1)\}\\ &= \mu(x * (x * y)).\end{aligned}$$

Hence $\mu(x * y) \geq \mu(x * (x * y))$. Therefore $\mu(x * y) = \mu(x * (x * y))$.
(2) $\Rightarrow$ (3) : Let $x, y, z \in X$. Since X is transitive, we get $y * z \leq (x * y) * (x * z)$ and hence we get that $x * (y * z) \leq x * ((x * y) * (x * z))$. Hence $\mu(x * (y * z)) \leq \mu(x * ((x * y) * (x * z)))$. Now, we get

$$\begin{aligned}\mu((x * y) * (x * z)) &= \mu(x * ((x * y) * z))\\ &= \mu(x * (x * ((x * y) * z)) && \text{by (2)}\\ &= \mu(x * ((x * y) * (x * z))\\ &\geq \mu(x * (y * z)).\end{aligned}$$

(3) $\Rightarrow$ (1) : Assume the condition (3). Let $x, y, z \in X$. Then $\mu(x * z) \geq \min\{\mu((x * y) * (x * z)), \mu(x * y)\} \geq \min\{\mu(x * (y * z)), \mu(x * y)\}$. Therefore μ is a fuzzy implicative filter in X. □

Proposition 14.3.6 *A fuzzy set μ of a BE-algebra X is a fuzzy implicative filter in X if and only if for each $\alpha \in [0, 1]$, the level subset μ_α is an implicative filter in X, when $\mu_\alpha \neq \emptyset$.*

Proof Assume that μ is a fuzzy implicative filter of X. Then $\mu(1) \geq \mu(x)$ for all $x \in X$. In particular, $\mu(1) \geq \mu(x) \geq \alpha$ for all $x \in \mu_\alpha$. Hence $1 \in \mu_\alpha$. Let $x * (y * z), x * y \in \mu_\alpha$. Then $\mu(x * (y * z)) \geq \alpha$ and $\mu(x * y) \geq \alpha$. Since μ is a fuzzy implicative filter, we get $\mu(x * z) \geq \min\{\mu(x * (y * z)), \mu(x * y)\} \geq \alpha$. Thus $x * z \in \mu_\alpha$. Therefore μ_α is an implicative filter in X.

Conversely, assume that μ_α is an implicative filter of X for each $\alpha \in [0, 1]$ with $\mu_\alpha \neq \emptyset$. Suppose there exists $x_0 \in X$ such that $\mu(1) < \mu(x_0)$. Again, let $\alpha_0 = \frac{1}{2}(\mu(1) + \mu(x_0))$. Then $\mu(1) < \alpha_0$ and $0 \leq \alpha_0 < \mu(x_0) \leq 1$. Hence $x_0 \in \mu_{\alpha_0}$ and $\mu_{\alpha_0} \neq \emptyset$. Since μ_{α_0} is an implicative filter in X, we get $1 \in \mu_{\alpha_0}$ and hence $\mu(1) \geq \alpha_0$, which is a contradiction. Therefore $\mu(1) \geq \mu(x)$ for all $x \in X$. Let $x, y, z \in X$ be such that $\mu(x * (y * z)) = \alpha_1$ and $\mu(x * y) = \alpha_2$. Then $x * (y * z) \in \mu_{\alpha_1}$ and $x * y \in \mu_{\alpha_2}$. Without loss of generality, assume that $\alpha_1 \leq \alpha_2$. Clearly $\mu_{\alpha_2} \subseteq \mu_{\alpha_1}$. Hence $x * y \in \mu_{\alpha_1}$. Since μ_{α_1} is implicative, we get $x * z \in \mu_{\alpha_1}$. Thus $\mu(x * z) \geq \alpha_1 = \min\{\alpha_1, \alpha_2\} = \min\{\mu(x * (y * z)), \mu(x * y)\}$. Therefore μ is a fuzzy implicative filter of X. □

Theorem 14.3.7 *Let F be an implicative filter of a BE-algebra. Then there exists a fuzzy implicative filter μ of X such that $\mu_\alpha = F$ for some $\alpha \in (0, 1)$.*

Proof Let μ be a fuzzy set in a BE-algebra X defined by

$$\mu(x) = \begin{cases} \alpha & \text{if } x \in F \\ 0 & \text{otherwise} \end{cases}$$

where α is a fixed number ($0 < \alpha < 1$). Since $1 \in F$, we get $\mu(1) = \alpha \geq \mu(x)$ for all $x \in X$. Let $x, y, z \in X$. Suppose $x * (y * z), x * y \in F$. Since F is an implicative filter, we get $x * z \in F$. Then $\mu(x * y) = \mu(x * (y * z)) = \mu(x * z) = \alpha$. Hence $\mu(x * z) \geq \min\{\mu(x * (y * z)), \mu(x * y)\}$. Suppose $x * (y * z) \notin F$ and $x * y \notin F$. Then $\mu(x * y) = \mu(x * (y * z)) = 0$. Hence $\mu(x * z) \geq \min\{\mu(x * (y * z)), \mu(x * y)\}$. If exactly one of $x * (y * z)$ and $x * y$ is in F, then exactly one of $\mu(x * (y * z))$ and $\mu(x * y)$ is equal to 0. Hence $\mu(x * z) \geq \min\{\mu(x * (y * z)), \mu(x * y)\}$. By summarizing the above results, we get $\mu(x * z) \geq \min\{\mu(x * (y * z)), \mu(x * y)\}$ for all $x, y, z \in X$. Therefore μ is a fuzzy implicative filter of X. Clearly $\mu_\alpha = F$. □

Theorem 14.3.8 *Let μ be a fuzzy implicative filter of a BE-algebra X. Then two level implicative filters μ_{α_1} and μ_{α_2}(with $\alpha_1 < \alpha_2$) of μ are equal if and only if there is no $x \in X$ such that $\alpha_1 \leq \mu(x) < \alpha_2$.*

Proof Assume that $\mu_{\alpha_1} = \mu_{\alpha_2}$ for $\alpha_1 < \alpha_2$. Suppose there exists some $x \in X$ such that $\alpha_1 \leq \mu(x) < \alpha_2$. Then μ_{α_2} is a proper subset of μ_{α_1}, which is impossible. Conversely, assume that there is no $x \in X$ such that $\alpha_1 \leq \mu(x) < \alpha_2$. Since $\alpha_1 < \alpha_2$, we get $\mu_{\alpha_2} \subseteq \mu_{\alpha_1}$. If $x \in \mu_{\alpha_1}$, then $\mu(x) \geq \alpha_1$. Hence by assumed condition, we get $\mu(x) \geq \alpha_2$. Hence $x \in \mu_{\alpha_2}$ and so $\mu_{\alpha_1} \subseteq \mu_{\alpha_2}$. Therefore $\mu_{\alpha_1} = \mu_{\alpha_2}$. □

Theorem 14.3.9 *Let μ be a fuzzy implicative filter of X with $Im(\mu) = \{\alpha_i \mid i \in \Delta\}$ and $\mathcal{F} = \{\mu_{\alpha_i} \mid i \in \Delta\}$ where Δ is an arbitrary indexed set. If μ attains its infimum on all implicative filters of X, then $\mathcal{F}$ contains all level implicative filters of μ.*

Proof Suppose μ attains its infimum on all implicative filters of X. Let μ_α be a level implicative filter of μ. If $\alpha = \alpha_i$ for some $i \in \Delta$, then clearly $\mu_\alpha \in \mathcal{F}$. Assume that $\alpha \neq \alpha_i$ for all $i \in \Delta$. Then there exists no $x \in X$ such that $\mu(x) = \alpha$. Let $F = \{x \in X \mid \mu(x) > \alpha\}$. Clearly $1 \in F$. Let $x, y, z \in X$ be such that $x * y \in F$ and $x * (y * z) \in F$. Then $\mu(x * y) > \alpha$ and $\mu(x * (y * z)) > \alpha$. Since μ is a fuzzy implicative filter in X, we get $\mu(x * z) \geq \min\{\mu(x * (y * z)), \mu(x * y)\} > \alpha$. Hence $\mu(x * z) > \alpha$, which implies that $x * z \in F$. Therefore F is an implicative filter of X. By the hypothesis, there exists $y \in F$ such that $\mu(y) = \inf\{\mu(x) \mid x \in X\}$. Hence $\mu(y) \in Im(\mu)$, which yields that $\mu(y) = \alpha_i$ for some $i \in \Delta$. It is clear that $\alpha_i \geq \alpha$. Hence, by assumption, we get $\alpha_i > \alpha$. Thus there exists no $x \in X$ such that $\alpha \leq \mu(x) < \alpha_i$. Hence by above theorem, we get $\mu_\alpha = \mu_{\alpha_i}$. Therefore $\mu_\alpha \in \mathcal{F}$. This completes the proof. □

Theorem 14.3.10 (Extension property) *Let μ and ν be two fuzzy filters of a BE-algebra X with $\mu \leq \nu$ and $\mu(1) = \nu(1)$. If μ is a fuzzy implicative filter of X, then so is ν.*

Proof Assume that μ and ν are two fuzzy filters of a BE-algebra X such that $\mu \leq \nu$ and $\mu(1) = \nu(1)$. Suppose μ is a fuzzy implicative filter of X. Let $x, y, z \in X$. Since $\mu \subseteq \nu$ and μ is a fuzzy implicative filter, by Theorem 14.3.4, we can obtain the following consequence.

$$\begin{aligned}
\nu((x * y) * (x * ((x * (y * z)) * z))) &\geq \mu((x * y) * (x * ((x * (y * z)) * z))) \\
&\geq \mu(x * (y * ((x * (y * z)) * z))) \\
&= \mu(x * ((x * (y * z)) * (y * z))) \\
&= \mu((x * (y * z)) * (x * (y * z))) \\
&= \mu(1) \\
&= \nu(1).
\end{aligned}$$

whence $\nu((x * (y * z)) * ((x * y) * (x * z))) = \nu((x * y) * ((x * (y * z)) * (x * z))) = \nu((x * y) * (x * ((x * (y * z)) * z))) \geq \nu(1)$. Hence $\nu((x * (y * z)) * ((x * y) * (x * z))) = \nu(1)$. Since ν is a fuzzy filter, by Lemma 13.1.3, we get that $\nu((x * y) * (x * z)) \geq \nu(x * (y * z))$. By Theorem 14.3.4, it yields that ν is a fuzzy implicative filter in X. □

Definition 14.3.11 Let $f : X \longrightarrow Y$ be a homomorphism of BE-algebras, and μ is a fuzzy set in Y. Then define a mapping $\mu^f : X \longrightarrow [0, 1]$ such that $\mu^f(x) = \mu(f(x))$ for all $x \in X$.

Clearly the above mapping μ^f is well-defined and a fuzzy set in X.

Theorem 14.3.12 *Let $f : X \longrightarrow Y$ be an onto homomorphism and μ a fuzzy set in Y. Then μ is a fuzzy implicative filter in Y if and only if μ^f is a fuzzy implicative filter in X.*

Proof Assume that μ is a fuzzy implicative filter of Y. For any $x \in X$, we have $\mu^f(1) = \mu(f(1)) = \mu(1') \geq \mu(f(x)) = \mu^f(x)$. Let $x, y, z \in X$. Then

$$\begin{aligned}
\mu^f(x * z) &= \mu(f(x * z)) \\
&= \mu(f(x) * f(z)) \\
&\geq \min\{\mu(f(x) * (f(y) * f(z))), \mu(f(x) * f(y))\} \\
&= \min\{\mu(f(x * (y * z))), \mu(f(x * y))\} \\
&= \min\{\mu^f(x * (y * z)), \mu^f(x * y)\}
\end{aligned}$$

Hence μ^f is a fuzzy implicative filter of X. Conversely, assume that μ^f is a fuzzy implicative filter of X. Let $x \in Y$. Since f is onto, there exists $y \in X$ such that $f(y) = x$. Then $\mu(1') = \mu(f(1)) = \mu^f(1) \geq \mu^f(y) = \mu(f(y)) = \mu(x)$. Let $x, y, z \in Y$. Then there exists $a, b, c \in X$ such that $f(a) = x$, $f(b) = y$, and $f(c) = z$. Hence we get

$$\begin{aligned}
\mu(x * z) &= \mu(f(a) * f(c)) \\
&= \mu(f(a * c)) \\
&= \mu^f(a * c) \\
&\geq \min\{\mu^f(a * (b * c)), \mu^f(a * b)\} \\
&= \min\{\mu(f(a) * (f(b) * f(c)), \mu(f(a) * f(b))\} \\
&= \min\{\mu(x * (y * z)), \mu(x * y)\}
\end{aligned}$$

Therefore μ is a fuzzy implicative filter in Y. □

14.4 Generalized Fuzzy Weak Implicative Filters

In this section, the notion of ε-generalized weak implicative filters is introduced in BE-algebras. Some properties of generalized weak implicative filters are investigated.

Definition 14.4.1 Let $\varepsilon \in \mathbb{R}^+$. A fuzzy subset μ of a BE-algebra X is said to be an *ε-generalized fuzzy weak implicative filter* of X if it satisfies the following:

(GFWIF1) $\mu(1) \geq \min\{\mu(x), \varepsilon\}$ for all $x \in X$,
(GFWIF2) $\mu(x * (x * z)) \geq \min\{\mu(x * (y * z)), \mu(x * y), \varepsilon\}$ for all $x, y \in X$.

If $\varepsilon \geq 1$, then every ε-generalized fuzzy weak implicative filter of a BE-algebra X is a fuzzy weak implicative filter. If $\varepsilon < 0$, then the above definition can be modified as given below and named as ε-generalized anti fuzzy weak implicative filter of X.

Definition 14.4.2 Let $\varepsilon \in \mathbb{R}^-$. A fuzzy subset μ of a BE-algebra X is said to be an *ε-generalized fuzzy weak implicative filter* of X if it satisfies the following:

(GAFWIF1) $\mu(1) \leq \max\{\mu(x), \varepsilon\}$ for all $x \in X$,
(GAFWIF2) $\mu(x * (x * z)) \leq \max\{\mu(x * (y * z)), \mu(x * y), \varepsilon\}$ for all $x, y \in X$.

Proposition 14.4.3 *Let $\varepsilon, \beta \in (0, 1]$ and $\beta \leq \varepsilon$. Then every ε-generalized fuzzy weak implicative filter of a BE-algebra X is a β-generalized fuzzy weak implicative filter.*

Proof Let $\varepsilon, \beta \in (0, 1]$ be such that $\beta \leq \varepsilon$. Suppose μ is a ε-generalized fuzzy weak implicative filter of X. Then $\mu(1) \geq \min\{\mu(x), \varepsilon\} \geq \min\{\mu(x), \beta\}$ for all $x \in X$. Let $x, y \in X$. Since μ is a β-generalized fuzzy weak implicative filter of X, we get

$$\begin{aligned}\mu(x * (x * z)) &\geq \min\{\mu(x * (y * z)), \mu(x * y), \varepsilon\} \\ &\geq \min\{\mu(x * (y * z)), \mu(x * y), \beta\}\end{aligned}$$

Therefore μ is a β-generalized fuzzy weak implicative filter of X. □

Theorem 14.4.4 *Let μ be an ε-generalized fuzzy weak implicative filter of a BE-algebra X and $\varepsilon \in (0, 1]$. Then $x \leq y$ implies* $\min\{\mu(x), \varepsilon\} \leq \mu(y)$ *for all $x, y \in X$.*

Proof Let $x, y \in X$. Suppose $x \leq y$. Then $x * y = 1$. Since μ is an ε-generalized fuzzy weak implicative filter, we get

$$\begin{aligned}\mu(y) &= \mu(1 * (1 * y)) \\ &\geq \min\{\mu(1 * (x * y)), \mu(1 * x), \varepsilon\} \\ &= \min\{\mu(x * y), \mu(x), \varepsilon\} \\ &= \min\{\mu(1), \mu(x), \varepsilon\} \\ &= \min\{\mu(x), \varepsilon\}.\end{aligned}$$

Hence $x \leq y$ implies $\min\{\mu(x), \varepsilon\} \leq \mu(y)$ for all $x, y \in X$. □

Theorem 14.4.5 *Let $\varepsilon, \beta \in (0, 1]$. A fuzzy subset μ of a BE-algebra X is an ε-generalized fuzzy weak implicative filter of X if it satisfies (GFWIF1) and any one of the following conditions:*

(1) $\mu(z) \geq \min\{\mu(x * (y * z)), \mu(x * y), \varepsilon\}$ *for all $x, y \in X$,*
(2) $\mu(x * z) \geq \min\{\mu(x * y), \mu(x * (y * z)), \varepsilon\}$ *for all $x, y \in X$,*
(3) $\mu(x * ((x * y) * (x * z))) \geq \min\{\mu(x * (y * z)), \varepsilon\}$ *for all $x, y \in X$.*

Proof (1) Let $x, y, z \in X$. Clearly $z \leq x * (x * z)$. Then by (1) and by the above theorem, we get

$$\begin{aligned}\mu(x * (x * z)) &\geq \min\{\mu(z), \varepsilon\}\\ &\geq \min\{\min\{\mu(x * (y * z)), \mu(x * y), \varepsilon\}, \varepsilon\}\\ &= \min\{\mu(x * (y * z)), \mu(x * y), \varepsilon\}.\end{aligned}$$

Therefore μ is an ε-generalized fuzzy weak implicative filter of X.
(2) Let $x, y, z \in X$. Since $x * z \leq x * (x * z)$, we get

$$\begin{aligned}\mu(x * (x * z)) &\geq \min\{\mu(x * z), \varepsilon\}\\ &\geq \min\{\min\{\mu(x * y), \mu(x * (y * z)), \varepsilon\}, \varepsilon\}\\ &= \min\{\mu(x * y), \mu(x * (y * z)), \varepsilon\}\end{aligned}$$

Therefore μ is an ε-generalized fuzzy weak implicative filter of X.
(3) Let $x, y, z \in X$. Assume that condition (3) holds. Then

$$\begin{aligned}\mu(x * (x * z)) &\geq \mu(1 * (1 * (x * (x * z))))\\ &\geq \min\{\mu(1 * (x * y)), \mu(1 * ((x * y) * (x * (x * z)))), \varepsilon\}\\ &= \min\{\mu(x * y), \mu(x * ((x * y) * (x * z))), \varepsilon\}\\ &= \min\{\mu(x * y), \min\{\mu(x * (y * z)), \varepsilon\}, \varepsilon\}\\ &= \min\{\mu(x * y), \mu(x * (y * z)), \varepsilon\}.\end{aligned}$$

Therefore μ is an ε-generalized fuzzy weak implicative filter of X. □

Definition 14.4.6 Let $\varepsilon \in (0, 1]$ and μ be a fuzzy subset of a BE-algebra X. For any $\lambda \in (0, 1]$, define the level subset $U(\mu, \lambda)$ of X as $U(\mu, \lambda) = \{x \in X \mid \mu(x) \geq \lambda\}$

It is clear that $U(\mu, \lambda) \subseteq U(\mu, \varepsilon)$ whenever $\lambda \geq \varepsilon$.

Theorem 14.4.7 *Let $\varepsilon \in (0, 1]$ and μ be a fuzzy subset of a BE-algebra X. Then μ is an ε-generalized fuzzy weak implicative filter of X if and only if the level set $U(\mu, \lambda)$ is a weak implicative filter of X for each $\lambda \in (0, 1]$ with $\lambda \leq \varepsilon$.*

Proof Assume that μ is an ε-generalized fuzzy weak implicative filter of X. Let $\lambda \in (0, 1]$ with $\lambda \leq \varepsilon$. Clearly $\mu(1) \geq \min\{\mu(x), \varepsilon\}$ for all $x \in X$. In particular, for $x \in U(\mu, \lambda)$, we get $\mu(1) \geq \min\{\mu(x), \varepsilon\} \geq \min\{\lambda, \varepsilon\} = \lambda$. Hence $1 \in U(\mu, \lambda)$. Let $x, y, z \in X$. Suppose $x * (y * z) \in U(\mu, \lambda)$ and $x * y \in U(\mu, \lambda)$. Then $\mu(x * (y * z)) \geq \lambda$ and $\mu(x * y) \geq \lambda$. Since μ is an ε-generalized fuzzy weak implicative filter of X, we get $\mu(x * (x * z)) \geq \min\{\mu(x * (y * z)), \mu(x * y), \varepsilon\} \geq \min\{\lambda, \lambda, \varepsilon\} = \lambda$. Hence $x * (x * z) \in U(\mu, \lambda)$. Therefore $U(\mu, \lambda)$ is a weak implicative filter of X.

Conversely, assume that $U(\mu, \lambda)$ is a weak implicative filter of X for all $\lambda \in (0, 1]$ with $\lambda \leq \varepsilon$. Suppose there exists $x_0 \in X$ such that $\mu(1) < \min\{\mu(x_0), \varepsilon\}$. Choose $\alpha_0 = \frac{1}{2}(\mu(1) + \min\{\mu(x_0), \varepsilon\})$. Then clearly $\mu(1) < \min\{\alpha_0, \varepsilon\}$ and $0 \leq$

$\alpha_0 \leq \min\{\mu(x_0), \varepsilon\} \leq \mu(x_0) \leq 1$. Hence $x_0 \in U(\mu, \alpha_0)$. Hence $U(\mu, \alpha_0) \neq \emptyset$. Since $U(\mu, \alpha_0)$ is a weak implicative filter, we get $1 \in U(\mu, \alpha_0)$. Hence $\mu(1) \geq \alpha_0$, which is a contradiction. Therefore $\mu(1) \geq \min\{\mu(x), \varepsilon\}$ for all $x \in X$. Let $x, y, z \in X$ be such that $\mu(x * (y * z)) = \lambda_1$ and $\mu(x * y) = \lambda_2$. Then $x * (y * z) \in U(\mu, \lambda_1)$ and $x * y \in U(\mu, \lambda_2)$. Without loss of generality, assume that $\lambda_1 \leq \lambda_2 \leq \varepsilon$. Then $U(\mu, \lambda_2) \subseteq U(\mu, \lambda_1)$. Hence $x * y \in U(\mu, \lambda_1)$. Since $U(\mu, \lambda_1)$ is a weak implicative filter, we get $x * (x * z) \in U(\mu, \lambda_1)$. Hence $\mu(x * (x * z)) \geq \lambda_1 = \min\{\lambda_1, \lambda_2, \varepsilon\} = \min\{\mu(x * (x * y)), \mu(x * y), \varepsilon\}$. Therefore μ is an ε-generalized fuzzy weak implicative filter of X. □

Theorem 14.4.8 *Let $\{\mu_{\varepsilon_i}\}_{i \in \Delta}$ be an indexed family of ε_i-generalized fuzzy weak implicative filters of a BE-algebra X and $\varepsilon_i \in (0, 1]$ for all $i \in \Delta$. Then their intersection $\bigcap\limits_{i \in \Delta} \mu_{\varepsilon_i}$ is an ε-generalized fuzzy weak implicative filter of X where $\varepsilon = \inf\{\varepsilon_i \mid i \in \Delta\}$.*

Proof Let $\{\mu_{\varepsilon_i}\}_{i \in \Delta}$ be an indexed family of ε_i-generalized fuzzy weak implicative filters of a BE-algebra X and $\varepsilon_i \in (0, 1]$ for all $i \in \Delta$. Take $\varepsilon = \inf\{\varepsilon_i \mid i \in \Delta\}$. Then

$$\begin{aligned}
\Big(\bigcap_{i \in \Delta} \mu_{\varepsilon_i}\Big)(1) &= \inf\{\mu_{\varepsilon_i}(1) \mid i \in \Delta\} \\
&\geq \inf\{\min\{\mu_{\varepsilon_i}(x), \varepsilon_i\} \mid i \in \Delta\} \\
&= \min\{\inf\{\mu_{\varepsilon_i}(x) \mid i \in \Delta\}, \inf\{\varepsilon_i \mid i \in \Delta\}\} \\
&= \min\{\Big(\bigcap_{i \in \Delta} \mu_{\varepsilon_i}\Big)(x), \varepsilon\}.
\end{aligned}$$

For brevity of expression, set $a = x * (y * z)$, $b = x * y$ and $c = x * (x * z)$. Then

$$\begin{aligned}
\Big(\bigcap_{i \in \Delta} \mu_{\varepsilon_i}\Big)(c) &= \inf\{\mu_{\varepsilon_i}(c) \mid i \in \Delta\} \\
&\geq \inf\{\min\{\mu_{\varepsilon_i}(a), \mu_{\varepsilon_i}(b), \varepsilon_i\} \mid i \in \Delta\} \\
&= \min\{\inf\{\mu_{\varepsilon_i}(a) \mid i \in \Delta\}, \inf\{\mu_{\varepsilon_i}(b) \mid i \in \Delta\}, \inf\{\varepsilon_i \mid i \in \Delta\}\} \\
&= \min\{\Big(\bigcap_{i \in \Delta} \mu_{\varepsilon_i}\Big)(a), \Big(\bigcap_{i \in \Delta} \mu_{\varepsilon_i}\Big)(b), \varepsilon\}
\end{aligned}$$

Therefore $\bigcap\limits_{i \in \Delta} \mu_{\varepsilon_i}$ is an ε-generalized fuzzy weak implicative filter of X. □

Theorem 14.4.9 *Let $\varepsilon \in (0, 1]$ and μ be an ε-generalized fuzzy weak implicative filter of a BE-algebra X. Then $X_\mu = \{x \in X \mid \mu(x) \geq \min\{\mu(1), \varepsilon\}\}$ is a weak implicative filter of X.*

Proof By the definition of X_μ, we get $\mu(1) \geq \min\{\mu(1), \varepsilon\}$. Hence $1 \in X_\mu$. Let $x, y, z \in X$ be such that $x * (y * z) \in X_\mu$ and $x * y \in X_\mu$. Then $\mu(x * (y * z)) \geq \min\{\mu(1), \varepsilon\}$ and $\mu(x * y) \geq \min\{\mu(1), \varepsilon\}$. Since μ is an ε-generalized fuzzy weak implicative filter, we get $\mu(x * (x * z)) \geq \min\{\mu(x * (y * z)), \mu(x * y), \varepsilon\} \geq \min\{\mu(1), \mu(1), \varepsilon\} = \min\{\mu(1), \varepsilon\}$. Hence $x * (x * z) \in X_\mu$. Therefore X_μ is a weak implicative filter of X. □

Definition 14.4.10 Let $\varepsilon \in (0, 1]$ and f is a homomorphism on a BE-algebra X. Define a mapping $\mu_f : X \to [0, 1]$ by $\mu_f(x) = \min\{\mu(f(x)), \varepsilon\}$ for all $x \in X$.

Theorem 14.4.11 *Let $\varepsilon \in (0, 1]$ and f is a homomorphism on a BE-algebra X. If μ is an ε-generalized fuzzy weak implicative filter of X, then μ_f is also an ε-generalized fuzzy weak implicative filter of X.*

Proof Assume that μ is an ε-generalized fuzzy weak implicative filter of X for $\varepsilon \in (0, 1]$. Let $x \in X$. Then $\mu_f(1) = \mu(f(1)) = \mu(1) \geq \min\{\mu(x), \varepsilon\} = \mu_f(x)$. Let $x, y, z \in X$. Since μ is an ε-generalized fuzzy weak implicative filter of X, we get

$$\begin{aligned}
\mu_f(x * (x * z)) &= \mu(f(x * (x * z))) \\
&= \mu(f(x) * (f(x) * f(z))) \\
&\geq \min\{\mu(f(x * (y * z))), \mu(f(x * y)), \varepsilon\} \\
&= \min\{\mu_f(x * (y * z)), \mu_f(x * y), \varepsilon\}
\end{aligned}$$

Therefore μ_f is an ε-generalized fuzzy weak implicative filter of X. □

14.5 Cartesian Products of Fuzzy Implicative Filters

In this section, we discuss some properties of the Cartesian products of fuzzy implicative filters of BE-algebras. The notion of fuzzy relations is extended to the case of fuzzy implicative filters of BE-algebras.

Definition 14.5.1 A fuzzy relation on a set S is a fuzzy set $\mu : S \times S \longrightarrow [0, 1]$.

Definition 14.5.2 Let μ be a fuzzy relation on a set S and ν a fuzzy set in S. Then μ is a *fuzzy relation* on ν if $\mu(x, y) \leq \min\{\nu(x), \nu(y)\}$ for all $x, y \in S$.

Definition 14.5.3 Let μ and ν be two fuzzy sets in a BE-algebra X. The *Cartesian product* of μ and ν is defined by $(\mu \times \nu)(x, y) = \min\{\nu(x), \nu(y)\}$ for all $x, y \in X$.

Theorem 14.5.4 *Let μ and ν be two fuzzy implicative filters of a BE-algebra X. Then $\mu \times \nu$ is a fuzzy implicative filter in $X \times X$.*

Proof Let $(x, y) \in X \times X$. Since μ, ν are fuzzy implicative filters in X, we get $(\mu \times \nu)(1, 1) = \min\{\mu(1), \nu(1)\} \geq \min\{\mu(x), \nu(y)\} = (\mu \times \nu)(x, y)$. Let (x, x'), (y, y'), $(z, z') \in X \times X$. Put $t = x * (y * z)$ and $t' = x' * (y' * z')$. Clearly $(t, t') = (x, x') * ((y, y') * (z, z'))$. Since μ and ν are fuzzy implicative filters in X, we can obtain the following consequence.

$$
\begin{aligned}
(\mu \times \nu)((x, x') * (z, z')) &= (\mu \times \nu)(x * z, x' * z') \\
&= \min\{\mu(x * z), \nu(x' * z')\} \\
&\geq \min\{\min\{\mu(x * y), \mu(t)\}, \min\{\nu(x' * y'), \nu(t')\}\} \\
&= \min\{\min\{\mu(x * y), \nu(x' * y')\}, \min\{\mu(t), \nu(t')\}\} \\
&= \min\{(\mu \times \nu)(x * y, x' * y), (\mu \times \nu)(t, t')\} \\
&= \min\{(\mu \times \nu)((x, x') * (y, y')), (\mu \times \nu)(t, t')\}.
\end{aligned}
$$

Therefore $\mu \times \nu$ is a fuzzy implicative filter in $X \times X$. □

Theorem 14.5.5 *Let μ and ν be two fuzzy sets in a BE-algebra X such that $\mu \times \nu$ is a fuzzy filter of $X \times X$. Then we have the following:*

(1) *either $\mu(x) \leq \mu(1)$ or $\nu(x) \leq \nu(1)$ for all $x \in X$,*
(2) *if $\mu(x) \leq \mu(1)$ for all $x \in X$, then either $\mu(x) \leq \nu(1)$ or $\nu(x) \leq \nu(1)$,*
(3) *if $\nu(x) \leq \nu(1)$ for all $x \in X$, then either $\mu(x) \leq \mu(1)$ or $\mu(x) \leq \mu(1)$,*
(4) *either μ or ν is a fuzzy implicative filter of X.*

Proof (1) Suppose that $\mu(x) > \mu(1)$ and $\nu(y) > \nu(1)$ for some $x, y \in X$. Then we get $(\mu \times \nu)(x, y) = \min\{\mu(x), \nu(y)\} > \min\{\mu(1), \nu(1)\} = (\mu \times \nu)(1, 1)$, which is a contradiction. Hence either $\mu(x) \leq \mu(1)$ or $\nu(x) \leq \nu(1)$ for all $x \in X$.
(2) Assume that $\mu(x) \leq \mu(1)$ for all $x \in X$. Suppose $\mu(x) > \mu(1)$ and $\nu(y) > \nu(1)$ for some $x, y \in X$. Then $(\mu \times \nu)(1, 1) = \min\{\mu(1), \nu(1)\} = \nu(1)$. Hence $(\mu \times \nu)(x, y) = \min\{\mu(x), \nu(y)\} > \nu(1) = (\mu \times \nu)(1, 1)$. which is a contradiction. Therefore (2) holds.
(3) It can be obtained in a similar fashion.
(4) By (1), either $\mu(x) \leq \mu(1)$ or $\nu(x) \leq \nu(1)$ for all $x \in X$. Without loss of generality, we may assume that $\mu(x) \leq \mu(1)$ for all $x \in X$. From (2), we can get either $\mu(x) \leq \nu(1)$ or $\nu(x) \leq \nu(1)$ for all $x \in X$.

Case. I: Suppose $\mu(x) \leq \nu(1)$ for all $x \in X$. Then $(\mu \times \nu)(x, 1) = \min\{\mu(x), \nu(1)\} = \mu(x)$ for all $x \in X$. Since $\mu \times \nu$ is a fuzzy implicative filter in $X \times X$, we get $\mu(x) = \min\{\mu(x), \nu(1)\} = (\mu \times \nu)(x, 1) \leq (\mu \times \nu)(1, 1) = \mu(1)$. Also

$$
\begin{aligned}
\mu(x * z) &= \min\{\mu(x * z), \nu(1)\} \\
&= (\mu \times \nu)(x * z, 1) \\
&= (\mu \times \nu)(x * z, x * 1) \\
&= (\mu \times \nu)((x, x) * (z, 1)) \\
&\geq \min\{(\mu \times \nu)((x, x) * (y, 1)), (\mu \times \nu)((x, x) * ((y, 1) * (z, 1)))\} \\
&= \min\{(\mu \times \nu)(x * x, y * 1), (\mu \times \nu)(x * (y * z), x * (1 * 1))\} \\
&= \min\{\min\{\mu(x * y), \nu(y * 1)\}, \min\{\mu(x * (y * z)), \nu(x * (1 * 1))\}\} \\
&= \min\{\min\{\mu(x * y), \nu(1)\}, \min\{\mu(x * (y * z)), \nu(1)\}\} \\
&= \min\{\mu(x * y), \mu(x * (y * z))\}.
\end{aligned}
$$

Therefore μ is a fuzzy implicative filter in X.

Case. II: Suppose $\nu(x) \le \nu(1)$ for all $x \in X$. Suppose $\mu(x) \le \nu(1)$ for all $x \in X$. Then it leads to case I. Suppose $\mu(t) > \nu(1)$ for some $t \in X$. Then $\mu(1) \ge \mu(t) > \nu(1)$. Since $\nu(x) \le \nu(1)$ for all $x \in X$, it yields that $\mu(1) > \nu(x)$ for all $x \in X$. Hence

$$\begin{aligned}
\nu(x * z) &= \min\{\mu(1), \nu(x * z)\} \\
&= (\mu \times \nu)(1, x * z) \\
&= (\mu \times \nu)(x * 1, x * z) \\
&= (\mu \times \nu)((x, x) * (1, z)) \\
&\ge \min\{(\mu \times \nu)((x, x) * (1, y)), (\mu \times \nu)((x, x) * ((1, y) * (1, z)))\} \\
&= \min\{(\mu \times \nu)(x * 1, x * y), (\mu \times \nu)(x * (1 * 1), x * (y * z))\} \\
&= \min\{(\mu \times \nu)(1, x * y), (\mu \times \nu)(1, x * (y * z))\} \\
&= \min\{\min\{\mu(1), \nu(x * y)\}, \min\{\mu(1), \nu(x * (y * z))\}\} \\
&= \min\{\nu(x * y), \nu(x * (y * z))\}.
\end{aligned}$$

Therefore ν is a fuzzy implicative filter in X. □

In the following, we present an example to show that if $\mu \times \nu$ is a fuzzy implicative filter of the product algebra $X \times X$, then μ and ν both need not be fuzzy implicative filters of X.

Example 14.5.6 Let X be a BE-algebra with $|X| > 2$, and let $\alpha, \beta \in [0, 1]$ be such that $0 \le \alpha \le \beta < 1$. Define fuzzy sets μ and $\nu : X \to [0, 1]$ by $\mu(x) = \alpha$ and

$$\nu(x) = \begin{cases} \beta \text{ if } x = 1 \\ 1 \text{ if } x \neq 1 \end{cases}$$

Then $(\mu \times \nu)(x, y) = \min\{\mu(x), \nu(y)\} = \alpha$ for all $(x, y) \in X \times X$. Hence $\mu \times \nu : X \times X \to [0, 1]$ is a constant function. Thus $\mu \times \nu$ is a fuzzy implicative filter of $X \times X$. Now μ is a fuzzy implicative filter of X, but ν is not a fuzzy implicative filter of X because ν does satisfy $(F1)$.

Definition 14.5.7 Let ν be a fuzzy set in a BE-algebra X. Then the *strongest fuzzy relation* μ_ν is a fuzzy relation on X defined by $\mu_\nu(x, y) = \min\{\nu(x), \nu(y)\}$ for all $x, y \in X$.

Theorem 14.5.8 *Let ν be a fuzzy set in X and μ_ν be the strongest fuzzy relation on X. Then ν is a fuzzy implicative filter of X if and only if μ_ν is a fuzzy implicative filter of $X \times X$.*

Proof Assume that ν is a fuzzy implicative filter of X. Then for any $(x, y) \in X \times X$, we have $\mu_\nu(x, y) = \min\{\nu(x), \nu(y)\} \le \min\{\nu(1), \nu(1)\} = \mu_\nu(1, 1)$. Let $(x, x'), (y, y')$ and $(z, z') \in X \times X$. Then we have the following:

$$\begin{aligned}
\mu_\nu((x, x') * (z, z')) &= \mu_\nu(x * z, x' * z') \\
&= \min\{\nu(x * z), \nu(x' * z')\} \\
&\geq \min\{\min\{\nu(x * y), \nu(t)\}, \min\{\nu(x' * y'), \nu(t')\}\} \\
&\qquad \text{where } t = x * (y * z) \text{and } t' = x' * (y' * z') \\
&= \min\{\min\{\nu(x * y), \nu(x' * y')\}, \min\{\nu(t), \nu(t')\}\} \\
&= \min\{\mu_\nu(x * y, x' * y'), \mu_\nu(x * (y * z), x' * (y' * z'))\} \\
&= \min\{\mu_\nu((x, x') * (y, y')), \mu_\nu((x, x') * ((y, y') * (z, z')))\}.
\end{aligned}$$

Therefore μ_ν is a fuzzy implicative filter of $X \times X$. Conversely, assume that μ_ν is a fuzzy implicative filter of $X \times X$. Then $\nu(1) = \min\{\nu(1), \nu(1)\} = \mu_\nu(1, 1) \geq \mu_\nu(x, y) = \min\{\nu(x), \nu(y)\}$ for all $x, y \in X$. Hence $\nu(x) \leq \nu(1)$ for all $x \in X$. Let $x, y, z \in X$. Then

$$\begin{aligned}
\nu(x * z) &= \min\{\nu(x * z), \nu(1)\} \\
&= \mu_\nu(x * z, 1) \\
&= \mu_\nu(x * z, z * 1) \\
&= \mu_\nu((x, z) * (z, 1)) \\
&\geq \min\{\mu_\nu((x, z) * (y, 1)), \mu_\nu((x, z) * ((y, 1) * (z, 1)))\} \\
&= \min\{\mu_\nu(x * y, z * 1), \mu_\nu(x * (y * z), z * (1 * 1))\} \\
&= \min\{\mu_\nu(x * y, 1), \mu_\nu(x * (y * z), 1)\} \\
&= \min\{\min\{\nu(x * y), \nu(1)\}, \min\{\nu(x * (y * z), \nu(1)\}\} \\
&= \min\{\nu(x * y), \nu(x * (y * z))\}.
\end{aligned}$$

Therefore ν is a fuzzy implicative filter of X. □

14.6 Vague Sets Applied to Weak Implicative Filters

In this section, the notion of vague sets is applied to weak implicative filters of BE-algebras. Some characterizations of vague weak implicative filters are derived. Interconnections among vague filters, vague associative filters, and vague weak implicative filters are investigated.

Definition 14.6.1 ([14]) A vague set A in the universe of discourse U is characterized by two membership functions given by:

(1) A truth-membership function $t_A : U \to [0, 1]$ and
(2) A false-membership function $f_A : U \to [0, 1]$.

where $t_A(u)$ is a lower bound of the grade of membership of u derived from the "evidence for u", and $f_A(u)$ is a lower bound on the negation of u derived from the "evidence against u", and $t_A(u) + f_A(u) \leq 1$.

Thus the grade of membership of u in the vague set A is bounded by a subinterval $[t_A(u), 1 - f_A(u)]$ of $[0, 1]$. This indicates that if the actual grade of membership is $\mu(u)$, then $t_A(u) \leq \mu(u) \leq 1 - f_A(u)$. The vague set A is written as

$$A = \{(u, [t_A(u), f_A(u)]) \mid u \in U\}$$

where the interval $[t_A(u), 1 - f_A(u)]$ is called the *vague value* of u in A and is denoted by $V_A(u)$.

It is worth to mention here that the interval-valued fuzzy sets (i-v fuzzy sets) are not vague sets. In the (i-v fuzzy sets), an interval-valued membership value is assigned to each element of the universe considering the evidence for u without considering the evidence against u. In vague sets, both values are independent proposed by the decision maker. This makes a major difference in the judgment about the grade of membership.

Definition 14.6.2 ([14]) A vague set A of a set U is said to be

(1) the *zero vague set* of U if $t_A(u) = 0$ and $f_A(u) = l$ for all $u \in U$,
(2) the *unit vague set* of U if $t_A(u) = 1$ and $f_A(u) = 0$ for all $u \in U$,
(3) the *α-vague set* of U if $t_A(u) = \alpha$ and $f_A(u) = 1 - \alpha$ for all $u \in U$ where $\alpha \in (0, 1)$.

Definition 14.6.3 ([14]) For $\alpha, \beta \in [0, 1]$. Suppose A is a vague set of a universe X with the true-membership function t_A and the false-membership function f_A. The (α, β)-cut of the vague set A is a crisp subset $A_{(\alpha,\beta)}$ of the set X given by

$$A_{(\alpha,\beta)} = \{x \in X \mid V_A(x) \geq [\alpha, \beta]\},$$

where $\alpha \leq \beta$. Clearly $A_{(0,0)} = X$. The (α, β)-cuts are also called vague cuts of the vague set A.

Definition 14.6.4 ([14]) The α-cut of the vague set A is a crisp subset A_α of the set X given by $A_\alpha = A_{(\alpha,\alpha)}$.

Note that $A_0 = X$ and if $\alpha \geq \beta$, then $A_\alpha \subseteq A_\beta$ and $A(\alpha, \beta) = A_\alpha$. Equivalently, we can define the α-cut as

$$A_\alpha = \{x \in X \mid t_A(x) \geq \alpha\}$$

For our discussion, we shall use the following notations which were given in [14] on interval arithmetic.

Notation. Let $I[0, 1]$ denote the family of all closed subintervals of $[0, 1]$. If $I_1 = [a_1, b_1]$ and $I_2 = [a_2, b_2]$ are two elements of $I[0, 1]$, we call $I_1 > I_2$ if $a_1 \geq a_2$ and $b_1 \geq b_2$. Similarly, we understand the relations $I_1 \leq I_2$ and $I_1 = I_2$. Clearly the relation $I_1 \geq I_2$ does not necessarily imply that $I_1 \supseteq I_2$ and conversely. We define the term "imax" to mean the maximum of two intervals

$$\text{imax}(I_1, I_2) = [\max(a_1, a_2), \max(b_1, b_2)]$$

Similarly, we define "imin". The concept of imax and imin could be extended to define "isup" and "iinf" of infinite number of elements of $I[0, 1]$. It is obvious that $L = \{I[0, 1], \text{isup}, \text{iinf}, \leq\}$ is a lattice with universal bounds $[0, 0]$ and $[1, 1]$.

Definition 14.6.5 ([4]) A vague set A of a BE-algebra X is said to be a *vague filter* of X if the following conditions are true:

(VF1) $V_A(1) \geq V_A(x)$ for all $x \in X$,
(VF2) $V_A(y) \geq \text{imin}\ \{V_A(x), V_A(x * y)\}$ for all $x, y \in X$.

Every vague filter of a BE-algebra X satisfies the following property:

$$x \leq y \text{implies } V_A(x) \leq V_A(y) \text{ for all } x, y \in X.$$

Definition 14.6.6 A vague set A of a BE-algebra X is said to be a *vague weak implicative filter* of X if the following conditions are true:

(VWIF1) $V_A(1) \geq V_A(x)$ for all $x \in X$,
(VWIF2) $V_A(x * (x * z)) \geq \text{imin}\ \{V_A(x * y), V_A(x * (y * z))\}$ for all $x, y \in X$.
That is

$$t_A(1) \geq t_A(x), 1 - f_A(1) \geq 1 - f_A(x) \quad \text{and}$$

$$t_A(x * (x * z)) \geq \min\{t_A(x * y), t_A(x * (y * z))\}$$

$$1 - f_A(x * (x * z)) \geq \min\{1 - f_A(x * y), 1 - f_A(x * (y * z))\}.$$

Definition 14.6.7 A vague set A of a BE-algebra X is said to be a *vague implicative filter* of X if the following conditions are true:

(VIF1) $V_A(1) \geq V_A(x)$ for all $x \in X$,
(VIF2) $V_A(x * z) \geq \text{imin}\ \{V_A(x * y), V_A(x * (y * z))\}$ for all $x, y \in X$.

Definition 14.6.8 A vague set A of a BE-algebra X is said to be a *vague associative filter* of X if the following conditions are true:

(VAF1) $V_A(1) \geq V_A(x)$ for all $x \in X$,
(VAF2) $V_A(z) \geq \text{imin}\ \{V_A(x * y), V_A(x * (y * z))\}$ for all $x, y \in X$.

Example 14.6.9 Let $X = \{1, a, b\}$. Define a binary operation $*$ on X as follows:

$*$	1	a	b
1	1	a	b
a	1	1	b
b	1	1	1

Clearly $(X, *, 1)$ is a BE-algebra. Let A be a vague set defined in X as follows:

$$A = \{(1, [0.7, 0.2]), (a, [0.5, 0.3]), (b, [0.4, 0.4])\}$$

It can be easily verified that A is vague weak implicative filter of X.

Proposition 14.6.10 *Every vague weak implicative filter of a BE-algebra is a vague filter.*

Proof By taking $x = 1$ in (VWIF2), it is clear. □

Theorem 14.6.11 *Let A be a vague weak implicative filter of a BE-algebra X. Then for any $\alpha, \beta \in [0, 1]$, the vague cut $A_{(\alpha,\beta)}$ is a weak implicative filter of X.*

Proof Clearly $1 \in A_{(\alpha,\beta)}$. Let $x, y \in X$. Suppose $x * y \in A_{(\alpha,\beta)}$ and $x * (y * z) \in A_{(\alpha,\beta)}$. Then $V_A(x * y) \geq [\alpha, \beta]$ and $V_A(x * (y * z)) \geq [\alpha, \beta]$. Since $V_A(x * y) \geq [\alpha, \beta]$, we get $t_A(x * y) \geq \alpha$ and $1 - f_A(x * y) \geq \beta$. Since $V_A(x * (y * z)) \geq [\alpha, \beta]$, we get $t_A(x * (y * z)) \geq \alpha$ and $1 - f_A(x * (y * z)) \geq \beta$. Since A is a vague weak implicative filter, we get

$$t_A(x * (x * z)) \geq \min\{t_A(x * y), t_A(x * (y * z))\} \geq \min\{\alpha, \alpha\} = \alpha$$

$$1 - f_A(x * (x * z)) \geq \min\{1 - f_A(x * y), 1 - f_A(x * (y * z))\} \geq \min\{\beta, \beta\} = \beta.$$

Hence $V_A(x * (x * z)) \geq [\alpha, \beta]$. Thus $x * (x * z) \in A_{(\alpha,\beta)}$. Therefore the vague cut $A_{(\alpha,\beta)}$ is a weak implicative filter of X. □

Theorem 14.6.12 *If X is transitive, every vague filter is a vague weak implicative filter.*

Proof Let X be a transitive BE-algebra and A be a vague filter of X. Then $V_A(1) \geq V_A(x)$ for all $x \in X$. Let $x, y, z \in X$. Since X is transitive, we get $x * (y * z) \leq x * ((x * y) * (x * z)) = (x * y) * (x * (x * z))$. Since A is a vague filter, we get $V_A(x * (x * z)) \geq$ imin $\{V_A(x * y), v_A((x * y) * (x * (x * z))\} \geq$ imin $\{V_A(x * y), v_A(x * (y * z))\}$. Therefore A is vague weak implicative filter. □

Proposition 14.6.13 *Every vague implicative filter of X is a vague weak implicative filter.*

Proof Let A be a vague implicative filter of X. Putting $x = 1$ in (VIF2), we get that A is a vague filter of X. Hence $V_A(1) \geq V_A(x)$ for all $x \in X$. Let $x, y, z \in X$. Then

$$\begin{aligned} V_A(x * (x * z)) &\geq V_A(x * z) \\ &\geq \text{imin}\{V_A(x * y), V_A(x * (y * z))\} \end{aligned}$$

Therefore A is a vague weak implicative filter of X. □

Theorem 14.6.14 *A vague weak implicative filter of a BE-algebra X is a vague implicative filter if and only if $V_A(x * (x * y)) = V_A(x * y)$ for all $x, y \in X$.*

Proof Let A be vague weak implicative filter of X. Assume that A is a vague implicative filter. Then A is a vague filter. Let $x, y \in X$. Since $x * y \leq x * (x * y)$, we get $V_A(x * y) \leq V_A(x * (x * y))$. Again, since A is a vague implicative filter, we get

$$\begin{aligned} V_A(x * y) &\geq \text{imin}\{V_A(x * x), V_A(x * (x * y))\} \\ &= \text{imin}\{V_A(1), V_A(x * (x * y))\} \\ &= V_A(x * (x * y)) \end{aligned}$$

Therefore $V_A(x * (x * y)) = V_A(x * y)$ for all $x, y \in X$.

Conversely, assume that the condition holds in X. Since A is a vague weak implicative filter, we get $V_A(x) \geq V_A(x)$ for all $x \in X$. Let $x, y, z \in X$. Then, by the assumed condition, we get $V_A(x * y) = V_A(x * (x * z)) \geq \text{imin}\,\{V_A(x * y), V_A(x * (y * z))\}$. Therefore A is a vague implicative filter of X. □

Theorem 14.6.15 *Every vague associative filter of a BE-algebra is a vague implicative filter.*

Proof Let A be a vague associative filter of X. By taking $x = 1$ in (VAF2), we get that A is a vague filter of X. Hence $V_A(1) \geq V_A(x)$ for all $x \in X$. Let $x, y, z \in X$. Since $z \leq x * z$, we get $V_A(z) \leq V_A(x * z)$. Since A is a vague associative filter, we get

$$\begin{aligned} V_A(x * z) &\geq V_A(z) \\ &\geq \text{imin}\{V_A(x * y), V_(x * (y * z))\} \end{aligned}$$

Therefore A is a vague implicative filter of X. □

Corollary 14.6.16 *Every vague associative filter of a BE-algebra is a vague weak implicative filter.*

Theorem 14.6.17 *A vague weak implicative filter of a BE-algebra X is a vague associative filter if and only if $V_A(x * (x * y)) = V_A(y)$ for all $x, y \in X$.*

Proof Let A be a vague weak implicative filter of X. Assume that A is a vague associative filter of X. Hence A is a vague filter of X. Let $x, y \in X$. Since $y \leq x * (x * y)$, we get $V_A(y) \leq V_A(x * (x * y))$. Since A is vague associative, we get

$$\begin{aligned} V_A(y) &\geq \text{imin}\{V_A(x * (x * y)), V_A(x * x)\} \\ &= \text{imin}\{V_A(x * (x * y)), V_A(1)\} \\ &= V_A(x * (x * y)). \end{aligned}$$

Therefore $V_A(x * (x * y)) = V_A(y)$ for all $x, y \in X$.

Conversely, assume that the condition holds in X. Clearly $V_A(1) \geq V_A(x)$ for all $x \in X$. Let $x, y, z \in X$. Since A is a vague weak implicative filter, we get $V_A(z) = V_A(x * (x * z)) \geq \text{imin}\,\{V_A(x * y), V_A(x * (y * z))\}$. Hence A is vague associative. □

Theorem 14.6.18 *A vague filter A of a BE-algebra X is a vague weak implicative filter if it satisfies the following condition for all $x, y, z \in X$:*

$$(VWIF3)\ V_A(x * (y * z)) \leq V_A(x * ((x * y) * (x * z))).$$

Proof Let A be a vague filter of X. Assume that A is satisfying the given condition. Let $x, y, z \in X$. Since A is a vague filter, we get

$$\begin{aligned} V_A(x * (x * z)) &\geq \text{imin}\{V_A(x * y), V_A((x * y) * (x * (x * z)))\} \\ &= \text{imin}\{V_A(x * y), V_A(x * ((x * y) * (x * z)))\} \\ &\geq \text{imin}\{V_A(x * y), V_A(x * (y * z))\} \end{aligned}$$

Therefore A is a vague weak implicative filter of X. □

Theorem 14.6.19 *Let A and B be two vague filters of a BE-algebra X such that $V_A(1) = V_B(1)$ and $V_A(x) \leq V_B(x)$ for all $x \in X$. If A is a vague weak implicative filter, then B is also a vague weak implicative filter.*

Proof Assume that A and B are two vague filters of a BE-algebra X such that $V_A(1) = V_B(1)$ and $V_A(x) \leq V_B(x)$ for all $x \in X$. Suppose A is a vague weak implicative filter of X. Let $x, y, z \in X$. For brevity, we set $x * (y * z) = a$. Since $V_A(x) \leq V_B(x)$ and A is a vague weak implicative filter, by the above theorem, we get

$$\begin{aligned} V_B(x * ((x * y) * (x * (a * z)))) &\geq V_A(x * ((x * y) * (x * (a * z)))) \\ &\geq V_A(x * (y * (a * z))) \\ &= V_A(x * (a * (y * z))) \\ &= V_A(a * (x * (y * z))) \\ &= V_A(a * a) \\ &= V_A(1) \\ &= V_B(1). \end{aligned}$$

whence

$$\begin{aligned} V_B(a * (x * ((x * y) * (x * z)))) &= V_B(x * (a * ((x * y) * (x * z)))) \\ &= V_B(x * ((x * y) * (a * (x * z)))) \\ &= V_B(x * ((x * y) * (x * (a * z)))) \\ &\geq V_B(1). \end{aligned}$$

Hence $V_B((x * (y * z)) * (x * (((x * y) * (x * z)))) = V_B(1)$. Since V_B is a vague filter, we get that $V_B(x * ((x * y) * (x * z))) \geq V_B(x * (y * z))$. Therefore, by the above theorem, B is a vague weak implicative filter of X. □

14.7 Falling Fuzzy Implicative Filters of BE-algebras

In this section, the notion of falling fuzzy shadows is applied to the case of implicative filters of BE-algebras. Relation between falling fuzzy filters and falling fuzzy implicative filters is studied. Relation between fuzzy implicative filters and falling fuzzy implicative filters is observed.

Definition 14.7.1 ([199]) Let $(\Omega, \mathcal{A}, P)$ be a probability space, and let $\xi : \Omega \to \mathcal{P}(X)$ be a random set. If $\xi(\omega)$ is a filter of X for any $\omega \in \Omega$, then the falling shadow $\tilde{H}$ of the random set ξ, that is,

$$\tilde{H}(x) = P(\omega \mid x \in \xi(\omega))$$

is said to be a *falling fuzzy filter* of X.

Definition 14.7.2 Let $(\Omega, \mathcal{A}, P)$ be a probability space, and let $\xi : \Omega \to \mathcal{P}(X)$ be a random set. If $\xi(\omega)$ is an implicative filter (resp. weak implicative filter) of X for any $\omega \in \Omega$, then the falling shadow $\tilde{H}$ of the random set ξ, that is,

$$\tilde{H}(x) = P(\omega \mid x \in \xi(\omega))$$

is said to be a *falling fuzzy implicative filter* (resp. *falling fuzzy weak implicative filter*) of X.

Theorem 14.7.3 *Let $(\Omega, \mathcal{A}, P)$ be a probability space, and let $\xi : \Omega \to \mathcal{P}(X)$ be a random set. Then every fuzzy implicative filter of X is a falling fuzzy implicative filter of X.*

Proof Let $\tilde{H}$ be a fuzzy implicative filter of X. Then the t-cut $\tilde{H}_t = \{u \in X \mid \tilde{H}(u) \geq t\}$ of $\tilde{H}$ is an implicative filter of X for all $t \in [0, 1]$. Let $\xi : [0, 1] \to \mathcal{P}(X)$ be a random set and $\xi(t) = \tilde{H}_t$. Therefore $\tilde{H}$ is a falling fuzzy implicative filter of X. □

Theorem 14.7.4 *Let $(\Omega, \mathcal{A}, P)$ be a probability space, and let $\xi : \Omega \to \mathcal{P}(X)$ be a random set. Then the falling shadow $\tilde{H}$ of the random set ξ is a falling fuzzy implicative filter of X if and only if it satisfies the following for each $\omega \in \Omega$:*

(1) $1 \in \xi(\omega)$,
(2) $(\forall (x * (y * z)) \in \xi(\omega))(\forall (x * z) \in X \setminus \xi(\omega))(x * y \in X \setminus \xi(\omega))$.

Proof Assume that $\tilde{H}$ is a falling fuzzy implicative filter of X. Then $\xi(\omega)$ is an implicative filter of X for all $\omega \in \Omega$. Let $x, y, z \in X$ be such that $x * (y * z) \in \xi(\omega)$ and $x * z \in X \setminus \xi(\omega)$. Suppose $x * y \in \xi(\omega)$. Since $\xi(\omega)$ is an implicative filter and $x * (x * y) \in \xi(\omega)$, we get $x * z \in \xi(\omega)$, which is a contradiction. Hence $x * y \notin \xi(\omega)$, which means $x * y \in X \setminus \xi(\omega)$.

Conversely, assume that the falling filter $\tilde{H}$ satisfies the given condition. Then $\xi(\omega)$ is a filter of X for each $\omega \in \Omega$. Hence $1 \in \xi(\omega)$. Let $x, y, z \in X$ be such that $x * y \in \xi(\omega)$ and $x * (y * z) \in \xi(\omega)$. Suppose $x * z \notin \xi(\omega)$, which means $x * z \in X \setminus \xi(\omega)$.

Since $x * (y * z) \in \xi(\omega)$, by the assumed condition, we get $x * y \in X \setminus \xi(\omega)$, which is a contradiction. Hence $x * z$ must be in $\xi(\omega)$. Thus $\xi(\omega)$ is an implicative filter of X. Therefore $\tilde{H}$ is a falling fuzzy implicative filter of X. □

For any subset S of X and $f \in F(X)$, let $S_f = \{\omega \in \Omega \mid f(\omega) \in S\}$,

$$\xi : \Omega \to \mathcal{P}(F(X)), \omega \mapsto \{f \in F(X) \mid f(\omega) \in S\}.$$

Then $S_f \in X$.

Example 14.7.5 Let $X = \{1, a, b, c, d\}$. Define an operation $*$ on X which is given in the following Cayley table:

$*$	1	a	b	c	d
1	1	a	b	c	d
a	1	1	b	c	d
b	1	1	1	c	d
c	1	1	1	1	d
d	1	1	b	c	1

Then $(X, *, 1)$ is a BE-algebra. Let $(\Omega, \mathcal{A}, P) = ([0, 1], \mathcal{A}, m)$, and let $\xi : [0, 1] \to \mathcal{P}(X)$ be defined by

$$\xi(t) = \begin{cases} \{1, a\} & \text{if } t \in [0, 0.2) \\ \{1, a, b\} & \text{if } t \in [0.2, 0.7) \\ \{1, a, d\} & \text{if } t \in [0.7, 1] \end{cases}$$

Then $\xi(t)$ is an implicative filter of X for all $t \in [0, 1]$. Hence $\tilde{H}(x) = P(t \mid x \in \xi(t))$ is a falling fuzzy implicative filter of X, and $\tilde{H}$ is represented as follows:

$$\tilde{H}(x) = \begin{cases} 1 & \text{if } x = 1, a \\ 0 & \text{if } x = c \\ 0.5 & \text{if } x = b \\ 0.3 & \text{if } x = d \end{cases}$$

Theorem 14.7.6 *If F is an implicative filter of X, then $\xi(\omega) = \{f \in F(X) \mid f(\omega) \in S\}$ is an implicative filter of $F(X)$.*

Proof Assume that F is an implicative filter of X, and let $\omega \in \Omega$. Since $\theta(\omega) = 1 \in F$, we get that $\theta \in \xi(\omega)$. Let $f, g, h \in F(X)$ be such that $f \circledast (g \circledast h) \in \xi(\omega)$ and $f \circledast g \in \xi(\omega)$. Then $f(\omega) \circledast (g(\omega) \circledast h(\omega)) = (f \circledast (g \circledast h))(\omega) \in F$ and $f(\omega) \circledast g(\omega) = (f \circledast g)(\omega) \in F$. Since F is an implicative filter of X, we get that $(f \circledast h)(\omega) = f(\omega) \circledast h(\omega) \in F$. Hence $f \circledast h \in \xi(\omega)$. Therefore $\xi(\omega)$ is an implicative filter of $F(X)$. □

Remark 14.7.7 Since $\xi^{-1}(f) = \{\omega \in \Omega \mid f \in \xi(\omega)\} = \{\omega \in \Omega \mid f(\omega) \in F\} = F_f \in \mathcal{A}$, we get that ξ is a random set on $F(X)$. Let $\tilde{H}(f) = P(\omega \mid f(\omega) \in F)$. Then $\tilde{H}$ is a falling fuzzy implicative filter of X.

Theorem 14.7.8 *Every falling fuzzy implicative filter is a falling fuzzy filter.*

Proof Let $\tilde{H}$ be a falling fuzzy implicative filter of X. Then $\xi(\omega)$ is an implicative filter of X, and hence, it is a filter of X. Therefore $\tilde{H}$ is a falling fuzzy filter of X. □

The converse of the above theorem is not true in general as shown by the following example.

Example 14.7.9 Let $X = \{1, a, b, c, d\}$. Define an operation $*$ on X which is given in the following Cayley table:

$*$	1	a	b	c	d
1	1	a	b	c	d
a	1	1	a	c	d
b	1	1	1	c	d
c	1	a	c	1	d
d	1	a	b	c	1

Then $(X, *, 1)$ is a BE-algebra. Let $(\Omega, \mathcal{A}, P) = ([0, 1], \mathcal{A}, m)$, and let $\xi : [0, 1] \to \mathcal{P}(X)$ be defined by

$$\xi(t) = \begin{cases} \{1, c\} & \text{if } t \in [0, 0.15) \\ \{1, d\} & \text{if } t \in [0.15, 0.45) \\ \{1, a, b\} & \text{if } t \in [0.45, 0.75) \\ \{1, c, d\} & \text{if } t \in [0.75, 1] \end{cases}$$

Then $\xi(t)$ is a filter of X for all $t \in [0, 1]$. Hence $\tilde{H}(x) = P(t \mid x \in \xi(t))$ is a falling fuzzy filter of X, and $\tilde{H}$ is represented as follows:

$$\tilde{H}(x) = \begin{cases} 1 & \text{if } x = 1 \\ 0.15 & \text{if } x = c \\ 0.45 & \text{if } x = d \\ 0.75 & \text{if } x = a, b \\ 1 & \text{if } x = c, d \end{cases}$$

In this case, we can easily check that $\tilde{H}$ is a fuzzy filter of X. If $t \in [0.75, 1]$, then $\tilde{H}_t = \{1, c, d\}$ is not an implicative filter of X since $a * (c * b) = a * c = c \in \tilde{H}_t$ and $a * c = c \in \tilde{H}_t$ but $a * b = a \notin \tilde{H}_t$. Therefore $\tilde{H}$ is not a falling fuzzy implicative filter of X.

Though a falling fuzzy filter is not a falling fuzzy implicative filter, we derive a set of equivalent conditions for a falling filter of a transitive BE-algebra to become a falling fuzzy implicative filter.

Theorem 14.7.10 *Let X be a transitive BE-algebra. Let $(\Omega, \mathcal{A}, P)$ be a probability space, and let $\xi : \Omega \to \mathcal{P}(X)$ be a random set. If the falling shadow $\tilde{H}$ is a filter of a random set $\xi : \Omega \to \mathcal{P}(X)$, then the following conditions are equivalent:*

(1) $\tilde{H}$ *is a falling fuzzy implicative filter of* X*;*
(2) *for all* $x, y \in X$, $\Omega(x * (x * y)) \subseteq \Omega(x * y)$*;*
(3) *for all* $x, y, z \in X$, $\Omega(x * (y * z) \subseteq \Omega((x * y) * (x * z))$.

Proof (1) $\Rightarrow$ (2): Assume that $\tilde{H}$ is a falling fuzzy implicative filter of X. Then $\xi(\omega)$ is an implicative filter for all $\omega \in [0, 1]$. Let $x, y \in X$ be such that $\omega \in \Omega(x * (x * y))$. Then $x * (x * y) \in \xi(\omega)$. Since $x * x = 1 \in \xi(\omega)$ and $\xi(\omega)$ is an implicative filter, we get $x * y \in \xi(\omega)$. Hence $\omega \in \Omega(x * y)$. Therefore $\Omega(x * (x * y)) \subseteq \Omega(x * y)$.
(2) $\Rightarrow$ (3): Assume that $\Omega(x * (x * y)) \subseteq \Omega(x * y)$ holds for all $x, y \in X$. Let $\omega \in \Omega((x * y) * z)$. Then $x * (y * z) \in \xi(\omega)$. Since X is transitive, we get $y * z \leq (x * y) * (x * z)$. Hence, we get the following consequence.

$$\begin{aligned} x * (y * z) &\leq x * ((x * y) * (x * z)) \\ &= x * (x * ((x * y) * z)) \end{aligned}$$

Since $\xi(\omega)$ is a filter of X, it immediately infers $x * (x * ((x * y) * z)) \in \xi(\omega)$. Hence, by the assumed condition, we get $(x * y) * (x * z) = x * ((x * y) * z) \in \xi(\omega)$. Thus the condition (3) holds.
(3) $\Rightarrow$ (1): Assume that the condition (3) holds. Let $x, y, z \in X$ be such that $x * (y * z) \in \xi(\omega)$ and $x * y \in \xi(\omega)$. Since $x * (y * z) \in \xi(\omega)$, we get $\omega \in \Omega(x * (y * z)) \subseteq \Omega((x * y) * (x * z))$. Since $\tilde{H}$ is a falling filter of X, it is clear that $\xi(\omega)$ is a filter of X. Since $x * y \in \xi(\omega)$, we get that $x * z \in \xi(\omega)$. Hence $\xi(\omega)$ is an implicative filter of X. Therefore $\tilde{H}$ is a falling fuzzy implicative filter of X. □

Theorem 14.7.11 *Let* $(\Omega, \mathcal{A}, P)$ *be a probability space and* $\xi : \Omega \to P(X)$ *be a random set. Suppose* $\tilde{H}$ *be a falling fuzzy implicative filter of a* BE*-algebra* X*. Then*

$$\tilde{H}(x * z) \geq T(\tilde{H}(x * (y * z)), \tilde{H}(x * y))$$

for all $x, y \in X$*, where* $T(s, t) = \min\{s + t - 1, 1\}$ *for any* $s, t \in [0, 1]$.

Proof From the definition of falling implicative filter, we get $\xi(\omega)$ is an implicative filter of X for any $\omega \in \Omega$. Let $x, y, z \in X$. Then by the above proposition, we get $\Omega(x * (y * z)) \cap \Omega(x * y) \subseteq \Omega(x * z)$. Hence

$$\begin{aligned} \tilde{H}(x * z) &= P(\omega \mid x * z \in \xi(\omega)) \\ &\geq P(\omega \mid x * (y * z) \in \xi(\omega)) \cap P(\omega \mid x * y \in \xi(\omega)) \\ &\geq P(\omega \mid x * (y * z) \in \xi(\omega)) + P(\omega \mid x * y \in \xi(\omega)) \\ &\qquad -P(\omega \mid x * (y * z) \in \xi(\omega) \text{ or } x * y \in \xi(\omega)) \\ &\geq \tilde{H}(x * (y * z)) + \tilde{H}(x * y) - 1 \end{aligned}$$

Hence $\tilde{H}(x * z) \geq \min\{\tilde{H}(x * (y * z)) + \tilde{H}(x * y) - 1, 1\} = T(\tilde{H}(x * (y * z)), \tilde{H}(x * y))$. □

Corollary 14.7.12 *Every falling fuzzy implicative filter is a T-fuzzy implicative filter.*

Exercise

1. For any two implicative filters of a BE-algebra, show that their direct product is also an implicative filter. On the other hand, show that every implicative filter of a product algebra can be represented as a product of implicative filters of the respective BE-algebras.
2. Show that the homomorphic image of an implicative filter of a BE-algebra is again an implicative filter, whenever the homomorphism is surjective.
3. Let F be a non-empty subset of a BE-algebra X. Define a fuzzy set $\mu_F : X \to [0, 1]$ as $\mu_F(x) = \alpha$ if $x \in F$ and $\mu_F(x) = 0$ if $x \notin F$ where $0 < \alpha < 1$ is fixed. If μ_F is a fuzzy weak implicative filter in X, then prove that F is a weak implicative filter in X.
4. Let F be a weak implicative filter of a BE-algebra X, and let $1 \neq \alpha < \beta$ be elements in $[0, 1]$. If $\mu : X \to [0, 1]$ is a fuzzy set defined by $\mu(x) = \alpha$ if $x \in F$ and $\mu(x) = \beta$ if $x \notin F$, then prove that μ is a fuzzy weak implicative filter of X.
5. For any fuzzy set ν in a BE-algebra X, define the strongest fuzzy relation μ_ν by $\mu_\nu(x, y) = \min\{\nu(x), \nu(y)\}$ for all $x, y \in X$. If ν is a fuzzy weak implicative filter in X, then prove that μ_ν is a fuzzy weak implicative filter of $X \times X$.
6. Prove that every weak implicative filter is a level weak implicative filter of an ε-generalized fuzzy weak of a BE-algebra for any $\varepsilon \in (0, 1]$.
7. Prove that every vague filter A of a BE-algebra X is a vague implicative filter if $V_A((x * y) * z) \geq V_A(x * (y * z))$ for all $x, y, z \in X$.
8. Let f be an onto homomorphism on a BE-algebra X and $\varepsilon \in (0, 1]$. Define a self-mapping μ_f on X by $m_f(x) = \min\{m(x), \varepsilon\}$. Prove that μ_f is an ε-generalized fuzzy weak implicative filter of X whenever μ is an ε-generalized weak implicative filter of X.
9. Let $\varepsilon \in (0, 1]$ and $f : X \to Y$ be an onto homomorphism of BE-algebras. Let μ be an ε-generalized fuzzy weak implicative filter of X. Define $f(\mu)$ by $f(\mu)(y) = \inf\{\mu(x) \mid f(x) = y$ for all $y \in Y\}$ and prove that $f(\mu)$ is an ε-generalized fuzzy weak implicative filter of Y.
10. Let μ and ν be two ε-generalized weak implicative filters of a BE-algebra X. Prove that the Cartesian product $\mu \times \nu$ is an ε-generalized weak implicative filter of $X \times X$.

Chapter 15
Transitive Filters

It is a general observation in a BE-algebra that the BE-ordering $\leq$ is reflexive but neither antisymmetric nor transitive. If the BE-algebra X is commutative, then the pair $(X, \leq)$ is a partially ordered set. If X is transitive, then $\leq$ satisfies only the transitive property. This is the exact reason to concentrate on the transitive property of the filters and to introduce the filters called transitive filters. It is also observed that every filter of a BE-algebra satisfies the transitive property whenever the BE-algebra is transitive. Along with the transitive property of filters, the distributive property is also studied in BE-algebras and introduced the notion of distributive and strong distributive filters in BE-algebras. It is also observed that every filter of a BE-algebra satisfies the distributive property whenever the BE-algebra is self-distributive.

It this chapter, the notion of transitive filters is introduced in BE-algebras. An equivalent condition is derived for every filter to become a transitive filter. The notions of distributive filters and strong distributive filters are introduced in BE-algebras and proved that the set of all distributive filters of a BE-algebra is properly contained in the set of all transitive filters as well as in the set of all implicative filters. The notion of semitransitive filters is introduced, and relations among semitransitive, transitive filters, weak implicative, associative filters, implicative filters are derived. The notion of fuzzy semitransitive filters is introduced in BE-algebras. Some properties of these fuzzy semitransitive filters are studied. Some sufficient conditions are derived for any fuzzy filter of a BE-algebra to become a fuzzy semitransitive filter. An extension property is derived for fuzzy semitransitive filters of BE-algebras.

Triangular normed fuzzification is applied to the semitransitive filters of BE-algebras. Some sufficient conditions are derived for every triangular normed fuzzy filter to become a triangular normed fuzzy semitransitive filter. Homomorphic images and inverse images of T-fuzzy semitransitive filters are studied using the imaginable property and the *sup* property of images.

S. R. Mukkamala, *A Course in BE-algebras*,
https://doi.org/10.1007/978-981-10-6838-6_15

15.1 Definition and Properties

In this section, the concepts of transitive filters, distributive filters, and strong distributive filters are introduced in BE-algebras and the properties of these filters are studied. An equivalent condition is obtained for every filter to become a transitive filter.

Definition 15.1.1 A non-empty subset F of a BE-algebra X is called a *transitive filter* if it satisfies the following properties, for all $x, y, z \in X$:

(TF1) $1 \in F$,
(TF2) $x * y \in F, y * z \in F$ imply that $x * z \in F$.

Obviously, the filter $\{1\}$ of a partially ordered BE-algebra is a transitive filter. For, consider $x, y, z \in X$ such that $x * y \in \{1\}$ and $y * z \in \{1\}$. Then $x \leq y$ and $y \leq z$. Hence $x \leq z$, which implies $x * z = 1 \in \{1\}$. Moreover, every filter of a transitive BE-algebra is a transitive filter.

Example 15.1.2 Let $X = \{a, b, c, d, 1\}$. Define an operation $*$ on X as follows:

$*$	a	b	c	d	1
a	1	b	c	d	1
b	a	1	c	c	1
c	1	b	1	b	1
d	1	1	1	1	1
1	a	b	c	d	1

Then clearly $(X, *, 1)$ is a BE-algebra. Now consider the set $F = \{b, 1\}$. It can be easily verified that F is a transitive filter of X.

Theorem 15.1.3 *Every implicative filter of a BE-algebra is transitive.*

Proof Let F be an implicative filter of X. Then F is a filter of X. Let $x, y, z \in X$ be such that $x * y \in F$ and $y * z \in F$. Since $y * z \in F$ and F is a filter, we get $x * (y * z) \in F$. Since F is implicative and $x * y \in F$, we get $x * z \in F$. Therefore F is a transitive filter of X. □

Proposition 15.1.4 *Let F be a transitive filter of X. Then the following hold:*

(1) *$x \in F, x \leq y$ imply that $y \in F$,*
(2) *$x \in F, x * y \in F$ imply that $y \in F$,*
(3) *Set intersection of transitive filters is again a transitive filter.*

Proof (1) Suppose that $x \in F$ and $x \leq y$. Then we get that $x * y = 1 \in F$. Since $1 * x \in F$ and $x * y \in F$, we get that $y = 1 * y \in F$.
(2) Let $x \in F$ and $x * y \in F$. Then $1 * x \in F$ and $x * y \in F$. Since F is transitive, we get $y \in F$.
(3) It is clear. □

From (1) and (2) of the above proposition, the following corollary is a direct consequence.

Corollary 15.1.5 *Every transitive filter of a BE-algebra is a filter.*

But the converse of the above corollary is not true. It can be observed in the following example.

Example 15.1.6 Let $X = \{a, b, c, d, 1\}$. Define a binary operation $*$ on X as follows:

$*$	a	b	c	d	1
a	1	b	c	b	1
b	a	1	b	a	1
c	a	1	1	a	1
d	1	1	b	1	1
1	a	b	c	d	1

Then $(X, *, 1)$ is a BE-algebra. It is easy to check that $F = \{a, c, 1\}$ is a filter of X but not a transitive filter, since $a * c = c \in F$ and $c * b = 1 \in F$ but $a * b = b \notin F$.

In the following theorem, a sufficient condition is derived for a filter of a BE-algebra to become a transitive filter.

Theorem 15.1.7 *Let F be a filter of X. Then F is a transitive filter if $x * (y * z) \in F$ implies $(x * y) * z \in F$ for all $x, y, z \in X$.*

Proof Let F be a filter of X which is satisfying the given condition. Let $x, y, z \in X$ be such that $x * y \in F$ and $y * z \in F$. Since $y * z \leq x * (y * z)$, we can get $x * (y * z) \in F$. By the assumed condition, we get $(x * y) * z \in F$. Since F is a filter and $x * y \in F$, it yields that $z \in F$. Since $z \leq x * z$, we get $x * z \in F$. Therefore F is a transitive filter of X. □

For any filter F, the relation θ_F on X defined by $(x, y) \in \theta_F$ if and only if $x * y \in F$ and $y * x \in F$ for all $x, y \in X$ is studied in Chap. 2. It is observed that θ_F is congruence whenever X is transitive. In the following, the relation between transitive filters and θ_F is studied.

Lemma 15.1.8 *If F is a transitive filter of X, then θ_F is an equivalence relation on X.*

Proof Since $x * x = 1 \in F$, we get $(x, x) \in \theta_F$ for any $x \in X$. Hence θ_F is reflexive. Clearly θ_F is symmetric. Again, let $(x, y) \in \theta_F$ and $(y, z) \in \theta_F$. Then $x * y, y * x, y * z, z * y \in F$. Since F is transitive, we get $x * z, z * x \in F$. Hence $(x, z) \in \theta_F$. Thus θ_F is an equivalence relation on X. □

In the following, the notion of *distributive filters* is introduced in BE-algebras.

Definition 15.1.9 A filter F of a BE-algebra X is said to be a *distributive filter* of X if it satisfies the following property for all $x, y, z \in X$;

(D1) $x * (y * z) \in F$ implies $(x * y) * (x * z) \in F$.

Proposition 15.1.10 *Every distributive filter of a BE-algebra is a transitive filter.*

Proof Let F be a distributive filter of a BE-algebra X. Let $x, y, z \in X$ be such that $x * y \in F$ and $y * z \in F$. Since $y * z \leq x * (y * z)$, we get $x * (y * z) \in F$. Since F is distributive, we get $(x * y) * (x * z) \in F$. Since $x * y \in F$, it gives that $x * z \in F$. Therefore F is transitive. □

Proposition 15.1.11 *Every distributive filter of a BE-algebra is an implicative filter.*

Proof Let F be distributive filter of a BE-algebra X. Clearly F is a filter of X. Let $x, y, z \in X$ be such that $x * (y * z) \in F$ and $x * y \in F$. Since F is distributive and $x * (y * z) \in F$, we get $(x * y) * (y * z) \in F$. Since F is a filter, it yields that $x * z \in F$. Therefore F is implicative. □

Theorem 15.1.12 *If F is a distributive filter of a BE-algebra X, then the relation $\theta_F = \{(x, y) \in X \times X \mid x * y \in F \text{ and } y * x \in F\}$ is a weak congruence on X.*

Proof In Lemma 15.1.8, it is proved that θ_F is an equivalence relation on X. Let $(x, y) \in \theta_F$. Then we get $x * y \in F$ and $y * x \in F$. Since F is a filter, we get $z * (x * y) \in F$ and $z * (y * x) \in F$. Since F is a distributive filter, we get $(z * x) * (z * y) \in F$ and $(z * y) * (z * x) \in F$. Hence $(z * x, z * y) \in \theta_F$. Therefore θ_F is a weak congruence on X. □

In the following, the notion of *strong distributive filters* is introduced in BE-algebras.

Definition 15.1.13 A filter F of a BE-algebra X is said to be a *strong distributive filter* if, for all $x, y, z \in X$, it satisfies the following condition:

(D2) $x * (y * z) \in F$ implies $(x * y) * (x * z) \in F$ and $(z * y) * (z * x) \in F$.

Clearly every strong distributive filter of a BE-algebra is a distributive filter.

Theorem 15.1.14 *If F is a strong distributive filter of a BE-algebra X, the relation θ_F is a congruence on X.*

Proof In Lemma 15.1.8, it is proved that θ_F is an equivalence relation on X. Let $(x, y) \in \theta_F$. Then by Theorem 15.1.12, we get that $(z * x, z * y) \in \theta_F$. Moreover, $(y * z) * (x * z) \in F$ and $(x * z) * (y * z) \in F$. Hence by the definition of θ_F, it yields that $(x * z, y * z) \in \theta_F$.

Now, let $(x, y) \in \theta_F$ and $(u, v) \in \theta_F$ for $x, y, u, v \in X$. Then by the above condition, we get that $(x * u, y * u) \in \theta_F$. Again, by the above condition, we get that $(y * u, y * v) \in \theta_F$. Since θ_F is transitive, we get $(x * u, y * v) \in \theta_F$. Therefore θ_F is a congruence on X. □

Corollary 15.1.15 *If F is a distributive filter, then $Ker(\theta_F)$ is a distributive filter.*

Proof Let F be a distributive filter of a BE-algebra X. Then clearly $Ker(\theta_F)$ is a filter of X. Let $x * (y * z) \in Ker(\theta_F)$. Then $(x * (y * z), 1) \in \theta_F$. Hence $x * (y * z) \in F$. Since F is distributive, it implies that $(x * y) * (x * z) \in F$. Thus $((x * y) * (x * z), 1) \in \theta_F$. Hence $(x * y) * (x * z) \in Ker(\theta_F)$. Therefore $Ker(\theta_F)$ is a distributive filter of X. □

Corollary 15.1.16 *If F is a strong distributive filter, then $Ker(\theta_F)$ is a strong distributive filter.*

Theorem 15.1.17 *If the smallest filter $\{1\}$ is strong distributive, then $Ker(\Delta_X)$ is strong distributive where Δ_X is the smallest congruence.*

Proof Assume that $\{1\}$ is strong distributive. Clearly $Ker(\Delta_X)$ is a filter of X. Let $x * (y * z) \in Ker(\Delta_X)$. Then $x * (y * z) = 1$. Since $\{1\}$ is strong distributive, we get $(x * y) * (x * z) \in \{1\}$ and $(z * y) * (z * x) \in \{1\}$. Hence $(x * y) * (x * z) = (z * y) * (z * x)$. Thus, it yields $(x * y) * (x * z), (z * y) * (z * x) \in Ker\ \Delta_X$. Therefore $Ker(\Delta_X)$ is strong distributive. □

15.2 Applications of $\mathcal{N}$-structure to Transitive Filters

In this section, the notion of $\mathcal{N}$-fuzzy transitive filters is introduced in BE-algebras. In [136], Kang and Jun introduced the notion of $\mathcal{N}$-ideals of BE-algebras and investigated some significant properties of these $\mathcal{N}$-ideals of BE-algebras. In this section, this $\mathcal{N}$-structure is applied to transitive filters and investigated some related properties.

Definition 15.2.1 ([136]) Let X be a non-empty set. A function $f : X \to [-1, 0]$ is called a negative-valued function (briefly $\mathcal{N}$-function or $\mathcal{N}$-fuzzy set) from X into $[-1, 0]$.

Let us denote the set of all $\mathcal{N}$-functions from the set X to $[1, 0]$ by $\mathcal{F}(X, [-1, 0])$. By an $\mathcal{N}$-structure, we mean an ordered pair (X, f) of X and an $\mathcal{N}$-function f on X. For any $t \in [-1, 0]$ the set $C(f, t) = \{x \in X \mid f(x) \leq t\}$ is called a closed (f, t)-cut of (X, f).

Definition 15.2.2 ([136]) Let f be an $\mathcal{N}$-function on a BE-algebra X. Then f is called an $\mathcal{N}$-filter of X if it satisfies the following property:

$$(\forall t \in [-1, 0])(C(f, t) \in \mathcal{J}(X) \cup \{\emptyset\}).$$

where $\mathcal{J}(X)$ is the set of all ideals of X.

Definition 15.2.3 ([136]) Let f be an $\mathcal{N}$-fuzzy set of a BE-algebra X. Then f is called an $\mathcal{N}$-fuzzy filter of X if it satisfies the following properties:

(NFF1) $f(1) \leq f(x)$ for all $x \in X$,
(NFF2) $f(y) \leq \max\{f(x), f(x * y)\}$ for all $x, y \in X$.

Every $\mathcal{N}$-filter is order reversing (i.e., $x, y \in X, x \leq y$ imply $f(y) \leq f(x)$).

Definition 15.2.4 Let f be an $\mathcal{N}$-fuzzy set of a BE-algebra X. Then f is called an $\mathcal{N}$-fuzzy transitive filter of X if it satisfies the following properties:

(NFT1) $f(1) \leq f(x)$ for all $x \in X$,
(NFT2) $f(x * z) \leq \max\{f(x * y), f(y * z)\}$ for all $x, y \in X$.

Example 15.2.5 Let $X = \{1, a, b, c, d\}$. Define a binary operation $*$ on X as follows:

$*$	1	a	b	c	d
1	1	a	b	c	d
a	1	1	b	c	d
b	1	a	1	c	c
c	1	1	b	1	b
d	1	1	1	1	1

Then $(X, *, 1)$ is a BE-algebra. Define an $\mathcal{N}$-fuzzy set $f : X \longrightarrow [-1, 0]$ as follows:

$$f(x) = \begin{cases} -0.8 & \text{if } x = 1, b \\ -0.3 & \text{otherwise} \end{cases}$$

for all $x \in X$. It can be easily verified that f is an $\mathcal{N}$-fuzzy transitive filter of X.

Proposition 15.2.6 *Every $\mathcal{N}$-fuzzy transitive filter of a BE-algebra is an $\mathcal{N}$-fuzzy filter.*

Proof Let f be an $\mathcal{N}$-fuzzy transitive filter of a BE-algebra X. Hence we get

$$\begin{aligned} f(y) &= f(1 * y) \\ &\leq \max\{f(1 * x), f(x * y)\} \\ &= \max\{f(x), f(x * y)\} \end{aligned}$$

Therefore f is an $\mathcal{N}$-fuzzy filter of X. □

The converse of the above proposition is not true. That is an $\mathcal{N}$-fuzzy filter of a BE-algebra need not be an $\mathcal{N}$-fuzzy transitive filter as shown in the following example.

Example 15.2.7 Let $X = \{1, a, b, c, d\}$. Define a binary operation $*$ on X as follows:

$*$	1	a	b	c	d
1	1	a	b	c	d
a	1	1	b	c	d
b	1	a	1	b	a
c	1	a	1	1	a
d	1	1	1	1	1

Clearly $(X, *, 1)$ is a BE-algebra. Define an $\mathcal{N}$-fuzzy set f on X as follows:

$$f(x) = \begin{cases} -0.6 & \text{if } x = 1 \\ -0.3 & \text{otherwise} \end{cases}$$

for all $x \in X$. Then clearly f is an $\mathcal{N}$-fuzzy filter of X, but f is not an $\mathcal{N}$-fuzzy transitive filter of X since $f(a * a) \nleq \max\{f(a * b), f(b * a)\}$.

We now derive some sufficient conditions for every $\mathcal{N}$-fuzzy filter of a BE-algebra to become an $\mathcal{N}$-fuzzy transitive filter.

Theorem 15.2.8 *Every $\mathcal{N}$-fuzzy filter f of a BE-algebra X is an $\mathcal{N}$-fuzzy transitive filter if it satisfies the following condition for all $x, y, z \in X$.*

(NFT3) $f(y) \leq \max\{f(y * z), f(x * (z * y))\}$

Proof Let f be an $\mathcal{N}$-fuzzy set of X with condition (NFT3). Then we get

$$\begin{aligned} f(y) &\leq \max\{f(x * (z * y)), f(y * z)\} \\ &\leq \max\{f(z * (x * y)), f(y * z)\} \\ &\leq \max\{f(x * y), f(y * z)\} \quad \text{since } x * y \leq z * (x * y) \end{aligned}$$

Since f is an $\mathcal{N}$-fuzzy filter, we get the following consequence:

$$\begin{aligned} f(x * z) &\leq \max\{f(y), f(y * (x * z))\} \\ &= \max\{f(y), f(x * (y * z))\} \\ &\leq \max\{\max\{f(x * y), f(y * z)\}, f(y * z)\} \\ &= \max\{f(x * y), f(y * z)\}. \end{aligned}$$

Therefore f is an $\mathcal{N}$-fuzzy transitive filter of X. □

Theorem 15.2.9 *Let f be an $\mathcal{N}$-fuzzy filter of a BE-algebra X. Then f is an $\mathcal{N}$-fuzzy transitive filter of X if it satisfies the following condition, for all $x, y, z \in X$:*

(NFT4) $f((x * y) * (x * z)) \leq \max\{f(x * y), f((x * y) * (y * z))\}$.

Proof Assume that f satisfies the condition (NFT4). Clearly $f(1) \leq f(x)$ for all $x \in X$. Let $x, y, z \in X$. Clearly $f(y * z) \geq f((x * y) * (y * z))$. Since μ is a fuzzy filter, by the assumed condition, we get the following:

$$\begin{aligned} f(x * z) &\leq \max\{f(x * y), f((x * y) * (x * z))\} \\ &\leq \max\{f(x * y), \max\{f(x * y), f((x * y) * (y * z))\}\} \\ &= \max\{f(x * y), f((x * y) * (y * z))\} \\ &\leq \max\{f(x * y), f(y * z)\} \end{aligned}$$

Therefore f is an $\mathcal{N}$-fuzzy transitive filter of X. □

Definition 15.2.10 Let f be an $\mathcal{N}$-fuzzy subset of a BE-algebra X. For any $\alpha \in [-1, 0]$, the set $f_\alpha = \{x \in X \mid f(x) \leq \alpha\}$ is called an $\mathcal{N}$-level subset of X.

Theorem 15.2.11 *Let f be an $\mathcal{N}$-fuzzy subset of a BE-algebra X. Then f is an $\mathcal{N}$-fuzzy transitive filter of X if and only if the $\mathcal{N}$-level subset $f_\alpha \neq \emptyset$ is a transitive filter of X for each $\alpha \in [-1, 0]$.*

Proof Assume that f is an $\mathcal{N}$-fuzzy transitive filter of X. Then $f(1) \leq f(x)$ for all $x \in X$. In particular, for $x \in f_\alpha$, we get $f(1) \leq f(x) \leq \alpha$. Hence $1 \in f_\alpha$. Let $x * y \in f_\alpha$ and $y * z \in f_\alpha$. Then $f(x * y) \leq \alpha$ and $f(y * z) \leq \alpha$. Since f is an $\mathcal{N}$-fuzzy transitive filter, we get $f(x * z) \leq \max\{f(x * y), f(y * z)\} \leq \max\{\alpha, \alpha\} = \alpha$. Hence $x * z \in f_\alpha$. Thus f_α is a transitive filter.

Conversely, assume that for each $\alpha \in [-1, 0]$, the $\mathcal{N}$-level subset $f_\alpha \neq \emptyset$ is a transitive filter of X. Suppose there exists $x_0 \in X$ such that $f(1) > f(x_0)$. Choose $\alpha_0 = \frac{1}{2}(f(1) + f(x_0))$. Then $f(1) > \alpha_0$ and $-1 \leq f(x_0) < \alpha_0 \leq 0$. Hence $x_0 \in f_{\alpha_0}$ and hence $f_{\alpha_0} \neq \emptyset$. Since f_{α_0} is a transitive filter of X, we get $1 \in f_{\alpha_0}$. Hence $f(1) \leq \alpha_0$, which is a contradiction. Therefore $f(1) \leq f(x)$ for all $x \in X$. Let $x, y, z \in X$ be such that $f(x * y) = \alpha_1$ and $f(y * z) = \alpha_2$. Then $x * y \in f_{\alpha_1}$ and $y * z \in f_{\alpha_2}$. Without loss of generality, assume that $\alpha_1 \leq \alpha_2$. Then clearly $f_{\alpha_1} \subseteq f_{\alpha_2}$. Hence $x * y \in f_{\alpha_2}$. Since f_{α_2} is an $\mathcal{N}$-fuzzy transitive filter, we get $x * z \in f_{\alpha_2}$. Thus $f(x * z) \leq \alpha_2 = \max\{\alpha_1, \alpha_2\} = \max\{f(x * y), f(y * z)\}$. Hence f is an $\mathcal{N}$-fuzzy transitive filter. □

Theorem 15.2.12 *Let f be a $\mathcal{N}$-fuzzy transitive filters of a BE-algebra X. Then two $\mathcal{N}$-level transitive filters f_{α_1} and f_{α_2} (with $\alpha_1 < \alpha_2$) of f are equal if and only if there is no $x \in X$ such that $\alpha_1 \leq f(x) < \alpha_2$.*

Proof Assume that $f_{\alpha_1} = f_{\alpha_2}$ for $\alpha_1 < \alpha_2$. Suppose there exists some $x \in X$ such that $\alpha_1 \leq f(x) < \alpha_2$. Then f_{α_2} is a proper subset of f_{α_1}, which is impossible. Conversely, assume that there is no $x \in X$ such that $\alpha_1 \leq f(x) < \alpha_2$. Since $\alpha_1 < \alpha_2$, we get $f_{\alpha_1} \subseteq f_{\alpha_2}$. If $x \in f_{\alpha_2}$, then $f(x) \leq \alpha_2$. Hence by assume condition, we get $f(x) \leq \alpha_1$. Hence $x \in f_{\alpha_1}$ and so $f_{\alpha_2} \subseteq f_{\alpha_1}$. Therefore $f_{\alpha_1} = f_{\alpha_1}$. □

Theorem 15.2.13 *Let F be a transitive filter of a BE-algebra X. Then there exists an $\mathcal{N}$-fuzzy transitive filter f such that $f_\alpha = F$ for some $\alpha \in [-1, 0]$.*

Proof Let F be a transitive filter of a BE-algebra X. Define an $\mathcal{N}$ fuzzy set $f : X \longrightarrow [-1, 0]$ as follows:

$$f(x) = \begin{cases} \alpha & \text{if } x \in F \\ 0 & \text{if } x \notin F \end{cases}$$

where $-1 < \alpha < 0$ is fixed. Since $1 \in F$, we get that $f(1) = \alpha \leq f(x)$ for all $x \in X$. Let $x, y \in X$. Then we can have the following cases:

Case I: Suppose $x * y, y * z \in F$. Since F is a transitive filter, we get $x * z \in F$. Then $f(x * y) = f(y * z) = f(x * z) = \alpha$. Hence $f(x * z) \leq \max\{f(x * y), f(y * z)\}$.

Case II: Suppose $x * y \notin F$ and $y * z \notin F$. Then $f(x * y) = f(y * z) = 0$. Hence $f(x * z) \leq \max\{f(x * y), f(y * z)\}$.

Case III: If exactly one of $x * y$ and $y * z$ is in F, then exactly one of $f(x * y)$ and $f(y * z)$ is equal to 0. Hence $f(x * z) \leq \max\{f(x * y), f(y * z)\}$.

Thus $f(x * z) \leq \max\{f(x * y), f(y * z)\}$ for all $x, y \in X$. Therefore f is an $\mathcal{N}$-fuzzy transitive filter of X. Clearly $f_\alpha = F$. Hence the proof is completed. □

Corollary 15.2.14 *Any transitive filter of a BE-algebra X can be realized as a level transitive filter of some fuzzy transitive filter of X.*

Definition 15.2.15 Let f and g be two $\mathcal{N}$-fuzzy sets of a BE-algebra X. The Cartesian product of f and g is defined by $(f \times g)(x, y) = \max\{f(x), g(y)\}$ for all $x, y \in X$.

Theorem 15.2.16 *Let f and g be two $\mathcal{N}$-fuzzy transitive filters of a BE-algebra X. Then $f \times g$ is an $\mathcal{N}$-fuzzy transitive filter of $X \times X$.*

Proof Let $(x, y) \in X \times X$. Since f and g are $\mathcal{N}$-fuzzy transitive filters of X, we get

$$\begin{aligned}(f \times g)(1, 1) &= \max\{f(1), g(1)\} \\ &\leq \max\{f(x), g(y)\} \qquad \text{for all } x, y \in X \\ &= (f \times g)(x, y)\end{aligned}$$

Let $(x, x'), (y, y'), (z, z') \in X \times X$. Since f and g are $\mathcal{N}$-fuzzy transitive filters of X, we can obtain the following consequence.

$$\begin{aligned}(f \times g)((x, x') * (z, z')) &= (f \times g)(x * z, x' * z') \\ &= \max\{f(x * z), g(x' * z')\} \\ &\leq \max\{\max\{f(x * y), f(y * z)\}, \max\{g(x' * y'), g(y' * z')\}\} \\ &= \max\{\max\{f(x * y), g(x' * y')\}, \max\{f(y * z), g(y' * z')\}\} \\ &= \max\{(f \times g)(x * y, x' * y), (f \times g)(y * z, y' * z')\} \\ &= \max\{(f \times g)((x, x') * (y, y')), (f \times g)(y * y', z * z')\}\end{aligned}$$

Therefore $f \times g$ is an $\mathcal{N}$-fuzzy transitive filter of $X \times X$. □

15.3 Transitive Hyper Filters of Hyper BE-algebra

In this section, the notion of transitive hyper filters is introduced in hyper BE-algebras. Relations among transitive hyper filters, implicative hyper filters, and hyper filters are studied. Properties of transitive hyper filters are studied.

Definition 15.3.1 ([197]) Let F be a non-empty subset of a hyper BE-algebra H and $1 \in F$. Then F is called

(1) a weak hyper filter of H if $x \circ y \subseteq F$ and $x \in F$ imply $y \in F$ for all $x, y \in H$;

(2) a hyper filter of H if $x \circ y \approx F$ and $x \in F$ imply $y \in F$, where $x \circ y \approx F$ means that $x \circ y \cap F \neq \emptyset$ for all $x, y \in H$.

Theorem 15.3.2 ([197]) *Let F be a hyper filter of a hyper BE-algebra H. For all $x, y \in H$, if $x \in F$ and $x < y$, then $y \in F$.*

Definition 15.3.3 Let F be a non-empty subset of a hyper BE-algebra H. Then F is said a transitive hyper filter of H if it satisfies the following condition for all $x, y, z \in H$:

$$x \circ y \cap F \neq \emptyset \text{ and } y \circ z \cap F \neq \emptyset \text{ imply that } x \circ z \cap F \neq \emptyset.$$

Example 15.3.4 Let $H = \{1, a, b\}$. Define the hyperoperation $\circ$ on X as follows:

$\circ$	1	a	b
1	$\{1\}$	$\{a\}$	$\{b\}$
a	$\{1\}$	$\{1\}$	$\{b\}$
b	$\{1\}$	$\{1, a\}$	$\{1\}$

It can be routinely verified that $(H, \circ, 1)$ is a hyper BE-algebra. Consider the set $F = \{1, a\}$. It can be easily seen that F is a transitive hyper filter of H.

In general, a transitive hyper filter of a hyper BE-algebra is not a hyper filter. For, consider the following example:

Example 15.3.5 Let $H = \{1, a, b\}$. Define the hyperoperation $\circ$ on X as follows:

$\circ$	1	a	b
1	$\{1\}$	$\{a, b\}$	$\{b\}$
a	$\{1\}$	$\{1, a\}$	$\{1, b\}$
b	$\{1\}$	$\{1, a, b\}$	$\{1\}$

It can be routinely verified that $(H, \circ, 1)$ is a hyper BE-algebra. Consider the set $F = \{1, b\}$. It can be easily seen that F is a transitive hyper filter of H but not a hyper filter of H because of $b \in F$ and $b \circ a \cap F = \{1, b\} \neq \emptyset$ but $a \notin F$.

Theorem 15.3.6 *Let $(H, \circ, 1)$ be an R-hyper BE-algebra and F is a non-empty subset of H. If F is a transitive hyper filter of H, then F is a hyper filter of H.*

Proof Let F be a transitive hyper filter of H. Let $x, y \in H$ be such that $x \in F$ and $x \circ y \approx F$. Hence $x \circ y \cap F \neq \emptyset$. Since H is R-hyper BE-algebra, we get $1 \circ x = x \in F$. Hence $1 \circ x \cap F \neq \emptyset$. Since F is transitive hyper filter, we get $1 \circ y \cap F \neq \emptyset$. Thus $y = 1 \circ y \in F$. Therefore F is a hyper filter of H. □

Theorem 15.3.7 *Let F be a non-empty subset of an R-hyper BE-algebra H. Then F is a transitive hyper filter of H if and only if*

(1) $1 \in F$,
(2) $F < x \circ y$ *and* $F < y \circ z$ *imply that* $F < x \circ z$ *for all* $x, y, z \in H$.

Proof Assume that F is a transitive hyper filter of H. Clearly (1) holds. Let $x, y, z \in H$ be such that $F < x \circ y$ and $F < y \circ z$. Then there exist $a, c \in F, b \in x \circ y$, and $d \in y \circ z$ such that $a < b$ and $c < d$. Since F is a hyper filter and $a, c \in F$, we get $b, d \in F$. Hence $b \in x \circ y \cap F$ and $d \in y \circ z \cap F$. Thus $x \circ y \cap F \neq \emptyset$ and $y \circ z \cap F \neq \emptyset$. Since F is transitive hyper filter, we get $x \circ z \cap F \neq \emptyset$. Therefore $F < x \circ z$.

Conversely assume that (1) and (2) hold. Let $x, y, z \in H$. Suppose $x \circ y \approx F$ and $y \circ z \approx F$, which means $x \circ y \cap F \neq \emptyset$ and $y \circ z \cap F \neq \emptyset$. Choose $a \in F$ and $a \in x \circ y$. Then clearly $F < x \circ y$. Similarly, we get $F < y \circ z$. By the condition (2), we get $F < x \circ z$. Then there exists $a \in F$ and $b \in x \circ z$ such that $a < b$. Since F is a hyper filter and $a < b$, we get $b \in F$. Hence $x \circ z \cap F \neq \emptyset$. Therefore F is a transitive hyper filter of H. □

In [45], X.Y. Cheng and X.L. X in introduced the notion of implicative hyper filters of hyper BE-algebras and the properties of implicative hyper filters are studied. In the following, we establish a relation between implicative hyper filters and transitive hyper filters of hyper BE-algebras.

Theorem 15.3.8 *Every implicative hyper filter of a hyper BE-algebra is a transitive hyper filter.*

Proof Let F be an implicative hyper filter of a hyper BE-algebra H. Clearly $1 \in F$. Let $x, y, z \in H$ be such that $x \circ y \cap F \neq \emptyset$ and $y \circ z \cap F \neq \emptyset$. Clearly $y \circ z \subseteq x \circ (y \circ z)$. Since $y \circ z \cap F \neq \emptyset$, we get $x \circ (y \circ z) \cap F \neq \emptyset$. Since F is hyper implicative and $x \circ y \cap F \neq \emptyset$, we get $x \circ z \cap F \neq \emptyset$. Therefore F is a transitive hyper filter of H.□

Theorem 15.3.9 *Let* $\{F_i \mid i \in \Delta\}$ *be an indexed family of transitive hyper filters of an R-hyper BE-algebra* H. *Then*

(1) $\bigcap_{i \in \Delta} F_i$ *is a transitive hyper filter of* H,
(2) *if* $\{F_i \mid i \in \Delta\}$ *is a chain of transitive hyper filters of* H *for all* $i \in \Delta$, *then* $\bigcup_{i \in \Delta} F_i$ *is a transitive hyper filter of* H.

Proof (1) Let F_i be a transitive hyper filter of H for each $i \in \Delta$. Since H is R-hyper, we get that F_i is a hyper filter of H for each $i \in \Delta$. Then $\bigcap_{i \in \Delta} F_i$ is a hyper filter of H. Clearly $1 \in \bigcap_{i \in \Delta} F_i$. Let $x, y, z \in H$. Suppose $x \circ y \cap \left(\bigcap_{i \in \Delta} F_i\right) \neq \emptyset$ and $y \circ z \cap \left(\bigcap_{i \in \Delta} F_i\right) \neq \emptyset$. Choose $a \in x \circ y \cap \left(\bigcap_{i \in \Delta} F_i\right)$ and $b \in y \circ z \cap \left(\bigcap_{i \in \Delta} F_i\right)$. Since $a \in F_i$ for each $i \in \Delta$, we get $F_i < x \circ y$ for each $i \in \Delta$. Similarly, we get $F_i < y \circ z$ for each $i \in \Delta$. Since each F_i is a transitive hyper filter of H, we get $F_i < x \circ z$ for each $i \in \Delta$. Hence $\bigcap_{i \in \Delta} F_i < x \circ z$. Since $\bigcap_{i \in \Delta} F_i$ is a hyper filter, we get $x \circ z \cap \left(\bigcap_{i \in \Delta} F_i\right) \neq \emptyset$. Therefore $\bigcap_{i \in \Delta} F_i$ is a transitive hyper filter of H.

(2) Suppose $\{F_i \mid i \in \Delta\}$ is a chain of transitive hyper filters of H for all $i \in \Delta$. Since H is R-hyper, we get that F_i is a hyper filter of H for each $i \in \Delta$. Hence $\bigcup_{i\in\Delta} F_i$ is a hyper filter of H. Clearly $1 \in \bigcup_{i\in\Delta} F_i$. Let $x, y, z \in H$. Suppose $x \circ y \cap \left(\bigcup_{i\in\Delta} F_i\right) \neq \emptyset$ and $y \circ z \cap \left(\bigcup_{i\in\Delta} F_i\right) \neq \emptyset$. Then there exists $m, n \in \Delta$ such that $x \circ y \cap F_m \neq \emptyset$ and $y \circ z \cap F_n \neq \emptyset$. Since $\{F_i \mid i \in \Delta\}$ is a chain, we get either $x \circ y, y \circ z \cap F_m \neq \emptyset$ and $x \circ y, y \circ z \cap F_n \neq \emptyset$. Since each F_i is a transitive hyper filter, we get either $x \circ z \cap F_m \neq \emptyset$ or $x \circ z \cap F_n \neq \emptyset$. Hence $x \circ z \cap \left(\bigcup_{i\in\Delta} F_i\right) \neq \emptyset$. Therefore $\bigcup_{i\in\Delta} F_i$ is a transitive hyper filter of H. □

For any two hyper BE-algebras $(H_1, \circ_1, 1_1)$ and $(H_2, \circ_2, 1_2)$ with $H = H_1 \times H_2$, define a hyperoperation $\circ$ on H by

$$(x_1, y_1) \circ (x_2, y_2) = (x_1 \circ x_2, y_1 \circ y_2)$$

for $(x_1, y_1), (x_2, y_2) \in H$. Then it is proved in [45] that $(H, \circ, 1)$ is a hyper BE-algebra with the greatest element $(1_1, 1_2)$. The above hyper BE-algebra is called a product hyper BE-algebra of H_1 and H_2. In the following, we prove the result for product of transitive hyper filters.

Theorem 15.3.10 *Let $(H_1, \circ_1, 1_1)$ and $(H_2, \circ_2, 1_2)$ be two R-hyper BE-algebras. If F_1 and F_2 are two transitive hyper filters of H_1 and H_2, respectively, then $F_1 \times F_2$ is a transitive hyper filter of the product hyper algebra $H_1 \times H_2$.*

Proof Assume that F_1 and F_2 are two transitive hyper filters of H_1 and H_2, respectively. Since H is R-hyper, we get F_1 and F_2 are hyper filters of H_1 and H_2, respectively. Hence $1_1 \in F_1$ and $1_2 \in F_2$. Thus $(1_1, 1_2) \in F_1 \times F_2$. Let $(x_1, x_2), (x_2, y_2), (x_3, y_3) \in H_1 \times H_2$. Let $F_1 \times F_2 < (x_1, y_1) \circ (x_2, y_2)$ and $F_1 \times F_2 < (x_2, y_2) \circ (x_3, y_3)$. Then $F_1 < x_1 \circ y_1$ and $F_1 < x_2 \circ y_2$. Since F_1 is transitive hyper filter, we get $F_1 < x_1 \circ x_3$. Similarly, we get $F_2 < y_1 \circ y_3$. Hence $F_1 \times F_2 < (x_1, y_1) \circ (x_3, y_3)$. Therefore $F_1 \times F_2$ is transitive hyper filter of $H_1 \times H_2$. □

15.4 Semitransitive Filters of BE-algebras

In this section, the concept of semitransitive filters is introduced in BE-algebras. Some sufficient conditions are derived for every filter to become a semitransitive filter. Interconnections among weak implicative filters, associative filters, semitransitive filters are observed. A necessary and sufficient condition is derived for every semitransitive filter to become implicative.

Definition 15.4.1 A non-empty subset F of a BE-algebra X is said to be a *semitransitive filter* of X if it satisfies the following properties, for all $x, y, z \in X$:

(STF1) $1 \in F$,
(STF2) $x * y \in F, y * z \in F$ imply that $x * (x * z) \in F$.

Proposition 15.4.2 *Every semitransitive filter of a BE-algebra is a filter.*

Proof Let F be a semitransitive filter of a BE-algebra X. Let $x, y \in X$. Suppose $x \in F$ and $x * y \in F$. Then we get $1 * x = x \in F$ and $x * y \in F$. Since F is a semitransitive filter, we get $y = 1 * y = 1 * (1 * y) \in F$. Therefore F is a filter of X. □

In a general BE-algebra, every filter need not be semitransitive. It can be seen in the BE-algebra of Example 15.1.6. Consider $F = \{a, c, 1\}$. It is easy to check that F is a filter of X. Take $a, b, c \in X$. Clearly $a * c = c \in F$ and $c * b = 1 \in F$ but $a * (a * b) = a * b = b \notin F$. Therefore F is not a semitransitive filter. However, in the following theorem, a set of sufficient conditions is derived for every filter of a BE-algebra to become semitransitive.

Theorem 15.4.3 *A filter F of a BE-algebra X is semitransitive if it satisfies one of the following:*

(1) $x * (y * z) \in F$ *implies* $x * ((x * y) * (x * z)) \in F$,
(2) $(x * z) * (y * z) \in F$ *implies* $(x * y) * (x * (x * z)) \in F$,

(3) $x * ((z * y) * y) \in F$ *implies* $(y * z) * (x * (x * z)) \in F$.

Proof (1) Assume that F satisfies the condition (1). Let $x, y, z \in X$. Suppose $x * y \in F$ and $y * z \in F$. Clearly $y * z \leq x * (y * z)$. Since $y * z \in F$ and F is a filter, we get $x * (y * z) \in F$. By the given condition, we get $(x * y) * (x * (x * z)) = x * ((x * y) * (x * z)) \in F$. Since $x * y \in F$ and F is a filter, we get $x * (x * z) \in F$. Therefore F is semitransitive.
(2) Assume that F satisfies the condition (2). Let $x, y, z \in X$. Suppose $x * y \in F$ and $y * z \in F$. Clearly $y * z \leq (x * z) * (y * z)$. Since $y * z \in F$ and F is a filter, we get that $(x * z) * (y * z) \in F$. By the given condition, we get $(x * y) * (x * (x * z)) \in F$. Since $x * y \in F$ and F is a filter, we get $x * (x * z) \in F$. Therefore F is semitransitive.
(3) Assume that F satisfies the condition (3). Let $x, y, z \in X$. Suppose $x * y \in F$ and $y * z \in F$. Clearly $x * y \leq (z * y) * (x * y)$. Since $x * y \in F$ and F is a filter, we get $x * ((z * y) * y) = (z * y) * (x * y) \in F$. By the given condition, we get $(y * z) * (x * (x * z)) \in F$. Since $y * z \in F$ and F is a filter, we get $x * (x * z) \in F$. Therefore F is semitransitive. □

The following two propositions are trivial due to structures of respective BE-algebras.

Proposition 15.4.4 *Every filter of a transitive BE-algebra is semitransitive.*

Proposition 15.4.5 *Every filter of a simple BE-algebra is semitransitive.*

Proposition 15.4.6 *Every transitive filter of a BE-algebra is a semitransitive filter.*

Proof Let $(X, *, 1)$ be a BE-algebra and F be a transitive filter of X. Then clearly F is a filter of X. Let $x, y, z \in X$ be such that $x * y \in F$ and $y * z \in F$. Since F is a transitive filter, we get that $x * z \in F$. Since F is a filter of X and $x * z \leq x * (x * z)$, it yields that $x * z \in F$. Therefore F is a semitransitive filter of X. □

Proposition 15.4.7 *Every associative filter of a BE-algebra is semitransitive.*

Proof Let F be an associative filter of a BE-algebra X. Clearly F is a filter of X. Let $x, y, z \in X$ be such that $x * y \in F$ and $y * z \in F$. Since $y * z \leq x * (y * z)$ and F is a filter, we get $x * (y * z) \in F$. Since $x * (y * z) \in F$; $x * y \in$ and F is associative, we get $z \in F$. Since F is a filter, we get $x * (x * z) \in F$. Therefore F is semitransitive. □

Proposition 15.4.8 *Every weak implicative filter of a BE-algebra is semitransitive.*

Proof Let F be a weak implicative filter of a BE-algebra X. Clearly F is a filter. Let $x, y, z \in X$ be such that $x * y \in F$ and $y * z \in F$. Since $y * z \leq x * (y * z)$ and F is a filter, we get $x * (y * z) \in F$. Since F is weak implicative, we get $x * (x * z) \in F$. Therefore F is semitransitive. □

Corollary 15.4.9 *Every implicative filter of a BE-algebra is semitransitive.*

Though the converse of the above corollary is not true in general, we derive a necessary and sufficient condition for a semitransitive filter of a BE-algebra to become implicative.

Theorem 15.4.10 *A semitransitive filter of a BE-algebra X is an implicative filter if and only if it satisfies the condition: $x * (x * (x * y)) \in F$ implies $x * y \in F$ for all $x, y \in X$.*

Proof Let F be a semitransitive filter of X. Clearly F is a filter. Assume that F is an implicative filter. Let $x, y \in X$ be such that $x * (x * (x * y)) \in F$. Since $x * x = 1 \in F$ and F is implicative, we get $x * (x * y) \in F$. Since $x * x = 1 \in F$ and F is implicative, we get $x * y \in F$.

Conversely, assume that F satisfies the given condition. Let $x * (y * z) \in F$ and $x * y \in F$ for all $x, y, z \in X$. Then we get $y * (x * z) \in F$ and $x * y \in F$. Since F is a semitransitive filter, it gives that $x * (x * (x * z)) \in F$. Hence, from the assumed condition, we get that $x * z \in F$. Therefore F is an implicative filter of X. □

15.5 Fuzzification of Semitransitive Filters

In this section, the notion of fuzzy semitransitive filters is introduced in BE-algebras. Some properties of these fuzzy semitransitive filters including the extension property are studied. Some sufficient conditions are derived for any fuzzy filter to become a fuzzy semitransitive filter. An extension property for fuzzy semitransitive filters of BE-algebras is derived.

Definition 15.5.1 A fuzzy set μ of a BE-algebra X is said to be a *fuzzy semitransitive filter* of X if it satisfies the following properties for all $x, y, z \in X$.

(FST1) $\mu(1) \geq \mu(x)$,
(FST2) $\mu(x * (x * z)) \geq \min\{\mu(x * y), \mu(y * z)\}$.

Example 15.5.2 Let $X = \{1, a, b, c\}$. Define a binary operation $*$ on X as follows:

$*$	1	a	b	c
1	1	a	b	c
a	1	1	a	a
b	1	1	1	a
c	1	1	a	1

Then $(X, *, 1)$ is a BE-algebra. Define a fuzzy set $\mu : X \longrightarrow [0, 1]$ as follows:

$$\mu(x) = \begin{cases} 1 & \text{if } x = 1 \\ 0 & \text{otherwise} \end{cases}$$

Then it can be easily verified that μ is a fuzzy semitransitive filter of X.

Proposition 15.5.3 *Every fuzzy semitransitive filter of a BE-algebra is a fuzzy filter.*

Proof Let μ be a fuzzy semitransitive filter of a BE-algebra X. Hence we get $\mu(y) = \mu(1 * (1 * y)) \geq \min\{\mu(1 * x), \mu(x * y)\} = \min\{\mu(x), \mu(x * y)\}$. Therefore μ is a fuzzy filter of X. $\square$

In the following theorem, we derive some sufficient conditions for every fuzzy filter of a BE-algebra to become a fuzzy semitransitive filter.

Theorem 15.5.4 *Every fuzzy filter μ of a BE-algebra X is a fuzzy semitransitive filter of X if it satisfies the following condition for all $x, y, z \in X$:*

$$\mu(y) \geq \min\{\mu(x * y), \mu(y * z)\}.$$

Proof Let μ be a fuzzy filter of X such that the above condition holds for all $x, y, z \in X$. Since μ is a fuzzy filter, we get the following consequence:

$$\begin{aligned} \mu(x * (x * z)) &\geq \min\{\mu(y), \mu(y * (x * (x * z)))\} \\ &= \min\{\mu(y), \mu(x * (x * (y * z))) \\ &\geq \min\{\min\{\mu(x * y), \mu(y * z)\}, \mu(x * (y * z))\} \\ &\geq \min\{\min\{\mu(x * y), \mu(y * z)\}, \mu(y * z)\} \\ &= \min\{\mu(x * y), \mu(y * z)\}. \end{aligned}$$

Therefore μ is a fuzzy semitransitive filter of X. $\square$

Theorem 15.5.5 *Every fuzzy filter μ of a BE-algebra X is a fuzzy semitransitive filter of X if it satisfies the following condition for all $x, y, z \in X$:*

$$\mu((x * y) * (x * (x * z))) \geq \min\{\mu(x * y), \mu(x * (y * z))\}.$$

Proof Let μ be a fuzzy filter of X such that the above condition holds for all $x, y, z \in X$. Since μ is a fuzzy filter, we get the following consequence:

$$\begin{aligned}\mu(x * (x * z)) &\geq \min\{\mu(x * y), \mu((x * y) * (x * (x * z)))\}\\ &\geq \min\{\mu(x * y), \min\{\mu(x * y), \mu(x * (y * z))\}\}\\ &= \min\{\mu(x * y), \mu(x * (y * z))\}\\ &\geq \min\{\mu(x * y), \mu(y * z)\}\}.\end{aligned}$$

Therefore μ is a fuzzy semitransitive filter of X. □

Theorem 15.5.6 *Let μ be a fuzzy filter of a BE-algebra X which satisfies $\mu((x * y) * ((x * y) * (x * z))) \geq \min\{\mu(x * y), \mu((x * y) * (y * z))\}$ for all $x, y, z \in X$. Then μ is a fuzzy semitransitive filter of X.*

Proof Assume that the condition holds in X. Clearly $\mu(1) \geq \mu(x)$ for all $x \in X$. Let $x, y, z \in X$. Clearly $\mu(y * z) \leq \mu((x * y) * (y * z))$. Since μ is a fuzzy filter, by the assumed condition, we get

$$\begin{aligned}\mu(x * (x * z)) &\geq \min\{\mu(x * y), \mu((x * y) * (x * (x * z)))\}\\ &= \min\{\mu(x * y), \mu(x * ((x * y) * (x * z)))\}\\ &\geq \min\{\mu(x * y), \mu((x * y) * (x * z))\}\\ &\geq \min\{\mu(x * y), \min\{\mu(x * y), \mu((x * y) * (y * z))\}\}\\ &= \min\{\mu(x * y), \mu((x * y) * (y * z))\}\\ &\geq \min\{\mu(x * y), \mu(y * z)\}.\end{aligned}$$

Therefore μ is a fuzzy semitransitive filter of X. □

As a converse of the above theorem, it can be easily observed that a fuzzy transitive filter of a BE-algebra cannot satisfy the condition of the above theorem. We now derive a sufficient condition for a transitive filter of a BE-algebra to satisfy the condition of the above theorem.

Theorem 15.5.7 *Let X be a BE-algebra with $(z * y) * (x * y) \leq (y * z) * (x * z)$ for all $x, y, z \in X$. A fuzzy semitransitive filter μ of X satisfies the following property for all $x, y, z \in X$:*

$$\mu((x * y) * ((x * y) * (x * z))) \geq \min\{\mu(x * y), \mu((x * y) * (y * z))\}$$

Proof Let μ be a fuzzy transitive filter of X. Hence μ is a fuzzy filter of X. Let $x, y, z \in X$. Then by the given condition, we get $\mu((x * y) * ((z * y) * (x * y))) \leq \mu((x * y) * ((y * z) * (x * z)))$. Hence

$$\begin{aligned}
\mu((x * y) * ((x * y) * ((x * y) * (x * z)))) \\
&\geq \min\{\mu((x * y) * (y * z)), \mu((y * z) * ((x * y) * (x * z)))\} \\
&= \min\{\mu((x * y) * (y * z)), \mu((x * y) * ((y * z) * (x * z)))\} \\
&\geq \min\{\mu((x * y) * (y * z)), \mu((x * y) * ((z * y) * (x * y)))\} \\
&= \min\{\mu((x * y) * (y * z)), \mu((z * y) * ((x * y) * (x * y)))\} \\
&= \min\{\mu((x * y) * (y * z)), \mu((z * y) * 1))\} \\
&= \min\{\mu((x * y) * (y * z)), \mu(1)\} \\
&= \mu((x * y) * (y * z)).
\end{aligned}$$

Since μ is a fuzzy filter and from the above observation, we get $\mu((x * y) * ((x * y) * (x * z))) \geq \min\{\mu(x * y), \mu((x * y) * ((x * y) * ((x * y) * (x * z))))\} \geq \min\{\mu(x * y), \mu((x * y) * (y * z))\}$. □

In the following theorem, an extension property for fuzzy semitransitive filters is obtained.

Theorem 15.5.8 (Extension property for fuzzy semitransitive filters) *Let X be a BE-algebra with $(z * y) * (x * y) \leq (y * z) * (x * z)$ for all $x, y, z \in X$. Suppose μ and ν are two fuzzy filters of X such that $\mu(1) = \nu(1)$ and $\mu \subseteq \nu$ (i.e., $\mu(x) \leq \nu(x)$ for all $x \in X$). If μ is a fuzzy semitransitive filter, then so is ν.*

Proof Assume that μ is a fuzzy semitransitive filter of X. Let $x, y, z \in X$. Then

$$\begin{aligned}
\nu((y * z) * ((x * y) * (x * z))) &\geq \mu((y * z) * ((x * y) * (x * z))) \\
&= \mu((x * y) * ((y * z) * (x * z))) \\
&\geq \mu((x * y) * ((z * y) * (x * y))) \\
&= \mu((z * y) * ((x * y) * (x * y))) \\
&= \mu((z * y) * 1) \\
&= \mu(1) \\
&= \nu(1).
\end{aligned}$$

Hence $\nu((y * z) * ((x * y) * (x * z))) = \nu(1)$. Since ν is a fuzzy filter, we get

$$\begin{aligned}
\nu((x * y) * ((x * y) * (x * z))) &\geq \min\{\nu(y * z), \nu((y * z) * ((x * y) * (x * z)))\} \\
&= \min\{\nu(y * z), \nu(1)\} \\
&= \nu(y * z) \\
&\geq \min\{\nu(x * y), \nu((x * y) * (y * z))\}.
\end{aligned}$$

Hence by Theorem 15.5.6, we get ν is a fuzzy semitransitive filter in X. □

15.6 Triangulation of Semitransitive Filters

It this section, the notion of triangular normed fuzzy semitransitive filters is introduced in BE-algebras. Some sufficient conditions are derived for every triangular normed fuzzy filter of a BE-algebra to become a triangular normed fuzzy semitransitive filter.

Definition 15.6.1 A fuzzy set μ of a BE-algebra X is said to be a *fuzzy semitransitive filter* with respect to a t-norm T (simply called T-fuzzy semitransitive filter) if it satisfies the properties

(1) $\mu(1) \geq \mu(x)$ for all $x \in X$,
(2) $\mu(x * (x * z)) \geq T(\mu(x * y), \mu(y * z))$ for all $x, y, z \in X$.

Proposition 15.6.2 *Every T-fuzzy semitransitive filter of a BE-algebra is a T-fuzzy filter.*

Proof Let μ be a T-fuzzy semitransitive filter of F. Let $x, y \in X$. Then we get $\mu(y) = \mu(1 * (1 * y)) \geq T(\mu(1 * x), \mu(x * y)) = T(\mu(x), \mu(x * y))$. Therefore μ is a T-fuzzy filter of X. □

In general, the converse of the above proposition is not true. However, some sufficient conditions are derived for every T-fuzzy filter to become a T-fuzzy semitransitive filter.

Theorem 15.6.3 *Every T-fuzzy filter μ of a BE-algebra X is a T-fuzzy transitive filter if it satisfies the following condition for all $x, y \in X$.*

(TFT1) $\mu((x * y) * (x * z)) \geq mu(x * (y * z))$.

Proof Let μ be a T-fuzzy filter of X such that the condition (TFT1) holds for all $x, y \in X$. Let $x, y, z \in X$. Then we get the following:

$$\begin{aligned}
\mu(x * (x * z)) &\geq T(\mu(x * y), \mu((x * y) * (x * (x * z)))) \\
&= T(\mu(x * y), \mu(x * ((x * y) * (x * z)))) \\
&\geq T(\mu(x * y), \mu((x * y) * (x * z))) \\
&\geq T(\mu(x * y), \mu(x * (y * z))) \\
&\geq T(\mu(x * y), \mu(y * z)).
\end{aligned}$$

Therefore μ is a T-fuzzy semitransitive filter of X. □

Theorem 15.6.4 *Every T-fuzzy filter μ of a BE-algebra X is a T-fuzzy semitransitive filter if it satisfies the following condition for all $x, y, z \in X$.*

(TFT2) $\mu((x * y) * z) \geq \mu(x * (y * z))$

Proof Let μ be a T-fuzzy filter of X such that the condition (TF2) holds for all $x, y, z \in X$. Let $x, y, z \in X$. Then we get the following consequence:

$$\begin{aligned}\mu(x*(x*z)) &\geq \mu(x*z)\\ &\geq \mu(z)\\ &= T(\mu(x*y), \mu((x*y)*z))\\ &\geq T(\mu(x*y), \mu(x*(y*z)))\\ &\geq T(\mu(x*y), \mu(y*z)).\end{aligned}$$

Therefore μ is a T-fuzzy semitransitive filter of X. □

Lemma 15.6.5 *Every imaginable T-fuzzy semitransitive filter is order preserving.*

Proof Let μ be a T-fuzzy semitransitive filter of a BE-algebra X. Let $x, y \in X$ be such that $x \leq y$. Then we get $x * y = 1$. Hence $\mu(y) = \mu(1 * y) = \mu(1 * (1 * y)) \geq T(\mu(1 * x), \mu(x * y)) = T(\mu(x), \mu(1)) \geq T(\mu(x), \mu(x)) = \mu(x)$. Therefore μ is order preserving. □

Proposition 15.6.6 *Every fuzzy semitransitive filter is a T-fuzzy semitransitive filter.*

Proof Let μ be a fuzzy semitransitive filter of a BE-algebra X. For $x, y, z \in X$, we have

$$\mu(x*(x*z)) \geq \min\{\mu(x*y), \mu(y*z)\} \geq T(\mu(x*y), \mu(y*z)).$$

Therefore μ is a T-fuzzy semitransitive filter in X. □

The converse of the above proposition is not true. However, we derive a sufficient condition for every T-fuzzy semitransitive filter to become a fuzzy semitransitive filter.

Theorem 15.6.7 *Every imaginable T-fuzzy semitransitive filter of BE-algebra is fuzzy semitransitive.*

Proof Let μ be an imaginable T-fuzzy semitransitive filter of a BE-algebra X. Let $x, y, z \in X$. Then clearly $\mu(x * (x * z)) \geq T(\mu(x * y), \mu(y * z))$. Since μ is imaginable and $\min\{\mu(x * y), \mu(y * z)\} \leq \mu(x * y), \mu(y * z)$, we get

$$\begin{aligned}\min\{\mu(x*y), \mu(y*z)\} &= T(\min\{\mu(x*y), \mu(y*z)\}, \min\{\mu(x*y), \mu(y*z)\})\\ &\leq T(\min\{\mu(x*y), \mu(y*z)\}, \mu(y*z))\\ &\leq T(\mu(x*y), \mu(y*z)\})\\ &\leq \min\{\mu(x*y), \mu(y*z)\}.\end{aligned}$$

Hence $T(\mu(x * y), \mu(y * z)) = \min\{\mu(x * y), \mu(y * z)\}$. Thus $\mu(x * (x * z)) \geq T(\mu(x * y), \mu(y * z)) = \min\{\mu(x * y), \mu(y * z)\}$. Therefore μ is a fuzzy semitransitive filter in X. □

Definition 15.6.8 Let X and X' be any two sets and $f : X \to X'$ be a function. If ν is a fuzzy set in $f(X)$, then the fuzzy set μ in X defined for all $x \in X$ by $\mu(x) = \nu(f(x))$ is said to be the *pre-image of* ν *under* f and is denoted by $f^{-1}(\nu)$. Clearly $f^{-1}(\nu) = \nu \circ f$.

Theorem 15.6.9 *Let $f : X \to Y$ be an onto homomorphism of BE-algebras. If ν is a T-fuzzy semitransitive filter of Y, then $f^{-1}(\nu)$ is a T-fuzzy semitransitive filter of X. Moreover, if ν satisfies the imaginable property, then so does $f^{-1}(\nu)$.*

Proof For any $x \in X$, $f^{-1}(\nu)(1) = \nu(f(1)) = \nu(1) \geq \nu(f(x)) = f^{-1}(\nu)(x)$. Let $x, y, z \in X$. Then $f^{-1}(\nu)(x * (x * z)) = \nu(f(x * (x * z))) = \nu(f(x) * (f(x) * f(z))) \geq T(\nu(f(x) * a), \nu(a * f(z)))$ for some $a \in Y$. Since f is onto, there exists $y_a \in X$ such that $f(y_a) = a$. Now

$$\begin{aligned} f^{-1}(\nu)(x * (x * z)) &\geq T(\nu(f(x) * a), \nu(a * f(z))) \\ &= T(\nu(f(x) * f(y_a)), \nu(f(y_a) * f(z))) \\ &= T(\nu(f(x * y_a)), \nu(f(y_a * z))) \\ &= T(f^{-1}(\upsilon)(x * y_a), f^{-1}(\nu)(y_a * z)). \end{aligned}$$

Since a is arbitrary, this inequality holds for all $y \in X$. Hence $f^{-1}(\nu)(x * (x * z)) \geq T(f^{-1}(\upsilon)(x * y), f^{-1}(\nu)(y * z))$. Therefore $f^{-1}(\nu)$ is a T-fuzzy semitransitive filter of X. Suppose ν satisfies the imaginable property. Then $T(f^{-1}(\upsilon)(x), f^{-1}(\upsilon)(x)) = T(\nu(f(x)), \nu(f(x))) = \nu(f(x)) = f^{-1}(\nu)(x)$. Therefore $f^{-1}(\nu)$ satisfies the imaginable property. □

Definition 15.6.10 Let X and X' be any two sets and $f : X \to X'$ be any function. If μ is a fuzzy set in X, then the fuzzy set ν in X' defined for all $x \in X'$ by $\nu(x) = \sup_{t \in f^{-1}(x)} \mu(t)$ is said to be the *image of* μ *under* f and is denoted by $f(\mu)$.

We say that a fuzzy set μ in X has the *sup* property if, for any subset A of X, there exists $a_0 \in A$ such that $\mu(a_0) = \sup_{a \in A} \mu(a)$.

Theorem 15.6.11 *Let $f : X \to Y$ be a homomorphism of a BE-algebra X onto a BE-algebra Y. Let μ be a T-fuzzy semitransitive filter of X which has the sup property. Then the image of μ under f is a T-fuzzy semitransitive filter of Y.*

Proof Since $1 \in f^{-1}(1)$, we get $f(\mu)(1) = \sup_{t \in f^{-1}(1)} \mu(t) = \mu(1) \geq \mu(x)$ for all $x \in X$. Hence $f(\mu)(1) \geq \sup_{t \in f^{-1}(a)} \mu(t) = f(\mu)(a)$ for all $a \in Y$. For any $a, b, c \in Y$, let $x_a \in f^{-1}(a)$, $x_b \in f^{-1}(b)$, and $x_c \in f^{-1}(c)$ be such that $\mu(x_a * (x_a * x_c)) = \sup_{t \in f^{-1}(a*(a*c))} \mu(t)$, $\mu(x_b * x_c) = \sup_{t \in f^{-1}(b*c)} \mu(t)$ and $\mu(x_a * x_b) = \sup_{t \in f^{-1}(a*b)} \mu(t)$. Then we get the following consequence:

$$\begin{aligned}
f(\mu)(a*(a*c)) &= \sup_{t \in f^{-1}(a*(a*c))} \mu(t) \\
&= \mu(x_a * (x_a * x_c)) \\
&\geq T(\mu(x_a * x_b), \mu(x_b * x_c)) \\
&= T(\sup_{t \in f^{-1}(a*b)} \mu(t), \sup_{t \in f^{-1}(b*c)} \mu(t)) \\
&= T(f(\mu)(a*b), f(\mu)(b*c)).
\end{aligned}$$

Therefore $f(\mu)$ is a T-fuzzy semitransitive filter of Y. □

Definition 15.6.12 Let μ and ν be two fuzzy sets in a BE-algebra X. Then the T-product of μ and ν is defined by $(\mu \times \nu)_T(x) = T(\mu(x), \nu(x))$ for all $x \in X$.

Definition 15.6.13 Let T and S be two t-norms on $I = [0, 1]$. Then the t-norm S is said to dominate the t-norm T if for all $\alpha, \beta, \gamma, \delta \in [0, 1]$, the following satisfies:

$$S(T(\alpha, \gamma), T(\beta, \delta)) \geq T(S(\alpha, \beta), S(\gamma, \delta)).$$

Theorem 15.6.14 *Let μ and ν be T-fuzzy semitransitive filters of a BE-algebra X. If a t-norm S dominates T, then the produce $(\mu \times \nu)_S$ is a T-fuzzy semitransitive filter of X.*

Proof For any $x \in X$, we can get that $(\mu \times \nu)_S(1) = S(\mu(1), \nu(1)) \geq S(\mu(x), \nu(x)) = (\mu \times \nu)_S(x)$. Let $x, y, z \in X$. Then

$$\begin{aligned}
(\mu \times \nu)_S(x*(x*z)) &= S(\mu(x*(x*z)), \nu(x*(x*z))) \\
&\geq S(T(\mu(x*y), \mu(y*z)), T(\nu(x*y), \nu(y*z))) \\
&\geq T(S(\mu(x*y), \nu(x*y)), S(\mu(y*z), \nu(y*z))) \\
&= T((\mu \times \nu)_S(x*y), (\mu \times \nu)_S(y*z)).
\end{aligned}$$

Therefore $(\mu \times \nu)_S$ is a T-fuzzy semitransitive filter of X. □

Exercise

1. Show that every implicative filter of a BE-algebra is a transitive filter and derive a sufficient condition for a transitive filter of a BE-algebra to become an implicative filter.
2. Let $f : X \to Y$ be an onto homomorphism of BE-algebras and μ is a fuzzy set in Y. Define $\mu^f : X \to [0, 1]$ such that $\mu^f(x) = \mu(f(x))$ for all $x \in X$. Then prove that μ is a fuzzy transitive filter of Y if and only if μ^f is a fuzzy transitive filter of X.
3. Let μ and ν be two fuzzy transitive filters of a BE-algebra X such that μ and ν have the finite images and have the identical family of level filters. If $Im(\mu) = \{\alpha_1, \alpha_2, \ldots, \alpha_m$ and $Im(\nu) = \{\beta_1, \beta_2, \ldots, \beta_n\}$, where $\alpha_1 > \alpha_2 > \cdots > \alpha_m$ and $\beta_1 > \beta_2 > \cdots > \beta_n$, then prove that $m = n$. Moreover, prove that $\mu_{\alpha_i} = \mu_{\beta_i}$ for $i = 1, 2, \ldots, m$.

4. Let μ be a fuzzy transitive filter of a BE-algebra X. Then prove that two level filters μ_{α_1} and μ_{α_2}(with $\alpha_1 < \alpha_2$) of μ are equal if and only if there is no $x \in X$ such that $\alpha_1 \leq \mu(x) < \alpha_2$.
5. If μ and ν are two fuzzy semitransitive filters of a BE-algebra X, then prove that their direct product $\mu \times \nu$ is also a fuzzy semitransitive filter in $X \times X$.
6. Let $f : X \to Y$ be an onto homomorphism of BE-algebras. If ν is a fuzzy semitransitive filter of Y, then prove that the inverse image $f^{-1}(\nu)$ is a fuzzy semitransitive filter of X.
7. Let f be a fuzzy set in a BE-algebra X and μ_f the strongest fuzzy relation on X defined by $\mu_f(x, y) = (f(x), f(y))$ for all $x, y \in X$. Then prove that f is an $\mathcal{N}$-fuzzy transitive filter of X if and only if μ_f is an $\mathcal{N}$-fuzzy transitive filter of $X \times X$.
8. Let f be an $\mathcal{N}$-fuzzy transitive filter of a BE-algebra X. Suppose $Im(f) = \{\alpha_1, \alpha_2, \ldots, \alpha_n\}$, where $\alpha_1 < \alpha_2 < \cdots < \alpha_n$. Then prove that the family $f_{\alpha_i}, i = 1, 2, \ldots, n$ constitutes all the level filters of f.
9. For any chain $\mathcal{F}_0 \supset \mathcal{F}_1 \supset \mathcal{F}_2 \supset \ldots \supset \mathcal{F}_n = X$ of transitive filters of X, prove that there exists an $\mathcal{N}$-fuzzy transitive filter f of X such that the level subsets of f coincide with the chain.
10. If every $\mathcal{N}$-fuzzy transitive filter of a BE-algebra X has a finite image, then prove that every ascending chain of transitive filters of X terminates after a finite number of steps.
11. For any non-empty subset F of a BE-algebra X, define a fuzzy set $\mu_F : X \to [0, 1]$ as $\mu_F(x) = \alpha$ if $x \in F$ and $\mu_F(x) = 0$ if $x \notin F$ where $0 < \alpha < 1$ is fixed. Then prove that μ_F is a fuzzy distributive filter in X if and only if F is a distributive filter in X.
12. Prove that a fuzzy set μ of a BE-algebra X is a fuzzy distributive filter if and only if for each $\alpha \in [0, 1]$, the level subset μ_α is a distributive filter in X, when $\mu_\alpha \neq \emptyset$.
13. Let $\Phi : X \to Y$ be a homomorphism and f an $\mathcal{N}$-fuzzy subset of X. Define a mapping $f_\Phi : X \to [0, 1]$ by $f_\Phi(x) = f(\Phi(x))$ for all $x \in X$. Prove that f is an $\mathcal{N}$-fuzzy transitive filter of X if and only if f_Φ is an $\mathcal{N}$-fuzzy transitive filter of X.

Bibliography

1. J.C. Abbott, Implicational algebras. Bull. Math. Soc. Sci. Math. R. S. Roumanie **11**(59), 3–23 (1967)
2. H.A.S. Abujabal, M.A. Obaid, M. Aslam, On annihilators of BCK-algebras. Czech. Math. J. **45**(120), 727–735 (1995)
3. S.S. Ahn, J.S. Han, On fuzzy positive implicative filters in BE-algebras. Sci. World J. **929162**, 1–5 (2014)
4. S.S. Ahn, J.M. Ko, On vague filters in BE-algebras. Commun. Korean Math. Soc. **26**(3), 417–425 (2011)
5. S.S. Ahn, K.S. So, On ideals and upper sets in BE-algebras. Sci. Math. Jpn. **68**(2), 279–285 (2008)
6. S.S. Ahn, K.S. So, On generalized upper sets in BE-algebras. Bull. Korean Math. Soc. **46**(2), 281–287 (2009)
7. S.S. Ahn, H.S. Kim, H.D. Lee, R-maps and L-maps Q-algebras. Int. J. Pure Appl. Math. **12**(4), 419–425 (2004)
8. S.S. Ahn, Y.H. Kim, J.M. Ko, Filters in commutative BE-algebras. Bull. Korean Math. Soc. **27**(2), 233–242 (2012)
9. J. Ahsan, E.Y. Deeba, A.B. Thaheem, On prime ideal of BCK-algebra. Math. Jpn. **36**, 875–882 (1991)
10. H. Aktas, N. Cagman, Soft sets and soft groups. Inform. Sci. **177**, 2726–2735 (2007)
11. C.K. Bae, Some properties of homomorphisms in BCK-algebras. J. Korea Soc. Math. Edu. **24**(1), 7–10 (1985)
12. R. Balbes, Ph Dwinger, *Distributive Lattices* (University of Missouri Press, Columbia, 1974)
13. P. Bhattacharya, N.P. Mukherjee, Fuzzy relations and fuzzy groups. Inform. Sci. **36**, 267–282 (1985)
14. R. Biswas, Vague groups. Int. J. Comput. Cognit. **4**, 20–23 (2006)
15. A. Borumand Saeid, N. Mohtashamnia, Stabilizers in residuated lattices. U.P.B. Sci. Bull. Ser. A **74**(2), 65–74 (2012)
16. A. Borumand Saeid, M. Haveshki, H. Babaei, Uniform topology on Hilbert algebras. Kyungpook Math. J. **45**, 405–411 (2005)
17. R.A. Borzooei, Stabilizer topology of hoops. Algebr. Struct. Appl. **1**(1), 35–48 (2014)
18. R.A. Borzooei, M. Bakhshi, Lattice structure on ideals of a BCK-algebra. J. Multiple-Valued Logic Soft Comput. **18**(1–2), 387–399 (2012)
19. R.A. Borzooei, S. Khosravi Shoar, Implication algebras are equivalent to the dual implicative BCK-algebras. Scientiae Mathematicae Japonicae **63**, 429–431 (2006)

S. R. Mukkamala, *A Course in BE-algebras*,
https://doi.org/10.1007/978-981-10-6838-6

20. R.A. Borzooei, A. Paad, Some new types of stabilizers in BL-algebras and their applications. Indian J. Sci. Technol. **5**(1), 1910–1915 (2012)
21. R.A. Borzooei, O. Zahiri, Prime ideals in BCI and BCK-algebras. Ann. Univ. Craiova, Math. Comput. Sci. Ser. **39**(2), 266–276 (2012)
22. R.A. Borzooei, A. Dvurecenskij, O. Zahiri, *State BCK-Algebras and State-Morphism BCK-Algebras* 25 Apr 2013, arXiv:1304.6963v1, [math.AC]
23. R.A. Borzooei, A. Borumand Saeid, R. Ameri, A. Rezaei, Involutory BE-algebras. J. Math. Appl. **37**, 13–26 (2014)
24. R.A. Borzooei, A. Borumand Saeid, A. Rezaei, R. Ameri, States on BE-algebras. Kochi. Math. **9**, 27–42 (2014)
25. B. Bosbach, Komplementare Halbgruppen. Axiomatic und Arithmetic. Fundamenta Mathematicae **64**, 257–287 (1969)
26. B. Bosbach, Komplementare Halbgruppen. Kongruenzen and Quotiente. Fundamenta Mathematicae **69**, 1–14 (1970)
27. D.J. Brown, R. Suszko, Abstract logics. Diss. Math. **102**, 4–42 (1973)
28. S. Burris, H.P. Sankappanavar, *Graduate Text in Math*, vol. 78, A course in Universal algebra (Springer, New York, 1981)
29. D. Busneag, A note on deductive systems of a Hilbert algebra. Kobe J. Math. **2**, 29–35 (1985)
30. D. Busneag, F-multipliers and the localization of Hilbert algebras. Zeitschr. F. Math. Logikund Grundlagen d. Math. Bd. **36**, 331–338 (1990)
31. D. Busneag, State-morphisms on Hilbert algebras. Ann. Univ. Craiova, Math. Comp. Sci. Ser. **37**(4), 58–64 (2010)
32. D. Busneag, D. Piciu, Localization of MV-algebras and lu-groups. Algebra Universalis **50**, 359–380 (2003)
33. D. Busneag, D. Piciu, On the lattice of deductive systems of a BL-algebra. Central Eur. J. Math. **1**(2), 221–238 (2003)
34. S. Celani, A note on homomorphisms of Hilbert algebras. Int. J. Math. Math. Sci. **29**(1), 55–61 (2002)
35. S. Celani, Deductive systems of BCK-algebras. Acta Univ Palac. Olomuc. Fac. Rer. Nat. Math. **43**(1), 27–32 (2004)
36. I. Chajda, R. Halaš, Congruences and ideals in Hilbert algebras. Kyungpook Math. J. **39**, 429–432 (1999)
37. I. Chajda, R. Halaš, Stabilizers in Hilbert algebras. Mult. Valued Logic **8**, 139–148 (2002)
38. I. Chajda, J. Kuhr, Algebraic structures derived from BCK-algebras. Miskolc Math. Notes **8**(1), 11–21 (2007)
39. I. Chajda, B. Zelinka, Tolarences relations on lattices. Časopis pro pěstování matematiky **99**(4), 394–399 (1974)
40. I. Chajda, R. Halaš, J. Zednik, Filters and annihilators in implication algebras. Acta Univ. Palack. Olomuc, Fac. Reum Natur. Math. **37**, 141–145 (1998)
41. I. Chajda, R. Halaš, Y.B. Jun, Annihilators and deductive systems in commutative Hilbert algebras. Comment. Math. Univ. Carolin. **43**(3), 407–417 (2002)
42. I. Chajda, R. Halaš, J. Kuhr, *Semilattice Structures* (Heldermann, Germany, 2007)
43. J. Chen, Y. Huang, Some results of self-maps in BCK-algebras. Scientiae Mathematicae **2**(1), 89–93 (1999)
44. Z.M. Chen, H.X. Wang, Closed ideals and congruences on BCI-algebras. Kobe J. Math. **8**, 1–9 (1991)
45. X.Y. Cheng, X.L. Xin, Filter theory on hyper BE-algebras. Italian J. Pure Appl. Math. **35**, 509–526 (2015)
46. Z. Chen, Y. Huang, E.H. Roh, $C(S)$ extensions of $S - I - BCK$-algebras. Comm. Korean Math. Soc. **10**(3), 499–518 (1995)
47. D. Chen, E.C.C. Tsang, D.S. Yeung, X. Wang, The parametrization reduction of soft sets and its applications. Comput. Math. Appl. **49**, 757–763 (2005)
48. Y.U. Cho, Y.B. Jun and E.H. Roh, *On sensible fuzzy subalgebras of BCK-algebras with respect to a s-norm*, Scientiae Mathematicae Japonicae Online, e-2005, 11-18

49. R. Cignoli, A. Torrens, Boolean products of MV-algebras: hypernormal MV-algebras. J. Math. Anal. Appl. **99**, 637–653 (1996)
50. R. Cignoli, A. Torrens, Glivenko like theorems in natural expansions of BCK-logics. Math. Log. Quart. **50**(2), 111125 (2004)
51. R. Cignoli, A. Torrens, Free algebras in varieties of BL-algebras with a Boolean retract. Algebra Universalis **48**(1), 55–79 (2002)
52. R. Cignoli, I.M.L. DOttaviano, D. Mundici, *Trends in Logic Studia Logica Library*, vol. 7, Algebraic foundations of many-valued reasoning (Kluwer Academic Publishers, Dordrecht, 2000)
53. Z. Ciloğlu and Y. Cevan, *Commutative and bounded BE-algebras*, Algebra, **2013**, 1-5
54. L.C. Ciunge, States on pseudo BCK-algebras. Math. Rep. **10(60)**(1), 17–36 (2008)
55. L.C. Ciungu, Bosbach and Riecan states on residuated lattices. Appl. Funct. Anal. **2**, 175–188 (2002)
56. L.C. Ciungu, A. Dvurecenskij, states and de Finetti maps on pseudo BCK-algebras. Fuzzy Sets Syst. **161**, 2870–2896 (2010)
57. L.C. Ciungu, A. Dvurecenskij, M. Hycko, State BL-algebras. Soft Comput. **15**, 619–634 (2011)
58. W.H. Cornish, The multiplier extension of a distributive lattice. J. Algebra **32**, 339–355 (1974)
59. W.H. Cornish, A multiplier approach to implicative BCK-algebras. Math. Semin. Notes Kobe Univ. **8**(1), 157–169 (1980)
60. W.H. Cornish, A large variety of BCK-algebras. Math. Jpn. **26**, 339–344 (1981)
61. K.H. Dar, B. Ahmad, Endomorphisms of BCK-algebras. Math. Jpn. **31**, 855–857 (1986)
62. K.H. Dar, M. Akram, On endomorphisms of BCH-algebras. Ann. Univ. Craiova Math. Comp. Sci. Ser. **33**, 227–234 (2006)
63. B.A. Davey, H. Priestley, *Introduction to Lattice and Order* (Camebridge University Press, U.S.A., 2002)
64. B. Davvaz, W.A. Dudek, Y.B. Jun, Intuitionistic fuzzy Hv-submodules. Inform. Sci. **176**, 285–300 (2006)
65. A. Di Nola, A. Dvurecenskij, State-morphism MV-algebras. Ann. Pure Appl. Logic. **161**, 161–173 (2009)
66. A. Di Nola, A. Dvurecenskij, On some classes of state-morphism MV-algebras. Math. Slovaca **59**, 517–534 (2009)
67. A. Di Nola, L. Leustean, Compact representations of BL-algebra. Arch. Math. Logic **42**, 737–761 (2003)
68. A. Di Nola, G. Georgescu, A. Iorgulescu, Pseudo BL-algebra: Part I. J. Mult. Val. Logic **8**(5–6), 673–714 (2002)
69. A. Diego, Sur les algbra de Hilbert. Coll. Logique Math. Ser. A **21**, 1–54 (1966)
70. W.A. Dudek, On BCC-algebras. Logique et Analyse **129**(130), 103–111 (1990)
71. W.A. Dudek, On fuzzification in Hilbert algebras, in *Proceedings of the Olomouc Conference and the Summer School 1998*. Contributions to General Algebra, vol. 11 (J. Heyn, Klagenfurt, 1999), pp. 77–83
72. W.A. Dudek, A new characterization of ideals in BCC-algebras. Novi Sad J. Math. **29**(1), 139–145 (1999)
73. W.A. Dudek, On ideals in Hilbert algebras. Acta Univ. Palacki. Olomouc. Fac. Rer. Nat. Math. **38**, 31–34 (1999)
74. W.A. Dudek, On embeddings of Hilbert algebras. Math. Moravica **3**, 25–28 (1999)
75. W.A. Dudek, Y.B. Jun, On fuzzy ideals in Hilbert algebras. Novi Sad J. Math. **29**(2), 193–207 (1999)
76. W.A. Dudek, Y.B. Jun, Normalizations of fuzzy BCC-ideals in BCC-algebras. Math. Moravica **3**, 17–24 (1999)
77. W.A. Dudek, Y.B. Jun, On fuzzy BCC-ideals over a t-norm. Math. Commun. **5**, 149–155 (2000)
78. W.A. Dudek, Y.B. Jun, Zero invariant and independent fuzzy BCC-subalgebras. Math. Slovaca **52**, 145–156 (2002)

79. W.A. Dudek, Y.B. Jun, Quasi p-ideals of quasi BCI-algebras. Quasigroups Relat. Syst. **11**, 25–38 (2004)
80. W.A. Dudek, Y.B. Jun, On multiplicative fuzzy BCC-algebras. J. Fuzzy Math. **13**, 929–939 (2005)
81. W.A. Dudek, Y.B. Jun, Poor and crazy filters of BCK-algebras. Int. J. Pure Appl. Math. Sci. **4**, 25–38 (2007)
82. W.A. Dudek, J. Thomys, On decompositions of BCH-algebras. Math. Jpn. **35**, 1131–1138 (1990)
83. W.A. Dudek, X.H. Zhang, On ideals and congruences in BCC-algebras. Czechoslovak Math. J. **48**(123), 21–29 (1998)
84. W.A. Dudek, X.H. Zhang, Initial segments in BCC-algebras. Math. Moravica **4**, 27–34 (2000)
85. W.A. Dudek, K.H. Kim, Y.B. Jun, Fuzzy BCC-subalgebras of BCC-algebras with respect to a t-norm. Scientiae Math. **3**, 99–106 (2000)
86. W.A. Dudek, Y.B. Jun, Z. Stojakovic, On fuzzy ideals in BCC-algebras. Fuzzy Sets Syst. **123**, 251–258 (2001)
87. W.A. Dudek, X.H. Zhang, Y.Q. Wang, Ideals and atoms of BZ-algebras. Math. Slovaca **59**, 387–404 (2009)
88. A. Dvurecenskij, Measures and states on BCK-algebras. Sem. Mat. Fis. Univ. Modena **47**, 511–528 (1999)
89. A. Dvurecenskij, Subdirectly irreducible state-morphism BL-algebras. Arch. Math. Logic **50**, 145–160 (2011)
90. A. Dvurecenskij, J. Rachunek, Probabilistic averaging in bounded commutative $R\ell$-monoids. Discret. Math. **306**, 1317–1326 (2006)
91. A. Dvurecenskij, J. Rachunek, On Riecan and Bosbach states for bounded non-commutative $R\ell$-monoids. Semigroup Forum **56**, 487–500 (2006)
92. A. Dvurecenskij, T. Kowalski, F. Montagna, State morphism MV-algebras. Int. J. Approx. Reason. **52**, 1215–1228 (2011)
93. A. Dvurecenskij, J. Rachunek, D. Salounova, State operators on generalizations of fuzzy structures. Fuzzy Sets Syst. **187**, 58–76 (2012)
94. T. Flaminio, F. Montagna, An algebraic approach to states on MV-algebras. Proc. EUSFLAT07 **2**, 201–206 (2007)
95. T. Flaminio, F. Montagna, MV-algebras with internal states and probabilistic fuzzy logics. Int. J. Approx. Reason. **50**, 138–152 (2009)
96. J.M. Font, A.J. Rodriguez, A. Torrens, Wajsberg algebras. Stochastica **8**, 5–31 (1984)
97. O. Frink, Pseudo-complements in semilattices. Duke Math. J. **29**, 505–514 (1962)
98. F.M. Garcia, A.J. Rodriguez, Linearization of BCK-logic. Stud. Log. **65**, 31–51 (2000)
99. Georgescu, Boshbatch states on fuzzy structures. Soft Comput. **8**, 217–230 (2004)
100. J. Gispert, A. Torrens, Bounded BCK-algebras and their generated variety. Math. Log. Q. **53**, 206–213 (2007)
101. J. Gispert, A. Torrens, Boolean representation of bounded BCK-algebras. Soft Comput. **12**, 941–954 (2008)
102. K. Goel, A.K. Arora, Characterization of semisimple BCK-algebras. Math. Jpn. **39**(2) (1994)
103. G. Gratzer, *General Lattice Theory* (Academic press, New York, 1978)
104. P. Hajek, *Metamathematics of Fuzzy Logic*, vol. 4 (Springer, Berlin, 1998)
105. P. Hajek, On very true. Fuzzy Sets Syst. **124**, 329–333 (2001)
106. R. Halaš, Remarks on commutative Hilbert algebras. Mathematica Bohemica **127**(4), 525–529 (2002)
107. R. Halaš, Annihilators in BCK-algebras. Czech. Math. J. **53**(4), 1001–1007 (2003)
108. A. Hasankhani, H. Saadatsome, Some Quotients on a BCK-algebra generated by a fuzzy set. Iran. J. Fuzzy Syst. **1**(2), 33–43 (2004)
109. S.M. Hong, Y.B. Jun, Fuzzy and level subalgebras of BCK/BCI-algebras. Pusan Kyongnam Math. J. **7**(2), 185–190 (1991)
110. S.M. Hong, Y.B. Jun, On deductive systems of Hilbert algebras. Comm. Korean Math. Soc. **11**(3), 595–600 (1996)

111. S.M. Hong, Y.B. Jun, Anti fuzzy ideals in BCK-algebras. Kyungpook Math. J. **38**(1), 145–150 (1998)
112. C.S. Hoo, Filters and ideals in BCI-algebras. Math. Jpn. **36**, 987–997 (1991)
113. Q.P. Hu, X. Li, On BCH-algebras. Kobe J. Math. **11**(2), 313–320 (1983)
114. Q.P. Hu, X. Li, On proper BCH-algebras. Math. Jpn. **30**(4), 659–661 (1985)
115. Y. Huang, *BCI-Algebras* (Science Press, Beijing, 2006)
116. Y.S. Hunge, Normal BCI-algebras. Scientiae Mathematicae Japonicae Online **9**, 321–331 (2003)
117. Y.S. Hunge, Z. Chen, Normal BCK-algebras. Math. Jpn. **45**(3), 541–546 (1997)
118. Y. Imai, K. Iśeki, On axiom systems of propositional Calculi. XIV Proc. Jpn. Acad. **42**, 19–22 (1966)
119. A. Iorgulescu, *Algebras of Logic as BCK-Algebras* (Editura ASE, Bucharest, 2008)
120. A. Iorgulescu, Classes of BCK-algebra, part III. Preprint Ser. Inst. Math. Roman. Acad., preprint nr **3**, 1–37 (2004)
121. K. Iśeki, An algebra related with a propositional calculus. Proc. Jpn. Acad. **42**, 26–29 (1966)
122. K. Iśeki, On some ideals in BCK-algebras. Math. Semin. Notes **3**, 65–70 (1975)
123. K. Iśeki, On BCI-algebras. Math. Sem. Notes Kobe Univ. **8**(1), 125–130 (1980)
124. K. Iśeki, S. Tanaka, An introduction to the theory of BCK-algebras. Math. Jpn. **23**(1), 1–26 (1979)
125. Y.B. Jun, Positive implicative ordered filters of implicative semigroups. Int. J. Math. Math. Sci. **23**(12), 801–806 (2000)
126. Y.B. Jun, Prime filter theorem of lattice implication algebras. Int. J. Math. Math. Sci. **25**(2), 115–118 (2000)
127. Y.B. Jun, Fantastic filter of lattice implication algebras. Int. J. Math. Math. Sci. **24**(4), 277–281 (2000)
128. Y.B. Jun, Fuzzy positive implicative and fuzzy associative filters of lattice implication algebras. Fuzzy Sets Syst. **121**, 353–357 (2001)
129. Y.B. Jun, Topological aspects of filters in lattice implication algebras. Bull. Korean Math. Soc. **43**(2), 227–233 (2006)
130. Y.B. Jun, Fuzzy quotient structures of BCK-algebras induced by fuzzy BCK-filters. Commun. Korean Math. Soc. **21**(1), 27–36 (2006)
131. Y.B. Jun, Soft BCK/BCI-algebras. Comput. Math. Appl. **56**, 1408–1413 (2008)
132. Y.B. Jun, S.S. Ahn, Fuzzy implicative filters with degrees in the interval (0, 1]. J. Comput. Anal. Appl. **15**, 1456–1466 (2013)
133. Y.B. Jun, W.A. Duduk, n-fold BCC-ideals of BCC-algebras. Scientiae Math. **3**, 171–178 (2000)
134. Y.B. Jun, W.A. Dudek, Intuitionistic fuzzy closed ideals in BCH-algebras. J. Fuzzy Math. **9**, 535–544 (2001)
135. Y.B. Jun, S.M. Hong, Fuzzy deductive systems of Hilbert algebras. Indian J. Pure Appl. Math. **27**(2), 141–151 (1996)
136. Y.B. Jun, M.S. Kang, Ideal theory of BE-algebras based on $\mathcal{N}$-structure. Haceteppe J. Math. Stat. **41**(4), 435–447 (2012)
137. Y.B. Jun, C.H. Park, Applications of soft sets in ideal theory of BCK/BCI-algebras. Inf. Sci. **178**, 2466–2475 (2008)
138. Y.B. Jun, C.H. Park, Falling shadows applied to subalgebras and ideals of BCK/BCI-algebras. Honam Math. J. **34**(2), 135–144 (2012)
139. Y.B. Jun, S.Z. Song, Generalized fuzzy interior ideals in semigroups. Inform. Sci. **176**, 3079–3093 (2006)
140. Y.B. Jun, X.L. Xin, Fuzzy prime ideals and invertible fuzzy ideals in BCK-algebras. Fuzzy Sets Syst. **117**, 471–476 (2001)
141. Y.B. Jun, X.L. Xin, On derivations of BCI- algebras. Inform. Sci. **159**, 167–176 (2004)
142. Y.B. Jun, S.M. Hong, E.H. Roh, Fuzzy characteristic subalgebras of a BCK-algebra. Pusan Kyongnam Math. J. **9**(1), 127–132 (1993)

143. Y.B. Jun, J.W. Nam, S.M. Hong, A note on Hilbert algebras. Pusan Kyongnam Math. J. **10**, 276–285 (1994)
144. Y.B. Jun, E.H. Roh, J. Meng, Annihilators in BCI-algebras. Math. Jpn. **43**, 559–562 (1996)
145. Y.B. Jun, J.U. Kim, H.S. Kim, Hilbert algebras inherited from the posets. Indian J. Pure Appl. Math. **28**(4), 471–475 (1997)
146. Y.B. Jun, Y. Xu, K. Qin, Positive implicative and Associate filters of lattice implication algebras. Bull. Korean Math. Soc. **35**(1), 53–61 (1998)
147. Y.B. Jun, E.H. Roh, H.S. Kim, On BH-algebras. Scientiae Mathematicae **1**(3), 347–354 (1998)
148. Y.B. Jun, S.M. Hong, S.J. Kim, S.Z. Song, Fuzzy ideals and fuzzy subalgebras of BCK-algebras. J. Fuzzy Math. **7**(2), 411–418 (1999)
149. Y.B. Jun, S.S. Ahn, H.S. Kim, Quotient structures of some implicative algebras via fuzzy implicative filters. Fuzzy Sets Syst. **121**, 325–332 (2001)
150. Y.B. Jun, E.H. Roh, H.S. Kim, On fuzzy B-algebras. Czechoslovak Math. J. **52**, 375–384 (2002)
151. Y.B. Jun, X. Xin, E.H. Roh, The role of atoms in BCI-algebras. Soochow. Math. **30**(4), 491–506 (2004)
152. K.H. Kim, A note on BE-algebras. Scientiae Mathematicae Japonicae **72**, 127–132 (2010)
153. H.S. Kim, Y.H. Kim, On BE-algebras, Sci. Math. Jpn. Online, 1299–1302 (e-2006)
154. C.B. Kim, H.S. Kim, On BM-algebras. Scientiae Mathematicae Japonicae **63**(3), 421–427 (2006)
155. H.S. Kim, K.J. Lee, Extended upper sets in BE-algebras. Bull. Malays. Math. Sci. Soc. **34**(3), 511–520 (2011)
156. K.H. Kim, Y.H. Yon, Dual BCK-algebra and MV-algebra. Scientiae Mathematicae Japonicae **66**, 247–253 (2007)
157. M. Kondo, Positive implicative BCK-algebra and its dual algebra. Math. Japon. **35**, 289–291 (1990)
158. M. Kondo, Some properties of left maps in BCK-algebras. Math. Japon. **36**, 173–174 (1991)
159. M. Kondo, Annihilators in BCK-algebras. Mem. Fac. Sci. Eng. Shimane Univ. Ser. B. Math. Sci. **31**, 21–25 (1998)
160. M. Kondo, W.A. Dudek, Characterization theorem of lattice implication algebras. Far East J. Math. Sci. **13**, 325–342 (2004)
161. M. Kondo, W.A. Dudek, On the transfer principle in fuzzy theory. Math-ware Soft Comput. **12**, 41–55 (2005)
162. M. Kondo, W.A. Dudek, Filter theory of BL-algebras. Soft Comput. **12**, 419–423 (2008)
163. J. Kuhr, Pseudo-BCK-algebras and related structures, Univerzite Palackeho Olomouci (2007)
164. K.J. Lee, Y.B. Jun, M.I. Doh, Fuzzy translations and fuzzy multiplications of BCK/BCI-algebras. Commun. Korean Math. Soc. **24**, 353360 (2009)
165. L. Liu, K. Li, Fuzzy filters of BL-algebras. Inf. Sci. **173**(13), 141–154 (2005)
166. L. Liu, K. Li, Fuzzy Boolean and positive implicative filters of BL-algebras. Fuzzy Sets Syst. **152**(2), 333–348 (2005)
167. X. Liu, Z. Wang, On very true operators and v-filters. WSEAS Trans. Math. **7**(10), 599–608 (2008)
168. X. Liu, Z. Wang, On v-filters of commutative residuated lattices with weak vt-operators, in *Proceedings of the 3rd WSEAS International Conference on Computer Engineering and Applications (CEA'09)* (2009), pp. 133–137
169. P.K. Maji, R. Biswas, A.R. Roy, Soft set theory. Comput. Math. Appl. **45**, 555–562 (2003)
170. P.K. Maji, A.R. Roy, R. Biswas, An application of soft sets in a decision making problem. Comput. Math. Appl. **44**, 1077–1083 (2002)
171. J. Meng, On ideals in BCK-algebras. Math. Japon. **40**(1), 143–154 (1994)
172. J. Meng, Implication algebras are dual to implicative BCK algebras. Soochow J. Math. **22**(4), 567–571 (1996)
173. B.L. Meng, On Filters in BE-algebras. Sci. Math. Japon, Online, 105–111 (e-2010)
174. J. Meng, Y.B. Jun, *BCK-Algebras* (Kyungmoon Sa Co, Seoul, 1994)

175. J. Meng, Y.B. Jun, Atomatic extensions of BCK-algebras to proper BCI-algebras. Math. Japon **44**(2), 313–315 (1996)
176. J. Meng, X.L. Xin, Characterization of atoms in BCI-algebras. Math. Japon **37**(2), 359–362 (1992)
177. B.L. Meng, X.L. Xin, On fuzzy ideals of BL-algebras. Sci. World J. **2014**, 12 (2014). Article ID 757382
178. J. Meng, Y.B. Jun, E.H. Roh, The role of $B(X)$ and $L(X)$ in the ideal theory of BCI-algebras. Indian J. Pure Appl. Math. **28**(6), 741–752 (1997)
179. J. Meng, Y.B. Jun, H.S. Kim, Fuzzy implicative ideals of BCK-algebras. Fuzzy Sets Syst. **89**, 243–248 (1997)
180. J. Meng, Y.B. Jun, X.L. Xin, Prime ideal in commutative BCK-algebras. Discussiones Mathematicae **18**, 5–15 (1998)
181. D. Molodtsov, Soft set theory - first results. Comput. Math. Appl. **37**, 19–31 (1999)
182. S.M. Mostafa, Fuzzy implicative ideals in BCK-algebras. Fuzzy Sets Syst. **87**, 361–368 (1997)
183. S. Motamed, L. Torkzadeh, A. Borumand Saeid, N. Mohtashmania, Radical of filters in BL-algebras. Math. Log. Quart. **57**(2), 166–179 (2011)
184. D. Mundici, Averaging the truth-value in Lukasiewicz. Studia Logica **55**, 113–127 (1995)
185. J.R. Munkres, *Topology a First Course* (Prentice-Hall, Upper Saddle River, 1975)
186. C. Muresan, Dense elements and classes of residuated lattices. Bull. Math. Soc. Sci. Math. Roumanie Tome **53**(1), 11–24 (2010)
187. R. Najafi, Pseudo-commutators in BCK-algebras. Pure Math. Sci. **2**(1), 29–32 (2013)
188. R. Najafi, A.B. Saeid, Solvable BCK-algebras. Cankaya Univ. J. Sci. Eng. **11**(2), 19–28 (2014)
189. A.S. Nasab, A.B. Saeid, Semi-maximal filters in Hilbert algebras. J. Intell. Fuzzy Syst. **30**(1), 7–15 (2016)
190. J. Neggers, H.S. Kim, On B-algebras. Matematichki Vesnik **54**(1–2), 21–29 (2002)
191. W.C. Nemitz, Implicative semilattices. Trans. Am. Math. Soc. **117**, 128–142 (1965)
192. M. Palasinski, On ideals and congruence lattices of BCK-algebras. Math. Semin. Notes **9**, 441–443 (1981)
193. M. Palasinski, Ideals in BCK-algebras which are lower semilattices. Math. Japonica **26**(2), 245–250 (1981)
194. D. Piciu, D.D. Tascau, The Localization of Commutative Bounded BCK-Algebras. Adv. Pure Math. **1**, 367–377 (2011)
195. J. Rachunek, D. Salounova, Truth values on generalization of some commutative fuzzy structures. Fuzzy Sets Syst. **157**, 3159–3168 (2006)
196. J. Rachunek, D. Salounova, State operators on GMV-algebras. Soft Comput. **15**, 327–334 (2011)
197. A. Radfar, A. Rezaei, A.B. Saeid, Hyper BE-algebras. Novi Sad J. Math. **44**(2), 137–147 (2014)
198. H. Rasiowa, *An Algebraic Approach to Non-classical Logics* (North Holland, Amsterdam, 1974)
199. B. Ravikumar, N. Rafi, B. Davvaz, Falling fuzzy filters in BE -algebra. J. Chungcheong Math. Soc. **30**(2), 201–211 (2017)
200. A. Rezaei, Congruencec relations on BE-algebras, 3th Math. Mashhad, Iran, Sci. Con. PNU, 57–64 (2010)
201. A. Rezaei, A. Borumand Saeid, Some results on BE-algebras. Analele Universitatii Oradea Fasc. Matematica, Tom XIX, 33–34 (2012)
202. A. Rezaei, A. Borumand Saeid, On fuzzy subalgebras of BE-algebras. Afrika Matematika **22**(2), 115–127 (2011)
203. A. Rezaei, A. Borumand Saeid, Commutative ideals in BE-algebras. Kyungpook Math. J **52**, 483–494 (2012)
204. A. Rezaei, A. Borumand Saeid, Generalized fuzzy filters(ideals) of BE-algebras. J. Uncertain Syst. **7**(2), 152–160 (2012)

205. A. Rezaei, A. Borumand, R.A. Borzooei, Relation between Hilbert algebras and BE-algebras. Appl. Appl. Math. **8**(2), 573–584 (2013)
206. A. Rosenfeld, Fuzzy groups. J. Math. Anal. Appl. **35**, 512–517 (1971)
207. T. Roudabri, L. Torkzadeh, A topology on BCK-algebras via left and right stabilizers. Iran. J. Math. Sci. Inf. **4**(2), 1–8 (2009)
208. A.B. Saeid, Smarandache weak BE-algebras. Commun. Korean Math. Soc. **27**(3), 489–496 (2012)
209. A.B. Saeid, A. Rezaei, R.A. Borzooei, Some type of filters in BE-algebras. Math. Comput. Sci. **7**, 341–352 (2013)
210. M. Sambasiva Rao, δ-ideals in pseudo-complemented distributive lattices. Archivum Mathematicum. **48**(2), 97–105 (2012)
211. M. Sambasiva Rao, On associative filters of lattice implication algebras. Int. J. Math. Archiv. **3**(8), 3118–3121 (2012)
212. M. Sambasiva Rao, Fuzzy filters of BE-algebras. Int. J. Math. Arch. **4**(6), 181–187 (2013)
213. M. Sambasiva Rao, Fuzzy implicative filters of BE-algebras. Ann. Fuzzy Math. Inf. **6**(3), 755–765 (2013)
214. M. Sambasiva Rao, On fuzzy filters of BE-algebras. Ann. Fuzzy Math. Inf. **7**(2), 229–238 (2014)
215. M. Sambasiva Rao, Fuzzy positive implicative filters of BE-algebras. Ann. Fuzzy Math. Inf. **7**(2), 263–273 (2014)
216. M. Sambasiva Rao, Fantastic filters and their fuzzification in BE-algebras. Ann. Fuzzy Math. Inf. **7**(4), 553–561 (2014)
217. M. Sambasiva Rao, Transitive and absorbent filters of lattice implication algebras. J. Appl. Math. Inf. **32**(3–4), 323–330 (2014)
218. M. Sambasiva Rao, Derivations of implicative semilattices. South. Asian Bull. Math. **39**(4), 555–566 (2015)
219. M. Sambasiva Rao, Simple filters and applications of soft sets to simple filters in BE-algebras. Int. J. Concep. Comput. Inf. Tech. **3**(2), 36–39 (2015)
220. M. Sambasiva Rao, Prime filters of commutative BE-algebras. J. Appl. Math. Inf. **33**(5–6), 579–591 (2015)
221. M. Sambasiva Rao, Filters of BE-algebras with respect to a congruence. J. Appl. Math. Inf. **34**(1–2), 1–7 (2016)
222. M. Sambasiva Rao, K.P. Shum, On filters of implicative n.p.o. semigroups. Asian Eur. J. Math. **5**(3), 1–10 (2012)
223. M. Sambasiva Rao, Multipliers and normal filters of BE-algebras. J. Adv. Res. Pure Math. **4**(3), 61–67 (2012)
224. M. Sambasiva Rao, V. Venkata Kumar, Dual annihilator filters of commutative BE-algebras. Asian Eur. J. Math. (to appear)
225. T. Senapati, M. Bhowmikb, M. Palc, Fuzzy dot subalgebras and fuzzy dot ideals of B-algebras. J. Uncertain Syst. **8**(1), 22–30 (2014)
226. B.T. Sims, *Fundamentals of Topology* (Macmillan Publishing Co. Inc, New York, 1976)
227. S.Z. Song, Y.B. Jun, K.J. Lee, Fuzzy ideals in BE-algebras. Bull. Malays. Math. Sci. Soc. **33**, 147–153 (2010)
228. M.H. Stone, A theory of representations for Boolean algebras. Tran. Am. Math. Soc. **40**, 37–111 (1936)
229. D. Sun, On atoms of BCK-algebras. Scientiae Mathematicae Japonicae Online **4**, 115–124 (2001)
230. U.M. Swamy, G. Suryanarayana Murti, Boolean centre of universal algebra. Algebra Universalis **13**, 202–205 (1981)
231. L. Torkzadeh, Dual stabilizers and dual normal BCK-algebras. 20th Seminar on Algebra Tarbiat Moallem University, 2-3 Ordibehesht, 1388 (22–23 Apr 2009), pp. 219-221
232. L. Torkzadeh, T. Roudbari, Dual normal BCK-algebras. Math. Sci. **3**(2), 111–132 (2009)
233. E. Turunen, Boolean deductive systems of BL-algebras. Arch. Maths. Logic. **40**, 467–473 (2001)

234. E. Turunen, J. Mertanen, States on semi-divisible residuated lattices. Fuzzy Sets Syst. **12**(4), 353–357 (2008)
235. E. Turunen, S. Sessa, BL-algebra and basic fuzzy logic. Math. Ware Soft Comput. **6**, 49–61 (1999)
236. K. Venkateswarlu, B.V.N. Murthy, Isomorphism theorems in Boolean like semirings. Int. J. Math. Sci. Appl. **1**(3), 1363–1369 (2011)
237. V. Vychodil, Truth-depressing hedges and BL-logic. Fuzzy Sets Syst. **157**, 2074–2090 (2006)
238. A. Walendziak, Some axiomatizations of B-algebras. Math. Slovaca **56**(3), 301–306 (2006)
239. A. Walendziak, On commutative *BE-algebras*, Sci. Math. Japon, Online, 585–588 (e-2008)
240. P.Z. Wang, *Fuzzy Sets and Falling Shadows of Random Sets* (Beijing Nnormal University Press, People's Republic of China, 1985). [In Chinese]
241. P.Z. Wang, E. Sanchez, in *Treating a Fuzzy Subset as a Projectable Random Set*, ed. by M.M. Gupta, E. Sanchez, Fuzzy Information and Decision (Pergamon, New York, 1982), pp. 212–219
242. A. Wronski, BCK-algebras do not form a variety. Math. Jpn. **28**, 211–213 (1983)
243. A. Wronski, Reflections and Distentiones of BCK-algebras. Math. Jpn. **28**, 215–225 (1983)
244. O.G. Xi, Fuzzy BCK-algebras. Math. Jpn. **36**(5), 935–942 (1991)
245. Y. Xu, K.Y. Qin, On filters of lattice implication algebras. J. Fuzzy Math. **1**, 251–260 (1993)
246. Y.H. Yon, S.M. Lee, K.H. Kim, On congruences and BE-relations in BE-algebras. Int. Math. Forum **5**, 2263–2270 (2010)
247. D.S. Yoon, H.S. Kim, Uniform structures in BCI-algebras. Commun. Korean Math. Soc. **17**(3), 403–408 (2002)
248. Y. Yu, J.N. Morderson, S.C. Cheng, Elements of L-algebra, Lecture Notes in Fuzzy Mathematics and Computer Science (Creighton University, Omaha, Nebraska 68178, USA, 1994)
249. L.A. Zadeh, From circuit theory to system theory. Proc. Inst. Radio Eng. **50**, 856–865 (1962)
250. L.A. Zadeh, Fuzzy sets. Inf. Control. **8**, 338–353 (1965)
251. L.A. Zadeh, Similarity relations and fuzzy ordering. Inf. Sci. **3**, 177–200 (1971)
252. L.A. Zadeh, Fuzzy logic and approximate reasoning. Synthesis **30**, 407–428 (1975)
253. L.A. Zadeh, Towards a generalized theory of uncertainty (GTU) - an outline. Inf. Sci. **172**, 1–40 (2005)
254. B. Zelinka, Tolerance in algebraic Structures. Czech. Math. J. **20**, 179–183 (1970)
255. X.H. Zhang, W.A. Dudek, Fuzzy $BIK+$-logic and non-commutative fuzzy logic. Fuzzy Syst. Math. **23**, 2–20 (2009)
256. X.H. Zhang, W.A. Dudek, Soft MTL-algebras based on fuzzy sets. Sci. Bull. Ser. A, Appl. Math. Phys. Politeh. Univ. Bucharest **74**, 41–56 (2012)
257. X.H. Zhang, K. Qin, W.A. Dudek, Ultra Li-ideals in lattice implication algebras and MTL-algebras. Czech. Math. J. **57**(132), 591–605 (2007)

Index

Symbols

A

B

S. R. Mukkamala, *A Course in BE-algebras*,
https://doi.org/10.1007/978-981-10-6838-6

Printed in the United States
By Bookmasters